SYNCHROTRON RADIATION APPLICATIONS

SYNCHROTRON RADIATION APPLICATIONS

Xinyi Zhang

Fudan University, China

World Scientific

NEW JERSEY • LONDON • SINGAPORE • BEIJING • SHANGHAI • HONG KONG • TAIPEI • CHENNAI • TOKYO

Published by

World Scientific Publishing Co. Pte. Ltd.
5 Toh Tuck Link, Singapore 596224
USA office: 27 Warren Street, Suite 401-402, Hackensack, NJ 07601
UK office: 57 Shelton Street, Covent Garden, London WC2H 9HE

Library of Congress Cataloging-in-Publication Data
Names: Zhang, Xinyi, 1942– editor.
Title: Synchrotron radiation applications / editor, Xinyi Zhang (Fudan University).
Description: Singapore ; Hackensack, NJ : World Scientific, [2018] |
Includes bibliographical references and index.
Identifiers: LCCN 2017043655| ISBN 9789813227668 (hardcover ; alk. paper) |
ISBN 9813227664 (hardcover ; alk. paper)
Subjects: LCSH: Synchrotron radiation--Industrial applications.
Classification: LCC QC793.5.E627 S85 2018 | DDC 539.7/35--dc23
LC record available at https://lccn.loc.gov/2017043655

British Library Cataloguing-in-Publication Data
A catalogue record for this book is available from the British Library.

For any available supplementary material, please visit
http://www.worldscientific.com/worldscibooks/10.1142/10643#t=suppl

Typeset by Stallion Press
Email: enquiries@stallionpress.com

Printed in Singapore

Preface

This is a review volume. It will tell our readers both the fundamentals and the advanced research results, covering the most important aspects of synchrotron radiation applications. Among the broad subjects of synchrotron radiation applications we choose subjects which cover applications in VUV, soft X-rays, hard X-rays, and XFEL (X-ray free electron laser), and contain important applications by various synchrotron-based techniques and methods, such as ARPES (angle-resolved photoemission spectroscopy), VUV photoionization spectroscopy, X-ray absorption/emission spectroscopy and X-ray absorption fine structure, X-ray diffraction, small angle X-ray scattering, X-ray excited optical luminescence, imaging, and high pressure techniques. Many interesting aspects with latest stirring results will be described, such as recent ARPES study on the iron based superconductors, the charge density wave materials, and the Mott-insulators; crystal structure of macromolecules in structural biology; medical imaging including clinical and pathological applications; study on various advanced functional materials; and the recent encouraging research achievements based on XFELs. One more thing that I would like to stress on is that developing new detectors matters as much as new experimental methods. Simply speaking, you can see nothing without the detectors, and it would be impossible to have top-notch experimental results. In this book, Chapter 5 of Hiroyuki

Oyanagi described the history of segmented X-ray detectors, including the *state-of-the-art* pixel detectors for fluorescence-detected synchrotron radiation X-ray spectroscopy experiments.

I did not expect it would cost me more than four years at the beginning when I accepted the writing job of this book. Tetsuya Ishikawa, director of Riken Harima Institute, once said to me, "The science with XFEL, which I plan to describe, has been expanding very rapidly. Actually, it is extremely difficult for anyone to compile the XFEL science in this rapidly changing time." He is right. In fact, not only XFEL but also various fields of synchrotron radiation applications have developed well and fast, new experimental methods and numerous research results keep on emerging. The team of Tetsuya Ishikawa finally offers our readers the wonderful Chapter 13 reviewing SACLA (Spring-8 Angstrom Compact Free-electron Laser) research achievements.

I kept thinking about collecting the latest research results about synchrotron radiation applications in this book. But it is too hard to do this for reasons that research objects and methods are differentiated in a wide range, experimental techniques are precise but complicated and development is so fast that sometimes it is beyond one's reach. Nevertheless, I tried to ask contributors of each chapter to include the latest synchrotron-based techniques and research results into their work time and time again. Authors were also asked to edit their manuscripts over and over so as to include new research developments. For example, Ando Masami, who wrote Chapter 8 of this book, had at least seven re-edited versions stored in my computer. In truth, many other authors did the similar things. Though this book took a long time, we are fortunate to enclose some new techniques quickly developed in recent years into this book. For example, at the beginning of writing this book, I once listed CDI (coherent diffractive imaging) in, but some others thought it was still under-developed to be described in this book. However, just two years later, CDI has finally succeeded using XFEL; hence we have Chapter 14 contributed by Andrew V. Martin and N. Duane Loh. Also, a short paragraph describing CDI was added in Tiqiao Xiao and Lanhong Xie's

Chapter 9. Many techniques mark great research developments in synchrotron radiation applications over these years, such as time and space resolved technique, especially dynamical research on the relationship between structure and function during operation, namely *operando* techniques; various new imaging techniques, and new X-ray diffractive technique for dynamical structural studies on increasingly complex systems.

This book can be a reference book for researchers and technicians taking up synchrotron radiation application researches and postgraduates majoring in this field. For it is finite in my knowledge, some inadequacies and imperfect points are inevitable; therefore I hope you feel free to enlighten me. Finally, I would like to thank all the authors who made great contributions to this book and also many others, who contributed as well but have not been listed is the chapter authors.

Xinyi Zhang
Department of Physics, Fudan University
xy-zhang@fudan.edu.cn
October 13, 2016, Shanghai

List of Abbreviations

2D : two-dimensional
3D : three-dimensional
AAV–NT-1 : adeno-associated viral vector–Netrin-1
ABI : analyzer-based imaging
ACA : anterior cerebral artery
ACF : absorption correction factor
AchA : anterior choroidal artery
AC-OSEM : Absorption-corrected ordered subsets expectation maximization
AD : Alzheimer's disease
AFM : atomic force microscopy
AMO : antimicrobial oligomers
ASAXS : anomalous small angle X-ray scattering
BL13W1 : X-ray imaging and biomedical applications beamline
BM : biological macromolecule
BN : boron nitride
BNSLs : binary NPSLs
BSRF : Beijing Synchrotron Radiation Facility
cc : corpus callosum
CCDs : Combustion chamber deposits
CDI : coherent diffraction imaging

CF : cystic fibrosis
CI : crystal interferometry
CMC : carboxymethyl cellulose
CMFs : cellulose microfibrils
CMMS : Calcium/magnesium-doped silica-based hierarchically mesoporous scaffolds
CNCD : computerized numerical control drilling
CPu : caudate putamen
CS : compressed sensing
CT : computed tomography
CTAC : cetyl trimethylammonium chloride
CT-DE : computed tomography-Diffraction enhancement
DA : decoupling approximation
DCM : data-constrained modeling
DEI : diffraction enhancement imaging
DLS : dynamic light scattering
DNA : deoxyribonucleic acid
DOSs : densities of states
DWBA : Distorted Wave Born Approximation
ECM : extracellular matrix
EDM : electrical discharge machining
EGFR : Epidermal Growth Factor Receptor
EM : electron microscopy
EPR : electron paramagnetic resonance
ERK2 : extracellular signal-regulated kinase 2
EROS : ensemble refinement of SAXS
EST : equally sloped tomography
FBP : filtered back projection
FEL : free electron laser
FOV : field of view
FT : Fourier transform
FWHM : Full width at half maximum
GAG : glycosaminoglycan
GDPC : Grating-based differential phase contrast imaging
GI : grating interferometry

GIFT : generalized indirect Fourier transformation
GISAXS : grazing incidence small angle X-ray scattering
GUI : graphical user interface
HePTP : hematopoietic tyrosine phosphatase
HNC : hypernetted chain
HTA : hypothalamic artery
ICH : intracranial hemorrhage
ICP-MS : Inductively coupled plasma mass spectroscopy
IFs : intermediate filaments
IL-PCI : in-line X-ray phase-contrast imaging
IL-XPμCT : in-line X-ray phase contrast micro-computed tomography
IL-XPCT : in-line X-ray phase-contrast CT
IMC : intermetallic compound
LMA : local monodisperse approximation
LV : lateral ventricle
MAP : mitogen-activated protein
MB : Methylene Blue
MBA : Modified Bronnikov algorithm
MCA : middle cerebral artery
MCAO : middle cerebral artery occlusion
MD : molecule dynamics
Micro-CT : X-ray microscopic computerized-tomography
MRI : magnetic resonance imaging
NAC : nascent polypeptide associated complex
NFTs : neurofibrillary tangles
NMR : nuclear magnetic resonance
NPs : nanoparticles
NPSL : nanoparticle superlattice
OA : osteoarthritis
ODDS : Osmotic drug delivery systems
OSEM : ordered subsets expectation maximization
OZ : Ornstein-Zernike
PAD : phase-attenuation duality
PBI : propagation-based phase-contrast imaging
PC : phosphatidylcholine

PCA : posterior cerebral artery
PC-CT : phase-contrast enhanced computed tomography
PCI : phase-contrast imaging
PCL : poly(ε-caprolactone)
PcomA : posterior communication arteries
PCXI : phase contrast X-ray imaging
PDB : protein data bank
PDDF : pair distance distribution function
$PeaT1$: protein elicitor from Alternaria tenuissima
PEG–PE : Poly(ethylene glycol)–phosphatidylethanolamine
PEO : poly-(ethylene oxide)
PET : positron emission tomography
PG : proteoglycan
PITRE : Phase-sensitive X-ray Image processing and Tomography Reconstruction
PLLA : Poly(L-lactide)
PPCI : X-ray propagation-based phase-contrast imaging
PPCT : X-ray propagation-based phase-contrast CT
PPI : phase propagation X-ray imaging
PPO : poly(propylene oxide)
PS-PVP : poly(styrene-b-2-vinylpyridine)
PVME : poly(vinyl methyl ether)
RC : rocking curve
rhBMP-2 : recombinant human bone morphogenetic protein-2
RNA : Ribonucleic acid
ROI : region of interesting
SAH : subarachnoid hemorrhage
SANS : small-angle neutron scattering
SAXS : small angle X-ray scattering
SCF/EP : short carbon fiber/epoxy
SD : Sprague-Dawley
SDD : sample-to-detector distance
SDF : size distribution function
SEM : scanning electron microscopy
SL : superlattice

SLS : static light scattering
SNSLs : single NPSLs
SPECT : single-photon emission computed tomography
SR-μCT : synchrotron based X-ray micro-CT
SR : synchrotron radiation
SR-CT : synchrotron radiation absorption-based CT
SSCA : size-spacing correlation approximation
SSRF : Shanghai Synchrotron Radiation Facility
SSRL : Stanford Synchrotron Radiation Lightsource
TBC : tetrabromocathecol
TBT : tangent by tangent
TCMs : traditional Chinese medicines
TE : tissue engineering
TEM : transmission electron microscopy
TIE : differential transport-of-intensity equation
TIEG1 : TGFβ inducible early gene 1
TMV : tobacco mosaic virus
t-SAXS : time-resolved small angle X-ray scattering
UBA : ubiquitin-associated
US : ultrasound
USAXS : ultra-small-angle X-ray scattering
UV : ultraviolet-visible
WAXS : wide-angle X-ray scattering
WHO : World Health Organization
XAFS : X-ray absorption fine structure
XANES : X-ray absorption near-edge structure
XDFI : X-ray dark field image
XFCT : X-ray fluorescence computed tomography
XPμCT : X-ray phase contrast micro-computed tomography
XPCI : X-ray phase-contrast imaging
XPCMT : X-ray phase contrast micro-tomography
XRD : X-ray diffraction
XRF : X-ray fluorescence

Program Names

ATSAS
PCG
DECON
GIFT
IsGISAXS
PRIMUS
GNOM
DAMMIN
DAMAVER

Contents

Chapter 1

Angle Resolved Photoemission Spectroscopy Study Utilizing the Synchrotron Radiation

Yan Zhang

International Center for Quantum Materials,
School of Physics, Peking University, Beijing 100871,
People's Republic of China

Dawei Shen

State Key Laboratory of Functional Materials for Informatics,
Shanghai Institute of Microsystem and Information Technology,
Chinese Academy of Sciences, Shanghai 200050,
People's Republic of China

Donglai Feng

State Key Laboratory of Surface Physics,
and Department of Physics, Fudan University,
Shanghai 200433, People's Republic of China

Abstract

Angle-resolved photoemission spectroscopy is a powerful technique that can directly probe the electronic structure of materials. By utilizing the synchrotron radiation, which can provide high-intense monochromatic photon beam with tunable polarization and photon energy, ARPES can resolve detailed electronic structure with ultra-high energy and momentum resolutions, detect three-dimensional momentum space, and probe electrons with different orbital characters selectively, etc. In this chapter, we will briefly review the recent progress obtained by ARPES using the synchrotron radiation. The focused materials will cover the iron-based high-T_C superconductors, the charge-density-wave materials, and Mott-insulators.

Keywords: band dispersion, charge-density-wave, density of states, energy gap, high-Tc superconductors, iron-based superconductors, Mott transition, photoemission spectroscopy, strongly correlated materials, transition metal dichalcogenides.

1.1. Introduction

Angle-resolved photoemission spectroscopy (ARPES) can provide rich information about the electronic structure of solids. In this section, we will first briefly introduce the ARPES technique, and then focus on three particular aspects which benefits from the synchrotron radiation: the ultra-high energy and momentum resolutions, the k_z-probing capability, and orbital-sensitive measurements.

1.1.1. *Angle-resolved photoemission spectroscopy*

The experimental setup of ARPES is shown in Fig. 1.1. Based on the photoelectric effect, when a beam of monochromatized radiation is shined on a sample, electrons are excited by absorbing photons, and then escape into vacuum [1, 2]. An energy conservation during this

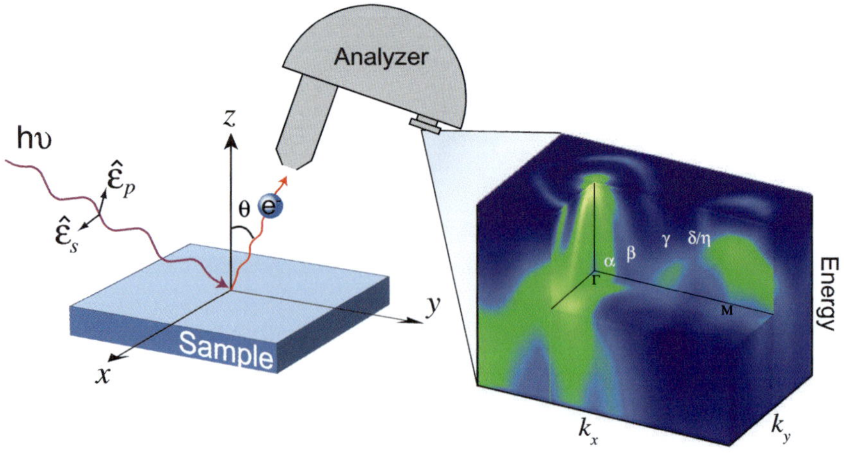

Fig. 1.1. The experimental setup of angle-resolved photoemission spectroscopy.

photoemission process can be described by:

$$E_{kin} = h\nu - |E_B| - \varphi,$$

where E_{kin} is the kinetic energy of the photoelectron, $h\nu$ is the photon energy, E_B is the electron binding energy, and φ is the work function of the solid. When electrons are excited from solids with crystalline order, there is a momentum conservation between the in-plane photoelectron momentum $\mathbf{p}_{\|}$ and the crystal momentum of electron $\hbar\mathbf{k}_{\|}$, which can be described by:

$$\mathbf{p}_{\|} = \hbar\mathbf{k}_{\|} = \sqrt{2mE_{kin}}\sin\theta.$$

By measuring the kinetic energy E_{kin} of the photoelectron and its emission angle θ with an electron analyzer, we can calculate the binding energy and crystal momentum of the electrons in the materials based on the energy and momentum conservations. The inset of Fig. 1.1 shows typical photoemission data taken on the iron-based superconductor LiFeAs. The distribution of electronic states in the $k_x - k_y - Energy$ space could be mapped out and viewed from the false color plot.

1.1.2. *Ultra-high energy and momentum resolutions*

After years of development, the typical energy and angular resolutions of ARPES can be better than 1 meV and 0.1 degree respectively, which enable us to distinguish the fine features of electronic structure. A synchrotron can provide high-quality monochromatic photon beam with tunable polarization and energy (Fig. 1.2). Electrons pass through an undulator and then generate an intense beam of synchrotron radiation. After reflected by grating, the beam pass through an exit slit and then is focused onto the sample. The general energy resolution of the beam varies from 1 to 10 meV depending on the selected photon energy. Such a energy scale is critical for studying the meV phenomena in solids, including superconductivity (SC), charge-density-wave (CDW), electron-phonon coupling, etc. The momentum resolution is determined by both the angular resolution

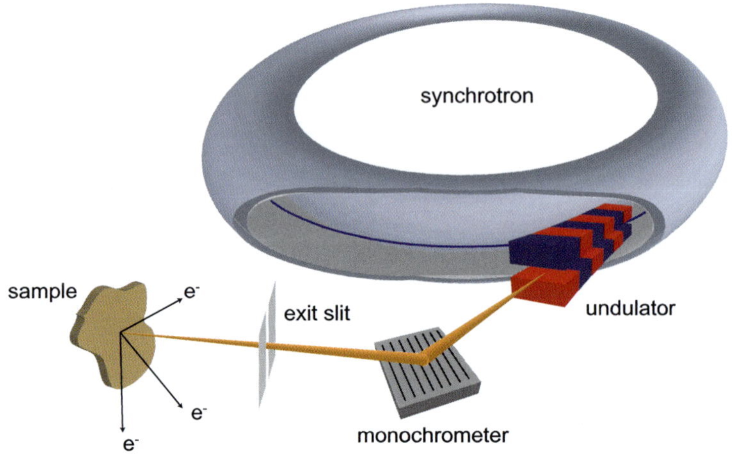

Fig. 1.2. The illustration of the synchrotron radiation and beamline setup.

and photon energy. Because of the electron optics of a electron analyzer, the angular resolution is limited by the beam spot size. The general spot size of the synchrotron radiation can be focused down to tens of microns, which could give us an angular resolution of better than 0.1 degree. On the other hand, the photon energy can be tuned from 5 to hundreds of electron volts. The momentum resolution is maximized at the lowest photon energy, while the detectable momentum range can be benefited at higher photon energy. As will be discussed in Section 1.2, the high energy and momentum resolutions are very important for the energy gap studies in CDW materials.

1.1.3. k_z *determination in ARPES*

Due to the lack of translational symmetry along the sample surface normal, the out-of-plane momentum $k_\perp$ (or k_z) is not conserved during the photoemission process. Such uncertainty in k_z is less relevant in the case of low-dimensional systems, such as the cuprates. However, for three-dimensional materials, the electronic structure often has considerable variation along the k_z direction. In ARPES experiments, although $k_\perp$ does not follow the simple law of momentum conservation, it could be approximately calculated by using "inner

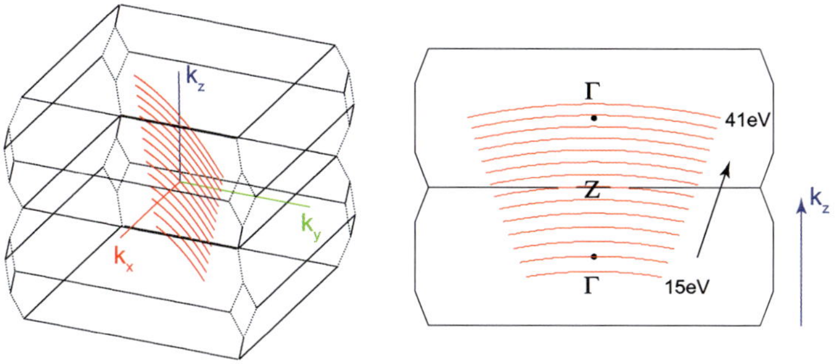

Fig. 1.3. The photon energy dependent measurement in ARPES. The black solid lines show the three-dimensional (3D) Brillouin zone of a body center lattice structure. The red solid lines illustrate the momentum cuts taken with the photon energies from 15 to 41 eV. The inner potential is set to be 15 eV.

potential" V_0 as following [1]:

$$k_{\perp} = \hbar^{-1}\sqrt{2m(E_{kin}\cos^2\theta + V_0)}.$$

In practice, by changing the incident photon energy, the E_{kin} will change accordingly. One can thus detect the electrons at different k_zs through the entire three-dimensional (3D) Brillouin zone (Fig. 1.3). The obtained band dispersion usually exhibits good periodicity along k_z, which could be used to estimate the value of inner potential. Therefore, conducting the photon energy dependent ARPES is an effective way to probe the 3D electronic structure. We will show an example in Section 1.3, the soft X-ray ARPES study on the 3D electronic structure of $NiS_{2-x}Se_x$. However, we should note that, for materials with moderate k_z dispersion, the k_z broadening and final states effects during the photoemission process could complicate the ARPES spectra and need to be carefully considered [1].

1.1.4. *Orbital-sensitive probe*

The polarization-sensitivity of orbitals in ARPES can be used to determine the orbital character of electronic structure. The photoemission intensity is proportional to the matrix element of photoemission process $I_0(\mathbf{k}, \mathbf{v}, \mathbf{A}) \propto |M_{f,i}^{\mathbf{k}}|^2$ [2, 3], which can be

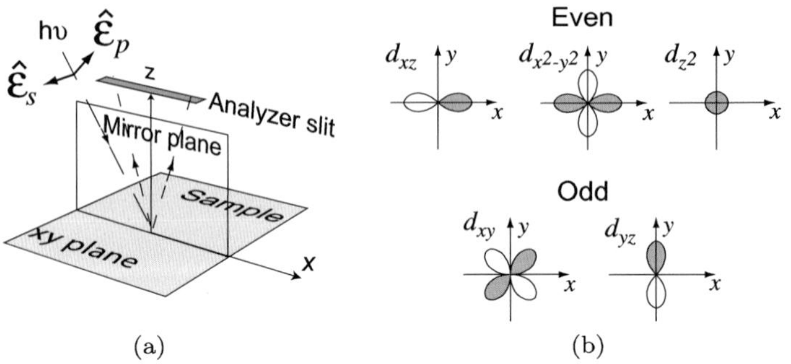

Fig. 1.4. (a) The experimental geometry of the polarization dependent ARPES. (b) The spatial symmetries of 3d orbitals with respect to the xz-plane.

described by:

$$|M^{\mathbf{k}}_{f,i}|^2 \propto |\langle\phi^{\mathbf{k}}_f|\hat{\varepsilon}\cdot\mathbf{r}|\phi^{\mathbf{k}}_{\mathbf{i}}\rangle|^{\mathbf{2}},$$

where $\hat{\varepsilon}$ is the unit vector of the electric field of light, and $\phi^{\mathbf{k}}_i$ ($\phi^{\mathbf{k}}_f$) is the initial-state (final-state) wave function. In order to have non-vanishing photoemission intensity, the whole integrand in the overlap integral must be an even function under a reflection with respect to a mirror plane. For high kinetic-energy photoelectrons, the final-state wave function $\phi^{\mathbf{k}}_f$ can be approximated by an even-parity plane-wave state $e^{i\mathbf{k}\cdot\mathbf{r}}$ with $\mathbf{k}$ in the mirror plane defined by the analyzer slit and the sample surface normal. In turn, this implies that $\hat{\varepsilon}\cdot\mathbf{r}|\phi^{\mathbf{k}}_{\mathbf{i}}\rangle$ must be even. For the p (or s) experimental geometry, where the electric field direction of the incident photons $\hat{\varepsilon}_{\mathbf{p}}$ (or $\hat{\varepsilon}_{\mathbf{s}}$) is parallel (or perpendicular) to the mirror plane (Fig. 1.4(a)), $\hat{\varepsilon}\cdot\mathbf{r}$ will be even (or odd). Therefore, only the even (or odd) parity initial states $\phi^{\mathbf{k}}_i$ will show finite photoemission signal and thus can be detected.

Such selection rules are very useful for studying the multi-orbital systems. Considering the spatial symmetry of 3d orbitals, when the analyzer slit is in the xz plane for example, the d_{xz}, d_{z^2}, and $d_{x^2-y^2}$ orbitals are even and the d_{xy} and d_{yz} orbitals are odd [3] (Fig. 1.4(b)). The even and odd orbitals could be only observed in the p and s geometries, respectively. By comparing the experimental results taken in different polarization geometries, we could then determine the orbital characters of bands experimentally. Note that, we use the atomic wave function of $3d$ electrons as the initial wave function for simplicity. To calculate the matrix element more strictly, the Bloch wave function needs to be used in the analysis [4]. In Section 1.4, we shall see that the polarization dependent ARPES plays an important role in determining the orbital character and observing orbital anisotropy in the iron-based superconductors.

1.2. High-Resolution ARPES Study on the Energy Gaps of Charge-Density-Wave Systems

CDW is one of the most common forms of order in solids, which has been widely observed in quasi-one dimensional compounds, high temperature superconductors, manganites and many other materials [5–8]. Conventionally, Fermi surface nesting is the dominant and textbook mechanism for CDW. In this typical picture of CDW, there are usually parallel Fermi surface sections, which guarantee a large number of electrons could be scattered from one parallel side to the other by softened phonons with a fixed momentum transfer corresponding to the ordering wave vector(s). As a result, the charge susceptibility at the ordering wave vector(s) will be enhanced, and thus CDW instability could be induced in collaboration with lattice. However, it has failed to explain the CDW in the $2H$-structured transition metal dichalcogenides ($2H$-TMDs), which is actually the very first two-dimensional CDW materials discovered around 40 years ago [9]. Even after three decades of quite intensive research on this subject, the CDW mechanism of $2H$-TMDs remains mysterious, and controversial.

To understand the unusual CDW transition, experimentally, one crucial bottleneck is the prolonged lack of the intimate view of the subtle low energy electronic structure changes upon the CDW transition. Combining the state-of-the-art electron analyzer and the intense photon beam from a modern synchrotron accelerator or high flux helium lamp, the ARPES apparatus is a direct probe of the electronic structure and can reach enough high energy and momentum resolutions for this issue as mentioned above.

In this section, we will discuss the electronic origin of the CDW in two prototype $2H$-TMDs in which the electron-phonon coupling is

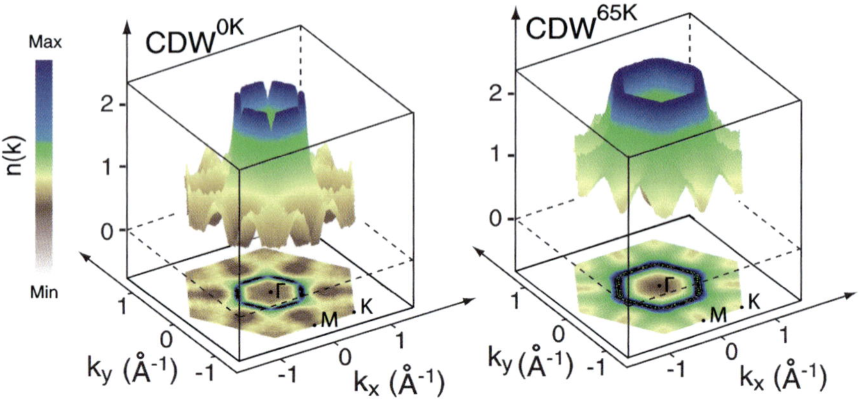

(a) (b)

Fig. 1.5. (a) and (b), The photoemission intensity integrated within 10 meV of E_F is shown for CDW^{0K} and CDW^{65K} respectively. Both data were taken at 15 K and the image was six-fold symmetrized [10].

extremely strong. We will first introduce the "Fermi patches" driven CDW mechanism in the 2H-Na_xTaS_2, and then present the major contributions to the CDW formation in $NbSe_2$ from these barely occupied states. Our findings not only resolve the long-standing puzzles, but also overthrow the conventional wisdom that CDW is dominated only by the regions around the Fermi surface with high density of states (DOS).

1.2.1. *The Fermi patches driven CDW transition in 2H-Na_xTaS_2*

2H-Na_xTaS_2 is advantageous for studying the CDW transition in 2H-TMDs, since one can control the strength of CDW upon adjusting the sodium doping level. For the systematic investigation purpose, 2H-Na_xTaS_2 with 2%, 5% and 10% Na concentrations were synthesized with CDW transition temperature T_{CDW}s of 68 K, 65 K and 0 K, respectively. The samples are labelled as CDW^{68K}, CDW^{65K}, and CDW^{0K} hereafter [10].

Figures 1.5(a) and (b) directly compare the photoemission intensity maps of samples with (CDW^{65K}) and without CDW (CDW^{0K}). The difference is striking. Although they have similar DOS at

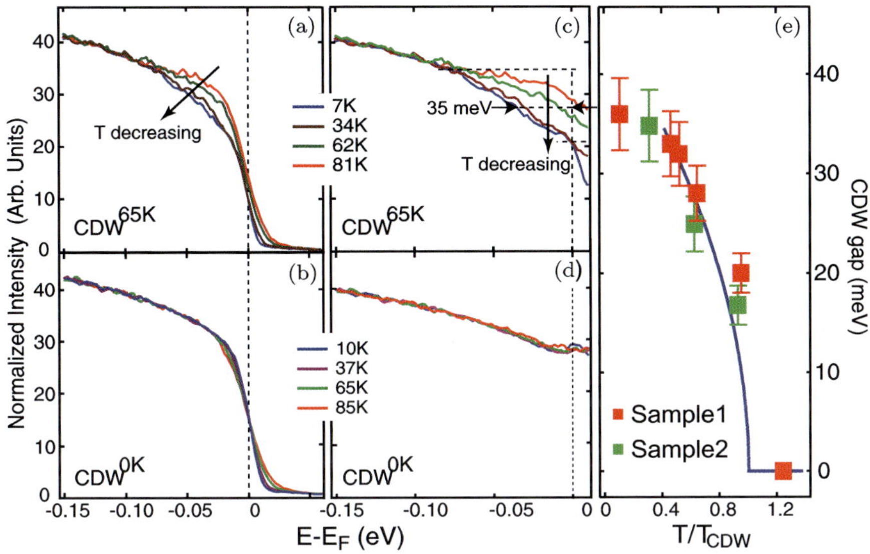

Fig. 1.6. The CDW gap measurements of 2H-Na_xTaS_2. ARPES spectra taken at **M** for different temperatures for (a), CDW^{65K} and (b), CDW^{0K}. (c)–(d), The spectra in (a) and (b) divided by the resolution convoluted Fermi-Dirac distribution at the corresponding temperatures. (e), The temperature dependence of the CDW gap. The solid line is the fit to a mean field formula, $\Delta_0\sqrt{1-\frac{T}{T_{CDW}}}$. Reprinted with permission from [10].

the Fermi crossing area, for momentum regions away from the Fermi surface, CDW^{65K} shows a much stronger spectral weight than CDW^{0K}, no matter whether it is inside the occupied regions (e.g. **M** point) or in the unoccupied regions (e.g., Γ and **K** points). This unconventional behavior of CDW^{65K} deviates much from the well-known Fermi liquid picture in the weakly correlated materials, indicating the strong electron-phonon coupling in this 5*d*-electron system. At the Fermi energy, the states locate not only at the conventional Fermi momenta, but also over other large regions in the Brillouin zone for the CDW samples, forming the so called Fermi patches.

Former searching effects for the CDW mechanism of 2H-TMD systems were mostly focused on the Fermi surfaces. Intrigued by the differences between CDW^{65K} and CDW^{0K} spectra in the Fermi-patch regions, Figures 1.6(a) and 1.6(b) examine spectra taken from **M** point at different temperatures for CDW^{65K} and CDW^{0K},

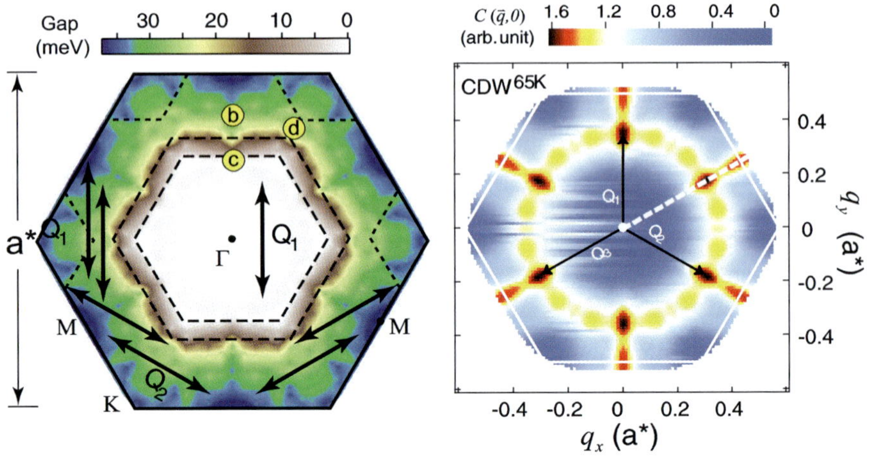

(a) (b)

Fig. 1.7. (a), The false color plots of the CDW gap in the first Brillouin zone of CDW^{65K}, where the dashed lines indicate the Fermi surfaces. States in the gapped region could be connected by the CDW wavevectors, $\mathbf{Q}_i$ ($i = 1,2,3$), as indicated by the double-head arrows. (b), The two-dimensional joint density of states result of the ARPES intensity in the normal state of CDW^{65K} [10].

respectively. While the CDW^{0K} spectra simply exhibit the clear Fermi crossing and thermal broadening, the CDW^{65K} spectra appear very anomalous. At low temperature, while the upper part of the spectrum is pushed to higher binding energy, the middle point of the leading edge of the lower part still matches E_F. To get rid of the complication caused by the thermal broadening, the spectra were divided with the corresponding finite temperature Fermi-Dirac distribution functions, as shown in Fig. 1.6(c). One can clearly observe the spectral weight suppression upon CDW transition for CDW^{65K}, and an energy gap of around 35 meV has opened on part of the states here. In contrast, there is no sign of gap opening for CDW^{0K} [Fig. 1.6(d)]. The temperature evolution of the gap size is shown in Fig. 1.6(e) for two different CDW^{65K} samples, which can be described well by the mean field formula.

Both temperature dependence and the doping dependence of the observed partial gap have unambiguously illustrated its CDW origin. Further experiment revealed that there is a clear CDW gap opening over the Fermi patches, just as illustrated by the Fig. 1.7(a). This finding undoubtedly indicates that the strong electron-phonon coupling in this system makes the electrons widely spread in the momentum space and the electronic states over the Fermi patches are actually responsible for the CDW.

To explain the 3×3 CDW ordering in this material, the joint density of states over the Fermi patches,

$$C(\vec{q},\omega) \equiv \int A(\vec{k},\omega)A(\vec{k}+\vec{q},\omega)d\vec{k},$$

where $A(\vec{k},\omega)$ is the spectral function, is calculated (Fig. 1.7(b)), as it describes the phase space for scattering of electrons from the state at $\vec{k}$ to the state at $\vec{k}+\vec{q}$ by certain modes with wavevector $\vec{q}$ [10–12]. Indeed, one could find that it peaks at exactly the ordering wavevector for $\omega = 0$. This result proves that such charge ordering is indeed favored by the collective behavior of the states at the Fermi patches. In this way, by means of high resolution ARPES, the CDW mechanism of this material is uncovered, and many existing controversies are resolved.

Because many complex materials bare similar electronic properties, such as Fermi patches and broad lineshapes, this Fermi-patch driven CDW mechanism may be a general theme for charge ordering in the strong electron-phonon coupling regime.

1.2.2. *Primary role of the barely occupied states in the CDW Formation of $NbSe_2$*

Niobium diselenide ($NbSe_2$) is another well-known $2H$-TMD for its prototypical SC and two dimensional CDW [13–18]. However, the very mechanism of the CDW in $NbSe_2$ itself has been mysterious and controversial for over three decades [16, 19–22], even though it is one of the very first two-dimensional CDW materials discovered [9].

In searching for the CDW mechanism of this material, previous researches mostly focused on the nesting of the momentum regions with high DOS at E_F in the Brillouin zone, e.g., Fermi surfaces or the so-called saddle band points. However, the CDW in $NbSe_2$

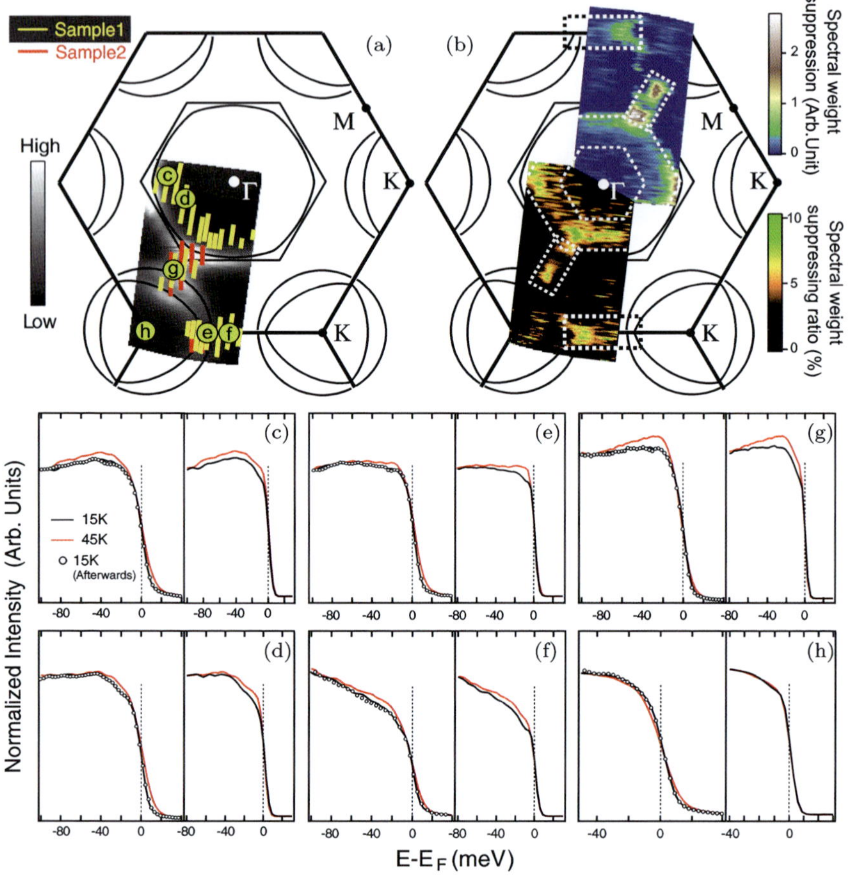

Fig. 1.8. (a) The illustration of the regions where gap opening occurs upon CDW transition in the first BZ. The gapped regions determined by comparison of the spectra below and above T_{CDW} one by one. (b) The spectral weight suppression map and its corresponding ratio map (see text for detail) are false color plotted, where the dashed lines contour the obviously suppressed regions and the black ones indicate the FSs. (c)–(h) Typical spectra (left) and the corresponding thermal effect removed (right) ones in the normal and CDW states at momenta marked in panel (a). Reprinted with permission from [23].

cannot be explained by any of these scenarios: neither can the CDW ordering wavevectors match the Fermi surface [16, 20, 21], nor match the saddle band points [20, 21]. The inapplicability of these common models for CDW in such a classic compound suggests that our current understanding of CDW need to be amended.

Inspired by the Fermi-patch driven CDW transition in 2H-Na_xTaS_2, more than 1,200 pairs of the normal and CDW state photoemission spectra were compared one by one over more than 1/6 of the whole Brillouin zone [23]. Some of the examples are given by Figs. 1.8(c–h), where the left panels show the raw spectra, and right panels show the corresponding spectra with thermal broadening removed. Eventually, by comparison the momentum regions where the spectral weight is suppressed can be determined accurately, as marked in Fig. 1.8(a) by bars. These regions are mainly composed of the large domains inside the Γ pockets, around the **M** points, and around the saddle band points. Note this distribution includes some "barely occupied" regions where the DOS at E_F is very weak. We emphasize that the weak spectral weight and its suppression are not due to the umklapp scattering of the CDW order, since it is not observed in the Fermi surface map or dispersion. Such findings have been confirmed by several samples, and the results based on two of them are shown here.

To further visualize such CDW gap distribution, the intensity map at 15 K (CDW state)[integrated over (E_F-80 meV, E_F)] is subtracted from that at 45 K (without CDW), and the result is shown in the upper half of Fig. 1.8(b). Note the thermal broadening effects have been removed from the spectra using the standard symmetrization method [24]. After normalization by the intensity map above the CDW transition temperature, the resulting spectral weight suppression ratio map [lower part of Fig. 1.8(b)] can further highlight the suppressed regions, which are consistent with those determined in Fig. 1.8(a). Note we have surveyed the entire Brillouin zone, and always observe such suppression only in the binding energy range of around (E_F–80 meV, E_F). Considering the conservation of the total electrons, the suppressed weight should be shifted to the higher binding energies. In this way, the electronic energetic gain for the CDW transition is naturally fulfilled.

The distribution of the CDW induced spectral weight suppression makes a broken-honeycomb-shaped structure in the Brillouin zone, as shown by Fig. 1.9(a). The suppression over the saddle band regions (surrounded by the dotted lines) only accounts for 43%

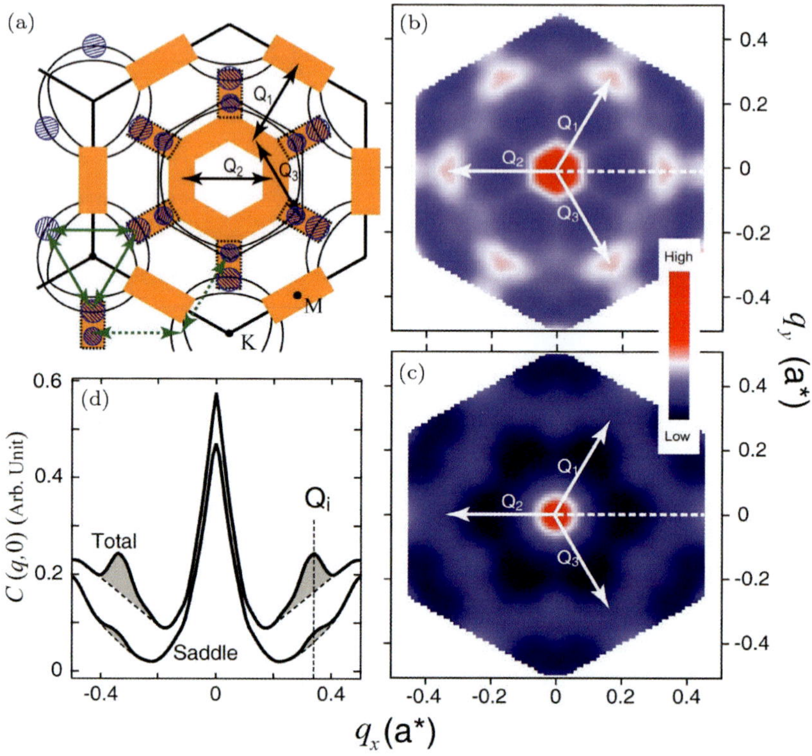

Fig. 1.9. (a) The CDW gap distribution over the entire BZ sketched from the spectral weight suppression map, which are marked by the orange broken-honeycomb-shaped structure. The gapped regions could be well connected by the CDW wavevectors, $\mathbf{Q}_i$s ($i = 1,2,3$), as indicated by the double-headed arrows. The gapped regions proposed by Kiss *et al.* and around the saddle band points are marked by the hashed circles and the dotted line surrounded rectangles. (b)–(c) Two dimensional joint DOS results on the honeycomb regions and the saddle band point regions, respectively. (d) Comparison of the corresponding joint DOS results along the dashed lines in panels (b) and (c). Reprinted with permission from [23].

of the total one. In other words, although the DOS is higher around the saddle band points, it only plays a secondary role in the CDW formation due to its relatively small momentum space. Moreover, the broken-honeycomb regions can be well connected by the CDW wavevector $\mathbf{Q}_i$'s [represented by the double-headed arrows in Fig. 1.9(a)]. In previous studies, since only a small portion of the weight-suppressed region was discovered (hashed circles), multiple scattering has to be invoked to barely make some of the regions connected by the two dashed arrows in Fig. 1.9(a) [22]. However, this multiple scattering assumption is not needed, once the large weight-suppressed region is recovered here.

In order to further assess the contribution to the CDW instability of the broken-honeycomb-shaped region, we calculated the weighted joint DOS in the similar way to 2H-Na_xTaS_2. Here, the matrix element is introduced assuming that only the states that directly couple (thus suppressed) with the softening phonon branch around $\mathbf{Q}_i$ [25], would contribute to the CDW instability. $C(\vec{q}, 0)$'s computed result on the honeycomb regions [Fig. 1.9(b)] and on the saddle band regions [Fig. 1.9(c)] are plotted in the same scale. Clearly, the peaks at $\mathbf{Q}_i$s are very prominent in Fig. 1.9(b), whereas they are not so pronounced in Fig. 1.9(c). This is further illustrated in Fig. 1.9(d), where the $C(q, 0)$'s along the **Γ-M** high symmetry direction are compared. The saddle band regions only contribute about 17% of the total $C(Q_i, 0)$ computed result over the whole broken-honeycomb region. Therefore, the nesting of states with low DOS at E_F indeed dominates the charge instability at $\mathbf{Q}_i$s in $NbSe_2$. This finding shows that collective contribution of the barely occupied states can play the primary role in CDW formation in some polaronic systems, and thus should not be overlooked.

1.3. The Three Dimensional Electronic Structure and Metal-Insulator Transition in $NiS_{2-x}Se_x$

As discussed in Section 1.1.3, when the escape depth of photoelectrons is small, the k_z is not well defined during the photoemission process. This is less relevant when the electronic structure is two-dimensional. However, for the materials with strongly three-dimensional electronic structure, poor k_z resolution would broaden the photoemission spectra significantly, making it hard to resolve the band structure. In this section, we will discuss the soft X-ray angle-resolved photoemission spectroscopy (SX-ARPES) study on the bulk electronic structure of $NiS_{2-x}Se_x$. We will first present the

3D electronic structure of $NiS_{2-x}Se_x$ observed by soft X-ray ARPES, and then discuss its doping dependence [35]. Our results provide a direct observation of bandwidth-control Mott transition (BCMT), and unveil its distinct characters in a multiband non-half-filled system.

1.3.1. *The three-dimensional electronic structure of $NiS_{2-x}Se_x$*

Mott transition is a metal-insulator transition (MIT) due to the presence of strong electron correlations. When the ratio between the onsite Coulomb repulsion U and the bandwidth W is larger than a certain critical value, the electron hopping between neighboring sites is prohibited [26]. The Mott transition could be realized by controlling either the filling or the bandwidth of the Hubbard bands [26]. In filling control, partially-filled Hubbard band is created by doping, as in the cuprate high temperature superconductors. In bandwidth control, W is tuned against U by applying physical or chemical pressure, as inferred indirectly from the transport properties of V_2O_3, $NiS_{2-x}Se_x$, etc [27, 28].

$NiS_{2-x}Se_x$ is believed to be a prototypical BCMT system, with the MIT at $x_c = 0.43$ [29–31]. However, the origin of its MIT is still under debates due to its non-half-filled multi-band nature. Previous ultraviolet ARPES (UV-ARPES) studies revealed a narrow band developing near the Fermi energy (E_F) in the metallic phase [32]. However, this narrow band persists in the insulating regime and its origin was attributed to possible carriering doping by Se/S vacancies or the surface effects [33, 34].

The photoemission data were taken with p-polarized light at the Advanced Resonant Spectroscopies (ADRESS) beam line of the Swiss Light Source (SLS) [36]. The Brillouin zone (BZ) corresponding to the face-centered-cubic (fcc) sublattice of Ni^{2+} in $NiS_{2-x}Se_x$ is shown in Fig. 1.10(a) [35]. Along the two high symmetric Γ-M-Z (pink) and Γ-X-Z (blue) planes, k_z dependent photoemission intensity distribution of $NiS_{0.33}Se_{1.67}$ at E_F are shown in Figs. 1.10(b) and (c), with the finite photon momentum corrected.

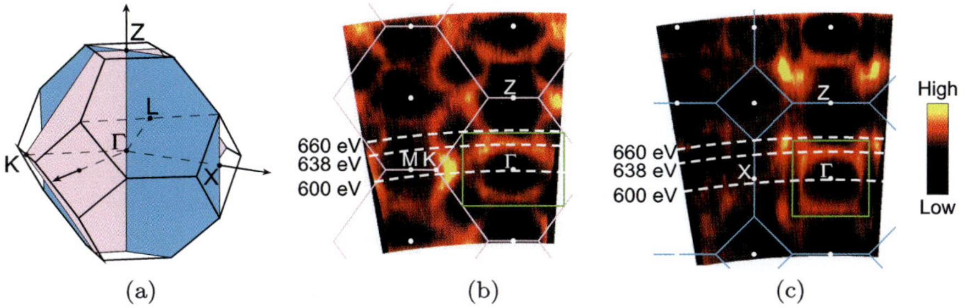

Fig. 1.10. (a) The Brillouin zone (BZ) for the face-centered-cubic sublattice of Ni^{2+} in $NiS_{2-x}Se_x$. (b), (c) The Fermi surface mapping in the Γ-M-Z plane, and the Γ-X-Z plane, respectively [35].

Such Fermi surface topology was resolved by ARPES for the first time. The periodicity of the Fermi surface could be clearly observed. The inner potential could then be determined to be 25 eV and the periodicity of the Fermi surface matches that of the BZ for the fcc sublattice very well. ARPES using ultraviolet light source is surface sensitive, since the escape depth of photoelectrons is below 10 Å if the photon energy is from 20 to 200 eV. Such surface confinement of the final states wave function makes the k_z ill-defined. However, if the photon energy is in the soft X-ray range, which is above 600 eV, the escape depth of photoelectron will increase significantly. Therefore, the soft X-ray ARPES provides much better k_z resolution and bulk sensitivity as its main advantages over the UV-ARPES.

Three cross-sections of the Fermi surface taken with different photon energies are shown in Fig. 1.11(a). The k_z dependence is notable. The Fermi surface crossings could be determined from the peak positions in the momentum distribution curves (MDC's) at the E_F [Figs. 1.11(b) and (c)]. Three sets of Fermi surface sheets could be observed in the Γ-X-M plane: 1. a hole pocket centered at Γ contributed by the band α, and a hole pocket centered at X noted as α' that is the counterpart of α in the 2nd simple cubic BZ; 2. an electron pocket contributed by the band β centered at M; 3. four small and weak electron pockets at the corner of simple cubic BZ, contributed by the band δ. The observed complex electronic structure contradicts

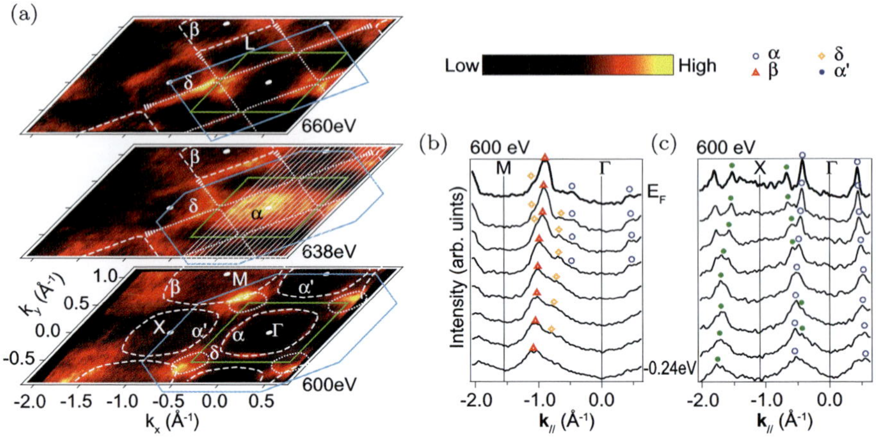

Fig. 1.11. (a) Fermi surface slices in the Γ-X-M plane ($h\nu = 600$ eV), the plane at the half BZ height (600 eV), and a intermediate plane (638 eV), respectively. (b), (c) Momentum distribution curves (MDC's) near E_F along Γ-M (b), and Γ-X (c), respectively [35].

the recent local-density approximation calculations, which give just one electron pocket around Γ in $NiSe_2$ [37].

1.3.2. *The metal-insulator transition in $NiS_{2-x}Se_x$*

In order to discuss the doping dependence of the electronic structure in $NiS_{2-x}Se_x$, we will focus our discussion on the data in the Γ-X-M plane, which represent the general electronic structure. The doping dependence of Fermi surface is shown in Figs. 1.12(b)–(f). In the metallic regime ($x > 0.5$), there is no detectable variation of Fermi surface topology. The Fermi crossings along Γ-X [Fig. 1.12(g)] and Γ-M [Fig. 1.12(h)] appear invariant. However, the intensity at E_F decreases gradually with decreasing doping, and the Fermi surfaces disappear in the insulating regime. The doping independent Fermi surface volume proves the isovalent nature of the S/Se substitution. Therefore, the decrease in carrier density near E_F close to the MIT is not due to the proposed Luttinger volume reduction [38, 39], but the diminishing spectral weight at E_F.

The spectral weight suppression at E_F and the invariance of Fermi surface topology suggest that the MIT in $NiS_{2-x}Se_x$ is controlled

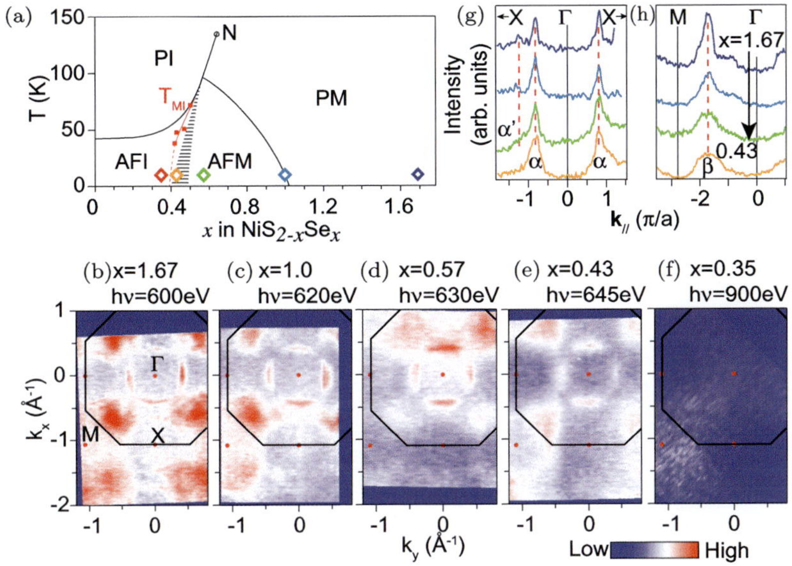

Fig. 1.12. (a) The phase diagram of $NiS_{2-x}Se_x$ reproduced from Ref. [31]. The measured dopings are marked by the diamonds. AFI, AFM, PI, and PM stand for antiferromagnetic insulator, antiferromagnetic metal, paramagnetic insulator, and paramagnetic metal, respectively. (b)–(f) The Fermi surface mapping for various dopings in the Γ-X-M plane. The intensities are integrated over $[E_F - 15$ meV, $E_F + 15$ meV$]$ and then normalized by the integrated intensity at -4 eV. (g), (h) The MDCs at E_F normalized by spectrum intensities for various dopings along Γ-X (g), and Γ-M (h), respectively. Reprinted with permission from [35].

by the bandwidth instead of band filling. In this case, the MIT is characterized by the disappearance of quasiparticle spectral weight and the divergence of effective mass. Figure 1.13 shows that the doping dependence of the photoemission intensity distribution along cut #3 in the first 0.4 eV below E_F. The spectra of the metallic samples are made of a large incoherent/dispersionless broad background and a weak dispersive feature, while that of the insulating sample is featureless.

To obtain band dispersion of the coherent feature, one needs to take the incoherent feature as a background and subtract it from the spectra. One could either take the broad spectrum at Γ (since one would not expect any feature near E_F there), or combine the momentum independent part at individual binding energies as

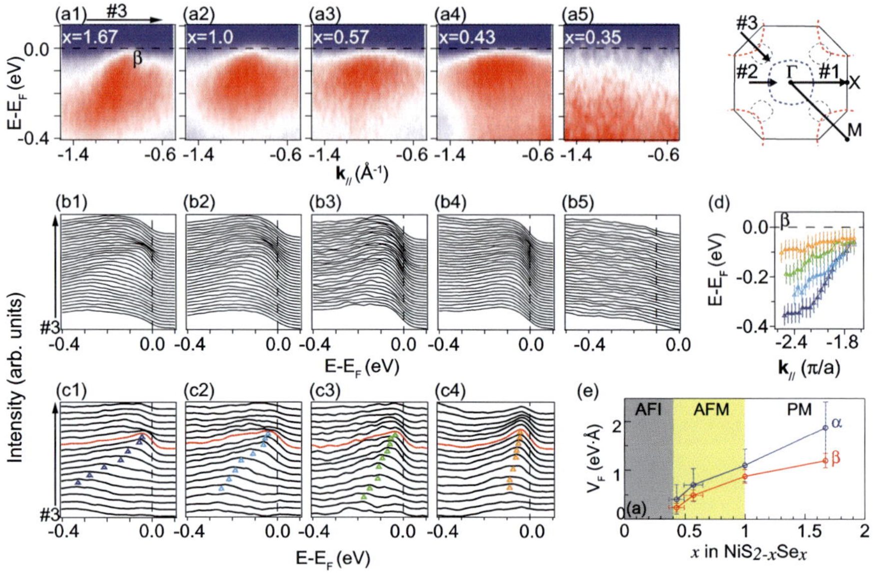

Fig. 1.13. (a1)–(a5) Intensity along cut #3 as indicated in the inset for doping $x = 1.67$, 1.0, 0.6, and 0.5, respectively. (b1)–(b5) Energy distribution curves (EDCs) corresponding to (a1)–(c5). (c1)–(c4) The EDCs of (a1)–(a4) after subtracting the incoherent background. (d) The doping dependence of the extracted dispersions. (e) The doping dependence of Fermi velocities (V_F). The data for the *alpha* band could be refer to [35].

the incoherent/dispersionless background. After subtraction, one can readily identify the Fermi velocities of the bands. The Fermi velocities of the quasiparticle dispersion is summarized in Fig. 1.13(e). They drop rapidly when approaching the Mott transition, and most likely diminish at the transition based on the trend presented here. Such a strong enhancement or even diverging quasiparticle mass near x_c is a direct demonstration of the BCMT picture in $NiS_{2-x}Se_x$.

1.3.3. *The incoherent spectral weight of $NiS_{2-x}Se_x$*

We note that the spectra of $NiS_{2-x}Se_x$ are dominated by strong incoherent feature near E_F which is not expected in a single band BCMT picture. The doping dependent of the incoherent feature is illustrated in Fig. 1.14(a). The dispersionless feature shifts towards

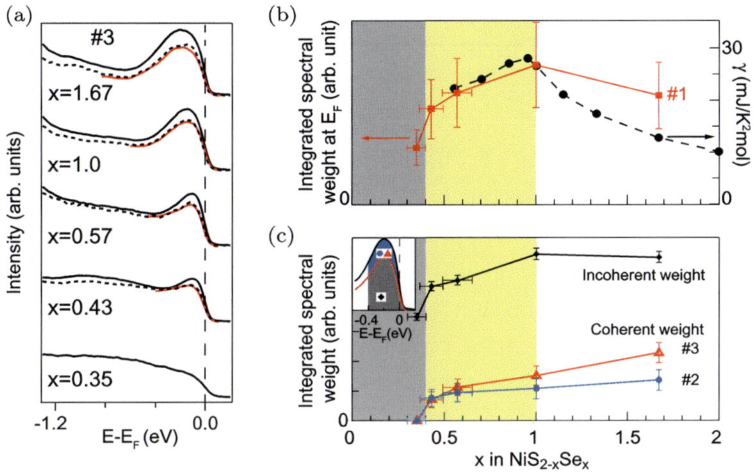

Fig. 1.14. (a) Integrated EDCs along cut #3 (black solid curves) plotted with EDCs at Γ (black dashed curves), and the incoherent/dispersionless background (red solid curves). (b) The doping dependence of spectral weight at E_F integrated over [E_F −40 meV, E_F +40 meV] along path #1 in Fig. 1.13, which resembles the specific heat linear coefficient γ [38]. (c) The doping dependence of the incoherent/dispersionless background (gray area illustrated in the inset) and the dispersive spectral weight along cuts #2 and #3 (blue area illustrated in the inset) over energy range [E_F −0.4 eV, E_F +0.1 eV] [35].

E_F with decreasing x, following the renormalization of the coherent bands, which implies that the incoherent weight could be some shake-off side bands from the coherent quasiparticles by strong correlations. However, the density of state at E_F does not exhibit any anomaly across the MIT, and it is still finite even for the insulator. As shown in Fig. 1.14(b), the integrated spectral weight at E_F shows similar doping dependence as the specific heat linear coefficient γ reproduced from [11]. Both quantities peak at $x = 1$ but smoothly evolve through the MIT. Such finite spectra weight at E_F in the insulating regime could originate from the incoherent part of the spectra. We plotted the doping dependence of the integrated spectra weight for the coherent and incoherent spectra separately in Fig. 1.14(c). The coherent part diminishes in the insulating regime, while the spectral weight of the incoherent part evolves smoothly through the MIT.

Intriguingly, the decreasing rate are accelerated near the MIT for both the coherent and incoherent weight.

It is clear that $NiS_{2-x}Se_x$ can be described neither by a simple band insulator picture, nor by the traditional single-band Hubbard model with the splitting of a half-filled band and a clean gap. Instead, the behavior of the coherent weight of the spectra could be described by the Mott physics in the picture of Brinkman and Rice [40], while the rest incoherent part gradually dominates the spectra weight near E_F due to rapidly increasing strong correlations. In this way, the system is converted from a bad metal to a bad insulator, as found by resistivity and optical measurements [41, 42]. Similar coherent-incoherent MIT characterized by finite but incoherent spectral weight at E_F in the insulating phase has been observed before in polaronic systems [43–45].

Our results directly demonstrate the electronic characters of the BCMT in $NiS_{2-x}Se_x$, highlighting the dominant role of the strong correlations. Beyond the simple BCMT picture, the observed incoherent/coherent spectra weight evolution suggests the complexity of the MIT in a non-half-filed multi-band system. The insulating phase is characterized by finite incoherent spectral weight at E_F without opening a charge gap. These results deepen our understanding of the MIT in strongly correlated materials. The high k_z resolution and bulk sensitivity of soft X-ray ARPES are critical for studying these 3D systems.

1.4. Multi-Orbital Nature of Iron-Based Superconductors

Iron-based superconductors belong to a new family of unconventional superconductors with the maximum superconducting transition temperature (T_C) up to 56 K in the bulk materials [46, 47]. It provides a new route to realize and understand the high-T_C superconductivity. Unlike the cuprates, where the Fermi surface is constructed just by the $d_{x^2-y^2}$ orbital [2], the iron-based superconductor exhibits multi-band and multi-orbital nature [48–52]. The band calculation shows

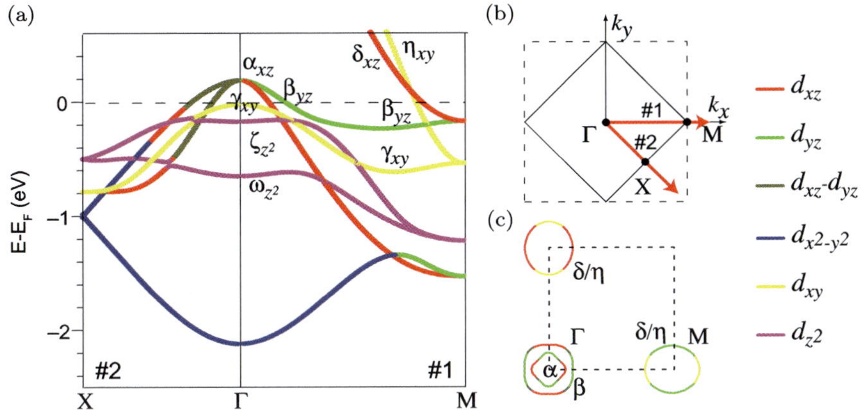

Fig. 1.15. (a) A typical orbital assignment of bands of iron-base superconductors as calculated in [52]. (b) Two-dimensional plot of the simplified Brillouin zone (solid line) and the unfolded Brillouin zone (dashed line). (c) The orbital distribution on the Fermi surface plotted in the unfold Brillouin zone.

that all five 3*d* orbitals of iron contribute to the low-lying electronic structure [Fig. 1.15(a)]. Determining the orbital character of the bands is thus very important. In this section, we will present the polarization dependent ARPES study on the band structure of iron-based superconductors. We will first show how to use the orbital sensitivity of ARPES to determine the orbital characters of electronic structure. We will then discuss the orbital anisotropic band reconstruction observed in the symmetry broken phases of iron-based superconductors.

1.4.1. *The orbital character of electronic structure*

As shown in Section 1.1.4, the orbital sensitivity in ARPES is a powerful tool to determine the orbital character of electronic structure. The Brillouin zone of iron-based superconductors shows four fold symmetry [Fig. 1.15(b)]. There are two high symmetry directions the Γ-M and Γ-X directions. Table 1.1 shows the possibility to detect 3*d* orbitals along these two high symmetry directions in *p* and *s* polarization geometry. For example, along the Γ-M direction, the d_{xz}, d_{z^2} and $d_{x^2-y^2}$ orbitals could be detected in *p*

Table 1.1. The possibility to detect 3d orbitals along two high symmetry directions in the p and s geometry by polarization-dependent APRES.

High-symmetry direction	Experimental geometry	3d orbitals				
		d_{xz}	$d_{x^2-y^2}$	d_{z^2}	d_{yz}	d_{xy}
#1 Γ(Z)-M(A)	p	√	√	√		
	s				√	√
#2 Γ(Z)-X(R)	p	√		√	√	√
	s	√	√		√	

polarization, while the d_{yz} and d_{xy} orbitals could be observed in s polarization.

Band calculation shows that the orbital character near E_F is dominated by three orbitals, the d_{xz}, d_{yz}, and d_{xy} [Fig. 1.15(c)]. They forms multiple disconnected Fermi surface sheets with hole pockets near the zone center and electron pockets near the zone corner. We will focus on the band structure near the zone center first. The polarization dependent photoemission data are shown in Fig. 1.16. Near E_F, three bands (α, β, and γ) could be identified (Fig. 1.16). α only shows up in the p geometry, while β only appears in the s geometry, indicating that the α and β bands show opposite spatial symmetries [Figs. 1.16(a)–(e)]. The band tops of both α and β degenerate at the Γ point. Such degeneracy is expected for the d_{xz} and d_{yz} bands due to the four-fold symmetry of the sample. Therefore, we attribute the α and β bands to be of the d_{xz} and d_{yz} orbitals, respectively in this direction. In addition, along the Γ-M direction, α disperses to the binding energy over 200 meV, and β disperses to about 150 meV and then bends over to lower binding energies toward the M point. Such dispersions are consistent with the prediction by theory for the α_{xz} and β_{yz} bands (Fig. 1.15(a)). Due to the four-fold symmetry, the d_{xz} component of α along the k_x axis has to become d_{yz} along the k_y-axis. Therefore, the equal mixing of these two orbitals along the Γ-X direction is expected. Strong polarization dependence could be observed for the α and β bands along Γ-X direction [Figs. 1.16(f)–(j)],

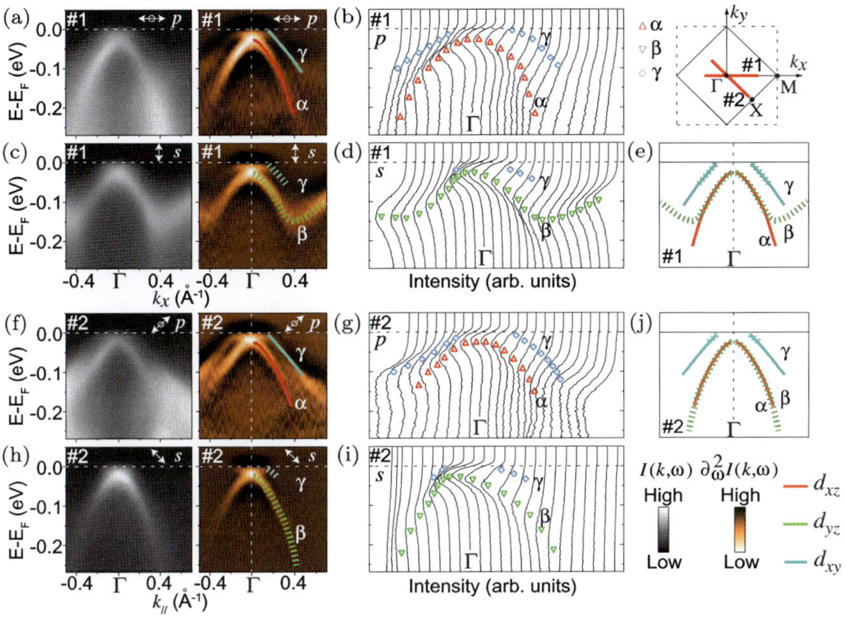

Fig. 1.16. (Color online) The polarization dependent APRES data around Γ taken with 47 eV photons. (a) $I(k,\omega)$ map and its second derivative with respect to energy $(\partial^2 I(k,\omega)/\partial\omega^2)$ taken with the p geometry along #1 direction as marked with a red thick solid line in the inset. (b) The EDCs for the data in panel a. (c) and (d) are the same as panels a and b, but taken with the s geometry. (e) The sketch of the band structure in panels a and c. (f)–(j) are the same as (a)–(e), but taken along cut #2 in the inset. The band dispersions are identified with the help of second derivative image and EDCs, and illustrated by the solid and dashed lines ploted on the second derivative image. The color scale of $I(k,\omega)$ and $\partial^2 I(k,\omega)/\partial\omega^2$ are shown in the inset. The solid and dashed sketches of the bands represent the even and odd components of the bands respectively. These color scales and notations are complied throughout the paper. Reprinted with permission from [3].

which reflects the hybridization of the d_{xz} and d_{yz} orbitals. For example, $\frac{d_{xz}+d_{yz}}{\sqrt{2}}$ and $\frac{d_{xz}-d_{yz}}{\sqrt{2}}$ are of even and odd spatial symmetry with respect to the Γ-X direction, respectively.

Besides the α and β bands, the γ band emerges in both s and p geometries along Γ-M direction indicating an orbital mixing. Along Γ-X direction, the γ band could be mostly observed in p geometry, which is consistent with the polarization dependence of the d_{xy}

orbital. We thus attribute the γ band to be the d_{xy} band, whose band top is below E_F in theoretical calculations.

Three hole-like bands could be observed near zone center, whose orbital character is consistent well with theoretical calculations. We will then turn to the electron pockets near the zone corner. Two electron pockets are predicted around the zone corner by theories. The outer one is constructed by the d_{xy} orbitals, and the inner one is constructed by the d_{xz} and d_{yz} orbitals. The polarization dependence of the spectral intensity at zone corner is more complicated than that around the zone center. Because one need to consider the symmetry of the Bloch wave function with finite momentum instead of only using the atomic wave function. The polarization dependent Fermi surface mappings are shown in Fig. 1.17. We could resolve two pockets, the δ and η pockets. The η pocket is square like and could be enhanced in p polarization, while the δ pocket is ellipse like and could be observed in s polarization. Based on the matrix element calculation, we attribute the orbital character of the η pocket to be the d_{xy} orbital and the δ pocket to be the d_{xz} and d_{yz} orbitals. The k_z dependence of the Fermi surface further supports the orbital assignment. The d_{xy} pocket is two-dimensional and its pocket size

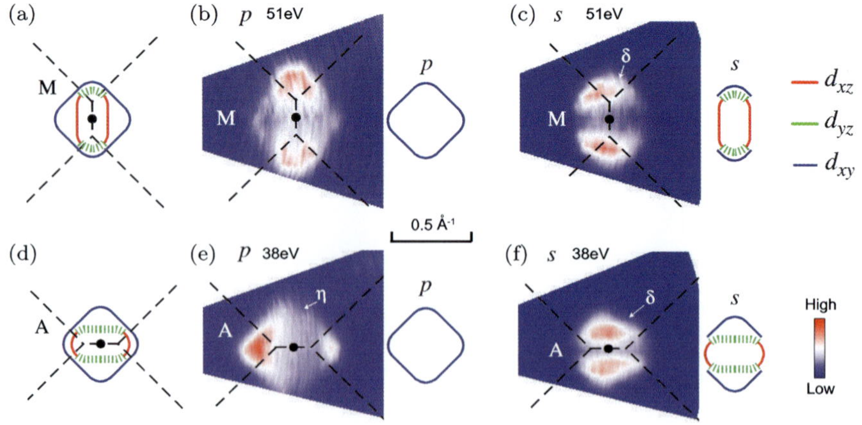

Fig. 1.17. (a) The Fermi surface and orbital character around the zone corner predicted by band calculations. (b), (c) The photoemission intensity maps taken with 51 eV photons around M in the p and s geometry respectively [3].

stays the same at the M and A points. In contrast to the d_{xy} orbital, the d_{xz} and d_{yz} orbitals is more three dimensional. As a result, the Fermi crossings of the δ pocket show strong k_z dependence, shrinking from M to A.

1.4.2. *The orbital anisotropic band reconstruction*

As shown above, polarization dependent ARPES is a powerful tool to detect the bands with different orbital characters selectively, which is important for the study of multi-orbital system. For iron-based superconductors, the system first breaks the C4 rotational symmetry through a tetragonal to orthorhombic structural transition and then form a collinear magnetic order at lower temperature [53, 54]. The understanding of these two transitions attracted intensive studies, since the driving force behind could be the effective interaction that mediates the superconducting pairing [50, 51, 55, 56].

We have studied the band structure evolution through the structural and magnetic transitions by polarization dependent ARPES [57]. The spectral intensity distribution taken above and below the transition temperatures are compared in Fig. 1.18. Two ellipse-like features observed in the p and s geometries originate from the d_{xz} and d_{yz} orbitals, respectively. They intertwine with each other forming the α and β hole pockets. The outmost feature is the γ pocket, whose intensity distribution is consistent with the matrix element distribution of the d_{xy} orbital. The electronic structure shows four-fold symmetry at 80 K, which is expected for the tetragonal phase above the structural transition temperature. However, when the sample is cooled down to 20 K, the symmetry breaking of electronic structure is obvious on the α and β hole pockets [Fig. 1.18(b)]. Both pockets extend along the AFM direction exhibiting C2 rotational symmetry. The d_{yz} portions is more anisotropic than the d_{xz} portions, indicating an orbital-dependent reconstruction of the electronic structure.

To further investigate the band reconstruction, we compared the photoemission spectral along k_x and k_y in Fig. 1.19. Remarkable band reconstruction was observed along k_x [Fig. 1.19(a)]. Near Γ, the β_x and

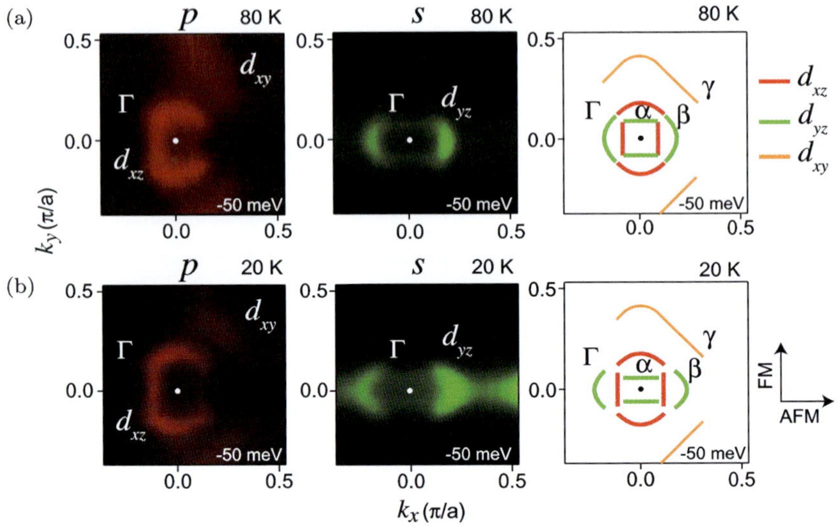

Fig. 1.18. (a) The left and middle panels are the photoemission intensity maps integrated over a 10 meV energy window at 50 meV below E_F taken at 80 K in the p and s geometries, respectively. The observed cross-sections of bands are summarized in the right panel. (b) Same as described for panel (a) but taken at 20 K. All the data were taken with 107-eV photons. Reprinted with permission from [57].

γ_x bands open a energy gap near E_F. Around the M point, the β_x band shifts up and its band top is pushed above E_F at 20 K. In contrast to the complex band reconstruction along k_x, the reconstruct of β band is negligible along the k_y direction [Fig. 1.19(b)]. The Fermi crossings of β_y are almost temperature independent. Around the M point, the β_y band shifts downward to about -50 meV at the lowest temperature. The different band behaviors along k_x and k_y direction further show that the electronic structure breaks the C4 rotational symmetry in the symmetry broken states. Furthermore, the β_y and β_x bands are constructed by the d_{xz} and d_{yz} orbitals respectively. The band reconstruction is thus strongly orbital dependent. The d_{yz} and d_{xy} bands reconstruct significantly while the d_{xz} bands were less affected.

So far, there are mainly two categories of theoretical proposals for understanding the origin of these transitions. One category of theories suggests that symmetry breaking could be induced by the orbital

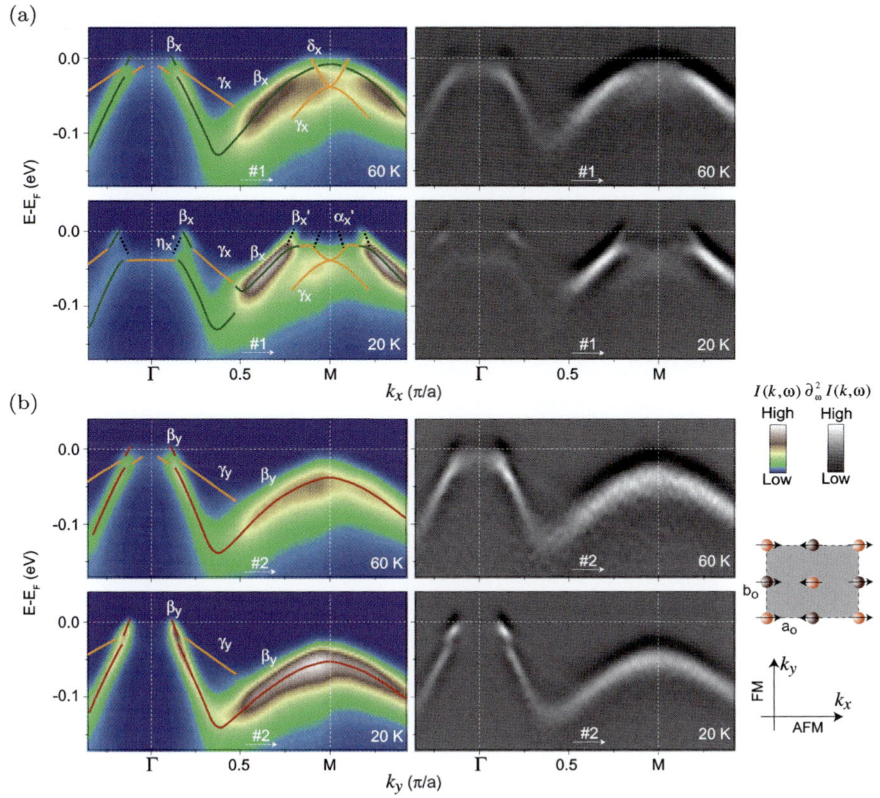

Fig. 1.19. (a) The temperature-dependent photoemission data along the AFM direction. The right panels are the second derivative images. (b) the the same as panel (a), but taken along the FM direction [57].

instability [58–60]. The structural transition is driven by the formation of ferro-orbital ordering between the d_{xz} and d_{yz} orbitals. The other category of theories suggests that the spin fluctuation would break the C4 rotational symmetry by forming a spin nematic state [61, 62]. Our results provide several important quantitative tests for these theoretical proposals. We checked the inequivalent occupation between the d_{xz} and d_{yz} orbitals, which is considered to be important for the formation of ferro-orbital ordering. The maximum change of occupation for the d_{xz} orbital is 0.04 electron per iron which is too small to drive the structural transition. The orbital dependent band reconstruction observed here suggests an important role of orbital degrees of freedom. Therefore, the strong spin fluctuation at high temperature and its coupling with the orbital-degree of freedom could be the driving force for these two transitions. The polarization dependent ARPES help us to study the bands with different orbital characters separately. The observed orbital-dependent band reconstruction provides solid experimental foundation for the microscopic understanding of the symmetry broken states in iron-based superconductors.

1.5. Summary

Utilizing the synchrotron radiation is one of the most important factors for the recent ARPES studies. The ultra-high beam resolution and small beam spot can help us to detect the detailed features of electronic structure, such as energy gap distributions. The photon energy can be tuned to the soft X-ray regime, which can be used to investigate the 3D materials. The variable polarization of the beam is critical for the study of multi-orbital system. With all these capability, synchrotron radiation makes ARPES an important tool to study the novel phenomena in condensed matter physics.

Acknowledgement

The authors gratefully acknowledge Prof. X. H. Chen, Prof. Minghu Fang, and Prof. H. H. Wen for providing high quality single crystals. The ARPES experiment support was provided by Dr. M. Arita, Prof. K. Shimada, Prof. H. Namatame, and Prof. M. Taniguchi at Hiroshima synchrotron radiation center (HiSOR), Dr. D. H. Lu, Dr. R. G. Moore, and Dr. M. Hashimoto at Stanford Synchrotron Radiation Lightsource (SSRL), Dr X. Y. Cui, Dr. M. Shi, Dr. V. N. Strocov, and Dr. M. Kobayashi at Swiss Light Source (SLS). The authors also thank the inspiring discussions with Prof. J. P. Hu, Prof. Chi-Cheng Lee, Prof. J. X. Li, Prof. T. Xiang, Prof. H. Q. Lin, Prof. Q. H. Wang, and Prof. J. Shi. This work is supported by the National Natural Science Foundation of China and Ministry of Science and Technology of China.

Bibliography

[1] Stefan Hüfner, *Photoelectron Spectroscopy*, 3nd edn. (Springer, New York, 2003), pp. 39–60.

[2] Andrea Damascelli, Zahid Hussain and Zhi-Xun Shen, *Rev. Mod. Phys.* **75**, 473–541, (2003).

[3] Y. Zhang, F. Chen, C. He, B. Zhou, B. P. Xie, C. Fang, W. F. Tsai, X. H. Chen, H. Hayashi, J. Jiang, H. Iwasawa, K. Shimada, H. Namatame, M. Taniguchi, J. P. Hu and D. L. Feng, Orbital characters of bands in the iron-based superconductor $BaFe_{1.85}Co_{0.15}As_2$, *Phys. Rev. B* **83**, 054510, (2011).

[4] V. Brouet, M. Fuglsang Jensen, Ping-Hui Lin, A. Taleb-Ibrahimi, P. Le Fvre, F. Bertran, Chia-Hui Lin, Wei Ku, D. Colson and A. Forget, Impact of the 2 Fe unit cell on the electronic structure measured by ARPES in iron pnictides, arXiv, 1205.4513 (2012).

[5] B. Dardel, D. Malterre, M. Grioni, P. Weibel and Y. Baer, Unusual photoemission spectral function of quasi-one-dimensional metals, *Phys. Rev. Lett.* **67**, 3144, (1991).

[6] B. Dardel, D. Malterre, M. Grioni, P. Weibel, Y. Baer, M. Schlenker and Y. Petroff, Temperature dependence of the spectral function through the peierls transition in quasi-one-dimensional compounds, *Europhys. Lett.* **19**, 525, (1992).

[7] K. M. Shen, F. Ronning, D. H. Lu, F. Baumberger, N. J. C. Ingle, W. S. Lee, W. Meevasana, Y. Kohsaka, M. Azuma, M. Takano, H. Takagi and Z.-X. Shen, Nodal quasiparticles and antinodal charge ordering in $Ca_{2-x}Na_xCuO_2Cl_2$, *Science* **307**, 901 (2005).

[8] T. Nakagawa, G. I. Boishin1, H. Fujioka, H. W. Yeom, I. Matsuda, N. Takagi, M. Nishijima and T. Aruga, Fermi surface nesting and structural transition on a metal surface: In /Cu(001), *Phys. Rev. Lett.* **86**, 854, (2001).

[9] J. A. Wilson, F. J. Di Salvo and S. Mahajan, Charge-density waves and superlattices in the metallic layered transition metal dichalcogenides, *Adv. Phys.* **24**, 117, (1975).

[10] D. W. Shen, B. P. Xie, J. F. Zhao, L. X. Yang, L. Fang, J. Shi, R. H. He, D. H. Lu, H. H. Wen and D. L. Feng, Novel mechanism of a charge density wave in a transition metal dichalcogenide, *Phys. Rev. Lett.* **99**, 216404, (2007).

[11] U. Chatterjee, M. Shi, A. Kaminski, A. Kanigel, H. M. Fretwell, K. Terashima, T. Takahashi, S. Rosenkranz, Z. Z. Li, H. Raffy, A. Santander-Syro, K. Kadowaki, M. R. Norman, M. Randeria and J. C. Campuzano, Nondispersive Fermi arcs and the absence of charge ordering in the pseudogap phase of $Bi_2Sr_2CaCu_2O_{8+\delta}$, *Phys. Rev. Lett.* **96**, 107006 (2006).

[12] K. McElroy, G.-H. Gweon, S. Y. Zhou, J. Graf, S. Uchida, H. Eisaki, H. Takagi, T. Sasagawa, D.-H. Lee and A. Lanzara, Elastic scattering susceptibility of the high temperature superconductor $Bi_2Sr_2CaCu_2O_{8+\delta}$: A comparison between real and momentum space photoemission spectroscopies, *Phys. Rev. Lett.* **96**, 067005, (2006).

[13] As documented in http://apps.isiknowledge.com/, there have been nearly 800 papers published solely on $NbSe_2$ so far.
[14] E. Revolinski, B. E. Brown, D. Beerntsen and C. H. Armitage, Selenide and Telluride Systems of Niobium and Tantalum, *J. Less-Common Met.* **8**, 63, (1965).
[15] J. M. E. Harper, T. H. Geballe and F. J. DiSalvo, Heat capacity of 2H-$NbSe_2$ at the charge density wave transition, *Phys. Lett.* **54A**, 27, (1975).
[16] Th. Straub, Th. Finteis, R. Claessen, P. Steiner, S. Hüfner, P. Blaha, C. S. Oglesby and E. Bucher, Charge-density-wave mechanism in 2H-$NbSe_2$: Photoemission results, *Phys. Rev. Lett.* **82**, 4504, (1999).
[17] T. Yokoya, T. Kiss, A. Chainani, S. Shin, M. Nohara and H. Takagi, Fermi surface sheet-dependent superconductivity in 2H-$NbSe_2$, *Science* **294**, 2518, (2001).
[18] T. Valla, A. V. Fedorov, P. D. Johnson, P-A. Glans, C. McGuinness, K. E. Smith, E. Y. Andrei and H. Berger, Quasiparticle spectra, charge-density waves, superconductivity and electron-phonon coupling in 2H-$NbSe_2$, *Phys. Rev. Lett.* **92**, 086401, (2004).
[19] T. M. Rice and G. K. Scott, New Mechanism for a Charge-Density-Wave Instability, *Phys. Rev. Lett.* **35**, 120, (1975).
[20] K. Rossnagel, O. Seifarth, L. Kipp, M. Skibowski, D. Voß, P. Krüger, A. Mazur and J. Pollmann, Fermi surface of 2H-$NbSe_2$ and its implications on the charge-density-wave mechanism, *Phys. Rev. B* **64**, 235119, (2001).
[21] M. D. Johannes, I. I. Mazin and C. A. Howells, Fermi-surface nesting and the origin of the charge-density wave in $NbSe_2$, *Phys. Rev. B* **73**, 205102, (2006).
[22] T. Kiss, T. Yokoya, A. Chainani, S. Shin, T. Hanaguri, M. Nohara and H. Takagi, Charge-order-maximized momentum-dependent superconductivity, *Nature Physics* **3**, 720, (2007).
[23] D. W. Shen, Y. Zhang, L. X. Yang, J. Wei, H. W. Ou, J. K. Dong, B. P. Xie, C. He, J. F. Zhao, B. Zhou, M. Arita, K. Shimada, H. Namatame, M. Taniguchi, J. Shi and D. L. Feng, Primary role of the barely occupied states in the charge density wave formation of $NbSe_2$, *Phys. Rev. Lett.* **101**, 226406, (2008).
[24] M. R. Norman, H. Ding, M. Randeria, J. C. Campuzano, T. Yokoya, T. Takeuchi, T. Takahashi, T. Mochiku, K. Kadowaki, P. Guptasarma and D. G. Hinks, Destruction of the Fermi surface in underdoped high-Tc superconductors, *Nature (London)* **392**, 157, (1998).
[25] D. E. Moncton, J. D. Axe and F. J. Di Salvo, Neutron scattering study of the charge-density wave transitions in 2H-$TaSe_2$ and 2H-$NbSe_2$, *Phys. Rev. B* **16**, 801, (1977).
[26] M. Imada, A. Fujimori and Y. Tokura, Metal-insulator transitions, *Rev. Mod. Phys.* 70, 1039, (1998).
[27] W. Bao, C. Broholm, S. A. Carter, T. F. Rosenbaum, G. Aeppli, S. F. Trevino, P. Metcalf, J. M. Honig and J. Spalek, Incommensurate spin density wave in metallic $V_{2-y}O_3$, *Phys. Rev. Lett.* **71**, 766, (1993).

[28] J. M. Honig and J. Spalek, Electronic properties of $NiS_{2-x}Se_x$ single crystals: From magnetic Mott-Hubbard insulators to normal metals, *Chem. Mater.* **10**, 2910, (1998).
[29] J. A. Wilson and G. D. Pitt, Metal-insulator transition in NiS_2, *Philos. Mag.* **23**, 1297, (1971).
[30] S. R. Krishnakumar and D. D. Sarma, X-ray photoemission study of $NiS_{2-x}Se_x$ ($x = 0.0 - 1.2$), *Phys. Rev. B*, 68, 155110 (2003).
[31] M. Matsuura, H. Hiraka, K. Yamada and Y. Endoh, Magnetic phase diagram and metal-insulator transition of $NiS_{2-x}Se_x$, *J. Phys. Soc. Jpn.* **69**, 1503 (2000).
[32] A. Y. Matsuura, Z. X. Shen, D. S. Dessau, C. H. Park, T. Thio, J. W. Bennett and O. Jepsen, Electronic structure and the metal-insulator transition in $NiS_{2-x}Se_x$, *Phys. Rev. B* **53**, R7584 (1996).
[33] A. Y. Matsuura, H. Watanabe, C. Kim, S. Doniach, Z.-X. Shen, T. Thio and J. W. Bennett, Metal-insulator transition in $NiS_{2-x}Se_x$ and the local impurity self-consistent approximation model, *Phys. Rev. B* **58**, 3690 (1998).
[34] D. D. Sarma, S. R. Krishnakumar, E. Weschke, C. Schüßler-Langeheine, Chandan Mazumdar, L. Kilian, G. Kaindl, K. Mamiya, S.-I. Fujimori, A. Fujimori and T. Miyadai, Metal-insulator crossover behavior at the surface of NiS_2, *Phys. Rev. B*, **67**, 155112 (2003).
[35] H. C. Xu, Y. Zhang, M. Xu, R. Peng, V. N. Strocov, M. Shi, M. Kobayashi, B. P. Xie and D. L. Feng, Direct observation of the bandwidth control Mott transition in the $NiS_{2-x}Se_x$ multiband system, arXiv. 1307.5941.
[36] V. N. Strocov, T. Schmitt, U. Flechsig, T. Schmidt, A. Imhof, Q. Chen, J. Raabe, R. Betemps, D. Zimoch, J. Krempasky, X. Wang, M. Grioni, A. Piazzalunga and L. Patthey, High-resolution soft X-ray beamline ADRESS at the Swiss Light Source for resonant inelastic X-ray scattering and angle-resolved photoelectron spectroscopies, *J. Synchrotron Rad.*, **17**, 631 (2010).
[37] J. Kunes, L. Baldassarre, B. Sch́achner *et al.*, Metal-insulator transition in $NiS_{2-x}Se_x$, *Phys. Rev. B*, **81**, 035122 (2010).
[38] S. Miyasaka, H. Takagi, Y. Sekine, H. Takahashi, N. Mori and R. J. Cava, Metal-insulator transition and itinerant antiferromagnetism in $NiS_{2-x}Se_x$ pyrite, *J. Phys. Soc. Jpn.* **69**, 3166 (2000).
[39] K. Iwaya, Y. Kohsaka, S. Satow *et al.*, Evolution of local electronic states from a metal to a correlated insulator in a $NiS_{2-x}Se_x$ solid solution, *Phys. Rev. B*, **70**, 161103 (2004).
[40] W. F. Brinkman and T. M. Rice, Application of Gutzwiller's variational method to the metal-insulator transition, *Phys. Rev. B*, **2**, 4302 (1970).
[41] R. L. Kautz, M. S. Dresselhaus, D. Adler and A. Linz, Electrical and optical properties of NiS_2, *Phys. Rev. B*, **6**, 2078 (1972).
[42] X. Yao, J. M. Honig, T. Hogan, C. Kannewurf and J. Sparłk, Electrical properties of $NiS_{2-x}Se_x$ single crystals: From Mott insulator to paramagnetic metal, *Phys. Rev. B*, **54**, 17469 (1996).
[43] P. Kwizera, M. S. Dresselhaus and D. Adler, Electrical properties of $NiS_{2-x}Se_x$, *Phys. Rev. B*, **21**, 2328 (1980).

[44] T. Valla, P. D. Johnson, Z. Yusof, B. Wells, Q. Li, S. M. Loureiro, R. J. Cava, M. Mikami, Y. Mori, M. Yoshimura and T. Sasaki, Coherenceincoherence and dimensional crossover in layered strongly correlated metals, *Nature*, **417**, 627 (2002).

[45] Y. Zhang, F. Chen, C. He, L. Yang, B. P. Xie, Y. L. Xie, X. H. Chen, Minghu Fang, M. Arita, K. Shimada, H. Namatame, M. Taniguchi, J. P. Hu and D. L. Feng, Strong correlations and spin-density-wave phase induced by a massive spectral weight redistribution in α-$Fe_1.06Te$, *Phys. Rev. B*, **82**, 165113 (2010).

[46] Yoichi Kamihara, Takumi Watanabe, Masahiro Hirano and Hideo Hosono, Iron-based layered superconductor $La[O_{1-x}F_x]FeAs$ (x = 0.05-0.12) with Tc = 26 K, *J. Am. Chem. Soc.* **130**, 3296–3297, (2008).

[47] X. H. Chen, T. Wu, G. Wu, R. H. Liu, H. Chen and D. F. Fang, Superconductivity at 43 K in $SmFeAsO_{1-x}F_x$, *Nature* **453**, 761–762, (2008).

[48] Fengjie Ma and Zhong-Yi Lu, Iron-based layered compound LaFeAsO is an antiferromagnetic semimetal, *Phys. Rev. B* **78**, 033111, (2008).

[49] D. J. Singh and M. H. Du, Density Functional Study of $LaFeAsO_{1-x}F_x$: A low carrier density superconductor near itinerant magnetism, *Phys. Rev. Lett.* **100**, 237003, (2008).

[50] Kazuhiko Kuroki, Seiichiro Onari, Ryotaro Arita, Hidetomo Usui, Yukio Tanaka, Hiroshi Kontani and Hideo Aoki, Unconventional pairing originating from the disconnected fermi surfaces of superconducting $LaFeAsO_{1-x}F_x$, *Phys. Rev. Lett.* **101**, 087004, (2008).

[51] I. I. Mazin, D. J. Singh, M. D. Johannes and M. H. Du, Unconventional superconductivity with a sign reversal in the order parameter of $LaFeAsO_{1-x}F_x$, *Phys. Rev. Lett.* **101**, 057003, (2008).

[52] S. Graser, T. A. Maier, P. J. Hirschfeld and D. J. Scalapino, Near-degeneracy of several pairing channels in multiorbital models for the Fe pnictides, *New. J. Phys.* **11**, 025016, (2009).

[53] Clarina de la Cruz, Q. Huang, J. W. Lynn, Jiying Li, W. Ratcliff Ii, J. L. Zarestky, H. A. Mook, G. F. Chen, J. L. Luo, N. L. Wang and Pengcheng Dai, Magnetic order close to superconductivity in the iron-based layered $LaO_{1-x}F_xFeAs$ systems, *Nature*, **453**, 899–902, (2008).

[54] Q. Huang, Y. Qiu, Wei Bao, M. A. Green, J. W. Lynn, Y. C. Gasparovic, T. Wu, G. Wu and X. H. Chen, Neutron-diffraction measurements of magnetic order and a structural transition in the parent $BaFe_2As_2$ compound of FeAs-based high-temperature superconductors, *Phys. Rev. Lett.* **101**, 257003, (2008).

[55] Jiangping Hu and Hong Ding, Local antiferromagnetic exchange and collaborative Fermi surface as key ingredients of high temperature superconductors, *Scientific Report* **2**, 381, (2012).

[56] Hiroshi Kontani and Seiichiro Onari, Orbital-fluctuation-mediated superconductivity in iron pnictides: Analysis of the five-orbital Hubbard-Holstein model, *Phys. Rev. Lett.* **104**, 157001, (2010).

[57] Y. Zhang, C. He, Z. R. Ye, J. Jiang, F. Chen, M. Xu, Q. Q. Ge, B. P. Xie, J. Wei, M. Aeschlimann, X. Y. Cui, M. Shi, J. P. Hu and D. L. Feng, Symmetry breaking via orbital-dependent reconstruction of electronic structure in detwinned NaFeAs, *Phys. Rev. B* **85**, 085121, (2012).
[58] Weicheng Lv, Jiansheng Wu and Philip Phillips, Orbital ordering induces structural phase transition and the resistivity anomaly in iron pnictides, *Phys. Rev. B* **80**, 224506, (2009).
[59] C.-C. Lee, W.-G. Yin and W. Ku, Ferro-orbital order and strong magnetic anisotropy in the parent compounds of iron-pnictide superconductors, *Phys. Rev. Lett.* **103**, 267001, (2009).
[60] C.-C. Chen, J. Maciejko, A. P. Sorini, B. Moritz, R. R. P. Singh and T. P. Devereaux, Orbital order and spontaneous orthorhombicity in iron pnictides, *Phys. Rev. B* **82**, 100504(R), (2010).
[61] C. Fang, H. Yao, W.-F. Tsai, J. P. Hu and S. A. Kivelson, *Phys. Rev. B* **77**, 224509 (2008).
[62] R. M. Fernandes, A. V. Chubukov, J. Knolle, I. Eremin and J. Schmalian, Preemptive nematic order, pseudogap, and orbital order in the iron pnictides, *Phys. Rev. B* **85**, 024534, (2012).

Chapter 2

Synchrotron-Based VUV Photoionization Mass Spectrometry in Combustion Chemistry Research

Nils Hansen

Combustion Research Facility,
Sandia National Laboratories,
Livermore, CA 94551, USA

Bin Yang

Center for Combustion Energy,
Department of Thermal Engineering,
Tsinghua University,
Beijing 100084, P. R. China

Tina Kasper

Faculty of Engineering,
Institute for Combustion and Gasdynamics,
University of Duisburg-Essen,
47048 Duisburg, Germany

Abstract

Since more than a decade, synchrotron-based methods play a major role in combustion chemistry research. This chapter focuses on the use of synchrotron-generated vacuum-ultraviolet (VUV) radiation as an ionization source in mass spectrometric applications. The use of tunable VUV radiation allows for the detection, identification, and quantification of different isomeric structures, thus adding a new dimension to mass spectrometric experiments which has not been previously accessible. In particular, this chapter highlights flame-sampling and kinetic experiments that provided new detailed insights into combustion chemistry processes. Examples are provided that are concerned with soot precursor formation and the combustion of prototypical biofuels.

Keywords: Combustion; laminar flames; reaction mechanism; mass spectrometry; vacuum ultraviolet (VUV); isomers; photoionization efficiency curve; soot formation; biodiesel.

2.1. Introduction

Today's global community faces the challenge of sustainably supplying its growing demand for energy, which is vital to modern day living in developed and developing countries. While the search for renewable energy sources and for efficient energy storage devices will continue, combustion processes are likely to remain in their role as the major energy supplier in the immediate and near-term future [1, 2]. It is predicted that modern societies will not overcome their dependence on combustion any time soon, largely because an astonishing 80% of today's global energy demand is supplied through combustion of fossil and biomass-derived renewable fuels. Associated with this ubiquitous utilization of combustion are a number of societal and political concerns, i.e. future supply shortages, environmental degradation, and human health effects [3–8].

From a scientific standpoint, combustion can be thought of as an intricate web of interrelated chemical reactions and fluid mechanical phenomena that liberates heat. Many free-radical species and reactive intermediates are connected via so-called chain-branching, chain-propagation, and termination reactions [9, 10]. Even with a single fuel, thousands of reactions involving hundreds of chemical species take place; a "real fuel", like gasoline, contains hundreds of single-fuel components. For most of the applications, the details of the combustion chemistry can be ignored and other aspects of the combustion process, like fuel vaporization or mixing of the components, are at the center of attention. When Otto and Diesel invented their internal combustion engines in the nineteenth century, they only needed to know that burning fuel gave heat. Today, for purposes such as increased combustion efficiency and pollution control, the chemical details of the combustion processes are important to know. Understanding the formation of airborne pollutants, including particulate matter, polycyclic aromatic hydrocarbons, nitrogen oxides, and unburned volatile hydrocarbons, requires accurate knowledge of the identity and concentration of many molecular precursor species that occur at the part-per-million level or below in the combustion mixture [11, 12]. Indeed, combustion chemists meet their greatest challenge in trying to predict the small quantities of pollutants that are formed as by-products of combustion.

To improve our detailed knowledge of combustion chemistry, many different experimental sources exist, including data from shock tubes, flow reactors, jet-stirred reactors, and laboratory flames in configurations that avoid the complex flow fields of turbulent combustion. The experimental observables, which reflect the chemical combustion network, can then be used in subsequent steps to develop a step-by-step understanding for the entire complicated combustion process by comparing the data with computer simulations that try to capture the chemical details in plausible reaction mechanisms. To verify the predictive capabilities of such combustion chemistry models, it is not sufficient to just test the predictions against experimentally measured global properties like ignition delay times and flame speeds, but it is advantageous to have many reliable quantitative species profiles from within laboratory flames available as validation targets. In subsequent steps, these detailed chemical reaction mechanisms are then reduced in order to be applicable to practical situations, i.e. internal combustion engines, etc.

Over the last decade, synchrotron-based studies have played a major role in the development of predictive combustion chemistry models [12–15]. Specifically, valence photoionization has been proven to be a very powerful tool where the most weakly bound electron is removed from a molecule using a photon with 6–20 eV energy. When combined with a mass spectrometer to separate the positive ions created by the photoionization process, this technique provides many advantages, including universal detection of all species simultaneously, high sensitivity and dynamic range, good selectivity between isomers, and control of molecular fragmentation. By now, many flames have been studied using a technique called flame-sampling photoionization mass spectrometry [13, 14, 16], which will be explained in more detail below.

In this chapter, we will give an overview of this technique and its applications to studies of the molecular level details of combustion processes in widely used laminar model flames. The flames studied covered a wide range of fuels, i.e. from hydrocarbons and oxygenated species such as alcohols, ethers, and esters, to nitrogen-containing compounds and fuel blends. The work on hydrocarbon flames has been motivated by the facts that hydrocarbons are the most dominant components of practical petroleum-derived fuels and the fundamental interest in the formation of polycyclic aromatic hydrocarbons (PAHs), soot, and their precursors. It is believed that a better understanding of the basic chemistry of soot formation can help to address future regulations for emissions from internal combustion engines and environmental and health concerns. Furthermore, many nations are motivated to utilize biomass-derived fuels, such as bio-alcohols or biodiesel as replacements or alternatives to conventional hydrocarbon fuels, because of anticipated future shortages in fossil fuel supplies, environmental degradation, and tightened regulations for emissions from internal combustion engines. While novel processes emerge, that can generate a variety of new fuels from biomass resources, the combustion chemistry of these fuels needs to be examined carefully, for example with respect to their pollutant emission potentials.

It is beyond the scope of this chapter to review every flame-sampling mass spectrometry study in detail, and we therefore refer any interested reader to the original literature which is cited throughout the following paragraphs. Instead, we will highlight a few insights on important areas of combustion research, including soot formation and oxidation chemistry of oxygenated, non-petroleum-based fuels, for which synchrotron radiation and its properties have been essential.

2.2. Flame Structures

To understand fundamental details of combustion processes, laminar premixed flames have usually been studied. See Fig. 2.1(a) as an example. In this configuration, the flames are stabilized on a porous-plug burner at a reduced pressure (15–30 Torr) to widen the reaction zone where many of the intermediate species are formed and consumed. Under appropriate conditions, such setups support

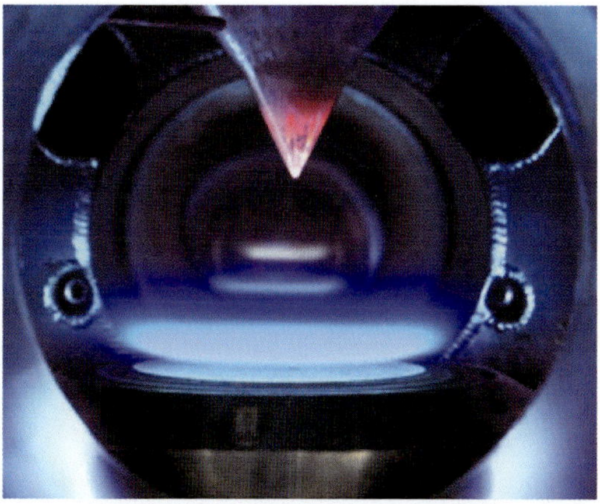

(a)

(b)

Fig. 2.1. Images of (a) a laminar premixed flame and (b) an opposed-flow diffusion flame both at a low pressure. Reprinted from [17], with permission from Elsevier.

one-dimensional flames in which the combustion process needs to be followed only in one dimension, i.e. the distance from the burner surface.

Although studying premixed low-pressure flat flames offers a lot of advantages, we need to realize that most practical combustion processes are non-premixed and that studying the chemical composition of non-premixed (diffusion) flames is of great interest as well. Consequently, non-premixed flames in the co-flow and opposed-flow configurations have been combined with molecular-beam mass spectrometry and VUV photoionization [18–20]. A photograph of an opposed-flow diffusion flame at low pressure is shown in Fig. 2.1(b). Such flames provide an interesting opportunity to examine both pyrolysis and oxidation chemistry under a wide range of conditions.

A crucial challenge in analyzing the chemical composition of such model flames lies in obtaining a representative sample from any given position within the flame followed by an accurate quantitative determination of the sampled gases. It is known that sampling from flames can cause several problems because of the steep gradients in the temperature and species' concentration profiles [21, 22]. For a meaningful interpretation of the underlying combustion chemistry in the targeted model flames, the disturbances through the sampling cone or microprobe must be kept to a minimum or care must be taken to quantify such influences [21–25]. If the aim of the study is to only identify combustion intermediates, flame disturbances through the sampling probes are likely to be not important.

In the next few paragraphs we describe how flame-sampling mass spectrometry with photoionization via synchrotron-generated vacuum-ultraviolet (VUV) radiation has been instrumental in achieving unprecedentedly detailed and accurate quantitative information about the chemical composition of the targeted model flames.

2.3. Vacuum Ultraviolet Photoionization and Isomer-Specificity

In synchrotron-based photoionization experiments, gas samples are withdrawn continuously from the targeted flame by a quartz cone or a microprobe, expanded into vacuum, and then quantitatively analyzed by time-of-flight mass spectrometry. The experiment is designed in a way that a complete mass spectrum can be taken at any point along the centerline of the flame. We employ synchrotron-generated vacuum ultraviolet (VUV) radiation only as a means to ionize the neutral atoms, radicals, and stable molecules that make up the complicated mix of species present in combustion processes [14, 26, 27], thus enabling separation of the species by their mass-to-charge (m/z) ratio.

The VUV region of the electromagnetic spectrum extends from about 180 nm (~6.8 eV) to about 30 nm (~41 eV) and this is the energy range needed to ionize all of the commonly expected combustion intermediates. With the development of modern synchrotron facilities, the difficulties of making an intense, high-duty cycle beam of VUV photon, whose energy is widely tunable but spectrally narrow, have been overcome. For example, the third-generation synchrotron at the Advanced Light Source (ALS) at the Lawrence Berkeley National Laboratory can generate VUV radiation in the range from 7.2 to 25 eV with a nominal flux of 10^{16} photons s^{-1} and a 2.5% bandwidth [28]. The spectral bandwidth can be improved to 10–50 meV with a 3-m off-plane Eagle monochromator at the expense of the photon flux (~10^{13}–10^{14} photons s^{-1}). Although the ALS radiation is pulsed with a 500 MHz repetition rate, the detection rate of the mass spectrometer is more than three orders of magnitude smaller, making the ALS-radiation appear to be quasi-continuous. Therefore, synchrotron-based flame experiments provide unprecedented data throughput.

These characteristics of the synchrotron-generated VUV photons combined with its tunability enable studies of the detailed combustion chemistry with high sensitivity and high selectivity when combined with mass spectrometry. Even structural isomers, which normally cannot be separated by mass spectrometry alone because of their identical m/z ratios, can be sorted out by virtue of their different photoionization efficiency (PIE) curves that generally have distinct ionization energies, shapes (as determined by the Franck–Condon overlap of the neutral with the cation state), and intensities (as determined by the absolute photoionization cross-section). Isomers usually have very different chemical behaviors and can react rather differently. As will be shown in this chapter, knowledge of concentrations of intermediates in their various isomeric forms is crucial to developing kinetic models for combustion.

An example on how synchrotron radiation is applied to separate isomeric contributions from a complex mixture is shown in Fig. 2.2. Representative data in the form of a flame-sampled time-of-flight mass spectrum is given in Fig. 2.2(a) and the peaks can easily be assigned to various common combustion intermediates. The

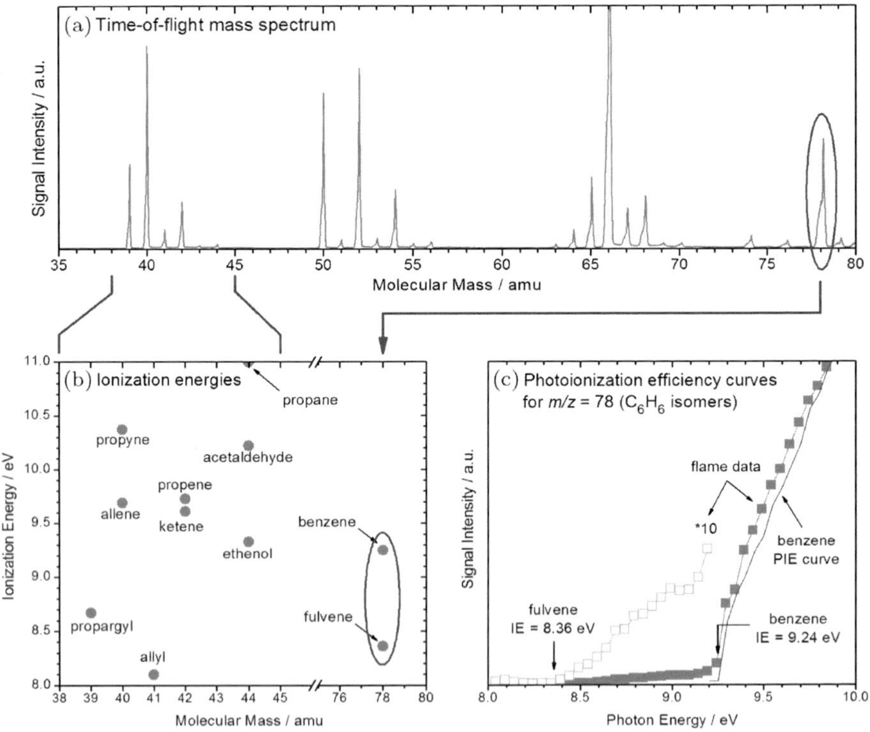

Fig. 2.2. (a) Flame-sampled time-of-flight mass spectrum between $m/z = 35$ and 80 from within a fuel-rich cyclopentene/O_2/Ar flame, (b) ionization energies of several conceivable combustion intermediates, and (c) flame-sampled photoionization efficiency (PIE) curve for $m/z = 78$ (C_6H_6 isomers) and its comparison to the cold-flow PIE curve of benzene.

ionization energies of the most likely intermediates are shown in Fig. 2.2(b) and it is obvious from this plot that an energy bandwidth of $\leq$50 meV of the ionizing photons is generally sufficient to separate the corresponding isomeric contributions. For example, by means of recording the so-called flame-sampled photoionization efficiency curve for $m/z = 78$ (C_6H_6), fulvene and benzene were found to contribute significantly [Fig. 2.2(c)]. Choosing a photon energy below the ionization energy of benzene allows for quantification of fulvene, while at photon energies above 9.24 eV, both fulvene and benzene contribute and the signal at $m/z = 78$ needs to be corrected for the known fulvene concentration to determine the benzene contributions.

Benefiting from the high energy resolution of the synchrotron-generated VUV photons, not only isomers can be separated, but also fragmentation interferences can largely be avoided by choosing the energy of the ionizing photons wisely. For example, a simple alkene like 1-hexene fragments significantly upon interactions with high energy photons (Fig. 2.3) [29]. However, fragmentation can be

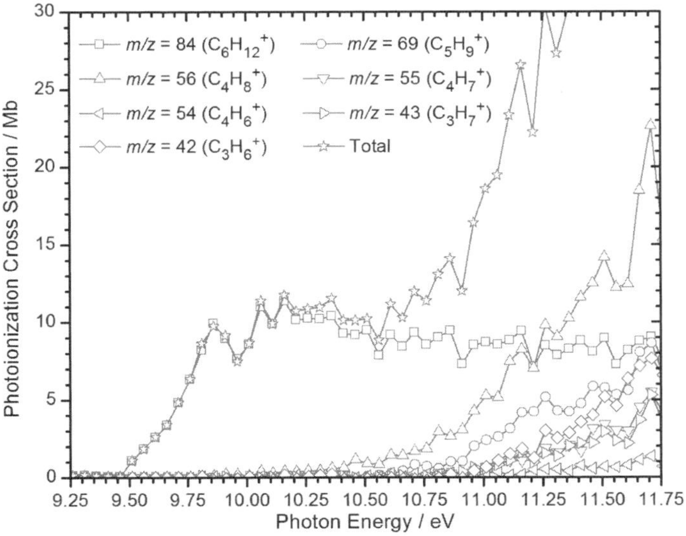

Fig. 2.3. Photoionization cross-sections for direct and dissociative photoionization of 1-hexene. Adapted from [29].

avoided when using photon energies below the appearance energies of the various dissociative ionization channels. The flexibility in tuning the energy of the synchrotron-generated photons together with the high energy resolution is beneficial for an accurate determination of a complex mixture composition as dissociative fragmentation can be prevented or quantitatively considered.

2.4. Combustion Chemistry

Using this VUV photoionization mass spectrometry technique, the exploration of isomer-selective combustion chemistry has only begun and a wealth of unprecedentedly detailed information must now be considered for the development of sophisticated combustion chemistry models. For example, this photoionization technique has been instrumental in the identification of many different isomeric species and has even enabled the discovery of a whole class of molecules, the enols, which has been absent from combustion chemistry models [31]. The photoionization efficiency curves for the $m/z = 58$ (C_3H_6O) and $m/z = 72$ (C_4H_8O) isomers in flames fueled by the butanol isomers are shown in Fig. 2.4 [30]. Ever since the discoveries of enol

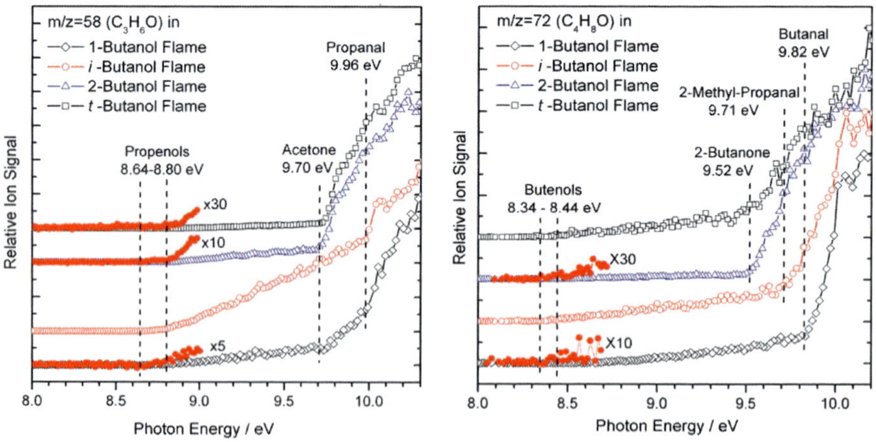

Fig. 2.4. Scaled PIE curves for $m/z = 58$ and 72 measured in the flames of the four butanol isomers. The known ionization energies of the various conceivable isomers are indicated. Reprinted from [30], with permission from Elsevier.

in flames, steps have been taken to consider their chemistry in combustion chemistry modeling [32]. It is being discussed [33, 34], that specifically in the combustion of alcohols, enols play a significant role and can be as abundant as their isomeric carbonyl compounds. Indeed, isomer-resolved studies have been crucial for developing models for alcohol combustion and significant progress has been made to predict the chemical structures of simple flames under the simplified conditions of premixed laminar flames [35–40].

By comparison, the chemistry and physics of soot formation processes are still relatively poorly understood. Current conventional wisdom has it that PAH's and soot are formed from a so-called "first aromatic ring", generally (but not always) benzene (C_6H_6) or phenyl (C_6H_5) [41, 42]. A significant amount of effort has been devoted to understanding how the first ring is formed because it then serves as a reactant for the formation of larger PAH's by-mechanisms that are currently still under investigations [43]. It is through providing species' identities and isomer-resolved mole fraction data from within model flames that synchrotron-based experiments have had a clear impact on our understanding of the early stages of soot formation.

From ALS-based experimental studies and in combination with modeling activities, it is now well accepted that under most combustion conditions, much of the phenyl/benzene is formed through the combination of two propargyl (C_3H_3) radicals [44]. Propargyl is a so-called resonantly stabilized radical and the unpaired electron is spread out over multiple sites, resulting in two resonant electronic structures. Resonance-stabilized radicals are energetically more stable than non-resonantly stabilized radicals and they can accumulate in higher concentrations in the hostile flame environments. For reasons which shall not be discussed here, the reactions of resonantly stabilized radicals are particularly relevant as cyclization steps. Other important resonantly stabilized radicals in combustion include C_3H_5 (allyl) and C_4H_5. The resonance structures of C_3H_3, C_3H_5, and C_4H_5 are shown in Fig. 2.5, including a total of three isomers for C_4H_5, the CH_2CHCCH_2 (so-called i-C_4H_5), CH_2CCCH_3 (so-called methyl-allenyl), and the CH_3CHCCH form. The n-isomer of C_4H_5

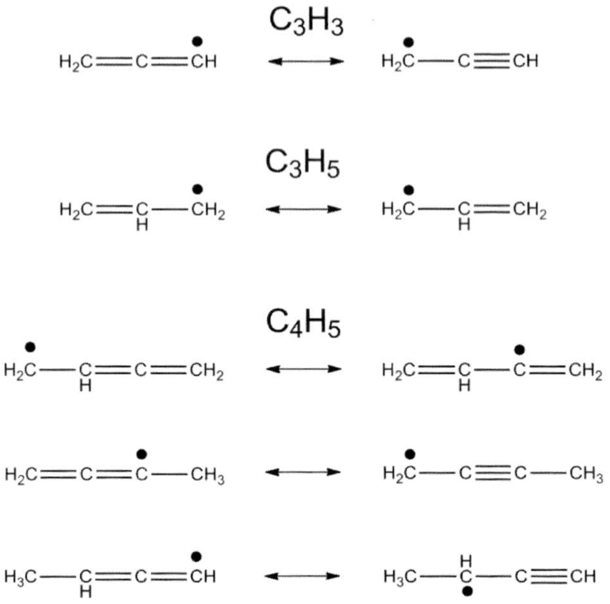

Fig. 2.5. Resonant structures of C_3H_3, C_3H_5, and C_4H_5 radicals. The bottom two isomers can be considered as methyl-substituted propargyl radicals.

($CH\text{-}CH\text{-}CH{=}CH_2$) is not resonance-stabilized and consequently is not shown.

Synchrotron-based studies have been essential for the identification and detection of the C_4H_5 isomers in flames and have helped to end a long-lasting debate over the significance of the n-C_4H_5+C_2H_2 *versus* the i-C_4H_5+C_2H_2 routes to benzene/fulvene formation [45]. The latter reaction actually forms dominantly fulvene+H [46], but in many combustion chemistry models the fulvene produced is subsequently converted to benzene by H-atom-assisted isomerization [47]. Based on their distinct ionization energies and on simulated photoionization efficiency curves, the resonantly stabilized C_4H_5 radicals, as shown in Fig. 2.5, were differentiated from the non-resonantly stabilized n-C_4H_5 radical. The corresponding flame-sampled photoionization efficiency curve is shown in Fig. 2.6 together with the simulated photoionization Franck–Condon envelopes for the three resonantly stabilized radicals. The observed photoionization

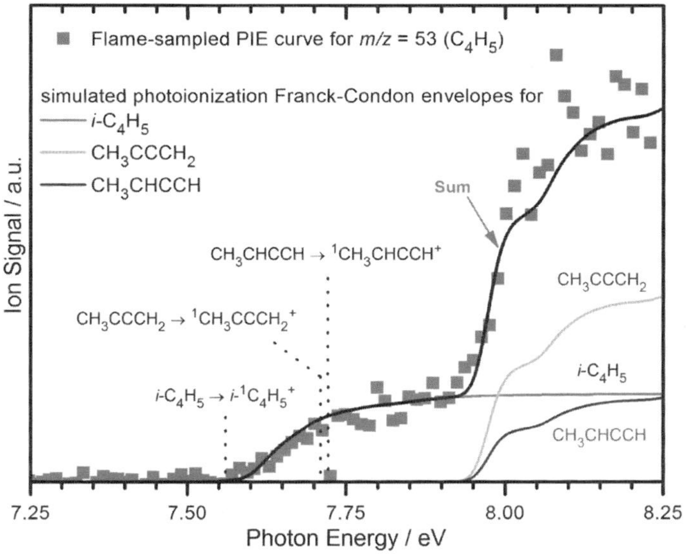

Fig. 2.6. Flame-sampled photoionization efficiency curve for m/z=53 (C_4H_5) and simulated photoionization Franck–Condon envelopes for three resonantly stabilized C_4H_5 radicals. Adapted from [45].

efficiency curve below 8 eV can be explained by the presence of the *i*-C_4H_5 isomer, while above 8 eV CH_3CCCH_2 and CH_3CHCCH can potentially contribute. The latter two isomers are hard to distinguish based on their ionization energy and simulated photoionization efficiency curves, and as explained in more detail by Hansen *et al.* [45], the presence of both these isomers is expected. The *n*-C_4H_5 radical was not found to be a likely candidate to explain the observed flame-sampled PIE curve for two reasons. First *n*-C_4H_5 is less likely to be present in high concentrations because of its thermodynamic instability compared to the resonantly stabilized C_4H_5 isomers, and secondly because of a poor Franck–Condon overlap between the neutral and the ion state.

Thanks to the theoretical work of Senosiain and Miller [46], it is now well established that the *n*-C_4H_5+C_2H_2 reaction is considerably faster than *i*-C_4H_5+C_2H_2, but based on the synchrotron-based identification of the isomeric forms, we now know that it is the latter reaction that plays the more important role. Indeed, for flames of 1,3-butadiene, in which i-C_4H_5 radicals are readily formed by H-abstraction reactions from the fuel, the modeling results indicated that the reaction of these radicals with C_2H_2 is of similar importance as the anticipated C_3H_3+C_3H_3 reaction [48]. The potential roles in cyclization reactions of the other identified C_4H_5 isomers require further investigations. It is worth pointing out that CH_3CCCH_2 and CH_3CHCCH can simply be seen as methyl-substituted propargyl radicals and that it is conceivable that they react with propargyl to form toluene or benzyl+H through a sequence of steps analogous to those by which the C_3H_3+C_3H_3 reaction forms benzene and phenyl+H. Similarly, they can react with themselves, leading ultimately to *o*-xylene. Other substituted propargyl radicals, which were detected in flames and identified through their distinct ionization energies and photoionization efficiency curves, are the *n*- and i-C_5H_3 isomers (HCCCHCCH and CH_2CCCCH) [49].

Once the identities of the intermediates are resolved, flame-sampling photoionization mass spectrometry allows for obtaining isomer-resolved mole fraction data. For example, in a study by Hansen *et al.* on the benzene formation routes in a 1-hexene flame, it was essential to separate the fulvene from the benzene contributions in order to validate the combustion chemistry model and to fully interpret the underlying chemistry [29]. The individual mole fraction profiles, determined as described above, are shown in Fig. 2.7 together with the modeling results. In order to correctly predict the experimentally observed levels of benzene in the flame of 1-hexene, it was necessary to include the appropriate fulvene chemistry in the combustion chemistry model. The ability to compare the modeling results to the actual fulvene mole fraction profile was the key component in the validation of the presented model. The key findings of this study were that in a flame of 1-hexene, benzene is formed dominantly via H-assisted isomerization of fulvene and that fulvene itself originates overwhelmingly from the C_3H_3+C_3H_5 reaction. The combination of two propargyl (C_3H_3) radicals was found to be only a small contributor to both the formation of fulvene and benzene. Furthermore, it was concluded that the importance of the C_3H_3+C_3H_5

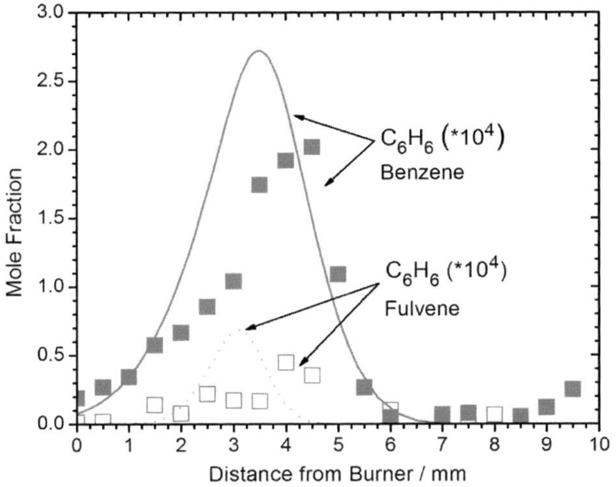

Fig. 2.7. Experimental (symbols) and modeled (lines) mole fraction profiles of fulvene and benzene in a fuel-rich 1-hexene/oxygen/argon flame. Adapted from [29].

pathway is a consequence of 1-hexene dominantly decomposing via unimolecular dissociation forming $C_3H_5 + n\text{-}C_3H_7$, thus providing a direct and vital route to C_3H_5. Similar conclusions were drawn by Nawdiyal *et al.* when modeling an extended set of 1-hexene flame data [50].

To summarize, synchrotron-based flame data in combination with modeling results have shed light on the chemistry of the early stages in soot formation. It is now accepted that besides the reaction of two C_3H_3 radicals, many other reactions can contribute and that their significance depends on how readily the precursors like allyl (C_3H_5), i-C_4H_5, etc. are formed through fuel decomposition [44, 51].

Isomer-resolved experiments at the ALS have also provided evidence that, besides unimolecular dissociation reactions, isomerization steps can be important fuel-consumption pathways as well. For example, Skeen *et al.* found experimental evidence for the unimolecular isomerization of methylcyclohexane (MCH) to heptenes and/or methyl-hexenes in the form of the flame-sampled photoionization efficiency curve at $m/z = 98$ (C_7H_{14}) as shown in Fig. 2.8 [52]. The signal below the ionization energy of MCH suggests the presence of

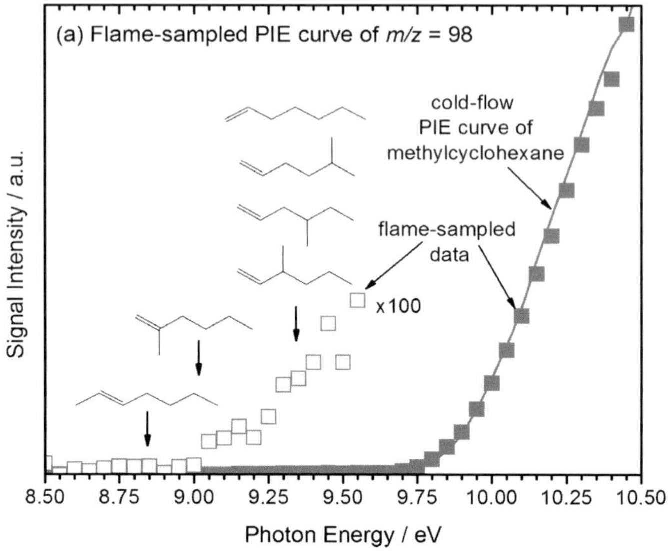

Fig. 2.8. Flame-sampled photoionization efficiency curve for $m/z = 98$ (C_7H_{14}) reveals evidence for the isomerization reaction of methylcyclohexane to heptenes and/or methyl-substituted hexene isomers.

the linear and branched C_7H_{14} isomers, which can readily be formed through stepwise or concerted ring-opening processes. The high-energy resolution of the ALS allowed for separation of the alkene-like isomers from the MCH and based on the observed species mole fractions and calculated potential energy surface, the authors concluded that as much as 15% of the fuel can be consumed through such an isomerization reaction.

Another kind of unimolecular reaction, a so-called complex fission during which multiple bonds break and form via 4-center and 6-centered transition states [Fig. 2.9(a) and (b)], has been identified to be important in ester flames [53–56]. For example, the flame sample PIE curve of $m/z = 56$ in methyl propanoate flame [Fig. 2.9(c)] suggests the presence of methyl ketene which may come from the 4-centered unimolecular dissociation of the fuel [Fig. 2.9(a)]. As illustrated in Fig. 2.9(b) and (d), acetic acid can be directly formed form the ethyl acetate through 6-centered molecular elimination reactions. No such pathway is feasible for methyl propanoate ($CH_3CH_2COOCH_3$).

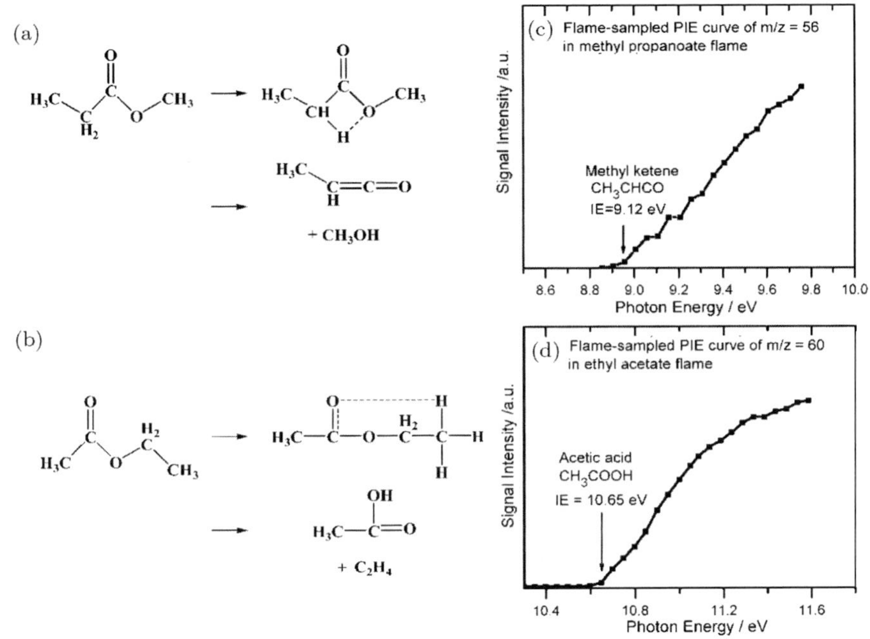

Fig. 2.9. Flame-sampled photoionization efficienty curves for $m/z = 56$ (methyl ketene) in methyl propanoate flame and for $m/z = 60$ (acetic acid) in ethyl acetate flame reveal evidence for the 4-centered and 6-centered unimolecular dissociation pathways in respective flame.

These fuel consumption pathways confirmed by synchrotron-based studies provide many of the clues to develop and refine the reaction sequences for the oxidation of these ester fuels. For example, the experimental mole fractions for acetic acid are helpful to determine the rates for related reactions.

Compared to the huge amounts of kinetic studies on hydrocarbon fuels, only few experimental or kinetic modeling studies have been carried out for alkyl ester molecules. This limited number of experimental data constrains the ability to develop accurate mechanisms that can successfully predict the ester's combustion properties including ignition delays, heat release rates, and amounts of pollutant emissions. Just in recent years, benefiting from the efficient output of synchrotron-derived data, the full intermediate pools for fuel-rich laminar premixed flames fueled by a family of methyl esters have been

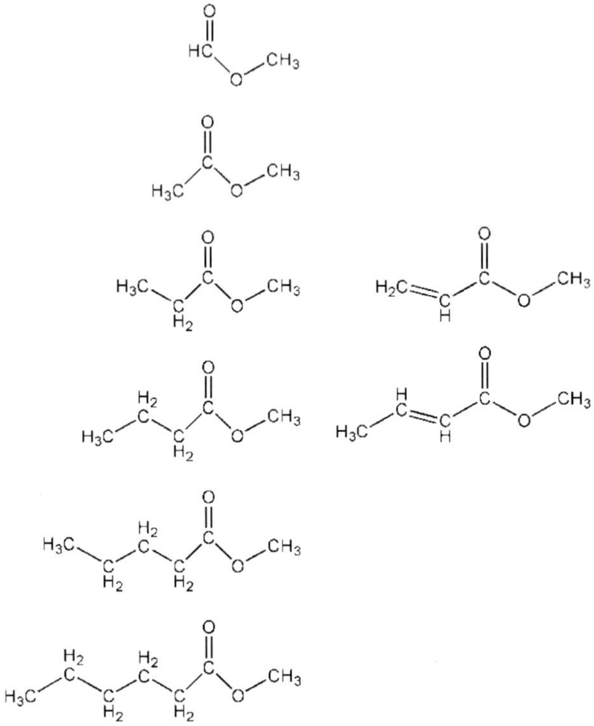

Fig. 2.10. A family of methyl esters have been studied through a hierarchical approach to develop a predictive combustion model for methyl esters.

mapped. This has played a major role in developments of predictive combustion chemistry models for methyl esters in a hierarchical manner. As shown in Fig. 2.10, the methyl esters from methyl formate ($m/z = 60$) to the isomers of $C_7H_{14}O_2$ ($m/z = 130$), as well as their unsaturated counterparts have been investigated [53–58], which help us to understand how the terminal methyl ester moiety present in biodiesel fuel molecules influences the laminar flame kinetics, such as lower soot production and early production of CO_2.

2.5. Other Combustion-Related Experiments Using Synchrotron Radiation

Following the pioneering work on laminar flames at the Advanced Light Source (Berkeley, USA) and the National Synchrotron

Radiation Laboratory (Hefei, China), more different combustion-relevant experiments have been developed to take advantage of synchrotron-based photoionization mass spectrometry with isomer-specificity. Recent investigations include jet-stirred and pyrolysis reactors [14, 59–63], flame-sampling aerosol mass spectrometry [64, 65], and kinetic studies in flow reactors and shock tubes [66–68]. A few highlights are presented in the following paragraphs. We should also mention – but not discussing in detail — small angle X-ray scattering (SAXS) experiments on combustion-generated soot particles in order to determine particle size distributions [69–71].

2.5.1. *Jet-stirred reactor*

Low-temperature oxidation chemistry of hydrocarbons plays an important role in ignition and in the initial reactions of the fuel in combustion processes. Jet-stirred reactors (JSR) have often been used to study the gas phase oxidation and the thermal decomposition of fuels in a less complex chemical environment than found in flames. A JSR is a continuously stirred tank reactor. Fresh, preheated gases are introduced into the reactor through typically four nozzles of turbulent jets ensuring good mixing of the gas phase and homogenous composition of the mixture in the reactor [72]. The conversion of reactants and the evolution of reaction products are studied as a function of reactor temperature, residence time, pressure and initial gas composition. Following sample extraction, the process gases can be analyzed with a variety of analytical techniques [72]. Coupling a jet-stirred reactor to a reflectron time-of-flight mass spectrometer with tunable synchrotron VUV photoionization using a molecular beam sampling system allows for particularly sensitive detection of reactive intermediates and has enabled the first detailed speciation study of the hydroperoxides likely to be involved in the gas-phase auto-ignition of hydrocarbons at low temperatures [59]. Further progress in understanding the low temperature oxidation chemistry and in particular the ignition chemistry of n-butane [73], propane [74], and n-heptane [75] was made using this technique. A similar system has been implemented at the Avanced Light Source in Berkeley and it has been successfully tested for studying the low temperature oxidation of dimethyl ether [76]. The tunable VUV synchrotron radiation was essential to provide experimental evidence for the anticipated Korcek decomposition channel of ketohydroperoxides.

2.5.2. *Pyrolysis reactors*

Pyrolysis reactions contribute to the fuel destruction in combustion processes and can be used in thermo-chemical conversion processes, e.g. for solid biomass conversion to lighter fuels [77, 78]. The fuel destruction and thermochemical processes in the absence of oxygen can be investigated in flow reactors.

For kinetic studies, the length and inner diameter of the tubular reactor and the gas flows are chosen so that a laminar plug flow results [79–82]. Under these conditions, the reaction progress depends linearly on the residence time in the reactor as well as initial mixture composition, temperature and pressure. Typically, residence times are in the order of milliseconds (ms), complementing the JSR experiments with residence times in the second (s) range. The pyrolysis reactor at the Hefei synchrotron consists of a flow tube made of inert aluminum oxide that is embedded in a high temperature furnace. The feed gases pass through a homogenously heated reaction zone in the flow tube and exit into an intermediate chamber at low or normal pressure. They are sampled as a molecular beam and analyzed by photoionization mass spectrometry [82]. Scans of the photon energy reveal the composition of the gas. Species profiles as a function of temperature reveal the decomposition temperature of the fuel and can be compared to simulations to gain insights into the decomposition pathways. Numerous fuels have been investigated — a summary can be found in [14] and references therein. The experimental data provide a validation base for chemistry models.

A similar flow reactor has been used to study the pyrolysis of biomass [83]. Solid biomass is introduced into the heated section of the flow reactor. Volatile pyrolysis products mix with a diluent gas flow and are transported to the exit of the reactor where they are sampled by a quartz probe. To improve sensitivity, the gases are ionized in the intermediate chamber and pass into a hybrid quadrupole-time-of-flight (QTOF) mass spectrometer offering higher mass resolution than the simple reflectron time-of-flight mass spectrometers used in other experiments. Mass spectra recorded as a function of photon energy, temperature and reaction time reveal the gas composition and sequence of pyrolysis product formation. Studies of different types of wood [60, 83] detected pyrolysis products from cellulose, hemicellulose, and lignin and could show that the earliest products are formed from lignin, followed by hemicellulose products, and then species from cellulose. These data provide the basis for the development of pyrolysis mechanisms that are needed for thermochemical conversion processes of biomass to desired products.

2.5.3. *Chemical kinetic photoionization mass spectrometer*

Combustion chemistry models require the input of kinetic data, e.g. the rate coefficients of the reactions that are considered in a certain mechanism. One instrument dedicated to the investigation of reaction kinetics was commissioned in 2008 at the ALS [66]. The previously described flame and reactor experiments all used stationary flow conditions and consequently the steady operation allowed averaging data at any position for the time required to obtain a good signal-to-noise (S/N) ratio. In experiments designed to follow the kinetics of elementary reactions, the concentration changes have to be measured as a function of time from the start of the reaction. Consequently, the kinetic experiments are not steady but have to be repeated in a reproducible manner until a required S/N ratio is reached. A good time-resolution is needed to follow the targeted elementary reactions.

The ALS-based experiment consists of a side-sampled quartz flow reactor, a flash photolysis system, and a photoionization chamber with a mass spectrometer. Reactants are diluted in a helium bath gas flowing through a heatable slow-flow quartz reactor and continuously sampled through an orifice in the side of the reactor to form a molecular beam. The molecular beam is skimmed and passes into a mass spectrometer. In the original set-up, a miniature double-focusing sector mass spectrometer of the Mattauch–Herzog geometry was used for detection and later upgraded to an orthogonal time-of-flight mass spectrometer. The gases are ionized with the quasi-continuous, tunable vacuum ultraviolet synchrotron radiation of the ALS. Photodissociation of suitable precursor molecules initiate radical reactions. Because the laser light is aligned parallel to the gas flow a uniform radical density is achieved along the reactor tube. Sequentially acquired mass spectra give the time evolution of the mixture composition for typically several hundred milliseconds with respect to the laser pulse as time zero. As described above, isomer-resolution is obtained by scanning the VUV-photon energy and comparing photoionization efficiency curves to calibration data.

The instrument has been used to directly measure illusive reaction intermediates. Most noted among the extensive work are the first experimental detection of Criegee intermediates important in the ozonoylsis of alkenes [84], their conformer dependent reactivity [85], and the controlled production of hydroperoxy alkyl radicals that potentially are important to autoignition processes [86, 87].

2.6. Extension of Detection Schemes: PEPICO Experiments at the SLS and SOLEIL

Recently, imaging photoelectron photoion coincidence spectrometry (iPEPICO) has been employed as an alternative detection technique to VUV photoionization mass spectrometry in flame experiments using the PEPICO instruments at the Swiss Light Source and SOLEIL [88–90]. The motivation of these efforts is to improve the isomer-resolving capabilities. In flame-sampling photoionization mass spectrometry, isomers are identified by their ionization energies and the form of the PIE curve as described above. Isomers can only be clearly identified if their ionization thresholds differ by more than the energy resolution of the ionizing photons and their PIE curves show distinctive non-overlapping features at good signal-to-noise. When several isomers with similar ionization properties are present at a certain mass-to-charge ratio, isomer separation becomes difficult [13, 24].

In PEPICO detection schemes, in addition to the mass spectra, a photoelectron spectrum is acquired for each mass. Close to threshold the photoelectrons produced during each ionization event have no extra kinetic energy so that they can be separated from electrons produced during other events. Mass-resolved photoelectron spectra show many vertical transitions which can be unambiguously assigned to a certain ionization event and provide a fingerprint for the identification of species. Initial studies illustrate the superior isomer identification abilities of PEPICO and demonstrate quantification strategies which result in high quality data sets for chemistry model validation [88].

2.7. Conclusion

Synchrotron radiation has been successfully applied to probe the details of the ongoing chemistry during combustion processes. Especially, synchrotron-based VUV has been used advantageously in combination with mass spectrometry to provide more insights into the early stages of soot formation chemistry in hydrocarbon flames and on how oxygenated, bio-derived fuels burn and produce potentially harmful byproducts. Because there exist so far no comparable analytical method, VUV single-photon ionization with mass spectrometry is not just used for the analysis of actual flames, but now also in combination with many different combustion applications, including flow reactors, pyrolysis reactors, and shock tubes. For all experiments, similar requirements apply to the detection of intermediates: high sensitivity, specificity, dynamic range and the capability of simultaneous detection of several chemically distinct species in a complex chemical gas mixture. Consequently, photoionization mass spectrometry is the tool of choice, that when combined with ionization by synchrotron radiation, allows the nearly fragmentation-free ionization of reactive and fragile intermediates with isomer-resolution.

Acknowledgments

We are thankful to our close collaborators Katharina Kohse-Höinghaus, Philipp R. Westmoreland, Fei Qi and the late Terrill

A. Cool. We also acknowledge our students and postdocs who have significantly contributed to the work and for their help with the data acquisition. The experiments at the Advanced Light Source (ALS) have profited from the expert technical assistance of Paul Fugazzi. The ALS is supported by the Director, Office of Science, Office of Basic Energy Sciences, of the U.S. Department of Energy under Contract No. DE-AC02-05CH11231. Sandia National Laboratories is a multimission laboratory managed and operated by National Technology and Engineering Solutions of Sandia, LLC., a wholly owned subsidiary of Honeywell International, Inc., for the U.S. Department of Energys National Nuclear Security Administration under contract DE-NA0003525. NH and BY have also been supported by the U.S. Department of Energy, Office of Basic Energy Sciences under the Energy Frontier Research Center for Combustion Science (Grant No. DE-SC0001198). BY is thankful for support through the "National 1000 Young Talents Program" of China. TK gratefully acknowledges financial support through the Ministry of Innovation, Science and Research of the state of North Rhine-Westfalia, Germany.

Bibliography

[1] International Energy Agency, World Energy Outlook 2012 Factsheet, http://www.worldenergyoutlook.org/media/weowebsite/2012/factsheets.pdf (04/26/2013).
[2] U.S. Energy Information Administration, International Energy Outlook 2011, http://www.eia.gov/forecasts/ieo/pdf/0484%282011%29.pdf (04/26/2013).
[3] C. J. Campbell and J. H. Laherrère, *Sci. Am.* **278**(3), 78–83, (1998).
[4] D. J. Wuebbles and A. K. Jain, *Fuel Process. Technol.* **71**(1–3), 99–119, (2001).
[5] M. D. Avakian, B. Dellinger, H. Fiedler, B. Gullet, C. Koshland, S. Marklund, G. Oberdorster, S. Safe, A. Sarofim, K. R. Smith, D. Schwartz and W. A. Suk, *Environ. Health Perspect.* **110**(11), 1155–1162, (2002).
[6] B. Nowack and T. D. Bucheli, *Environ. Pollut.* **150**(1), 5–22, (2007).
[7] J. M. Samet, S. L. Zeger, F. Dominici, F. Curriero, I. Coursac, D. W. Dockery, J. Schwartz and A. Zanobetti, in Samet, M. Jonathan, Scott L. Zeger, Francesca Dominici, Frank Curriero, Ivan Coursac, Douglas W. Dockery, Joel Schwartz and Antonella Zanobetti, *National Morbidity, Mortality, and Air Pollution Study, Part II: Morbidity and Mortality from Air Pollution*

in the United States 2000-01-01 edn. (Health Effects Institute, Boston, MA, USA, 2000).

[8] S. Solomon, D. Qin, M. Manning, Z. Chen, M. Marquis, K. B. Averyt, M. Tignor and H. L. Miller (Ed.), *Contribution of Working Group I to the Fourth Assessment Report of the Intergovernmental Panel on Climate Change* (Cambridge University Press, Cambridge, United Kingdom and New York, NY, USA, 2007).

[9] W. C. Gardiner Jr. (Ed.), *Gas-phase Combustion Chemistry* (Springer-Verlag, New York, Berlin, Heidelberg, 2000).

[10] I. Glassman, *Combustion* (Academic Press, New York, San Francisco, London, 1977).

[11] C. S. McEnally, L. D. Pfefferle, B. Atakan and K. Kohse-Höinghaus, *Progr. Energy Combust. Sci.* **32**(3), 247–294, (2006).

[12] K. Kohse-Höinghaus, P. Oswald, T. A. Cool, T. Kasper, N. Hansen, F. Qi, C. K. Westbrook and P. R. Westmoreland, *Angew. Chem. Int. Ed.* **49**(21), 3572–3597, (2010).

[13] N. Hansen, T. A. Cool, P. R. Westmoreland and K. Kohse-Höinghaus, *Progr. Energy Combust. Sci.* **35**(2), 168–191, (2009).

[14] F. Qi, *Proc. Combust. Inst.* **34**(1), 33–63, (2013).

[15] C. A. Taatjes, N. Hansen, D. L. Osborn, K. Kohse-Höinghaus, T. A. Cool and P. R. Westmoreland, *Phys. Chem. Chem. Phys.* **10**(1), 20–34, (2008).

[16] N. Hansen, S. A. Skeen, H. A. Michelsen, K. R. Wilson and K. Kohse-Höinghaus, *J. Vis. Exp.* **87**, e51369, (2014). doi:10.3791/51369 (2014).

[17] M. D. Smooke, *Proc. Combust. Inst.* **34**(1), 65–98, (2013).

[18] Y. Li and F. Qi, *Acc. Chem. Res.* **43**(1), 68–78, (2010).

[19] S. A. Skeen, B. Yang, H. A. Michelsen, J. A. Miller, A. Violi and N. Hansen, *Proc. Combust. Inst.* **34**, 1067–1075, (2013).

[20] A. Cuoci, A. Frassoldati, T. Faravelli, H. F. Jin, Y. Z. Wang, K. W. Zhang, P. Glarborg and F. Qi, *Proc. Combust. Inst.* **34**, 1811–1818, (2013).

[21] J. C. Biordi, *Progr. Energy Combust. Sci.* **3**(3), 151–173, (1977).

[22] J. C. Biordi, C. P. Lazzara and J. F. Papp, *Combust. Flame*, **23**(1), 73–82, (1974).

[23] U. Struckmeier, P. Oswald, T. Kasper, L. Böhling, M. Heusing, M. Köhler, A. Brockhinke and K. Kohse-Höinghaus, *Z. Phys. Chem.* **223**(4–5), 503–537, (2009).

[24] F. N. Egolfopoulos, N. Hansen, Y. Ju, K. Kohse-Höinghaus, C. K. Law and F. Qi, *Progr. Energy Combust. Sci.* **43**(1), 36–67, (2014).

[25] V. Gururajan, F. N. Egolfopoulos and K. Kohse-Höinghaus, *Proc. Combust. Inst.* **35**, 821–829, (2015).

[26] T. A. Cool, A. McIlroy, F. Qi, P. R. Westmoreland, L. Poisson, D. S. Peterka and M. Ahmed, *Rev. Sci. Instrum.* **76**(9), 094102, (2005).

[27] F. Qi, R. Yang, B. Yang, C. Huang, L. Wei, J. Wang, L. Sheng and Y. Zhang, *Rev. Sci. Instrum.* **77**(8), 084101, (2006).

[28] P. A. Heimann, M. Koike, C. W. Hsu, D. Blank, X. M. Yang, A. G. Suits, Y. T. Lee, M. Evans, C. Y. Ng, C. Flaim and H. A. Padmore, *Rev. Sci. Instrum.* **68**(5), 1945–1951, (1997).

[29] N. Hansen, W. Li, M. E. Law, T. Kasper, P. R. Westmoreland, B. Yang, T. A. Cool and A. Lucassen, *Phys. Chem. Chem. Phys.* **12**(38), 12112–12122, (2010).

[30] B. Yang, P. Oswald, Y. Y. Li, J. Wang, L. X. Wei, Z. Y. Tian, F. Qi and K. Kohse-Höinghaus, *Combust. Flame* **148**(4), 198–209, (2007).

[31] C. A. Taatjes, N. Hansen, A. McIlroy, J. A. Miller, J. P. Senosiain, S. J. Klippenstein, F. Qi, L. S. Sheng, Y. W. Zhang, T. A. Cool, J. Wang, P. R. Westmoreland, M. E. Law, T. Kasper and K. Kohse-Höinghaus, *Science* **308**(5730), 1887–1889, (2005).

[32] C. A. Taatjes, N. Hansen, J. A. Miller, T. A. Cool, J. Wang, P. R. Westmoreland, M. E. Law, T. Kasper and K. Kohse-Höinghaus, *J. Phys. Chem. A* **110**(9), 3254–3260, (2006).

[33] G. Black, H. J. Curran, S. Pichon, J. M. Simmie and V. Zhukov, *Combust. Flame* **157**(2), 363–373, (2010).

[34] J. M. Simmie and H. J. Curran, *J. Phys. Chem. A* **113**(27), 7834–7845, (2009).

[35] A. Frassoldati, R. Grana, T. Faravelli, E. Ranzi, P. Oswald and K. Kohse-Höinghaus, *Combust. Flame* **159**(7), 2295–2311, (2012).

[36] N. Hansen, M. R. Harper and W. H. Green, *Phys. Chem. Chem. Phys.* **13**, 20262–20274, (2011).

[37] S. M. Sarathy, M. J. Thomson, C. Togbe, P. Dagaut, F. Halter and C. Mounaim-Rousselle, *Combust. Flame* **156**(4), 852–864, (2009).

[38] S. M. Sarathy, S. Vranckx, K. Yasunaga, M. Mehl, P. Oswald, W. K. Metcalfe, C. K. Westbrook, W. J. Pitz, K. Kohse-Höinghaus, R. X. Fernandes and H. J. Curran, *Combust. Flame* **159**, 2028–2055, (2012).

[39] N. Hansen, S. S. Merchant, M. R. Harper and W. H. Green, *Combust. Flame* **160**(11), 2343–2351, (2013).

[40] S. M. Sarathy, P. Oswald, N. Hansen and K. Kohse-Höinghaus, *Progr. Energy Combust. Sci.* **44**, 40–102, (2014).

[41] J. A. Miller, M. J. Pilling and J. Troe, *Proc. Combust. Inst.* **30**, 43–88, (2005).

[42] H. Richter and J. B. Howard, *Progr. Energy Combust. Sci.* **26**(4–6), 565–608, (2000).

[43] H. Wang, *Proc. Combust. Inst.* **33**, 41–67, (2011).

[44] N. Hansen, J. A. Miller, S. J. Klippenstein, P. R. Westmoreland and K. Kohse-Höinghaus, *Combust. Expl. Shock Waves* **48**(5), 508–515, (2012).

[45] N. Hansen, S. J. Klippenstein, C. A. Taatjes, J. A. Miller, J. Wang, T. A. Cool, B. Yang, R. Yang, L. X. Wei, C. Q. Huang, F. Qi, M. E. Law and P. R. Westmoreland, *J. Phys Chem. A* **110**(10), 3670–3678, (2006).

[46] J. P. Senosiain and J. A. Miller, *J. Phys. Chem. A* **111**(19), 3740–3747, (2007).

[47] A. W. Jasper and N. Hansen, *Proc. Combust. Inst.* **34**, 279–287, (2013).

[48] N. Hansen, J. A. Miller, T. Kasper, K. Kohse-Höinghaus, P. R. Westmoreland, J. Wang and T. A. Cool, *Proc. Combust. Inst.* **32**, 623–630, (2009).

[49] N. Hansen, S. J. Klippenstein, J. A. Miller, J. Wang, T. A. Cool, M. E. Law, P. R. Westmoreland, T. Kasper and K. Kohse-Höinghaus, *J. Phys Chem. A* **110**(13), 4376–4388, (2006).
[50] A. Nawdiyal, N. Hansen, T. Zeuch, L. Seidel and F. Mauß, *Proc. Combust. Inst.* **35**, 325–332, (2015).
[51] N. Hansen, T. Kasper, B. Yang, T. A. Cool, W. Li, P. R. Westmoreland, P. Oswald and K. Kohse-Höinghaus, *Proc. Combust. Inst.* **33**, 585–592, (2011).
[52] S. A. Skeen, B. Yang, A. W. Jasper, W. J. Pitz and N. Hansen, *Energy Fuels* **25**(12), 5611–5625, (2011).
[53] B. Yang, C. K. Westbrook, T. A. Cool, N. Hansen and K. Kohse-Höinghaus, *Phys. Chem. Chem. Phys.* **13**(15), 6901–6913, (2011).
[54] B. Yang, C. K. Westbrook, T. A. Cool, N. Hansen and K. Kohse-Höinghaus, *Z. Phys. Chem.* **225**(11–12), 1293–1314, (2011).
[55] B. Yang, C. K. Westbrook, T. A. Cool, N. Hansen and K. Kohse-Hoeinghaus, *Proc. Combust. Inst.* **34**, 443–451, (2013).
[56] O. P. Korobeinichev, I. E. Gerasimov, D. A. Knyazkov, A. G. Shmakov, T. A. Bolshova, N. Hansen, C. K. Westbrook, G. Dayma and B. Yang, *Z. Phys. Chem.* **229**(5), 759–780, (2015).
[57] S. Dooley, F. L. Dryer, B. Yang, J. Wang, T. A. Cool, T. Kasper and N. Hansen, *Combust. Flame* **158**(4), 732–741, (2011).
[58] C. K. Westbrook, W. J. Pitz, P. R. Westmoreland, F. L. Dryer, M. Chaos, P. Oswald, K. Kohse-Höinghaus, T. A. Cool, J. Wang, B. Yang, N. Hansen and T. Kasper, *Proc. Combust. Inst.* **32** 221–228, (2009).
[59] F. Battin-Leclerc, O. Herbinet, P. A. Glaude, R. Fournet, Z. Y. Zhou, L. L. Deng, H. J. Guo, M. F. Xie and F. Qi, *Angew. Chem. Int. Ed.* **49**(18), 3169–3172, (2010).
[60] J. J. Weng, L. Y. Jia, S. B. Sun, Y. Wang, X. F. Tang, Z. Y. Zhou and F. Qi, *Anal. Bioanal. Chem.* **405**(22), 7097–7105, (2013).
[61] K. N. Urness, Q. Guan, A. Golan, J. W. Daily, M. R. Nimlos, J. F. Stanton, M. Ahmed and G. B. Ellison, *J. Chem. Phys.* **139**(12), (2013).
[62] G. T. Buckingham, T. K. Ormond, J. P. Porterfield, P. Hemberger, O. Kostko, M. Ahmed, D. J. Robichaud, M. R. Nimlos, J. W. Daily and G. B. Ellison, *J. Chem. Phys.* **142**(4), (2015).
[63] T. K. Ormond, A. M. Scheer, M. R. Nimlos, D. J. Robichaud, T. P. Troy, M. Ahmed, J. W. Daily, T. L. Nguyen, J. F. Stanton and G. B. Ellison, *J. Phys. Chem. A* **119**(28), 7222–7234, (2015).
[64] S. A. Skeen, H. A. Michelsen, K. R. Wilson, D. M. Popolan, A. Violi and N. Hansen, *J. Aerosol Sci.* **58**, 86–102, (2013).
[65] K. O. Johansson, J. Y. W. Lai, S. A. Skeen, D. M. Popolan-Vaida, K. R. Wilson, N. Hansen, A. Violi and H. A. Michelsen, *Proc. Combust. Inst.* **35**, 1819–1826, (2015).
[66] D. L. Osborn, P. Zou, H. Johnsen, C. C. Hayden, C. A. Taatjes, V. D. Knyazev, S. W. North, D. S. Peterka, M. Ahmed and S. R. Leone, *Rev. Sci. Instrum.* **79**(10), (2008).
[67] R. S. Tranter and P. T. Lynch, *Rev. Sci. Instrum.* **84**(9), (2013).

[68] P. T. Lynch, T. P. Troy, M. Ahmed and R. S. Tranter, *Anal. Chem.* **87**(4), 2345–2352, (2015).
[69] S. di Stasio, J. L. LeGarrec and J. B. A. Mitchell, *Energy Fuels* **25**(3), 916–925, (2011).
[70] J. B. A. Mitchell, S. di Stasio, J. L. LeGarrec and A. I. Florescu-Mitchell, *Nucl. Instrum. Meth. B* **268**(9), 1486–1491, (2010).
[71] J. B. A. Mitchell, J. L. LeGarrec, G. Saidani, F. Lefeuvre and S. di Stasio, *Energy Fuels* **27**(8), 4891–4898, (2013).
[72] O. Herbinet and G. Dayma, in *Cleaner Combustion: Developing Detailed Chemical Kinetic Models*, eds. F. Battin-Leclerc and J. M. Simmie and E. Blurock (Springer-Verlag: London, 2013), pp. 183–210.
[73] O. Herbinet, F. Battin-Leclerc, S. Bax, H. Le Gall, P. A. Glaude, R. Fournet, Z. Y. Zhou, L. L. Deng, H. J. Guo, M. F. Xie and F. Qi, *Phys. Chem. Chem. Phys.* **13**(1), 296–308, (2011).
[74] M. Cord, B. Husson, J. C. Lizardo Huerta, O. Herbinet, P. A. Glaude, R. Fournet, B. Sirjean, F. Battin-Leclerc, M. Ruiz-Lopez, Z. D. Wang, M. F. Xie, Z. J. Cheng and F. Qi, *J. Phys. Chem. A* **116**(50), 12214–12228, (2012).
[75] O. Herbinet, B. Husson, Z. Serinyel, M. Cord, V. Warth, R. Fournet, P. A. Glaude, B. Sirjean, F. Battin-Leclerc, Z. D. Wang, M. F. Xie, Z. J. Cheng and F. Qi, *Combust. Flame* **159**(12), 3455–3471, (2012).
[76] K. Moshammer, A. W. Jasper, D. M. Popolan-Vaida, A. Lucassen, P. Dievart, H. Selim, A. J. Eskola, C. A. Taatjes, S. R. Leone, S. M. Sarathy, Y. G. Ju, P. Dagaut, K. Kohse-Höinghaus and N. Hansen, *J. Phys. Chem. A* **119**(28), 7361–7374, (2015).
[77] D. Neves, H. Thunman, A. Matos, L. Tarelho and A. Gomez-Barea, *Progr. Energy Combust. Sci.* **37**(5), 611–630, (2011).
[78] D. Vamvuka, *Int. J. Energy Res.* **35**(10), 835–862, (2011).
[79] M. Cathonnet, J. C. Boettner and H. James, *Proc. Combust. Inst.* **18** 903–913, (1981).
[80] P. Glarborg, P. G. Kristensen, S. H. Jensen and K. Damjohansen, *Combust. Flame* **98**(3), 241–258, (1994).
[81] P. G. Kristensen, P. Glarborg and K. DamJohansen, *Combust. Flame* **107**(3), 211–222, (1996).
[82] L. Zhao, M. F. Xie, L. L. Ye, Z. J. Cheng, J. H. Cai, Y. Y. Li, F. Qi and L. D. Zhang, *Combust. Flame* **160**(10), 1958–1966, (2013).
[83] J. J. Weng, L. Y. Jia, Y. Wang, S. B. Sun, X. F. Tang, Z. Y. Zhou, K. Kohse-Höinghaus and F. Qi, *Proc. Combust. Inst.* **34**, 2347–2354, (2013).
[84] O. Welz, J. D. Savee, D. L. Osborn, S. S. Vasu, C. J. Percival, D. E. Shallcross and C. A. Taatjes, *Science* **335**(6065), 204–207, (2012).
[85] C. A. Taatjes, O. Welz, A. J. Eskola, J. D. Savee, A. M. Scheer, D. E. Shallcross, B. Rotavera, E. P. F. Lee, J. M. Dyke, D. K. W. Mok, D. L. Osborn and C. J. Percival, *Science* **340**(6129), 177–180, (2013).
[86] J. Zador, H. F. Huang, O. Welz, J. Zetterberg, D. L. Osborn and C. A. Taatjes, *Phys. Chem. Chem. Phys.* **15**(26), 10753–10760, (2013).

[87] J. D. Savee, E. Papajak, B. Rotavera, H. F. Huang, A. J. Eskola, O. Welz, L. Sheps, C. A. Taatjes, J. Zador and D. L. Osborn, *Science* **347**(6222), 643–646, (2015).

[88] P. Oswald, P. Hemberger, T. Bierkandt, E. Akyildiz, M. Köhler, A. Bodi, T. Gerber and T. Kasper, *Rev. Sci. Instrum.* **85**, 025101, (2014).

[89] J. Krüger, G. A. Garcia, D. Felsmann, K. Moshammer, A. Lackner, A. Brockhinke, L. Nahon and K. Kohse-Höinghaus, *Phys. Chem. Chem. Phys.* **16**(41), 22791–22804, (2014).

[90] D. Felsmann, K. Moshammer, J. Kruger, A. Lackner, A. Brockhinke, T. Kasper, T. Bierkandt, E. Akyildiz, N. Hansen, A. Lucassen, P. Oswald, M. Köhler, G. A. Garcia, L. Nahon, P. Hemberger, A. Bodi, T. Gerber and K. Kohse-Höinghaus, *Proc. Combust. Inst.* **35**, 779–786, (2015).

Chapter 3

Developments in Synchrotron X-ray Diffraction

Qiyun Xie and Xiaoshan Wu

School of Physics, Nanjing University, Nanjing 210093, China

Abstract

This chapter gives first a short introduction on the basic properties of X-ray diffraction as well as current synchrotron light sources. It then goes into the details of recent developments on three popular X-ray diffraction techniques based on synchrotron radiation source, namely powder diffraction, surface X-ray diffraction and 3D reciprocal space mapping. A further outlook on the new coming source of X-ray free electron laser is also given.

Keywords: Powder diffraction; Surface X-ray diffraction; Reciprocal space mapping (RSM); X-ray free-electron lasers.

3.1. Introduction

One century ago, the important pioneering works by scientists Max von Laue, William Henry Bragg and his son etc. have disclosed a fact that the atoms in the crystals can diffract X-rays [1]. The collected diffraction pattern could be used to back-calculate the structure information of crystals. Since then, this non-destructive method with high resolution, sensitivity to the inside of materials, and insensitivity to physical properties, has become the leading tool for studying the structure of matter at atomic and molecular level. More and

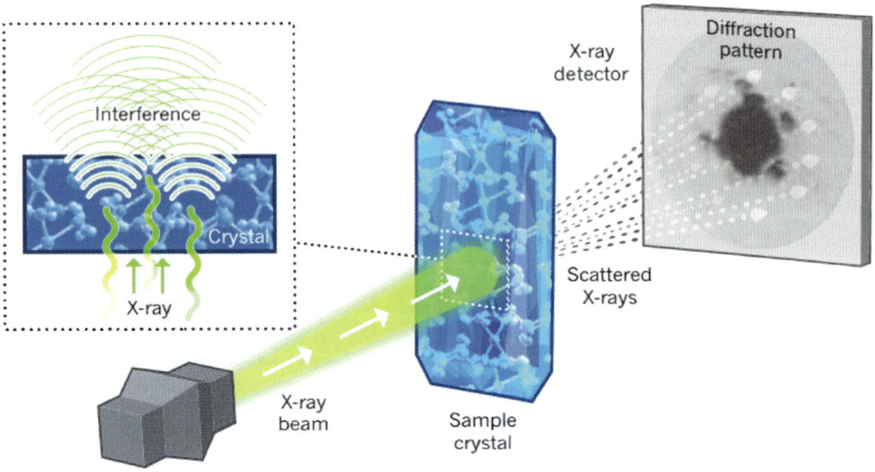

Fig. 3.1. The idea of X-ray diffraction proposed by Max von Laue [1].

more crystalline structures from simple minerals to complex molecular have been solved.

With the improvements in technology, especially with the advent of sophisticated synchrotron X-ray source, the pace of structure characterization has been accelerated and revolutionized. Five Nobel prizes have been awarded for work that depended on synchrotron X-ray studies [2]. When an electron is accelerated, it emits powerful X-rays. These short pulses, high flux, high intensity, well-collimated X-ray beams allow new experimental approaches and scientific discoveries. Currently, most of the sources around the world belong to the third generation. The fourth generation, which can produce flux densities a thousand to a million times higher than those current, are well operating. Up to now, ten US famous synchrotron light sources have developed. For example, Advance light Source (ALS), Advance Photo Source (APS), Linac coherent light Source (LCLS) etc. are included, which widely extends the understanding of matter at physical, chemical and biological research fields. Today, we can see that researchers in scientific disciplines ranging from biology to solid state physics are benefiting more from synchrotron facilities than ever before. It is time for us to take good charge of this technique.

As we know, in laboratory instruments, as long as we get the angles and intensities of the diffracted X-ray beam, the basic structure information may be revealed. But with aid of synchrotron radiation, the capability of conventional X-ray diffraction is undoubtedly expanded. In this chapter, we will focus on discussing the recent developments on three popular X-ray diffraction techniques based on synchrotron irradiation source, namely powder diffraction, surface X-ray diffraction and 3D strain mapping. We are optimistic that the synchrotron X-ray diffraction will always be in high demand in the future, which guarantees more challenging scientific works.

3.2. Synchrotron Powder X-ray Diffraction

Powder diffraction, which has been practiced for almost one century, is widely applied for phase identification and crystal structure determination on polycrystalline samples. However, the development in powder diffraction has not ever ceased. The Rieveld method [3, 4] promotes a big step on structure determination from the former simple high-symmetry system to complex low-symmetry system, allowing the local atomic coordination, lattice parameter and peak-shape parameters to be obtained. As shown in Fig. 3.2, the structure of the tetragonal $(1-x)\mathrm{PbTiO_3}-x\mathrm{BiInO_3}$ phase [5] has been investigated by synchrotron radiation powder diffraction (SPD) which is a powerful technique for resolving atomic positions.

The synchrotron radiation source enabled rapid collection of X-ray diffraction pattern with much higher intensity, higher resolution and lower background than that obtained in laboratory. Combining the high flux and narrow X-ray beam, it is even able to study the structure of individual constitute single crystals in magnetite Fe_3O_4 powders [6]. The reciprocal lattice synchrotron X-ray diffraction of (hkl) intensities in the $h = 50$ plane for magnetite microcrystal grains at 90 K is shown in Fig. 3.3.

Synchrotron beamlines have improved detecting and focusing capabilities. For solid state physics area, people are more and more concerned about how the structures evolve under the external disturbances such as temperature, pressure, X-ray irradiation and electrical

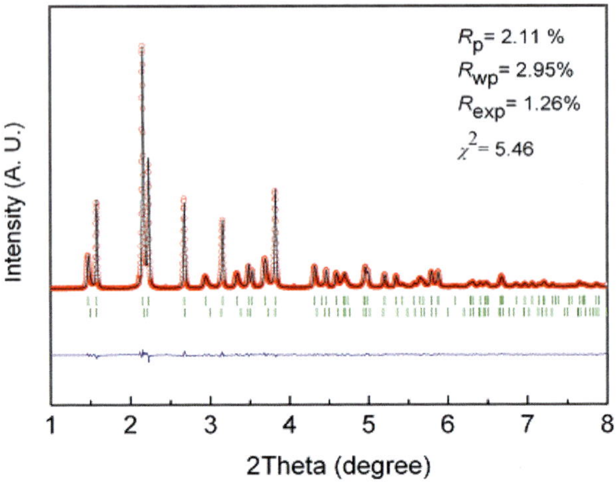

Fig. 3.2. Experimental, calculated and difference synchrotron radiation diffraction patterns of the $(1-x)\mathrm{PbTiO_3} - x\mathrm{BiInO_3}$ with $x = 0.15$ [5].

field. That is to say, *in situ* X-ray diffraction is of particular current interest. One of the representative works is to unravel how the main mineral of lower mantle $MgSiO_3$ transforms under extreme pressures and temperatures [7]. Intense synchrotron X-ray beams are essential for megabar crystallography, to target minute single crystals (of around 5 micrometers) between the diamond tips. Another typical work is to probe the structural transition of $BiMnO_3$ (BMO) by synchrotron X-ray diffraction [8]. By Rietveld refinement of synchrotron X-ray diffraction patterns under different pressures, it is found that BMO exhibits at above 37 GPa a polar monoclinic phase with an unprecedented giant elongation for a perovskite under high-pressure.

The transient state of the phase transition of a positive electrode material used for lithium-ion battery and the mechanism behind a material that can be used for fast-charge and fast-discharge batteries have been clarified by powder diffraction [9]. The phase transition between $LiFePO_4$ and $FePO_4$ during nonequilibrium battery operation was tracked in real time using time-resolved X-ray diffraction. This study is the first to report the experimental observation of a phase transition through a metastable phase during nonequilibrium battery operation. The formation of the metastable phase decreases

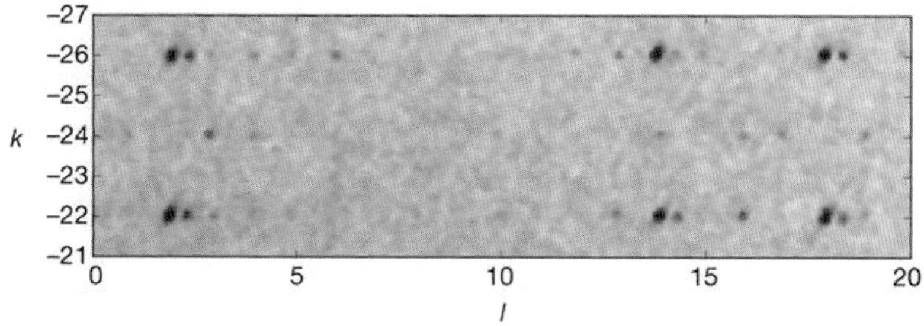

(a)

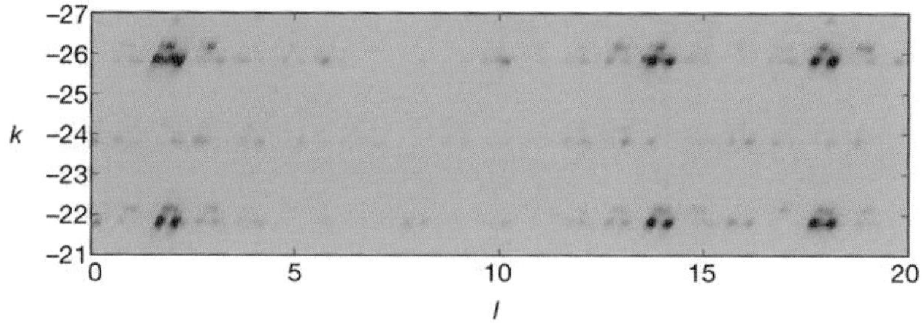

(b)

Fig. 3.3. Sections of (hkl) intensities in the $h = 50$ plane are shown. (a) Intensities for an approximately spherical grain of diameter 40 μm, which was used to determine the crystal structure. The microcrystal is almost single domain; faint spots to the right of the main intensities reveal the secondary twin. (b) Intensities from an irregular grain of dimension $\sim$100 μm showing multiple twin domains [6].

the lattice mismatch, dominates the nucleation energy. The phase transition path through the metastable phase contributes to the high-rate performance of the $LiFePO_4$ system. Use of the transient phase of two-phase reaction systems in LIBs electrodes, which enables high-rate charge–discharge, may pave the way for creating indispensable power sources for the future society.

In the above, we are talking about synchrotron X-ray powder diffraction on crystalline samples. While for the structures of systems, which are not long-range ordered, it is shown that the approach of pairwise distribution functions (PDFs) [10] plays an important role. The state-of-art 11-ID-B beamline at Argonne's Advanced Photon Source strongly maximizes the PDF quality and has pushed the experimental limitations of PDF studies approaching 0.1 Å. As far as synchrotron powder diffraction is concerned, subsequent works are still needed, including the further promotion of source intensity, high resolution, high sensitiveness of X-ray detectors and more powerful refinements codes etc. As the commentary published in Nature Materials this year indicated [11], powder X-ray diffraction has provided countless insights into the properties of materials. It will continue to do so in the future, but with an emphasis on elucidating how materials respond to external stimuli.

3.3. Surface X-ray Diffraction

The bulk system can be characterized by powder or single crystal diffraction. Whereas for the ultrathin film system where the thickness is reduced to several layers, obviously the structure information cannot be analyzed based on the positions and intensities of Bragg reflections. As we know, a flat surface is defined by a sudden termination of the crystal at a plane, and the Fourier transformation of such a termination results in a rod-shaped scattering intensity distribution around each Bragg reflection in the reciprocal space, i.e. crystal truncation rod (CTR) scattering. Due to the weak scattering of several layers, synchrotron sources are indispensable for this study. The CTR scattering profile is modulated by the surface structure. So the Synchrotron radiation surface X-ray diffraction is a surface-sensitive method. Once the scattering amplitude and phase are obtained, it is straightforward to determine the electron density distribution via a Fourier transformation. In reciprocal space, the diffraction pattern of a system with ultrathin thin layer plus the bulk can be seen as a superposition of two parts, namely, the surface (line) and the bulk (dot), as shown in Fig. 3.4.

Therefore, the observed CTR scattering amplitude can be written as [12]

$$F(\vec{Q}) = F_B(\vec{Q}) + F_S(\vec{Q}),$$

where $F_B(\vec{Q})$ represents the scattering amplitude from an ideal surface of a bulk structure and is easily calculated. The $F_S(\vec{Q})$

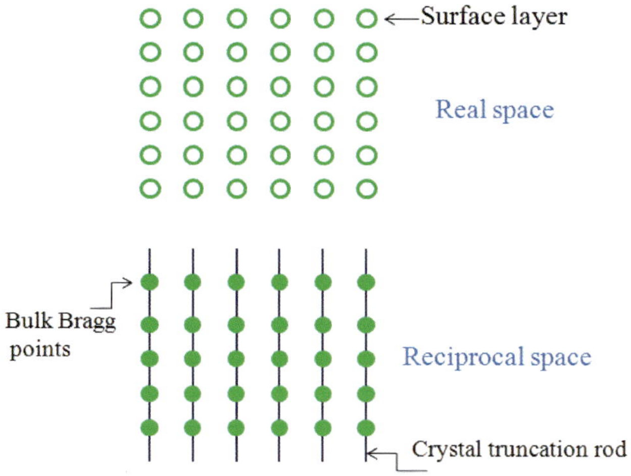

Fig. 3.4. Schematic diffraction pattern of the system of a very thin layer existing on top of the bulk.

represents the scattering amplitude from the top layer, which may have a structure of its own or a simple continuation of the bulk but with small modifications. These $F_S(\overrightarrow{Q})$ are complex numbers, and the observed intensity is the modulus square of $F(\overrightarrow{Q})$. A detailed description of the calculation of the CTR structure factors can be found elsewhere [13]. The simple form of I_{CTR} is given by:

$$I_{CTR} \propto |F(\overrightarrow{K})|^2 \frac{\sin^2(\pi N_1 \overrightarrow{K} \cdot \overrightarrow{a_1})}{\sin^2(\pi \overrightarrow{K} \cdot \overrightarrow{a_1})} \frac{\sin^2(\pi N_2 \overrightarrow{K} \cdot \overrightarrow{a_2})}{\sin^2(\pi \overrightarrow{K} \cdot \overrightarrow{a_2})} \frac{1}{2\sin^2(\pi \overrightarrow{K} \cdot \overrightarrow{a_3})},$$

where $\overrightarrow{a_1}$, $\overrightarrow{a_2}$ and $\overrightarrow{a_3}$ are lattice parameter directions, N_1 and N_2 are unit cell numbers along the $\overrightarrow{a_1}$ and $\overrightarrow{a_1}$ directions, $\overrightarrow{K}$ is the scattering vector and $F(\overrightarrow{K})$ is the structure factor. In the above formula, the I_{CTR} is approximated by considering the interference function with a sharp termination plane along the $\overrightarrow{a_3}$ direction. Namely, the $\frac{\sin^2(\pi N_3 \overrightarrow{K} \cdot \overrightarrow{a_3})}{\sin^2(\pi \overrightarrow{K} \cdot \overrightarrow{a_3})}$ term in 3D interference function has been replaced by $\frac{1}{2\sin^2(\pi \overrightarrow{K} \cdot \overrightarrow{a_3})}$.

When performing the surface X-ray diffraction experiments, the angle of the incoming X-ray beam is set slightly higher than the critical angle. Usually the crystallographic directions are chosen such that H and K lie in the surface plane, while L is the direction perpendicular to the surface. CTRs were collected by subsequent acquisition of rocking scans, i.e., by rotating the sample around the axis perpendicular to its surface at different L values for fixed (H, K) values. Finally, the possible presence of reconstructions was checked by scans along different H and K directions at small L values in reciprocal space.

CTR scattering has been found to be a very useful tool for surface studies of Si, Au, Ag and other transition metal oxide thin film systems [14–16]. A representative example is to use surface X-ray diffraction technique to investigate the atomic structure of two types of interface between $LaAlO_3$ and $SrTiO_3$ [16]. Measurements were made in air at room temperature with 12 keV X-rays. The corresponding results are shown in Fig. 3.5. The sharp peaks at the integer position are from the Bragg reflection of substrate. The broad intensity undulations come from the homogeneous thickness of the $LaAlO_3$ film. The lines in Fig. 3.5(c) and (f) show the depth dependences of the electron densities of two types of interfaces, which were performed by an electron density analysis known as coherent Bragg rod analysis (COBRA). With surface X-ray diffraction, authors argued that the $SrTiO_3$ in the sample with the n-type interface has a large polarized region, while that with the p-type interface has a limited polarized region. With the development of modern synchrotron source, surface X-ray diffraction technique will continue to allow us to study the near-surface electronic states of many more complex systems.

3.4. 3D reciprocal Space Mapping

For samples mounted in the "four-circle" diffractometer, the rotation around the sample will lead to different X-ray scan types as shown in Fig. 3.6 [17]. Here, ω is the angle for the incidence beam with respect to the sample surface, while 2θ is the angle for the scattered beam with respect to incident beam. By taking a series of

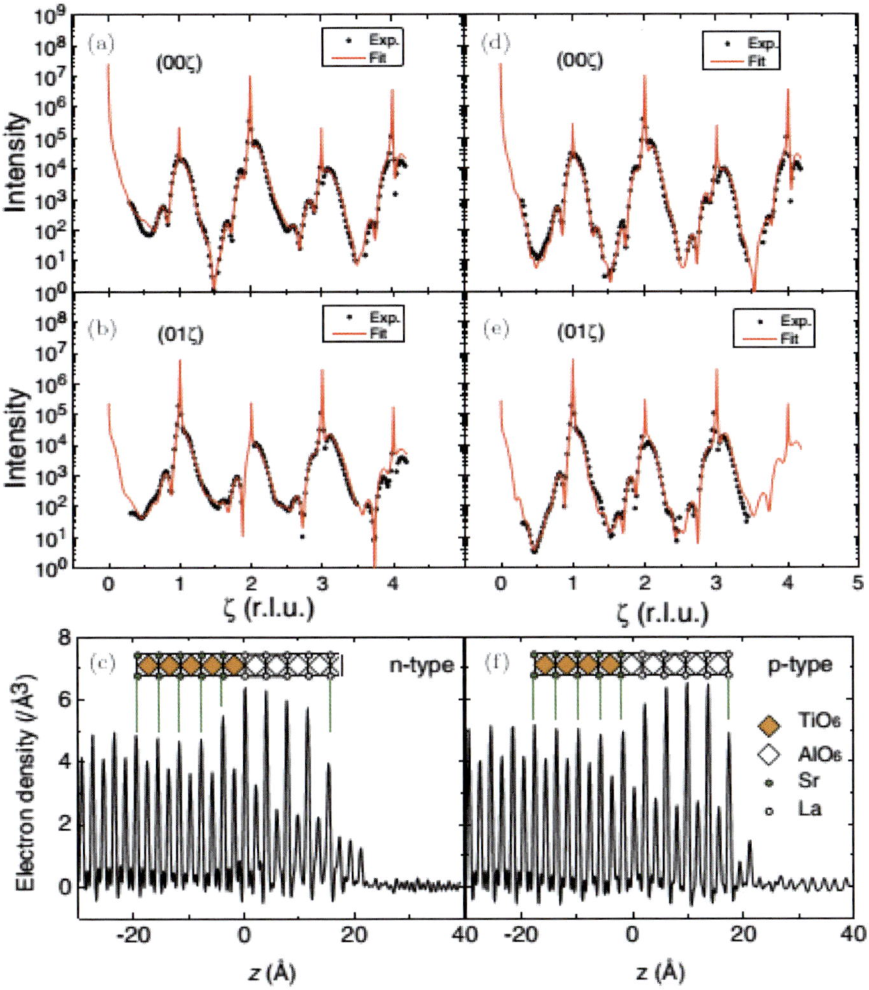

Fig. 3.5. X-ray scattering intensity profiles along (a) the (00ζ) line and (b) the (01ζ) line for the n-type sample. The depth profile of the electron density, obtained from the electron density analysis performed on the (00ζ) rod, is shown in (c). Panels (d)–(f) show those for the p-type sample [16].

$\omega - 2\theta$ scans at successive ω values and presenting the experimental results in the map form, we can get the so-called reciprocal space mapping (RSM). Imperfect materials benefit enormously more than perfect materials from these methods. The method has been applied

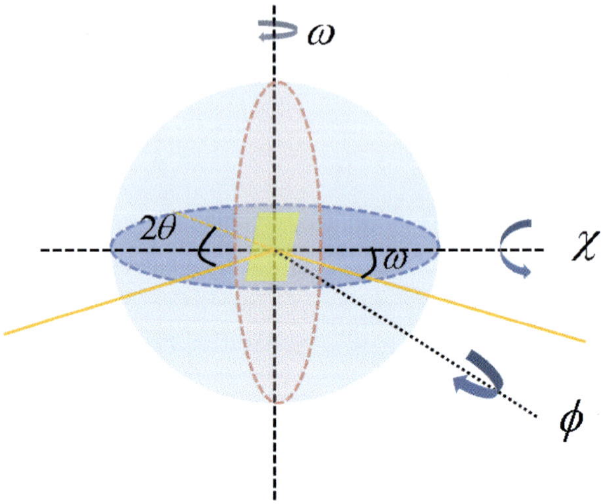

Fig. 3.6. Schematic of different X-ray scan types.

to semiconductors, protein crystals etc. A good review can be found elsewhere [18]. The conversion of angles used in measurement to the coordination in reciprocal space is given by

$$Q_{\parallel} = \frac{\lambda}{2}\sin\theta\sin(\theta-\omega) = \frac{1}{d_{\mathrm{hkl}}}\sin(\theta-\omega)$$

$$Q_{\perp} = \frac{\lambda}{2}\sin\theta\cos(\theta-\omega) = \frac{1}{d_{\mathrm{hkl}}}\cos(\theta-\omega)$$

The intensity shown in RSMs is normally a projection of the 3D intensity of the diffraction spots onto a 2D plane. By analyzing the RSM, the information of lattice parameter, lattice defects and strain relaxation can be deduced. Warusawithana *et al.* [19] used RSM to determine the strain relaxation of $SrTiO_3$ film grown on Si substrate. As we seen that the RSM of 202 $SrTiO_3$ peak for 6 molecular layers in Fig. 3.7(b), diffracted intensity mostly centered at $h = k = 2.00$ Si Relative Light Unit, whereas for the 8 ML sample (Fig. 3.7(d)), more spectral weight is observed at lower values of $h = k$ because of relaxation of the $SrTiO_3$.

2D RSM requires making assumptions about the sample because it is the projection of a plane in reciprocal space. Any sample that

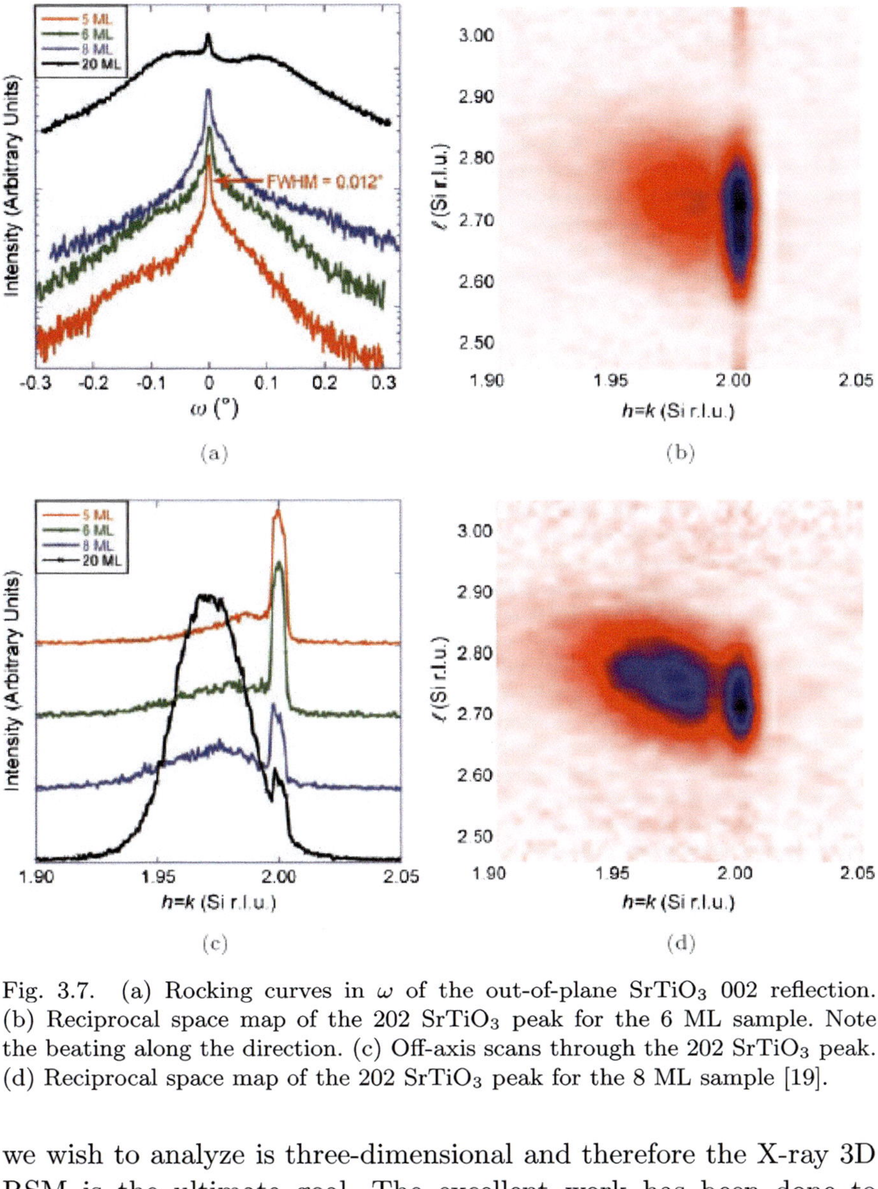

Fig. 3.7. (a) Rocking curves in ω of the out-of-plane $SrTiO_3$ 002 reflection. (b) Reciprocal space map of the 202 $SrTiO_3$ peak for the 6 ML sample. Note the beating along the direction. (c) Off-axis scans through the 202 $SrTiO_3$ peak. (d) Reciprocal space map of the 202 $SrTiO_3$ peak for the 8 ML sample [19].

we wish to analyze is three-dimensional and therefore the X-ray 3D RSM is the ultimate goal. The excellent work has been done to characterize polar nano-regions, morphology of crystallites and nano-particles strained *in situ* with X-ray 3D RSM. Luo *et al.* [20] have performed 3D-RSM experiments on $BiFiO_3$ film on BL14B beamline of the Shanghai Synchrotron Radiation Facility with an image-plate detector, and at the 33IDD beamline on Advanced Photon Source (APS) with a CCD detector. They demonstrate that 3D-RSM is an effective way to present a structural overview of the different BFO polymorphs, domain variants, and even the interfacial regions between coexisting triclinic phases. The corresponding experimental result is shown in Fig. 3.8.

One century later, Bragg's law is still the heart of the modern technique for us to peer into molecules and atoms. It is desirable

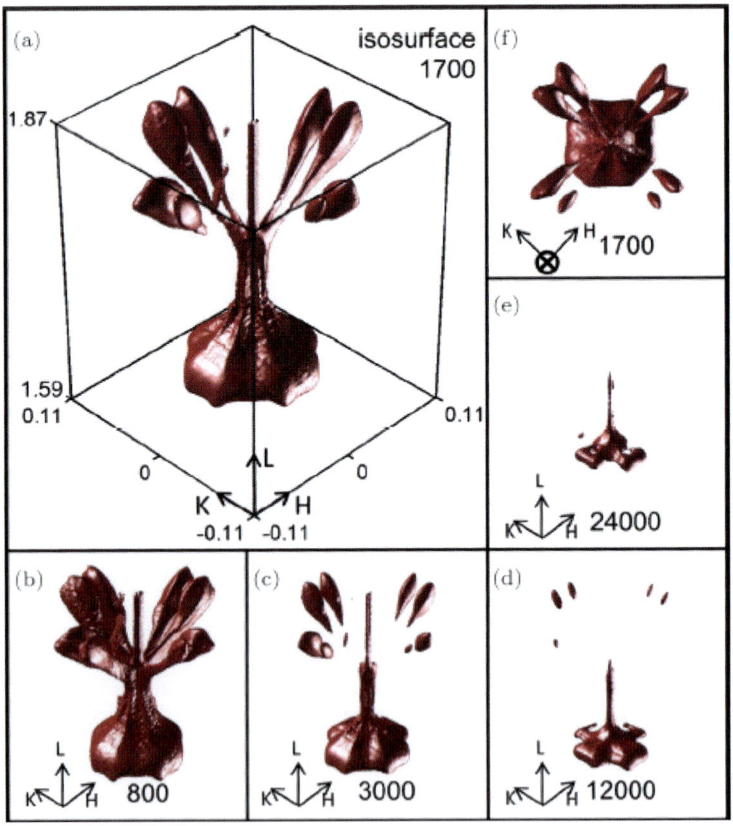

Fig. 3.8. 3D-RSMs of $BiFeO_3/LaAlO_3$ (LAO) film around LAO (002) Bragg spot, reconstructed with the help of Matlab and represented as isosurfaces in reciprocal lattice units of LAO [20].

that the X-ray pulses are short enough to avert radiation damage, but sufficiently intense to produce high-quality diffraction data. The revolutionary thing in US synchrotron facilities is the construction of X-ray free-electron lasers on LCLS. Synchrotrons have long been the preferred X-ray sources for crystallography, but competition has arrived with the advent of X-ray free-electron lasers, which has been called the fourth generation light source. The readers can refer to the review paper [21] for detailed description on the development of X-ray free-electron laser sources. Generally, the X-ray source uses special undulator magnets to wiggle an electron beam so that it emits intense, laser-like pulses of X-ray light. The angstrom-level wavelengths are so small and coherent enough to image individual atoms. Compared with the conventional static structure studies of samples, it is now possible to do dynamical structural studies on increasingly complex systems, which is really a big surprise especially for chemists and biologists. Now, Germany, Sweden, Japan and Brazil are all working on designs for synchrotrons that would compete with the US machines. Despite the many experimental and technical challenges that will exist at X-ray-free electron laser sources, it is definite that this new coming source will brighten the future of X-ray diffraction.

Bibliography

[1] N. Jones, *Nature* **505**, 602, (2014).

[2] S. Mcsweeney, *Nature* **505**, 620, (2014).

[3] H. M. Rietveld, *Acta Crystrallogr.* **22**, 151, (1967).

[4] H. M. Rietveld, *J. Appl. Crystallogr.* **2**, 65, (1969).

[5] F. F. Wang, L. L. Fan, Y. Ren, J. Chen and X. R. Xing, *Appl. Phys. Lett.* **104**, 252901, (2014).

[6] M. S. Senn, J. P. Wright and J. P. Attfield, *Nature* **481**, 173, (2011).

[7] M. Murakami, K. Hirose, K. Kawamura, N. Sata and Y. Ohishi, *Science* **304**, 855, (2004).

[8] Yuki Orikasa, Takehiro Maeda, Yukinori Koyama, Haruno Murayama, Katsutoshi Fukuda, Hajime Tanida, Hajime Arai, Eiichiro Matsubara, Yoshiharu Uchimoto and Zempachi Ogumi, *J. Am. Chem. Soc.* **135**, 5497–5500, (2013).

[9] M. Guennou, P. Bouvier, P. Toulemonde, C. Darie, C. Goujon, P. Bordet, M. Hanfland and J. Kreisel, *Phys. Rev. Lett.* **112**, 075501, (2014).

[10] B. E. Warren, *Phys. Rev.* **45**, 657, (1934).

[11] A. K. Cheetham and A. L. Goodwin, *Nat. Mater.* **13**, 760, (2014).
[12] Y. Wakabayashi, *J. Phys.: Condens. Matter* **23**, 483001, (2011).
[13] R. Feidenhansl, *Surf. Sci. Rep.* **10**, 105, (1989).
[14] V. Vonk, S. Konings, G. J. van Hummel, S. Harkema and H. Graafsma, *Surface Science* **595**, 183, (2005).
[15] A. Fragneto, G. M. D. Luca, R. D. Capua, U. S. Uccio, M. Salluzzo, X. Torrelles, T. L. Lee and J. Zegenhagen, *Appl. Phys. Lett.* **91**, 101910, (2007).
[16] R. Yamamoto, C. Bell, Y. Hikita, H. Y. Hwang, H. Nakamura, T. Kimura and Y. Wakabayashi, *Phys. Rev. Lett.* **107**, 036104, (2011).
[17] T. Lahaye, C. Menotti, L. Santos, M. Lewenstein and T. Pfau, *Rep. Prog. Phys.* **72**, 036502, (2009).
[18] P. F. Fewster, *Critical Reviews Rerirws in Solid State Materials Science* **22**, 69, (1997).
[19] M. P. Warusawithana, Cheng Cen, Charles R. Sleasman, J. C. Woicik, Y. L. Li, L. F. Kourkoutis, Jeffrey A. Klug, Hao Li, Philip Ryan, Li-Peng Wang, Michael Bedzyk, D. A. Muller, L. Q. Chen, J. Levy and D. G. Schlom, *Science* **324**, 367 (2009).
[20] Z. L. Luo, H. Huang, H. Zhou, Z. H. Chen, Y. Yang, L. Wu, C. Zhu, H. Wang, M. Yang, S. Hu, H. Wen, X. Zhang, Z. Zhang, L. Chen, D. D. Fong and C. Gao, *Appl. Phys. Lett.* **104**, 182901, (2014).
[21] R. J. D. Miller, *Science* **343**, 1108, (2014).

Chapter 4

Structural Biology and Synchrotron Radiation

Zihe Rao and Zhiyong Lou

Laboratory of Structural Biology,
School of Medicine, Tsinghua University,
Beijing, 100084, China

Abstract

In the past decades, X-ray crystallography and structural biology are rapidly growing up with the supports of impressive achievements of synchrotron radiation technologies. In recent years, XFEL advances X-ray structural biology in femtosecond time scale and in nano-crystal scale. In this chapter, we will summarize the progress of synchrotron radiation and its utilization in structural biology field.

Keywords: protein crystallography; macromolecule; protein structure; X-ray free electron laser; XFEL; MAD; SAD; anomalous; biosafety

4.1. Discovery, Development and Application of Synchrotron Radiation in Protein Crystallography

4.1.1. *The history and development of synchrotron radiation*

Synchrontron has become one of the most powerful instruments in biological research, but its invention and earlier use were mainly for high-energy and nuclear physics research. Until 1947, accompanied high intensity X-ray was first discovered and reported when electrons were forced to travel in a curved path or orbit by a magnetic field. Though initially regarded as a nuisance, the newly discovered X-ray radiation was soon widely used to boost the development of protein crystallography. At the very start, the protein crystallographers became "parasitic" users of synchrotron laboratory applications. Soon after, the dedicated second and third generations of synchrotron was specially built for crystallographic purposes.

The application of synchrotron radiation broke the bottleneck of crystallography, and formerly unusable small and weak-diffracting protein crystals, notoriously crystals of large protein complexes and membrane proteins, could now be routinely used for data collection and structure determination. Till today, protein crystallography remains the first choice for revealing the detailed atomic structures of macromolecular machineries.

Now, the powerful fourth generation of synchrotron radiation centers has enabled the forthcoming of space-and-time resolved measures of proteins and other important biological assemblies, which brought unprecedented opportunity for the classic crystallography world.

4.1.2. *Properties and advantages of synchrotron radiation over traditional in house X-ray*

Comparing to traditional rotating anode X-ray generator source, synchrotron offers several advantages for the acquisition of diffraction data from protein crystals. Revolutionarily, the properties of synchrotron beam make it able to determine multiple aspects of macromolecular structure. As previously mentioned, structures of dramatically large complexes have been solved, and these studies could hardly be done with conventional sources. Synchrotron provides extremely intense beams, with photon fluxes that are orders of magnitude stronger than those from conventional rotating anode sources. The current third-generation synchrotrons inserted devices into the main synchrotron ring to produce very small, highly directional X-ray beams. Synchrotron has been proven cost-effective despite the initial high construction cost per beamline. The booming of synchrotron radiation centers has yielded a thriving structural biology community that routinely uses these tunable, high-intensity X-ray sources for crystallographic experiments.

Unlike home X-ray sources that are limited to a few specific wavelengths corresponding to the K_α radiation from elements such as copper and chromium, synchrotrons provide access to a continuous range of X-ray energies or wavelengths. Therefore, a number of successful phasing experiments could be performed, and Fe, Hg, Yb, Br, W, Pt, Au and Se elements can be used as anomalous diffracting source for phase solving. The wavelength tunability makes it possible to use the technique of multiple wavelength anomalous dispersion for phase determination over a wide range of absorption edges. Commonly, X-rays of selected energies will scatter anomalously from the heavy atoms in macromolecule, and these can be used to solve the phase problem during structure determination. The currently available beamlines have been designed to meet the stringent experimental criteria of multi-wavelength anomalous dispersion (MAD) or single wavelength anomalous dispersion (SAD) method, such as the accurate and reproducible wavelength tunability and stability as well as the accurate measurement of diffraction intensities. The appearance of those dedicated beamlines, in particular the use of third generation sources has resulted in a rapid increase in the number of successful MAD/SAD.

4.1.3. *Synchrotron devices and radiation centers*

To improve X-ray beams, a series of devices such as undulators and wigglers are inserted to the synchrotron radiation facility. They include an array of closely spaced magnets of opposite polarity that deflect the electrons by small amounts from a linear path, and thus generate synchrotron radiation at each position of detection. In particular, the undulators make the X-ray beam highly collimated and deliver all photons into a small area, which provides a way to take the maximum advantage of intrinsic brightness of the synchrotron-radiation source. The magnetic structure of most widely-used undulators is an array of closely spaced, vertically oriented dipole magnets of alternating polarity. As the electron beam passes longitudinally through the array, its trajectory oscillates in the horizontal plane. Owing to the relatively weak field, the radiation cones emitted at each bend in the trajectory overlap, giving rise to a constructive interference effect that results in one or a few spectrally narrow peaks in a beam that is highly collimated in both horizontal and vertical directions. The wiggler is a series of magnets designed to periodically and laterally deflect a beam of charged particles inside a storage ring of a synchrotron. In an undulator, the detection is kept within the natural angle of emission so that the radiation from different poles is in-phase for a particular energy; the net intensity is then the sum of the squares of the individual amplitudes. By this way, undulators produce the X-ray sources with the highest brilliance.

The X-ray beam is often conditioned or altered in some way before it is used for a diffraction experiment. The X-ray wavelength bandwidth can be narrowed and extraneous radiation can be eliminated by a monochromator. Typically, Bragg diffraction is used for this purpose, and monochromator crystals can be various types such as a man-made "crystal" of pyrolytic graphite. The very parallel synchrotron beam can be monochromated by a perfect crystal, and silicon cut to use the (1,1,1) planes is the usual choice. The tight monochromatization means reduced backgrounds and decreased radiation damage for the same exposure. X-ray mirrors play an essential role at synchrotron radiation sources in rejecting high-order harmonic reflections from a monochromator. The combination of brighter, tighter optics makes synchrotron the best source for very large unit cells like viruses and macromolecular complexes like ribosome. Usually a pair of curved, very smooth low-angle reflection mirrors would focus the beam from a conventional source in the vertical direction first and then in the horizontal direction. Therefore the X-ray beam could be further focused on the production of tightly collimated high intensity beams.

Macromolecular crystals are quite sensitive to X-ray damage owing partly to the relatively non-covalent loose packing bonds between protein molecules, so the absorption of a high energy X-ray photon by macromolecular crystals will result in quick decay of diffraction power. Thus, for macromolecular crystallography, crystal cryocooling is very important and cryocoolers have been ubiquitous at synchrotron facilities to prolong the diffraction life and extend the diffraction limit of precious protein crystals.

A diffractometer apparatus is used to measure diffraction data. It is typically composed of some sort of "collimation" system that includes an aperture or slit pair to give a final limit to the size of the X-ray beam and a detector to record the diffracted X-rays. The aperture can hold the specimen crystal precisely in the beam and to move it in a controlled direction to sample diffraction space. The X-ray detector is the most important component of the diffractometer. It is usually a charge-coupled device (CCD) that reduces the read-out noise and provides rapid readout. It also exhibits high spatial resolution, large dynamic range and high linearity.

The European Synchrotron Radiation Facility (ESRF) in Grenoble was the first third-generation hard X-ray source to operate. It came on line for experiments by users with a 6-GeV storage ring and a partial complement of commissioned beamlines in 1994. Following ESRF, the Advanced Photon Source at Argonne National Laboratory (7 GeV) and the SPring-8 (8 GeV) in Harima Science Garden City in Japan opened in late 1996 and in late 1997, respectively. These machines are physically large (850–1440 meters in circumference) with a capability for 30 or more insertion devices, and a comparable number of bend-magnet beamlines. To date, there have been more than fifty synchrotrons worldwide supporting studies of the crystalline forms of proteins. Detailed introduction of each synchrotron radiation center is out of scope of this chapter and only a few typical synchrotron facilities would be covered to give a brief view.

4.1.4. *High throughput crystal diffraction data collection and structure determination at synchrotron radiation centers*

In most synchrotron centers, the oscillation method is still the first choice for crystal diffraction data collection. The crystal orienter is often a single axis rotated perpendicular to the X-ray beam, to enable the crystal oscillated through a small angle in the beam during data collection (Fig. 4.1). The synchrotron radiation source can not only provide highly brilliant X-rays, stable energy and high resolution, but also has fast read-out CCD detectors that can collect data from a single crystal in just a few minutes, compared with tens of hours of data collection time at home-source. The short exposure time required at beamlines also allowed the collection of time-resolved protein crystallographic data to reveal more detailed aspects of biochemical or molecular processes. Now, scientists have developed a number of time-resolved methods to observe the structural changes of macromolecules with active biological functions.

It can be expected that much higher resolution of crystal diffracts is achieved in this way, and people tend to harvest and make use of much weaker reflections at the edge of detector. For a high-resolution data collector, it may be required to collect two or more passes at different resolutions to avoid overloading. It became common sense

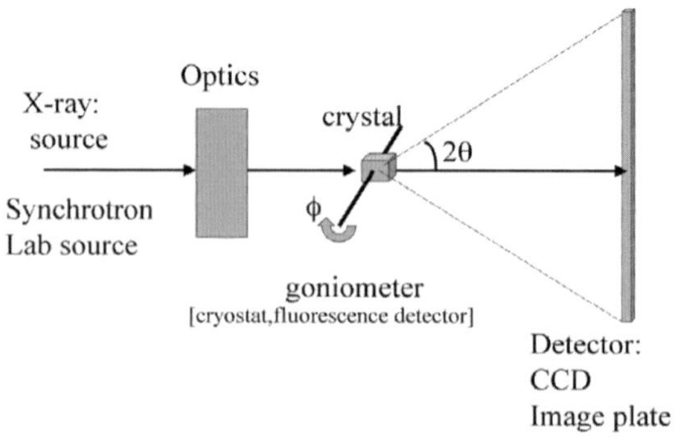

Fig. 4.1. Simple schematic view of datacollection at synchrotron radiation centers.

that taking extra time to optimize data quality is very important because good data can facilitate structure solution and avoid frustration or failure. It is thus worthwhile to collect data as complete as possible in the available time. Incomplete data, especially low-resolution data with systematically missing components, can result in failed structure solution or uninterpretable electron density maps.

After diffraction data have been measured, one must "reduce" the data to extract observed structure factor amplitudes from diffraction images. The steps involve assigning general indices (*hkl*) for each diffraction spot, measuring integrated intensity of diffraction spot, estimating the error of that measurement, and making all known geometric corrections to that intensity.

MAD method is still the gold standard for high-quality phases that are independent of any protein mode. But in the past years, SAD method that only depends on the anomalous scattering from heavy atoms in a crystal has become increasingly popular. In a MAD/SAD experiment, it is essential to maximize the anomalous scattering signal from an incorporated or existing heavy element. Therefore the measurement of an X-ray fluorescence spectrum is indispensable at most synchrotron beamlines for protein crystallography. With the assistance of highly brilliant and tunable synchrotron sources, in recent years it becomes more and more common to use an anomalous signal from sulfur atom in the protein structure to locate and refine these atoms. In the native proteins, sulfur ($Z = 16$) is the heaviest element and its K-shell resonance at 2.47 keV ($l = 5.02$ Å) is inaccessible to standard MAD experiments. Nevertheless, sulfur anomalous scattering can suffice for SAD phasing and resolve the phase ambiguity. Therefore, much trouble and agony could be avoided in comparison with the standard heavy atom soaking or selenomethionine derivative preparation. This will be particularly useful for structure determination of difficult protein targets for which heavy atom derivatives or selenomethionine substitution does not work and other standard phasing methods fail to give structure solutions. The sulfur SAD phasing method was firstly used to determine the structure of crambin [1].

Fig. 4.2. The *in situ* diffraction of protein crystals at microfocus I24 beamline of Diamond Light Source, Oxford.

If a crystal does not diffract when cryocooled, it is worthwhile to perform a room temperature test to find out whether the crystal has no intrinsic diffracting power or whether the order is destroyed by cryocooling protocol. Through recording diffraction *in situ* from crystals in their crystallization plates at room temperature, a number of problems associated with crystal handling and cryoprotection can be side-stepped. Using a dedicated goniometer installed on microfocus macromolecular crystallography beamline I24 at Diamond Light Source, crystals have been studied *in situ* with an intense and flexible microfocus beam, allowing weakly diffracting samples to be assessed without a manual crystal-handling step but with good signal/noise ratio, despite the background scattering from the plate. These demonstrate the potential for *in situ* data collection and structure solution with microbeams.

Currently, researchers are focusing on the automation of data collection steps previously described to develop high-throughput "pipelines" for structure solution. Robots that can transfer a crystal from a liquid nitrogen Dewar onto the goniometer head are now under test at most synchrotrons [2–5]. The crystals will be aligned under control of automated software for initial test, and then, the best crystals are selected based on various criteria in the diffraction image processing software and used for data collection. In the near future, these automatic pipelines will steer a structure solution to the point of producing an electron density map. Newly developed crystallographic software such as HKL3000 that integrates data collection, data reduction, phasing and model building significantly accelerates the process of structure determination and minimizes the number of data sets and synchrotron time required for structure solution. Although not routine yet, it has become usual at a SR source to look at an interpretable electron density map only a few hours after an uncharacterized crystal has been put into the X-ray beam.

4.1.5. *The future development of synchrotron radiation and its application in protein crystallography*

The race to develop a new generation of synchrotron radiation sources with significantly enhanced performance has already begun, even as the third-generation facilities have entered their prime and brought us to the future, namely, to the fourth generation. The candidate with the best scientific case for a fourth-generation source is the hard X-ray (wavelength less than 1 Å) free-electron laser (FEL) based on a very long undulator in a high-energy electron linear accelerator. Such a device would have a peak brightness with orders of magnitude in comparison to the third-generation sources, as well as pulse lengths of 100 fs or shorter, and it would be fully coherent. Many laboratories around the world are working hard to overcome several technical challenges.

4.2. Current Progress on Synchrotron Radiation Application of Life Sciences

4.2.1. *Microfocus beamline*

Microfocus Beamline is one of the latest synchrotron technologies in protein crystallography. Some challenging projects, such as membrane protein, virus, or super complex can only give tiny crystals, such that crystals of Enterovirus 71 have the dimension of several micrometers [6]. Those tiny crystals often have poor diffraction quality, which urges the construction of beamlines with so called microfocus X-ray beams.

Normal synchrotron beamlines designed for protein crystallography usually offer X-ray beam with diameter of 50–100 micrometers, while the beam size of microfocus beamline is only 5–10 micrometers. Thus, the microfocus beamline can be used to collect high-quality diffraction data from crystals with dimensions of only $5 \times 5 \times 5\,\mu$m. It could also be useful to collect data with large crystals. The small beamsize can be used to test the diffraction quality at different positions of the crystal, and choose the best one to collect data. It is extremely valueable to deal with the crystals when only part of the crystals have single lattice, when the data can be collected in microfocus X-ray beam to avoid the twinning part of crystal. The X-ray beam of microfocus beamline has very high intensity, which makes the crystal decay in minutes or even in seconds, but decay usually happens in the part of crystal exposed to X-ray beam, while the rest is still intact. So it is possible to collect a complete data set by shooting several different positions with focused X-ray beam.

Beamline ID23-2 in European Synchrotron Radiation Facility (ESRF) and I24 in Diamond Light Source (DLS) are two most famous microfocus beamlines in the world. Before constructing ID23-2, ESRF has conducted experiments in another microfocus beamline ID13 to confirm the importance of microfocus X-ray beam for protein crystallography. The beamline ID13 was constructed in 1994. It could offer an X-ray beam with the diameter of only $30\,\mu$m, which is only half of the X-ray beams generated by normal beamline. The microfocus X-ray beam of ID13 was used to collect data from very small protein crystals and solved a structure of bacteriorhodopsin [7]. However, the sample environment of beamline ID13 was not designed specifically for protein crystallography, and the demand of microfocus beamline by protein crystallography let ESRF decide in 2001 to construct a microfocus beamline only for macromolecular crystallography. This beamline was named ID23-2 and opened in 2005. ID23-2 could generate X-ray beam with the size of only $7.5 \times 7.5\,\mu$m in normal operating conditions.

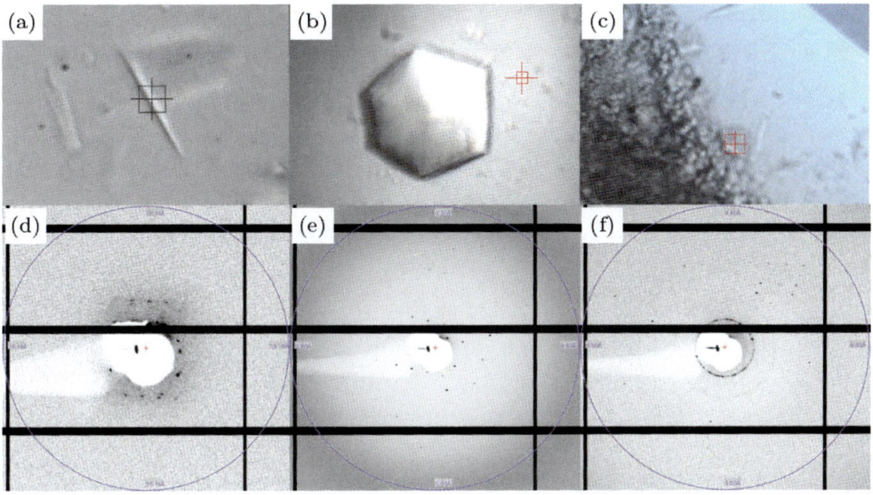

Fig. 4.3. On-axis microscope images of crystal hits and *in situ* diffraction patterns from three example membrane proteins in I24 beamline. (a, d) A G-protein-coupled receptor protein crystallized in lipid cubic phase. (b, e) A complex of two domains of a bacterial membrane protein (the apparent misalignment of the beam centre and crystal in this case is a consequence of large refraction effects induced by the use of curved bottomed wells). (c, f) A bacterial membrane-anchor cytochrome protein. In (a), (b) and (c) the square box and cross-hair represent the beam size and position, respectively. In each case the size was 10×10 mm.

The beamline I24 in Diamond Light Source opened in 2008. I24 is also a microfocus beamline specifically used for protein crystallography. It can offer an X-ray beam of only $5 \times 5 \mu m$, by which a crystal as small as $1.5 \mu m$ has been successfully measured. I24 is also equipped with the most cutting-edge data collection system and detector. In addition, the specially designed goniometer makes I24 able to collect data directly from the crystal plate in room temperature, and many virus and membrane protein structures were solved by this beamline (Fig. 4.3) [8].

4.2.2. *High-speed detector*

The most commonly used detector for protein crystallography in synchrotron is the CCD detector, and it has been equipped with nearly all beamlines in synchrotron. The typical readout time of CCD detector is about 1 second, which allows collecting a single data set of native crystal in several minutes to half an hour, or collecting a set of MAD data of heavy atom derivatives in several hours. The emerging of new generation high-speed detector not only makes the data collection faster, but also changes the data collection mode.

The readout time of the new generation high-speed detector, such as PILATUS series detector is only 3 microseconds [9]. (PILATUS is a silicon hybrid pixel detector system, operating in single-photon-counting mode, that has been developed at the Paul Scherrer Institut for the needs of macromolecular crystallography at the Swiss Light Source (SLS). A calibrated PILATUS module has been characterized with monochromatic synchrotron radiation.) This makes it possible to collect data in a continuous sample rotation mode, when the crystal is rotated by an angle of 0.05°–0.2° during each frame, and thus, a set data of 180° can cover thousands of frames. In this way, the crystal can also be rotated continuously in the beam, and the beam shutter does not have to open and close for each frame because the readout time is too short to cause any significant impact to the veracity of the data. Accuracy is the significant advantage of PILATUS since it does not add any noise to the data.

The new generation high-speed detector brings an evolution to the diffraction data collection in protein crystallography. The shutterless data collection mode and extremely short readout time not only dramatically reduces the time cost in data collection (e.g. collecting a set of data of 180° within several minutes), but also improves the data collection accuracy.

4.2.3. *Bio-protection*

Protein crystals brought to the synchrotron beamline are usually bio-safe. Proteins are normally expressed by *in vitro* system and crystalized, and the reagents used here are usually not harmful to the human body. But the situation is different for scientists who want to study the structure of infectious pathogens in synchrotron. Viruses are the most common pathogens in synchrotron. Some viruses, such as enterovirus 71 [6] and foot-mouth disease virus [10], should only be handled in biosafety level 2 or even level 3 laboratories. Those samples may cause serious and potentially lethal human diseases. In order to investigate the crystal structures of these important viruses, many synchrotrons have built beamlines with biosafety facilities. For example, beamline I23 and I24 in Diamond Light Source are equipped with biosafety level 2 facilities and both can be used to collect diffraction data of corresponding virus crystals. There is also a special department in Argonne National Laboratory, the BioCARS, which is an integral part of the multi-disciplinary Center for Advanced Radiation Sources (CARS) run by the University of Chicago, which manages BL-14-ID-B and BL14-BM-C. All areas and equipments in these two beamlines achieve biosafety level 3. The BioCARS has carried out the first diffraction experiment of biosafety level 3 sample in 2004, and plays an important role in study of infectious pathogen's structures safely.

4.3. Perspective on Next Generation Application for Synchrotron Application in Life Sciences

4.3.1. *X-ray free electron laser*

X-ray free electron laser (XFEL) produces X-rays of the highest intensity we can obtain now. The Linac Coherent Light Source (LCLS) of SLAC National Accelerator Laboratory can generate X-rays a billion times brighter than that produced by normal third generation synchrotron. The wavelength of X-rays generated by LCLS can also meet the protein crystallographers' requirement for as short as 1 angstrom, which could be used to analyze the protein structures at atomic resolution.

Collection protein crystal diffraction data in XFEL facility is quite different from using traditional synchrotron beamlines. The normal continuous exposure method used in traditional synchrotron is no longer suitable for XFEL facility due to the extremely high intensity of X-rays generated by XFEL. A single crystal could be destroyed in femtosecond time scale when exposed in the XFEL, thus collecting a complete data from one crystal also becomes impossible.

The newly developed method applied on XFEL is Serial Femtosecond Crystallography (SFX). X-ray generated by XFEL is not continuous but composed of a series of X-ray pulses which last from several femtoseconds to several hundred femtoseconds. The pulse frequency is usually between 30–120 Hz, and the pulse duration is strictly controlled to make sure that radiation damage will not impair diffraction. It has been shown that the radiation damage process could be mostly avoided before the ending diffraction process if the timescale of pulse is short enough [11]. The obvious movement of atomic nucleus happens only after 100 femtoseconds interaction between X-rays and crystal. Break of covalent bonds and change of crystal cell parameters occur in even longer timescale. Using an X-ray pulse shorter than 100 femtoseconds will not bring any notable effect of radiation damage to the diffraction data. For example, the data collected by XFEL of 70 femtoseconds pulse showed no apparent radiation damage, and the data collected by the 200 femtoseconds pulse is not affected by radiation damage within 25 angstroms. By this "diffraction before destruction" approach, we can take the advantage of the extraordinarily high intensity of XFEL without being puzzled by radiation damage.

In SFX, a complete data set is collected from a great mount of microcrystals. Instead of being fixed at goniometer and placed in liquid nitrogen stream, the single protein crystal is crack into thousands of microcrystals with size of only hundreds nanometers to several micrometers. These microcrystals are suspended in mother liquid solution and spouted at very high speed (10 m/s) in a continuous liquid water jet of only several micrometers in diameter [12, 13]. The liquid stream containing microcrystals flows across X-ray beams and the diffraction happens when X-ray pulse and a microcrystal coincidently meet at the cross point. So every diffraction frame collected by SFX is independent of other frames. New methods have been applied to scale such diffraction frames to a complete data set. Some

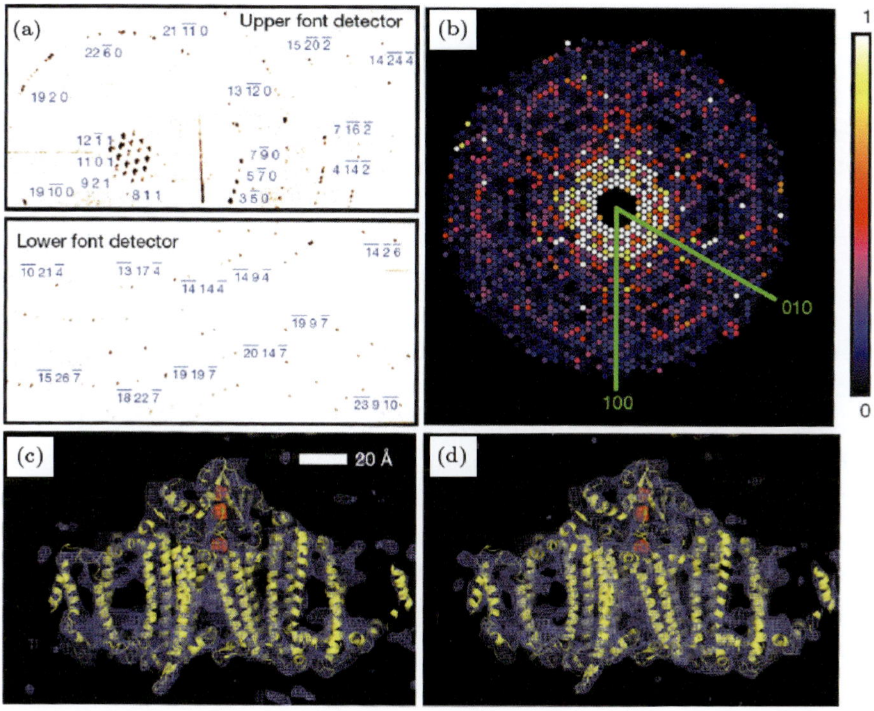

Fig. 4.4. Diffraction intensities and electron density of photosystem I. (a), Diffraction pattern recorded on the front pnCCDs with a single 70 fs pulse after background subtraction and correction of saturated pixels. Some peaks are labelled with their Miller indices. The resolution in the lower detector corner is 8.5 Å. (b), Precession-style pattern of the [001] zone for photosystem I, obtained from merging femtosecond nanocrystal data from over 15,000 nanocrystal patterns, displayed on the linear colour scale shown on the right. (c),(d) Region of the 2mFo-DFc electron density map at 1.0σ (purple mesh), calculated from (c) the 70 fs data and from (d) conventional synchrotron data truncated at a resolution of 8.5 Å and collected at a temperature of 100 K (Methods). The refined model is depicted in yellow.

softwares, such as Cheetah and CrystFEL, were developed to deal with diffraction frames collected by SFX and XFEL [14].

Several protein crystal structures have been solved through XFEL. The first structure solved by XFEL is the structure of photosystem I (Fig. 4.4) [12], which is one of the largest membrane protein complex solved by traditional X-ray crystallography. Although the X-ray wavelength used in this validation experiment is only 6.9 angstroms, and the resolution of final model is only 8.5 angstroms, it proves the capability of XFEL to solve even the most complicated protein structures. LCLS shortened their X-ray wavelength to about 1 angstrom later and successfully solved the first structure at atomic resolution by SFX [13]. The structure of *Tb*CatB is the first novel structure solved by XFEL [15].

Currently, LCLS is the only available XFEL facility used for protein crystallography. Europe, Japan and Korea have their own XFEL facilities in construction and will provide beamlines for protein crystallography in the near future. As the next generation of light source, XFEL will become one of the most important instruments for structure determination. It opens up an area for protein crystallographers to study important proteins that were previously hampered by limited size or poor diffraction quality. XFEL also sheds a light for exploring the potential for molecular imaging, which means to solve the protein structure without crystal in the future.

4.4. Synchrotron Radiation in China

4.4.1. *The first and the second generation: Beijing Synchrotron Radiation Facility (BSRF) and National Synchrotron Radiation Laboratory (NSRL)*

As an essential part of the Beijing Electron Positron Collider (BEPC) project, Beijing Synchrotron Radiation Facility (BSRF) was constructed almost in parallel with BEPC and opened to users since 1991. After upgrade to BEPCII, BSRF runs in 2.5 GeV full-energy injection and 250 mA beam current in the dedicated mode of synchrotron radiation. The intensity of hard X-rays has increased one magnitude and the stability was also significantly improved [16].

Currently, there are 3 experimental halls, 5 insertion devices, 14 beamlines and 15 experimental stations at BSRF. The synchrotron radiation light of BSRF covers the energy from vacuum ultraviolet to hard X-rays, and supports multiple experimental techniques, such as X-ray topography and X-ray imaging, small angle X-ray scattering, diffuse X-ray scattering, biological macromolecule structure, X-ray fluorescence analysis, X-ray absorption fine structure, photoelectron spectroscopy, circular dichroism spectra, soft X-ray calibration, mid-energy X-ray optics, high pressure structure, LIGA and X-ray lithography. These meet the requirements of many fundamental and applied researches, such as Condensed Matter Physics, High Pressure Physics, Chemistry, Material Science, Biology, Goescience, Environmental Science, Micro-electronics, Micromachining, Metrology, Optics and Probing Technique [17, 18].

The accumulated experimental time for the dedicated mode of synchrotron radiation is at least three months per year. Additionally, six beamlines (mid-energy X-ray station, soft X-ray absorption spectroscopy station, photoelectron spectroscopy station, vacuum ultraviolet station, small angle X-ray scattering station and macromolecular crystallography station) could also run while BEPCII is working for high energy physics, which is called coupling mode and supplies more experimental time to users. Totally, BSRF can supply beamtime to a maximal of 500 experiments for over 1000 users from more than 100 institutes and universities in China and abroad.

The design, establishment and usage of synchrotron radiation in China are initiated from Hefei Light Source (HLS), which is the first dedicated synchrotron light source, at National Synchrotron Radiation Laboratory (NSRL), located at the west campus of the University of Science and Technology of China (USTC), in Hefei, Anhui Province. It is actually a second generation synchrotron radiation source. As one of the key issues of China's investments on large-scale scientific research facilities, the Phase-I construction of HLS began in 1984 and was finished in 1991, including a 200 MeV LINAC, an 800 MeV storage ring with 160 nm.rad emitance and 200 mA beam current, and five beamlines open to users. After the fully functional running of eight years, the Phase-II Project was carried out to upgrade HLS from 1999 to 2004, to add eight new beamlines and to improve the stability and availability of HLS. From 2012 to

Fig. 4.5. Campus of Shanghai synchrotron radiation facility.

2014, HLS underwent its major upgrade and reconstruction, aiming at 38 nm.rad emittance, 800 MeV full energy injection, and 300 mA beam current. Up to date, HLS is a fully functional synchrotron radiation facility working in VUV and soft X-ray region with ten beamlines, meeting the requirements from fundamental and applied research areas, such as clean energy and environment, combustion and flame dynamics, soft X-ray image of biomaterials, catalysis and surface science, condensed matter physics, atomic and molecular physics. However, the usage of NSRL for structural biology is still very limited

4.4.2. *The third generation synchrotron: Shanghai Synchrotron Radiation Facility (SSRF)*

Shanghai Synchrotron Radiation Facility (SSRF) is a third-generation of synchrotron radiation light source, and would be an invaluable instrument for Chinese research and industry community. Up to now, SSRF is the biggest scientific platform for scientific research and technology development in China, and hundreds of scientists and engineers from universities, institutes and industries in domestic and even overseas can use it everyday.

The SSRF project was proposed and designed to meet the growing demand for Synchrotron radiation in China. From 1995 to 2001, this project went through a concept design study phase and an R&D program phase, and finally was approved by the central government in January 2004 [19]. There are 40 bending magnets, 16 standard straights (6.5 meters) and 4 long straights (12 meters) along the ring, so more than 60 beamlines could be installed in the ring. Among them, 26 are based on insertion devices (by installing two mini-gap undulators for several straights), 36 lines are based on bending magnets, and several infrared beamlines are also available. SSRF will also include 7 initial beamlines and experimental stations, which are used for macromolecular crystallography, XAFS, hard X-ray microfocus, X-ray imaging and biomedical application, soft X-ray spectromicroscopy, diffraction and small angle X-ray scattering respectively. The former five beamlines are based on insertion devices, and the other two are based on bending magnets [20–23].

In the high cost performance mode, the energy of the storage ring is 3.5 GeV, which is the highest value for medium-energy light source. Its performance is optimized in the widely used X-ray energy region. The advanced technology of insertion devices can generate high brilliance synchrotron radiation light with 1∼5 keV photon energy as well as high brilliance hard X-ray with 5∼20 keV photon energy, which has similar properties as the radiation generated by high-energy light source (6∼8 GeV). The brilliance is hundreds of millions of times higher than that of X-tube. The light brilliance of main spectra region is 1017–1020 photon/(S.mm2.mrad2.10-3bw). It offers high spatial resolution, high momentum resolution and fast time resolution condition to gain the out-breaking achievement. Up to date, more than 60 beamlines and more than 100 experimental stations will be set up. Therefore it will offer the consumer more than 5000 hours/year. Every day, it contacts hundreds of scientists and engineers from different fields and they can use the synchronic radiation light day and night.

With an electron beam energy of 3.5 GeV, just lower than APS, ESRF and SPring-8, storage ring emittance of 4 nm.rad, SSRF will be the fourth highest energy in the world. Its performance is better than the existing third generation of synchrotron radiation light source with the same energy region, and it is also one of the best light sources in the world. The size and cost of SSRF is in accord with the situation in China, and have a good cost performance ratio in a wide photon energy region. Its brilliance in the hard X-ray region (5$\sim$20 keV) is close to the 6$\sim$8 GeV light source, which is extremely large and expensive. From 1 keV to 5 keV photon energy region, its brilliance is on top of the world. Comparable with the third generation of synchrotron radiation light sources in the United States and Europe, SSRF, together with SPring-8 (8 GeV) in Japan, PLS in South Korea, TLS in Taiwan and Indus-II in India, will become the world-class synchrotron radiation light source platform, and cover all the photon energies from low to high energy in Asia.

As a third generation of synchrotron light source, SSRF's electron energy is 3.5 Gev, which is fourth in the world, only lower than SPring-8 of Japan (8 GeV), APS of America (7 GeV) and ESRF of European Community (6 GeV). SSRF consists of a 150 MeV LINAC, a booster that can increase the electron energy from 150 MeV to 3.5 GeV in 0.5 second, and a 3.5 GeV electron storage ring. The 20 cell, 4-fold structure is used to store an electron beam (average current 300 mA) with low emittance (minimum 4 nmrad) and long lifetime ($>$10 hours). With the advanced insertion device, synchrotron radiation light with high flux and high brilliance will be produced, and the range of the photon energy is from 0.1 to 40 keV, which have the call of the users. The brightness of the photon is higher than the 19th power of 10. There are 40 bending magnets, 16 standard straights (6.5 meters) and 4 long straights (12 meters) along the ring, so more than 60 beamlines could be installed in the ring, among which 26 are based on insertion devices (by installing two mini-gap undulators for several straights), 36 are based on bending magnets and several infrared beamlines are available. SSRF will also include 7 initial beamlines and experimental stations, which are used for macromolecular crystallography, XAFS, hard X-ray microfocus, X-ray imaging and biomedical application, soft X-ray spectromicroscopy, diffraction and small angle X-ray scattering respectively. The former five beamlines are based on insertion devices, and the other two are based on bending magnets.

SSRF is an complex project with many subsystems, most of which are dealt with advanced technology, such as superconductive RF and cryogenic, ultra-high vacuum, ultra-high precision digital power supply, high performance magnets and mechanical collimation, beam diagnosis, advanced control system, and advanced beamline. The difficulty of the system development and integration is very high, especially in how to keep the fault rate very low on the premise of ensuring all system's performance, in order to achieve the goal to store beam for tens of hours and provide synchrotron radiation light for more than 5000 hours per year.

Small emittance is necessary for the demand of high brilliance. The horizontal emittance of SSRF is about 4 nmrad, the beam size at the light source point is only about 150 μm horizontally and 10 μm vertically. But low emittance requirement will make the ring's dynamic aperture very small, which causes beam instability and shortens beam lifetime. So it is a challenging to optimize the dynamic property of light source.

To make the beam more stable, the vertical orbit stability must be less than 1 μm, which is one of the difficulties of SSRF. A lot of methods are used to satisfy this demand, such as control of groundsill sedimentation, distortion of the tunnel and floor strictly, restriction of the temperature variety of air and water, monitor and control of all vibration source, optimization of mechanical structure of the device, close off and damp of vibration, increase of power supply stability, and application of orbit feedback system, etc. All these technics provide world-class stable and highly efficient experimental condition and lead the production of dozens of important scientific advancements. For example, a very recent Nature paper reports the mechanism of recognition of strigolactone by a non-canonical hormone DWARF14, in which the crystallographic data is acquired from SSRF [24].

In the past decades, the development of synchrotron radiation significantly prompted the development of crystallography and single macromolecule structural biology. With the requirement to study huge biocomplex and other challenging proteins (it is difficult to obtain single crystals for traditional crystallographic study.) new technologies need to be developed. Fortunately, XFEL and other techniques based on synchrotron radiation with rapid improvement will greatly benefit crystallographic researchers.

Bibliography

[1] M. M. Teeter, J. A. Mazer and J. J. L'Italien, Primary structure of the hydrophobic plant protein crambin., *Biochemistry* **20**, 5437–5443, (1981).

[2] M. Y. Heidari Khajepour, X. Vernede, D. Cobessi, H. Lebrette, P. Rogues, M. Terrien, C. Berzin and J. L. Ferrer, REACH: Robotic Equipment for Automated Crystal Harvesting using a six-axis robot arm and a micro-gripper, *Acta. Crystallogr. D Biol. Crystallogr.* **69**, 381–387, (2013).

[3] M. Gelin, V. Delfosse, F. Allemand, F. Hoh, Y. Sallaz-Damaz, M. Pirocchi, W. Bourguet, J. L. Ferrer, G. Labesse and J. F. Guichou, Combining 'dry' co-crystallization and *in situ* diffraction to facilitate ligand screening by X-ray crystallography, *Acta. Crystallogr. D Biol. Crystallogr.* **71**, 1777–1787, (2015).

[4] A. le Maire, M. Gelin, S. Pochet, F. Hoh, M. Pirocchi, J. F. Guichou, J. L. Ferrer and G. Labesse, In-plate protein crystallization, *in situ* Ligand soaking and X-ray diffraction, *Acta. Crystallogr. D Biol. Crystallogr.* **67**, 747–755, (2011).

[5] M. Sugahara, Y. Asada, K. Shimizu, H. Yamamoto, N. K. Lokanath, H. Mizutani, B. Bagautdinov, Y. Matsuura, M. Taketa, Y. Kageyama, N. Ono, Y. Morikawa, Y. Tanaka, H. Shimada, T. Nakamoto, M. Yamamoto and N. Kunishima, High-throughput crystallization-to-structure pipeline at RIKEN SPring-8 Center, *J. Struct. Funct. Genomics* **9**, 21–28, (2008).

[6] X. X. Wang, W. Peng, J. S. Ren, Z. Y. Hu, J. W. Xu, Z. Y. Lou, X. M. Li, W. D. Yin, X. L. Shen, C. Porta, T. S. Walter, G. Evans, D. Axford, R. Owens, D. J. Rowlands, J. Z. Wang, D. I. Stuart, E. E. Fry and Z. H. Rao, A sensor-adaptor mechanism for enterovirus uncoating from structures of EV71, *Nat. Struct Mol. Biol.* **19**, 424–429, (2012).

[7] D. Flot, T. Mairs, T. Giraud, M. Guijarro, M. Lesourd, V. Rey, D. van Brussel, C. Morawe, C. Borel, O. Hignette, J. Chavanne, D. Nurizzo, S. McSweeney and E. Mitchell, The ID23-2 structural biology microfocus beamline at the ESRF *J Synchrotron Radiat.* **17**, 107–118, (2010).

[8] D. Axford, R. L. Owen, J. Aishima, J. Foadi, A. W. Morgan, J. I. Robinson, J. E. Nettleship, R. J. Owens, I. Moraes, E. E. Fry, J. M. Grimes, K. Harlos, A. Kotecha, J. Ren, G. Sutton, T. S. Walter, D. I. Stuart and G. Evans, In situ macromolecular crystallography using microbeams, *Acta. Crystallogr. D Biol. Crystallogr.* **68**, 592–600, (2012).

[9] B. Henrich, A. Bergamaschi, C. Broennimann, R. Dinapoli, E. F. Eikenberry, I. Johnson, M. Kobas, P. Kraft, A. Mozzanica and B. Schmitt, PILATUS: A single photon counting pixel detector for X-ray applications, *Nucl. Instrum. Meth. A* **607**, 247–249, (2009).
[10] D. Logan, R. Abughazaleh, W. Blakemore, S. Curry, T. Jackson, A. King, S. Lea, R. Lewis, J. Newman, N. Parry, D. Rowlands, D. Stuart and E. Fry, Structure of a major immunogenic site on foot-and-mouth-disease virus, *Nature* **362**, 566–568, (1993).
[11] R. Neutze, R. Wouts, D. van der Spoel, E. Weckert and J. Hajdu, Potential for biomolecular imaging with femtosecond X-ray pulses, *Nature* **406**, 752–757, (2000).
[12] H. N. Chapman, P. Fromme, A. Barty, T. A. White, R. A. Kirian, A. Aquila, M. S. Hunter, J. Schulz, D. P. DePonte, U. Weierstall, R. B. Doak, F. R. N. C. Maia, A. V. Martin, I. Schlichting, L. Lomb, N. Coppola, R. L. Shoeman, S. W. Epp, R. Hartmann, D. Rolles, A. Rudenko, L. Foucar, N. Kimmel, G. Weidenspointner, P. Holl, M. N. Liang, M. Barthelmess, C. Caleman, S. Boutet, M. J. Bogan, J. Krzywinski, C. Bostedt, S. Bajt, L. Gumprecht, B. Rudek, B. Erk, C. Schmidt, A. Homke, C. Reich, D. Pietschner, L. Struder, G. Hauser, H. Gorke, J. Ullrich, S. Herrmann, G. Schaller, F. Schopper, H. Soltau, K. U. Kuhnel, M. Messerschmidt, J. D. Bozek, S. P. Hau-Riege, M. Frank, C. Y. Hampton, R. G. Sierra, D. Starodub, G. J. Williams, J. Hajdu, N. Timneanu, M. M. Seibert, J. Andreasson, A. Rocker, O. Jonsson, M. Svenda, S. Stern, K. Nass, R. Andritschke, C. D. Schroter, F. Krasniqi, M. Bott, K. E. Schmidt, X. Y. Wang, I. Grotjohann, J. M. Holton, T. R. M. Barends, R. Neutze, S. Marchesini, R. Fromme, S. Schorb, D. Rupp, M. Adolph, T. Gorkhover, I. Andersson, H. Hirsemann, G. Potdevin, H. Graafsma, B. Nilsson and J. C. H. Spence, Femtosecond X-ray protein nanocrystallography, *Nature* **470**, 73–U81, (2011).
[13] S. Boutet, L. Lomb, G. J. Williams, T. R. M. Barends, A. Aquila, R. B. Doak, U. Weierstall, D. P. DePonte, J. Steinbrener, R. L. Shoeman, M. Messerschmidt, A. Barty, T. A. White, S. Kassemeyer, R. A. Kirian, M. M. Seibert, P. A. Montanez, C. Kenney, R. Herbst, P. Hart, J. Pines, G. Haller, S. M. Gruner, H. T. Philipp, M. W. Tate, M. Hromalik, L. J. Koerner, N. van Bakel, J. Morse, W. Ghonsalves, D. Arnlund, M. J. Bogan, C. Caleman, R. Fromme, C. Y. Hampton, M. S. Hunter, L. C. Johansson, G. Katona, C. Kupitz, M. N. Liang, A. V. Martin, K. Nass, L. Redecke, F. Stellato, N. Timneanu, D. J. Wang, N. A. Zatsepin, D. Schafer, J. Defever, R. Neutze, P. Fromme, J. C. H. Spence, H. N. Chapman and I. Schlichting, High-resolution protein structure determination by serial femtosecond crystallography, *Science* **337**, 362–364, (2012).
[14] T. A. White, A. Barty, F. Stellato, J. M. Holton, R. A. Kirian, N. A. Zatsepin and H. N. Chapman, Crystallographic data processing for free-electron laser sources, *Acta Crystallogr D* **69**, 1231–1240, (2013).
[15] L. Redecke, K. Nass, D. P. DePonte, T. A. White, D. Rehders, A. Barty, F. Stellato, M. N. Liang, T. R. M. Barends, S. Boutet, G. J. Williams,

M. Messerschmidt, M. M. Seibert, A. Aquila, D. Arnlund, S. Bajt, T. Barth, M. J. Bogan, C. Caleman, T. C. Chao, R. B. Doak, H. Fleckenstein, M. Frank, R. Fromme, L. Galli, I. Grotjohann, M. S. Hunter, L. C. Johansson, S. Kassemeyer, G. Katona, R. A. Kirian, R. Koopmann, C. Kupitz, L. Lomb, A. V. Martin, S. Mogk, R. Neutze, R. L. Shoeman, J. Steinbrener, N. Timneanu, D. J. Wang, U. Weierstall, N. A. Zatsepin, J. C. H. Spence, P. Fromme, I. Schlichting, M. Duszenko, C. Betzel and H. N. Chapman, Natively inhibited trypanosoma brucei cathepsin B structure determined by using an X-ray laser, *Science* **339**, 227–230, (2013).

[16] E. T. and D. X. Xiaoming Jiang, Beijing Synchrotron Radiation Facility, *Review of Scientific Instruments* **66**, (1995).

[17] W. A. Hendrickson, Synchrotron crystallography, *Trends Biochem Sci.* **25**, 637–643, (2000).

[18] W. Minor, D. Tomchick and Z. Otwinowski, Strategies for macromolecular synchrotron crystallography, *Structure* **8**, R105–110, (2000).

[19] Z. T. Z. H. J. Xu, SSRF: A 3.5 Gev synchrotron light source for China, *Proceedings of EPAC 2004* (2004).

[20] X. Wang, Y. Cao, H. Du and L. Yin, Dynamic performance of the beam position monitor support at the SSRF, *J. Synchrotron. Radiat.* **16**, 1–7, (2009).

[21] Y. Tao, Y. Huang, Z. Gao, H. Zhuang, A. Zhou, Y. Tan, D. Li and S. Sun, Developing VUV spectroscopy for protein folding and material luminescence on beamline 4B8 at the Beijing Synchrotron Radiation Facility, *J. Synchrotron. Radiat.* **16**, 857–863 (2009).

[22] X. Wang, L. Chen, Z. Yan, H. Du and L. Yin, Mechanical stability of the magnet girder assembly at the SSRF, *J. Synchrotron. Radiat.* **15**, 385–391, (2008).

[23] X. Wang, L. Chen, H. W. Du and L. X. Yin, Measurement of underground motion at the SSRF site, *J. Synchrotron. Radiat.* **15**, 350–354, (2008).

[24] R. Yao, Z. Ming, L. Yan, S. Li, F. Wang, S. Ma, C. Yu, M. Yang, L. Chen, L. Chen, Y. Li, C. Yan, D. Miao, Z. Sun, J. Yan, Y. Sun, L. Wang, J. Chu, S. Fan, W. He, H. Deng, F. Nan, J. Li, Z. Rao, Z. Lou and D. Xie, DWARF14 is a non-canonical hormone receptor for strigolactone, *Nature* 2016 Aug 25;536(7617):469–73.

Chapter 5

Fluorescence Detected XAS — Unconventional Applications

Hiroyuki Oyanagi

Research Institute of Instrumentation Frontier, AIST, Tsukuba, Ibaraki 305-8568, Japan

Abstract

With the advent of modern synchrotron radiation facilities, fluorescence-detected X-ray absorption spectroscopy (XAS) has proven its capability as a highly sensitive local structural tool. Previous applications have been limited to dilute systems and thin films where the absorption of incoming incidence X-ray and outgoing fluorescence X-ray can be neglected. In this chapter we describe unconventional applications beyond this criterion, which are made possible by improving detection efficiency by orders of magnitude combining a high flux insertion devices and highly efficient detectors. In the first application, the non-equilibrium local structure of photo-induced phase of Fe(II) spin crossover complex, $[Fe(2\text{-pic})_3]Cl_2EtOH$ (2-pic=2-aminomethyl pyridine), was investigated under laser illumination. The results indicated no symmetry breaking of FeN_6 clusters upon spin-state switching, establishing the origin of new Raman lines as unrelaxed "frozen-in" distorted outer ligand molecules. Secondly, we demonstrate that polarization-dependent XAS is obtained with a high quality for high temperature superconducting (HTSC) single crystals. The results of temperature-dependent local lattice study on high-quality single crystal of LaSrCuO revealed doping-induced local lattice distortion $T_d^{\max}$ that maximizes at the superconducting critical temperature $T_c^{\max}$, indicating the strong intimacy of HTSC with local lattice, casting doubts to the purely electronic pairing mechanisms that ignore the contribution of lattice. Lastly, we describe that, combining a microfluidic cell with a fluorescence detection, position-dependent XAS provides time-dependent

XAS for dilute systems. The application to CdSe nanocrystal formation demonstrated that such an approach would allow *in situ* studies on the structural evolution of intermediate state of nanocluster formation known as puzzling state, *i.e.* "monomers", in association with molecular orbital calculations. Because of a limited space, unconventional applications are briefly described in this chapter but it could be sufficient to expect more outputs from the 4th generation synchrotron facilities that would allow higher brilliance and a smaller focus size because the methodology described here is strongly dependent on synchrotron instrumentation.

Keywords: X-ray absorption spectroscopy (XAS); EXAFS and XANES; high temperature super conducting (HTSC); fluorescence X-ray; fluorescence detection; self absorption effect; Zn filter; surface-sensitive X-ray absorption fine structure; multipole wiggler (MPW); multi-element Si(Li) solid-state detector; submonolayer sensitivity; first nearest neighbor; linear polarization; solid angle; high energy resolution detector; Myoglobin (Mb); solid state detector (SSD); carboxyl hemeprotein; Mn filter; fluorescence yield; UHV; Si(Li) detector; pixel array detector (PAD); FET capacitance; PSC 741-N; CAMAC; digital signal processor (DSP); TSFZ method; Photo-induced phase transition; charge transfer (CT); light-induced excited spin state trapping (LIESST); LaFeAsO (LFAO); antiferromagnetic (AFM); DFT calculation; Hubbard U; polaron; colossal magnetoresistance (CMR); $La_{2-x}Sr_xCuO_4$ (LSCO); mean square relative displacement (MSRD); pseudo-JT mode; Q_2-type JT mode; nanoparticle (NP); microfluidics; KAPTON; TEFLON; Se-TOP; ATHENA; ARTEMIS; Silicon drift diode (SDD); ligand field; beat; small cluster (SC); IFEFFIT.

5.1. Introduction

X-ray absorption spectroscopy (XAS) is a local structure tool, which rapidly advanced as dedicated (second generation) synchrotron radiation facilities proliferated in the 1980's. XAS provides unique structural information around an excited atom without a long range order complementally to crystallography. The advantage of EXAFS and XANES is closely related to their origin, or energy-dependent fine structures observed in absorption coefficient as a consequence of the interference of photoelectron waves with outgoing ones (Fig. 5.1(a)). A conventional experimental setup of XAS is based on a transmission geometry where incident beam and transmitted beam intensities are measured as illustrated in Fig. 5.2(b), which generally gets difficult

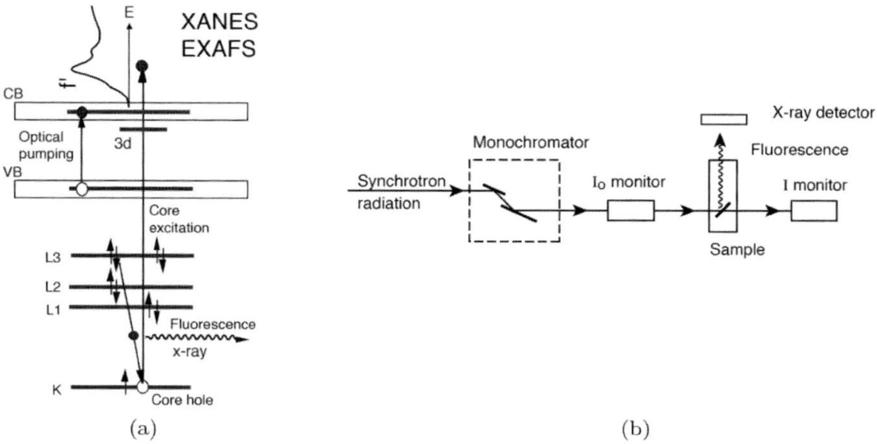

Fig. 5.1. (a) Energy diagram of core excitation and XAS principle and (b) fundamental experimental setup.

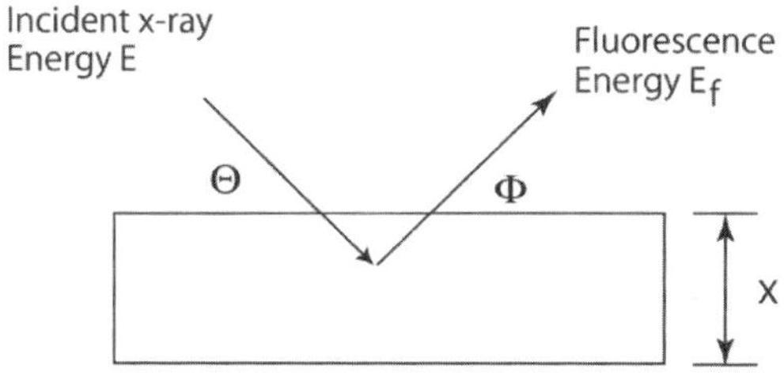

Fig. 5.2. Schematic geometry of fluorescence detection.

when one goes to specimens with low concentration or thin layers as they lack the sufficient number of absorbing atoms. The situation is similar to the relationship between photoluminescence and absorption in UV-visible spectroscopy. It is reasonable as an absorption is measured by quantity (attenuated beam intensity) that exponentially decays as thickness increases, while fluorescence intensity is linear to the concentration. Fluorescence detection [1] is an alternative, highly sensitive experimental means which has been widely used in dilute atom systems such as biological samples, impurities or thin films [2–5]. Although most of the fluorescence XAS applications were on dilute systems here we describe rather unconventional applications, *i.e.* fine polycrystalline specimen and single crystals with heavy elements both of which satisfies self-absorption limit. Typical examples of former and latter cases are photo-excited systems and high temperature super conducting (HTSC) single crystals described in later sections.

A simple schematic of fluorescence excitation and detection is illustrated in Fig. 5.2. Here we assume that the incident X-ray beam with the energy E irradiates a sample (thickness x) with an incidence angle Θ and the fluorescence X-ray (with the energy E_f) is detected by a detector with a detection angle of Φ and a solid angle Ω. The fluorescence intensity I_f (E) accepted by a detector with a solid angle of $\Omega/4\pi$ is described by the following equation. Here ε is the quantum efficiency of fluorescence excitation and I_o the incident beam intensity and $\mu_a(E)$ is the absorption coefficient for the excited atom component.

$$
\begin{aligned}
I_f(E) = I_0(\Omega/4\pi)\varepsilon\mu_T(E)/\sin\Theta \times \\
1 - \exp[-(\mu(E)/\sin\Theta + \mu(E_f)/\sin\Phi)x]/ \\
[\mu(E)/\sin\Theta + \mu(E_f)/\sin\Phi] \qquad (5.1)
\end{aligned}
$$

This reduces to a simple equation $I_f(E) = I_0(\Omega/4\pi)\varepsilon\mu_a(E)$ in the limit of dilution or thin film x *i.e.* $\mu_a(E) \ll 1$ or $x \ll 1$. In such systems (defined as conventional application of fluorescence detection), one can measure the XAS spectra without using a transmission experiment.

This scheme enables us to go beyond the dilute limit, making it possible to study impurities [6], biological systems [7], and thin films [8]. Sensitivity is defined as a signal to background ratio which should be larger than unity for which high resolution X-ray detection and additional equipment, X-ray filters are used. In order to increase surface sensitivity, a grazing incidence geometry is useful [9]. In a widely used surface-sensitive geometry, the signal fluorescence to background scattering ratio is drastically reduced when the incidence angle becomes smaller than the total reflection angle [10] by orders

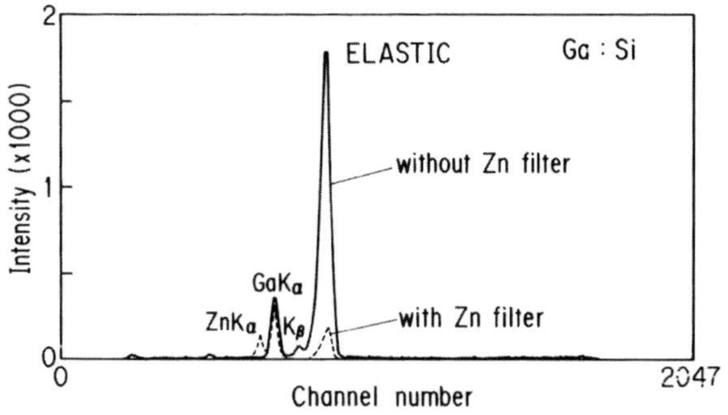

Fig. 5.3. MCA output of fluorescence X-ray detector (SSD) for Ga impurity atoms in Si.

of magnitude, while the signal intensity is suppressed as the effective thickness is lowered. In general, the self absorption effect causes distortion in spectra, and several correction methods and codes were proposed [11]. Here we describe some of the conventional fluorescence XAS applications, or dilute impurities embedded in semiconductor lattice. Figure 5.3 shows the energy spectrum for Ga impurity atoms (10^{19} atoms/cm^3) in silicon with and without a Zn filter to reduce the elastic peak intensity. An optimum thickness filter roughly reduces the elastic peak by an order of magnitude sacrificing the signal by half.

Surface-sensitive X-ray absorption fine structure (XAFS) with sub-monolayer sensitivity based on grazing-incidence fluorescence detection is described here. The efficiency of fluorescence detection increased by more than two orders of magnitude by combining a multipole wiggler (MPW) with a multi-element Si(Li) solid-state detector. The capability of the present technique for structural studies of surfaces and buried interfaces in the hard X-ray region was demonstrated by As K-edge XAFS studies of the InP(001) surface exposed to AsH_3 flow. The results indicated that ~0.1 monolayer As atoms are incorporated into the surface replacing the P atoms. Figure 5.4 shows the magnitude of Fourier transform of the As K-EXAFS oscillations for (a) 0.1 monolayers (MLs) on InAs surface (measured by fluorescence) capped by several InP layers for protection and (b) bulk

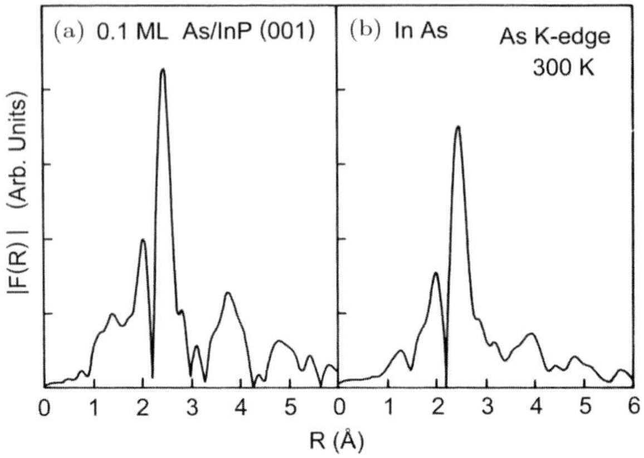

Fig. 5.4. (a) Fourier transform magnitudes for the As K-EXAFS oscillations of 0.1 MN In as on InP and (b) bulk InAs measured at 300 K.

InAs (measured by transmission) [12]. One can find that the grazing-incidence geometry has submonolayer sensitivity and the first nearest neighbor In coordination is essentially preserved. The FT peak in the surface experiment is enhanced because of polarization factor. For optimizing a grazing-incidence angle, the ionization chamber placed behind a sample is monitored as the angle of incidence is slowly varied and/or a fluorescence signal intensity is also informative on the critical angle of the total reflection.

The fluorescence detection mode is also useful in probing photo-excited specimen which is made "X-ray thin" in order to maximize efficiency of photo-excitation in a visible photon region. See Fig. 5.5 which illustrates the two cases of fundamental geometry of experiments. Figure 5.5(a) is a conventional setup for unpolarized powder specimen or linear polarization with the electrical vector $\boldsymbol{E}$ perpendicular to the single crystal plane while Fig. 5.5(b) is used for grazing incidence experiment or the linear polarization with $\boldsymbol{E}$ parallel to the surface.

As indicated by Eq. (5.1), at the strong absorption limit where self-absorption of fluorescent X-rays is strong and the outgoing signal is attenuated, fluescence yield is proportional to the solid angle of detection Ω. At the early stage of development, efforts were taken

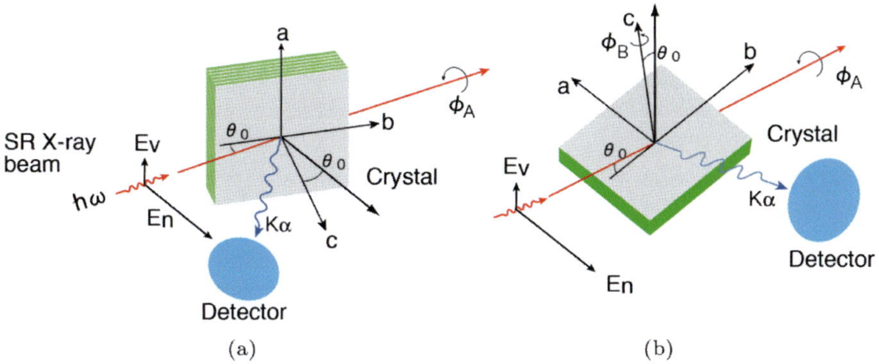

Fig. 5.5. Conventional fluorescence detection setup (**E** perpendicular to crystal plane) a and a setup for a grazing incidence geometry (**E** parallel to the crystal *ab* plane).

to increase Ω to provide high efficiency. For that purpose a standard approach was to use an array of scintillation detectors, which were later replaced with better energy resolution detectors [13, 14]. The fluorescence yield can replace absorption spectrum in the limit of strong absorption which holds in either thick dilute or thin film cases. Use of a grazing-incidence geometry has successfully improved surface sensitivity by suppressing the background scattering and unnecessary fluorescence signals originating from deeper regions such as substrate. Applications in research fields such as biological specimen in solution and thin films soon became popular and led to a number of successful research outputs. For fluorescence XAS applications, a high energy resolution detector is required in order to filter out the signal from elastic peaks and/or other fluorescence lines.

Because of the lack of energy resolution in scintillation X-ray detectors, concentration limit of fluorescence XAS is strongly limited. A typical setup of fluorescence XAS experiment with multiple NaI detectors with a focused X-ray beam from a bending magnet is shown in Fig. 5.6.

Myoglobin (Mb) is an oxygen storage metal protein with a molecular weight of 17, 500 around one heme iron which is one of the most thoroughly studied biological specimen. Planar four ligands are all nitrogen atoms of the heme molecule and the 5th ligand is linked

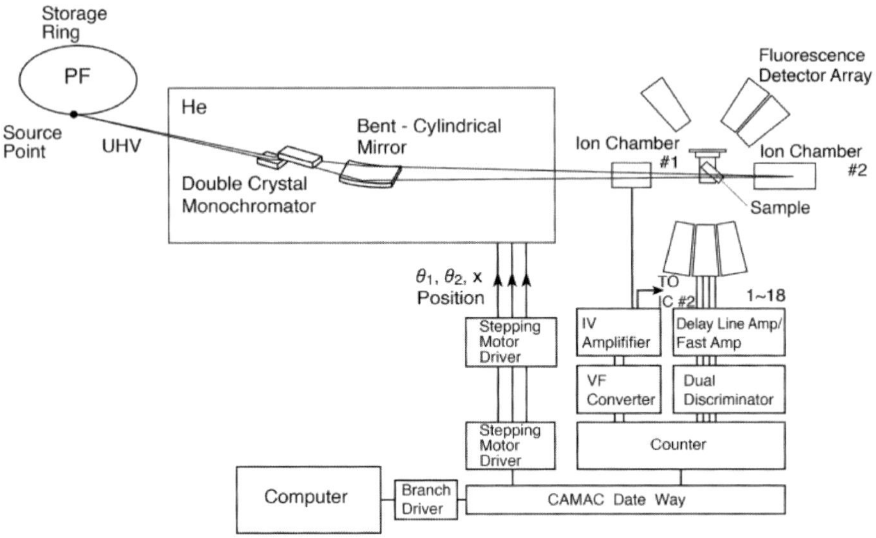

Fig. 5.6. Schematic diagram of a fluorescence XAS apparatus equipped with an array of NaI detectors.

to the protein through hystidine (Hy). Its biological functions are strongly related to its vacant ligand (6th ligand) which has affinity with various molecules such as O_2, CO and CN *etc.* The electronic structure is similar to a spin cross-over complex whose spin state is dependent on the environments (pressure, temperature or electronic excitation). In Fig. 5.7, the Fe K-XANES spectrum for MbOH (Fe^{2+}) taken at room temperature is shown as a function of energy, together with the derivative spectrum [7].

The energy diagram of the Fe $3d$ level is understood as the $3d$ orbitals which split into e_g and t_{2g} orbitals. The electron configuration takes either the high spin $S = 5/2$ state or the low spin $S = 3/2$ state depending on the temperature or pressure. In Fig. 5.7, the three specific energy regions indicated by arrows are spin-dependent features [7] which are described by a multiple scattering calculation of the near-edge structure [15]. It was found that the local structure of Fe^{2+} central ion is distorted in the high spin state, resulting in a large displacement from the heme plane by 0.4 Å as shown in Figs. 5.8 and 5.9.

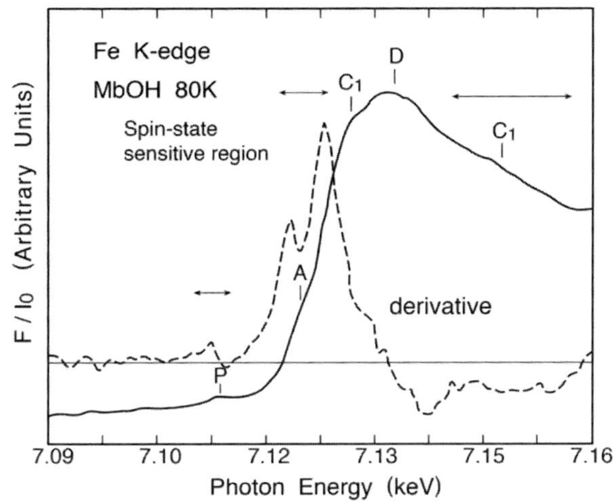

Fig. 5.7. Fe K-XANES for MbOH at 80 K. Dashed line indicates the derivative spectrum.

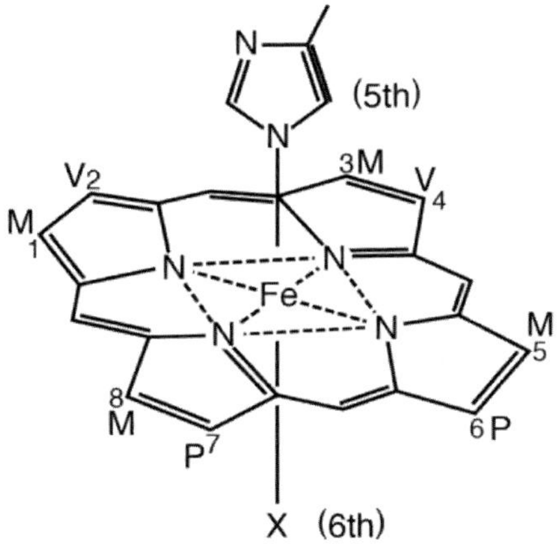

Fig. 5.8. Local structure of myoglobin (Mb) heme with a vacant 6th ligand site.

This displacement is unfavorable for adsorbing molecules and thus affinity is reduced because of steric hindrance effect. It is interesting to note that such a symmetry breaking also reduces the ligand field stabilizing the high spin configuration, indicating a close relationship between the structure and function.

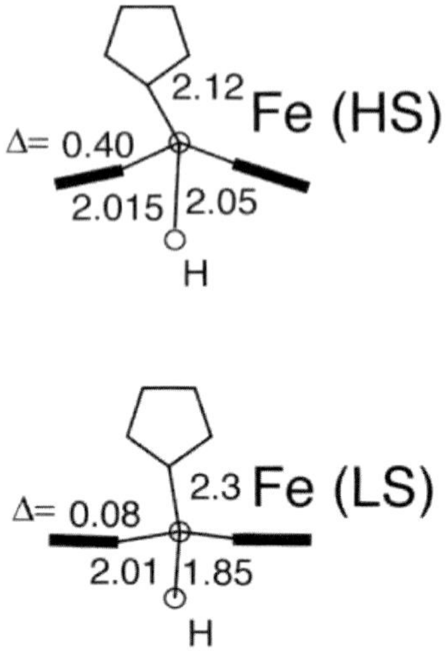

Fig. 5.9. Local structures of MbOH in the high spin state (HS) and in the low spin state (LS) derived by a multiple scattering calculation and fitting of the XANES shown in Fig. 5.7.

A standard X-ray detector beyond a scintillation detector in this respect is a solid state detector (SSD). For XAS applications covering a wide range of atomic number z, a high efficiency Ge detector over a wide energy range (4–60 keV) is required. Since the throughput of a single element SSD is limited by a linear amplifier dead time, a multi-element SSD with a high packing fraction is a practical approach [13, 14]. A monolithic pixel array detector [16] is an ultimate strategy for improving the throughput and packing ratio. Fluorescence XAS experiments on photo-excited state became feasible that allowed a time-resolved snapshot experiment for a carboxyl hemeprotein [17]. As far as low concentration limit is concerned, as illustrated in Fig. 5.10, even the SSD measurement is limited to the concentration of mM/l range while crystal monochromator is used to go beyond the limit.

In this case, even using a Mn filter to cut the background scattering, the signal-to-background ratio is about 0.3 and is close to the concentration limit (0.3 mM/l). One can use a crystal monochromator to go beyond the dilute limit which depends on the detector

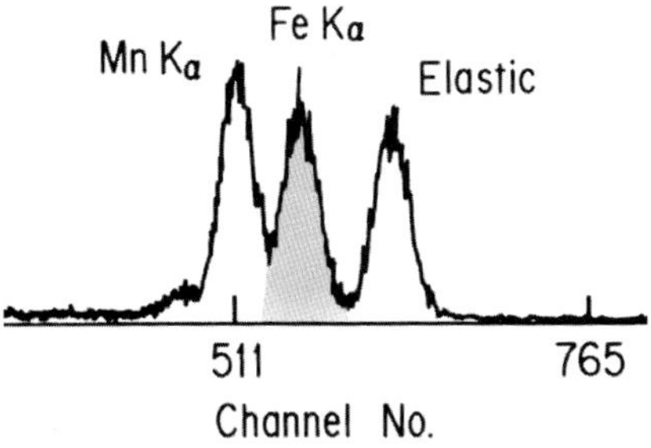

Fig. 5.10. Energy spectrum for 1 mM/l metmyoglobin above the Fe K-edge excitation with a Mn filter.

energy resolution. The disadvantage of crystal monochromator is the limited solid angle and an array of crystals is commonly used. The fluorescence yield spectrum is defined as the X-ray intensity with the energy window (shaded peak in Fig. 5.10) as a function of excitation energy normalized by the incident beam intensity.

5.2. Detector Development-History

Bulk sensitivity limit in fluorescence XAS depends on both the flux of incident X-ray beam and detector performance, *i.e.* energy resolution, efficiency and throughput. XAS is a two-wheel machine where a progress in one of the two wheels (detector) is inadequate and both sides have to be improved for achieving high quality reliable data taking. For surface studies where a typical density of adsorbed atoms is in the order of $10^{14}/cm^2$, a grazing-incidence geometry [9] can provide a means of surface-sensitive measurement. Owing to the recent advances of synchrotron radiation, *i.e.* insertion devices such as wigglers and undulators at third-generation synchrotron radiation facilities, rapid and sensitive measurements of surface XAS experiments became feasible. Effective throughput of a solid state detector (SSD) can be increased by assembling detector elements with a close packing geometry. Various types of multi-element SSD arrays have been designed for this purpose [13, 14, 16]. Fluorescence signal is seriously distorted by elastic or inelastic scattering. A large number of segmented X-ray detectors can control the *systematic (nonstatistical)*

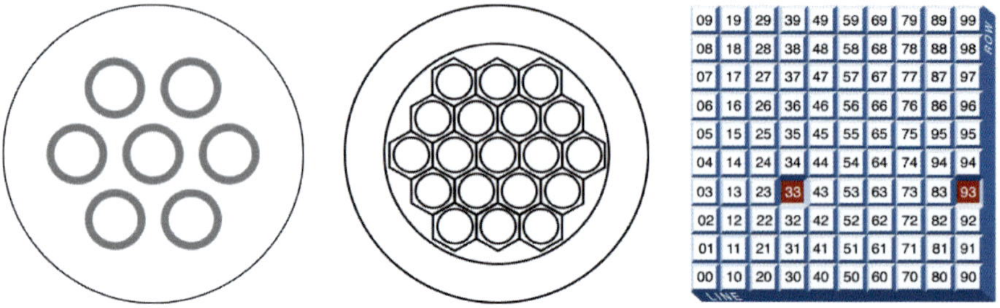

Fig. 5.11. Schematic representation of various segmented X-ray detectors. To the right, UHV-compatibel 7-element Si(Li) SSD. 19-element Ge SSD, and 100 pixel array detector.

error by eliminating the channels that suffer from strong scattering. A large-scale segmented X-ray detector is needed to achieve high quality XAS by minimizing both statistical and nonstatistical error.

For UHV *in situ* surface studies, a UHV-compatible 7-element Si(Li) detector was designed [18] where each Si(Li) segment has an independent Be window (Fig. 5.11 left column).

Since a Be window can separate the UHV and detector electronics sections in terms of vacuum, the vacuum chamber can be baked, which allows the operation in UHV chamber with better vacuum than 10^{-11} Torr. Combined use of the Si(Li) detector and UHV-compatible eight-axis goniometer were designed and built for surface XAS studies such as Ge submonolayers on clean silicon surface [18]. To cover higher energy region (>20 keV) with a better resolution and higher packing ratio, a germanium SSD array was developed [14].

5.3. Pixel Array Detector

Figure 5.11 illustrates the evolution of segmented detectors Si(Li) (left), pure Ge (center) and Ge pixel array (right). Our R&D of a challenging project (a pixel array detector) started in Strasbourg in mid 1990s took about seven years.

The pixel array detector (Fig. 5.11 (right)) can be called the first generation which uses a restive feedback front-end while the second generation based on a transister reset was developed by Gary Foran [19]. Details of pixel array detector kept continuous upgrading since, which led to the fifth generation standard in major synchrotron radiation facilities. We note that Silicon-Drift Diode (SDD) [20] is a state-of-the-art detector in a lower energy region (<15 kev) but its manufacturing process (ion implantation) limits active thickness and therefore high energy limit is not enough for the third generation light sources. Fabrication of Ge pixel array detector is similar to the MOS device in that it is based on a simple principle of pin junction on a single crystal. Fluorescence XAS experiments at high energy light sources, such as ESRF, APS and SPring-8 (6–8 GeV), require high detecting efficiency over a wide range of energy (5–60 keV). This energy range is hardly covered by SDD [20]. To meet these requirements, the first generation Ge pixel array detector was designed and built. Passivation of silicon, one of the key technologies of the MOS device process, could not be applied to germanium, as the germanium-oxygen bond is much weaker than silicon-oxygen.

Pixel array detectors we developed consist of 10×10 arrays. The two sets of pixel array detector have the same pixel length (y) and pitch (p) dimensions but different thickness (z), *i.e.* the first pixel array detector #1 has $z = 10$ mm while the second one (#2) has $z = 7.5$ mm. Pixel array detector #1 with thicker z was designed for use with a higher energy storage ring (8 GeV). Electronic noise of a semiconductor detector is proportional to the input capacitance at the FET, which is a sum of the detector (*variable*) and FET capacitance ($\sim$0.9 pF). Thus the strategy to maximize the energy resolution was to minimize the total capacitance which is dependent on p, x and z. Fluorescence signal taken over a cone-like limited solid angle away from the scattering plane gives the best signal-to-background ratio. From the geometrical consideration, we have first determined the pitch p as 5 mm. Capacitance sharply increases as a function of x below 300 micron and x was compromised. The pixel-to-ground capacitance also increases with thickness z. Energy resolution measured with 5.9 keV X-ray agreed with the theoretical calculation for the model structure.

The Ge pixel array is attached to contact pins and cooled by liquid nitrogen. Each output is fed into a separate analog preamplifier board. The transistor reset preamplifier was developed because of the following reason. In resistor feedback preamplifiers, the need for large feedback resistances (>10^9 ohms) to reduce the noise induced by the resistor imposes a limit on the counting rate. Pulsed-light feedback has drawbacks such as slow diffusion in the input FET of the charge generated by the infrared flash upon preamplifier reset. The PSC 741-N preamplifier consists of two components; an input circuit contained in the cryostat and cooled to a temperature slightly higher than that of 77 K, including the field-effect transistor, the reset transistor and the feedback capacitor and an external part containing a main board which processes the input pulses. To evaluate the performance, the analog-digital "hybrid" CAMAC module was developed [16]. The energy resolution was determined for each channel from the full width at half maximum (FWHM) of the Mn-K_α peak (5.9 keV) using a standard source (^{55}Fe). The best energy resolution was 215 eV at a low count rate using a shaping time of 6 μs. The average energy resolution evaluated from FWHM values of Mn K_α peak (5.9 keV) was 235 eV with $z = 7.5$ mm (#2).

The average FWHM value for $z = 10$ mm (#1) degraded to 245 eV due to the increase of pixel-to-ground capacitance. Amplifier dead time degrades the throughput due to the pulse pile-up. The total maximum count rate is evaluated to be maximum *ca.* 10 MHz. The present maximum throughput is limited by both analog (amplifier dead time) and digital (SCA and memory) circuits. If a digital signal processor (DSP) is used, the total throughput can be upgraded. Recent progress in DSP technology has made the maximum count rate of 800 kcps per channel [21]. Combined with such a rapid data acquisition system, potential of Ge PAD as a high throughput detector (*ca.* 0.1 GHz) would be feasible. Count rate for Ge SSD is much slower than silicon [22] which is already entering into the GHz range.

Apart from an advantage in counting rate, segmented fluorescence detection provides superior fluorescence signal purity which is monitored real time over a wide solid angle.

Segmented detection of fluorescence minimizes systematic noise arising from scattering or diffraction that cause nonlinear response

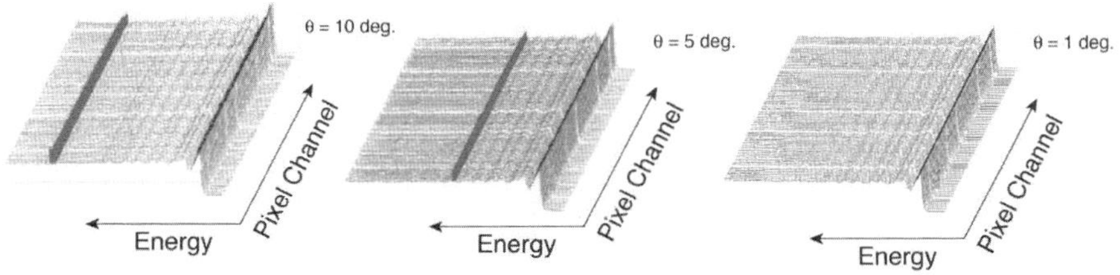

Fig. 5.12. Fluorescence yield spectra independently measured for 100 pixels. Sample is TSFZ grown *ab*-plane oriented LaSrCuO thin film single crystal grown by MBE on single crystal substrate. Anomalies caused by strong interference (scattering, diffraction and standing wave) is reduced by grazing incidence angle and is completely removed by optimizing the incidence angle. Taken from [11].

of fluorescence detection as illustrated in Fig. 5.12. By choosing a proper incidence angle (1 deg) corresponding to a film thickness and further adjustment of orientation, the substrate scattering effect is completely removed. A typical LSCO bulk single crystal grown by TSFZ method was 2 mm × 2 mm × 1 mm in dimension. Thin film single crystals (10 mm × 10 mm × 100 nm) were grown on single crystal substrates by molecular beam epitaxy (MBE) [23]. Decrease of incident angle (center) and *in-plane* rotation (right) remarkably remove energy dependent glitch due to strong substrate scattering. Figure 5.13 shows the normalized EXAFS spectra with various geometries. Strong anomaly data points are completely removed by optimized geometry.

Single crystal samples in our experiments are attached to an aluminum holder with a strain-free glue and cooled down using a closed-cycle He refrigerator. In EXAFS where the variation of absorption coefficients is several percent in magnitude, each data set must have photon statistics better than 0.1% or 10^6 photons. In a typical experiment, each data point collects an average of 3.6×10^7 photons making statistical (non-systematic) noise negligibly small. The 36-pixel Ge PAD (second generation) showed higher energy resolution [19]. At this moment, one can have third generation Ge pixel array detector with better energy resolution, FWHM@5.9 keV <130 eV (long shaping time) and less degraded value with high count rate <230 eV (0.5 μs).

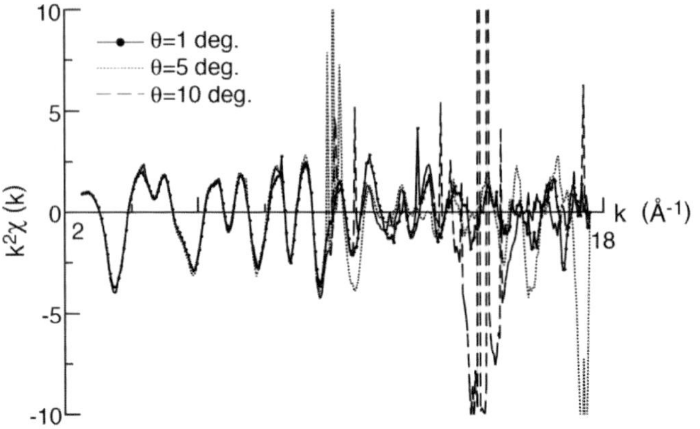

Fig. 5.13. Cu K-EXAFS spectra for 100 nm thick LSCO thin film single crystal grown by MBE on LaSrAlO substrate, corresponding to raw XAS spectra shown in Fig. 5.20. By choosing a right incidence angle ($\theta = 1$ deg.), artifacts are completely removed. EXAFS curves in closed circle, dot and dash denote spectra with $\theta = 1$ deg., 5 deg. and 10 deg., respectively.

5.4. Photo-induced Phase Transition

Photo-irradiation stimulates a cooperative phase transition between a ground state and a metastable state which may not be the same phase with that of thermally excited phase transition [24]. Photo-induced phase transitions such as ionic-neutral phase transitions in organic charge transfer (CT) crystals [25], magnetic phase transitions in spin crossover complexes [26], and photo-induced charge, spin and orbital order/disorder transitions in perovskite manganites [27] form rich class of multi-stable ground state materials where photo-stimulation eventually results in a global phase transition with pronounced changes in macroscopic physical properties.

[Fe(II)(2-pic)$_3$]Cl$_2$EtOH (2-pic=2-aminomethyl-pyridine), hereafter abbreviated as Fe-pic, is a typical spin-crossover complex which shows a thermally induced first-order phase transition from a low-spin (LS, S=0) to a high spin (HS, S=2) state as shown in Fig. 5.14.

Photo-induced phase transition is understood in terms of trapping of metastable spin state, *i.e.* light-induced excited spin state trapping (LIESST) [28]. A structural relaxation in an excited state may proceed with symmetric and non-symmetric normal coordinates

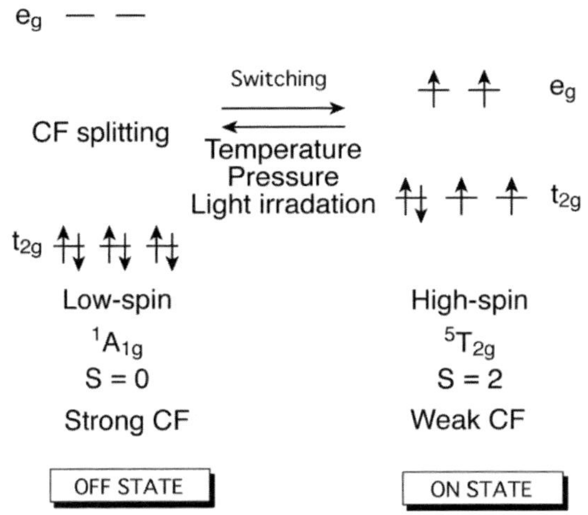

Fig. 5.14. Energy diagram and spin configurations of [Fe(II)(2-pic)$_3$]Cl$_2$EtOH (2-pic=2-aminomethyl-pyridine). Structural phase transition is dependent on stability of spin configuration against ligand field.

and therefore symmetry breaking could be observed. The excited state trapped by a metastable state may also relax to another *"false"* ground state which may not be the same with the original one and the nature of photo-induced phase is still unclear. Details of photo-excitation dynamics have been extensively studied by ultra-fast time-resolved synchrotron experiments [29, 30] and electronic aspects are becoming more clear. However, the nature of phase transition is closely linked with structure that causes nonlinear phase transition which is the central topic of investigation here. As chemical modification strongly affects phase transition, researchers are eager to synthesize modified versions of photo-sensitive materials. Naturally, quick inspection of irradiation effects on powder samples are required for which fluorescence detection is convenient.

Tayagaki *et al.* recently investigated the symmetry of photo-induced phase by Raman spectroscopy and observed possible broken symmetry in the photo-induced phase which is distinguished from the thermally induced phase [31]. The information on the local structure is essential to understand the nature of photo-induced high spin (PIHS) phase, *i.e.* symmetry breaking. The Fe K-edge X-ray

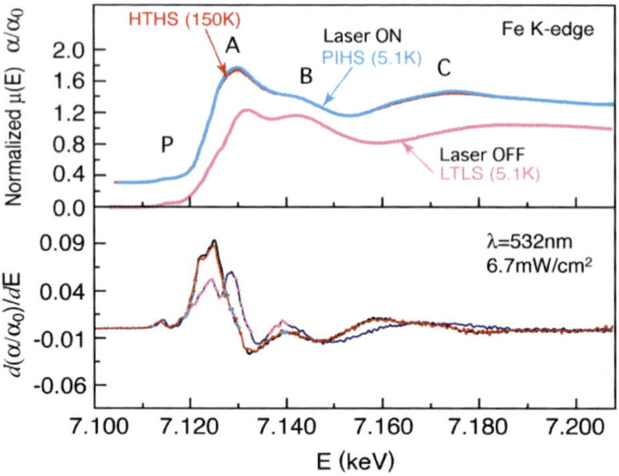

Fig. 5.15. Upper column, Fe K-XANES spectra for [Fe(II)(2-pic)$_3$]Cl$_2$EtOH (2-pic=2-aminomethyl-pyridine). Photo-induced high spin state (PIHS), thermally induced high temperature high spin state (HTHS) and low temperature low spin state (LTLS) are shown as blue, red and pink curves, respectively. Lower column indicates the derivative spectra.

near-edge absorption structure (XANES) spectra and the difference spectra taking the low temperature low spin (LTLS) phase as a standard are shown in Fig. 5.15. Energy shift of features B and C to higher energy upon laser light (532 nm) illumination is due to the $\mathrm{S}=0 \rightarrow \mathrm{S}=2$ spin transformation.

Thermally induced first order $S = 0 \rightarrow S = 2$ phase transition occurs at 120 K. Photo-induced spin conversion induces stretching of $R_{\text{Fe-N}}$ on going from a low spin state (strong ligand field) to a high spin state (weak ligand field). Fine powder sample (less than 1 mg) was embedded into carbon tape in order to achieve good thermal contact. Use of PAD allowed us to easily collect more than 10^6 photons per data giving 0.1% statistics. The local structure around Fe ions was studied by Fourier transform (FT) of the Fe K-EXAFS oscillations multiplied by k^3. The FT magnitude plots clearly show difference between LTLS and HTHS, between PIHS and LTLS, but surprisingly, similarity between PIHS and HTHS phases. Prominent peaks around 1.5–1.7 Å are due to the nearest neighbor (NN) nitrogen atoms.

Correcting the phase shift for the Fe-N contributions using FEFF code [32], the $R_{\text{Fe-N}}$(LTLS) and $R_{\text{Fe-N}}$(HTHS) values are determined as 2.01±0.01 Å and 2.20±0.01 Å, respectively. $\Delta R_{\text{Fe-N}} \sim 0.20$ Å is in agreement with crystallographic data of Mikami *et al.* [33]. The monoclinic space group $P2_1/c$ is unchanged upon the thermal phase transition. Recently, dynamics of photo-induced spin transition showed nonlinear characteristics such as the threshold light intensity, incubation period and phase separation. Although nonlinear properties are explained by a simple two-step model with a symmetrical lattice vibration mode with a long-range interaction, the recent Raman experiment raised a puzzling question, *i.e.* possible symmetry breaking in the near neighbor coordination.

Difference between the photo- and thermally induced HS phases is seen in the next-nearest neighbor (NNN) peaks due to Fe-C distances. The nearest Fe-N peak is almost the same for the two phases except for thermal broadening in the high temperature phase. Symmetry breaking during photo-induced spin conversion has been an open question. The local structure (FeN_6 octahedron and coordinated ligands) of Fe-pic is schematically illustrated in Fig. 5.16. The XANES and EXAFS results suggest that the octahedron expands by *ca.* 0.2 Å, maintaining the nearly octahedral (O_h) geometry. Nonlinear characteristics of phase transition indicates degraded cooperativity [34]. Let us find why this occurs. Three ligand molecules in a unit cell are linked by hydrogen bonds and compressive strain caused by local expansion propagates *via* a long-range intermolecular interaction. The LS→HS conversion in one molecule is associated with the local expansion of FeN_6 cluster by more than 30% *local* volume change which gives a compressive strain to the nearest neighbor three molecules.

As the Fe-C (NNN) distances expand roughly by the same amount, the FeN_6 deformation leads to isotropic expansion of ligand molecules as illustrated in Fig. 5.16. On the other hand, the PIHS phase is formed at low temperature where thermal relaxations does not take place. Local strains give rise to distant bond bending distortions observed as inversion symmetry breaking. A subtle change

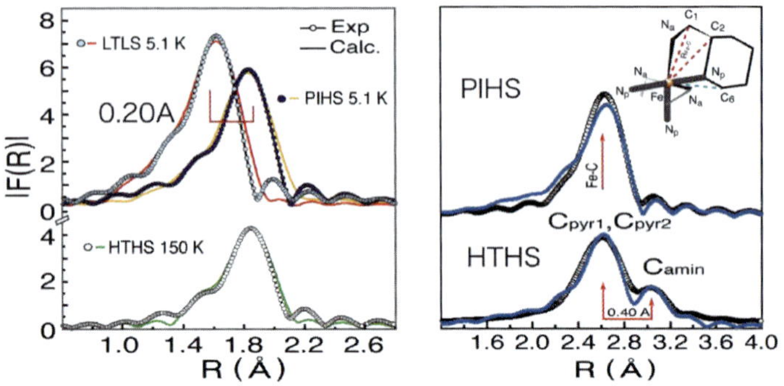

Fig. 5.16. Left column, nearest neighbor peak of LTLS and PIHS phases of $[Fe(II)(2\text{-pic})_3]Cl_2EtOH$ (2-pic=2-aminomethyl-pyridine) compared with HTHS phase. Experimental data and theoretical calculations are plotted in open circle and solid line, respectively. Right column, next nearest neighbor peak for PIHS and HTHS Fe-pic. The inset shows distorted geometry of ligand molecules in PIHS.

in the NNN Fe-C peak between the PIHS and HTHS phases indicate distortions of ligand molecules which would introduce inhomogeneity in strain distribution. Fe K-edge XANES and EXAFS results showed that the variation of nearest-neighbor local structure upon photo-induced $S = 0 \rightarrow S = 2$ spin transition is essentially an expansion of octahedral coordination (O_h) without lowering symmetry. The nearest neighbor Fe-N distance ($R_{\text{Fe-N}}$) increases by *ca.* 0.20 Å on going from LS to HS state, resulting in an increase of *10Dq* by 60%. Symmetry of the nearest neighbor FeN_6 octahedron in the photo-induced phase is unchanged on spin conversion by laser light illumination.

EtOH molecules that are disordered in orientation at room temperature are frozen in the photo-induced PIHS state. Because of hydrogen bonding, this solvent has played as a synapse of elastic force. Combining together with other experiments (XRD and Raman), we concluded that newly observed Raman lines are caused by disordered ligand molecules, while lack of homogeneity and frozen synapse disorder (solvent molecule) suppress cooperativity [34]. Photo-induced phase transition is like playing dominoes on a *non-flat* plane. We demonstrated *in situ* XAS studies with PAD lead to finding unusual trapped metastable states with unique properties.

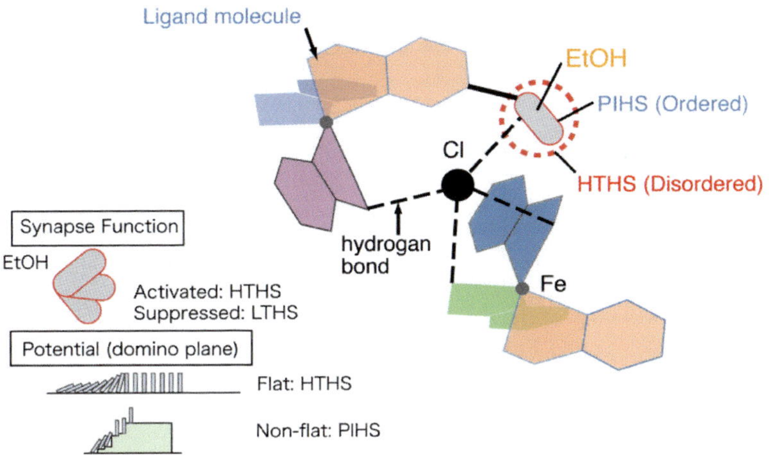

Fig. 5.17. Schematic representation of Fe molecular units linked by hydrogen bonds in [Fe(II)(2-pic)$_3$]Cl$_2$EtOH (2-pic=2-aminomethyl-pyridine). Each subunit with Fe ions is connected by hydrogen bond via EtOH molecule which acts as a synapse deciding direction of elastic force. At room temperature, EtOH is disordered in orientation but at low temperature, its orientation is frozen and synapse function switching elastic force is suppressed.

Photo-excitation creates spin conversion which locally expands the coordinated nitrogen atoms keeping symmetry. Local expansion is transferred to outer ligand molecules, which propagates elastic force to nearby units which are linked by hydrogen bonds *via* solvent molecules. Incomplete relaxation of ligand molecules causes inhomogeneous strain distribution.

Further, elastic force synapse function of solvent molecule is suppressed because of frozen orientation of molecules as schematically shown in Fig. 5.17 [35].

5.5. Lattice Effects in High Temperature Superconductors

While the microscopic mechanism of high temperature superconductivity (HTSC) is still a mystery more than 24 years after the discovery [36], recently reported superconductivity in fluorine-doped LaFeAsO (LFAO) [37] (Fig. 5.18) has revived interests in HTSC.

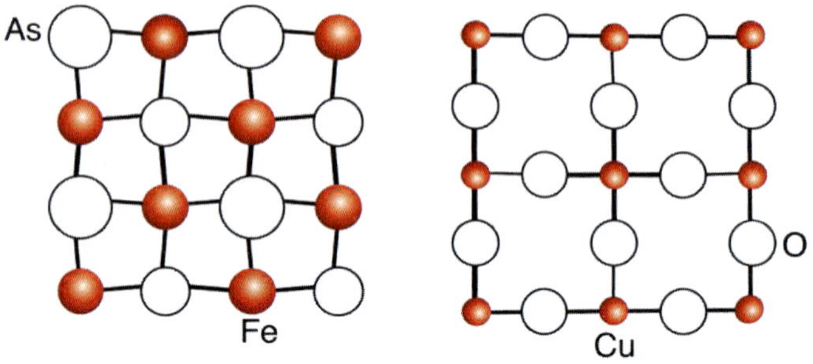

Fig. 5.18. Planar view of the conducting plane in iron pnictide (left) and cuprates (right). Four-fold Cu and Fe ions form the layered structure which is locally modified by polaron formation. Antiferromagnetic (AFM) spin configuration phase is adjacent with superconducting phase in both systems.

In LFAO, either replacing oxygen with fluorine or oxygen vacancy introduces charge carriers (electrons) which are transferred from the La-O(F) "charge reservoir" layer to the Fe-As conducting layer. Superconductivity emerges as the F-doping concentration exceeds about 5%, away from nondoped antiferromagnetic (AFM) parent phase. Inspite of similarity in the phase diagram with that of cuprates, *i.e.* superconducting phase near the AFM phase, the mechanism of superconductivity is still unclear, allowing diversity and contradictions in theoretical models. A conventional phonon mechanism is unlikely as the DFT calculations indicated a weak electron–phonon coupling, yet, the purely electronic mechanism [38] is also less likely than the case of cuprates as the Hubbard U is not large ($U < W$).

Let us recall that Cu K-EXAFS for cuprates detected a small displacement of oxygen atoms for doped specimen which is absent in non-doped samples [39–42]. Here we revisit this issue and describe the phenomena in terms of polaron resonance. In case of $La_{2-x}Sr_xCuO_4$ (LSCO), a typical cuprate superconductor, after doping holes by replacing La with Sr ions, local distortion occurs below a specific temperature (T^*) which corresponds to opening of a pseudogap. Polaron is formed when an electron is strongly coupled to nearby atoms in a crystal. Review article on polaronic mechanisms of superconductivity is found elsewhere [43].

Compared to manganites where coherent polaron condensation is the driving mechanism of colossal magnetoresistance (CMR) phenomena, less knowledge is accumulated for cuprates, although strong coupling between electrons and lattice in HTSC cuprates was indicated by angle-resolved photoemission spectroscopy (ARPES) [43]. Here we describe lattice effects in a typical cuprate $La_{2-x}Sr_xCuO_4$ (LSCO) focusing on polaron effects.

As the magnitude of EXAFS oscillations is in the order of 10^{-2} and anomalous lattice contribution is even smaller by an order, minimizing statistical error and systematic noise in FY measurement is essential. As structure of cuprates is strongly anisotropic (layered), polarization experiments to independently probe local structure along *in-plane* and *out-of-plane* directions are crucial. Polarized XAS experiments on single crystalline cuprates were performed by transmission technique on "sliced" single crystals or "oriented" powder specimen, which limits polarization geometry due to exponentially increasing absorption as the incidence angle is lowered. We found that unconventional use of FY for bulk single crystals can provide reliable polarized EXAFS [44]. This is based on the fact that "thin film approximation" of fluorescence detection was valid because of strong absorption by heavy ingredient elements such as La and Sr.

Difficulty in polarized FY experiments on *as-grown* single crystals (strong diffractions/scattering and standing wave excitations) was overcome by high-precision alignment of sample and segmented fluorescence detection. This becomes even more intensified for thin film single crystals because of strong interference of substrate. Use of segmented FY monitoring to discriminate noise and optimizing the incidence and *in-plane* orientation angles are crucial.

In the upper column of Fig. 5.19, the *in-plane* and *out-of-plane* polarized Cu K-EXAFS oscillations for LSCO ($x = 0.15$) taken at 20 K and 300 K are shown. The Fourier transform (FT) magnitude functions are shown in the lower column. The oxygen ion displacement is described by the mean square relative displacement (MSRD) which is plotted as a function of time. As seen in the plot, anomalous temperature dependence is seen in (1) "upturn" (negative

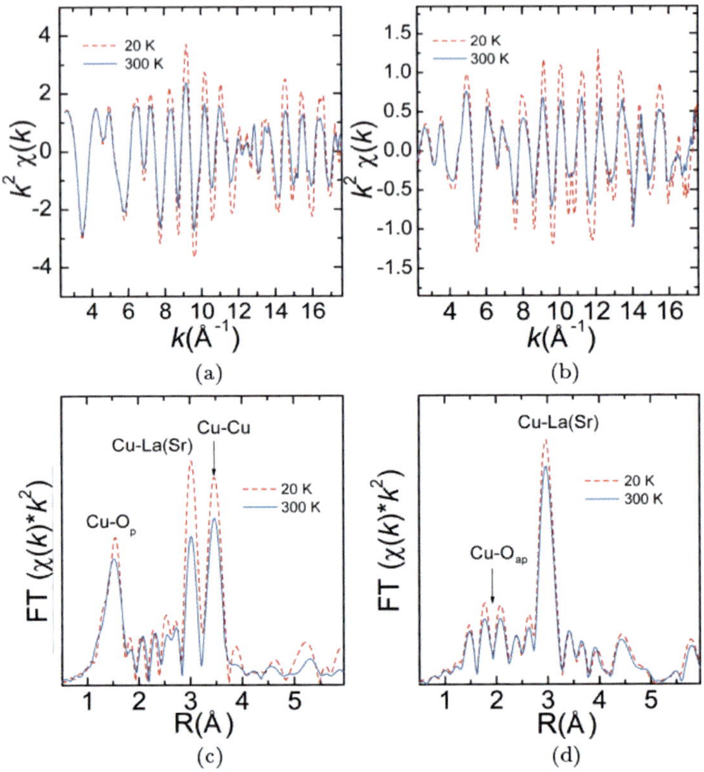

Fig. 5.19. Left upper column, *in-plane* polarized Cu K-edge EXAFS oscillations weighted by k^2 for $La_{1.85}Sr_{0.15}CuO_4$ at 20 K and 300 K. Left lower column, magnitude of the complex FT of k^2 times $\chi(k)$ for $La_{1.85}Sr_{0.15}CuO_4$ at 20 and 300 K. Right upper column, *out-of-plane* polarized Cu K-edge EXAFS oscillations weighted by k^2 for $La_{1.85}Sr_{0.15}CuO_4$ at 20 K and 300 K. Right lower column, magnitude of the complex FT of k^2 times $\chi(k)$ for $La_{1.85}Sr_{0.15}CuO_4$ at 20 K and 300 K.

temperature dependence) trend below T^* and (2) a sharp drop at the critical temperature as shown in Fig. 5.20. These anomalies indicate that local lattice around copper ions becomes distorted as a result of carrier doping. Here we note that the apical oxygen displacement has larger uncertainty because it is derived by a curve fit analysis together with La contribution but as can be seen "relative" error bar among sequential measurement as a function of time is much smaller as seen in small scattered data points.

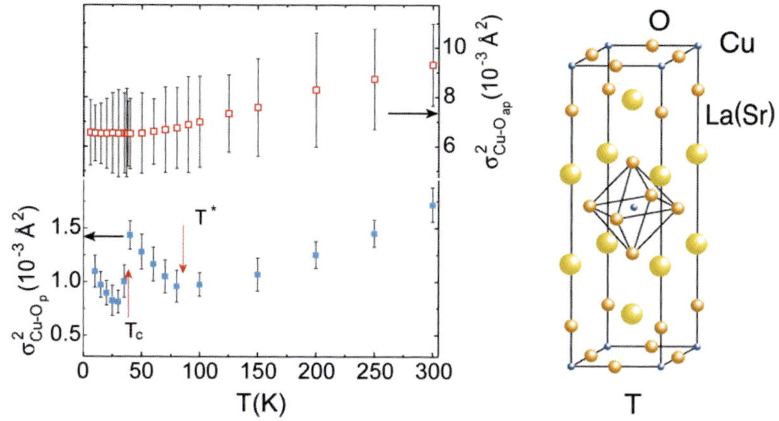

Fig. 5.20. Left, mean-square relative displacement (MSRD) of the Cu-O bond in superconducting $La_{1-x}Sr_xCuO_4$ ($x = 0.15$) with the *in-plane* ($E//ab$) and *out-of-plane* ($E//c$) geometries as a function of temperature. Figures are taken from [34]. Right, crystal structure of $La_{1-x}Sr_xCuO_4$.

Let us examine possible distortion modes that are consistent with the observed lattice distortion (bond splitting). Among possible distortion models having stretched bonds, models which have correlated displacement of apical oxygen atoms such as low-temperature-tetragonal distortion are omitted because of irrelevance of apical oxygen ions judged from smooth normal phononic behavior. In contrast, two *in-plane* distorted pairs in either pseudo-JT mode or Q_2-type JT mode are considered.

Correlated distortions allow long-range propagation with minimized elastic energy increase. Whether the nature of paired distortions is dynamic or not is unclear as the time scale of EXAFS experiments (10^{-15} s) is orders of magnitude faster than lattice motions (10^{-12} s). Before discussing the nature of lattice distortion mode further, we show that the anomaly observed in apparent increase of MSRD is due to splitting bond lengths, rather than disorder as the displacement is small (~0.1 Å), and could be doubted as experimental error. It is well known that two bond lengths that are separated by a small distance give "beat" in the EXAFS oscillations as illustrated in Fig. 5.21 (right column). Indeed, the appearance of beat feature at k ~ 12 $Å^{-1}$ corresponds to the bond length difference

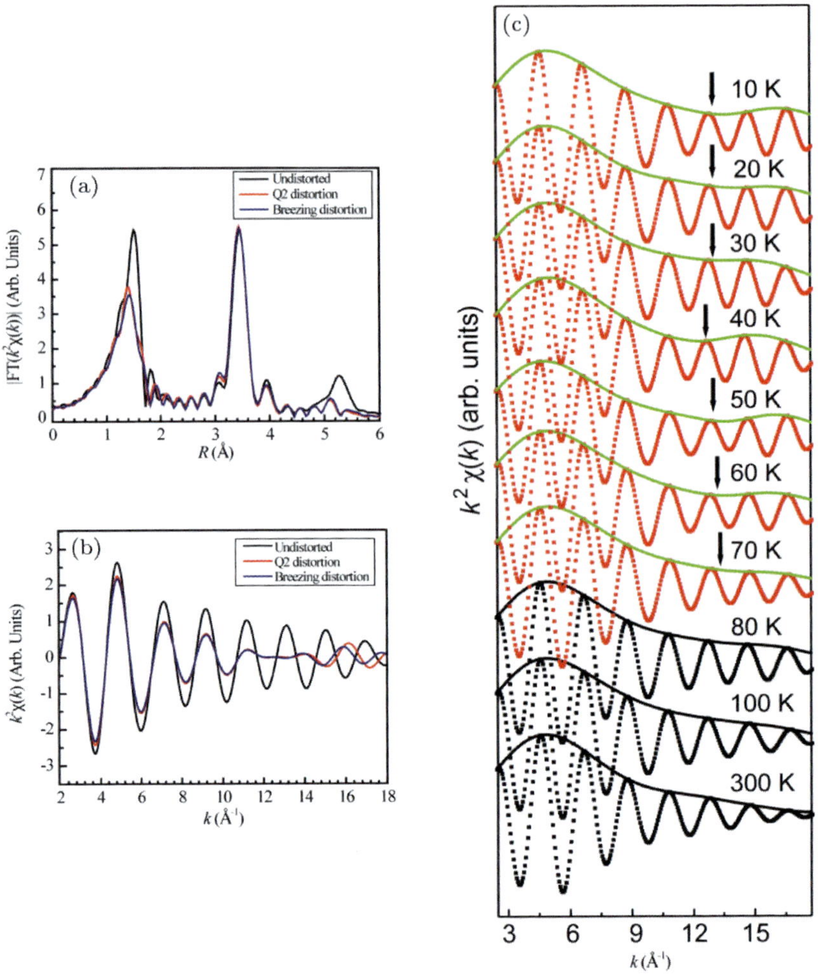

Fig. 5.21. (a) Fourier transform magnitude curves of simulated E//*ab* EXAFS oscillations for undistorted (solid line) and distorted CuO_4-CuO_4 units (red curve: JT Q_2 mode, purple curve: pseudo-JT (breezing) mode distortions). (b) Nearest neighbour contribution of simulated EXAFS oscillations for undistorted and distorted models. Interference of long and short Cu-O bonds give rise to beat features at $k \sim 12$ Å^{-1}. (c) Experimental nearest neighbor Cu-O EXFS curves measured at various temperatures. Beat feature similar to (b) is observed at $k = 14$ Å^{-1} near T_c (40 K).

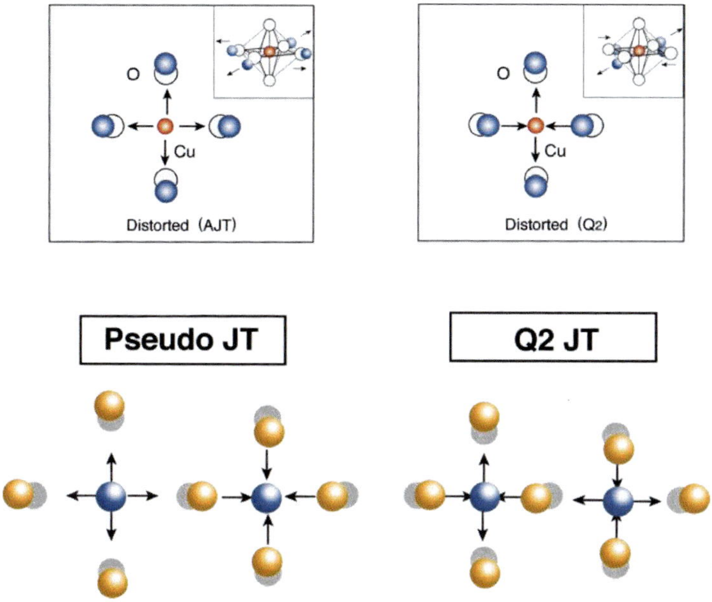

Fig. 5.22. Upper column, schematic local lattice distortions which result in expansion of *in-plane* Cu-O bonds. Lower column, correlated deformation schemes with *in-plane* atomic displacements. Two linked units are distorted in an opposite direction. In case of pseudo JT distortion model, oxygen atoms are distorted in a similar fashion with a breezing mode where all Cu-O bonds for each Cu are either elongated or shortened. In the case of Q_2 type JT distortion, half of Cu-O bonds are either elongated or shortened.

of ~0.1 Å. Assuming that these two bond lengths come from a single site due to Anti-Jahn Teller (JT) distortion such as Q_2 JT or pseudo JT distortions in the crystal structure (Fig. 5.22), model EXAFS oscillations were calculated. The results show that the anomaly we see in the *in-plane* EXAFS is indeed due to split bonds separated by ~0.1 Å^{-1}.

Let us discuss about some of the implications. Our view is that observed lattice anomaly reflects polarons formed at low temperatures which may play a significant role in pairing mechanism. Polarons are often considered as localized and regarded as hardly contributing to conduction but resonance [45] (tunneling [41] or percolation [46]) may contribute to metallic conduction within coherent homogeneous neighbors.

Here we argue that the two polaron models with different types of *in-plane* modes (pseudo JT [47] *versus* Q_2-type JT [48]) that are consistent with the experimental observation. The former model (pseudo JT distortions with the two CuO_6 units) gives $R = 1.82$ Å and 1.96 Å in good agreement with the EXAFS results [42] (Fig. 5.22).

The effect of superconducting coherence probed by MSRD observed as a sharp drop at T_c (proportional to polaron density) corresponds to the inflection point of resistivity $\partial r/\partial T$. This suggests that the onset of superconductivity is driven when localized polarons approach, exceed the critical value and resonate. Let us consider a simple model where the distance between polaronic domains L and the superconducting coherence length z_0 are key parameters.

As doping proceeds, reduced Coulomb repulsion (U) would promote integration into extended domains with a larger size. The average distance between extended domains L decreases allowing resonating (tunneling) over insulating domains. In contrast, suppressed resonating due to strain field such as tensile strain or impurity doping into copper sites decreases fraction of superconductivity (superfluid density), and hence the critical temperature.

In contrast, polaron may also create perturbations in spin configuration so that spin vortices are formed that grows into a macroscopic supercurrent [49]. The observed magnitude of oxygen displacement is in good agreement with the charged cluster calculation on the CuO_6 deformed pair [47] (Fig. 5.22). Whether polaron works to perturb spin configurations leading to spin vortices growing into spin loop current [49] or it enhances strong coupling of spin, charge and lattice is an open question which would be a key to understanding a microscopic mechanism of HTSC.

5.6. Initial Growth Studies of Nanoparticles and Clusters

The past two decades have witnessed an explosive expansion of research on semiconductor nanoparticles (NPs), due to their size-dependent properties and applications in a variety of fields such as solar cells, light emitting diodes, and biological imaging *etc.* To synthesize these NPs with desired size and shape and thus to tailor their structural and physical properties, understanding of the nucleation and growth processes is of essential importance. In most cases, the nucleation is described by the classical nucleation theory which assumes the same structure of nuclei as that of the stable NPs. NPs grown from these nuclei can be monitored by transmission electron microscopy (TEM) or optical spectra, such as UV-*vis* absorption and photoluminescence (PL). However, optical spectra provide little information during the time region when no optical absorption features are observed or the initial nuclei are not in ordered state, making UV-*vis* spectroscopy even more difficult to estimate the size. Since the nucleation strongly influences the subsequent growth process varying from system to system, *in-situ* structural information on nucleation is essential to optimize the properties.

The time-dependent X-ray absorption spectroscopy measured with a microfluidics is advantageous in sensitivity and key information on NPs (size and density) is available in addition to local structure (bond length, coordination number and relative displacement). The use of fluorescence detection is essential for dilute samples ($<3\,\mathrm{mM}$) but a conventional time dependence with a sealed cell allows a limited accumulation of fluorescence photons while statistics requires millions of photons accumulated per data point that limits a practical time window or time resolution. On the contrary, a microfluidics setup with a fluorescence detector allows one to take as much time as one needs for accumulating photons which overcomes the limitation for dilute specimen.

Recently, wet synthesis using microfluidics has attracted attention in various fields such as organic and inorganic chemistry, biology, and medicine which is a powerful synthesis method for NPs with respect to kinetic control. A special feature of microfluidics is controllability of reaction parameters (*e.g.* reaction time, temperature, and concentration) because of compact reaction volume. Following the first application of microfluidic cell to synchrotron radiation research

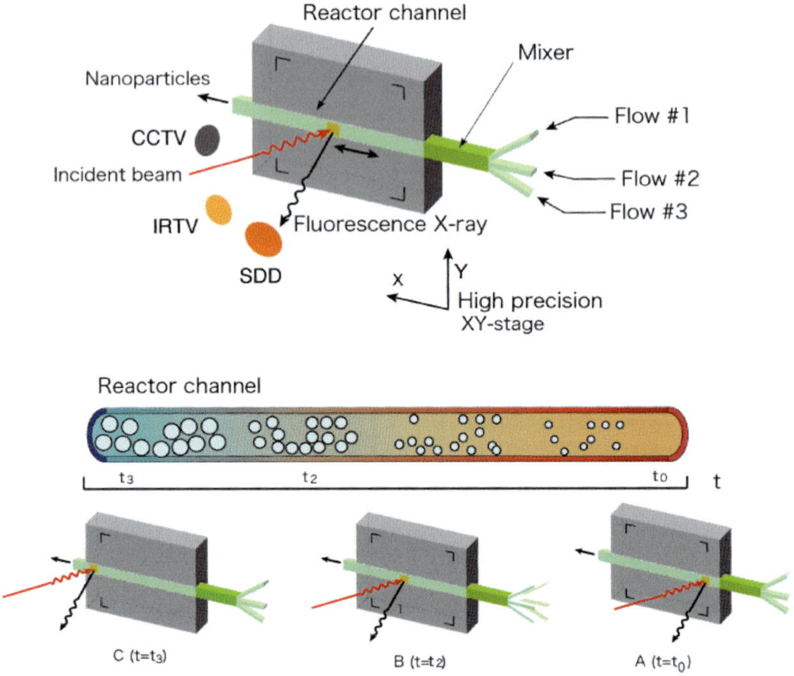

Fig. 5.23. Schematic principle of *in situ* XAFS experimental setup where time-dependence is obtained by probing a specific point along a microfluidic cell mounted on a high-precision XY stage (top column). Along a reactor channel (middle column), the initial stage of the growth of nanoparticles is observed by changing the position (bottom column).

[50] using a micro-beam to observe a cation exchange reaction, we performed the *in situ* XAS experiment using a focused undulator X-ray beam which matches a typical size of a flow cell (hundred microns in inner diameter) designed for initial growth study of CdSe NPs [51].

Figure 5.23 schematically illustrates the principle of the experimental setup for time-dependence of XAS during the nucleation and growth of NPs. The basic idea is to measure fluorescence XAS by means of position-dependence recorded at different positions along a laminar flow of a reactor channel. A constant velocity flow conveys a small volume of dilute solutions mixed prior to the injection to the reactor which creates nuclei upon heating that grows into the NPs as a function of time. Since the incident beam is fixed in position, a reactor channel needs to be controlled in position so that the beam always impinges on the region of interests. This can be achieved by varying the position of a microfluidic cell on a computer-controlled high-precision XY-stage. One can then record position-dependent XAS spectra in a fluorescence detection mode. Temperature distribution of the reactor channel is monitored real-time by an IR-TV camera.

An abrupt temperature rise is observed near the flow inlet ($t = t_0$) and a constant temperature distribution extending along the rest of channel. *In situ* EXAFS measurement at the Se K-edge was performed at beamline BL13B1 of Photon Factory (PF) and NW2 of Advanced Ring for Pulsed X-rays, (PF-AR) of High Energy Accelerator Research Organization, Japan. Reactor *KAPTON* tube is embedded in a thermal insulator and mounted on an aluminum holder as illustrated in Fig. 5.23. Both glass and KAPTON materials were resistive to radiation damage in the hard X-ray region (~11 keV) while TEFLON quickly deteriorates by high flux (NW2 focused undulator) beam, although the latter has superior characteristic against precipitation on the tube wall. Three raw material stock solutions (Se source, Cd source and dodecylamine (DDA) ligand solutions) were prepared. The selenium stock solution (TOP-Se) was prepared by dissolving selenium powder in trioctylphosphine (TOP) and then diluted in octadecene (ODE), a non-coordinating solution. The cadmium stock solution was prepared by heating a mixture of $Cd(CH_3COO)_2{\cdot}2H_2O$, oleic acid and ODE at 180°C under Ar gas flow. The DDA solution was prepared by dissolving DDA in ODE. These stock solutions were loaded into glass syringes separately and then mixed in a mixer connected to a capillary KAPTON tube with an inner diameter of 0.5 mm. Several reactor channel materials such as TEFLON, KAPTON and glass were tested and compared in terms of radiation damage and precipitation on the tube surface.

A length of 70 cm of the capillary tube was attached to the heating unit, where the mixed solution injected at a constant velocity of 7.6 mm/s was heated to a target temperature of 240°C.

In the mixed solution, the Se and Cd concentrations were kept at 30 and 12 mM, respectively, and the DDA concentration varied from 5 to 10 wt.%. The linear microfluidic cell is mounted on a stepping-motor-controlled XY-stage which rotates around the goniometer whose rotating axis is perpendicular to the stage normal, allowing either horizontal or vertical arrangements of a reactor. Our standard reactor mounting is horizontal while vertical mounting is preferable for moving reactor channel to minimize precipitation. For each DDA concentration, a series of EXAFS data, corresponding to the reaction time of 0.0, 1.1, 1.4, 1.8, 2.4, 3.0, 5.0 and 8.1 s, were collected. Under the same reaction conditions as in the *in situ* EXAFS experiment, *ex situ* UV-*vis* absorption and PL spectra were measured using a UV-*vis* spectrophotometer. PL spectra were measured at an excitation wavelength of 365 nm. The optical spectrophotometers were attached to a glass tube with an inner diameter of 0.5 mm connected to the outlet of the reactor tube. While a typical intensity of a fluorescence signal for Se-TOP exceeds 50–60 kcps above the Se K-edge the intensity of fluorescence X-ray was kept within at a safe limit to minimize dead time correction.

The temporal evolutions of the UV-vis absorption and PL spectra of the CdSe NPs clearly indicate the wavelength shift intrinsic to the size of NP in nanometer regime. Evidently, the absorption and luminescence curves are featureless within the first 2.4 s of the reaction showing difficulty in the application to the initial growth. As the reaction time approached to 2.4 s or longer, excitonic absorption and PL peaks appear, with the intensities and peak positions increasing rapidly with time. The appearance of these absorption and emission peaks indicates the formation of crystalline CdSe NPs, and the gradual red-shift of the peaks displays their size growth. The full-width at half-maximum (FWHM) of the PL peak in the presence of 5 wt.% DDA decreases continuously from 53 nm at 2.4 s to 35 nm at 8.1 s, indicating the size focusing of the crystalline particles accompanied with the growth. However, no obvious size focusing of the crystalline NPs was inferred for the 10 wt.% DDA case. The absence of any visible UV-vis absorption peak in the wavelength range of 350–450 nm

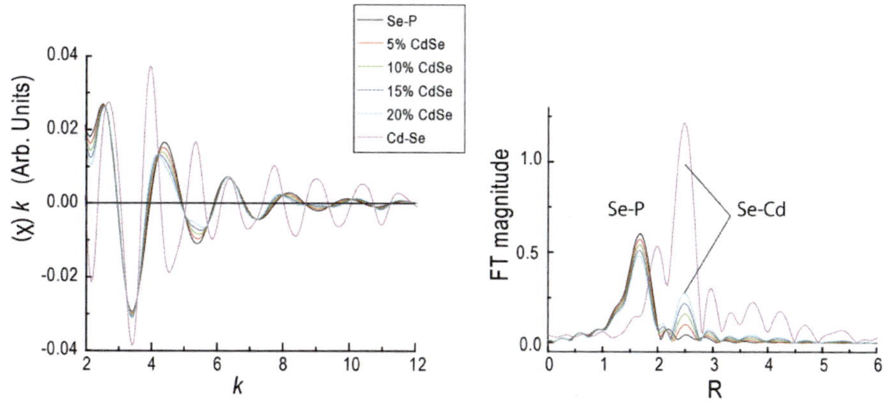

Fig. 5.24. First-shell Se K-EXAFS oscillations for Se-TOP containing various concentration CdSe simulating the formation of CdSe nanoparticles (left column). Fourier transform of the Se K-EXAFS oscillations for the mixture of Se-P and Se-Cd bonds.

rules out the possibility of existence of magic-size clusters, which are known to be stable only at relatively low temperatures of 90–150°C. Using the absorbance intensity and the FWHM of the PL spectrum, the reaction yield of Se, the average size and concentration of the NCs can be estimated. As illustrated in Fig. 5.24 which shows the simulated EXAFS oscillations and the Fourier transform on the Se K-edge, the important point is that it is necessary to observe the change of modulation in the EXAFS spectra with good statistics. Accumulating millions of photons (fluorescence signal) is required to observe the breaking of the Se-P bonds and the formation of Se-Cd bonds.

The Cu NCs were synthesized by a wet-chemical reduction method in organic solution. The copper stock solution, denoted as Cu-$(hfac)_2$ was prepared by dissolving Cu(II) hexafluoroacetylacetonate [Cu$(hfac)_2$] with diethylene glycol diethyl ether (DEGDEE). The Cu-$(hfac)_2$ solution was reacted with oleylamine (OLA) as a ligand and finally mixed with a reductant, diphenylsilane (DPS). The Cu K-XANES spectra were sequentially taken at each step, *i.e.* (i) after the dissolving with a solvent, (ii) after a ligand exchange

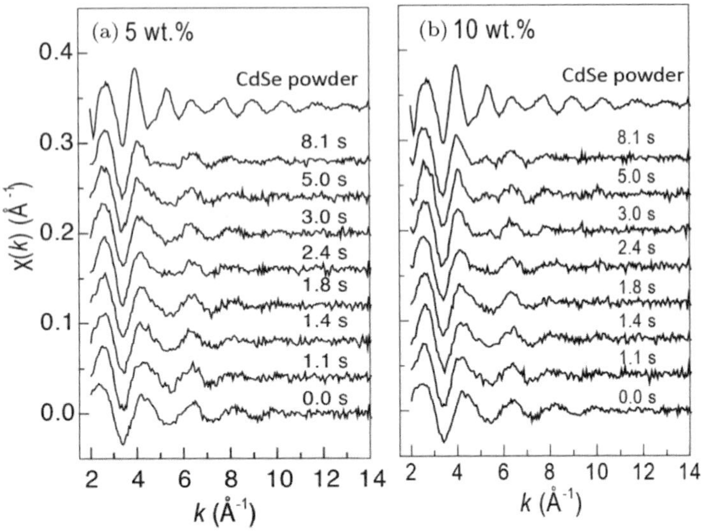

Fig. 5.25. (a) The Se K-edge EXAFS oscillation function $\chi(k)$ with the addition of 5 wt.% dodecylamine (DDA). (b) Magnitude of the Fourier transform (FT) of the k^2-weighted EXAFS data. The solid lines represent the experimental data and the open circles show the fitting results.

reaction and (iii) after adding DPS. We also measured reference compounds CuO (Cu^{2+}) and Cu metal (Cu^0) in both transmission and fluorescence modes.

The Fourier transform shows a prominent peak at 1.6 Å (the Se-P bond of TOP-Se) which gradually decreases its intensity as a fraction of CdSe increases associated with the emerging of the Se-Cd bond. The Fourier transform of the EXAFS oscillations are shown in Fig. 5.25, respectively. The oscillation functions were extracted from the raw experimental spectra using the ATHENA module [52], and the least-squares parameter fitting was performed using the ARTEMIS module [53]. Both modules are implemented in the IFEFFIT package [54]. Solid lines in Fig. 5.26 indicate the Fourier transform magnitudes of the k^2-weighted $\chi(k)$ functions using a Hanning window and a k-range of 2.5–10.5 Å^{-1} (shorter k-range of 2.5–8.5 Å^{-1} was used for the 3.0 and 5.0 s data for 10 wt.% DDA due to worse statics of the data). At the starting time $t = 0.0$ s, only a prominent peak at 1.6 Å ascribed to the Se-P bond of TOP-Se is visible

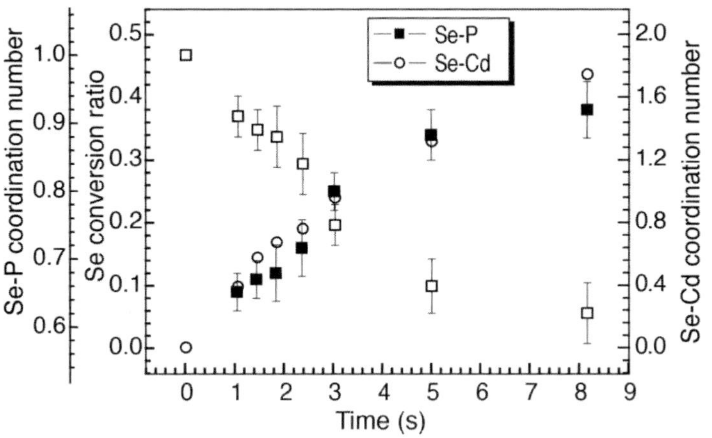

Fig. 5.26. Time-dependence of the Se-P coordination number, Se conversion ratio, and Se-Cd coordination number estimated from the analysis of the Se-P and Se-Cd bonds for the TOP-Se solution.

for all DDA concentrations, which resembles the spectral shape of a starting material (TOP-Se). After that a peak at 2.5 Å attributed to the Se-Cd bonds becomes apparent even at the reaction time as short as 1.1 s, indicating a quick formation of CdSe particles. This observation is remarkably different from those by optical spectra, which indicated the formation of CdSe NCs only after 2.4 s of reaction.

Upon further increasing the reaction time, the Se-Cd peak intensity is enhanced, and at the same time, the Se-P peak intensity decreases gradually in Fig. 5.26. Furthermore, the Se-Cd peak intensity increase (or the decrease of the Se-P peak) is quicker when the DDA is added, consistent with the fact that a surfactant (DDA) contributed to the conversion of TOP-Se into CdSe particles. Least-squares parameter fitting was performed to extract quantitative results using the same fitting strategy as described in detail in our previous paper.

The obtained bond lengths ($R_{\text{Se-P}} = 2.11 - 2.13 \pm 0.02$ Å, $R_{\text{Se-Cd}} = 2.59 - 2.63 \pm 0.04$ Å) did not show obvious time-dependence within the error bars, indicating that the size growth does not change the local structure in the growth regime in contrast with the small clusters (SCs) that depend strongly on size. The formation of CdSe particles observed by EXAFS at 1.1 s is not consistent with the results of UV-*vis* absorption spectroscopy which gives the formation after 2.4 s. Such discrepancy is explained by the formation of amorphous NPs for which the current XAS study is clearly advantageous over UV-*vis* spectroscopy. The NC size increase with time demonstrates the simultaneous occurrence of growth and crystallization of CdSe NPs under current reaction conditions. In the presence of 10 wt.% DDA, the percentage of crystalline particles is below 10% within the first 3 s of reaction, implying the dominant existence of amorphous nuclei in the initial reaction stage. This observation is in agreement with previous theoretical and experimental studies that the initial nuclei are disordered, contrary to the CNT assuming the same structure of nuclei as that of the stable NPs.

The plot clearly indicates that modest surfactant concentration (upto 5 wt.%) contributes to a rapid growth of size during the initial stage (2–5 s) but higher concentration (10 wt.%) found that the addition of surfactant ineffective at all (See Fig. 5.27). This surprising result may come from too much surfactant on the surface negatively interacting due to Coulomb repulsion and steric hindrance, causing NPs apart or prohibiting the growth. Thus without a surfactant or too much of them lead to a rather constant size (2.2–2.3 nm) during

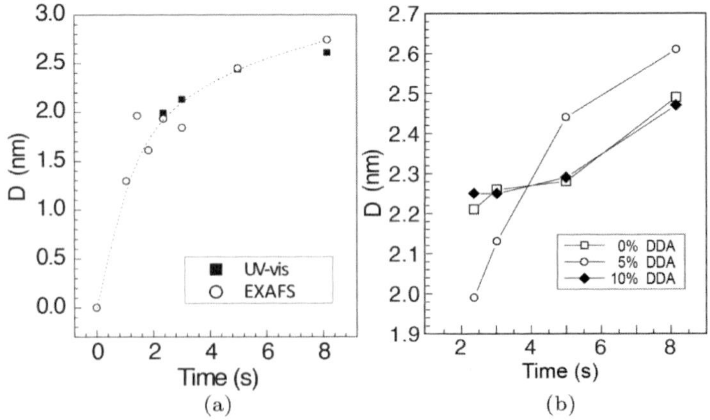

Fig. 5.27. (a) Temporal evolution of the partical size for CdSe nanocrystals compared to the UV-*vis* results. (b) Variation of the particle size estimated from the bond formation kinetics in the presence of 5 and 10 wt % DDA.

the initial period (2–5 s) slowing down of the further growth. The results clearly demonstrate the effect of surfactant that is *nonlinear* to the concentration.

Although theoretical studies of the role of surfactant are necessary, the present result establishes the superiority of XAS as a means of study of surfactants in the growth of NPs. In the case of SCs consisting of 10–20 atoms, the XANES spectra becomes dependent on the details of cluster such as the symmetry and surface effect. In Fig. 5.28, a model structure of $CdSe_{13}$ cluster is shown in the left column. The MS calculation of XANES spectra for each paths compared in the right column show that how the average spectrum is formed by the sum of separate paths. By analyzing the experimental data compared with the MS calculations for model clusters, one can evaluate the model and symmetries in some cases.

In situ XAS technique was applied to the Cu SCs prepared by a reduction reaction of Cu(II) in an organic solution. Cu K-XANES spectra were recorded as the copper stock solution was prepared by dissolving Cu(II) hexafluoroacetylacetonate [$Cu(hfac)_2$]

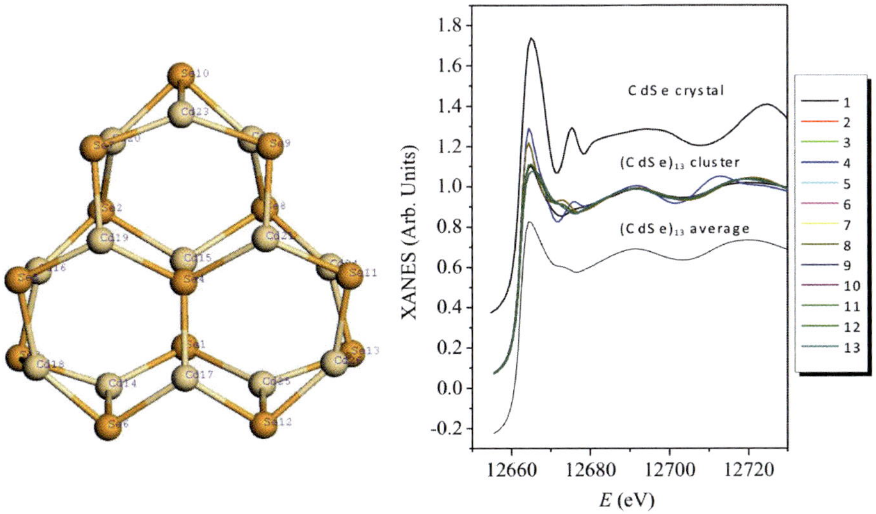

Fig. 5.28. $CdSe_{13}$ model structure (left column) and the experimental and theoretical calculations of the Se K-XANES spectra by the MS calculation for the model structure (right column).

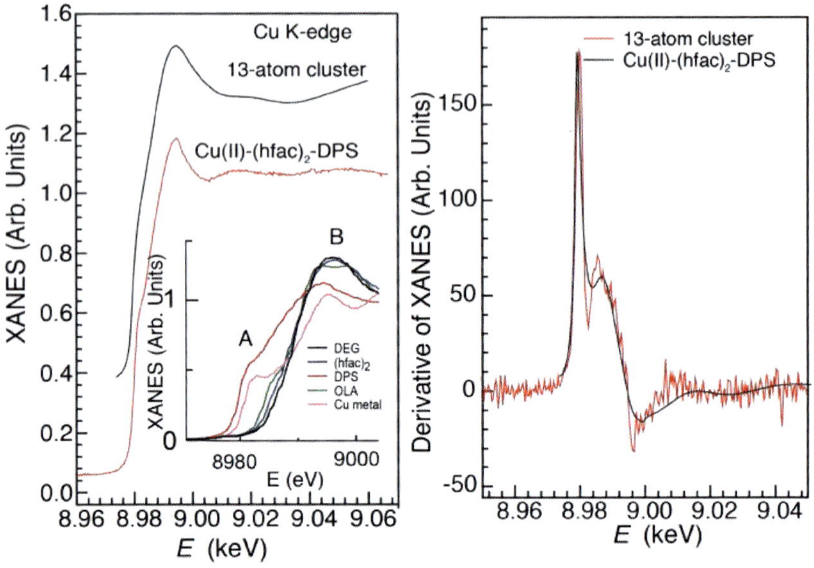

(a) (b)

Fig. 5.29. Comparison of the experimental and theoretical XANES spectra for a 13-atom copper cluster in (a) raw and (b) derivative spectra, respectively. The experimental XANES data (denoted as Cu-$(hfac)_2$-DPS in Fig. 5.1) was taken after the reduction by diphenylsilane (DPS).

with diethylene glycol diethyl ether (DEGDEE). The Cu-$(hfac)_2$ solution was reacted with oleylamine (OLA) as a ligand and finally mixed with a reductant, diphenylsilane (DPS). The near-edge spectra in the inset of Fig. 5.29 sequentially taken at each step; (i) after the dissolving with a solvent, (ii) after a ligand exchange reaction and (iii) after adding DPS clearly indicate the formation of the Cu(0) state in the final stage ($Cu^{2+} \rightarrow Cu^0$). A sharp absorption-edge shoulder seen for the Cu^{2+} data has been assigned to a "shakedown" process involving a Cu^{2+} $1s \rightarrow 4p$ transition simultaneous with ligand-to-metal charge transfer which is strongly polarized to the axial ligand direction. Although not shown in this figure, the valence state Cu^+ is associated with a sharp $1s$–$4p$ transition and the near edge features can estimate the Cu valency between 0 and +2. The near-edge features are also dependent on the local structure and the data for fcc Cu metal are characterized by the features at around 9000 eV.

The location of the threshold of peak A (8980.3 eV) is the Fermi level of the Cu metal (fcc) while the presence of B and C (not shown) indicates a larger cluster than $n = 43$ [50].

The experimental data shown in Fig. 5.29(a) is well reproduced by a SC with $n = 13$. The MS calculation showed that the characteristic double-peak near-edge features (B, C) around 9000 eV appear only when n exceeds 43. As SCs ($n \leq 19$) have a single peak at this energy, the first peak feature around 9000 eV can serve as a signature of a SC ($n \leq 19$). Figure 5.29(b) compares experimental XANES for the Cu SC with the theoretical calculation on a 13-atom model cluster, where (a) and (b) correspond to raw and derivative spectra, respectively. In case of a 13-atom SC which separates large optical gap small clusters, only the central atom is coordinated by the near neighbor copper atoms and other 12 atoms are on the surface, strongly interacting with ligand atoms or solvent molecules. As the surface of SCs in solution may be coordinated with amine molecules as a surfactant, the interaction with ligands (OLA) and amine (DPS) must be taken into account to better describe the local structure. Focusing on a 13-atom cluster, there are three possible models with different symmetries, cubo-octahedron (fcc), hcp and icosahedron. Our XANES calculation was based on a cubo-octahedron which smoothly connects to a fcc structure neglecting the size-dependent bond length. Although other polymorphs (hcp and icosahedron) are less likely in terms of total energy and packing fraction, conclusion on symmetry must be reserved.

The advantage of a microfuidic cell is the sensitivity with respect to volume and concentration and a wide dynamic range in elapsed time (depending on probe beam, flow rate and channel length). Most importantly for XAS experiments, signal (fluorescence) photons are accumulated without limitation of time window of a conventional Q-scan allowing one to move back and forth in time scale by positioning the incident beam along a reactor channel. Our *in situ* EXAFS study on CdSe NPs found discrepancy with the optical spectroscopy that implies that at the initial stage of the nucleation and growth, most of the particles are amorphous, contrary to hypothesis of the classical nucleation theory. These amorphous nuclei would crystallize rapidly over time, with the rate depending on the reaction conditions such as growth temperature, reaction time, surface ligand, and so on. The effect of surfactant (amine) was studied through a comparison of the growth and crystallization processes in the presence of DDA with different concentrations. The results show a strong non-linear effect in surfactant concentration. The results indicate that surfactants play in the beginning competitive roles: accelerating the formation of amorphous nuclei, and eventually hindering the subsequent crystallization and growth when the ligand concentration exceeds a critical limit. The local structure of copper small clusters grown by a chemical reaction in organic solution was investigated by XANES and MS calculation. The Cu K-edge XANES confirmed the reduction to monovalency as SCs are formed upon a reduction reaction with a reductant. Comparing with the size-dependent XANES calculation based on a multiple scattering calculation for n-atom model clusters, it was found that the experimental XANES data is well-reproduced by a 13-atom model with a cubo-octahedron symmetry.

Acknowledgements

The author would like to express his thanks to the Research and Development team, C. Fonne, D. Gutknecht, P. Dressler, R. Henck and M.-O. Lampert, S. Ogawa and K. Kasai. Early stage of segmented detector R&D efforts by M. Martini and M. Saito are also acknowledged. The author would like to express his thanks to T. Tayagaki, K. Tanaka, N. L. Saini, A. Tsukada, M. Naito, Y. Kamihara, H. Hosono, Z. Sun and C. Zhang for collaborations. Encouragements by A. Bianconi and K. A. Müller are also greatly acknowledged. The author also thanks A. Bussmann-Holder, H. Koizumi, J. Mustre de Leon, J. Ranninger, H. Kamimura, S. Conradson and A. Bishop for invaluable discussions. The author greatly expresses his thanks to X. Zhang for encouragement throughout the work.

Bibliography

[1] J. Jaklevic, J. A. Kirby, M. P. Klein, A. S. Robertson, G. S. Brown and P. Eisenberger, *Solid State Commun.* **23**, 679, (1977).
[2] B. Chance, R. Fishetti and L. Powers, *Biochem.* **22**, 3820, (1983).
[3] P. H. Citrin, P. Eisenberger and R. C. Hewitt, *Phys. Rev. Lett.* **41**, 309, (1978).
[4] P. Eisenberger, R. G. Shulman, G. S. Brown and S. Ogawa, *Proc. Nat. Acad. Sci. USA* **72**, 4003, (1975).
[5] J. Stöhr, L. Johansson, L. Lindau and P. Pianetta, *Phys. Rev.* **B20**, 664, (1979).
[6] J. Ye, Y. Jiang, Q. Liu, T. Yao, Z. Pan, H. Oyanagi, Z. Sun, W. Yan and S. Wei, *J. Appl. Physics* **106**, 103517, (2009).
[7] Oyanagi *et al.*, *J. Phys. Soc. Jpn.* **56**, 3381, (1987).
[8] H. Oyanagi, Y. Takeda, T. Matsushita, T. Ishiguro, T. Yao and A. Sasaki, *Solid State Commun.* **67**, 453, (1988).
[9] S. M. Heald, E. Keller and E. A. Stern, *Phys. Lett.* **A103**, 155, (1984).
[10] R. S. Becker, J. A. Golvchenko and J. R. Patel, *Phys. Rev. Lett.* **50**, 153, (1983).
[11] L. Troger, D. Arvanitis, K. Baberschke, H. Michaelis, U. Grimm and E. Zschech, *Phys. Rev. B* **46**, 3283, (1992).
[12] H. Oyanagi, K. Sakamoto, R. Shioda, Y. Kuwahara and K. Haga, *Phys. Rev. B* **52**, 5824, (1995).
[13] S. P. Cramer, O. Tench, M. Yocum and G. N. George, *Nucl. Inst. Meth.* **A266**, 586, (1988).
[14] H. Oyanagi, M. Martini and M. Saito, *Nucl. Inst. Meth.* **A403**, 58, (1998).
[15] S. Della Longa *et al. J. Biol. Chem.* **272**, 21025, (1996).
[16] H. Oyanagia, C. Fonne, D. Gutknechtb, P. Dressler, R. Henck, M.-O. Lampert, S. Ogawa, K. Kasai and S. B. Mohamed, *Nucl. Inst. Meth.* **A513**, 340, (2003).
[17] D. M. Mills, A. Lewis, A. Harootunian, J. Huang and B. Smith, *Science* **223**, 811, (1984).
[18] H. Oyanagi, I. Owen, M. Grimshaw, P. Head, M. Martni and M. Saito, *Rev. Sci. Instrum.* **66**, 5477, (1995).
[19] G. Foran *et al.*, unpublished.
[20] C. Gauthier *et al.*, Proc. of 9th Int. Conf. on X-ray Absorption Fine Structure, Grenoble, 1996.
[21] W. Warburton *et al.*, Proc. 9th Int. Conf. on X-ray Absorption Fine Structure, Grenoble, 1996.
[22] K.-C. Cheung, private commun.; R. J. Cernik *et al.*, Synchrotron Radiation Detector Roadmap Ver. 1.6.
[23] H. Oyanagi, A. Tsukada, M. Naito and N. L. Saini, *Phys. Rev. B*, 024511 (2007).
[24] K. Nasu, *Photoinduced Phase Transitions*, Phase Transition Vol. 74 (Gordon and Breach Publishers, Malaysia, 2001).

[25] S. Koshihara, Y. Tokura, T. Mitani, G. Saito and T. Koda, *Phys. Rev.* **B42**, 6853, (1990).
[26] P. Gutlich, A. Hauser and H. Spiering, *Angew. Chem.* **33**, 2024, (1994); G. A. Renovitch and W. A. Baker Jr., *J. Am. Chem. Soc.* **89**, 6377, (1967).
[27] V. Kiryukhin, D. Casa, J. P. Hill, B. Keimer, A. Vigliante, Y. Tomioka and Y. Tokura., *Nature (London)* **386**, 813, (1997).
[28] A. Hauser, *J. Chem. Phys.* **94**, 2741, (1991); S. Decurtins, P. Gütlich, K. M. Hasselbach, A. Hauser and H. Spiering, *Inorg. Chem. Cryst.* **24**, 2174, (1985).
[29] W. Gawelda *et al.*, *Phys. Rev. Lett.* **98**, 057401, (2007).
[30] Ch. Bressler *et al.*, *Science* **323**, 489, (2009).
[31] T. Tayagaki and K. Tanaka, *Phys. Rev. Lett.* **86** (2001).
[32] J. Rehr, S. I. Zabrinsky and R. C. Albers, *Phys. Rev. Lett.* **69**, 3397, (1992).
[33] M. Mikami, M. Konno and Y. Saito, *Acta Cryst.* **36**, 275, (1980).
[34] Y. Ogawa, S. Koshihara, K. Koshino, T. Ogawa, C. Urano and H. Takagi, *Phys. Rev. Lett.* **84**, 3181, (2000).
[35] H. Oyanagi, T. Tayagaki and K. Tanaka, *J. Phys. Chem. Solids* **65**, 1485, (2004).
[36] J. G. Bednorz and K. A. Müller, *Z. Phys. B* **64**, 189, (1986).
[37] Y. Kamihara, T. Watanabe, M. Hirano and H. Hosono, *J. Am. Chem. Soc.* **130**, 3296, (2008).
[38] P. W. Anderson, *The Theory of Superconductivity in the Cuprates* (Princeton University Press, Princeton, 1997).
[39] A. Bianconi, N. L. Saini, A. Lanzara, M. Missori, T. Rossetti, H. Oyanagi, H. Yamaguchi, K. Oka and T. Ito, *Phys. Rev. Lett.* **76**, 3412, (1996).
[40] N. L. Saini, A. Lanzara, H. Oyanagi, H. Yamaguchi, K. Oka, T. Ito and A. Bianconi, *Phys. Rev. B* **55**, 12759, (1997).
[41] J. Mustre de Leon, S. D. Conradson, I. Batistic and A. R. Bishop, *Phys. Rev. Lett.* **65**, 1675, (1990).
[42] C. J. Zhang and H. Oyanagi, *Phys. Rev. B* **79**, 064521, (2009).
[43] A. Alexandrov and J. Devereese, Advances in polaron physics, in Springer Series in Solid-State Sciences, Vol. 159 (2010), http://www.springer.com/materials/book/978-3-642-01895-4.
[44] A. Lanzara *et al.*, *Nature* **412**, 510, (2001); D. Reznik *et al.*, *Nature* **440**, 1170, (2006).
[45] J. Ranninger and A. Romano, *Phys. Rev. B* **78**, 054527, (2008).
[46] D. Mihailovic, V. V. Kabanov and K. A. Müller, *Europhys. Lett.* **57**, 254, (2002).
[47] S. Miyaki, K. Makoshi and H. Koizumi, *J. Phys. Soc. Jpn.* **77**, 034702, (2008).
[48] A. Bussmann-Holder and H. Keller, *Eur. Phys. J. B* **44**, 487, (2005); B. I. Kochelaev, J. Sichelschmidt, B. Elschner, W. Lemor and A. Loidl, *Phys. Rev. Lett.* **79**, 4274, (1997).
[49] H. Koizumi, *J. Phys. Soc. Jpn.* **77**, 034712, (2008).

[50] E. M. Chan, M. Marcus, S. Fakara, M. ElNaggar, R. A. Mathies and A. P. Alivisatos, *J. Phys. Chem. A* **111**, 12210, (2007).
[51] Z. H. Sun, H. Oyanagi, M. Uehara, H. Nakamura, K. Yamashita, A. Fukano and H. Maeda, *J. Phys. Chem. C* **114**, 10126, (2010).
[52] B. Ravel and M. Newville, *J. Synchrotron Rad.* **12**, 537, (2005).
[53] M. Newville, *J. Synchrotron Rad.* **8**, 322, (2001).
[54] H. Oyanagi, Z. H. Sun, Y. Jiang, M. Uehara, H. Nakamura, K. Yamashita, Y. Orimoto, L. Zhang, C. Lee, A. Fukano and H. Maeda, *J. Appl. Phys.* **111**, 084315, (2012).

Chapter 6

The Application of X-ray Absorption Fine Structure Spectroscopy in Functional Materials

Zhihu Sun and Shiqiang Wei

National Synchrotron Radiation Laboratory,
University of Science and Technology of China,
Hefei 230029, China

Xinyi Zhang

Department of Physics,
Surface Physics Laboratory
(National Key Laboratory),
and Synchrotron Radiation,
Research Center, Fudan University,
220 Handan Road, Shanghai 200433, China

Abstract

X-ray absorption fine structure (XAFS) spectroscopy has been widely used for decades in a wide range of scientific fields, including physics, chemistry, biology, materials, environmental sciences, and so on. In this chapter, we introduce the XAFS principles, including its basic theory, data analysis and experiment, from the view point of practical use. To show its strength as a local structure probe, applications of XAFS in various functional materials are introduced, covering nanoparticles and catalysts, magnetic semiconductors, thin film materials, complex compounds, *in situ* probing of the nucleation and growth processes of nanomaterials, as well as operando study of catalysts under working conditions.

In addition, we also briefly introduce some relatively new XAFS-related techniques, such as time-resolved and space-resolved XAFS techniques.

Keywords: XAFS; EXAFS; XANES; absorption edge; local structure; short-range order; Fourier transform, single-scattering; multiple-scattering; transmission mode; fluorescence mode; electronic structure; nanoparticles; catalysts; diluted magnetic semiconductors; thin films; *in situ* measurements; time-resolved; space-resolved; kinetics; nucleation and growth; complex compounds; operando.

6.1. Introduction

The last four decades has witnessed the rapid development of X-ray absorption fine structure (XAFS) spectroscopy as an important probing technique of the local structure around selected atomic species in solids, liquids, and even gases. Its most outstanding feature is the sensitivity to short-range order and element-specificity, which permits us to selectively probe the environments surrounding a specific element in a sample. Although this phenomenon has been long known even since the 1920s, it did not become a practical tool of structure characterization until the development of an effective scattering theory to express the essential physics into a standard XAFS equation, and the availability of synchrotron radiation light sources that emits tunable and brilliant X-ray photons. As a consequence of these developments and some others, the number of XAFS experiments performed has grown exponentially since the 1970s and XAFS has been routinely used as a local structure probe nowadays. This section is to provide an overview of XAFS and to provide a brief survey of the history of this powerful technique.

6.1.1. *Overview of XAFS*

In the X-ray spectral region, the impinging photons interact with matter through the dominant photoelectric absorption. The basic physical quantity measured in XAFS is the X-ray absorption coefficient $\mu(E)$, which describes how the X-rays absorption changes as a function of the incident X-ray photons with energy E. Generally, as the X-ray energy increases, the absorption coefficient $\mu(E)$ decreases,

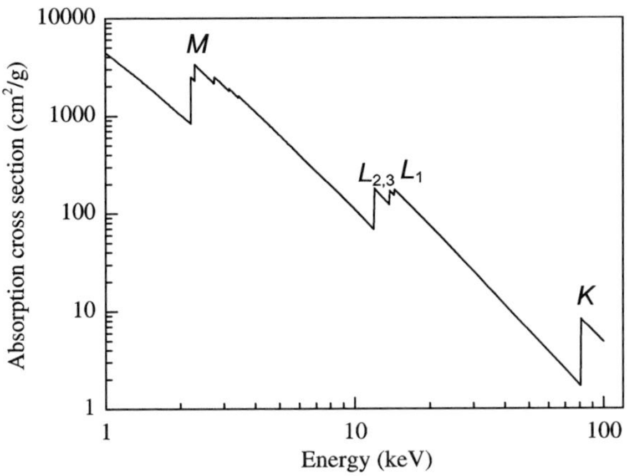

Fig. 6.1. Log-Log plot of the X-ray absorption cross-section of gold metal as a function of X-ray energy, demonstrating the K, L and M edges.

roughly proportional to $1/E^3$. However, at specific energies, the X-ray photon has sufficient energy to liberate electrons from the low-energy bound states in the absorbing atoms, and causes a sudden increase in the absorption coefficient. These energies are called X-ray absorption edges, which are element-specific. Excitation of electrons from $1s, 2s, 2p_{1/2}$, and $2p_{3/2}$ correspond to the $K, L_{1,2}$, and L_3 edges, and so on. Figure 6.1 displays the X-ray absorption cross-section, which is proportional to the X-ray absorption coefficient, of metal gold across the K, L and M edges.

Across the absorption edge, the X-ray absorption coefficient $\mu(E)$ exhibits a jump, followed by oscillatory structure at higher energies [1]. The relative magnitude of the oscillation could be as high as ~10% of the edge jump. XAFS spectrum, as exemplified by the Au L_3-edge spectrum of gold foil in Fig. 6.2, describes the details of X-ray absorption at energies near and above the absorption edge. It is usually separated into two regions: the extended and the near-edge region. The oscillatory structure observed over a wide energy range above the edge, roughly covering typically 30~50 eV until 1000 eV or more is generally referred to as extended X-ray absorption fine structure (EXAFS). The spectral region extending the energy range from the absorption edge (or a little below) to 30~50 eV above

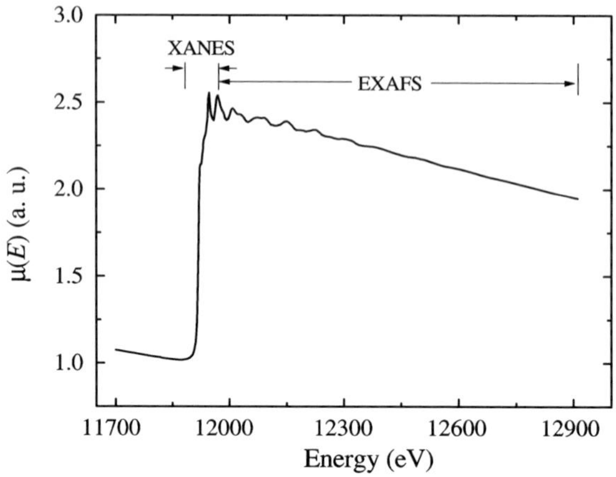

Fig. 6.2. Au L_3-edge XAFS spectrum for gold foil highlighting the EXAFS and XANES spectral region.

it, is commonly denoted as X-ray absorption near-edge structure (XANES). The distinction between XANES and EXAFS is somewhat arbitrary and there is no fundamental difference in the physics giving rise to the fine structures. However, the complexity and the analysis method for the EXAFS and XANES spectra are quite different. In the EXAFS region, the excited photoelectrons have high kinetic energy and are weakly affected by the neighboring atoms' potential; therefore, the single-scattering processes of the photoelectrons by the neighbors are dominant in most cases. In contrast, in the XANES region, the photoelectron energy is close to the absorption edge and the neighboring atoms' potential has significant influence on the photoelectron scattering; consequently, the multiple-scattering processes give a great contribution to the fine structure.

In the last 40 years, XAFS has evolved into a powerful structural characterization tool. It possesses striking advantages such as element-specificity, sensitivity to the short-range order (typically several Å) and chemical state [1]. It can provide quantitative structural information at an atomic scale about the local environment around the absorbing atom, i.e., near-neighbor species and distance, symmetry, coordination number and fluctuation in bond distance. More specifically, EXAFS is sensitive to the distance, coordination number and species of the atoms immediately surrounding the absorbing atoms. The EXAFS distance determination can be accurate to about 0.01 Å or better. XANES can distinguish the electronic state and the three-dimensional coordination geometry (including symmetry, atomic types, bond angles and distances). The sensitivity to the short-range order makes XAFS capable of probing the local structure for a diversity of systems, no matter the studied sample is in solid (crystalline or amorphous), liquid or even gas state. Thanks to the development of high-brilliance synchrotron radiation (SR) light sources, many kinds of measurement methods have been developed to obtain high quality XAFS data, such as transmission XAFS for usual bulk material, fluorescence XAFS for dilute impurity and thin film, grazing-incidence XAFS for surface research, magnetic XAFS for spin study, *in situ* (high-temperature, high-pressure, real reaction environment) XAFS for state or stress transformation investigation, time-resolved XAFS for kinetics studies and even space-resolved XAFS for microzone studies. Nowadays, XAFS has been routinely used in a wide range of scientific fields, including physics, chemistry, biology, materials sciences, environmental sciences, and so on.

6.1.2. *History of XAFS*

XAFS has a long history of more than 90 years since it was first observed by Friche and Hertz in 1920 [2, 3]. However, this phenomenon has long been plagued by controversial physical interpretations. In 1931, Kronig originally interpreted the XAFS oscillations in crystals as a density-of-states effect in terms of the so-called long-range order (LRO) theory [4]. But attempts to match EXAFS peaks with predictions of LRO theories proved unsatisfactory, and afterwards in 1932, Kronig presented an alternative theory for small molecules [5, 6], in which the oscillatory structure in EXAFS was attributed to the influence of neighboring atoms on the transition matrix element in the golden rule, i.e., a short range-order (SRO) effect. The debate between the LRO and SRO theories remained for 40 years until in 1971 Sayers, Stern, and Lytle performed the k-space Fourier transform to separate peaks corresponding to various atomic shells [7]. This is a milestone in the development story of XAFS. After that it is widely recognized that the SRO theory is more appropriate in interpreting the origin of XAFS. Since then it has been widely recognized that the oscillations in the X-ray absorption spectra, more than just a physical phenomenon, could be used as a quantitative structure-determination technique. There are many book chapters or excellent review articles [1, 8, 9] devoted to the early history of XAFS for interested readers.

It should be remarked that the controversy between the early SRO and LRO XAFS theory actually lies in the energy-dependent competition between scattering strength and inelastic losses. The excited high-energy photoelectron state must decay as a function of time and distance, which makes it not infinitely long lived and hence not able to probe long-range effects. This decay is due to both the intrinsic lifetime of the core-hole state (intrinsic losses) and the inelastic losses (extrinsic losses) as the photoelectron propagates in the material. Therefore, the excited outgoing photoelectron wave dies away as it moves farther away from the absorbing atom and becomes too weak to significantly reflect any waves scattered by distant atoms. The net effect is that XAFS can only measure the local atomic structure within a range limited by the effective mean free path of the excited photoelectron, typically in the order of ten angstroms.

Another key development in the XAFS theory is how to treat the multiple-scattering effects. Taking the advantage of Fourier transform in the EXAFS data analysis, in 1975, Lee and Pendry first proposed the theoretical treatment of multiple-scattering EXAFS in terms of effective path length [10]. Later, the path-by-path XAFS theory was greatly developed by Rehr and Albers who proposed a separable representation of the Green's-function propagators [11] to overcome all the computational difficulties. So far, various advancements in the theoretical models have been obtained such as full multiple-scattering (FMS) [12], self-consistent field and fully relativistic calculations [13] especially for a quantitative analysis of the near-edge spectroscopy. The EXAFS theory is implemented in various software packages or codes like CONTINUUM [14], EXCURV

[15], FEFF [16], GNXAS [17], WIEN2k [18], the Munich SPRKKR package [19], NSRL-XAFS [20] and others [21–23].

The decisive factor contributing to the development of XAFS experiment is the availability of brilliant synchrotron radiation light sources allowing for acquisition of better resolved experimental spectra [8, 14, 24–27]. From the 1930s to 1960, XAFS were measured in transmission with low-power conventional X-ray tubes. The last four decades has witnessed a rapid development of the XAFS experiment technique, thanks to the application of synchrotron radiation (SR) light sources that possess a number of excellent properties, such as high brilliance, wide and continuous wavelength range, high polarizability and low emittance. The usefulness of SR for XAFS measurements was first demonstrated in 1975 by Kincaid and Eisenberger [28], who used the first dedicated synchrotron radiation sources of the Stanford Synchrotron Radiation Project facility to greatly improve the data quality and the acquisition time for the XAFS measurements. This signaled the beginning of rapid development in measurement techniques. A transmission technique using a channel-cut Si(220) double-crystal monochromator and ion chamber was developed, followed shortly by the fluorescence technique based on a solid-state detector for dilute systems [29], an energy-dispersive technique for time-resolved experiments [30] and an electron detection technique for surface-sensitive measurement [31]. Furthermore, more attention was paid to the dynamics, as, for example, to the anisotropy of the surface Debye-Waller factors [32, 33]. In the mid-1990s when the third generation synchrotron radiation facilities (like ALS, APS, BESSY-II, ESRF, SPring-8) came into operation, insertion devices like wigglers and undulators started to be used routinely. These allowed for the measurement of XAFS with excellent statistics and high energy resolution. It is nowadays possible to record magnetic EXAFS (normally in the range of $10^{-2} - 10^{-3}$ of the isotropic absorption), nearly free of noise with a detailed fine structure. Because of the high spectral purity, even small oscillatory fine structures in the X-ray absorption coefficient as, for example, the so-called atomic EXAFS can be identified by polarization-dependent surface EXAFS measurements. Regarding the progresses made in the experimental side of XAFS techniques in the past three decades, there are some of the review papers and books for interested readers [34, 35].

6.2. Principles of XAFS

In this section, a simple physical description of the XAFS process and the basic EXAFS formula in the frame of single-scattering approximation will be presented from the view point of practical use. Various treatments of disorder problems will be described in some detail, followed by a brief introduction of the multiple-scattering EXAFS theory. An outline of the typical EXAFS data processing and analysis procedure will be presented. Finally, a brief introduction of XANES in the framework of multiple-scattering theory will be introduced.

6.2.1. *Single-scattering theory of EXAFS*

In the X-ray absorption process, when an incident X-ray photon with energy higher than the absorption edge of a specific atom A (as shown in Fig. 6.3) is absorbed, the X-ray photon has sufficient energy to emit a core-level electron with a kinetic energy $T = E - E_0$ given by the Einstein equation. (E is the incident photon energy and E_0

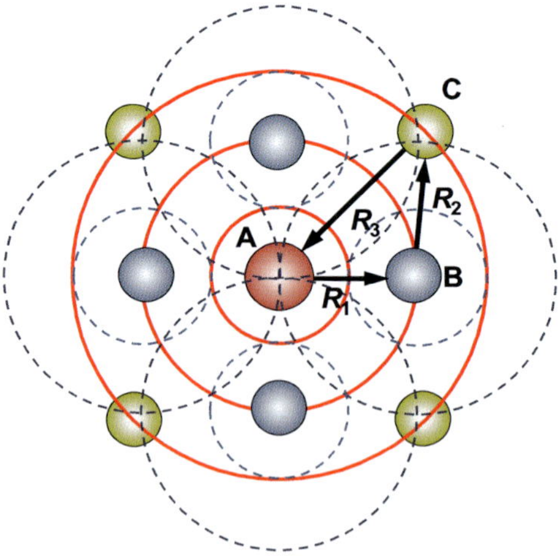

Fig. 6.3. Pictorial view of the scattering of an outgoing photoelectron wave by neighboring atoms of the absorbing atom A. The solid circles stand for the outgoing photoelectron waves, and the dashed lines stand for the scattered waves.

is the absorption edge energy.) This excited photoelectron with the wavenumber $k = [2m(E - E_0)/\hbar^2]^{1/2}$ (where $\hbar$ is the Planck's constant and m is the mass of an electron) outgoes away from atom A. If there is a neighboring atom B as schematically described in Fig. 6.3, the excited photoelectron wave will be scattered back to the absorber by the potential barrier of electron shells of this neighbor. The interference between the outgoing and scattered photoelectron waves thus changes the value of the X-ray absorption coefficient of the sample, giving rise to the oscillation in the X-ray absorption coefficient: this is the origin of EXAFS. The maxima in the EXAFS spectrum mean the constructive interference of the incoming and outgoing waves, and the minima correspond to destructive interference. For this reason, the EXAFS spectrum of a free atom has no oscillatory structures above the absorption edges and only shows a monotonic decrease in absorption coefficient with increasing X-ray photon energy.

Like any interference effect, the positions of the maxima and minima of the EXAFS spectrum depends on the wavelength of the scattered field and the geometry of the scattering objects. On the geometrical dependence side, the constructive or destructive interference depends on the phase difference $2kR$ between the two waves, where R is the interatomic distance and $2R$ the path difference. In the framework of single-scattering which assumes that the outgoing wave is scattered back only once by the neighboring atoms (e.g., the single-scattering path A $\rightarrow$ B $\rightarrow$ A in Fig. 6.3), the oscillation part of the absorption coefficient $\Delta\mu(E)$, normalized to the atomic background $\mu_0(E)$, is given by Eq. (6.1):

$$\begin{aligned}\chi(k) &= \frac{\mu(E) - \mu_0(E)}{\mu_0(E)} \\ &= \sum_j (3\cos^2\theta_j) \frac{S_0^2 f_j(k)}{k} \int_0^\infty \frac{g(R)}{R_j^2} \\ &\quad \times e^{-2R_j/\lambda(k)} \sin[(2kR_j + \delta_j(k) + 2\varphi_C(k))]dR. \qquad (6.1)\end{aligned}$$

Here S_0^2 is the amplitude reduction factor, R is the distance between the absorbing and scattering atoms, $g(R)$ is the pair distribution function of the scattering atom, $\lambda(k)$ is the mean free path of the excited photoelectron, $f(k)$ and $\delta(k)$ are the backscattering amplitude and scattering phase shift of the scattering atom, respectively, $\varphi_C(k)$ is the phase-shift of the absorbing atom, θ is the angle between the electric field vector of the incident X-ray and the radial direction of the scattering atom. The X-ray of synchrotron radiation is highly polarized, with the electric field vector in the horizontal plane. Thus the relevance of EXAFS function to the angle θ indicates that it is polarization-dependent and provides a means to determine the anisotropic geometry of atoms and electronic states in single crystals. This character is beneficial for studying the epitaxially grown thin films that will be covered in this Chapter. The amplitude reduction factor S_0^2 accounts for the reduction in single-electron excitation, which originates from intrinsic multi-electron excitations including "shake-up" and "shake-off" processes [36]. S_0^2 is weakly dependent on the absorber elements and is typically in the range of 0.7–1.0. The $e^{-2R/\lambda}$ term is due to the limited lifetime of the core hole (~1 fs) and the limited range (within 10 Å) of the excited photoelectron within the matrix, both of which have the effect of damping the oscillation amplitude and are the origin of the local sensitivity of EXAFS.

For randomly oriented polycrystalline or solution samples, an isotropic average over θ_i angles must be performed. The integral of the $3\cos^2\theta_i$ term in Eq. (6.1) over θ_i is unity. Therefore, the contributions from atoms of the same atomic number and at similar distances from the absorbing atom may not be resolvable from each other. Around the central absorbing atom, the neighboring atoms could be grouped into various coordination shells, consisting of N_j identical atoms at the same interatomic distance R_j within each j-shell. Then Eq. (6.1) is reduced to

$$\chi(k) = \sum_j \frac{N_j S_0^2 f_j(k)}{k} \times \int_0^\infty \frac{g(R)}{R_j^2} e^{-2R_j/\lambda(k)} \sin[(2kR_j + \delta_j(k) + 2\varphi_C(k))]dR. \quad (6.2)$$

Equation (6.2) is the general EXAFS formula which is applicable to all systems with disorder degree from low to high. Especially for low disorder systems, the $g(R)$ function can be represented by a Gaussian distribution,

$$g(R) = (2\pi\sigma^2)^{-\frac{1}{2}} \exp\left[-\frac{(R-R_j)^2}{2\sigma^2}\right]. \tag{6.3}$$

Substituting Eq. (6.3) into Eq. (6.2) gives

$$\chi(k) = \sum_j \frac{N_j S_0^2 f_j(k)}{kR_j^2} e^{-2k^2\sigma_j^2} e^{-2R_j/\lambda(k)} \sin[2kR_j + \delta_j(k) + 2\varphi_C(k)], \tag{6.4}$$

which is commonly called the standard EXAFS formula. The σ^2 term, known as the Debye-Waller factor, is the mean square variation in the interatomic distance R. The $e^{-2k^2\sigma_j^2}$ term accounts for the effects of dynamic (thermal) vibration and configuration (structural) disorders which smear out the EXAFS oscillations at high k region.

For systems with moderately higher disorder, like high-temperature samples and surfaces, the Gaussian distribution fails to represent the pair distribution function, and the asymmetry in the pair distribution must be considered. For instance, thermal expansion cannot be described within a harmonic oscillator but requires anharmonic interatomic potential. These systems have much larger vibrational amplitude and structural disorder, and surface atoms have more significant anharmonicity compared with bulk atoms. In such a case, a widely used model-independent method is cumulant expansion that takes into account of the inharmonic pair distributions in the EXAFS analysis [37, 38]. A thermal average of the EXAFS function $\chi(k)$ in a single shell is then

$$\langle\chi(k)\rangle = \frac{NS_0^2}{k}\left\langle \frac{f(k)}{R^2} e^{-2R/\lambda} \operatorname{Im}\{e^{i(2kR+\delta(k))}\}\right\rangle. \tag{6.5}$$

Compared with the e^{i2kR} term, the dependence of $e^{-2R/\lambda}$, $f(k)$, $1/R^2$ and $\delta(k)$ on the interatomic distance R is much less and could be neglected. Then the thermal average in Eq. (6.5) is simplified to

$$\langle \chi(k) \rangle = \frac{NS_0^2 f(k)}{kR_0^2} e^{-2R_0/\lambda} [\mathrm{Im} e^{i\delta(k)} \langle e^{i2kR} \rangle], \tag{6.6}$$

where R_0 stands for the average distance $R_0 = \langle R \rangle$. Let $r = R - R_0$, the thermal average could be expressed in terms of cumulants:

$$\langle e^{i2kR} \rangle = \exp\left[i2kR_0 + \sum_{n=1} \left(\frac{(2ik)^n}{n!} C_n \right) \right], \tag{6.7}$$

where $C_n = \langle (r - \langle r \rangle)^n \rangle$ is the n-the cumulant moment.

The above equation indicates that for Gaussian distribution, all the cumulant moments higher than C_2 are zero. When the anharmonic distribution is not too large, the high-order cumulants such as C_3 and C_4 have finite values and the EXAFS equation becomes

$$\chi(k) = \sum_j \frac{N_j S_0^2 f_j(k)}{kR_j^2} e^{-2k^2\sigma_j^2 + \frac{2}{3}k^4 C_{4,j}} e^{-2R_j/\lambda(k)}$$
$$\times \sin\left[2kR_j - \frac{4}{3}k^3 C_{3,j} + \delta_j(k) + 2\varphi_C(k) \right]. \tag{6.8}$$

Seen from Eq. (6.8), the C_3 term appears in the phase which shifts the peak position of the radial distribution function. Therefore, care must be taken that in the presence of considerable asymmetric contribution, the approximation of Gaussian distribution would lead to mistakenly bond contraction [39]. This should be kept in mind when interpreting the EXAFS results. It is also worthy of note that Eq. (6.8) is derived by considering only a thermal average, it is still valid to systems with structural (or static) disorder as long as the total disorder σ_T^2 ($\sigma_T^2 = \sigma_t^2 + \sigma_s^2$, where σ_t^2 is the thermal disorder and σ_s^2 is the structural disorder) could be described using a few cumulant moments.

In systems with even higher disorder, such as liquid or amorphous materials, many high-order cumulants are required to ensure the convergence of the cumulant expansion in Eq. (6.7); this makes the cumulant expansion method practically insufficient. In this case, some other model-dependent approximation has been considered, for example, a weighted exponential distribution function could be used to account for the structural disorder [40, 41]. In this case, the $g(R)$ function is modeled as a convolution $P_{\mathrm{G}} \otimes P_{\mathrm{E}}$, where P_G is a normalized Gaussian function centered at $R = 0$, and P_{E} is a weighted exponential distribution function:

$$P_E(R) = \begin{cases} \dfrac{1}{\sigma_s^{m+1}}(R - R_0)^m & (m = 0, 1, 2), \quad R \geq R_0 \\ \quad \times \exp[-(R - R_0)/\sigma_s]; & \\ \quad 0. & R < R_0 \end{cases} \tag{6.9}$$

The EXAFS function Eq. (6.4) is then expressed as:

$$\begin{aligned} \chi(k) = \sum_j \frac{N_j S_0^2 f_j(k)}{k R_{0,j}^2} & \exp(-2k^2 \sigma_{t,j}^2) \\ & \times [1 + (2k\sigma_{s,j})^2]^{-(1+m)/2} e^{-2R_{0,j}/\lambda(k)} \\ & \times \sin[2kR_{0,j} + \delta_j(k) + 2\varphi_C(k) + (m+1)\mathrm{ac}\tan(2k\sigma_s)]. \end{aligned} \tag{6.10}$$

Here R_0 is the distance of central atom to the closely packed atom, and the average interatomic distance is $R = R_0 + (m + 1)\sigma_s$. The difference between Eqs. (6.4) and (6.10) is that the $e^{-2k^2\sigma_j^2}$ term in Eq. (6.4) is converted to $e^{-2k^2\sigma_t^2}[1 + (2k\sigma_s)^2]^{-(1+m)/2}$ in amplitude and a term $(m + 1)\mathrm{actant}(2k\sigma_s)$ is added in the phase shift.

6.2.2. *Multiple-scattering theory of EXAFS*

The EXAFS equation (6.1) is derived based on the single-scattering approximation, which assumed that the outgoing photoelectron wave is only scattered once by the neighboring atoms before returning to the absorbing atoms (single-scattering). But in reality the photoelectron is scattered by more than one neighboring atom (multiple-scattering) and these events also contribute to the interference phenomenon. For instance, in Fig. 6.3 the outgoing photoelectron emitted from atom A could be scattered by atoms B and C before coming back to atom A. If the multiple-scattering contributions are small, the single-scattering approximation would be sufficient to describe the EXAFS process. Sometimes the addition of further scatterings such as scattering from one neighbor to the next before interfering with the outgoing wave is important if the scattering is strong, especially in the low-k region for a low-Z nearest neighbor. In collinear configuration where three or more atoms are aligned or almost aligned, the multiple-scattering EXAFS contributions can even be larger than the high shell single-scattering contribution because of the shadowing or focusing effect. Therefore, in order to avoid poor or misleading results in extracting the structural information beyond the first shell, attention to the multiple-scattering events involving different shell atoms should be paid.

Historically, the theoretical treatment of multiple-scattering EXAFS was first proposed by Lee and Pendry who classified the multiple-scattering contribution in terms of effective path length [10]. This approach takes the advantage of the Fourier transform in the EXAFS data analysis: since we are ultimately interested in Fourier transforming the absorption coefficient to position space, at a given radial distance we can only consider those multiple-scattering effects which contribute to the Fourier transform at this particular radius and neglect all the others. Its main advantages are its computational efficiency and accuracy; moreover, it provides a geometrical parametrization and interpretation of EXAFS, which is ideal for analyzing interatomic distances and other structural quantities. For example, the double-scattering path A $\rightarrow$ B $\rightarrow$ C $\rightarrow$ A in Fig. 6.3 has an effective path length $R_{\text{eff}} = (R_1 + R_2 + R_3)/2$. Principally, the number of multiple-scattering paths is infinite, but in practice, the signals for long path lengths and high order scattering are rapidly damped. In k-space the multiple-scattering events give rise to high frequency oscillations which tend to cancel out. The multiple-scattering becomes important only in the low-k region where the electron free path becomes very long.

A major advance in the path-by-path EXAFS theory is the separable representation of the Green's-function propagators developed by Rehr and Albers [11], which overcomes all the major computational difficulties of the multiple-scattering expansion. For an N-leg path Γ with scatterers at $\boldsymbol{R}_1, \boldsymbol{R}_2, \ldots, \boldsymbol{R}_N = \mathbf{R}_0$, the result for the EXAFS amplitude is:

$$\chi_\Gamma(p) = \mathrm{Im}\left\{ S_0^2 \frac{e^{i(\rho_1+\rho_2+\rho_N+\cdots+2\delta_1)}}{\rho_1\rho_2\cdots\rho_N} e^{-2\sigma_\Gamma^2 p^2/2} \mathrm{Tr}[M_l F^N \cdots F^2 F^1] \right\}. \tag{6.11}$$

Here $\boldsymbol{\rho}_i = p(\boldsymbol{R}_i - \boldsymbol{R}_{i-1}), p = (E - V_{\mathrm{mt}})^{1/2}$ is the photoelectron momentum measured with respect to the muffin-tin zero, F^i is the scattering matrix at site i, M_l is the termination matrix for the final state of angular momentum l. Finally, when one substitutes $k = (p^2 - k_F^2)^{1/2}$ into Eq. (6.11), the result for χ_Γ can be recast exactly in the form of the standard XAFS Eq. (6.3), but with an effective scattering amplitude f_{eff},

$$\chi_\Gamma(k) = \mathrm{Im}\left[\frac{f_{\mathrm{eff}}(k)}{kR_{eff}^2} e^{i(2kR_{eff}+\delta_l(k))} e^{-2k^2\sigma^2} \right]. \tag{6.12}$$

For more details on the multiple-scattering theory, the reader is referred to the excellent review article written by Rehr and Albers [9]. The implementation of these algorithms into the *ab initio* EXAFS code known as FEFF — named after the effective scattering amplitude f_{eff} in the theory — has made accurate calculations of high-order multiple-scattering EXAFS in general materials routine [16]. By using the FEFF code, the effective amplitude function f_{eff}, phase shift function $\delta_l(k)$ and the mean free path of electrons $\lambda(k)$ can be calculated for both single- and multiple-scattering paths.

In order to extract structural parameters from EXAFS analysis, amplitude and phase-shift functions are very important. They were obtained in two different ways before the development of FEFF. The first way is to extract these functions from experimental spectra standard materials with known structure, and the second way is to use the tabulated backscattering amplitudes and phase shifts by Teo and Lee based on a plane wave approximation [24, 42], or those later improved by McKale considering the spherical wave theory [43]. But these two methods suffer from many disadvantages, e.g., in many cases it is hard to find the proper standard material, and the distance determination is not very accurate. These disadvantages have been well overcome by the FEFF code that provides a reliable and convenient theoretical standard for most applications. Koningsberger conducted detailed investigations on whether these different theoretical parameters were acceptable for accurate structure determination [44], and the Committee on Standard and Criteria in EXAFS spectroscopy strongly discouraged the use of Teo and Lee's and McKale's tables.

6.2.3. *EXAFS data analysis*

The aim of EXAFS data processing and analysis is to extract structure parameters such as coordination numbers, interatomic distances, disorder degrees and sometimes, types of atoms, in various coordination shells for the "unknown" sample [5]. The accuracy of EXAFS in obtaining these structural information depends heavily on the standard phase-shift and amplitude functions which were derived from either experimental reference compounds or theoretical calculations. During the early stage of EXAFS development, experimental standards were more reliable than theoretical ones. However, experimental standards suffer from several limitations, including the difficulty in obtaining pure well-characterized compounds, unknown Debye-Waller factors for different scattering paths, and errors introduced by background removal methods in different experimental spectra. With the steady development of theoretical standards, especially the accurate *ab initio* FEFF code, it has become increasing feasible for EXAFS to determine accurate structural information, which is not limited to the first coordination shell around the absorbing atoms. Now, the general EXAFS data analysis procedure has been established, which consists of the following steps: deglitching, pre-edge background removal; normalization; conversion to k-space; Fourier transform to R-space and inverse Fourier transform. The order of these steps is not critical and not all the steps are necessary, and great care in the data analysis must be taken to draw a reliable conclusion. In the following, we will take the EXAFS spectrum of Cu foil as an example to show this general procedure in some detail, addressing some points that need special caution.

(1) Deglitching

The raw EXAFS spectra sometimes contain abnormally sharp data points called "glitches". Their origin might be diffractions from the crystal monochromator or from the single crystal samples (including crystal substrates), or multi-electron excitations. These glitches, intrinsic to the sample, might be many times larger than the real EXAFS signal can cause "spectral leakage" in the Fourier transform curves, and hence they must be removed before further processing. The common deglitching technique is smooth interpolation of the data on both sides of the glitch. There are also other experimental faults causing glitches, like bad linearity of the detector and electronics, pinholes of the samples, or higher harmonics of the beam. These faults make the observed EXAFS unreliable and must be carefully eliminated in experiments.

(2) Pre-edge background removal

The most interesting region in the EXAFS data is the part above the absorption edge, and the pre-edge background absorption should be removed. This generally involves a fitting of the data before the edge to some functional form and extrapolating this function into the post-edge data. A traditional function used for the fitting is the Victoreen empirical formula $\mu_\nu(E) = aE^{-3} + bE^{-4}$ developed for an in-laboratory EXAFS apparatus. However, modern EXAFS spectra are mostly recorded at synchrotron radiation source. For the data collected in transmission mode, one should be cautious to use this formula. This is because in calculating the absorption coefficient $\mu = \ln(I/I_0)$, only a fraction ($\sim$20%) of I_0 is measured by the ion chamber. In order to compensate for this distortion, a modified Victoreen formula $\mu_\nu(E) = aE^{-3} + bE^{-4} + c$ could be used instead that could give a better fitting to the pre-edge region, as exemplified in Fig. 6.4(a).

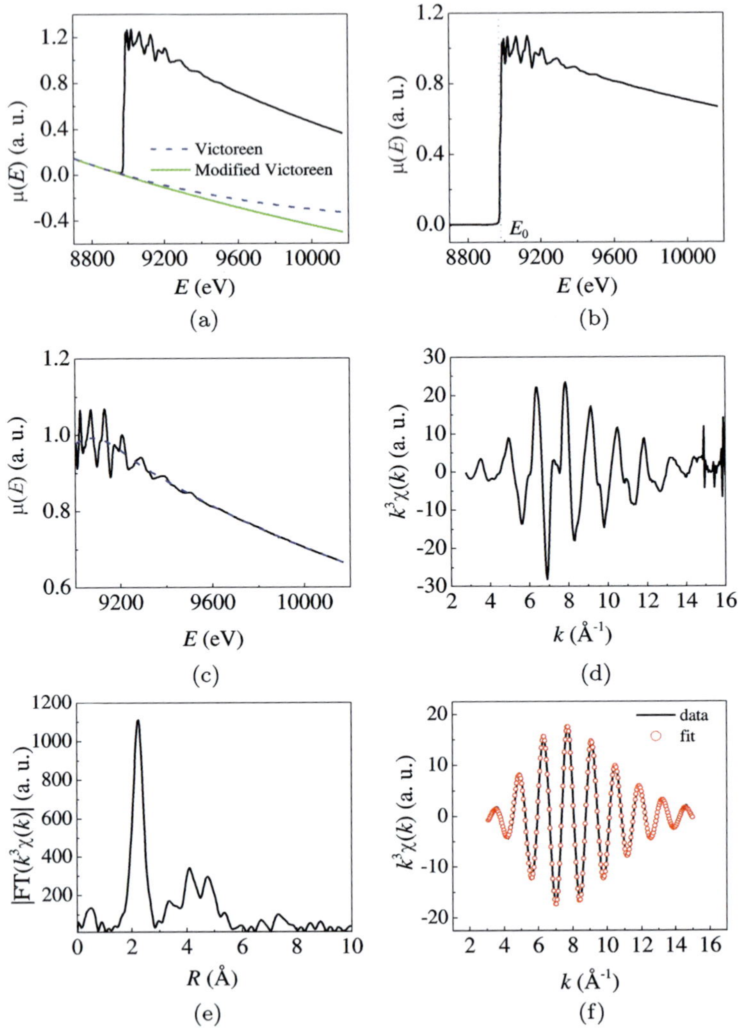

Fig. 6.4. Typical EXAFS data processing and analysis procedure: (a) pre-edge background removal, (b) normalization, (c) curve fitting of μ_0, (d) conversion to k-space and weighting, (e) Fourier transformation to R-space, and (f) reverse Fourier transform and fitting in the k-space.

(3) Data normalization

This step is taken to normalize the absorption coefficient to that of one single atom, for annihilating the difference in samples. The most widely used normalization method is to use the relation,

$$\chi(E) = \frac{\mu(E) - \mu_0(E)}{\Delta\mu(E_0)}. \tag{6.12}$$

Here $\mu_0(E)$ is the atomic part of the absorption, $\Delta\mu(E_0)$ is height of the edge-step, and E_0 is the threshold energy at which the photoelectron is excited as a free electron. This E_0 is close to, but somewhat different from the absorption edge of the element under investigation, as it is affected by the chemical state or valence. There is no general way to determine this threshold energy, and in practice it is chosen as an inflection point (where the $d\mu(E)/dE$ reaches the maximum) or the middle point of the edge jump. The errors in the determination of E_0 could further be compensated in the curve-fitting procedure, where a parameter ΔE will be optimized.

A common method of determining the step is to fit the data within a few hundred eV below and above the absorption edge with low order polynomials, then extrapolate them to the E_0 point (Fig. 6.4(b)). The difference between these two curves at E_0 is the step $\Delta\mu(E_0)$, and the resulting EXFAS spectrum is then normalized to an edge jump of 1.

It should be remarked that the atomic background absorption $\mu_0(E)$ is energy- dependent, therefore, the normalization by the edge-step introduces an energy-dependent attenuation to the amplitude of $\chi(E)$. Such an attenuation might be negligible for heavy elements, but can be important for light elements. This energy dependence of the edge normalization could be further corrected using the X-ray cross-sections of free atoms tabulated by McMaster *et al.* [45].

(4) Curve fitting of $\mu_0(E)$

Till now, the normalized $\mu(E)$ data still contain the atomic absorption $\mu_0(E)$ that must be subtracted out. This is a crucial and most difficult stage in EXAFS data processing because in general $\mu_0(E)$ is neither measurable nor can be calculated theoretically. Practically, a least-squares cubic spline functions fitting is used to obtain $\mu_0(E)$ (Fig. 6.4(c)). In doing so, the data are divided into several regions (the divided points are called nodes), and separate cubic polynomials are fit to the data over each region. The cubics are constrained so that the curves in each region are connected continuously (the values, the first and second derivatives match at the junctures). The atomic background absorption is then removed to reveal the EXAFS oscillatory curve $\chi(E)$. It is important during the background removal not to remove the EXAFS oscillations nor to leave low-frequency components in the extracted data.

(5) Conversion to k-space and weighting

For performing the subsequent Fourier transform (FT), the E-space EXAFS data $\chi(E)$ must be converted to $\chi(k)$ in the k-space. This is easily done according to $k = [2m(E - E_0)/\hbar^2]^{1/2}$. At the same time, the $\chi(k)$ data are generally weighted with some power of $k^n(n = 1, 2, 3)$ to compensate for the decay of the waves with extended k (Fig. 6.4(d)) as well as to diminish the ambiguity of E_0. The choice of the k-weighing power is arbitrary, but there are still some general comments. The k^3-weight is most frequently used for analysis of EXAFS data contributed by heavy atoms coordination, because these atoms have a large scattering amplitude in the high k-region and high power of k-weighting emphasizes their contribution. In comparison, the k- or k^2-weight is more suitable for lighter coordination atoms in order to enhance their contributions maximized in the low k-region.

(6) Fourier transform to R-space

The next step in the analysis is Fourier transform (FT) of the $k^n\chi(k)$ function to the R-space (Fig. 6.4(e)). This is a very important step in EXAFS analysis, since the EXAFS oscillations are expressed by the sum of different sinusoidal functions of many shells as in Eq. (6.3) and the Fourier transform technique could separate their contributions. In practice, the FT requires a uniform grid in k-space; before the transform one should interpolate the experimental data in an even grid$k$-space, typically $\Delta k = 0.05$ Å^{-1} in EXAFS analysis. It must be addressed that in the FT curve, the peak position is not identical to the interatomic distance as the real Radial Distribution Function (RDF) does, but shifts to the lower R-side by the phase-shift in the EXAFS equation. This shift is typically by 0.3–0.5 Å for the first shell atoms.

One should be cautious of the FT that a real EXAFS data set has a limited k-range. The resolution of the FT function increases with larger k-range, and the $k_{\max}$ should be as large as possible. Anyway the $k_{\max}$ is not infinite but truncated at a finite value, which inevitably produces some wiggles (called "ghost peaks") in the FT curves. To reduce such extra peaks, window functions are necessary in the FT. The most popular window function is the Hanning window:

$$W(k) = \begin{cases} \sin^2\left[\dfrac{\pi(k-k_{\min})}{2d_1}\right], & k_{\min} < k < k_{\min} + d_1 \\ 1, & k_{\min} + d_1 \le k \le k_{\max} - d_2 \\ \sin^2\left[\dfrac{\pi(k_{\max}-k)}{2d_2}\right]. & k_{\max} - d_2 < k < k_{\max} \end{cases} \tag{6.13}$$

It is worthy to note that for heavy coordination atoms in the fifth or sixth row of the Periodic Table of the Elements, a side lobe peak also appears at the left side of the main peak, due to the nonlinear behavior of the phase-shift functions of the heavy atoms.

(7) Reverse Fourier transform

The following step is to inversely transform the R-space peak corresponding to a single shell to k-space (Fig. 6.4(f)). Then the oscillation of this specific shell is filtered from the overall spectrum, which allows extracting the structural information of that shell through a curve-fitting procedure. We note that this step is not necessary, since nowadays the curve-fittings are more commonly done in the R-space.

(8) Fitting procedure

Once the single-shell oscillation has been isolated, the quantitative structural parameters can be obtained through a curve-fitting procedure according to the EXAFS standard formula Eq. (6.4). Besides the structural parameters N, R, σ^2 that appear explicitly in the EXAFS equation, there is another parameter ΔE_0 to be optimized. This parameter at least plays dual roles of correcting the errors in determining the threshold of absorption energy and compensating the inaccuracy of phase-shift. Before curve-fitting the amplitude and phase-shift functions should be known. As mentioned before, these functions can be determined experimentally from model compound through either fitting or ratio method, from tabulated values or from theoretical calculations. Thanks to the development of the accurate *ab initio* calculations of FEFF code, nowadays the theoretical amplitude and phase-shift are more routinely used than the experimental ones.

There are various algorithms for fitting the filtered experimental EXAFS function, among which the least-squares algorithm is most commonly used. However, the least-squares algorithm suffers from some drawbacks, for example, many local minima in the sum of squares function may exist, and the obtained structural parameters are often physically impossible. To overcome this problem, some new algorithms such as simulated annealing method have been developed. We have developed an USTCXAFS3.0 software package [20] which implements the simulated annealing method. Shown in Table 6.1 are the structural parameters of Cu foil extracted from EXAFS fitting with USTCXAFS3.0 software package as an example.

For an EXAFS spectrum with a good signal/noise ratio, the accuracy of the structural parameter determination can be: interatomic distance R, ± 0.01 Å or better for the first shell; coordination number N, typically within $\pm 10\%$; and Debye-Waller factor σ^2 in the

Table 6.1. The structure parameters of Cu foil from a curve fitting.

Parameter	Real value	Fitting result
R (Å)	2.55	2.55 ± 0.01
N	12	12.0 ± 0.4
σ^2 (Å)		0.0086 ± 0.0003
ΔE_0		-4.4 ± 0.5

Gaussian approximation, typically within ±0.0005 Å^2. In practice, one problem in the multi-shell EXAFS curve-fitting is the correlation among various parameters. For instance, both the coordination number and the Debye-Waller factor are related to the amplitude of the oscillation and hence are strongly correlated; if both are treated as free parameters, the accuracy would be decreased. It often occurs that when too many parameters are to be fitted, the correlations may force the fitting results to converge on values with unphysical meaning. Therefore, reasonable constraints between parameters are necessary, and the international XAFS working group has extensively discussed this issue and published standards and criteria in EXAFS data analysis [46].

6.2.4. *Multiple-scattering theory of XANES*

XANES is also sometimes called near-edge X-ray absorption fine structure (NEXAFS). It is associated with the excitation process of a core-level electron to bound states (located below the ionization threshold) and quasi-bound states (located above the threshold) that interact with the continuum. XANES contains information on the electronic structure of the absorbing atom as well as the three-dimensional structural information around it. Within the framework of single-electron approximation, an excited electron from the core-level behaves as an active electron and the remaining other electrons are in a passive state affected by multi-electron interactions. There are also cases where the one-electron picture is invalid, interactions between active and passive electrons must be taken into account which make the interpretation of XANES rather complicated. The most challenging part in interpreting XANES is how to get the final state wave function. Several complementary ways have been developed, including quantum chemistry, band-structure, and multiple-scattering theories.

For condensed matter that will be covered in this chapter, the multiple-scattering method is the most widely used and we give a brief introduction to it. In the single-electron and dipole approximations, the transition probability of a core-level electron state $|i\rangle$ of the absorbing atom into the final unoccupied state $|f\rangle$ is described by Fermi's Golden Rule:

$$\sigma(\omega) \propto \sum_{j} |\langle i|\vec{r}\cdot\vec{\varepsilon}\,|\rangle|^2 \rho(E_f), \tag{6.14}$$

where $\vec{r}\cdot\vec{\varepsilon}$ is the dipole operator for the incident electromagnetic wave interacting with the material and the sum of index f covers all the final states with energies compatible with the energy conservation as $E_f = E_0 + \hbar\omega$ and the $\rho(E_f)$ is the density of the final state. The deep-core state is typically a $1s, 2s$, or $2p$ state that corresponds to the K, L_1 and $L_{2,3}$ edge XAFS spectrum, respectively. As a consequence of the dipole approximation, there are selection rules which determine what final states are allowed for a given initial state. For a free atom the selection rules are $\Delta l = \pm 1, \Delta m = 0, \pm 1$, where l is the orbital angular momentum quantum number and m is its projection on the quantization axis. Therefore, for the K and L_1 edge absorption, the final state could only be p states, and for $L_{2,3}$ edge absorption, the transition could be to d and s final states. But the matrix element for the $p \to d$ transition is larger than the $p \to s$ transition, therefore, the $L_{2,3}$ edge absorption spectrum mainly probes the unoccupied d states.

In a real material system, the presence of neighboring atoms around the absorbing atom modifies the final of the transition. The calculation of the absorption cross-section could be done by a multiple scattering theory, which is based on the muffin-tin approximation of the scattering potential. This includes a spherical average of the potential around each atom and a constant interstitial potential in between. Under this assumption, Eq. (6.14) could be rewritten using a Green's function: [47]

$$\mu(E) \propto -\frac{1}{\pi}\mathrm{Im}\langle i|\vec{r}\cdot\vec{\varepsilon}^{\,*}\mathbf{G}(r,r';E)\,\vec{r}'\cdot\vec{\varepsilon}\,|\rangle\Theta(E - E_f). \tag{6.15}$$

Here, Θ is the Heaviside step function assuring that the cross-section is non-zero only above the Fermi energy E_F. $\mathbf{G}$ is the function that describes all possible ways for a photoelectron to interact with the surrounding atoms in the presence of the scattering potential. It could be expressed in a series by the Dyson equation in terms of the free electron propagator G^0 and atomic scattering matrix T:

$$\mathbf{G} = (1 - G^0T)^{-1}G^0 \tag{6.16}$$

$$= G^0 + G^0TG^0 + G^0TG^0TG^0 + G^0TG^0TG^0TG^0 + \cdots \tag{6.17}$$

G^0 describes how an electron propagates between two points in space and T describes how a photo-electron scatters from a neighboring atom.

In the EXAFS energy region where the X-ray photon has sufficiently high energy above the absorption edge, the formal expansion of $\mathbf{G}$ in Eq. (6.17) is convergent very quickly. Hence the multiple-scattering series of the propagator as a sum of limited number of terms is very effective, corresponding to the path-by-path theory. However, in the XANES region, the matrix expansion in Eq. (6.16) is not convergent. Instead, it is necessary to solve the inversion of the matrix $(1 - G^0T)^{-1}G^0$ that is the most computationally expensive part in the XANES calculation. This actually is equal to expand the propagator $\mathbf{G}$ by considering an infinite number of scattering paths and therefore this method is called full multiple scattering (FMS) technique. FMS calculation techniques with different details of implementation have been developed separately by several groups [14, 16, 48].

Besides, time-dependent density-functional theory (TDDFT) has also been successfully applied to X-ray absorption spectra at the $L_{2,3}$ edges for the $3d$ transmission metal (TM) series [49, 50]. However, because of the strong electron correlation effects in the L-edge spectra of $3d$ TMs, atomic multiplet theory is more commonly used to interpret the spectra [51]. The interested readers are referred to the review paper [52] written by de Groot for more details.

At present a fully quantitative treatment of the near-edge structures is still challenging, due to various complex many-body effects, such as the photoelectron core-hole interaction, multiplet effects, the photoelectron self-energy and inelastic losses. Nevertheless, it still finds wide applications in many different fields, often used as a "fingerprint" to identify an unknown sample by comparing it with those of a number of reference compounds with known structures. A variety of quantitative tools are also available for interpreting XANES data, such as edge positioning, peak fitting, linear combination fitting, and principle components analysis. In combination with theoretical simulations or interpretations, analysis of XANES spectrum could also provide valuable information on the site symmetry, oxidation states, bond length and so forth.

6.3. Detection Schemes and Experimental Setups of XAFS

The objective of a XAFS measurement is to record an energy-dependent quantity that is directly proportional to the absorption coefficient $\mu(E)$ of the atom of interest. Since the magnitude of the XAFS oscillatory function is typically in the order of 10^{-2} of the edge-jump and decreases with increasing energy, a fairly precise and accurate measurement scheme is required to collect high quality data. The main experimental key points are the availability of a tunable X-ray source with high photon flux and high energy resolution, and high performance detectors for recording the XAFS data. The modern XAFS measurements are overwhelmingly performed on synchrotron radiation light sources, which deliver highly brilliant X-rays with a wide range of X-ray wavelengths. In order to select monochromatized photons from the incident beam, a monochromator is needed. The most commonly used monochromator is the double-crystal monochromator, which, according to the Bragg's equation, selects the energy of the incident beam E to the Bragg angle θ_B as

$$E_{\text{hkl}} = \frac{hc}{\lambda_{\text{hkl}}} = \frac{12.39852}{\lambda_{\text{hkl}}} = \frac{6.19926}{d_{hkl}\sin(\theta_B)}(\text{keV}). \tag{6.18}$$

Here h is the Planck constant, c is the light velocity in vacuum, d_{hkl} is the interplanar spacing (in Å) of the *hkl* plane of the crystal, and λ_{hkl} and E_{hkl} are the wavelength (in Å) and energy (in keV) of the monochromatic X-ray photons, respectively. The most commonly used crystal planes are Si(111) and Si(311) with d values of 3.1356 and 1.6375 Å, respectively.

Depending on different schemes of detecting $\mu(E)$, various XAFS experimental modes have been developed. Among them the transmission, fluorescence and total electron yield modes are the most frequently used and will be briefly described. More details on the measurement requirements for getting good XAFS data could be found in many available review papers or book chapters [1, 35, 53].

6.3.1. *Transmission mode*

The transmission mode is the most convenient method for collecting XAFS data. It relies on the measurement of monochromatic photons intensities incident on (I_0) and transmitted by (I_t) a sample of thickness d. The incident and transmitted beam intensities I_0 and I_t are related by $I_t = I_0 e^{-\mu(E)d}$, then the absorption coefficient is calculated as

$$\mu(E)d = -\ln(I_{\mathrm{t}}/I_0) = \ln(I_0/I_t). \tag{6.19}$$

A typical transmission experimental setup is schematically shown in Fig. 6.5. The wide-band hard X-ray comes out from a synchrotron radiation beamline and is monochromatized by a double-crystal monochromator to select a particular energy. The intensities of I_0 and I_{t} are usually monitored by using ion chambers. Care must be taken to make sure that the two ion chambers are working in the linear region, and the choice of their lengths and gas compositions are essential. The higher-harmonic contamination should also be minimized, otherwise the spectrum will be distorted by serious systematic

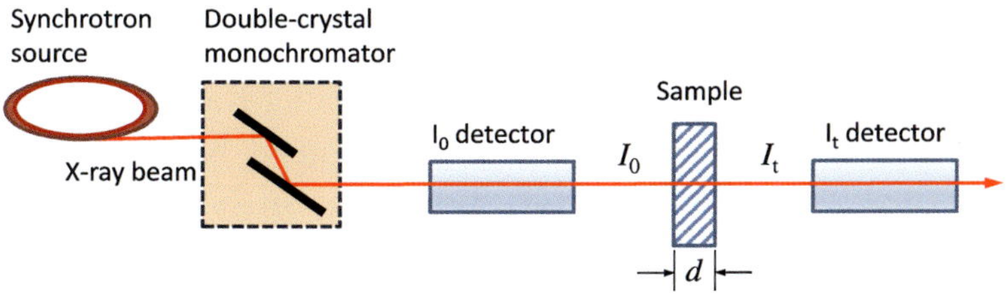

Fig. 6.5. The schematic XAFS experimental setup in a transmission mode.

error. This can easily be done by either using a mirror or detuning the two crystals of the monochromator.

XAFS measurement in transmission mode is the most widely used and has the best data quality. But it is mostly suitable for concentrated samples in which the element of interest is the major component (>10%). A crucial requirement for the sample is that it must be uniform and pinhole-free. Typically, the sample thickness d should be optimized so that the edge jump $\Delta\mu d \approx 1$. Under this condition, the "thickness effects" could be avoided and the data could have a good signal/noise ratio [54]. This requires that the grain sizes for powder samples be smaller than or comparable to the absorption length of the material. For a single substance, the absorption coefficient μ is related to the absorption cross σ by $\mu = \rho\sigma$, where ρ is the density. If the unit of ρ is g/cm^3, the cross-section is in cm^2/g and μ is in cm^{-1}. For a compound consisting of different elements, the total absorption coefficient μ is given by:

$$\mu \approx \sum_i \rho_i\sigma_i = \rho_M \sum_i \frac{m_i}{M}\sigma_i = \rho_N \sum_i \frac{n_i}{N}\sigma_i, \tag{6.20}$$

where ρ_M is the mass density of the material as a whole, ρ_N is the number density of the material as a whole, and m_i/M and n_i/N are the mass fraction and number fraction of element i, respectively. Sources of the energy-dependent cross-section data can be found in literatures [55, 56].

Here are some examples of the optimal thickness for different materials. For Fe metal, $\rho = 7.86\,\text{g/cm}^3$, the absorption cross-sections exactly before and after the absorption edge (7112 eV) are 407.6 and 53.3 cm^2/g; therefore across the absorption edge $\Delta\mu = \rho\Delta\sigma = 7.86\times(407.6-53.3) = 2784.8\ \text{cm}^{-1}$. The requirement of $\Delta\mu d \approx 1$ then gives the optimal thickness $d = 1/2784.8 = 0.00036\ \text{cm} = 3.6\,\mu\text{m}$. For iron oxide, such as Fe_3O_4, calculations using Eq. (6.19) gives the optimal thickness of 7.5 μm. For heavier element like Ag metal, the optimal thickness is much larger, 21 μm at Ag K-edge. For dilute solutions, d is typically in the millimeter range. In the soft X-ray region, the optimal thickness is much smaller. For example, at Al K-edge the optimal thickness for Al metal is only 1.0 μm. Due to the too small thickness required, it is practically very hard to make a uniform sample and the transmission technique usually cannot be applied.

6.3.2. *Fluorescence yield mode*

In a transmission experiment for a dilute sample, like the lightly doped samples such as diluted magnetic semiconductors and thin films grown on substrates, the thickness of the sample has to be adjusted so that the edge jump $\Delta\mu d \approx 1$. Under such circumstance, the background absorption due to the matrix or substrate would be very large and the transmitted beam intensity would be exponentially attenuated, resulting in very bad quality of the data. To reduce the background absorption, selectively recording the atomic-specific signals is a good choice. Among them, monitoring the X-ray fluorescence is the most preferred, since the fluorescence yield is proportional to the absorption efficient under proper conditions (see Eqs. (6.22) and (6.23) below).

To make good fluorescence XAFS measurements, one desires to collect the available signal as much as possible. In a fluorescence XAFS measurement, the X-rays emitted from the sample will include the fluorescence line of interest, fluorescence lines from other elements in the sample, and both elastic and inelastic (Compton) scattering X-rays. Although in general the fluorescence is emitted isotropically, the scatter is not emitted isotropically because the X-rays from a synchrotron are polarized in the plane of the synchrotron. This means that elastic scatter is greatly suppressed at 90° to the incident beam in the horizontal plane. Therefore, fluorescence detectors are normally placed in the horizontal plane at a right angle to the incident beam, and the sample is placed at 45° to both the incident beam and the fluorescence detector, as shown schematically in Fig. 6.6. In this geometry, the fluorescence intensity $I_f(E)$ accepted by a detector with a solid angle of $\Omega/4\pi$ is:

$$I_f(E) = I_0(\Omega/4\pi)\gamma\frac{\mu_A(E)}{\mu_{tot}(E)+\mu(E_f)}\{1-\exp[-(\mu_{tot}(E)+\mu(E_f))d]\}. \tag{6.21}$$

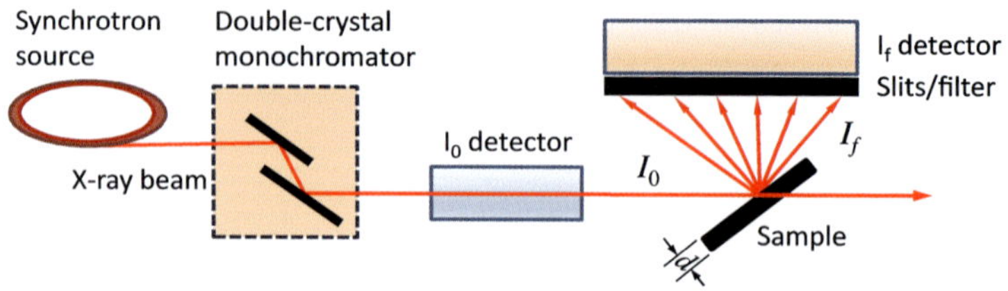

Fig. 6.6. The schematic XAFS measurement in a fluorescence mode.

Here γ is the fluorescence efficiency, d is the sample thickness, E_f is the energy of the fluorescent photons, $\mu_A(E)$ is the absorption from the element of interest, and $\mu_{tot}(E)$ is the total absorption coefficient that is the sum of the contributions $\mu_A(E)$ from the element of interest and $\mu_B(E)$ from the background absorption coefficient of all the other elements in the matrix. Because $\mu_{\text{tot}}(E)$ depends on $\mu_{\text{A}}(E)$, the measured fluorescence intensity is in general not proportional to the absorption coefficient $\mu_{\text{A}}(E)$ that one wants to measure. However, there are two limiting cases where this proportion holds:

(1) Thin samples (typically below 300 nm), in which the series expansion of the exponential in Eq. (6.21) can be truncated at the term linear in d, i.e., $1 - \exp[-(\mu_{\text{tot}}(E) + \mu(E_f))d] \approx (\mu_{\text{tot}}(E) + \mu(E_f))d$, and then

$$I_f(E) = I_0(\Omega/4\pi)\gamma d\mu_{\text{A}}(E) \propto I_0\mu_{\text{A}}(E). \tag{6.22}$$

(2) Thick samples with a dilute concentration of the element of interest, in which $\mu_{\text{tot}}(E) \gg \mu_{\text{A}}(E)$ and $1 - \exp[-(\mu_{\text{tot}}(E) + \mu(E_f))d] \approx 1$. Then Eq. (6.21) reduces to

$$I_f(E) = I_0(\Omega/4\pi)\gamma\frac{\mu_{\text{A}}(E)}{\mu_{\text{tot}}(E)} \propto I_0\mu_{\text{A}}(E). \tag{6.23}$$

If neither of the above two conditions could be met, a strong suppression of features in the spectrum could be observed. Hence, use of a thick, concentrated sample results in severely distorted EXAFS amplitudes and XANES spectra. Such a phenomenon, known as the "self-absorption" effect of the fluorescence data, should be corrected before further processing, and numerous correction procedures have been proposed [57, 58].

In the fluorescence mode one desires to collect only the fluorescence line of interest, and completely suppress the scatter peak and other fluorescence lines. This would greatly decrease the background intensity, and increase the signal/noise ratio. One can reduce the intensity of the scattering by an order of magnitude by inserting an X-ray filter that has the absorption edge between the fluorescence line of interest and the elastic scattering peak. A thin metal foil or a homogeneous powder containing the $Z-1$ element (Z is the atomic number of the element of interest) is conventionally used as a filter. For example, to measure Co K-edge fluorescence XAFS data, a 3- or 6-μm thick Fe_2O_3 filter is usually used. Besides, to avoid re-radiation from the filter itself, a set of slits (Soller slits) is often used to preferentially collect emission from the sample and block the emission from the filter.

To collect fluorescence X-ray, an ionization chamber–Lytle detector [59] could be used, but it cannot discriminate the fluorescence of interest from the background. To suppress the background intensity, detectors with energy-discriminating ability, such as high-purity Ge or Si solid-state detector (SSD) and silicon drift diode (SDD) detector, have been developed and routinely used. The energy resolution of these detector is typically in the range of 150–200 eV at 6 keV, as a result of a balance between the counting rate and background removal. The maximum counting rate of these detectors is limited by the detector electronics, which can be enhanced by an order of magnitude by assembling multiple detector elements with a close packing geometry. The dilute limit of the fluorescence XAFS measurements is determined by the detector counting rate and incident beam intensity. A state-of-the-art Ge pixel array detector (PAD) with 100 segments has been developed to gain high throughput measurement of fluorescence yield by Oyanagi [60], who describes more detail about the design of the PAD in Section 5.3 of this book. Because both the incident and detected signals in the fluorescence mode are penetrating X-rays, fluorescence XAFS is the most suitable choice to study bulk samples. It could also be a surface-sensitive method by using a grazing-incidence geometry [61]. Nowadays, samples with interested element concentration of several to tens of ppm or thin films with thickness equivalent to 0.1–1 monolayers are measurable by XAFS with adequate signal/noise ratio [61, 62].

6.3.3. *Electron yield detection mode*

When the core level electron absorbs an X-ray photon with energy above the absorption edge, the core hole left behind could decay by emitting fluorescence and Auger electron. Analogous to the fluorescence mode, detection of the Auger electrons provides another way to collect the XAFS signal. Depending on which electrons are detected, there are various electron detection yield detection modes. Among them, the total electron yield (TEY) detection mode is experimentally the simplest, which measures the drain current using an electrometer under ultra-high vacuum (Fig. 6.7) [63]. In this mode, the sample should be made electrically conductive. The TEY mode is especially important for collecting the XAFS data in the soft X-ray region, like O or N K-edge and 3d TM $L_{2,3}$-edge spectra [64]. A variation of TEY mode is the partial electron yield (PEY) mode in which only electrons of a given energy range are collected with retarding grids and electron detectors. There is also the conceptually simplest Auger electron yield (AEY) detection mode, in which the number of the Auger electrons is recorded by using an electron analyzer operating in ultrahigh-vacuum.

The short escape-length of the electrons (tens to hundreds of angstroms, depending on the energy of the original Auger electrons) makes the electron yield mode sensitive to near-surface regions

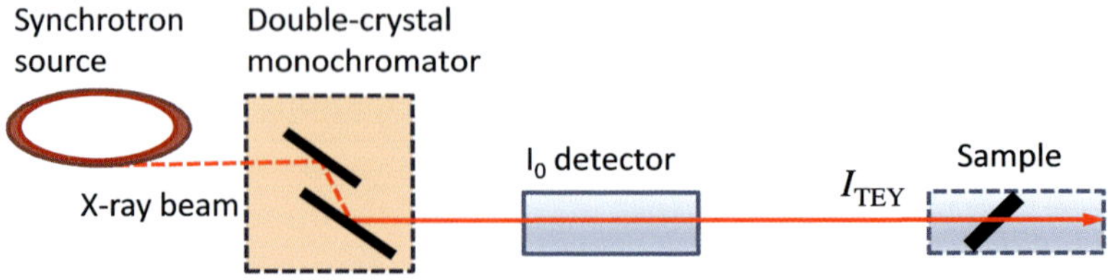

Fig. 6.7. The schematic XAFS measurement in a total electron detection mode.

or interfaces, in contrast to the bulk-sensitivity of the fluorescence mode. This method works efficiently for thick and concentrated specimens that cannot be measured by other XAFS modes. It also suffers less from the "self-absorption" effect occurring in the fluorescence mode, because the electron absorption has much higher efficiency than the photon absorption.

6.4. XAFS Applications in Functional Materials

The following subsections will be attributed to the applications of XAFS to the study of various functional materials. These materials range from catalysts, magnetic semiconductors, thin films, to complex compounds (high-temperature superconductivity and giant magnetoresistance materials). The application of XAFS in the *in situ* probing of nucleation and growth processes of nanomaterials will also be introduced, followed by a brief description of the recently developed time-resolved and space-resolved XAFS techniques. The content will be arranged according to the materials systems, with focus on the XAFS-revealed atomic and electronic structural features that play important roles in understanding the structure-functionality relationships.

6.4.1. *Nanoparticles and catalysts*

Monodisperse nanoparticles have fascinating size-dependent optical, electronic, catalytic and chemical properties which make them very promising in the development of new generations of catalysts, sensors, and optoelectronic devices [65–68]. To understand the interplay between the size and the surface effects that determines the structure and the electronic properties of gold nanoparticles (NP), Zhang and Sham conducted a study on a series of alkanethiol-capped Au NPs prepared by colloidal synthesis techniques in the presence of thiol [69]. The size of the thiol-capped NPs were determined by the ratio of the Au precursor ($HAuCl_4$) and the thiol; Au:S ratios of 1:3, 1:1, and 6:1 yielded NPs of 1.6, 2.4, and 4.0 nm, respectively. The Au L_3-edge EXAFS data show a size-dependent lattice contraction, as apparent in the back-transform of the radial distribution of the first

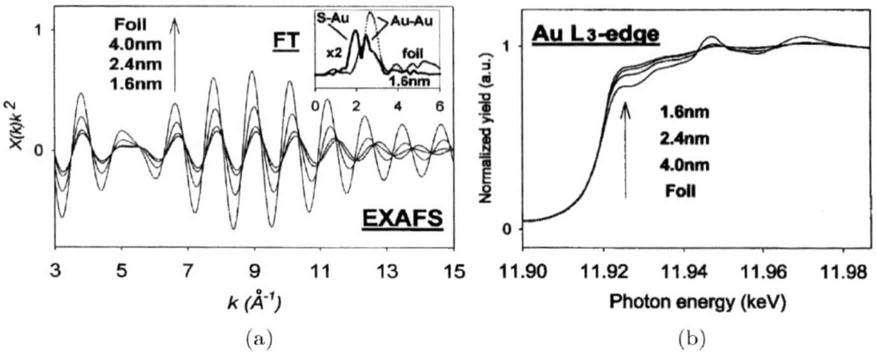

Fig. 6.8. (a) Fourier back-transform of the first Au-Au shell in R space of the three Au NPs samples and the Au foil. Inset shows the Fourier transformed $k^2\chi(k)$ EXAFS of the 1.6 nm NPs compared with Au foil. (b) Au L_3-edge XANES of the NPs samples and Au foil.

Au-Au shell in the k space where the oscillation becomes progressively farther apart as the Au NP size decreases (Fig. 6.8(a)). Quantitative results indicate the contractions of 0.7%, 1.1%, and 1.4% relative to the bulk for the 4.0, 2.4, and 1.6 nm Au NPs, respectively. The EXAFS data also show a reduction in coordination number and increasing disorder (less intense and broad oscillations). The normalized Au L_3-edge XANES of the three NPs and Au metal (bulk) show some interesting features (Fig. 6.8(b)). First, all NP spectra exhibit the same resonance pattern as that of Au metal albeit with a significant broadening, indicating the Au-like environment but with a significant disorder. Second, the whiteline intensity increases in the order of bulk < 4:0 nm NP < 2:4 nm NP < 1:6 nm NP, suggesting an increase in d-hole population (7.2%, 9.0%, and 11.2% relative to the Au metal) with decreased particle size, in good agreement with electronegativity considerations. The implication of these observations is discussed in terms of the interplay of quantum-size and surface effect. It appears that controlling the balance of the size and the nature of the capping molecules could afford some degree of tuning the electronic properties of Au NPs.

Supported metal nanoparticles are widely used as catalysts, whose activity and selectivity are determined by the chemical and

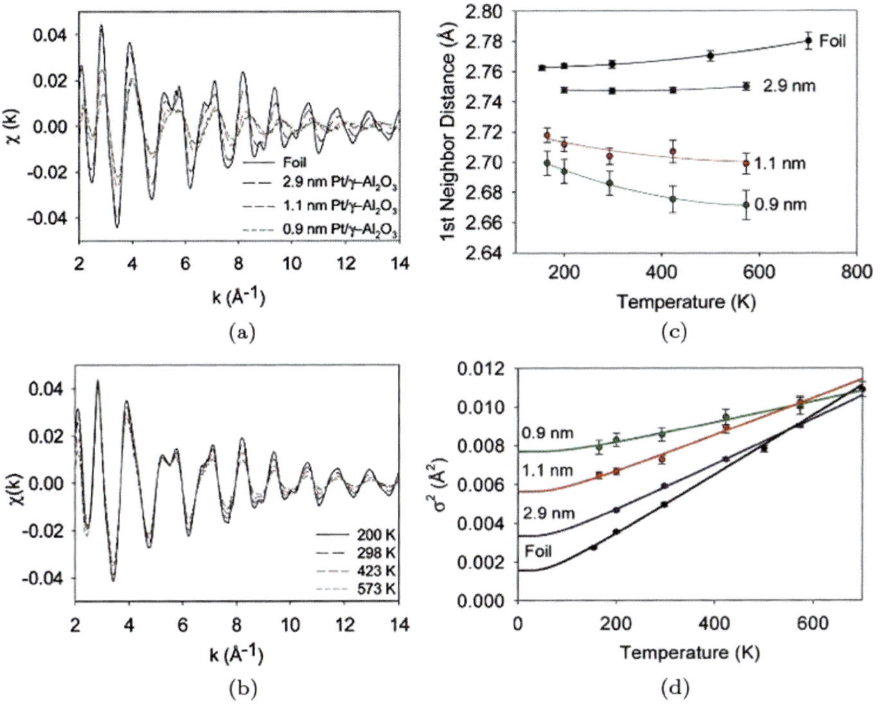

Fig. 6.9. (a) EXAFS data in k-space illustrating the effect of cluster size on the oscillatory pattern of different Pt/γ-Al_2O_3 samples, (b) progressive changes in intensity of the EXAFS data in k-space for a 2.9 nm Pt/γ-Al_2O_3 sample in response to temperature variation. (c) Thermal dependence of the 1NN Pt-Pt bond distances for the different sized Pt/γ-Al_2O_3 nanoparticles, and (d) temperature-dependence of the Debye-Waller factors for the various sized Pt/γ- Al_2O_3 nanoparticles and Pt foil plotted with their respective Einstein models (lines).

structural nature of catalytically active surface sites. To addresses the structural, dynamic, and electronic features exhibited in a highly prototypical catalytic system, Sanchez and coworkers studied γ-Al_2O_3 supported Pt NPs [70]. By collecting EXAFS data for different-sized Pt NPs samples at the same temperature (Fig. 6.9(a)) and the data for a given sample at different temperatures (Fig. 6.9(b)), they could discriminate the effects of temperature and particle size on the structural dynamics. The sensitivity to cluster size therefore permits the extraction of quantitative information about particle sizes based on the obtained coordination numbers, which is consistent with the average particle sizes obtained by TEM.

A full comparison of the temperature dependences of the 1NN Pt-Pt bond lengths for all three samples measured under He (Fig. 6.9(c)) reveals that the subnanometer Pt/γ-Al_2O_3 NPs exhibit a dramatically nonbulk-like nature: it possesses marked relaxations of the Pt-Pt bond distances, negative thermal expansion (NTE) with an average linear thermal expansion coefficient of $(-2.4 \pm 0.4) \times 10^{-5}\,K^{-1}$. The temperature-dependence of the EXAFS Debye-Waller factors σ^2 in Fig. 6.9(d) indicates that the static disorder σ_s^2 and the Einstein temperature increase rapidly as the nanoparticle size decreases. Adsorption of hydrogen on the sub-nanometer clusters leads to an increase of the Pt-Pt bond lengths and significant attenuation of the disorder. Oxidation of these same clusters have the opposite effect, leading to an increase in Pt-Pt bond strain and subsequent enhancement in nonbulk-like thermal properties. This study highlights that the charge exchange due to both metal-support and metal-adsorbate interactions plays an important role in mediating the structural dynamics of supported nanoscale metal clusters catalysts. Furthermore, the same group studied the influence of adsorbates on the electronic structure, bond strain of the ~1.0 nm Pt/γ-Al_2O_3 catalyst [71]. They demonstrated that the Pt-Pt bonding in these small clusters is responsive to the presence of adsorbates, exhibiting pronounced coverage-dependent strains, with concomitant modifications of their electronic structures. Moreover, the Pt-Pt bond strains within the clusters could be relaxed by H_2 and CO adsorbates.

Noble bimetallic nanostructures have unique catalytic, electronic, and magnetic properties different from their monometallic counterparts. In addition to size, shape, and chemical composition, the performance of bimetallic nanostructures are also largely associated with their structural diversities. Bimetallic nanostructures with non-random metal atoms distribution are an important type of such materials. Teng and coworkers reported the formation of Pd/Au bimetallic nanostructures (2.4 ± 0.2 nm in width, over 30 nm in length) from Pd nanowires via galvanic replacement reaction between Pd ultrathin nanowires and $AuCl_3$ in toluene [72]. Continuous changes of chemical composition and crystalline structure from Pd nanowires to $Pd_{68}Au_{32}$ and $Pd_{45}Au_{55}$ alloys, and to Au nanoparticles were observed. The statistical distributions of Pd and Au atoms inside the nanoparticles, which are very crucial to the catalytic performance of Au/Pd catalysts, were determined by a combined EXAFS/EDS analysis of the intra-particle structure and composition of resultant Pd/Au compounds.

EXAFS data measured at Pd K-edge (Fig. 6.10(a,b)) indicate some degree of alloying for the $Pd_{68}Au_{32}$ and $Pd_{45}Au_{55}$ samples,

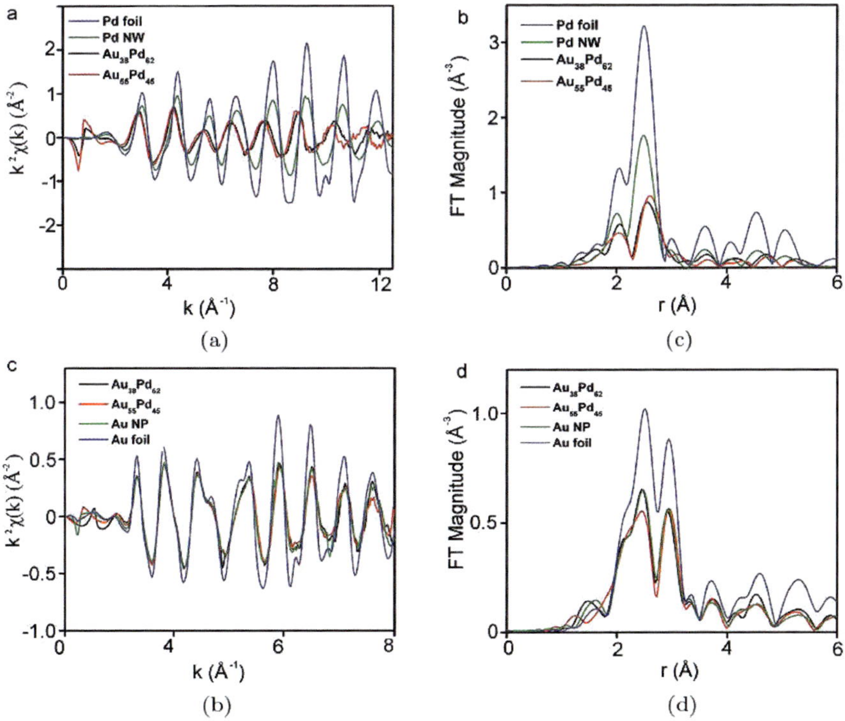

Fig. 6.10. EXAFS data for as-made Pd nanowires, $Pd_{68}Au_{32}$, $Pd_{45}Au_{55}$, and Au nanomaterials: (a) $k^2\chi(k)$ EXAFS signal and (b) Fourier transform magnitudes for Pd K-edge EXAFS; (c) $k^2\chi(k)$ EXAFS signal and (d) Fourier transform magnitudes for Au L_3-edge EXAFS.

while the Au L_3-edge data (Fig. 6.10(c,d)) show no phase shift relative to the bulk Au. The quantitative structural parameters suggest the following sequence of local structural transformations in the sample. For the $Pd_{68}Au_{32}$ nanostructure, the $N_{\text{Au-Au}}/N_{\text{Au-Pd}}$ ratio (~14) is much greater than the ratio of the compositions $X_{\text{Au}}/X_{\text{Pd}}$ (~0.47), indicating the formation of core-shell-like non-random Pd/Au alloy during the initial stage. With continued reaction, the $N_{\text{Au-Au}}/N_{\text{Au-Pd}}$ (1.58) became comparable to $X_{\text{Au}}/X_{\text{Pd}}$ (1.22) and implies the preference of random alloying in the resultant $Pd_{45}Au_{55}$ nanostructures. After the 10 h replacement reaction, all Pd was replaced by Au^{3+}, and only Au nanoparticles were left in the system. For more detailed studies on applications of EXAFS spectroscopy to bimetallic nanoparticle catalysts (covering both its power and limitations), interested readers are referred to an excellent review article written by Frenkel [73].

6.4.2. *Magnetic semiconductors*

The rapidly developing field of spintronics requires the simultaneous manipulation of spin and charge [74, 75], which could provide enhanced functionality and performance in traditional charge-based electronics. One of the promising material candidates for semiconductor spintronics is diluted magnetic semiconductor (DMS) which is refered to as semiconductor substitutionally incorporated by transition-metal (TM) magnetic ions [76]. A magnetically doped semiconductor could be called a "true" DMS material only when the dopants occupy the substitutional positions in the host, as suggested by Dietl *et al.*'s prediction [77]. Mn-doped GaAs is the prototypical DMS system, on which many studies have been performed. Shioda *et al.* [78] reported a study of the local structures around Mn in $Ga_{1-x}Mn_xAs$ (x = 0.005 and 0.074) films grown on GaAs(001) substrates by low-temperature molecular-beam epitaxy (LT-MBE) at 250–300°C under As-rich conditions. A comparison of the EXAFS spectra of $Ga_{1-x}Mn_xAs$ with that of zinc-blende structured GaAs shows that their nearest-neighbor peaks are similar in both position and shape. This indicates that the nearest-neighbor atoms should be the same species and reveals the complete incorporation of Mn dopants at the substitutional Ga sites in the GaAs lattice. Soo and coworkers [79] compared the local environment surrounding ferromagnetically ordered Mn in Mn/GaAs digital layers and (Ga, Mn)As random alloys. The observed Mn L_3-edge XMCD spectra and Mn K-edge EXAFS spectra are very similar for both samples, but different from those for the MnAs film. Therefore, the Mn atoms in the Mn/GaAs digital layers are exclusively substitutional for Ga sites in the GaAs sites, just like in the (Ga, Mn)As random alloy.

XANES has also made a significant contribution to the identification of the occupation sites of TM dopants in DMS materials. Wei and colleagues carried out investigations on Mn-doped GaN system by comparison of experimental and theoretical spectra of XANES [80]. Theoretical XANES spectra of representative structure models of Mn in GaN lattice were calculated, and compared with the experimental data of 300 nm thick $Ga_{1-x}Mn_xN$ ($x = 0.010$, 0.025, 0.10) thin films grown on GaAs(001) substrates using plasma-assisted MBE (Fig. 6.11(a)). It is shown that the substitutional Mn (Mn_{Ga}) in GaN is characterized by a pre-edge peak at 2.0 eV and a post-edge multiple-scattering peak at 29.1 eV. The peaks are shifted in position and dropped in intensity dramatically for the interstitial Mn_I and Mn_{Ga}-Mn_I dimer, and then disappear completely for metallic Mn clusters. The comparison between the experimental and theoretical XANES spectra indicate that the majority of Mn atoms locate at the Mn_{Ga} sites in $Ga_{1-x}Mn_xN$ with the lowest Mn content (0.010). Upon increasing the Mn content to 0.025, a significant fraction of Mn dopants exists in the form of Mn_{Ga}-Mn_I dimers. At the highest Mn doping concentration (0.100), a significant amount of metallic Mn clusters were precipitated.

Kunisu *et al.* [81] also reported a Mn K-edge XANES study on the local environment of Mn dopants in ZnO polycrystals. The spectra as shown in Fig. 6.11(b) at the Mn concentrations between 0.01 and 5 at.% indicate almost identical features, which could be well reproduced by first-principles spin-polarized density functional theory (DFT) calculations for substitutional Mn at Zn sites in ZnO. The XANES at 10 at.% Mn is also very similar to the ones with lower

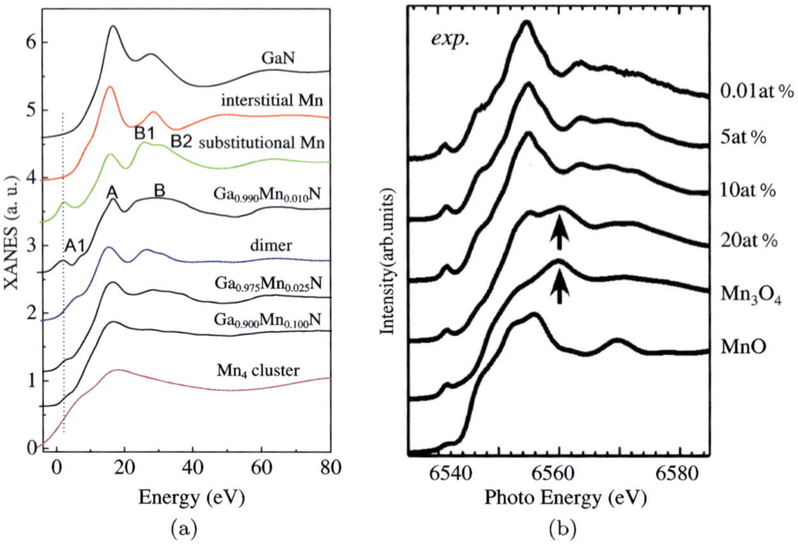

Fig. 6.11. (a) Comparison of the experimental XANES spectra for $Ga_{1-x}Mn_xN$ with various Mn concentration with the calculated Mn K-edge XANES spectra for representative structure models. (b) XANES spectra at Mn K-edge for polycrystalline Mn-doped ZnO polycrystals with Mn concentrations of 0.01, 5, 10, and 20 at.%.

concentrations, whereas a minor portion of Mn exists in the form of Mn_3O_4 as found in XRD pattern. At a higher Mn concentration of 20 at.%, change in the XANES shape is clearly recognized. The Co and O K-edges spectra were also combined to investigate the Co occupation sites and distributions in $Zn_{1-x}Co_xO$ thin films with low Co concentrations ($x = 0.005$, 0.02) [82]. Co K-edge EXAFS analysis indicates unambiguously that the doped Co ions are substantially incorporated into the ZnO host lattice. However, strikingly different features in the O K-edge XANES features for the $Zn_{1-x}Co_xO$ films than those for ZnO bulk are observed, which can be reproduced by the assembling of substitutional Co ions via intervening oxygen atoms. This provides experimental evidence for theoretical predications [83, 84] that the substitutional Co ions in ZnO tend to gather together, rather than distribute randomly in the whole lattice.

Interests in the Co-doped ZnO system have also been extended to zero-dimensional and one-dimensional nanostructures. Especially,

Co-doped ZnO nanowires (NWs) have attracted considerable attention. Yuhas and colleagues [85] synthesized single crystalline, single domain, and single phase Co:ZnO NWs with uniform diameters of 35 ± 5 nm and lengths ranging from 2 to 6 μm using a solution-based synthetic route. The Co concentrations, ranging from 5.35 to 15.50 at.%, are far beyond the equilibrium solubility of Co in bulk ZnO, whereas, the doped Co ions reside in a uniform subsitutional environment with a small degree of disorder. A similar phenomenon was also observed by Yao and coworkers on $Zn_{0.98}Co_{0.02}O$ NWs [86]. The substitutional incorporation of Mn ions up to 10% into CdSe quantum nano-ribbons was also readily achieved by Yu and coworkers via a nucleation-controlled doping process [87].

Due to the low equilibrium solubility of TM in most of the semiconductor hosts, the doped ions are not always substitutionally incorporated into the host lattice, but may form various secondary phases or metal clusters, whose presence should be clearly identified if the magnetic properties of the material are to be fully understood. Zhang and coworkers [88] carried out Mn K-edge EXAFS measurements for two cubic zinc-blende structured p-type $Ga_{1-x}Mn_xN$ films ($x = 0.025$ for sample A and $x = 0.10$ for sample B). Figure 6.12(a) shows their

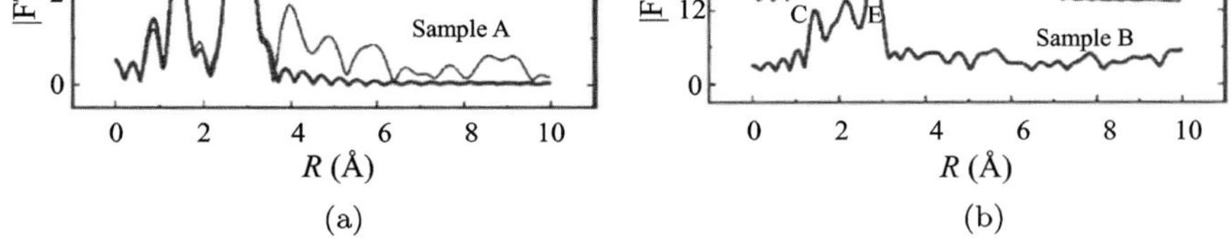

Fig. 6.12. (a) Magnitude of Fourier transformed EXAFS $k^3\chi(k)$ functions at Mn K-edge for $Ga_{1-x}Mn_xN$ samples A and B. (b) Comparison of the Fourier transformed $k^3\chi(k)$ curve of sample B with the simulation results for various structure models.

Fourier transformed EXAFS $k^3\chi(k)$ functions. For sample A, peak A corresponds to the nitrogen nearest neighbors surrounding Mn impurities with a coordination number 3.4 at a distance of 2.00 Å. Peak B consists of the contributions of 12 Ga and 12 N neighbors at distances of 3.22 and 3.93 Å, respectively. Compared with the Ga-N bond length (1.93 Å) in cubic GaN, the Mn-N bond in sample A is considerably elongated, possibly due to the existence of a large number of nitrogen defects. For sample B, an additional peak D that is absent for sample A is obviously apparent in the R-range of 2.00–3.22 Å. This indicates that some other Mn related structure more than substitutional Mn is formed in sample B. The authors simulated the EXAFS spectra for kinds of model structures, and found that only the model containing a mixture of substitutional Mn, interstitial Mn and Mn clusters could best reproduce the features of sample B (Fig. 6.12(b)). They concluded that in heavily doped $Ga_{1-x}Mn_xN$ ($x = 0.10$) films the Mn atoms tended to form Mn interstitial and Mn clusters; a consistent conclusion is also drawn from XANES analysis [80].

When the dopant concentration in DMSs is beyond its solubility in the host, other precipitates or compounds more than metallic clusters could also be formed. Shi *et al.* [89] using the sol-gel method synthesized $Zn_{1-x}Co_xO$ ($x = 0.02 - 0.25$) nano-composites that showed paramagnetic behavior in the temperature range of 10–400 K. XRD measurements indicated the single phase of wurtzite ZnO when Co concentration x was below 0.15 and the formation of an additional Co_3O_4 phase at $x > 0.15$. The EXAFS data shown in Fig. 6.13 indicate that at low Co concentrations ($x \leq 0.05$), the Co atoms are substitutionally incorporated into the ZnO lattice. At higher Co concentrations of $x \geq 0.10$, the formation of a Co_3O_4 secondary phase could be readily observed. Correlating the magnetic behavior with the structural properties of the $Zn_{1-x}Co_xO$ nano-composites, it is concluded that the paramagnetism is the intrinsic nature of Co-doped ZnO at low Co doping level if no other defects (such as oxygen vacancies) or significant structural distortion are present. The purely paramagnetic behavior for isolated or paired Co dopant atoms

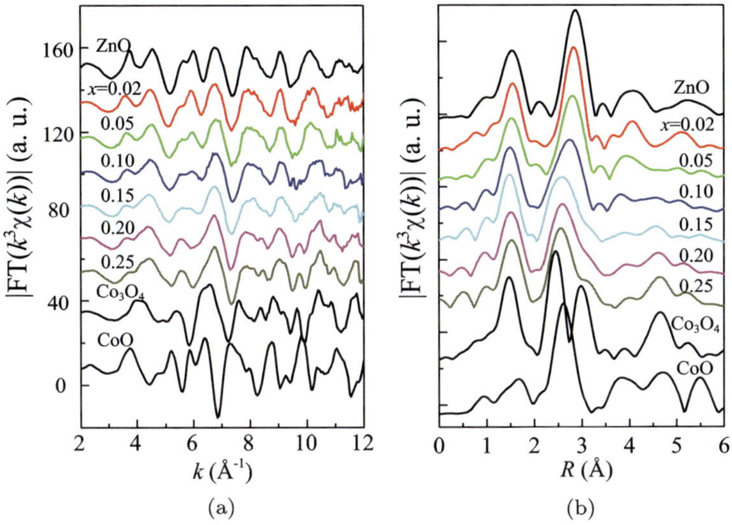

Fig. 6.13. (a) The Co K-edge EXAFS $k^3\chi(k)$ spectra and (b) the corresponding Fourrier transformed curves for the $Zn_{1-x}Co_xO$ nano-composites with various Co concentrations and reference samples.

in Co:ZnO films was also confirmed by Ney *et al.* [90] using XMCD and conventional magnetometry.

The valence of the TM ions in DMS materials is an important parameter because it is relevant to the TM band position and correlates with the hybridization strength between TM 3*d* states and the impurity bands. Yan and coworkers anticipated that magnetism tuning of oxide DMSs might be achieved via doping magnetic TM ions with variable valences, and proposed a concept of valence state dependent ferromagnetism. This idea is based on the bound magnetic polaron (BMP) model [91], where the magnetic behaviors of the oxide DMSs, correlated with strength of the hybridization between the TM ions and the impurity bands, are strongly dependent on the position of the TM 3*d* bands within the bandgap. Using Mn-doped NiO as a prototype material, they experimentally verified that the magnetic properties of the material could be fine-tuned by the intentional variation of Mn^{3+}/Mn^{2+} ratio via *p*-type Li-codoping. The Mn K-edge EXAFS analysis indicates the substitution of Mn for

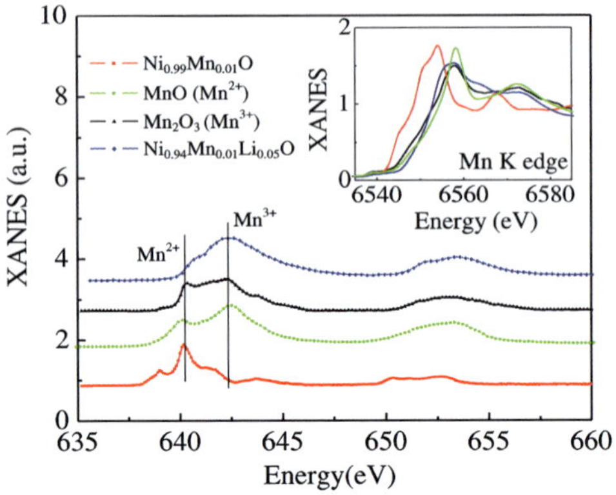

Fig. 6.14. The Mn $L_{2,3}$-edge XANES spectra of the $Ni_{0.99}Mn_{0.01}O$, $Ni_{0.94}Mn_{0.01}Li_{0.05}O$, MnO, and Mn_2O_3. The inset displays their Mn K-edge XANES spectra.

Ni sites in both $Ni_{0.99}Mn_{0.01}O$ and $Ni_{0.94}Mn_{0.01}Li_{0.05}O$ thin films. The Mn $L_{2,3}$-edge XANES spectra as plotted in Fig. 6.14 are used to infer the valence state of Mn ions. The Mn $L_{2,3}$-edge XANES of $Ni_{0.99}Mn_{0.01}O$ presents two peaks at 640 and 642 eV corresponding to Mn^{2+} and Mn^{3+}, respectively. Upon Li codoping in $Ni_{0.99}Mn_{0.01}O$, the Mn^{3+}/Mn^{2+} ratio is enhanced as suggested by the significantly intensified peak at 642 eV relative to the peak at 640 eV, presumably due to the Li^+ induced holes that facilitates the transformation of adjacent Mn^{2+} ions to Mn^{3+}. This picture is also supported by the Mn K-edge XANES (inset of Fig. 6.14), where Li-codoping significantly up-shifts the Mn absorption edge of $Ni_{0.99}Mn_{0.01}O$ to the higher-energy side. First-principles calculations based on density functional theory predicted that Mn^{3+} ions at substitution position interacted ferromagnetically through BMP mechanism while the Mn^{2+} parts contributed to antiferromagnetism via superexchange interactions.

6.4.3. *Thin film materials*

Artificially ordered Ge/Si thin films and superlattices possess exciting electrical and optical properties that open a doorway to band-structure engineering by formation of strained epitaxial layer of

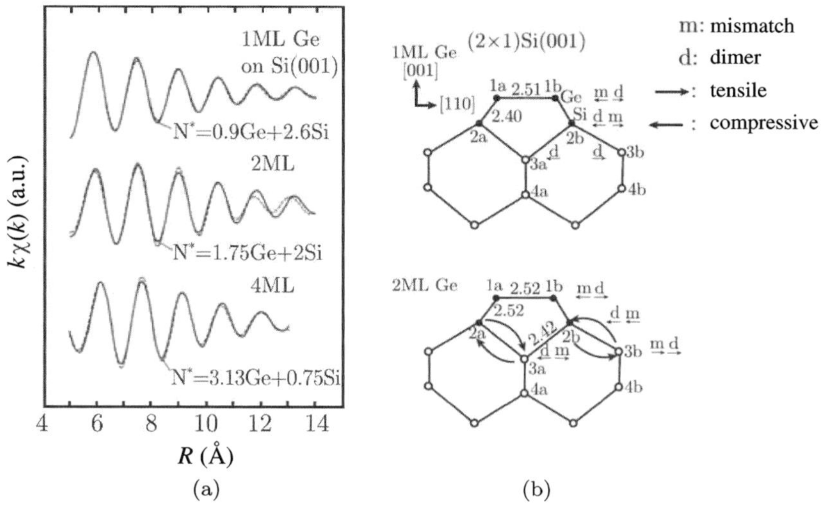

Fig. 6.15. (a) The first-shell EXAFS oscillations and the calculated curves of model structure for 1–4 ML Ge on Si(001). (b) The schematic surface strain in 1 and 2 ML Ge on Si(001).

Si and Ge. A clean Si(001) surface shows a (2×1) reconstruction and shortened interatomic distance than the counterpart in bulk Si. For Ge overlayers grown on Si(001), the formation of the dimmer and the interatomic distance are more complicated, depending on the number of Ge_n overlayers. Oyanagi and colleagues studied the local structure of Ge epitaxial overlayers with Ge thickness from 1 to 4 monolayers (ML) by fluorescence XAFS in the grazing incidence geometry [61]. The Fourier filtered first-shell Ge K-edge EXAFS oscillations are displayed in Fig. 6.15. The obtained Ge-Ge distance was 2.51 ± 0.01 Å and Ge-Si distance was 2.40 ± 0.01 Å for 1 ML Ge on Si(001), indicating that 1 ML Ge overlayer on Si(001) forms an elongated dimmer structure. It is interesting that the Ge-Ge distance was significantly longer than the interatomic distance 2.45 Å in Ge bulk, in spite of the uniaxial strain due to lattice mismatch which should shorten the bond length within a simple elastic deformation model. These bond length values remained almost unchanged within ± 0.01 Å for 2 ML Ge on Si(001). However, upon increasing the number of Ge_n overlayers to $n = 4$, the Ge-Ge distance was reduced to 2.47 ± 0.01 Å and the Ge-Si distance to 2.38 ± 0.01 Å. Figure 6.15(b) shows the surface strain in 1 ML and 2 ML Ge on Si(001) schematically. Stress components are indicated by directions of arrows. Strains are induced by surface reconstruction (dimmer strain) and atomic size effect (mismatch starin). A possible channel of atomic migrations is indicated by arrows in the bottom figure.

A $(Ge_4Si_4)_5$ monolayer strained-layer superlattice grown on Si(001) substrate shows strong optical transitions at 0.75, 1.25, 2.31 and eV unique to the superlattice period, found neither in constituent crystals nor in the $Ge_{0.5}Si_{0.5}$ alloy [92]. Using grazing-incidence fluorescence EXAFS, Wei and colleagues comprehensively studied the local structure of $(Ge_4Si_4)_5$ monolayer strained-layer superlattice [41]. EXAFS spectra for both $(Ge_4Si_4)_5$ and $Ge_{0.5}Si_{0.5}$/Si(001) have an intermediate profile suggesting that these two samples are characterized by an almost equal amount of Ge and Si scatterers (Fig. 6.16(a)). The least-squares curve-fitting results indicate

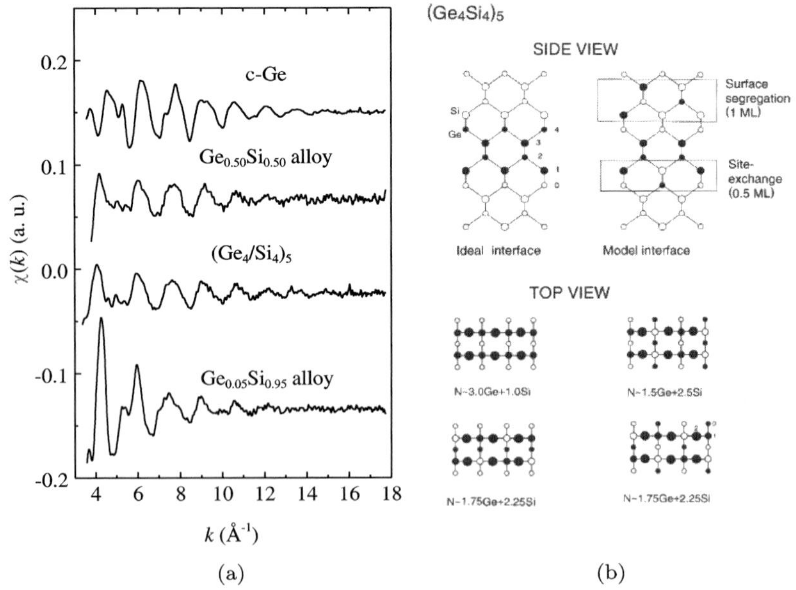

Fig. 6.16. (a) The EXAFS $\chi(k)$ function for $(Ge_4Si_4)_5$, crystalline Ge (c-Ge) and Ge_xSi_{1-x} alloys. (b) The schematic structures for $(Ge_4Si_4)_5$ with an ideal Ge/Si interface (left) and with an interface mixing (right).

that the bond lengths for $(Ge_4Si_4)_5$ are $R_{\text{Ge-Ge}} = 2.42 \pm 0.01$ Å, $R_{\text{Ge-Si}} = 2.38 \pm 0.01$ Å, and the coordination numbers are $N_{\text{Ge}} = 1.8$ and $N_{\text{Si}} = 2.2$, respectively. $R_{\text{Ge-Si}} = 2.38$ Å in $(Ge_4Si_4)_5$ is identical with the value for the dilute Ge atoms doped into Si crystal, suggesting that the bond-length compression takes the upper limit of isotropic deformation. The observed $R_{\text{Ge-Ge}}$ and $R_{\text{Ge-Si}}$ values take slightly larger values than the calculated values for model $(Ge_4Si_4)_5$ superlattice. The overestimation of bond-length compression may occur since most of model structures do not take interface mixing and nonuniform longitudinal displacement into account. Based on these results, it may be concluded that the lattice matching is achieved by tetragonal deformation via bond bending but the Ge layer is highly compressed. Figure 6.16(b) illustrates the schematic structures for $(Ge_4Si_4)_5$ with an ideal Ge/Si interface (left) and with an interface mixing (right). In the latter model, the Ge overlayer is assumed to have 0.5 ML site exchange upon deposition and 1 ML segregation upon Si overlayer growth.

XAFS spectra of multi-edges were used to study structural and electronic property evolution of nickel and nickel silicide thin films [93]. A series of nickel and nickel silicide thin films prepared by magnetron sputtering nickel on Si(100) substrates and sequential annealing procedures. XAFS at the Ni K edge, Si K edge, and Si $L_{3.2}$ edge were combined to monitor the structure and bonding systematics at different stages of the silicidation process. Ni and Si K-edge data were measured in the TEY mode in order to address the structural information in the vicinity of the Ni-Si interface. The Ni K-edge XAFS recorded in the TEY mode for the as-deposited Ni-Si(1) film looks the same as that of the Ni foil while progressive change in both the XANES and EXAFS are evident upon annealing (Fig. 6.17(a)). In contrast, the Si K-edge results display significantly different features from that of Si(100) substrate, indicating evident Ni-Si intermixing in the vicinity of the interface of the as-deposited film. The Ni K-edge EXAFS for the film after modest temperature annealing (300°C, Ni-Si(2)) shows a noticeable reduction ($\sim$40%) in amplitude and a broadening of the oscillations (Fig. 6.17(b)). Annealing

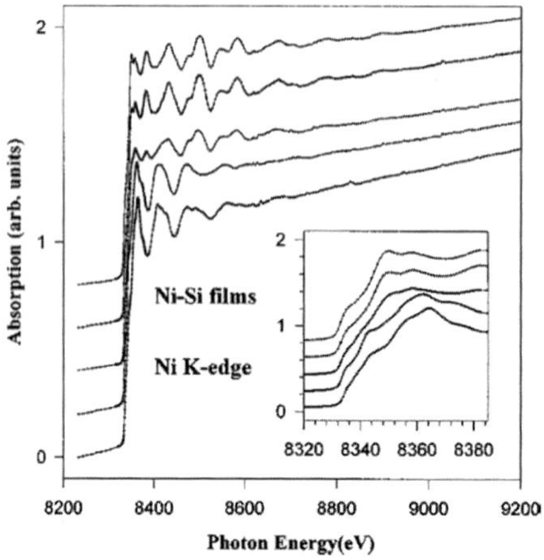

(a)

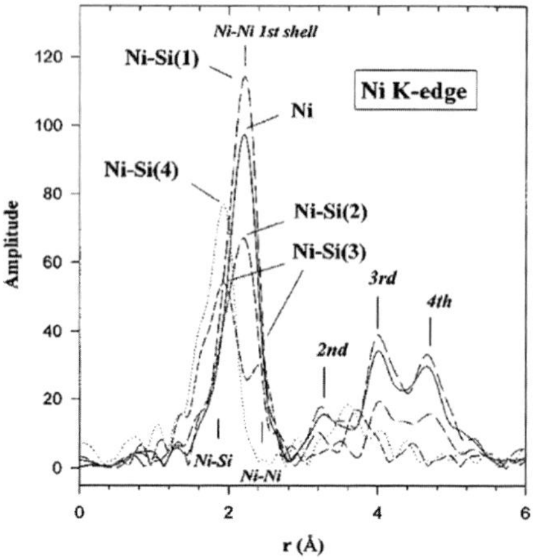

(b)

Fig. 6.17. (a) Ni K-edge XAFS (TEY mode) for a Ni foil and the films. From top to bottom: Ni-Si(1), Ni-Si(2), Ni-Si(3), and Ni-Si(4). The XANES are also shown in the same order in the inset. (b) Fourier transform of Ni K-edge EXAFS $k^2\chi(k)$ data.

at higher temperature induces significant changes in the spectra. For Ni-Si(3) annealed at ~500°C, the *fcc*-like oscillations are replaced by an entirely different pattern that is characteristic of a single Ni-Si phase. Further annealing induces still another change as seen in the spectrum for Ni-Si(4) (~850°C annealing), which reproduces the spectrum of a $NiSi_2$ phase. This is also confirmed by Si K-edge, $L_{2,3}$-edge data, XRD and Auger spectroscopy. Therefore, the as-deposited film exhibits noticeable intermixing at the Ni-Si interface at room temperature, leading to the formation of a Ni-rich silicide in the vicinity of the interface; as the annealing temperature increases, predominantly NiSi and $NiSi_2$ phases were sequentially formed.

Inorganic graphene analogues are a conceptually new class of materials with attractive applications in next-generation flexible and transparent nanodevices. Sun and colleagues reported the fabrication of large-area freestanding single layers of non-layered ZnSe with four-atomic thickness [94], by virtue of lamellar hybrid (Zn_2Se_2)(pa) (pa stands for n-propylamine) intermediate, in which hybrid ZnSe-pa single layers were initially exfoliated and then pa molecules were

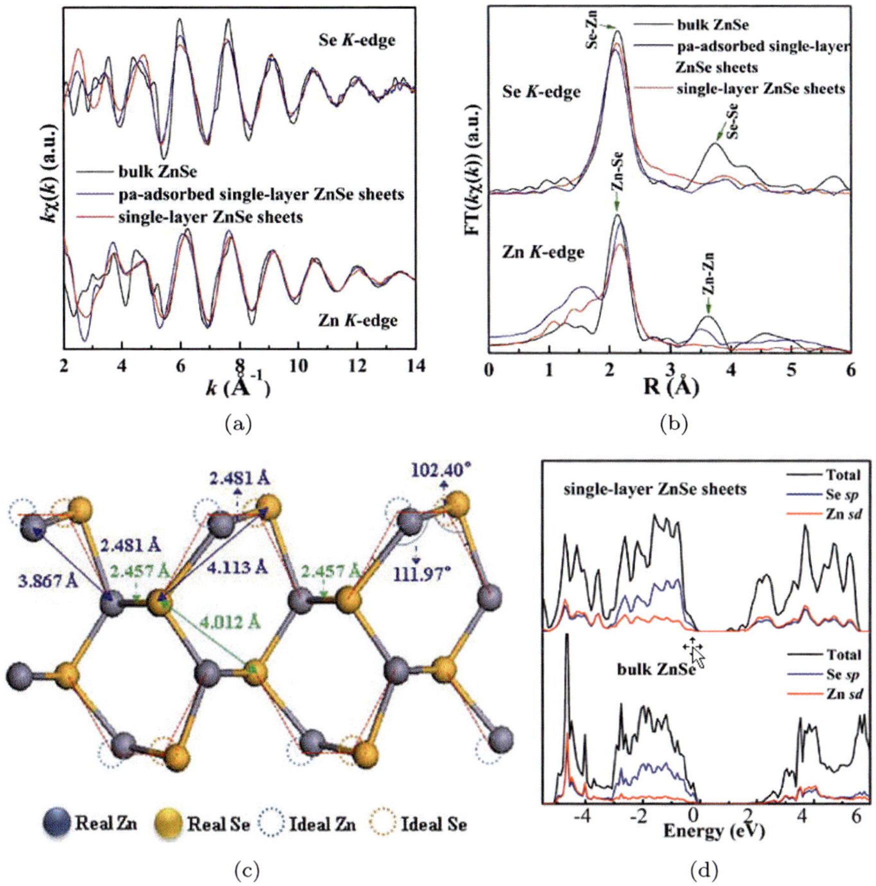

Fig. 6.18. (a) Zn and Se K-edge extended XAFS oscillation function $k\chi(k)$ and (b) the corresponding Fourier transforms; (c) structural model of single-layer ZnSe sheets (d) the calculated DOS.

removed by heat treatment at 140°C for 2 h. To interpret the stability of the ZnSe single layers, EXAFS data at both Zn and Se K-edges were measured (Fig. 6.18(a,b)). The Zn K-edge $k\chi(k)$ oscillation curves and their FT curves exhibit different local atomic arrangements of clean ZnSe single layers with their bulk counterpart and exfoliated ZnSe-pa single layers, as also confirmed by their Se K-edge data, where the Se-Se peak position is shifted to higher-R side and the peak intensity is greatly reduced. Quantitative results obtained from least-squares curve-fitting suggest that the coordination between Zn atoms and pa molecules result in slight expansion of the surface Zn-Se bond lengths, but the next nearest Zn-Zn distances (3.85 Å) are contracted and the Se-Se distances (4.11 Å) are elongated considerably relative to the bulk value (4.012 Å). Such a noticeable distortion on the surface of exfoliated ZnSe-pa single layers (Fig. 6.18(c)), as revealed by first-principles theoretical calculations, gives rise to a unique electronic structure and an excellent structural stability (Fig. 6.18(d)), resulting in an enhanced solar water splitting efficiency and photostability (195 times higher photocurrent density than that of bulk counterpart). The same group also studied pits-confined in ultrathin CeO_2 nanosheets for studying catalytic centers in carbon monoxide oxidation [95]. EXAFS analysis revealed the average coordination numbers of 4.6 for the artificial three-atomic-layer thin CeO_2 nanosheets that have approximately 20% pits occupancy. Density-functional calculations disclose that the four- and five-fold coordinated pit-surrounding cerium sites assume their respective role in carbon monoxide adsorption and oxygen activation, which lowers the activation barrier and avoids catalytic poisoning and leads to reduced conversion temperature and enhanced carbon monoxide catalytic ability.

6.4.4. *Complex compounds: High-temperature superconductivity and giant magnetoresistance materials*

After the discovery of high-temperature superconductivity (HTSC), the mechanism of the pairing has been a subject of intensive investigations. One of the striking features of HTSC is that their superconducting properties are highly sensitive to the microstructures of the materials. Bianconi and colleagues employed polarization-dependent EXAFS to determine the local lattice distortions in the CuO_2 plane of $La_{1.85}Sr_{0.15}CuO_4$ [96]. They measured the Cu-O distances by using polarized Cu K-edge EXAFS in $La_{1.85}Sr_{0.15}CuO_4$ crystal. The Cu-O (planar) bond lengths have been measured by $E//ab$ polarized spectra while the Cu-O (apical) bond lengths by $E//c$ polarized spectra.

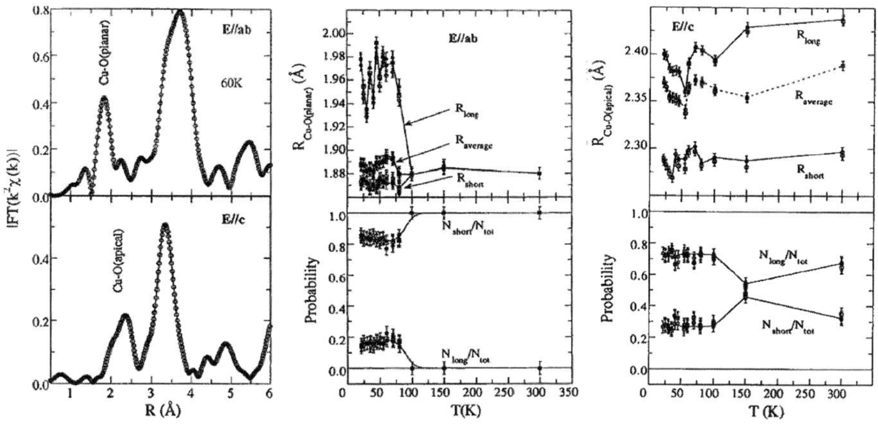

Fig. 6.19. (a) Magnitude of the Fourier transforms of the Cu K-edge $k^2\chi(k)$ EXAFS recorded at 60 K in the $E//ab$ (upper panel) and $E//c$ (bottom panel) geometry of the measurements. (b) Temperature dependence of the Cu-O (planar) distances (upper panel) and their relative probability (bottom panel) generated from the fits of the filtered $E//ab$ EXAFS spectra. (c) Temperature dependence of the Cu-O (apical) distances (upper panel) and their relative probability (bottom panel) generated from the fits of the filtered $E//c$ EXAFS spectra.

Figure 6.19(a) indicates that the signal of the Cu-O (planar) shell is completely suppressed in the $E//c$ spectra and the Cu-O (apical) signal is suppressed in $E//ab$. The EXAFS signals of the Cu-O bonds in the two spectra are well separated from the longer bond contributions. The temperature dependence of the two Cu-O (planar) distances (R_{short} and R_{long}) and their probabilities ($N_{\text{short}}/N_{\text{tot}}$ and $N_{\text{long}}/N_{\text{tot}}$) obtained from $E//ab$ are shown in Fig. 6.19(b), and those for the apical Cu-O bonds obtained from $E//c$ are shown in Fig. 6.19(c). The results in Fig. 6.19(c) show that the two apical Cu-O distances, R_{short} and R_{long}, are separated by 0.08 to 0.14 Å while the average crystallographic structure predicts a single distance at 2.405 Å. The anomalous 2.3 Å short Cu-O (apical) and 1.96 Å long Cu-O (planar) bond lengths are associated with distorted Cu octahedral, and could be assigned to two types of stripes with different lattice. This experiment supports a model of "two components" spatially separated in a superlattice of quantum stripes for the

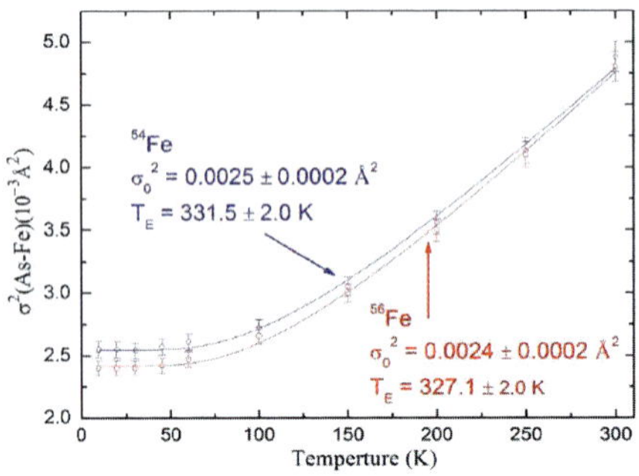

(a)

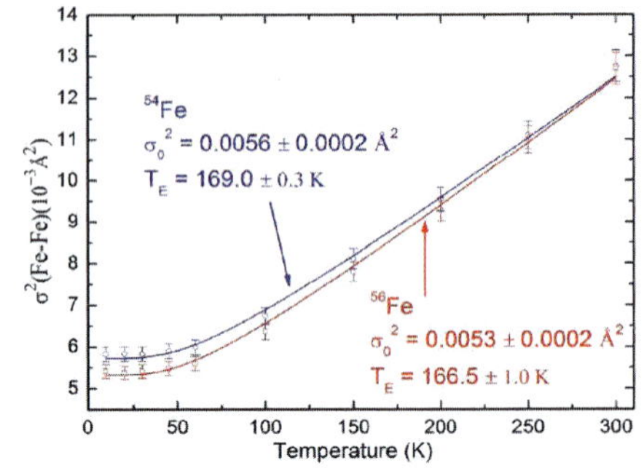

(b)

Fig. 6.20. Temperature dependent Debye-Waller factors σ^2 for the bond of the ^{54}Fe- and the ^{56}Fe-samples: (a) As-Fe bond; (b) Fe-Fe bond. Symbols are experimental data extracted from the EXAFS fits while solid lines are simulations based on the Einstein model.

anomalous properties of cuprate superconductors. For more studies and details using polarization-dependent XAFS on HTSC, the readers are referred to Chapter 5 of this book written by Oyanagi.

Chu and coworkers performed an investigation of the Fe isotope effect in the (Ba,K)Fe_2As_2 superconductor by temperature-dependent EXAFS [97]. They used the Einstein model to separate the static structural isotope contribution and the local lattice isotope contribution to the Debye-Waller factors. For both Fe-As (Fig. 6.20(a)) and Fe-Fe (Fig. 6.20(b)) pairs, the isotope substitution has a negligible effect on inter-atomic distances. On the contrary, both the local vibrations of the Fe-As and the Fe-Fe pairs show strong iron isotope dependence, marking the "intrinsic" isotope effect. Based on the iron isotope contribution to the local lattice dynamics, the authors suggest that the local phonons play an important role in the (Ba,K)Fe_2As_2 superconductor and a three-dimensional electron-phonon interaction should be considered to understand the superconducting mechanism of iron-based superconductors.

A temperature-dependent EXAFS investigation of $La_{1.2}Sr_{1.8}Mn_2O_7$(LSMO) was performed in order to understand the colossal magnetoresistance (CMR) from the viewpoint of local structure dynamics [98]. XAFS measurements with X-rays polarized parallel

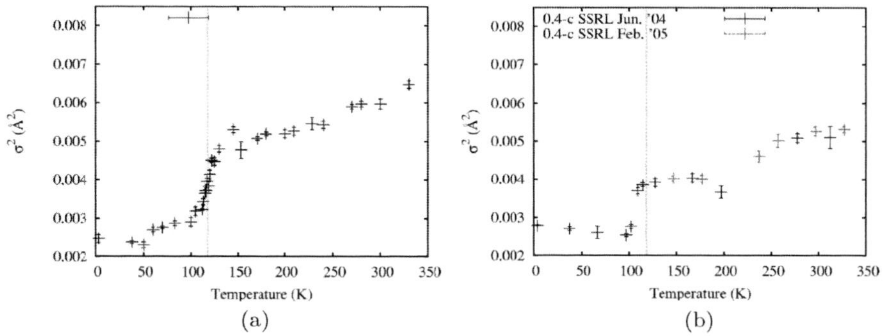

Fig. 6.21. Temperature-dependence of Debye-Waller factors σ^2 for $La_{1.2}Sr_{1.8}$ Mn_2O_7: (a) $E//ab$, (b) $E//c$.

to the *c*-axis and in the *ab*-plane were performed over a temperature range of 3–300 K covering the ferromagnetic transition temperature ($T_C \sim 125$ K). Figure 6.21 shows the temperature-dependence of σ^2 for both the *ab*-plane and the *c*-axis. The change in σ^2 is extremely sharp at T_C and approximately follows the magnetization data for both polarizations. This jump could be attributed to a large increase in the Jahn-Teller distortions about the Mn site, associated with dimerons and some isolated JT distorted sites, as the temperature is increased through T_C. The magnitude of this jump is about 65% larger for the *ab*-plane data than for the *c*-axis results, indicating more long bonds developed in the *ab*-plane than along the *c*-axis. Detailed analysis suggests that the changes in σ^2 at $T < T_C$ is linearly correlated with the sample magnetization, and that there is a break in the slope of this correlation when the sample is ~80% magnetized, consistent with the dimeron model. A JT distortion clearly exists well above T_C and some distortions still remain for a range of temperatures below T_C.

$LaCoO_3$ is a nonmagnetic, small gap semiconductor at low T, but exhibits a wealth of unusual magnetic properties associated with Co sites at increased temperature and particularly with Sr doping on the La sites to form the $La_{1-x}Sr_xCoO_3$ (LSCO) compound. With increasing Sr doping level x, the rhombohedral structural distortion gradually decreases and the crystal approaches the cubic perovskite structure [99]. Jiang and coworkers performed a temperature-dependent Co K-edge XAFS investigation of LSCO over a wide doping concentration range, $0 \leq x \leq 0.35$. XANES spectra indicate there is essentially no edge shift (≤ 0.15 eV) with increasing Sr concentration x to 0.3. This is in stark contrast to other Co systems and many manganites, for which the edge shift is roughly 3 eV per valence unit [100]. EXAFS results show that the Co-O bond length is nearly independent of x, indicating no change of the Co $3d^6$ electron configuration.

To address the issue of where the Sr-induced holes go, Medling and coauthors used the XAFS spectroscopy and X-ray magnetic circular dicrhoism (XMCD) at the O K-edge that help establish the connection between local structure and changes in electronic and magnetic states [101]. The XANES fluorescence data at the O K-edge are shown in Fig. 6.22(a), which indicate a shift of the leading part of the O K-edge to lower energy and an increase in absorption as x increases, corresponding to the addition of empty states. The magnetic state on O is probed by XMCD as shown in the lower part of Fig. 6.22(b). A significant signal is only observed over a short energy range ($\sim$2 eV) at the beginning of the O K-edge, indicating a nonzero orbital moment on the O site. The amplitude of the XMCD signal increases with x and is surprisingly large above $x = 0.19$ compared to measurements in other systems. The formation of the ferromagnetic

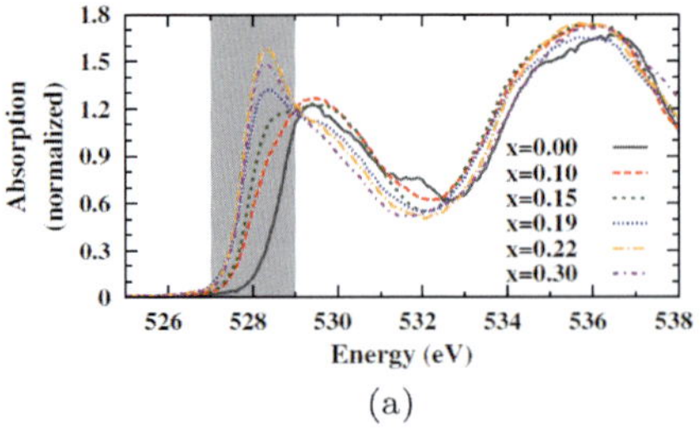

(a)

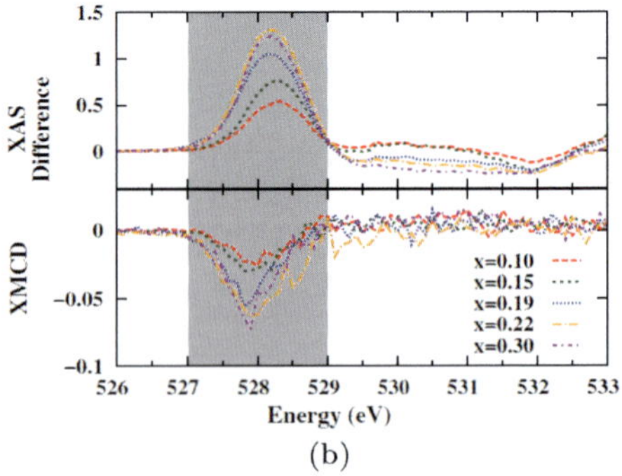

(b)

Fig. 6.22. (a) O K-edge XANES spectra for LSCO. (b) Difference in the O K-edge XANES (top) and XMCD (bottom). A significant XMCD signal is only observed over this same small energy range (shaded region).

metallic state appears to arise from the decreasing rhombohedral distortion with x. The Sr-induced change in XANES defined as XANES(x)–XANES($x = 0$) is shown in the top of Fig. 6.22(b). These results are consistent with the formation of ferromagnetic clusters occurring within a nonmagnetic insulating matrix at small x. With increasing Sr doping, Sr-induced magnetic O-hole states form just above E_F and grow. These nonlinear experimental spectral changes with x could be correlated with the overall magnetism in the system, which changes from glasslike with FM clusters below 18% to metallic FM above 18%.

6.4.5. *In situ probing the nucleation and growth processes of nanomaterials*

Quick-scan XAFS (QXAFS) or energy-dispersive XAFS (EDXAFS) techniques that have been developed for and widely applied in the *in situ* study of catalysts have also found extensive applications in the study of nucleation and growth processes of nanoparticles, which is a fundamentally important issue in the controllable synthesis of nanometerials. Wei and coworkers developed a continuous-flow *in situ* QXAFS technique to investigate the nucleation of colloidal Au nanocrystals reduced by citrate in water in the presence of polyvinylpyrrolidone (PVP) as surfactant [102]. The reactive solution prepared by the procedure was continuously circulated along the tubes by peristaltic pump during the acquisition of XAFS spectra and flowed into Teflon cell sealed with Kapton windows. Because the static reacted vessel is not suitable for the *in situ* measurement directly, recirculation system was used instead to ensure that the samples at the spot were newly formed from the reacted solution to achieve *in situ* detection. The recirculation system also has additional advantages that it could avoid the precipitation of particles after a long time reaction and the possible influence of X-ray irradiation in XAFS measurements.

The *in situ* XAFS spectra clearly show the whole process by reduction of the precursor $AuCl_4^-$ and the formation of Au nanocrystals (Fig. 6.23(a,b)). Interestingly, the expansion of the Au-Au bond

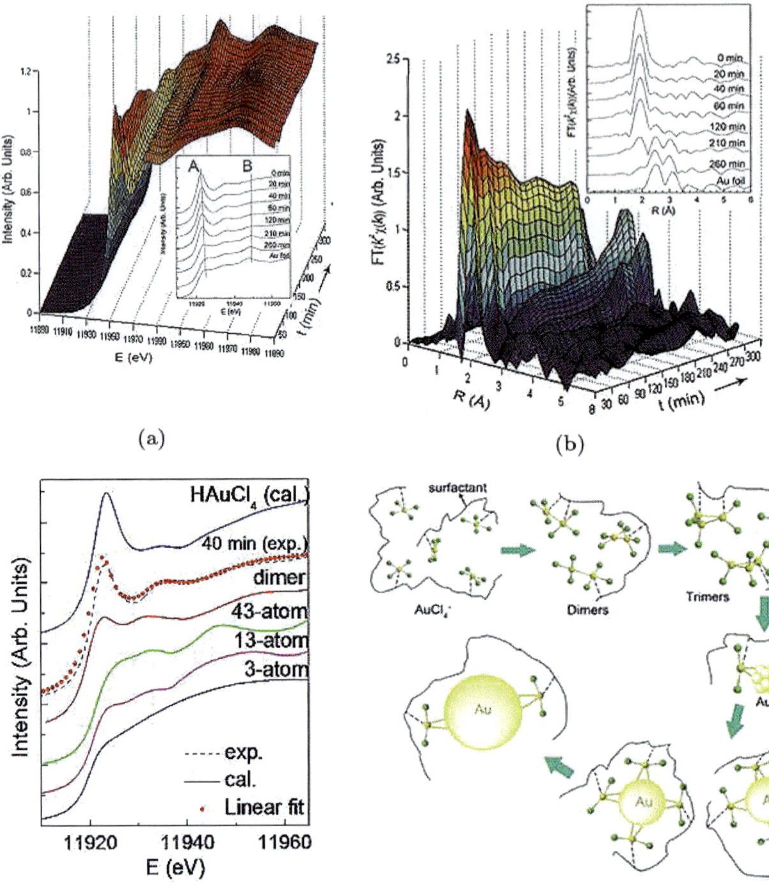

(a) (b) (c) (d)

Fig. 6.23. Series normalized Au L_3-edge *in situ* (a) XANES, and (b) k^2-weighted EXAFS Fourier transforms (FTs) spectra as a function of time reaction time. The insets show the typical Au L_3-edge XAFS spectra at different reaction time. (c) Comparison of the experimental XANES spectrum at 40 min with calculations for various Au clusters.

length for the small nuclei at the initial stage implied the formation of novel Au-contained complex which was further confirmed to be the 'Cl_3^-Au-$AuCl_3^-$' dimer cluster by a simulation of the XANES spectra for various cluster configurations in Fig. 6.23(c). The experimental XANES spectra during the first 60 min, for example, the spectrum of

40 min, are more close to that of dimer configuration, and thus can be well fitted (closed squares) by a linear combination of the spectra of the initial state ($HAuCl_4$) and intermediate state ('Cl_3^-Au-$AuCl_3^-$' dimers) with the partially reduced $AuCl_3^-$ ions. As the subsequent reduction continued, higher order 'Au_nCl_{n+x}' oligomers were formed in the nucleation stage, and then the growth was accelerated by the coalescence of small clusters. According to the temporal evolutions of the fitted results, the whole reaction could be classified into three stages: including the initial nucleation, slow growth, and eventual coalescence (Fig. 6.23(d)).

Combing *in situ* XAFS with other spectroscopy techniques (such as UV-vis) permits the simultaneous determination of the structural information and species evolution during nanocrystal formation. Weckhuysen and co-workers have designed a combined *in situ* UV-vis/XANES setup to probe the influence of X-rays on aqueous homogeneous copper solutions [103, 104]. They found that that X-rays had effect on the nucleation behavior of nanoparticles due to the radiolysis of water. This effect is mainly localized in a small sample volume within the spot size, with the time scale of several seconds.

Wei and coworkers developed an *in situ* measurement combining QXAFS and UV-vis absorption spectroscopies to monitor the tracks of the nucleation pathways in the solution synthesis of Pt nanocrystals [105]. Figure 6.24(a) and (b) indicate that the resultant morphologies of Pt nanocrystals are strongly dependently on the strength of reductants: a wire-like shape for weak reductant of ethylene glycol (EG) and a spherical shape for strong reductant of citric acid (CA). *In situ* XAFS and UV-vis probing (Fig. 6.24(c)–(f)) reveals that two different nucleation events involving the formation of one-dimensional "Pt_nCl_x" complexes from the polymerization of linear "$Cl_3Pt-PtCl_3$" dimers and spherical "Pt_n^0" clusters from the aggregation of Pt^0 atoms occur for the cases of EG and CA reductants, which then lead to the resultant morphologies of nanowires and nanospheres. This suggests that the strength of reductant can be used to switch the nucleation between the two different pathways, which consequently has a great influence on the postgrowth dynamics and the final shapes of the obtained nanocrystals.

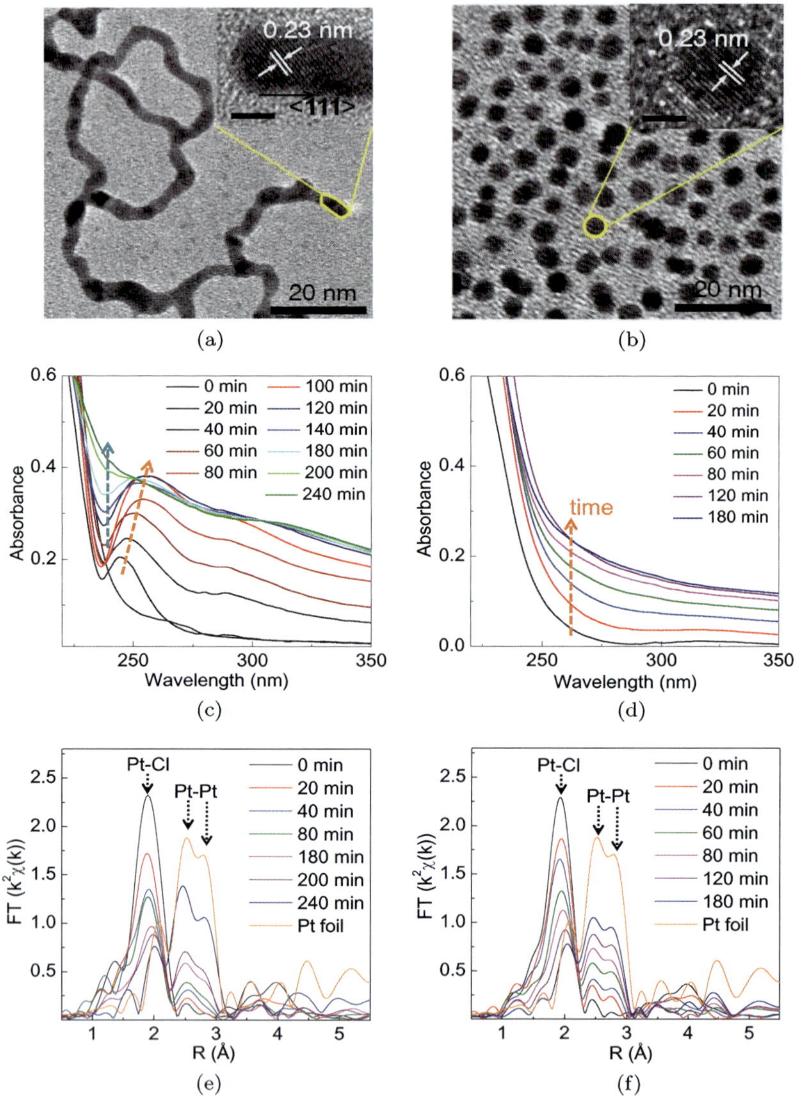

Fig. 6.24. Two representative mophologies, i.e., (a) nanowires and (b) nanospheres synthesized by using ethylene glycol and citric acid, respectively. Temporal evolution of UV-Vis absorption spectra of Pt colliod solutions for the K_2PtCl_4 reduced by (c) ethylene glycol and (d) citric acid. (e) and (f) show the time variations in k^2-weighted EXAFS Fourier transforms spectra during the reaction processes of the K_2PtCl_4 reduced by ethylene glycol and citric acid, respectively.

Tanaka and coauthors also reported the formation mechanism of monolayer-protected gold nanoparticles with time-resolved QXAFS spectroscopy [106–108]. For Au nanoparticles with sizes of 2–4 nm synthesized by the one-pot reduction of $HAuCl_4$ in dimethyl formamide (DMF) solution, TEM images showed that the resulting nanoparticles in the presence of porphyrin-cored tetradentate surfactant (**1**) were smaller than those prepared using the same S/Au molar ratio with a monodentate (**2**) surfactant. Figure 6.25(a) shows the time-resolved Au L_3-edge XANES spectral changes of the reaction solution upon adding $NaBH_4$ with **1** as the surfactant. The white line decreased in intensity and completely disappeares at 5.6 s, which corresponds to the reduction of the Au ion and the generation of a very small Au cluster, i.e., Au nuclei. After 5.6 s, the XANES spectrum is

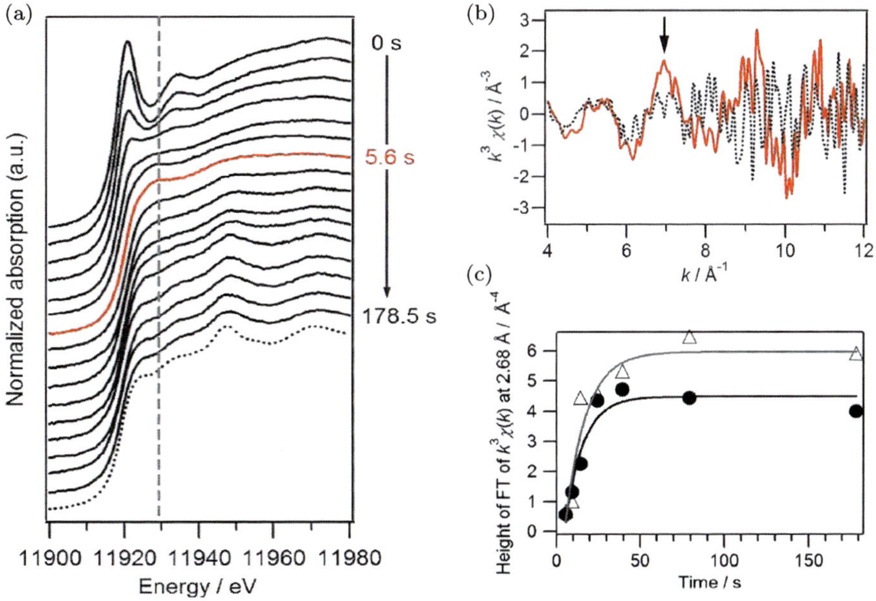

Fig. 6.25. Au L_3-edge XANES spectra during $HAuCl_4$ reduction in the presence surfactant **1** (solid line), together with the spectrum of the Au foil (dotted line). The position of the small peak at 5.6 s is indicated by dashed line. (b) EXAFS spectra of Au nuclei generated with **1** (solid line) and **2** (dotted line) at 5.6 s. (c) Au-Au peak amplitude of the FT-EXAFS spectra against reaction time, with the presence of **1** and **2** after the formation of Au nuclei.

converted to the spectrum similar to gold foil, showing two peaks in the energy range 11940–11980 eV. From these results, it is concluded that the change in the spectral feature from 5.6 s corresponds to the particle growth by aggregation of Au nuclei. Figure 6.25(b) shows the k^3-weighted EXAFS spectra of Au nuclei generated using **1** or **2**, indicating the slightly higher peak intensity of the Au nuclei with **1** at around 6.9 (marked by an arrow) than that with **2**. Furthermore, comparison of the formation kinetics obtained from analysis of the Au-Au peak amplitudes in the FT-EXAFS curves (Fig. 6.25(c)) with surfactants **1** and **2** shows that in both cases all of Au ions in solution were reduced to compose small Au clusters, i.e., nuclei, just after the $NaBH_4$ reduction of $HAuCl_4$, but their size varied with the initial S/Au molar ratios and structure of the surfactants. The surfactant **1** could stabilize smaller gold nanoparticles, 2–4 nm in size, but it was less efficient in trapping the Au nuclei formed at a very early stage, in comparison to the surfactant **2**.

Harada and co-workers performed a series of studies on the formation mechanism of different metal nanoparticles through photoreduction of the corresponding metal precursors. The investigated systems include the formation process of Au from $AuCl_4^-$ [109], Pd from $PdCl_2$ [110], Rh from $RhCl_3$ [111], and Pt from $PtCl_4^{2-}$ [112], in the presence of PVP as surfactant. Shown in Fig. 6.26(a) and (b) are the time evolution of the Pt L_3-edge XANES spectra and their Fourier transforms (FT) magnitudes of the EXAFS $k^3\chi(k)$ functions for Pt nanoparticles, respectively. Seen from Fig. 6.26(a), the intensity of the white line for precursor H_2PtCl_6 solution (before photoirradiation) decreases during the first 10 min and then gradually decreases in course of the reduction time. The isosbestic point shifts from 11 569 to 11 568 eV accordingly, especially in the reduction time between 10 and 40 min, indicating the stepwise transformation of one species to another ($Pt^{4+} \rightarrow Pt^{2+} \rightarrow Pt^0$). Meanwhile, the maximum peak position (11590 eV) attributed to metallic Pt-Pt bond increases with the increase of reduction time, suggesting the accumulation of small Pt nanoparticles. The EXAFS spectra indicate that when the photoirradiation time increases, the Pt-Cl peak located at

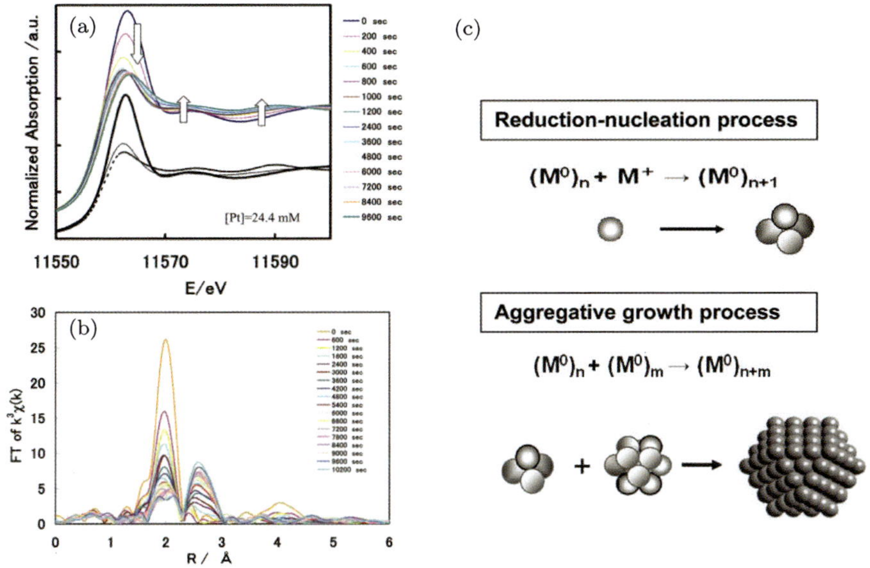

Fig. 6.26. Evolution as a function of the reduction time for (a) Pt L_3-edge XANES spectra and (b) Fourier transforms of $k^3\chi(k)$ EXAFS spectra for Pt nanoparticles prepared by the photoreduction in the presence of PVP. (c) Schematic presentation of the formation process of Pt nanoparticles involving the reduction-nucleation and aggregative growth progresses.

1.96 Å gradually decreases in intensity, and the other peak at around 2.57 Å assigned to a metallic Pt-Pt bond becomes more and more intensive. Based on these spectral observations as well as TEM and UV-vis absorption spectra, the authors proposed a general process involving the initial formation of Pt nuclei (i.e., $(Pt^0)_m$ nucleates approximately $m = 4$), followed by the aggregative particle growth with the autocatalytic reduction of Pt ionic species on the surface of Pt nuclei to produce Pt nanoparticles (Fig. 6.26(c)). Comparison of the different photoreduction reactions for various nanaoparticles indicates that the reductions of precursor metal species with high oxidation state possessed an induction period before the onset of metal particle formation. In these photoreduction reactions, the onset of metal-metal bond signal occurs after the fully reduction of the metal ionic species into zero-valent atoms.

6.4.6. *In situ XAFS under operando conditions*

Iwasawa and coworkers developed a novel time-gating quick XAFS (TG-QXAFS) technique with 1 s time resolution and an energy-dispersive XAFS (EDXAFS) system with 4 ms time resolution [113]. Using these techniques, they first conducted a study on the *in situ* time-resolved dynamic surface events on the Pt/C cathode in a fuel cell under operando conditions. Figure 6.27(a,b) shows a series of TG-QEXAFS oscillations at a Pt L_3 edge for a Pt/C cathode catalyst during a voltage-stepped process from 1.0 to 0.4 V and the Fourier transforms from 0.4 to 1.0 V. All of the TG-QEXAFS spectra are resolved enough to use for EXAFS curve-fitting analysis in order to evaluate the numbers and lengths of chemical bonds on the Pt cathode catalyst every 1 s. Based on the parameters (electrochemical charges, numbers and lengths of Pt-Pt and Pt-O bonds,

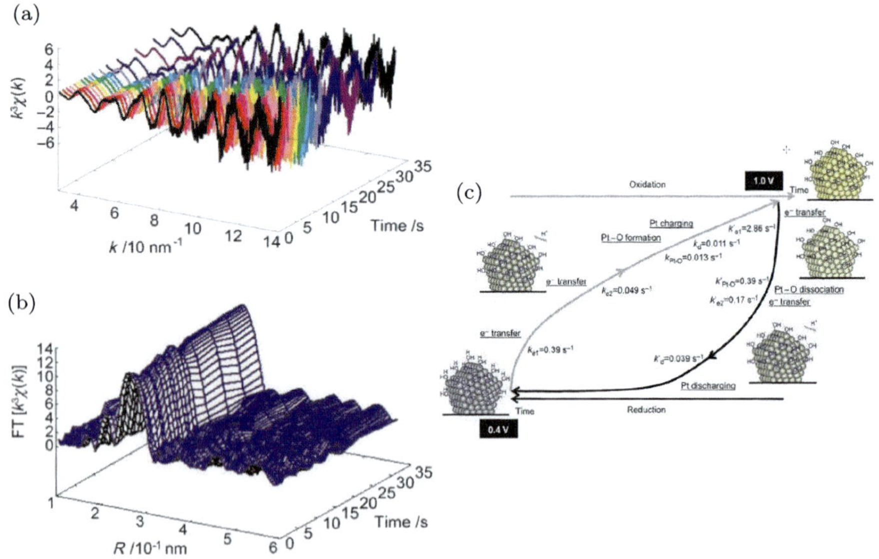

Fig. 6.27. (a) A series of TG-QEXAFS oscillations of a reduction process (1.0 → 0.4 V), and (b) the Fourier transforms of an oxidation process (0.4 → 1.0 V) at Pt L_3-edge in a Pt/C fuel cell measured at 333 K. (c) Reaction mechanism for changes in the Pt/C fuel-cell catalysts in cell voltages of 0.4/1.0 V under H_2 (anode) and N_2 (cathode).

electron density of Pt d orbitals) yielded by time-resolved analyses, they succeeded in determining the eight rate constants for the oxidation and reduction processes, respectively. The eight rate constants (Fig. 6.27(c)), significantly different from each other, are related to the electron transfer, structural change, and *d*-electron density change at the Pt/C cathode catalyst. A hysteresis loop for the structural changes in the Pt cathode particles in the oxidation and reduction processes is observed.

Using the established TG-QXAFS method, the Iwasawa group also studied dynamic structural changes and their kinetics of a Re_{10} cluster catalyst in the direct phenol synthesis from benzene and O_2 [114]. They also studied the reductive formation and oxidative phase separation of Pt_3Sn alloy nanoparticles supported on γ-Al_2O_3 in the reduction and oxidation processes at 673 K [115]. The mechanisms of the alloy formation and core-shell phase separation were discussed on the basis of the structural kinetics and structural parameters of the Pt_3Sn bimetallic nanoparticles on γ-Al_2O_3. Very recently, the Iwasawa group and coworkers used the time-resolved XAFS with a 0.5 s time resolution to study the structural kinetics of surface events on a Pt_3Co/C cathode catalyst in a polymer electrolyte fuel cell (PEFC) under operando conditions [116].

Frenkel and cooperators combined operando XAFS, X-ray diffraction (XRD), and Raman spectroscopy to study the structural and compositional changes of the partially reduced Fe_2O_3 and 3% chromium oxide-modified Fe_2O_3 (3% Cr_2O_3/Fe_2O_3) catalysts before, during, and after the water gas shift (WGS) reaction. The use of these methods is advantage that they could provide complementary information on the structure and valence state of Fe_2O_3 during the reaction. Figure 6.28 shows a series of (a) XANES, and (b) EXAFS spectra at Fe *K*-edge collected after exposing Fe_2O_3 to O_2 at RT and 400°C, during the WGS reaction at 400°C and in the WGM flow condition at RT. The XANES spectra in Fig. 6.28(a) show that the Fe *K*-edge positions measured in O_2 flow at RT and 400°C are similar, indicating the preservation of the charge state of Fe. However, in WGM conditions at 400°C, the edge is shifted to the lower

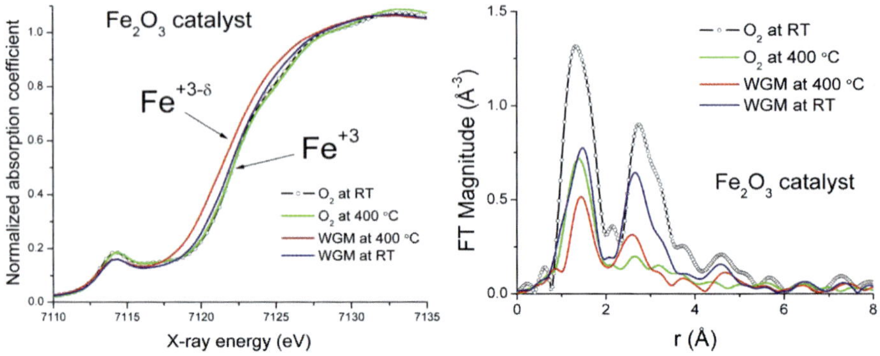

(a) (b)

Fig. 6.28. (a) Normalized XANES spectra, and (b) the corresponding EXAFS data at Fe K-edge collected after exposing Fe_2O_3 to O_2 at RT and 400°C, during the WGS reaction at 400°C and in the WGM flow condition at RT.

energy compared with its oxidized state prior to the WGS reaction, consistent with the partial reduction of Fe from Fe_2O_3 to Fe_3O_4. The corresponding EXAFS data in Fig. 6.28(b) demonstrate significant changes in the local structural environment of Fe in both catalysts, and the changes are markedly different between the two systems. The first nearest Fe-O peak located between 1–2.0 Å decreases evidently after the WGS reaction. This reduction can be interpreted either by the decreased Fe-O coordination number (CN) or the increased bond length disorder or both, but the former factor cannot be too significant due to the similar Fe-O coordination numbers in purely octahedrally coordinated α-Fe_2O_3 (CN = 6) to a mixture of two-thirds Fe_2O_3 and one-third tetrahedrally coordinated Fe_3O_4 (CN = 5.34). The second peak in Fig. 6.28(b) corresponds to the contributions of Fe-Fe bonds, and the reduction of the peak intensity for the 3% Cr_2O_3/Fe_2O_3 sample compared with the Fe_2O_3 sample indicates a more heterogeneous mixture of different Fe oxide phases with different Fe-Fe distances. Moreover, compared with the data under the WGM conditions, the Fe-Fe peak intensity shows a most notable decrease under O_2 conditions at the same temperature of 400°C. Such behavior is consistent with the large disorder in Fe environment under O_2 prior to the onset of the WGS reaction. The interpretation of the EXAFS data is also supported by *in situ* XRD results and Raman spectroscopy. Combining multiple techniques reveals the role of Cr in stabilizing the low-temperature γ-Fe_2O_3 phase, the nature of the disordered phase in the active state of the catalysts, and the possible deactivation mechanism.

6.4.7. *Time-resolved and space-resolved XAFS techniques*

Time-resolved XAFS technique using pump/probe scheme has been a very promising technique in detecting rapid processes, due to the time structure of storage ring of synchrotron radiation sources [117]. In 2001, Chen and coworkers were the first to conduct nanosecond-scale time-resolved laser pump/X-ray probe XAFS studies using X-ray pulses from a third-generation synchrotron source [118]. The experiments were conducted at a wiggler beamline at APS with 100 ps at full width at half-maximum of the duration of the X-ray pulses. This time-scale is suitable for capturing intermediate structures with sub-nanosecond resolution, and permits one to take "snapshots" of molecules when the population of the laser-generated transient species reaches a maximum. The transient molecular structure of a reaction intermediate produced by photoexcitation of $NiTPPL_2$ (NiTPP, nickeltetraphenylporphyrin; L, piperidine) in solution was determined. Shown in Fig. 6.29 is the Ni K-edge XANES spectra for the laser pumped $NiTPPL_2$, along with those of the ground states $NiTPPL_2$ and NiTPP. A distinct shoulder at 8.340 keV, which is attributed to the 1s to $4p_z$ transition in a square planar transition metal complex, appears for the ground state NiTPP and the laser-pumped sample, but with a lower amplitude, and disappeared when the pump laser beam was blocked. This suggests that the laser-irradiation created a transient square planar NiTPP-like structure, whose XANES spectrum could be reconstructed by mixing the ground state $NiTPPL_2$ and NiTPP-like spectra with a ratio of $\sim$7:3. Fourier-transformed EXAFS spectra (Fig. 6.29(b)) clearly show that the nearest Ni-N distances could be best described by two distances of 2.10 ± 0.02 Å and 1.92 ± 0.02 Å, with a relative ratio of 3.2:1.

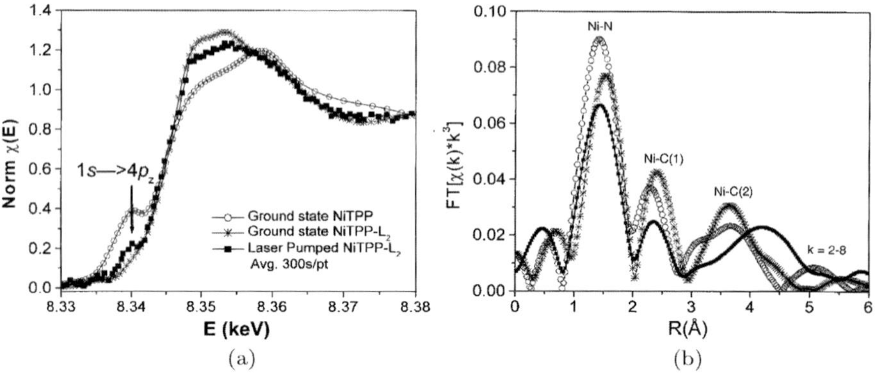

Fig. 6.29. (a) Normalized Ni K-edge XANES spectra for the laser pumped $NiTPPL_2$, and the ground states $NiTPPL_2$. The spectrum for the laser-pumped sample can be approximated by mixing the ground state spectra of $NiTPPL_2$ and NiTPP in the ratio of 7:3. (b) Fourier-transformed $k^3\chi(k)$ EXAFS spectra, with labeled neighboring atoms from the central Ni atom.

In contrast, the spectra of ground state NiTPP and $NiTPPL_2$ could be best fit with a single Ni-N distance (1.91 and 2.09 Å for NiTPP and $NiTPPL_2$, respectively.) The XAFS measurements then confirmed rapid removal by photoexcitation of both axial ligands to produce a transient square-planar intermediate, NiTPP, with a lifetime of 28 nanoseconds. Based on these results, it is proposed that the transient structure of the photodissociated intermediate adopts the same structure as the ground state in a noncoordinating solvent, before it recombines with two ligands to form the more stable octahedrally coordinated $NiTPPL_2$.

Starting from this pioneering attempt, X-ray transient absorption (XTA) spectroscopy has been rapidly developed, which uses a laser pulse as a pump to initiate a chemical reaction synchronously among different molecules and an X-ray pulse as a probe to detect transient structures associated with the reaction dynamics as a function of the delay time between the pump and probe. After over a decade of development, XTA has become an increasingly important and common tool in following molecular structures in chemical reactions and to map out structural factors that could influence reaction pathways [119–121].

Currently, the time-resolution of *in situ* XAFS covers a large range, from seconds to picoseconds. Nozawa and coworkers [122] used a Ti:Sapphire femtosecond laser to excite the low-spin (LS) ground state to high-spin (HS) of aqueous $[FeII(phen)_3]^{2+}$, and used picosecond X-ray pulse as a probe for direct observation of the transient spin-state dynamics coupled with electronic and structural modifications. The modifications in the electric state and molecular structure are observed from a Fe K-edge XAFS spectrum of the LS and HS analogy ground states. By observing the evolution of the Fe(II) $1s$-$3d$ transition, the spin crossover transition from the $^1A_1$ LS state to $^5T_2$ HS state has been inferred from the changes of the XANES features before and after pulse excitations (Fig. 6.30(a)). The EXAFS spectra of the LS, HS analog, and photoexcited HS at 50 ps (as shown in Fig. 6.30(b)) indicate the first nearest-neighbor (NN) in the Fe-N shell and the second NN carbon atoms (Fe-C) in the phenanthroline ligands. Quantitatively, the Fe-N distance of the LS state is 1.98 ± 0.01 Å, and expanded to 2.15 ± 0.02 Å after 50 ps of photoexcitation, in good agreement with the value in the HS analogue (2.15 ± 0.01 Å). The Fe-C shell is also expanded from 2.88 ± 0.08 Å

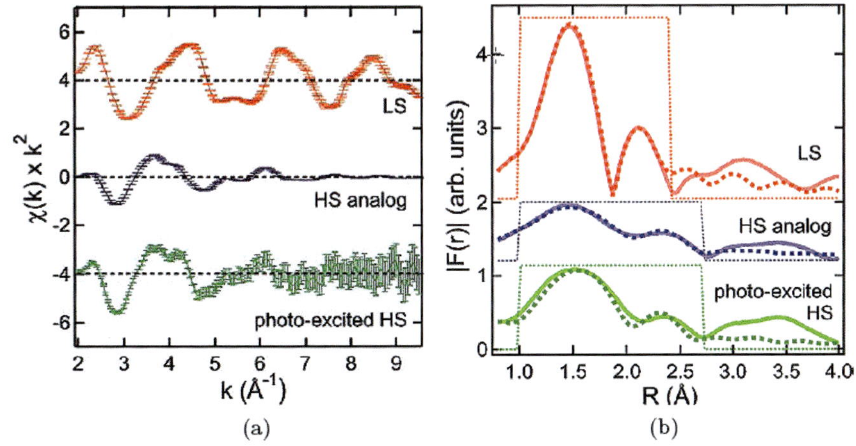

Fig. 6.30. (a) EXAFS spectra $\chi(k)$ weighted by k^2. (b) Fourier transformed EXAFS spectra (solid line). The window functions indicated by the dashed line show the R-range employed in the curve-fitting analysis. The dotted lines correspond to the fits of the data in R space.

to 3.01 ± 0.05 Å in the photoexcited HS state, corresponding to the photoinduced movement of the phenanthroline ligands. This study exemplifies that the picosecond time-resolved XAFS technique is well suited to capturing ultrafast photoinduced reactions in spin crossover transitions as well as in magnetic photocatalysts, nanomagnets, and many other disordered magnetic systems.

Besides the time-resolved XAFS technique, space-resolved XAFS has also been developed and applied to studies of functional materials. A good example is the work by Martínez-Criado *et al.* who detected metallic Mn in GaN based on XANES imaging with micron resolution [123]. This valuable instrument is capable of quantifying and mapping mixed valence state samples with high spatial resolution, elemental sensitivity, and chemical speciation. The GaN samples with Mn concentrations ranging from 0.05% up to 11% were grown by low temperature plasma-assisted MBE. Scanning μ-X-ray fluorescence (XRF) and XANES measurements were carried out at the ID21 and ID22 X-ray microprobe beamlines of the European Synchrotron Radiation Facility (ESRF). The monochromatic beam of ID21 beamline was focused on the sample to a spot of $0.6 \times 0.9\,\mu m^2$ size with a focus depth of ~0.4 mm by means of Au Fresnel zone plates as focusing lenses. At the ID22 beamline, the hard X-ray monochromatic beam was focused on the sample to a spot of $1.5 \times 3.5\,\mu m^2$ size with a focus depth of 300 μm by Kirkpatrick-Baez mirrors. In order to study in the micrometer scale the distribution and partitioning of Mn atoms in specific valence states, oxidation state mapping was performed. Two-dimensional X-ray fluorescence maps were taken at preferential monochromatic energies, which excited different Mn oxidation states. The XRF images were obtained by measuring the Mn K_α line intensities at different excitation energies. It is shown that for the sample with lowest Mn content (0.05%), Mn element is distributed homogeneously on the micrometer scale for all the chosen energies. On the contrary, for the highest Mn concentration (11%), different Mn patterns appear as a function of the incident energy. The Mn oxidation state maps show that for GaN samples with lower Mn content, a preferential homogeneous

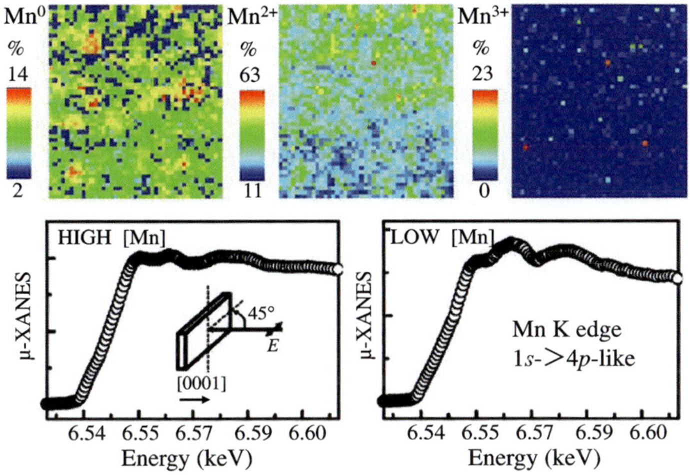

Fig. 6.31. Mn oxidation state maps of the highest Mn-doped GaN sample containing Mn^0, Mn^{2+}, and Mn^{3+} centers. The typical Mn K-edge micro-XANES data taken on different regions (low and high Mn concentrated) are included in the lower part.

distribution of mixed Mn^{2+} and Mn^{3+} states was observed rather than metallic Mn centers. For the highest Mn-doped GaN sample, the intensity scales clearly show Mn^{2+} as the dominant center, accompanied with Mn^0 in a lesser degree, and almost no signal from Mn^{3+} (Fig. 6.31). The Mn^0 pattern suggests the presence of specific cluster-like features, indicating surface segregation of metallic Mn centers.

On the same beamline of ID22 of the ESRF, Segura-Ruiz and coworkers used a hard X-ray nanoprobe to investigate the local structure of single Co-implanted ZnO NWs with a Co concentration of 0.6 at.% relative to Zn [124]. The monochromatic X-ray beam was focused to $100 \times 100\,\text{nm}^2$ spot size with $\sim 10^{11}$ ph/s using a pair of Kirkpatrick-Baez Si mirrors. The uniform Zn and Co distributions along the wires are confirmed by X-ray fluorescence maps. EXAFS analysis reveals the substitution of Co^{2+} ions in Zn sites and the high structural order in the host lattice. The polarization-dependent Zn K-edge XANES spectra at different points along the NW shown in Fig. 6.32 reveal the high structural order of the host lattice along both radial and axial directions. Specifically, the XANES data with

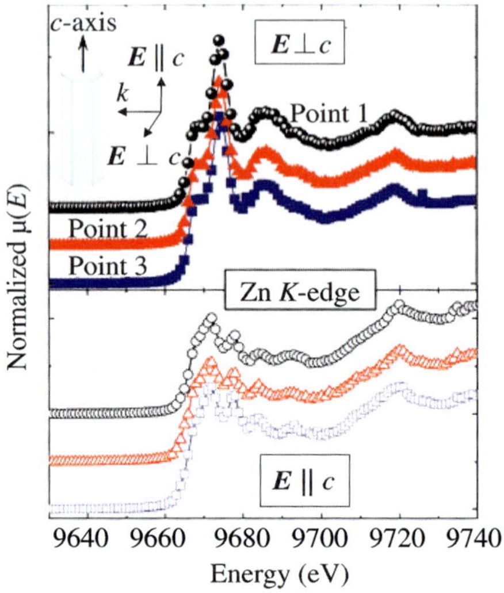

Fig. 6.32. Polarization-dependent Zn K-edge XANES spectra of Co-implanted ZnO NWs.

the electric field of the absorbed photons perpendicular to the c-axis ($E \perp c$) probe the symmetry of the ab-plane, namely p_{xy} conduction band states. The spectral features in Fig. 6.32(a) reflect the typical wurtzite structure of the ZnO lattice, and signatures of Zn metallic precipitates after the ion implantation process are observed. No structural disorder in the radial direction along the implanted NW is suggested according to the line shapes and peaks energies. For the axial direction ($E//c$), the XANES data probe the symmetry of the c-plane, namely p_z conduction band states. Again, the spectra exhibit the peaks associated to the hexagonal structure, without any evidence of lattice damage along the NW (Fig. 6.32(b)).

6.5. Outlook and Perspective

XAFS has been developed into a mature technique for studies on the local atomic environment and electronic structure around a specific type of atoms. The recent significant advances in theory, experimental methods, and data analysis have made this technique more and more widely and extensively used. With new generations of synchrotron radiation light sources and more advanced XAFS beamline (especially, space-resolved and time-resolved XAFS) becoming available, there is no doubt that the range of XAFS applications will continue extending. In the future XAFS will continue to be a powerful tool in a large diversity of areas such as physics, chemistry, materials science, biology, environmental science, and so forth.

6.6. Acknowledgements

The authors are grateful to the National Natural Science Foundation of China (Grant Nos. 11135008, 11175184, 11205158 and 11079004), the National Basic Research Program of China (2012CB825800), and the Foundation for Innovative Research Groups of the National Natural Science Foundation of China (11321503) for financial support. We would like to express our gratitude to Prof. Kunquan Lu, Prof. Ziyu Wu, and Prof. Tiandou Hu for valuable discussions and encouragements. We also express our thanks to Dr. Wensheng Yan, Dr. Zhiyun Pan, Dr. Qinghua Liu, Dr. Tao Yao, Dr. Yong Jiang, Dr. Yongfu Sun, Dr. Jingfu He, Mr. Zhi Xie, Mr. Bo He and Miss Fengchun Hu for fruitful collaborations. We appreciate National Synchrotron Radiation Laboratory (NSRL), Beijing Synchrotron Radiation Facility (BSRF), and Shanghai Synchrotron Radiation Facility (SSRF) for the valuable beamtime.

Bibliography

[1] D. C. Koningsberger and R. Prins, *X-ray Absorption: Principles, Applications, Techniques of EXAFS, SEXAFS and XANES* (Wiley, New York, 1988).
[2] H. Fricke, *Physical Review* **16**, 202, (1920).
[3] G. Z. Hertz, *Phys.* **3**, 19, (1920).
[4] R. D. Kronig, *Zeitschrift Fur Physik* **70**, 317, (1931).
[5] R. D. Kronig, *Zeitschrift Fur Physik* **75**, 468, (1932).
[6] R. D. Kronig, *Zeitschrift Fur Physik* **75**, 191, (1932).
[7] D. E. Sayers, E. A. Stern and F. W. Lytle, *Phys. Rev. Lett.* **27**, 1204, (1971).
[8] P. A. Lee, P. H. Citrin, P. Eisenberger and B. M. Kincaid, *Rev. Mod. Phys.* **53**, 769, (1981).

[9] J. J. Rehr and R. C. Albers, *Rev. Mod. Phys.* **72**, 621, (2000).
[10] P. A. Lee and J. B. Pendry, *Phys. Rev. B* **11**, 2795, (1975).
[11] J. J. Rehr and R. C. Albers, *Phys. Rev. B* **41**, 8139, (1990).
[12] T. Fujikawa and N. Yiwata, *Surf. Sci.* **357**, 60, (1996).
[13] A. L. Ankudinov, B. Ravel, J. J. Rehr and S. D. Conradson, *Phys. Rev. B* **58**, 7565, (1998).
[14] C. R. Natoli, D. K. Misemer, S. Doniach and F. W. Kutzler, *Physical Review A* **22**, 1104, (1980).
[15] N. Binsted, (1998).
[16] J. J. Rehr, J. M. Deleon, S. I. Zabinsky and R. C. Albers, *J. Am. Chem. Soc.* **113**, 5135, (1991).
[17] A. Filipponi, A. Di Cicco and C. R. Natoli, *Phys. Rev. B* **52**, 15122, (1995).
[18] P. Blaha, K. Schwarz, G. K. H. Madsen, D. Kvasnicka and J. Luitz (Karlheinz Schwarz, Universität Wien, Austria, 2001).
[19] H. Ebert, in *Lecture Notes in Physics*, 2000), Vol. 535, p. 191.
[20] W. Zhong and S. Wei, *J. of University of Science and Technology of China* **31**, 228, (2001).
[21] K. Baberschke, *Physica B* **158**, 19, (1989).
[22] H. Magnan, D. Chandesris, G. Rossi, G. Jezequel, K. Hricovini and J. Lecante, *Phys. Rev. B* **40**, 9989, (1989).
[23] O. Heckmann, H. Magnan, P. Lefevre, D. Chandesris and J. J. Rehr, *Surf. Sci.* **312**, 62, (1994).
[24] P. A. Lee and G. Beni, *Phys. Rev. B* **15**, 2862, (1977).
[25] J. E. Müller and W. L. Schaich, *Phys. Rev. B* **27**, 6489, (1983).
[26] D. Norman, J. Stöhr, R. Jaeger, P. J. Durham and J. B. Pendry, *Phys. Rev. Lett.* **51**, 2052, (1983).
[27] J. Stöhr and K. R. Bauchspiess, *Phys. Rev. Lett.* **67**, 3376, (1991).
[28] B. M. Kincaid and P. Eisenberger, *Phys. Rev. Lett.* **34**, 1361, (1975).
[29] J. Jaklevic, J. A. Kirby, M. P. Klein, A. S. Robertson, G. S. Brown and P. Eisenberger, *Solid State Commun.* **23**, 679, (1977).
[30] U. Kaminaga, T. Matsushita and K. Kohra, *Japanese Journal of Applied Physics* **20**, L355, (1981).
[31] J. Stohr, R. Jaeger and T. Kendelewicz, *Phys. Rev. Lett.* **49**, 142 (1982).
[32] P. Roubin, D. Chandesris, G. Rossi, J. Lecante, M. C. Desjonquères and G. Tréglia, *Phys. Rev. Lett.* **56**, 1272, (1986).
[33] F. Sette, C. T. Chen, J. E. Rowe and P. H. Citrin, *Phys. Rev. Lett.* **59**, 311, (1987).
[34] F. W. Lytle, *J. Synchrotron Rad.* **6**, 123, (1999).
[35] G. Bunker, (Cambridge University Press, Cambridge, 2010).
[36] J. J. Rehr, E. A. Stern, R. L. Martin and E. R. Davidson, *Phys. Rev. B* **17**, 560, (1978).
[37] G. Bunker, *Nuclear Instruments & Methods in Physics Research* **207**, 437, (1983).
[38] J. M. Tranquada and R. Ingalls, *Phys. Rev. B* **28**, 3520, (1983).
[39] P. Eisenberger and G. S. Brown, *Solid State Commun.* **29**, 481, (1979).

[40] E. D. Crozier and A. J. Seary, *Can. J. Phys.* **58**, 1388, (1980).
[41] S. Q. Wei, H. Oyanagi, W. H. Liu, T. D. Hu, S. L. Yin and G. Z. Bian, *J. Non-Cryst. Solids* **275**, 160, (2000).
[42] B. K. Teo and P. A. Lee, *J. Am. Chem. Soc.* **101**, 2815, (1979).
[43] A. G. McKale, G. S. Knapp and S. K. Chan, *Phys. Rev. B* **33**, 841, (1986).
[44] D. C. Koningsberger, *Japanese Journal of Applied Physics Part 1-Regular Papers Short Notes & Review Papers* **32**, 877, (1993).
[45] W. H. McMaster, N. Kerr-Del Grande, J. H. Mallett and J. H. Hubbell, (National Bureau of Standards, Springfield, VA, 1969).
[46] F. W. Lytle, D. E. Sayers and E. A. Stern, *Physica B* **158**, 701, (1989).
[47] B. Ravel, *J. Alloys Compd.* **401**, 118, (2005).
[48] T. Fujikawa, *J. Phys. Soc. Jpn.* **50**, 1321, (1981).
[49] J. Schwitalla and H. Ebert, *Phys. Rev. Lett.* **80**, 4586, (1998).
[50] A. L. Ankudinov, A. I. Nesvizhskii and J. J. Rehr, *Phys. Rev. B* **67**, 115120, (2003).
[51] F. M. F. Degroot, J. C. Fuggle, B. T. Thole and G. A. Sawatzky, *Phys. Rev. B* **42**, 5459, (1990).
[52] F. de Groot, *Coord. Chem. Rev.* **249**, 31, (2005).
[53] S. Diaz-Moreno, *J. Synchrotron Rad.* **19**, 863, (2012).
[54] E. A. Stern and K. Kim, *Phys. Rev. B* **23**, 3781, (1981).
[55] B. L. Henke, E. M. Gullikson and J. C. Davis, *At. Data Nucl. Data Tables* **54**, 181, (1993).
[56] C. T. Chantler, *J. Phys. Chem. Ref. Data* **24**, 71, (1995).
[57] L. Troger, D. Arvanitis, K. Baberschke, H. Michaelis, U. Grimm and E. Zschech, *Phys. Rev. B* **46**, 3283, (1992).
[58] C. H. Booth and F. Bridges, *Phys. Scr.* **T115**, 202, (2005).
[59] F. W. Lytle, R. B. Greegor, D. R. Sandstrom, E. C. Marques, J. Wong, C. L. Spiro, G. P. Huffman and F. E. Huggins, *Nuclear Instruments & Methods in Physics Research Section a-Accelerators Spectrometers Detectors and Associated Equipment* **226**, 542, (1984).
[60] H. Oyanagi, C. Fonne, D. Gutknecht, P. Dressler, R. Henck, M. O. Lampert, S. Ogawa, K. Kasai and S. B. Mohamed, *Nuclear Instruments & Methods in Physics Research Section a-Accelerators Spectrometers Detectors and Associated Equipment* **513**, 340, (2003).
[61] H. Oyanagi, K. Sakamoto, R. Shioda, Y. Kuwahara and K. Haga, *Phys. Rev. B* **52**, 5824, (1995).
[62] Y. Takeda, H. Ofuchi, H. Kyouzu, R. Takahashi and M. Tabuchi, *J. Synchrotron Rad.* **12**, 494, (2005).
[63] W. Gudat and C. Kunz, *Phys. Rev. Lett.* **29**, 169, (1972).
[64] J. Stöhr, *NEXAFS Spectroscopy: Springer Series in Surface Sciences* (Springer, Berlin Heidelberg, 1992).
[65] Y. Yin and A. P. Alivisatos, *Nature* **437**, 664, (2005).
[66] C. D. Donega, *Chem. Soc. Rev.* **40**, 1512, (2011).
[67] C. Vericat, M. E. Vela, G. Benitez, P. Carro and R. C. Salvarezza, *Chem. Soc. Rev.* **39**, 1805, (2010).

[68] Y. N. Xia, C. M. Cobley, J. Y. Chen, E. C. Cho and L. V. Wang, *Chem. Soc. Rev.* **40**, 44, (2011).
[69] P. Zhang and T. K. Sham, *Phys. Rev. Lett.* **90**, 245502, (2003).
[70] S. I. Sanchez, L. D. Menard, A. Bram, J. H. Kang, M. W. Small, R. G. Nuzzo and A. I. Frenkel, *J. Am. Chem. Soc.* **131**, 7040, (2009).
[71] M. W. Small, S. I. Sanchez, N. S. Marinkovic, A. I. Frenkel and R. G. Nuzzo, *Acs Nano* **6**, 5583, (2012).
[72] X. W. Teng, Q. Wang, P. Liu, W. Han, A. Frenkel, W. Wen, N. Marinkovic, J. C. Hanson and J. A. Rodriguez, *J. Am. Chem. Soc.* **130**, 1093, (2008).
[73] A. I. Frenkel, *Chem. Soc. Rev.* **41**, 8163, (2012).
[74] S. A. Wolf, D. D. Awschalom, R. A. Buhrman, J. M. Daughton, S. von Molnar, M. L. Roukes, A. Y. Chtchelkanova and D. M. Treger, *Science* **294**, 1488, (2001).
[75] I. Zutic, J. Fabian and S. Das Sarma, *Rev. Mod. Phys.* **76**, 323, (2004).
[76] H. Ohno, *Science* **281**, 951, (1998).
[77] T. Dietl, H. Ohno, F. Matsukura, J. Cibert and D. Ferrand, *Science* **287**, 1019, (2000).
[78] R. Shioda, K. Ando, T. Hayashi and M. Tanaka, *Phys. Rev. B* **58**, 1100 (1998).
[79] Y. L. Soo, G. Kioseoglou, S. Kim, X. Chen, H. Luo, Y. H. Kao, H. J. Lin, H. H. Hsieh, T. Y. Hou, C. T. Chen, Y. Sasaki, X. Liu and J. K. Furdyna, *Phys. Rev. B* **67**, (2003).
[80] S. Q. Wei, W. S. Yan, Z. H. Sun, Q. H. Liu, W. J. Zhong, X. Y. Zhang, H. Oyanagi and Z. Y. Wu, *Appl. Phys. Lett.* **89**, 121901, (2006).
[81] M. Kunisu, F. Oba, H. Ikeno, I. Tanaka and T. Yamamoto, *Appl. Phys. Lett.* **86**, (2005).
[82] Z. Sun, W. Yan, G. Zhang, H. Oyanagi, Z. Wu, Q. Liu, W. Wu, T. Shi, Z. Pan, P. Xu and S. Wei, *Phys. Rev. B* **77**, 242508, (2008).
[83] N. A. Spaldin, *Phys. Rev. B* **69**, 125201, (2004).
[84] P. Gopal and N. A. Spaldin, *Phys. Rev. B* **74**, 094418, (2006).
[85] B. D. Yuhas, S. Fakra, M. A. Marcus and P. D. Yang, *Nano Lett.* **7**, 905, (2007).
[86] T. Yao, W. S. Yan, Z. H. Sun, Z. Y. Pan, Y. Xie, Y. Jiang, J. Ye, F. C. Hu and S. Q. Wei, *J. Phys. Chem. C* **113**, 14114, (2009).
[87] J. H. Yu, X. Y. Liu, K. E. Kweon, J. Joo, J. Park, K. T. Ko, D. Lee, S. P. Shen, K. Tivakornsasithorn, J. S. Son, J. H. Park, Y. W. Kim, G. S. Hwang, M. Dobrowolska, J. K. Furdyna and T. Hyeon, *Nat. Mater.* **9**, 47, (2010).
[88] B. He, X. Y. Zhang, S. Q. Wei, H. Oyanagi, S. V. Novikov, K. W. Edmonds, C. T. Foxon, G. E. Zhou and Y. B. Jia, *Appl. Phys. Lett.* **88**, (2006).
[89] T. F. Shi, S. Y. Zhu, Z. H. Sun, S. Q. Wei and W. H. Liu, *Appl. Phys. Lett.* **90**, 102108, (2007).
[90] A. Ney, K. Ollefs, S. Ye, T. Kammermeier, V. Ney, T. Kaspar, S. Chambers, F. Wilhelm and A. Rogalev, *Phys. Rev. Lett.* **100**, 157201, (2008).

[91] J. M. D. Coey, M. Venkatesan and C. B. Fitzgerald, *Nat. Mater.* **4**, 173, (2005).
[92] T. P. Pearsall, J. Bevk, L. C. Feldman, J. M. Bonar, J. P. Mannaerts and A. Ourmazd, *Phys. Rev. Lett.* **58**, 729, (1987).
[93] S. J. Naftel, I. Coulthard, T. K. Sham, S. R. Das and D. X. Xu, *Phys. Rev. B* **57**, 9179, (1998).
[94] Y. F. Sun, Z. H. Sun, S. Gao, H. Cheng, Q. H. Liu, J. Y. Piao, T. Yao, C. Z. Wu, S. L. Hu, S. Q. Wei and Y. Xie, *Nature Communications* **3**, (2012).
[95] Y. F. Sun, Q. H. Liu, S. Gao, H. Cheng, F. C. Lei, Z. H. Sun, Y. Jiang, H. B. Su, S. Q. Wei and Y. Xie, *Nat Commun* **4**, 2899, (2013).
[96] A. Bianconi, N. L. Saini, A. Lanzara, M. Missori, T. Rossetti, H. Oyanagi, H. Yamaguchi, K. Oka and T. Ito, *Phys. Rev. Lett.* **76**, 3412, (1996).
[97] W. S. Chu, J. Cheng, S. Q. Chu, T. D. Hu, A. Marcelli, X. H. Chen and Z. Y. Wu, *Scientific Reports* **3**, (2013).
[98] T. A. O'Brien, F. Bridges, L. Downward, J. F. Mitchell and H. Zheng, *Phys. Rev. B* **75**, (2007).
[99] M. Kriener, M. Braden, H. Kierspel, D. Senff, O. Zabara, C. Zobel and T. Lorenz, *Phys. Rev. B* **79**, 224104, (2009).
[100] M. Sikora, C. Kapusta, K. Knížek, Z. Jirák, C. Autret, M. Borowiec, C. J. Oates, V. Procházka, D. Rybicki and D. Zajac, *Phys. Rev. B* **73**, 094426, (2006).
[101] S. Medling, Y. Lee, H. Zheng, J. F. Mitchell, J. W. Freeland, B. N. Harmon and F. Bridges, *Phys. Rev. Lett.* **109**, 5, (2012).
[102] T. Yao, Z. H. Sun, Y. Y. Li, Z. Y. Pan, H. Wei, Y. Xie, M. Nomura, Y. Niwa, W. S. Yan, Z. Y. Wu, Y. Jiang, Q. H. Liu and S. Q. Wei, *J. Am. Chem. Soc.* **132**, 7696, (2010).
[103] J. G. Mesu, A. M. J. van der Eerden, F. M. F. de Groot and B. M. Weckhuysen, *J. Phys. Chem. B* **109**, 4042, (2005).
[104] J. G. Mesu, A. M. Beale, F. M. F. de Groot and B. M. Weckhuysen, *J. Phys. Chem. B* **110**, 17671, (2006).
[105] T. Yao, S. J. Liu, Z. H. Sun, Y. Y. Li, S. He, H. Cheng, Y. Xie, Q. H. Liu, Y. Jiang, Z. Y. Wu, Z. Y. Pan, W. S. Yan and S. Q. Wei, in *J. Am. Chem. Soc.* **134**, 9410, (2012).
[106] J. Ohyama, K. Teramura, Y. Higuchi, T. Shishido, Y. Hitomi, K. Aoki, T. Funabiki, M. Kodera, K. Kato, H. Tanida, T. Uruga and T. Tanaka, *Phys. Chem. Chem. Phys.* **13**, 11128, (2011).
[107] J. Ohyama, K. Teramura, Y. Higuchi, T. Shishido, Y. Hitomi, K. Kato, H. Tanida, T. Uruga and T. Tanaka, *Chemphyschem* **12**, 127, (2011).
[108] T. Tanaka, J. Ohyama, K. Teramura and Y. Hitomi, *Catal. Today* **183**, 108, (2012).
[109] M. Harada and H. Einaga, *Langmuir* **23**, 6536, (2007).
[110] M. Harada and Y. Inada, *Langmuir* **25**, 6049, (2009).
[111] M. Harada, N. Tamura and M. Takenaka, *J. Phys. Chem. C* **115**, 14081, (2011).
[112] M. Harada and Y. Kamigaito, *Langmuir* **28**, 2415, (2012).

[113] M. Tada, S. Murata, T. Asakoka, K. Hiroshima, K. Okumura, H. Tanida, T. Uruga, H. Nakanishi, S. Matsumoto, Y. Inada, M. Nomura and Y. Iwasawa, *Angew. Chem. Int. Ed.* **46**, 4310, (2007).
[114] M. Tada, Y. Uemura, R. Bal, Y. Inada, M. Nomura and Y. Iwasawa, *Phys. Chem. Chem. Phys.* **12**, 5701, (2010).
[115] Y. Uemura, Y. Inada, K. K. Bando, T. Sasaki, N. Kamiuchi, K. Eguchi, A. Yagishita, M. Nomura, M. Tada and Y. Iwasawa, *J. Phys. Chem. C* **115**, 5823, (2011).
[116] N. Ishiguro, T. Saida, T. Uruga, S. Nagamatsu, O. Sekizawa, K. Nitta, T. Yamamoto, S. Ohkoshi, Y. Iwasawa, T. Yokoyama and M. Tada, *Acs Catalysis* **2**, 1319, (2012).
[117] C. Bressler and M. Chergui, *Chem. Rev.* **104**, 1781, (2004).
[118] L. X. Chen, W. J. H. Jager, G. Jennings, D. J. Gosztola, A. Munkholm and J. P. Hessler, *Science* **292**, 262, (2001).
[119] M. Newcomb, J. A. Halgrimson, J. H. Horner, E. C. Wasinger, L. X. Chen and S. G. Sligar, *Proc. Natl. Acad. Sci. U. S. A.* **105**, 8179, (2008).
[120] M. R. Harpham, S. C. Nguyen, Z. R. Hou, J. C. Grossman, C. B. Harris, M. W. Mara, A. B. Stickrath, Y. Kanai, A. M. Kolpak, D. Lee, D. J. Liu, J. P. Lomont, K. Moth-Poulsen, N. Vinokurov, L. X. Chen and K. P. C. Vollhardt, *Angew. Chem. Int. Ed.* **51**, 7692, (2012).
[121] L. X. Chen and X. Y. Zhang, *Journal of Physical Chemistry Letters* **4**, 4000, (2013).
[122] S. Nozawa, T. Sato, M. Chollet, K. Ichiyanagi, A. Tomita, H. Fujii, S. Adachi and S. Koshihara, *J. Am. Chem. Soc.* **132**, 61, (2010).
[123] G. Martinez-Criado, A. Somogyi, A. Homs, R. Tucoulou and J. Susini, *Appl. Phys. Lett.* **87**, (2005).
[124] J. Segura-Ruiz, G. Martinez-Criado, M. H. Chu, S. Geburt and C. Ronning, *Nano Lett.* **11**, 5322, (2011).

Chapter 7

Small Angle X-ray Scattering and Its Applications

Zhonghua Wu and Xueqing Xing

Beijing Synchrotron Radiation Facility, Institute of High Energy Physics, Chinese Academy of Sciences, 19B Yu-Quan Road, Shi-Jing-Shan District, Beijing 100049, China

Abstract

This chapter is divided into three sections: introduction, SAXS technique and SAXS applications. The introduction gives a brief description of the X-ray scattering basics, which includes the classification of scattering systems, the SAXS intensity, and the gyration radii of some scatters with specific shape. In the second section, the SAXS techniques including the normal SAXS, WAXS, and USAXS with transmission mode, the GISAXS with reflection mode, and the ASAXS with energy-scanning mode are simply mentioned. Some typical application examples of the SAXS techniques are described in the third section. Five main areas of SAXS applications including biological macromolecule, phospholipids and micelles, polymer and related materials, nanomaterial and superlattice, as well as natural biomaterials are discussed. Shape analysis of scatterers, complex and assembly, *in-situ* processes, size distribution of scatterers, and hierarchical structure are emphasized in the above five application areas, respectively. Finally, the methods and applications of ASAXS and GISAXS techniques are also presented briefly in the third section.

Keywords: SAXS, ASAXS, biological macromolecules, conformation, form factor, structure factor, GISAXS, gyration radius, Kratky plot, molecular weight, monodisperse system, nanomaterials, particle shape,

Porod's law, scattering intensity, scattering vector, shape restoration, size distribution, t-SAXS, USAXS, WAXS

7.1. Introduction

When an object is irradiated by an X-ray beam, the bonding electrons in the object will generate forced vibrations under the electromagnetic field (E_0) of the incident X-ray photons. This forced vibration of electron will accelerate its velocity of movement and radiate a new X-ray. When the X-ray radiation has the same energy as the incident X-ray beam, this X-ray radiation is called the Thomson scattering. It is well known that the Thomson scattering intensity decreases with the increase of scattering angle (2θ), where the scattering angle was defined as the separation angle between the incident and the scattering X-ray beams as shown in Fig. 7.1.

The total X-ray scattering intensity of an object is the sum of scattering intensities from all electrons irradiated by the incident X-ray. For an object with homogeneous electron density (ρ), approximately, the scattering angle $2\theta \approx \lambda/D$, where D is the average diameter of the object. λ is the incident X-ray wavelength. If there are nanoscale regions (or particles) with electron density (ρ) embedded in a sample and the particles have different electron density ($\Delta\rho$) from the matrix (ρ_0), then the X-ray scattering from these particles will locate at small angle extent. This is the so-called small angle X-ray scattering (SAXS). Usually, the small angle X-ray scattering is located within 5° around the incident X-ray beam. Actually, the SAXS intensity distribution in angular space depends on the incident X-ray wavelength and the particle sizes. When the particle size

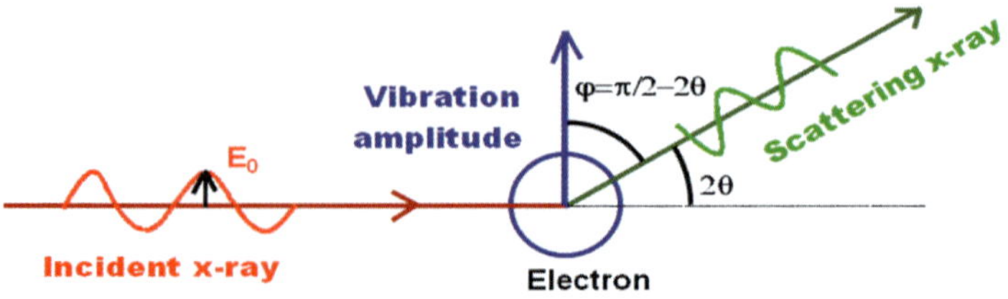

Fig. 7.1. Sketch map of X-ray scattering of an electron.

is extremely large or the incident X-ray wavelength is very short, the SAXS intensity will be condensed into the direct beam. In this case, the SAXS signal is very difficult to be experimentally separated from the direct beam. When the particle size is extremely small or the incident X-ray wavelength is very long, the SAXS intensity will be expanded to a quite wide angular extent. In such a case, the quick decay of SAXS intensity with the increase of the scattering angle causes much larger noise in the high-angle region, so that the high-angle SAXS intensity is too weak to be available. Therefore, the effective SAXS intensity is mainly distributed within $\sim 5°$, which reflects the nanoscale electron density fluctuation ($\Delta\rho = \rho - \rho_0$) in a sample. In other words, there are nanoscale regions with electron density (ρ) distributed in another homogeneous medium (ρ_0). These nanoscale regions are also called scatterers. Usually, the detectable particle size from SAXS is in a range of $2 \sim 100$ nm. Evidently, SAXS technique is a powerful tool to investigate nanoscale structures. In the following text, "electron density" is designed to the electron density fluctuation relative to the medium or matrix if no specification.

According to the status or the distribution of scatterers in samples, the SAXS samples can be divided into seven categories, i.e., monodisperse system, polydisperse system, orientated system, inhomogeneous system, dense system, periodic system, and arbitrary system. A schematic diagram for the sample categories is shown in Fig. 7.2. In a monodisperse system, all the scatterers (including nanoparticles or nanopores) have the same shape and electron density. And these scatterers are sparsely distributed in a sample. Therefore, the interference effect among the scatterings of different

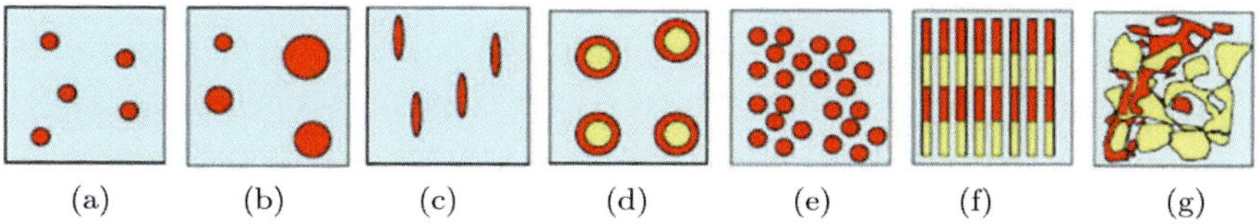

Fig. 7.2. A schematic diagram of the SAXS sample categories. (a) Monodisperse system; (b) Polydisperse system; (c) Orientated system; (d) Inhomogeneous system; (e) Dense system; (f) Periodic system. (h) Arbitrary system.

scatterers is neglectable in a monodisperse system. A sort of typical monodisperse sample is the purified and diluted protein solution. For such a monodisperse system, extracting the shape of the scatterers is the main purpose by SAXS study. When the scatterers with different particle sizes are sparsely and randomly distributed in a sample, it is a polydisperse system. In this case, the particle size distribution is the main research purpose by SAXS. Most of diluted nanoparticle solutions can be attributed to the polydisperse system. In fact, the polydispersity of particle size and shape of scatterers is ubiquitous. In an orientated system, the scatterers have anisotropic shape, same electron density, sparse distribution, and preferred orientation in the sample. A sort of typical orientated system is the fibrous sample. For example, the needle-like nanopores are sparsely distributed in carbon fibers. For an orientated sample, SAXS study is often used to extract the orientation degree and the size distribution of the scatterers. Although the scatterers are still distributed sparsely in the sample, the electron density of scatterers is not uniform or there is a transition layer between the scatterers and the matrix, such a sample can be attributed to the inhomogeneous system. A typical inhomogeneous system is the diluted solution of Au@SiO_2 nanoparticles with core-shell structure. In this system, the interfacial structure or the thickness of the transition layer is often the main issue studied by SAXS. Sparse distribution of scatterers in samples is the common character in the above four sample systems. However, when the scatterer distribution is not sparse any more, the interference effect of scatterings from different scatterers cannot be ignored. If the scatterers are still distributed randomly in the sample, such a sample can be attributed to the dense system. In this case, the SAXS intensity is the joint contribution of the particle scattering and the particle interference. Nanoparticle powder sample belongs frequently to the dense system. When the scatterers in a dense system are arranged into an ordered structure, the interference effect among different scatterers will appear in the form of diffraction peaks. This is quite similar to the normal X-ray diffraction (XRD). Because the d-spacings of this ordered structure are in nanometer range, it can be also ascribed to the research category of SAXS. Different from the normal XRD, the scattering unit is already not an atom but a particle. The particle shape with a certain size needs often to be considered in the profile analysis of the scattering signals. Many mesoscopic samples with assembly of nanoparticles and perhaps the porous materials can be attributed to the periodic system. There are always some samples with more complicated and nanoscale structures. These samples cannot be attributed to any one of the above mentioned systems. Even their nanoscale structures are difficult to be described. These samples are attributed to the arbitrary system. Some natural materials can be classified into the arbitrary system, which often present hierarchical structures. Even if the research interest focuses on the nanoscopic structures, the nanoscopic structures in a natural sample are still difficult to be attributed to an ideal sample system. Anyway, the size distribution, particle shape, and periodic structure of scatterers in a sample are the main issues studied by SAXS. Generally, the size distribution and the particle shape of scatterers cannot be resolved independently. If the sample is a polydisperse system and the scatterers have the same shape, SAXS is mainly used to determine the particle size distribution. If the sample is a monodisperse system, i.e., the scatterers have the same shape and size, then not only the size but also the shape of the scatterers can be determined by SAXS.

For a single particle with a volume V, its SAXS intensity can be expressed as Eq. (7.1):

$$\begin{aligned} I(q) &= I_e \int_V \int_V \rho(\vec{r}_1)\rho(\vec{r}_2) e^{-i\vec{q}\cdot(\vec{r}_1-\vec{r}_2)} dv_1 dv_2 \\ &= I_e \left| \int_V \rho(\vec{r}) e^{-i\vec{q}\cdot\vec{r}} dr \right|^2 \\ &= I_e \rho^2 V^2 F(\vec{q}). \end{aligned} \tag{7.1}$$

Here, I_e is the scattering intensity of an electron. $\rho(\vec{r}_1)$ and $\rho(\vec{r}_2)$ are the electron densities of the particle at $\vec{r}_1$ and $\vec{r}_2$. $\vec{r} = \vec{r}_1 - \vec{r}_2$ is the distance vector between two points in the particle. $\vec{q}$ is the scattering vector and $q = 4\pi \cdot \sin\theta/\lambda$. λ is the incident X-ray wavelength.

2θ is the scattering angle. $\rho V = n$, is the total electron number in the particle. $F(\vec{q})$ is the form factor of the particle.

For a monodisperse system, after averaging over all orientations of the particles, the SAXS intensity can be expressed as Eq. (7.2):

$$\begin{aligned} I(q) &= I_e \int_V \int_V \rho(\vec{r}_1)\rho(\vec{r}_2)\frac{\sin qr}{qr} dv_1 dv_2 \\ &= 4\pi \int_0^\infty p(r)\frac{\sin qr}{qr} dr. \end{aligned} \tag{7.2}$$

When $0 \leq r \leq D_{\max}, p(r) \geq 0$; and when $r \geq D_{\max}, p(r) = 0$. $D_{\max}$ is the maximum dimension of the particle. $p(r)$ is the pair distance distribution function (PDDF) restricted in one particle. On the basis of Eq. (7.2), the PDDF $p(r)$ can be expressed as the Fourier transformation of the scattering intensity $I(q)$ as shown in Eq. (7.3):

$$p(r) = \frac{1}{2\pi^2} \int_0^\infty I(q) \cdot qr \cdot \sin(qr) \cdot dq \tag{7.3}$$

In principle, the PDDF $p(r)$ contains the same information as the scattering intensity, but it is a real-space function. Therefore, PDDF is much easily understood as comparing to the scattering intensity in reciprocal space. Usually, $p(r)$ is first calculated from the Fourier transform of the scattering intensity $I(q)$, and then the particle shape and structural information are extracted from the PDDF. [1] Globular particle has a bell-shaped $p(r)$ function with its maximum at about $D_{\max}/2$, here $D_{\max}$ is the diameter of the Globular particle. Long-rod particle has a skewed $p(r)$ function with its maximum corresponding to the radius of the cross-section of the long-rod particle. Flat-disk particle presents a bell-like $p(r)$ function, but its profile is broadened and the maximum is obviously shifted to smaller distance than $D_{\max}/2$, here $D_{\max}$ is the diameter of the cross-section of the flat-disk particle. Core-shell spherical particle with lower electron density in the core-part and higher electron density in shell-part has a triangle-like $p(r)$ function with a maximum shifting towards larger distance than $D_{\max}/2$. Reversely, if a core-shell particle has higher electron density in core-part and lower electron density in shell-part, its $p(r)$ function will show a maximum shifting to lower distance than $D_{\mathrm{max}}/2$. Dumbbell-like particle consists of two well-separated sub-spheres, two extremums appear simultaneously at its $p(r)$ function. The lower-r and higher-r extremums are, respectively, corresponding to the radius of the sub-sphere and the interdistance between the two sub-spheres. Multiple extrema of $p(r)$ function are the feature of particle with well-separated subunits. Due to the distinguishable PDDF profiles of particles with different shapes, frequently the SAXS intensity of monodisperse particles are first Fourier-transformed into the $p(r)$ function in the SAXS data analysis. Especially for macromolecular or micellar samples, extracting the shape of the particles is the main purpose of the SAXS data analysis. Svergun and Koch [1] reviewed the SAXS studies of biological macromolecules in solution. An extensive-use of SAXS software (ATSAS) for protein shape analysis can be found from the website: http://www.embl-hamburg.de/biosaxs/software.html. The PCG is another software package [2–4] often used to analyze the SAXS data of micelles.

Early in 1939, Guinier gave an approximate SAXS expression for a single particle as shown in Eq. (7.4):

$$I(q) = I_e n^2 \left(1 - \frac{q^2}{3} R_g^2 + \cdots\right) \approx I_e n^2 e^{-q^2 R_g^2/3} \tag{7.4}$$

Here, $n = \rho V$ is the total electron number in the particle. R_{g} is the gyration radius of the particle. Equation (7.4) is the well-known "Guinier approximation". Because the Guinier approximation is a power expansion of the SAXS intensity at $qr \ll 1$, generally, it is only suitable for the low-angle region of the SAXS intensity. However, the Guinier approximation is independent on the particle shape, which means that the gyration radius R_{g} can be directly obtained from the slope of a linear fitting to the $\ln[I(q)]$ versus q^2 curve (or the Guinier plot). Regardless of the particle shape, the gyration radius of a particle can be easily obtained from the Guinier plot. But the gyration radius of a particle is related to the particle shape. For some regular geometrical shapes, the relation between the shape parameters and the gyration radius can be calculated as listed in Table 7.1.

Table 7.1. Relationship between the gyration radius R_g and the shape parameters for some particles with regular shape.

Regular particle	Shape parameter	Gyration radius R_g
Sphere	Radius: R	$\sqrt{3/5}R$
Hollow sphere	Outer radius: R, Inner radius: CR	$\sqrt{3/5}R\sqrt{(1-C^5)/(1-C^3)}$
Cylinder	Height: $2H$, Radius: R	$\sqrt{R^2/2+H^2/3}$
Fiber	Length: $2H$	$H/\sqrt{3}$
Ellipsoid	Half axes: abc	$\sqrt{(a^2+b^2+c^2)/5}$
Thin ellipse disc	Half axes:a and b	$\sqrt{(a^2+b^2)/2}$
Thin disc	Radius: R	$R/\sqrt{2}$
Cuboid	Lengths of side: $2a \times 2b \times 2c$	$\sqrt{(a^2+b^2+c^2)/3}$
Cube	Length of side: 2a	a
Elliptic cylinder	Height: h, Half axes: a and b	$\sqrt{(a^2+b^2+h^2/3)/2}$
Ellipsoid of revolution	Radii: $a, a, \omega a$	$a\sqrt{(2+\omega^2)/5}$
Hollow cylinder	Height: h, Outer radius: r_1 Inner radius: r_2	$\sqrt{(r_1^2+r_2^2)/2+h^2/12}$

For a polydisperse system, extracting the particle size distribution or particle number distribution is the essential. The X-ray scattering intensity of a polydisperse system with Guinier approximation can be described by Eq. (7.5).

$$I(q) = I_e \int_0^\infty N(R_g)\rho^2(R_g)V^2(R_g)F(R_g)dR_g$$
$$\approx I_e \int_0^\infty N(R_g)n^2(R_g)e^{-q^2R_g^2/3}dR_g, \tag{7.5}$$

where R_g is the radius of gyration, and $N(R_g)$ is the number of particles or the particle number distribution function with R_g. $V(R_g)$ is the volume of a particle with R_g. $\rho(R_g)$ is the electron density in a particle with R_g. $n(R_g)$ is the electron number in a particle with gyration radius R_g. On the basis of Eq. (7.5), by assuming the particle shapes (or the form factors) and fitting the SAXS intensity, the particle number distribution function or the particle size distribution can be tentatively obtained. A model-independent tangent-by-tangent method can be also used to extract the particle number distribution or the particle size distribution. For polydisperse systems, especially more complicated dense and arbitrary systems, their SAXS data analysis methods are still expected to be developed and improved further.

7.2. SAXS Technique

Since synchrotron radiation (SR) source was applied to scientific researches, SAXS experimental techniques had developed very fast. According to the experimental set-up, SAXS measurements can be classified into two basic types, i.e., the transmission SAXS mode and the reflection SAXS mode. The transmission SAXS mode contains SAXS, wide-angle X-ray scattering (WAXS), combing SAXS/WAXS, and ultra-small-angle X-ray scattering (USAXS). The reflection mode contains the grazing incidence SAXS (GISAXS). If the incident X-ray beam is fixed at a constant energy far away from the absorption edge of any elements in the sample, each of the above-mentioned SAXS techniques can be classified into normal SAXS-related techniques, otherwise it is classified into anomalous SAXS (ASAXS). In addition, if the real-time change of SAXS signal needs to recorded, the corresponding SAXS technique is called time-resolved SAXS (t-SAXS). In principle, if the SAXS signal at some transient state of nanoscale matter needs to be probed, a pump-probe SAXS technique could also be developed in future. But now there is no report about the transient SAXS measurement. Of course, if a SAXS change with the environmental condition of sample is recorded, the related SAXS measurement is called the *in situ* SAXS.

7.2.1. *Normal SAXS*

It is used to probe the mesoscopic lengths in an object. The incident X-ray beam with constant X-ray wavelength is often defined by multiple slits and penetrates directly into the sample. The X-ray scattering from the scatterers in the sample is recorded by an area detector. The detectable length scale of scatterers depends on the SAXS camera length (i.e. the sample-to-detector distance), the incident X-ray wavelength, and the active area of the detector. For SR-based SAXS experimental stations, the normal SAXS covers roughly a scattering vector q-range of $0.02 - 5.0\,\text{nm}^{-1}$. The lower limit determines the measurable maximum particle size, $D_{\text{max}} = \pi/q_{\text{min}}$. The upper limit determines the SAXS resolution, $D_{\text{min}} = 2\pi/q_{\text{max}}$. On the one hand, in order to reduce the lower limit of q-value and detect larger scatterer size, USAXS techniques have been developed. When the camera length increases and/or the incident X-ray wavelength increases, the lower limit of detectable q-value can reach approximately to $0.005\ \text{nm}^{-1}$. With an advanced-design Bonse-Hart camera with integrated fixed length, USAXS often can cover a q-range of 0.0005 to $1\ \text{nm}^{-1}$, which can be even used to detect micronsized structures. On the other hand, the upper limit of q-value has to increase for detection of smaller scatterer sizes. For this purpose, WAXS techniques were developed by reducing the camera length and/or the incident X-ray wavelength. Usually, WAXS can cover a q-range of $1-50\,\text{nm}^{-1}$, which is frequently used to detect the ordered structures with nanoscale spacings, even the crystal structures with interatomic spacings. The detectable particle-size ranges and the corresponding q-regions for SR-SAXS techniques are schematically shown in Fig. 7.3. In order to collect simultaneously the SAXS signal

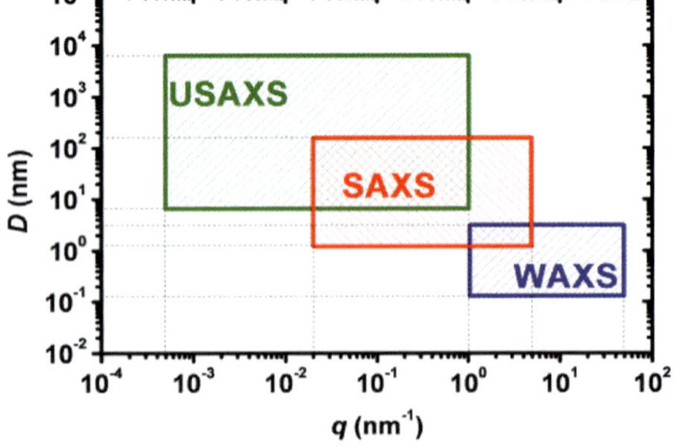

Fig. 7.3. Sketch map of the detectable particle size and the q-value region of USAXS, SAXS, and WAXS.

from random-distributed nanoscale structures and the WAXS pattern from ordered structures of scatterers in a sample, SAXS/WAXS combining techniques have been developed. Generally, one detector near the sample is used to collect the WAXS pattern, while another detector away from the sample is used to collect the SAXS pattern.

7.2.2. *GISAXS*

When some scatterers are included in a film or a near-surface layer which is supported by a substrate, transmission mode cannot be used to detect the nanoscale density correlations and/or the shape of the scatterers due to the restriction of the substrate thickness. Joanna Levine and coworkers introduced the GISAXS technique to study these nanostructures from the nanoscopic objects at surface, at buried interface, or in the film. Because the incident X-ray beam irradiates onto the surface in a grazing angle and an area detector is used to collect the GISAXS patterns, X-ray scattering and X-ray reflection signals are coupling into the GISAXS pattern. A specific software IsGISAXS (website: http://ln-www.insp.upmc.fr/axe4/Oxydes/IsGISAXS/figures/doc/manual.html) dedicated to GISAXS data analysis has been developed. The successful simulation to GISAXS patterns and the reasonable resolution for nanostructure systems not only bring flourishing GISAXS studies, but also make the GISAXS technique to be a versatile tool.

7.2.3. *ASAXS*

Utilizing the element selectivity, ASAXS can be used to distinguish the contribution of the specific elements to the nanoscale structures from a multi-element nanostructure system. The experimental setup of ASAXS is almost the same as the others except that the monochromator can be continuously tuned to select the wanted X-ray energy. Usually, several X-ray energies are chosen to repeat the SAXS measurements, but one of them must coincide with or be very close to the absorption-edge of the specific element, while another must be far away from the absorption-edge, and the others can be chosen around the absorption-edge. By resolving the SAXS signals measured with different X-ray energies, the contribution of the interested element to the nanostructures could be isolated. It is well known that SAXS depends only on the electron density fluctuation between the scatterers and the matrix or medium. If two scatterers have the same sizes and shapes, but one has a positive electron density fluctuation ($+\Delta\rho$), another has a negative electron density fluctuation ($-\Delta\rho$), for instance, a real particle and a nanopore, then both have the same SAXS patterns and the two scattering systems are complementary. Therefore, the scattering contributions from both complementary systems cannot be distinguished by the normal SAXS techniques. It becomes even a weakness of normal SAXS techniques. However, the element selectivity of ASAXS technique is an obvious advantage in distinguishing the real particles and the nanopores.

7.3. SAXS Application

In the past decades, with the application of SR sources and the progress of data analysis methods, SAXS has become a powerful tool to study the nanostructure systems. Various samples (solid, liquid, and even gas) are suitable for SAXS studies as long as there is nanoscale electron density fluctuation in the sample. Therefore, SAXS techniques have been widely applied in material science, physics, chemistry, biology, biomedicine, and so on. For example, *in situ* SAXS technique was applied to study the nanostructure changes induced by applied field (temperature, stress, voltage, electricity, magnetism, etc). Real-time SAXS technique was used to monitor the nanostructure changes of dynamic process (nucleation, growth, reaction, phase transition, folding and unfolding, coiling and uncoiling, etc). Now, SAXS has come to full flourish for the study of nanoscale structures. Some application examples of SAXS techniques are summarized as follows.

7.3.1. *Biological macromolecule*

The SAXS application in biological macromolecules (BMs) is an important field. Many BMs, such as membrane proteins, are difficult to crystallize or the crystal size is too small to be used for X-ray diffraction (XRD) analysis. However, their structural information is necessary for understanding the biological functions. SAXS provides not only the particle size and aggregation information, but also the shape and conformation information. Furthermore, the structure and function of BMs are closely related to the environment where BMs are located. Thus, obtaining the structures and structural evolution of BMs in physiological environment is essential to understand its structural mechanism of function. SAXS technique can offer the complementary information about the macromolecular folding-unfolding [5–11], aggregation [12,13], conformation [14], conformation changes [15–19], shape [20–23], shape changes [24–26], and assembly state [27–29] in solution near a real physiological environment. As compared with nuclear magnetic resonance (NMR), electron microscopy (EM), and XRD techniques, SAXS is more effective for diluted non-crystal samples. Although the SAXS resolution of about 1–3 nm is not so high, SAXS can provide the global shape of protein molecules in solution. Combining the known or predictive component structures, SAXS can provide not only more-detailed three-dimensional conformation, but also the primary phase for more accurate structure determination.

The shape determination of molecules or complex in solution is very important in order to understand their biological functions. Usually, the studied protein solution should be a monodisperse system. However, for most protein solutions, the electron density fluctuation between the protein molecules and the solvent is quite small. As a result, the scattering signal from the protein molecules is very weak for a diluted protein solution. Although a higher concentration of protein will enhance the scattering intensity and improve the signal-to-noise ratio, the possible interference effect will also distort the scattering profile, so that the form factor of the protein cannot be extracted well and truly. Usually, the solution sample of protein is required to be monodisperse for the shape determination. If the right SAXS patterns cannot be obtained from the protein solution with a felicitous concentration, it is an appropriate choice that the low-q SAXS intensity from one lower-concentration solution and the high-q SAXS intensity from another higher-concentration solution are merged together for protein shape analysis.

Recently, the progress of SAXS techniques enables high-throughput collection of SAXS signals of BMs. The development and improvement of SAXS data analysis enable the three-dimensional shape restoration of protein molecules from the one-dimensional SAXS curve. ATSAS software is a successful program package used for SAXS data analysis of BMs and consists of more than 30 programs, including PRIMUS [30], GNOM [31], DAMMIN [32], DAMAVER [33] and so on. PRIMUS is used to manipulate the SAXS data files for averaging, background-subtraction, merging, concentration-extrapolation and curve-fitting, as well as for the parameter evaluation, such as the radius of gyration (for globular, flat and rod-type particles), the Porod's volume, the zero intensity, and the molecular weight. GNOM is used to evaluate the PDDF $p(r)$ for monodisperse systems or to calculate the size distribution function (SDF) D(R) for polydisperse systems. DAMMIN is based on the simulated annealing algorithm and is used to determine the protein shape constructed by single-phase dummy atoms. DAMAVER is used to get the average shape of BMs in the solution from a dozen independent dummy-atom models.

7.3.1.1. *Macromolecule conformation*

The biological function of BMs is tightly related to the molecule conformation, while a detained molecule structure is often obtained from the BM crystal by using XRD. However, maybe the macromolecule conformations are not the same in the BM crystal as in its physiological environment. Therefore, it is needful to clarify the conformation difference between the BM solution and the crystal as the relationship between structure and function is investigated. Trempe

J. F. studied the structure of parkin, revealing the mechanism of ubiquitin ligase activation [14]. SAXS data were obtained for the full-length and R0-RBR (amino acids 141 to 465) parkins in solution. These SAXS data were fitted to single chains in each crystal structure, demonstrating that the experimental SAXS curves can be fitted well by the full-length and R0-RBR parkins. The calculated SAXS data from respective molecule chains in crystal structures were in good agreement with the experimental SAXS curves [14]. SAXS analysis identified that the full-length and R0-RBR parkins adopted the same conformation in the solution and in the crystals. In most cases, the BM conformations are the same in crystal and in solution, but there are also some exceptions [34].

7.3.1.2. *Shape and shape evolution*

SAXS technique can be used to monitor the structural response of a biological system to a perturbation in the physical or chemical environment. The shape evolution of a thermotolerant protein *PeaT*1 with temperature was studied [24] by using SAXS technique at Beijing Synchrotron Radiation Facility (BSRF). The solution of *PeaT*1 protein was heated to 25, 55, and 85°C, then it was cooled to 55 and 25°C. Five SAXS patterns were collected at the above five temperatures. In order to eliminate the concentration effect, the protein solutions with concentrations of 3.75, 7.5, and 30 mg/ml were used in the SAXS measurements. After normalizing the protein concentration and subtracting the background scattering, the low-q SAXS intensity with concentration of 7.5 mg/ml and the high-q SAXS intensity with concentration of 30 mg/ml were merged into the final SAXS curve at each temperature. The PDDFs of the protein at the above five temperatures were calculated from the merged SAXS data by using GNOM program as shown in Fig. 7.4(a). The protein shapes of *PeaT*1 at the five temperatures were restored from the SAXS data by using program DAMMIN and DAMAVER. Four structural domains F, NAC, T, and UBA were confirmed to be included in a single *PeaT*1 molecule. The *PeaT*1 protein was determined existing in the solution in the form of homodimer. The SAXS shapes with

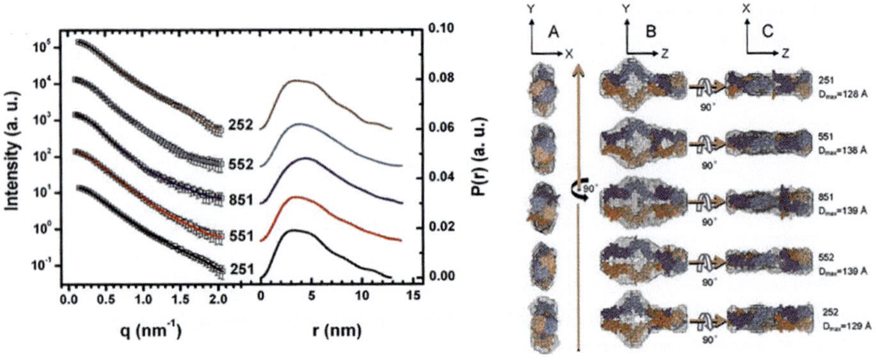

(a) (b)

Fig. 7.4. (a) SAXS intensities and $p(r)$ profiles of *PeaT*1 protein at different temperatures, as well as (b) the shape change of *PeaT*1 homodimer in solution with temperature. Reproduced from [24] with permission from John Wiley and Sons.

the docked domains were shown in Fig. 7.4(b). The SAXS results illustrate that the shape change of *PeaT*1 homodimer in the solution is approximately reversible with temperature change. With the temperature increasing, the two molecules in a *PeaT*1 homodimer are pushed away from each other. With the temperature decreasing, the two molecules are pulled towards each other to close up. During a heating–cooling cycle, the two NAC domains hug together face to face, and their relatively stable structure plays a crucial role of frame in the thermotolerance of *PeaT*1 protein.

For some biologically important complexes, their structures tightly control their activities. For example, in the complex of extracellular signal-regulated kinase 2 (ERK2) and the hematopoietic tyrosine phosphatase (HePTP), the HePTP negatively regulates T cell activation in lymphocytes via the dephosphorylation of the mitogen-activated protein (MAP) kinase ERK2. However, available structural information about the complex of ERK2:HePTP is very limited. Utilizing SAXS technique combined with the ensemble refinement of SAXS (EROS) [35], Francis *et al.* [26] characterized the structures of the resting and active states of ERK2:HePTP complexes. HePTP monomer and ERK2 monomer were found forming

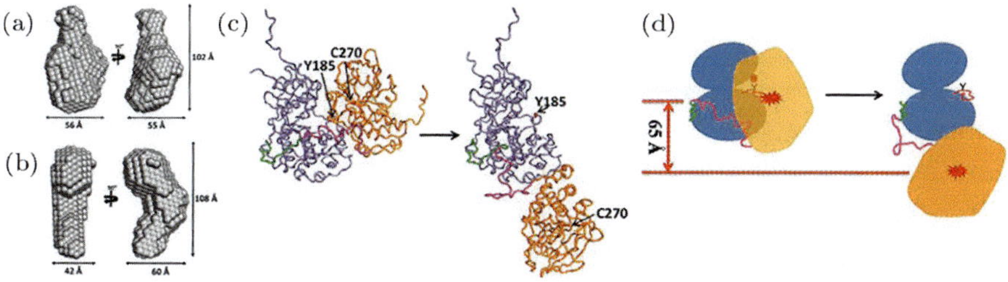

Fig. 7.5. SAXS analysis of the ERK2:HePTP complex. (a) The molecular shape of the active ERK2:HePTP complex; (b) The molecular shape of the resting ERK2:HePTP complex; (c) Ribbon diagram of the active (left) and resting (right) states of the ERK2:HePTP complex; (d) Cartoon representation of the active (left) and resting (right) states of the ERK2:HePTP complex. ERK2 is blue, HePTP-KIM is green, HePTP-KIS is magenta. The locations of C270 (HePTP) and Y185 (ERK2) are indicated in (c). The HePTP catalytic domain rotates by nearly 180° and the HePTP active site cysteine moves by ~65 Å between the active and resting state complexes. Reproduced from [26] with permission from American Chemical Society.

heterodimer in solution. Through the SAXS data analysis, they obtained the shapes of the complex as well as the respective orientations of ERK2 and HePTP in the resting and active states as shown in Figs. 7.5(a)–(c). The results demonstrated that the active state has more compact conformation than the resting state. A cartoon representation of the active (left) and resting (right) states of the ERK2:HePTP complex is shown in Fig. 7.5(d). Figure 7.5 shows that the active state of ERK2:HePTP complex is more globular and HePTP is perfectly positioned to interact with the phosphorylated tyrosine on the ERK2 activation loop. Accompanying the dephosphorylation, the HePTP catalytic domain rotates by nearly 180° and moves away from the ERK2 activation loop, which results in an extended resting state of the complex. The HePTP active site cysteine moves by ~65 Å between the resting and the active states of the complexes. Therefore, the resting state of the ERK2:HePTP complex adopts a highly extended, dynamic conformation, while the active state adopts a compact and ordered conformation. This study demonstrates that SAXS technique can be used to experimentally characterize the dynamic structural changes of complexes in solution and provide the first structural insight into an active state of complexes.

7.3.1.3. *Folding and unfolding*

BMs must be in some folding states to fulfill their biological functions. The misfolding of BMs is related to some diseases. SAXS of BMs can be used to study the folded, partially folded, and unfolded conformations of BMs. This conformational change of BMs can be evaluated from the profile of Kratky plot ($q^2I(q)$ versus q). Generally, the SAXS data of globular macromolecules follows the Porod's law, and its Kratky plot is a bell-shaped curve. While, the Kratky plots [36] of partially folded and unfolded BMs (such as the unfolded peptides) have an obvious increment or present a plateau in the high-q end as shown in Fig. 7.6.

Two different models were all used to explain the unfolding effect of urea over the past 50 years. One invoked the disruptive effects on water structure; another proposed that urea binds directly to multiple sites on the protein backbone. Both models would stabilize the unfolded state in the presence of high concentrations of urea. Combining SAXS and small-angle neutron scattering (SANS), Gabel *et al.* [10] studied the urea binding to unfolded ubiquitin. The SAXS curves of ubiquitin in 8 M urea/H_2O at pH 6.5 and 2.5 with protein concentration of 1.7 mM, as well as the calculated scattering curve of ubiquitin (PDB 1UBQ) using the program CRYSOL [37] are compared in Fig. 7.7(a). The comparability in the entire q-range between the pH-6.5 experimental SAXS data and the calculated one indicates that

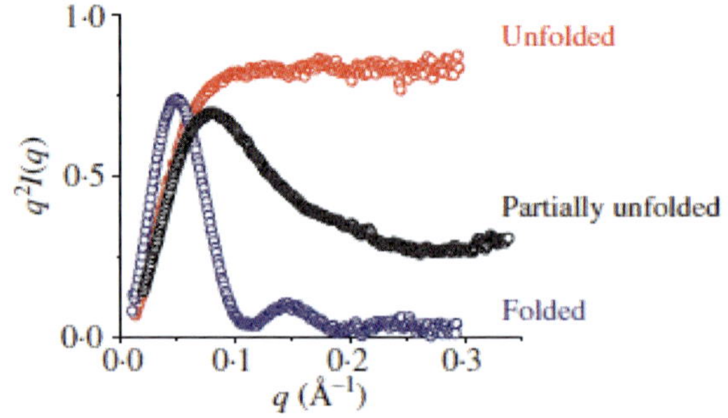

Fig. 7.6. Kratky plots of the SAXS data of partially folded and unfolded BMs. Reproduced from [36] with permission from Cambridge University Press.

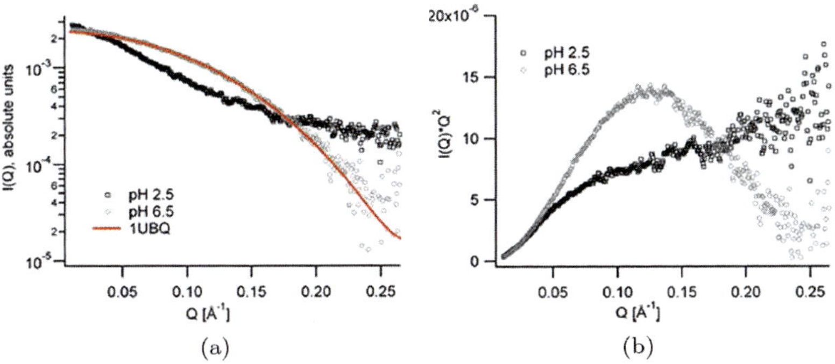

Fig. 7.7. (a) SAXS scattering curves of ubiquitin in 8 M urea/H_2O at pH 6.5 and 2.5 with protein concentration of 1.7 mM. The solid line represents the calculated scattering curve of ubiquitin (PDB 1UBQ) by using CRYSOL. (b) Kratky plot of the same data. Reproduced from [10] with permission from American Chemical Society.

the protein is in its native conformation even when it was dissolved in the high-concentration denaturant. However, when the pH value was lowered to 2.5, the SAXS curve had a significant increase in the high-q end, suggesting that the protein was unfolded into a disordered conformation. This finding was corroborated by the Kratky plot [10] as shown in Fig. 7.7(b), showing a transition from native folding state into unfolding one. Through the model-free analysis, they found that a pH change from 6.5 to 2.5 triggers a recruitment of ~20 urea molecules from the bulk solution per ubiquitin molecule during the unfolding process. This information qualitatively supports the model of direct binding of urea to the protein backbone.

Ribozymes must fold into compact, native structures to function properly in the cell. However, the formation of RNA tertiary structure includes the neutralization of the phosphate charge and the collapse of the unfolded molecules. In order to understand the forces that enable specific collapse in RNA, Roh *et al.* [9] used stopped-flow time-resolved SAXS technique to observe the multistage collapse of a bacterial ribozyme. A schematic view of the stopped-flow mixer (SFM400) was shown in Fig. 7.8(a). The Kratky plots of the real-time folding data in 1.5 mM $MgCl_2$ were shown in Fig. 7.8(b). Their results demonstrates that the preferred collapse mechanism depends on the

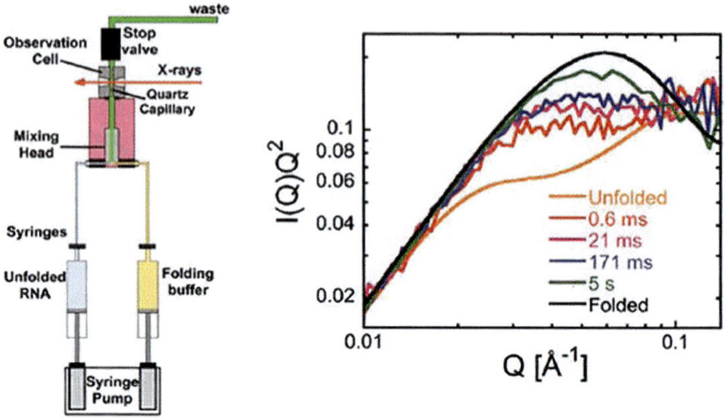

(a) (b)

Fig. 7.8. (a) Schematic view of the stopped-flow mixer for SAXS measurements, as well (b) as the Kratky plots of the real-time folding SAXS data of the 1.5-mM $MgCl_2$. Reproduced from [9] with permission from American Chemical Society.

extent of Mg^{2+}-dependent charge neutralization and that non-native interactions within the unfolded ensemble contribute to the heterogeneity of the ribozyme folding pathways at the very earliest stages of tertiary structure formation. These observations demonstrate the importance of electrostatic interactions for the nucleation of long-range tertiary structure in RNA.

7.3.1.4. *Aggregation*

Tau protein is an intrinsically disordered protein that is abundant in neuronal axons where it promotes and stabilizes microtubule assembly. Alzheimer's disease (AD) is related to the intracellular neurofibrillary tangles (NFTs) consisting of Tau protein. With progression of AD, Tau aggregates and accumulates into NFTs. Therefore, studying the inhibitors of tau aggregation and their mode of action plays an important role in therapeutic approaches for Alzheimer's disease. Akoury *et al.* [12, 13] studied the mechanistic basis of phenothiazine-driven inhibition of Tau aggregation, in which the aggregated state of Tau protein at different ratios of Methylene Blue

(MB), its derivatives azure A and B was observed by using SAXS. They concluded that MB and its derivatives azure A and B retain Tau in a monomeric aggregation incompetent state. Besides, the organic compound tetrasulfonate (PcTS) can also inhibit the aggregation of Tau protein. Combining nuclear magnetic resonance (NMR) spectroscopy, electron paramagnetic resonance (EPR), and SAXS, Akoury *et al.*[12] found that PcTS stimulated the assembly of soluble Tau oligomers having a dynamic core with diameters $30 \sim 40\,\mathrm{nm}$, which inhibited the Tau protein aggregation. The specific modulation of the conformation of tau is a viable strategy for reduction of pathogenic tau deposits.

7.3.2. *Phospholipids and micelles*

Phospholipids are the main components of biological membranes. Because they are amphoteric molecules, the SAXS application of phospholipids is discussed separately as a special case. Phospholipids and micelles are the important application fields of SAXS technique. The size distribution, electron density distribution, internal structure, phase transition, interaction, specific surface, and the interface structure are often the main research contents of SAXS in phospholipids and micelles.

Cell membrane is mainly composed of phospholipid bilayer, proteins and polysaccharides, which is usually displaced by self-assembly lipid-bilayers in a stimulation experiment. Phospholipid membrane is an impermeable barrier for most water-soluble molecules. Under some special conditions, for instance light illumination, heating process [38], as well as contact with chemical substances [39] and nanomaterials [40], the phospholipid membrane could behave as different phases [41]. Due to the spatial thickness of lipid layers is in nanoscale, the X-ray reflections from the crystalline phases of phospholipid membrane are often appearing on a small-angle region. Therefore, SAXS technique is very suitable for the study of the phase change of cell membrane. From the relative q-spacings of diffraction peaks, the crystalline phase can be easily determined. For example, lamellar phase: $1{:}2{:}3{:}4{:}\cdots\cdots$; hexagonal phase: $1 : \sqrt{3} : 2 : \sqrt{7} : \cdots\cdots$; and cubic phase: $1 : \sqrt{4/3} : \sqrt{8/3} : \sqrt{11/3} : \cdots\cdots$; or other complicated phases.

Yaghmur *et al.* [38] investigated the lamellar and hexagonal phase transformation of vesicles of N-methylated dioleoylphosphatidylethanolamine interacted with gold nanoparticles. By using SAXS technique, Yang *et al.* [42] studied the synthetic antimicrobial oligomers (AMO), by inducing a composition-dependent topological transition in membranes. They found that the antibacterial activity correlates with the AMO-induced topological transition from small unilamellar vesicles into inverted hexagonal phase. In this phase change, hexagonal arrays of 3.4 nm water channels defined by lipid tubes are formed. The 1D and 2D electron density functions of the inverted hexagonal phases were reconstructed from the SAXS data. The details about the density calculation can be found in literatures [43–45]. Wang *et al.* [46] studied the structure phase transformation of an organogel composed of a simple surfactant molecule-sodium laurate. Two diffraction peaks with position ratio of $1 : \sqrt{2}$ indicated that the cylindrical micelles of sodium laurate had formed a square lattice with lattice parameter of 3.2 nm. Wang *et al.* [47] also prepared a viscoelastic vesicle gel by mixing nonionic surfactant, tetraethylene glycol monododecyl ether ($C_{12}EO_4$), anionic dye, and sodium 4-phenylazobenzoic acid (AzoNa). SAXS study revealed the structural ordering, the interlamellar spacing, and the microstructural change with AzoNa concentrations and pH value in the vesicle gel. In a study [48] of nonaqueous lyotropic liquid crystal fabricated by a polyoxyethylene amphiphile in protic ionic liquid, SAXS results identified that hexagonal liquid crystalline phase can only be formed in the Brij 97/EAN binary system. And the structural ordering decreased with increasing temperature, and the lattice spacings shrank with reduction of ionic liquid content at room temperature. Martiel *et al.* [49] reported for the first time the formation of a reverse micellar mesophase of face-centered cubic symmetry in a ternary mixture of phosphatidylcholine (PC)/water/organic solvent. SAXS was used to identify the $Fm\bar{3}m$ space group and detect the scattering information of pure micelles. By fitting the form factors in the pure L_2 phase and in the $Fm\bar{3}m - L_2$ coexistence region, the PC interface rigidity was quantitatively estimated.

Micelles composed of amphiphilic block copolymers are of particular interest in drug delivery systems. These nano-sized drug carriers have many advantages, including easy preparation, biodegradability, controllable drug release, long systemic circulation time and enhanced accumulation in tumors via the enhanced permeability and retention (EPR) effect [50]. Poly(ethylene glycol)–phosphatidylethanolamine (PEG–PE) is a kind of biodegradable amphiphilic diblock copolymer, which can self-assemble in to spherical micelles [51, 52] with core-shell structure in aqueous solution. It was reported [52, 53] that cationic amphiphilic drugs can be easily encapsulated into PEG_{2000}-PE micelles by self-assembly method. This drug encapsulation shows high efficiency, controllable drug release and better micellar stability. Usually, the density difference between micelle and solvent (or water) is quite small, and the micelle concentration in solution is relatively high (25mg/ml or more) in practice. In other words, the interference effect to SAXS signal from the higher concentration is often not negligible. Therefore, the SAXS intensity of micelle solution has to contain the joint contributions of form factor and structure factor as shown in Eq. (7.6).

$$I(q) = nP(q)S(q), \tag{7.6}$$

where, n is the particle number density, $P(q)$ is the averaged form factor, and $S(q)$ is the static structure factor. It is the contribution of $S(q)$ that makes the SAXS intensity appearing interference effect as shown in low-q region of Fig. 7.9(a,d). In Fig. 7.9, "M" refers to the empty micelle of PEG_{2000}-PE; "M-Dox1/2" and "M-Dox1/1" refer to PEG_{2000}-PE micelles with loading doxorubicin with mole ratios of PEG_{2000}-PE molecules to doxorubicin molecules 2:1 and 1:1; "M-Vnr1/6" and "M-Vnr1/4" refer to PEG_{2000}-PE micelles with loading vinorelbine with mole ratios of PEG_{2000}-PE molecules to vinorelbine molecules 6:1 and 4:1. The static structure factor $S(q)$ reflects the interaction-potential information between particles and can be described as the Fourier transform of total correlation

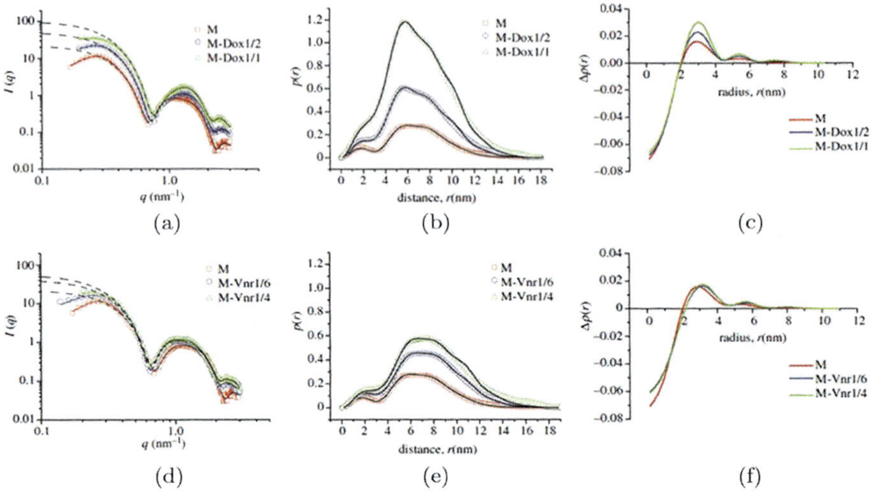

Fig. 7.9. (a) SAXS data (circle line) of PEG-PE micelle without loading drug (marked as M) and with loading doxorubicin in mole ratios of PEG-PE/doxorubicin 2:1 (M-Dox1/2) and 1:1 (M-Dox1/1); (b) and (c) are, respectively, the calculated $p(r)$ and radial densities of the three samples in (a); (d) SAXS data (circle line) of PEG-PE micelle without loading drug and with loading vinorelbine in mole ratios of PEG-PE/doxorubicin 6:1 (M-Vnr1/6) and 1:4 (M-Vnr1/4); (e) and (f) are the calculated $p(r)$ and radial densities of the three samples in (d). Solid lines are the fitting curves. Dash lines are the calculated P(q). Reproduced from [52] with permission from The Royal Society.

function, $h(r) = g(r) - 1$, as shown in Eq. (7.7).

$$S(q) = 1 + 4\pi n \int_0^{\infty} (g(r) - 1) r^2 \frac{\sin qr}{qr} dr, \tag{7.7}$$

where $g(r)$ is the pair-correlation function. For an infinite dilute solution, $S(q) = 1$. For a higher concentration micelle system, the software GIFT [51, 54, 55] can be used for its SAXS data analysis. At first, a model-free form factor $P(q)$ is evaluated from the SAXS intensity as ignoring the structural factor $S(q)$. Then, a model-dependent $S(q)$ is calculated based on a selected model. After optimizing the initial $P(q), S(q)$, and relative parameters, the final $P(q)$ can be obtained. Simultaneously, the PDDF $p(r)$ can be calculated by using the generalized indirect Fourier transformation (GIFT) technique, as shown in Fig. 7.9(b,e). The PEG_{2000}-PE molecules and the cationic drugs are all charged, so the charged monodisperse sphere structure factor model was chosen in GIFT calculations. The screened Coulomb (Yukawa) potential and the hypernetted chain (HNC) closure relation [2] for the Ornstein-Zernike (OZ) equation were used in the $S(q)$ calculation. More detailed theory about this technique can be found elsewhere [3, 56, 57]. For sphere scattering objects, the scattering length density profile is simply the function of radial position. The radial electron density profile is related to the PDDF $p(r)$ as in Eq. (7.8).

$$p(r) = r^2 \Delta\tilde{\rho}^2(r) \tag{7.8}$$

Program DECON [4, 58], a numerical convolution square root technique, can be used to calculate the radial electron density distribution $\Delta\rho(r)$, and the results were shown in Fig. 7.9(e,f).

The SAXS results [52] demonstrated that the PEG_{2000}-PE empty micelle is a core-shell spherical nanoparticle. Its outer radius ($D_{\max}/2$) is about 9 nm and its inside radius (or the core size, r_c) is about 2 nm. The main electron density is distributed in the region of 2–4 nm with a maximum occurring at 3 nm ($r_{\max}$), where phosphorus atoms are located. However, loading doxorubicin into the PEG_{2000}-PE micelles did not change their core size and the outer radius, but loading vinorelbine increased slightly the core size owing to larger molecular volume of vinorelbine. These SAXS results for the empty and doxorubicin-loading micelles had been also confirmed by a molecule dynamics (MD) calculation. Based on the SAXS analysis and the MD calculations, the sketch diagrams for the empty and the doxorubicin-loading micelles are shown in Fig. 7.10. From this study [52], it can be concluded that doxorubicin is distributed at the core-shell interface of PEG_{2000}-PE micelles because of the interaction between the positively charged amine from doxorubicin and the negatively charged phosphate group from PEG_{2000}-PE. And, the anthracycline ring of doxorubicin is inserted into the hydrophobic PE core of the PEG_{2000}-PE micelles.

Low molecule-weight triblock copolymer consisting of poly(ethylene oxide) (PEO) and poly(propylene oxide) (PPO) is nonionic

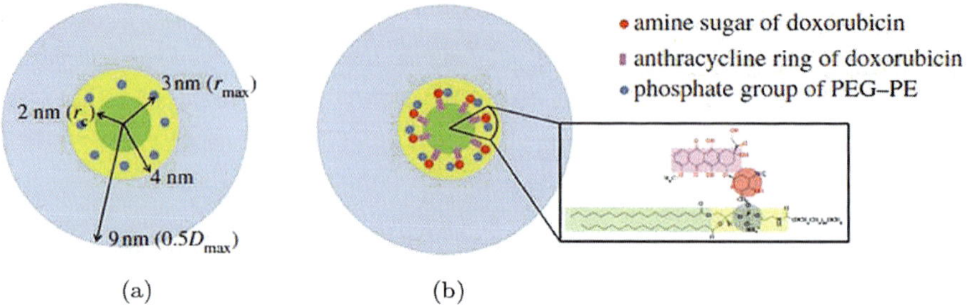

Fig. 7.10. Sketch diagrams of PEG-PE micelles (a) without and (b) with doxorubicin-loading. Reproduced from [52] with permission from The Royal Society.

and water soluble. Jasson *et al.* [59] studied PEO-PPO-PEO triblock copolymer/cationic surfactant complexes in aqueous solution using SAXS, NMR, dynamic light scattering (DLS) and static light scattering (SLS), getting the structure and composition of the complexes. For simplifying, the triblock copolymer PEO-PPO-PEO ($EO_{20}PO_{68}EO_{20}$) is shortened to P123, and the cationic surfactant hexadecyltrimethylammonium chloride is shortened to CTAC. In previous light scattering and calorimetry study [60] for the P123-CTAC system, three different surfactant concentration regimes was observed. In order to investigate the structures and compositions of the formed complexes as well as the P123 micelle-CTAC surfactant interaction, the SAXS data were analyzed by GIFT to separate the form factor and the structure factor. The internal electron density distribution was also calculated by using DECON. The apparent molecule weight was also obtained from the SAXS intensity on absolute scale. It was found that the P123-CTAC complexes exhibit a rather small decrease in size and apparent molecular weight under low CTAC concentrations. During the evaluation of SAXS curves, the interparticle interaction was chosen as the Percus-Yevick closure (excluded volume interaction). However, only CTAC-rich complexes are present when the CTAC concentration is high. And the CTAC-rich complex has slightly smaller particle size than the pure CTAC micelle, as well as a lower apparent molar mass. In this case, the hypernetted-chain closure was selected to describe the repulsive interparticle interaction for this highly charged complex.

For cylindrical colloids, assuming a constant electron density along the cylinder axis, then the distance distribution function of the cross-session $p_c(r)$ and the radial electron density of the cross-session $\rho_c(r)$ can be also calculated by GIFT and DECON from SAXS experimental data. Utilizing SAXS technique, Aramaki group studied widely self-assembled rod/cylindrical micelle systems with nonionic surfactants, such as the nonionic fluorinated micelles [61] in aqueous system, the short haired wormlike micelles in mixed nonionic fluorocarbon surfactants [62], the diglycerol monolaurate reverse micelles in nonpolar oils cyclohexane and octane [63], and the nonionic surfactant (glycerol α-monomyristate) micelles in organic solvents [64]. All these micelles or reverse micelles can form cylindrical shape under some suitable conditions. Their self-organization behaviors are highly influenced by the alkyl chain length of the solvent, the surfactant concentration, the extra adding oligomers, and the temperature etc. The micelle shape, size and internal structures were determined by SAXS technique from the calculated distance distribution function $p_c(r)$ and the radial electron density $\rho_c(r)$.

7.3.3. *Polymer and related materials*

Polymers are of great importance due to their peculiar chemical-physical properties and wide applications. In view of the complicated character in polymer physics researches, the polymer researches either in science or in engineering all rely heavily on SR-based experimental techniques. Specially, SAXS enables structural studies of polymers with molecular weight from kDa to GDa under extreme experimental environments. *In situ* and time-resolved SAXS techniques promise to provide the unique information [65] about the kinetics and dynamics of polymer processes The SAXS applications on polymers include: (1) Determining the shape and size [66, 67] of polymers in solution; (2) Determining the shape, size and distribution [68] of polymer colloidal particles; (3) Detecting the deformation mechanism [69, 70], mesostructure [71], crystalline grains, cavities or cracks; (4) Determining the crystalline structures, lamellae and long spacing in polymers, as well as phase transition;

(5) Measuring the average molecular weight of polymers; (6) Estimating the fractal structure in polymers and so on. For example, Jiang [72] group focuses on the relationship research between structure and mechanical property of polymers by using SAXS and WAXS techniques in recent years. They investigated quantitatively the lamellar orientation of Poly(L-lactide) (PLLA) as a function of shear rate. Orientation parameter f, proposed by Hermans was adopted to characterize the orientation degree of sample as defined by Eq. (7.9):

$$f = (3 < \cos^2 \varphi > -1)/2, \tag{7.9}$$

where φ is the angle between the chain axis and the reference axis (stretching direction), and $\langle \cos^2 \varphi \rangle$ is defined as Eq. (7.10):

$$\cos^2 \varphi = \frac{\int_0^{\pi/2} I(\varphi) \sin \varphi \cos^2 \varphi d\varphi}{\int_0^{\pi/2} I(\varphi) \sin \varphi d\varphi} \tag{7.10}$$

where $I(\varphi)$ is the scattered intensity along the angle φ. Different shear rates were applied to the sample, yielding a change of the orientation degree of PLLA. The azimuthal profiles extracted from 2D SAXS patterns of PLLA are shown in Fig. 7.11(a). The orientation degree of PLLA as a function of shear rate is shown in Fig. 7.11(b). The SAXS results demonstrate that even a weak shear flow can also

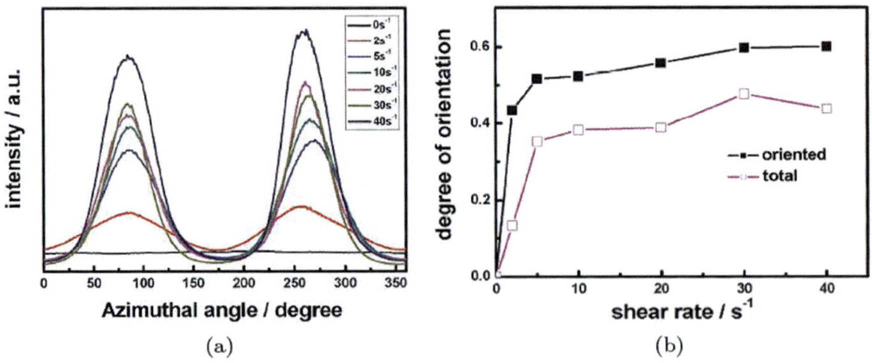

Fig. 7.11. (a) Azimuthal profiles of 2D SAXS patterns of PLLA crystallized after steady shear with different shear rates, (b) degree of orientation of PLLA as a function of shear rate. Reproduced from [72] with permission from The Royal Society of Chemistry.

increase dramatically the degree of orientation. When the shear rate is larger than 5 s^{-1}, the degree of orientation increases slightly to a constant. It was concluded that the amount of isotropic structure played an important role in the variation trend of orientation degree with shear rate.

Men [73] group is also dedicated to the SAXS application on polymers. Utilizing *in situ* SAXS and scanning SAXS techniques, they investigated the tensile deformation of oriented poly(ε-caprolactone) (PCL) and its miscible blends with poly(vinyl methyl ether) (PVME). The SAXS or WAXS patterns of PCL/PVME blends or pure PCL micromolded at different injection velocities were measured as a function of sample position or thickness, or true strain. The average thicknesses of the crystalline lamellae and amorphous layers measured along the stretching direction were derived from the electron density correlation function $K(z)$ as in Eq. (7.11).

$$K(z) = \frac{\int_0^\infty I(q)q^2 \cos(qz)dq}{\int_0^\infty I(q)q^2 dq}, \tag{7.11}$$

where $I(q)$ is the experimentally SAXS intensity distribution and z is along the drawing direction. The obtained one-dimensional correlation functions for pure PCL are shown Fig. 7.12(a). The long spacing and the average thicknesses of the crystalline lamellae and the amorphous layers were evaluated as shown in Fig. 7.12(b) for the micromolded PCL/PVME blends.

From these SAXS studies, the structural evolution of micromolded PCL and its miscible blends with PVME at the nanoscale were obtained as a function of deformation ratio and blend composition At small deformations, the crystal block slips within the lamellae. Then a stress induced fragmentation and recrystallization process occurs along the drawing direction at a critical strain where the average thickness of the crystalline lamellae remains essentially constant during stretching. The critical strain is dependent on the amount of the amorphous component or PVME incorporated in the blends, corresponding to a transition from lamellae to fibrils Beyond the critical strain the slippage of the fibrils (stacks of newly formed

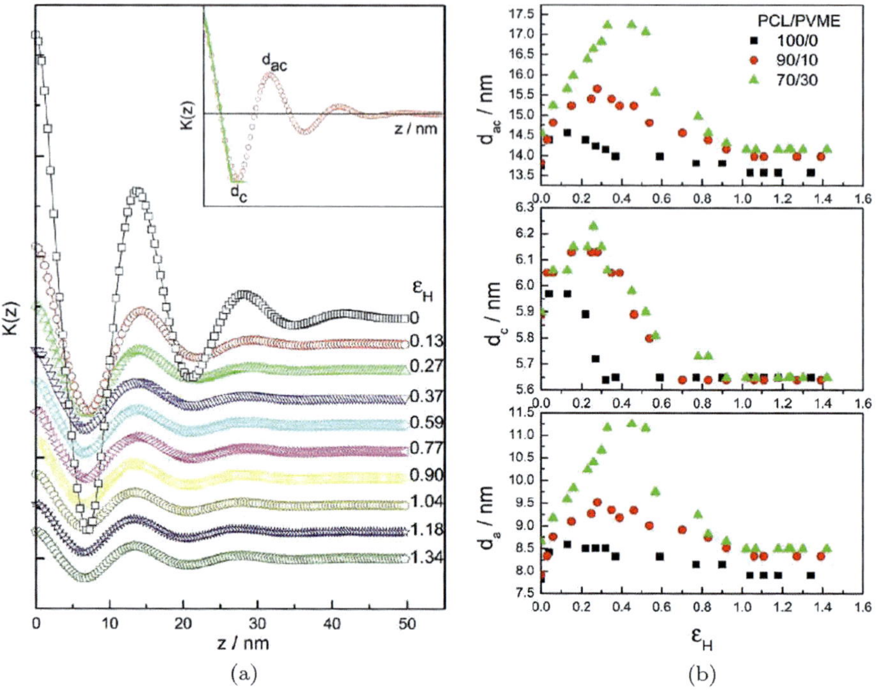

Fig. 7.12. (a) One-dimensional correlation functions for pure PCL taken at different deformations. The long spacing (d_{ac}) together with the average thicknesses of crystalline lamellae (d_c) and amorphous layers (d_a) can be obtained from the correlation function as illustrated in the inset diagram. (b) Evolution of the long spacing (d_{ac}) (top) together with the average thicknesses of the crystalline lamellae (d_c) (middle) and the amorphous layers (d_a) (bottom) as a function of strain measured for the micromolded PCL/PVME blends. Reproduced from [73] with permission from American Chemical Society.

lamellae) results in a relaxation of stretched interlamellar amorphous chains. As a consequence of phase separation induced by tensile deformation in the miscible PCL/PVME blends, the resulting jammed packing of the fibrils makes it more difficult for the lamellae stacks to reconfigure themselves at large strains compared to the pure PCL.

Copolymer is a polymer derived from two (or more) monomeric species. Block copolymer is made up of blocks of different polymerized monomers. Modern synthetic methods afford access to a broad portfolio of multiblock molecular architectures. SR-based

SAXS techniques have been extensively applied to study the nanoscale structures in block copolymers, complexes, or more complicated compounds. For example, drug carrier [52], single-chain polymeric nanoparticles mimicking secondary-structuring elements in proteins [74], as well as biocompatible thermoresponsive [75], photovoltaic [76], Semiconducting [77], supercapacitor and electrocatalyst [78] materials and so on. Besides, block copolymer melts [79], synthetic methods for novel nanoporous materials [78], and ordered diblock copolymer/ polyoxometalate nanocomposite films [80], have been also widely studied.

7.3.4. *Nanomaterial and superlattice*

Due to the essential nanoscale structures in nanomaterials, SAXS techniques are well-suited for the structural characterization of nanomaterials. Even compared with visual transmission electron microscopy (TEM), scanning electron microscopy (SEM), atomic force microscopy (AFM), and X-ray imaging techniques, SAXS techniques still have three outstanding advantages in the nanomaterial researches. One is the higher statistical property for scatterers in sample; another is global property within the gauge volume irradiated by X-ray and is not only limited to the sample surface. Besides, the spacious space and the low-vacuum or atmosphere requirement for SAXS sample environment make SAXS techniques especially suitable for nanomaterial characterization under *in situ* measurement and real condition. The main applications of SAXS techniques on nanomaterials include determining the superlattice structure consisted of nanoparticles (NPs); the NP size and distribution; as well as the nucleation and growth processes.

Superlattice

For realization and application of some functions, people are very interested in colloidal superlattices (SLs) self-assembled from inorganic nanoparticles or organic molecules [81–84], BMs like proteins [85], DNA [86, 87] and so on. Also SAXS techniques are often used to detect the structures of self-assembled SLs.

Wang and his coworkers reported [81, 82] the self-assembly of CdSe-CdS core-shell semiconductor nanorods, which was mediated by shape and structural anisotropy. TEM and SAXS measurements confirmed that the CdSe-CdS core-shell semiconductor nanorods were self-assembled to mesoscopic colloidal SLs, having multiple well-defined supercrystalline domains. The supercrystalline structure of double-domed cylinders adopt both 1D and 2D supercrystalline orders along the directions parallel and perpendicular to the long axis of constituent nanorods, respectively. These findings demonstrate that nano-objects could be self-assembled into anisotropic shapes, and multiple anisotropic nano-objects could be self-assembled into well-defined mesoscopic and macroscopic complex architectures. Colloidal nanoparticle superlattice (NPSL) metamaterials hold the promise of novel and tunable collective properties due to the periodic NP arrangement.

In situ real-time SAXS technique [83] was also used to monitor the growth of 3D colloidal NPSLs. A binary solution of 5.9 nm and 12.4 nm iron-oxide NPs was drop-casted on a Kapton substrate to achieve successful SL formation. After solvent evaporation, the AlB_2 stacking was expected to be formed in the NPSLs. The growth of the NPSLs during the drying processes was real-time monitored through the evolution of the radially-integrated structure-factor peaks in their SAXS patterns. The results demonstrate that when two different sized NPs were used to form the binary NPSLs (BNSLs), these excess NPs relative to the final BNSL stoichiometry were expelled during the BNSL formation. And these expelled NPs formed single NPSLs (SNSLs). The long-range ordering of these SNSLs and the BNSLs can occur faster than the NP expulsion. The bonding force of the NPSLs can be attributed to the van der Waals interactions between magnetite cores. This study is very helpful to design rationally multicomponent NPSLs by controlling NP self-assembly.

Li *et al.* studied [85] a biomolecular assembly of thermoresponsive superlattices of the tobacco mosaic virus (TMV) with large tunable interparticle distances. The rodlike TMV was chosen as a model 1D building block and sodium carboxymethyl cellulose (CMC), a representative rigid polysaccharide, was used as depletant. Without CMC, the TMV solution was clear and its SAXS profile showed no structure factor peaks. With CMC at high concentration, the TMV solution became opaque and its SAXS profile showed seven sharp diffraction peaks perfectly corresponding to a 2D hexagonal lattice of TMV bundles. In this study, SAXS was mainly used to detect the superlattice and the change of interparticle distances with the change of condition. The lattice parameter, d, of the superlattice of rodlike particles can be tuned from approximately 1 to 5 times the diameter, D, of the rodlike particles. The results demonstrate that the interparticle distance can be controlled reversibly in response to external stimuli, such as temperature, when rigid gel-forming polymers are used.

The concept "nanoparticle superlattice engineering" was proposed by Macfarlane *et al.* [86] A current limitation in nanoparticle superlattice engineering is that the identities of the particles being assembled often determine the structures that can be synthesized. Specific crystallographic symmetries or lattice parameters can only be achieved using specific nanoparticles as building blocks (and vice versa). Referring to the limitation in nanoparticle superlattice engineering, Macfarlane *et al.* gave six design rules for the deliberately preparation of nine distinct colloidal crystal structures, and outlined a strategy to independently adjust each of the relevant crystallographic parameters, including particle size, periodicity, and interparticle distance. SAXS technique was used to characterize all the lattices for *in situ* analysis of highly solvated materials. Jones *et al.* [87] also studied the superlattices related with DNA. A way to bypass the limitation of classical chemical purification with DNA-programmable nanoparticle recrystallization was found. The DNA-programmable nanoparticles crystallization was capable of maximizing the thermodynamic selectivity for one particle shape to form an ordered phase in preference to another. In this investigation, SAXS technique was used to detect the formation and types of superlattices of nanoprisms or nanospheres at three representative temperatures during the cooling process.

Size and Distribution

For a common preparation of NPs, the obtained particle sizes are often polydisperse. To estimate the polydispersity of particle sizes, it is necessary to measure the particle size and distribution. For *in situ* SAXS measurements of nanomaterials changed with applied field, the main research purpose is to trace the change of particle size and distribution. Wang *et al.* [88] studied the size changes of platinum nanoparticles with temperature by using *in situ* SAXS technique. The particle size distributions were obtained from the *in situ* SAXS patterns by using the tangent by tangent (TBT) method as shown in Fig. 7.13(a). The high-temperature XRD patterns of the same platinum nanoparticles were also collected at the same temperatures as in the SAXS samples. Scherrer formula was used to evaluate the crystallite sizes in the samples from the full width at half maximum (FWHM) of the Pt(111) reflections. By comparing the average sizes of Pt particles and Pt crystallites obtained, respectively, from the SAXS and XRD measurements, it was found that the average particle sizes were always larger than the average crystallite sizes as shown in Fig. 7.13(b). These results suggest that Pt NPs growth experienced an initial Oswald

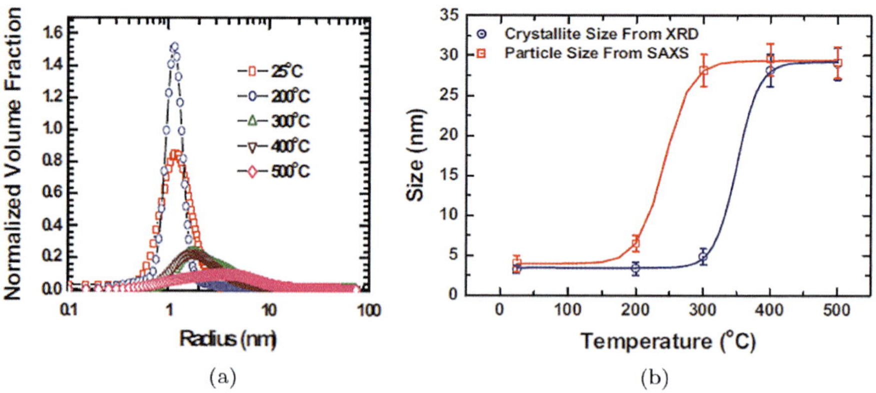

Fig. 7.13. (a) Pt nanoparticle size distribution, showing an increase of the average particle size with temperatures. (b) Average size change of Pt nanoparticles and Pt crystallites with temperature. Reproduced from [88] with permission from Springer.

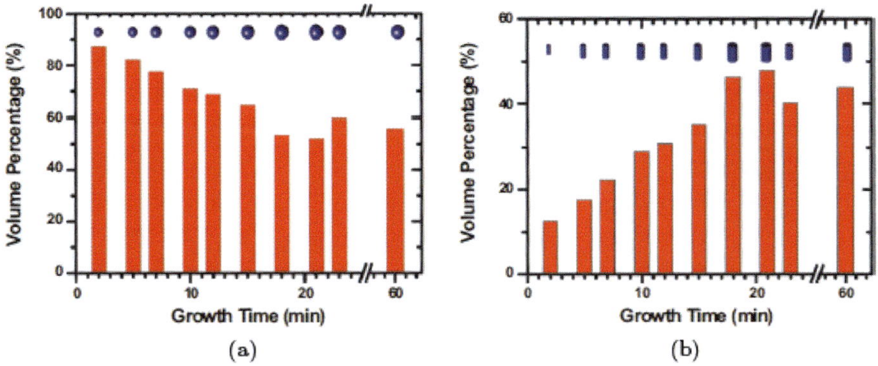

Fig. 7.14. Evolution of average size and volume percentage of the gold nanospheres (left) and nanorods (right) in the particle suspension with growth time. Reproduced from [89] with permission from Springer.

ripening, a middle aggregation growth, and a final evaporation and agglomeration growth stages during the heating process.

Real-time SAXS and ultraviolet-visible (UV) spectra were also used to study size and shape evolution [89] of gold nanoparticles in aqueous solution where gold nanospheres and nanorods coexisted. By combining SAXS measurement and UV-vis spectra, the contributions to SAXS from gold nanorods and nanospheres were successfully separated. The size and shape evolution of both gold nanoparticles as well as their volume fractions were obtained as shown in Fig. 7.14. Although all the nanoparticle sizes increased with reaction time, the volume fractions of nanospheres and nanorods presented different change trend. The results demonstrate that a mutual competitive growth occurred between the gold nanorods and nanospheres within the first 18 min, and then the growth was almost stopped. The aspect ratio of gold nanorods in the particle suspension was found to follow an exponential decay change after the initial 5 min growth.

7.3.4.1. *Nucleation and Growth*

Controllable preparation and functional design of materials depend highly on the knowledge of nucleation and growth of materials. Combining high-flux SR source and quick-response detector, SAXS has been developed to be a very powerful technique to monitor the nucleation and growth process of materials. Utilizing t-SAXS technique, Plech *et al.* [90] researched the X-ray induced synthesis process of gold nanoparticles from a mixed solution of gold hydrochlorate and tris-sodium citrate. The reactant was filled in a capillary, and the images were taken at a fixed position on the capillary in intervals between 1 and 60 s to cover a reaction time up to 2 h. The probed volume was also co-irradiated by UV light. However, the SAXS data both with and without UV light are almost the same, illustrating that the reducing power of the X-ray beam itself by far exceeds the UV light. The SAXS results demonstrate that smaller and larger particles coexist in the solution during growth. The electrostatic stabilization defines the final particle sizes and, probably, shapes.

Abécassis *et al.* [91] investigated the effect of two kinds of capping agents (alkanoic acid and alkylamine) on the growth of Au nanoparticles in organic solvent using a stopped-flow technique. Time-resolved SAXS/WAXS data with time resolution of 200 ms and the UV-visible spectra were collected as showed in Fig 7.15. The size distribution and particle numbers were obtained by best fitting the SAXS data using a Levenberg-Marquardt algorithm. For the acid case, the total number of particles had a rapid increase, but the average radius remained around 1 nm in the first second. After 1.5 s, the particle number remained almost constant while the radius increased by consumption of the remaining atoms in solution. For the amine case, a burst of nucleation with nearly no growth regime was observed. The nucleation rate was evaluated by dividing the number density of particles at the end of the nucleation period by the time required to generate these stable nuclei. The final particle radius was 3.5 nm for the acid ligand or 1.4 nm for the amine ligand. The ligands determine final size of gold nanoparticles by controlling their nucleation rate.

Polte *et al.* [92–94] also studied the nucleation and growth process of gold nanoparticles with t-SAXS experiment. The early reduction information [92] of the tetrachloroauric acid by trisodium citrate was obtained by coupled *in situ* SAXS and X-ray absorption near-edge spectroscopy (XANES) measurements. The time resolution

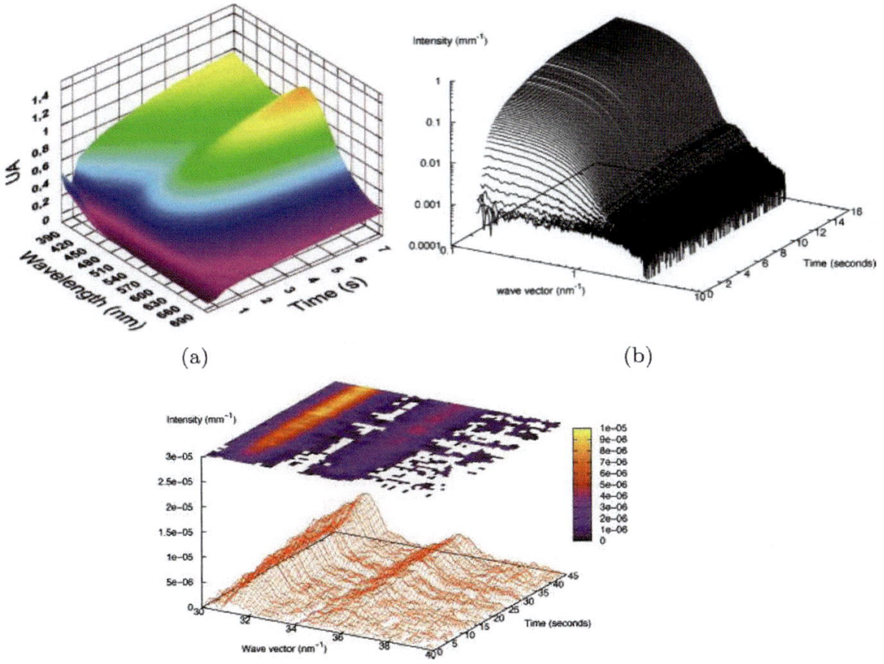

Fig. 7.15. Raw experiment patterns of the in situ UV-visible spectroscopy and the time-resolved SAXS/WAXS data. Reproduced from [91] with permission from American Chemical Society.

of the measured combined SAXS/XANES data amounts to ca. three minutes. A four-step formation mechanism of the gold nanoparticles was proposed, i.e., the fast initial nucleation, the nuclei coalescence, the slow particle growth, and the fast reduction of the remaining precursor species. A schematic illustration for the deduced process is showed in Fig. 7.16. They also developed a free-jet setup [94] used for SAXS experiments to detect the nucleation and growth of colloidal nanoparticles. This technique offers a unique possibility to follow the complete nanoparticle growth process *in situ* right from the beginning for hours Its time resolution depends only on the photon flux and the acquisition time of the detector. Moreover, it also prevents the fresh sample from the contamination of the aged sample absorbed on the wall of sample cell. This free-jet setup was also used to provide

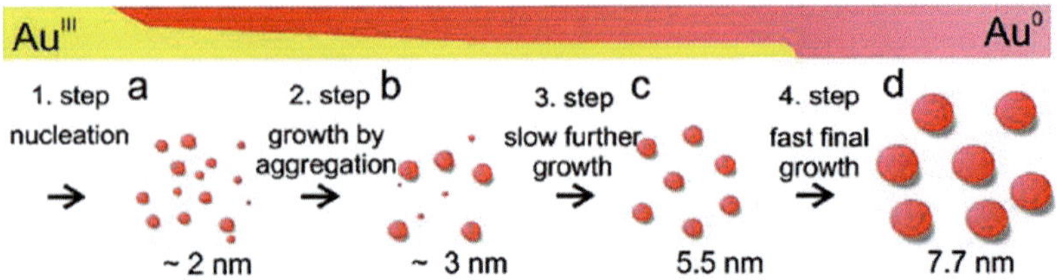

Fig. 7.16. Schematic map of the deduced process of gold nanoparticles formation. Reproduced from [92] with permission from American Chemical Society.

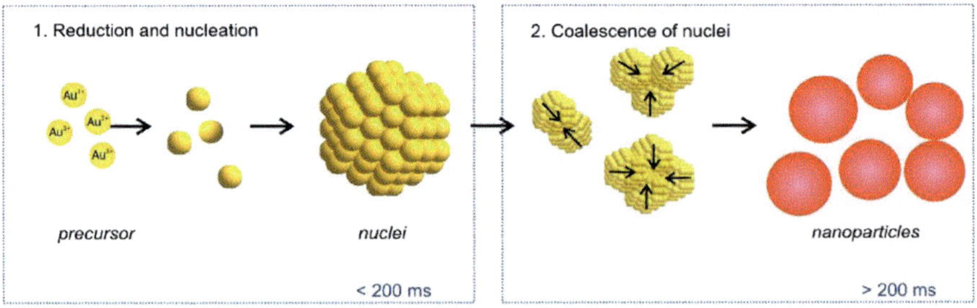

(a) (b)

Fig. 7.17. Schematic illustration for the deduced process of gold nanoparticle formation. Reproduced from [93] with permission from American Chemical Society.

time-resolved quantitative SAXS data for the synthesis reactions of gold nanoparticles via the reduction of aqueous $HAuCl_4$ by trisodium citrate. The obtained SAXS results refined the previous four-step model. Previously not available information about the initial phase of the synthesis could be gained, which provides new insights into the nucleation process.

To get a higher time resolution, a continuous-flow system was used to characterize the reduction [93] of tetrachloroauric acid by sodium borohydride in a mixed solution. The corresponding addressable reaction times ranged from 100 ms to about 136 s. Combining the SAXS results with XANES, SEM, and UVvis spectroscopy measurements, a two-step (i.e. nucleation and growth) formation mechanism of gold nanoparticle was deduced as shown in Fig. 7.17. In the nucleation stage, an initial rapid and complete conversion of the gold precursor into Au(0) in the form of gold nuclei was finished within less than 200 ms. In the following growth stage, particle growth proceeded exclusively via coalescence of the nuclei into bigger particles.

7.3.5. *Natural biomaterials*

Natural biomaterial is a mine of materials. Because of its excellent performances and the applications of bionics, natural biomaterials have attracted more and more attentions. The main research interests for natural biomaterials focus on the hierarchical structures, the formation process, and the relationship between structure and performance. From these researches, people expect to extract the structural models and synthesize novel biomimetic or biomedical materials with excellent preferences. Hierarchical structure is almost a character of natural biomaterials, in which abundant nanoscale structures are especially suitable for SAXS study. Here, several examples of SAXS studies on natural biomaterials are given.

7.3.5.1. *Teeth and jaws*

Biomineralization is quite a universal phenomenon in nature, which is frequently found in the claws, shell and teeth of invertebrates and in the bones of vertebrates. Biominerals are widely exploited to harden or stiffen tissues in living organisms. The most common biominerals are Ca-, Si-, and Fe-based ones. Undoubtedly, teeth and jaws are most biomineralized in both invertebrates and vertebrates. An unusual Cu-based biomineral was found by Lichtenegger *et al.* [95] in the jaws of the marine bloodworm Glycera dibranchiate. A combination of position-resolved XRD, SAXS, electron microscopy, and nanoindentation techniques was used to identify the biomineral, structural function, and mechanical properties of the jaw. The XRD of Glycera jaws revealed strong Bragg reflections in the tip of the jaw and the diffraction peaks were agreed well with the calculated pattern of atacamite [$Cu_2(OH)_3Cl$]. TEM results showed that atacamite is organized in fibers in the jaw tip and these mineralized fibers are arranged in layers under the very surface and not in the inner parts. Energy-dispersive X-ray analysis in a single fiber and between fibers showed that most of the copper is concentrated in the fibers. SAXS studies showed that the mean fiber orientation throughout the tip is roughly parallel to its outer shape. Nanoindentation experiments and microprobe element analysis were carried out on the same sample, and it was found that the Glycera jaw has excellent resistance to wear and abrasion, significantly exceeding that of vertebrate dentin and approaching that of tooth enamel Cricket is a truculent insect with its stiff and sharp teeth as fighting weapon. The hierarchical structures and the possible biomineralization [96] in cricket teeth have been also studied by using SAXS and WAXS techniques. The element distribution was revealed by X-ray fluorescence (XRF) analysis. SEM was used to observe the nanoscale morphograph. WAXS was used to probe the crystalline structures of possible biominerals, and SAXS was used to evaluate the size distribution of scatterers in cricket teeth. It was found that the surface was different from the interior of the tooth, having a different crystalline structure. WAXS patterns confirm that the most appropriate candidates of the crystalline phases at the surface and in the interior are, respectively, $ZnFe_2(AsO_4)_2(OH)_2(H_2O)_4$ and $ZnCl_2$. The tooth cusp was found to consist of microfibrils with their axial direction pointing at the top of tooth cusp. The size distribution of these microfibrils was obtained from the SAXS data by TBT analysis [88]. The microfibrils were randomly aggregated into intermediate filaments, forming a hierarchical structure in the tooth cusp. It was the biomineralization of the teeth to $ZnCl_2$ with fibrous and hierarchical structures at the tooth cusp that strengthen the hardness and sharpness of the cricket teeth.

7.3.5.2. *Human hair*

Mercury is one of the most hazardous environmental pollutants. Mercury pollution can cause brain, kidney, and neurological damage to human body. Human hairs [97–110] have been extensively studied for the structures and the correlation with some diseases. A structural change [111] of human hair induced by mercury exposure has been reported. SAXS technique was used to compare the structural difference between the normal and the Hg-exposure hairs.

Inductively coupled plasma mass spectroscopy (ICP-MS) was used to detect the element contents. It was found that the 4.5 nm diffraction ring disappeared in the Hg-exposure hairs, which was identified to originate from the ordered fibrillar structure of glycosaminoglycan (GAG) in proteoglycan (PG) composed the extracellular matrix (ECM). Besides, the structural parameters and the arrangement of the intermediate filaments (IFs) in hairs were also obtained. This study confirms that the increase of Ca content makes the regular oriented fibrillar structure of GAG transform to a random oriented one, and broadens the extent of the azimuthal angle of the reflection with 4.5 nm spacing from a diffraction arc to a diffraction ring. However, the Hg is bonded to the cysteines at different sites of the core protein to form S-Hg-S bonding, which makes the core proteins coiled and destroys the ordered arrangements of fibrillar GAG attached to core protein. A sketch map of the structure model of IFs (A), ECM (B), and mercury effect on PG (C) in human hair is shown in Fig. 7.18. Consequently, an overdose Hg will result in the disappearance of the 4.5 nm reflection, which can be used as bio-indicator of Hg-contamination to human body.

7.3.5.3. *Wood materials*

Wood is a highly sophisticated and hierarchical material. Its swelling behavior and the hierarchical structures are common interesting

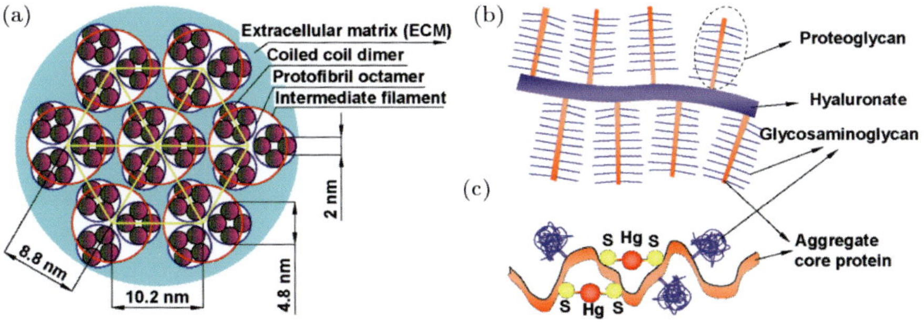

Fig. 7.18. (a) Sketch map of the structure model of IFs, (b) ECM, and (c) mercury effect on PG in human hair. Reproduced from [111] with permission from American Chemical Society.

issues. The nanostructural change [112] with absorbed water content was compared for the red pine, American pine, and white ash specimens by using SAXS technique. The normalized volume fractions of scatterers in the three wood materials were calculated from SAXS data by TBT analysis [88]. The results show that there are a sharp component and a wide component in the nanostructures of all the three wood specimens. The main contribution to the sharp component comes from cellulose microfibrils (CMFs) with constant size ($\sim$3 nm) during the whole water-absorbing process. The wide component includes the voids, microcracks, and CMF aggregates in the woods. However, the average size changes of the wide component are different between the softwood (Red pine and American pine) and hardwood (White ash). The average sizes of the wide component increase with the increasing water content for the softwood specimens, but decreases for the hardwood specimen. This result demonstrates different swelling behaviors between hardwood and softwood. As a consequence, it can be concluded that the absorbed water exists mainly in the open lumens in softwood, and in the amorphous hemicellulose and lignin surrounding CMFs in hardwood. A sketch map to describe the absorbed water in softwood and hardwood is shown in Fig. 7.19. A fractal analysis for the SAXS data was also performed. The results demonstrate that mass and surface fractals coexist in all the three samples. The mass fractal dimensions reveal that the structural looseness of the scatterers increases with the increase of the volume percentage of moisture in all the three wood species. But the surface fractal dimensions reveal that the surface or interface of the scatterers is coarser in hardwood (white ash) than in softwood (red pine and American pine). Such a coarse surface of scatterers compensates partially the swelling caused by the absorbed water.

7.3.5.4. *Bombyx mori silk*

Some natural silks, for example, spider silk and Bombyx mori silk, attract wide attention due to their excellent mechanical properties, bio-compatibility, and thermal stability. To make natural silk fiber artificially by the technique of genetic engineering so

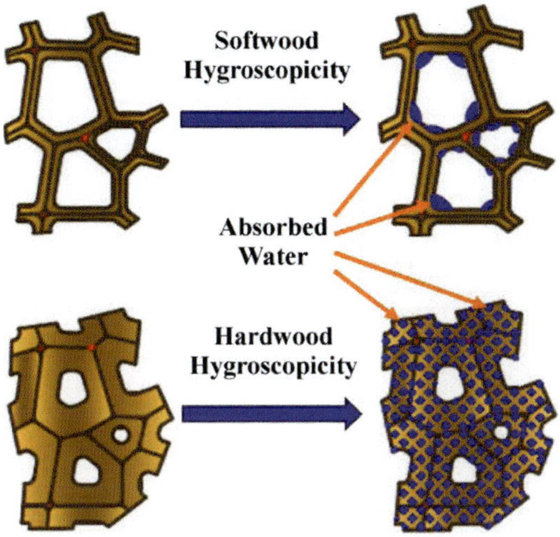

Fig. 7.19. Sketch map of softwood and hardwood hygroscopicity.

as to explore novel biomaterial with unique mechanical properties and good biocompatibility, people expect to understand well the relationships between the spinning conditions and the molecular structure, mechanical properties of the natural silks. By analyzing the biological spinning mechanisms, the molecular conformation change in the spinning process and the effects of spinning conditions on the molecular structure are often the main research contents. SAXS and WAXS techniques can be used to analyze the nanoscale structures in natural silks. A microstructural change for degummed Bombyx mori silk was reported [113] by using in-situ stretching WAXS technique. Bombyx mori silk has been used as textile material since time immemorial in China. In modern complex materials, Bombyx mori silk is also used as scaffold in cell and tissue cultures *in vivo* or *in vitro*, drug controlled release material, human wound surgical sutures and artificial blood vessels. In this study, WAXS patterns confirmed that the crystalline and amorphous regions coexisted in the silk fibers. The crystalline phase was determined to have an orthorhombic unit cell with lattice parameters: $a = 9.10$ Å, $b = 9.71$ Å and $c = 6.80$ Å. Its crystallite size, crystallite orientation and crystallinity as well as their changes with stress-strain can be estimated based on the WAXS data. With the increase of the stretching strain, the crystallite size (~2.5 nm) was unchanged, but the unit-cell orientation degree increased first then kept unchanged. Meanwhile, the random coil molecular chains in the amorphous region were aligned through the unfolding and rotating of the molecule chains, causing a linear increase of the crystallinity and the reduction of the amorphous region. The microstructure of the degummed silk fibers can be well described by an oriented β-sheet structure with banded feature.

7.3.6. *ASAXS application*

The electron density difference between scatterers and the surrounding matrix determines the SAXS contrast. In order to make the scatterers easy-detectable or monodisperse, the scattering contrast could be varied by manipulating the electron density of the matrix. For example, boron nitride (BN) powder is often used as the matrix to support other scatterers in a solid system, or it is used as a thinner for a condensed system. Glycerol or sucrose solvents can be also used for aqueous dispersion in a liquid system. This method to vary the SAXS contrast is only amenable for very limited systems. For nanocomposites consisting of multiple components, normal SAXS techniques hordly distinguish the contributions of different components if these components have similar particle sizes, shapes and distribution. However, when one or more of the multiple components are composed by heavier elements with accessible absorption edge by incident X-ray beam, ASAXS technique can be used to strengthen the contribution of one or more components to SAXS so that the interested component can be separated from the composites. A typical ASAXS method and theory [114] have been reported for the structural characterization of bulk block copolymer/nanoparticle composites. In this study, ASAXS was used to investigate the morphology of the polymer phases and the nature of dispersion of PSSH (thiol-terminated PS) tethered Au nanoparticles in bulk poly(styrene-b-2-vinylpyridine) (PS-PVP) diblock copolymer composites. Here, Au was used as the anomalous scattering atom. In the vicinity of the L_3 absorption edge of Au element, the atomic scattering factor strongly depends on the

X-ray energy (E) as shown in Eq. (7.12).

$$f_{Au}(E) = f_{0,Au} + f'_{Au}(E) + if''_{Au}(E) = f_{1,Au}(E) + if''_{Au}(E), \tag{7.12}$$

where f_0 is the Thomson scattering term, f' and f'' are the energy-related real and imaginary parts of anomalous dispersion. In the sample, the electron density of particles or matrix at an X-ray energy E can be expressed as Eq. (7.13).

$$\rho_e(E) = \sum_i^N n_i f_i(E), \tag{7.13}$$

where N is the types of atoms in the particles or in the matrix, n_i is number density of the ith atom in such a domain. The SAXS signal originates the square of the electron density difference between the particles and the matrix. For this studied PVP-PS/Au system, two kinds of electron density difference have to be considered, that is the electron density difference between Au and the surrounding PS $\rho_\mathrm{g}(\mathrm{E})$, and that between PS and PVP in the block copolymer ρ_b. In order to decompose partial scattering functions from the ASAXS data, they developed a single-subtraction method and a double-subtraction method to analyze the ASAXS data.

For a correlated system, the SAXS data should be collected at three different X-ray energies (E_1, E_i, and E_2). Energy E_1 is far away from the absorption edge and energy E_2 is near the edge in order have the maximum contrast. If the nonresonant component is needed to be extracted, the intermediate energy E_i is chosen to satisfy Eq. (7.14),

$$\frac{1}{\rho_g(E_1)} - \frac{1}{\rho_g(E_i)} = \frac{1}{\rho_g(E_i)} - \frac{1}{\rho_g(E_2)}. \tag{7.14}$$

Also a double-subtraction method with normalized contrasts is performed as in Eq. (7.15).

$$\begin{aligned} I_{ds,1}(q) &\equiv \left[\frac{I(q,E_1)}{\rho_g^2(E_1)} - \frac{I(q,E_i)}{\rho_g^2(E_i)}\right] - \left[\frac{I(q,E_i)}{\rho_g^2(E_i)} - \frac{I(q,E_2)}{\rho_g^2(E_2)}\right] \\ &= \rho_\alpha S_b(q) F_b^2(q) \end{aligned} \tag{7.15}$$

where $q = 4\pi \sin\theta/\lambda$, is the scattering vector $S_{\mathrm{b}}(q)$ is the structural factor of the block copolymer, $\rho_\alpha = \rho_b^2(E_1)/\rho_g^2(E_1) + \rho_b^2(E_2)/\rho_g^2(E_2) - 2\rho_b^2(E_i)/\rho_g^2(E_i)$ and $F_b(q)$ is the scattering amplitude of the copolymer. If the resonant component is needed to be extracted, the intermediate energy E_i is chosen to satisfies Eq. (7.16),

$$\rho_g(E_1) - \rho_g(E_i) = \rho_g(E_i) - \rho_g(E_2). \tag{7.16}$$

Also a double-subtraction method without normalized contrasts is performed as in Eq. (7.17).

$$I_{ds,2}(q) \equiv [I(q,E_1) - I(q,E_i)] - [I(q,E_i) - I(q,E_2)] = 2S_b(q)\rho_s^2 F_g^2(q), \tag{7.17}$$

where $\rho_s = \rho_g(E_1) - \rho_g(E_i) = \rho_g(E_i) - \rho_g(E_2)$, and $F_g(q)$ is the scattering amplitude of the dispersed Au nanoparticles.

For an uncorrelated or macrophase-separated system, the SAXS data measured at two different X-ray energies (E_1 and E_2) is enough. Energy E_1 is far away from the absorption edge and energy E_2 is near the edge in order have the maximum contrast. A single-subtraction method with normalized contrasts can be used to extract the non-resonant component as in Eq. (7.18).

$$I_{s,1}(q) = \frac{(\rho_g^2(E_2) - \rho_g^2(E_1))\rho_{b,0}^2}{\rho_g^2(E_1)\rho_g^2(E_2)} S_b(q) F_b^2(q), \tag{7.18}$$

where $\rho_{b,0} = \rho_{e,\widehat{PVP}} - \rho_{e,\widehat{PS}}$. A single-subtraction method without normalized contrasts can be used to extract the resonant component as in Eq. (7.19).

$$I_{s,2}(q) = (\rho_g^2(E_1) - \rho_g^2(E_2)) F_g^2(q). \tag{7.19}$$

In general, the double-subtraction method is not necessary for a completely phase-separated system. If the intensities from the single- and double-subtraction methods are different, it implies that the studied system is more complicated than a completely phase-separated system, perhaps including the correlated and uncorrelated structural domains in the system.

On the basis of the above ASAXS theory, Lee *et al.* [114] studied the PVP-PS/Au systems with Au concentrations at 8.8%

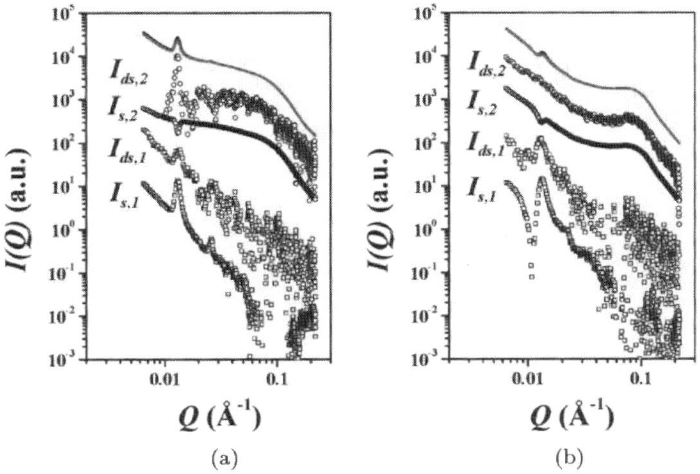

Fig. 7.20. Scattering intensities normalized by the m values ($= \rho_g^2(E)/\rho_g^2(E_L)$) at E = 11.919 (red) and 11.758 keV (black) and single and double subtracted intensities (open squares) for the (a) 8.8% and (b) 27.0% Au composites. Reproduced from [114] with permission from American Chemical Society.

and 27.0%. SAXS data were collected at the three X-ray energies: $E_1 = 11.758, E_2 = 11.917$ and $E_\mathrm{i} = 11.893$ keV. The normalized SAXS intensities, and the single-subtraction ($I_{\mathrm{s},1}, I_{\mathrm{s},2}$) and double-subtraction ($I_{\mathrm{ds},1}, I_{\mathrm{ds},2}$) intensities are shown in Fig. 7.20. It can be that $I_{\mathrm{ds},1}(q)$ and $I_{\mathrm{ds},2}(q)$ are not proportional to $I_{\mathrm{s},1}(\mathrm{q})$ and $I_{\mathrm{s},2}(q)$ respectively, which indicate that both composites contain correlated domains. The structure factor of $S_\mathrm{b}(q)$ is clearly visible in $I_{\mathrm{ds},2}(q)$ for the block copolymer with 8.8% of Au as shown in Fig. 7.20(a), Although the structure factor is less visible in the $I_{\mathrm{ds},2}(q)$ for the 27.0% Au composite, the presence of a dip in the $I_{\mathrm{s},2}(\mathrm{q})$ in Fig. 7.20(b) is a clear indication of the effect of $S_\mathrm{b}(q)$ convolution.

By fitting the $I_{\mathrm{ds},2}$ data for both samples using a combination of a power law and the polydisperse spherical particles with a hard-sphere structure factor, the average radius of Au particles is about 1.0 nm. The average distance between Au particles is about 6.3 nm and the Au-particle volume fraction is about 11.8% in the PS domain for the 27.0% Au composite. The Au nanoparticles are well dispersed in the middle region of the lamellar PS domain in the 8.8%

Au composite. But the morphology of PS-PVP is transformed into hex structure and the aggregated Au nanoparticles are located at the inter-cylinder domains constituted by the PS in the 27.0% composite.

For highly negatively charged DNA or RNA, it is important to know the number and spatial distribution of small positively charged ions around them. This information can provide entropic contributions to the free energy of binding in the cell. Among all techniques that can characterize the ion atmosphere around nucleic acids, ASAXS is the only one that can provide information on the spatial distribution of ions. Pabit *et al.* [115] used ASAXS highlighting difference ion distributions around comparably sequenced double-stranded DNA and RNA helices. ASAXS was used to count the number of ions around nucleic acids while measuring the ion-nucleic acid spatial distribution [116]. SAXS profiles at different X-ray energies were normalized, and were calibrated by using water. The numbers of ions around the 25 bp DNA samples were obtained to be 34 ± 3 for monovalent ions and 19 ± 2 for divalent ions.

PEG-P[Asp(Bzl)] as block copolymer can form polymeric micelles where PEG, i.e. poly(ethylene glycol), is often used as the shell part of a drug delivery micelle. P[Asp(Bzl)], i.e. poly(partially benzyl-esterified aspartic acid), is hydrophobic in the micelle building block. Sanada *et al.* [117] investigated the distribution information of the hydrophobic compound tetrabromocathecol (TBC) drugs in the micelles of PEG-P[Asp(Bzl)] using ASAXS technique. Bromine atoms contained in TBC were used as the resonant atoms. SAXS data were collected at five different X-ray energies. In which, four X-ray energies (13.386, 13.436, 13.446, and 13.471 keV) were near the K-absorption edge of Bromine (13.486 keV). One X-ray energy (12.4 keV) was far away from the absorption edge and was noted as noted as E_0. When the X-ray energy is far away from the absorption edge of elements, the anomalous scattering from the resonant atoms is negligible. In the ASAXS data analysis, an assumption that the scattering objects are monodisperse and centro-symmetric was adopted. Therefore, the SAXS intensity can be expressed as the product of the form factor $P(q, E)$ and the structure factor $S(q)$.

The scattering intensity from the bromine atoms was extracted and the distribution of TBC drugs in the PEG-P[Asp(Bzl)] micelles was calculated. The result demonstrated that when the amount of TBC is less than 1 wt%, TBC was dispersed only in the hydrophobic core. With the amount of TBC increasing to 1.5–4.0 wt %, TBC started to spread out the core. When the TBC amount was greater than 4.0 wt%, it had been beyond the capacity of the polymeric micelle.

7.3.7. *GISAXS application*

Classical SAXS experiments align the samples in transmission geometry. However, this transmission geometry is unsuitable for thicker samples due to the drastic attenuation of the transmission X-ray intensity by absorption. Especially for a thin film sample deposited on a thick substrate, grazing incidence geometry has to be used for the collection of SAXS data. Levine *et al.* [118] named first this SAXS experiment with sample in grazing-incidence geometry as grazing incidence small angle X-ray scattering (GISAXS). Two-dimensional (2D) area detector is often used to collect the 2D-GISAXS pattern which combines the classical SAXS and 2D-diffraction with grazing-incidence geometry. A sketch map of GISAXS measurements is shown in Fig. 7.21.

GISAXS intensity is the product of form factor $F(q)$ and structural factor $S(q)$. Form factor $F(q)$ is the Fourier transform of the particle shape, and structural factor $S(q)$ is the Fourier transform of the particle-particle pair correlation function. GISAXS pattern is the reciprocal image of the real-space scatterers. Therefore, the average particle lateral size d is inversely proportional to the spread of the scattering parallel to sample surface. The average particle height h is inversely proportional to the spread of the scattering perpendicular to the sample surface, while the average particle spacing D is inversely proportional to the distance between the main spots along the parallel direction.

Distorted Wave Born Approximation (DWBA) was developed to treat the form factor in GISAXS for buried density contrast or surface roughness [119], for nanostructure [120], and for particle size

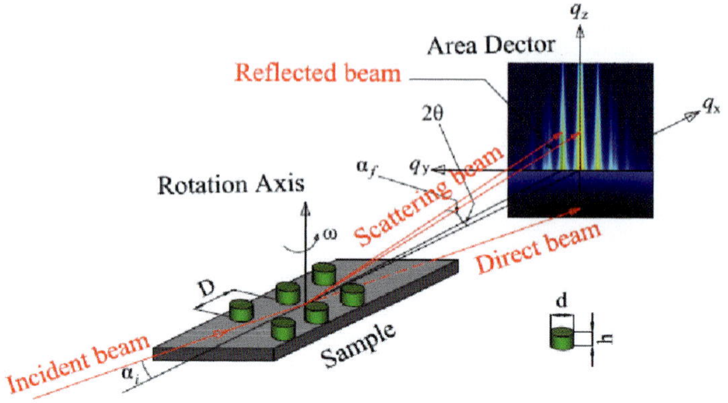

Fig. 7.21. A sketch map of GISAXS measurement. The angles $a_{\rm i}$ and $a_{\rm f}$, and 2θ are the grazing incidence angle, the out-of-plane and in-plane scattering angles. ω is the rotation angle of sample around its surface normal, which defines the orientation of the incident x-ray beam with respect to the in-plane crystallographic directions. The scatters on or in the substrate have a diameter (d) and a height (h) with a interdistance (D). A beam stop protects the detector from the direct and reflected beams (not shown in this figure).

and spacing [121]. A software "*IsGISAXS*" [75] dedicated to the simulation and analysis of GISAXS from nanostructures has been developed. The scattering cross-section is decomposed in terms of interference function and particle form factor. Various kinds of simple geometrical shapes are also available with a full account of size and shape distributions in the decoupling approximation (DA), in the local monodisperse approximation (LMA) and in the size-spacing correlation approximation (SSCA). Based on these progress of experiment, theory, and data analysis, GISAXS has become a sensitive technique to the surface morphology of the scattering entities either deposited on a flat substrate or buried in a thin layer [122, 123] It can give the in-plane and out-of-plane structural information of the thin films and is often used to probe the morphology of quantum dots or nanoparticles on or in the sample surface.

GISAXS technique can be used to monitor in real-time the growth of nanoparticles. A typical example [124] of GISAXS application is the real-time monitoring of the Volmer-Weber growth of Pd on MgO(001) surface and the self-organized growth of Co nanodots on the herring-bone reconstructed Au(111) surface. The GISAXS experiment was performed at the European Synchrotron Radiation Facility (ESRF) through a dedicated ultrahigh-vacuum GISAXS setup.

The *in situ* GISAXS patterns during the growth of Pd films from 0.1 to 2.8 nm deposited on MgO(001) at 650 K were collected. The shrinkage of the scattered intensity toward the origin of the reciprocal space with increasing film thickness revealed continuous island growth in all real-space directions. Except the specular reflectivity close to $q_y = 0$ was not calculated, all the features in the experimental GISAXS patterns were highly reproduced in the simulated GISAXS ones. A rod of scattering at 54.7° with respect to the surface normal appears when the incident beam was along the $\langle 110 \rangle$ direction, but disappears when the beam was along the $\langle 100 \rangle$ direction. This difference of GISAXS patterns between the different incident X-ray beam directions ($\langle 110 \rangle$ and $\langle 100 \rangle$) demonstrate the existence of $\{111\}$ facets on the Pd islands. The appearance of the scattering peaks perpendicular to the surface also indicated flat Pd(001) terraces. These features pointed to a truncated, octahedron-like, average particle shape with $\{100\}$ and $\{111\}$ facets. The best fit with a truncated octahedron for the 1.5 nm thick Pd deposit on MgO(001) at 650 K yielded that $d = 12.6 \pm 0.4$ nm with a log-normal distribution of 6 nm FWHM, $D = 21 \pm 0.2$ nm, $h_{(001)} = 5.8 \pm 0.1$ nm, and a total height of $h = 7.9 \pm 0.1$ nm with a distribution of 0.8 ± 0.4 nm FWHM, as well as a constant ratio $h_{(001)}/d = 0.46$. The obtained structural factor $S(q)$ indicated a disordered arrangement of the particles.

Bibliography

[1] D. I. Svergun and M. H. J. Koch, Small-angle scattering studies of biological macromolecules in solution, *Rep. Prog. Phys.* **66**, 1735–1782, (2003).

[2] G. Fritz and O. Glatter, Structure and interaction in dense colloidal systems: Evaluation of scattering data by the generalized indirect Fourier transformation method, *J. Phys.: Condens. Matter* **18**, S2403–S2419, (2006).

[3] G. Fritz and A. Bergmann, Interpretation of small-angle scattering data of inhomogeous ellipsoids, *J. Appl. Cryst.* **37**, 815–822, (2004).

[4] L. K. Shrestha, O. Glatter and K. Aramaki, Structure of nonionic surfactant (Glycerol r-Monomyristate) micelles in organic solvents: A SAXS study, *J. Phys. Chem. B* **113**, 6290–6298, (2009).

[5] A. E. Brown, R. I. Litvinov, D. E. Discher, P. K. Purohit and J. W. Weisel, Multiscale mechanics of fibrin polymer: Gel stretching with protein unfolding and loss of water, *Science* **325**, 741–744, (2009).

[6] D. E. Spratt, R. J. Martinez-Torres, Y. J. Noh, P. Mercier, N. Manczyk, K. R. Barber, J. D. Aguirre, L. Burchell, A. Purkiss, H. Walden and G. S. Shaw, A molecular explanation for the recessive nature of parkin-linked Parkinson's disease, *Nat. Commun.* **4**, 1983, (2013).

[7] D. Kilburn, J. H. Roh, R. Behrouzi, R. M. Briber and S. A. Woodson, Crowders perturb the entropy of RNA energy landscapes to favour folding, *J. Am. Chem. Soc.* **135**, 10055–10063, (2013).

[8] J. H. Roh, M. Tyagi, R. M. Briber, S. A. Woodson and A. P. Sokolov, The dynamics of unfolded versus folded tRNA: The role of electrostatic interactions, *J. Am. Chem. Soc.* **133**, 16406–16409, (2011).

[9] J. H. Roh, L. Guo, J. D. Kilburn, R. M. Briber, T. Irving and S. A. Woodson, Multistage collapse of a bacterial ribozyme observed by time-resolved small-angle X-ray scattering, *J. Am. Chem. Soc.* **132**, 10148–10154, (2010).

[10] F. Gabel, M. R. Jensen, G. Zaccaï and M. Blackledge, Quantitative model-free analysis of urea binding to unfolded ubiquitin using a combination of small angle X-ray and neutron scattering, *J. Am. Chem. Soc.* **131**, 8769–8771, (2009).

[11] Y. J. Shiu, U. S. Jeng, Y. S. Huang, Y. H. Lai, H. F. Lu, C. T. Liang, I. J. Hsu, C. H. Su, C. Su, I. Chao, A. C. Su and S. H. Lin, Global and local structural changes of cytochrome c and lysozyme characterized by a multigroup unfolding process, *Biophys. J.* **94**, 4828–36, (2008).

[12] E. Akoury, M. Pickhardt, M. Gajda, J. Biernat, E. Mandelkow and M. Zweckstetter, Mechanistic basis of phenothiazine-driven inhibition of tau aggregation, *Angew. Chem. Int. Ed.* **52**, 3511–3515, (2013).

[13] E. Akoury, M. Gajda, M. Pickhardt, J. Biernat, P. Soraya, C. Griesinger, E. Mandelkow and M. Zweckstetter, Inhibition of tau filament formation by conformational modulation, *J. Am. Chem. Soc.* **135**, 2853–2862, (2013).

[14] J. F. Trempe, V. Sauve, K. Grenier, M. Seirafi, M. Y. Tang, M. Menade, S. Al-Abdul-Wahid, J. Krett, K. Wong, G. Kozlov, B. Nagar, E. A. Fon and K. Gehring, Structure of parkin reveals mechanisms for ubiquitin ligase activation, *Science* **340**, 1451–1455, (2013).

[15] W. T. Heller, D. Vigil, S. Brown, D. K. Blumenthal, S. S. Taylor, J. C. Trewhella, C subunits binding to the protein kinase A RIα dimer induce a large conformational change, *J. Biol. Chem.* **279**, 19084–19090, (2004).

[16] D. Vigil, D. K. Blumenthal, S. S. Taylor, J. Trewhella, Solution scattering reveals large differences in the global structures of type II protein kinase A isoforms, *J. Mol. Biol.* **357**, 880–889, (2006).

[17] E. C. Dueber, A. J. Schoeffler, A. Lingel, J. M. Elliott, A. V. Fedorova, A. M. Giannetti, K. Zobel, B. Maurer, E. Varfolomeev, P. Wu, H. J. Wallweber, S. G. Hymowitz, K. Deshayes, D. Vucic and W. J. Fairbrother, Antagonists induce a conformational change in cIAP1 that promotes autoubiquitination, *Science* **334**, 376–380, (2011).

[18] N. Nishimura, K. Hitomi, A. S. Arvai, R. P. Rambo, C. Hitomi, S. R. Cutler, J. I. Schroeder and E. D. Getzoff, Structural mechanism of abscisic acid binding and signaling by dimeric PYR1, *Science* **326**, 1373–1379, (2009).

[19] S. R. Akabayov, B. Akabayov, C. C. Richardson and G. Wagner, Molecular crowding enhanced ATPase activity of the RNA helicase eIF4A correlates with compaction of its quaternary structure and association with eIF4G, *J. Am. Chem. Soc.* **135**, 10040–10047, (2013).

[20] N. Ando, Y. Kung, M. Can, G. Bender, S. W. Ragsdale and C. L. Drennan, Transient B_{12}-dependent methyltransferase complexes revealed by small-angle X-ray scattering, *J. Am. Chem. Soc.* **134**, 17945–17954, (2012).

[21] C. D. Schwieters, J.-Y. Suh, A. Grishaev, R. Ghirlando, Y. Takayama and G. M. Clore, Solution structure of the 128 kDa enzyme I dimer from escherichia coli and its 146 kDa complex with HPr using residual dipolar couplings and small- and wide-angle X-ray scattering, *J. Am. Chem. Soc.* **132**, 13026–13045, (2010).

[22] J. Wang, X. Zuo, P. Yu, I.-J. L. Byeon, J. Jung, X. Wang, M. Dyba, S. Seifert, C. D. Schwieters, J. Qin, A. M. Gronenborn and Y.-X. Wang, Determination of multicomponent protein structures in solution using global orientation and shape restraints, *J. Am. Chem. Soc.* **131**, 10507–10515, (2009).

[23] J. Blobel, P. Bernadó, D. I. Svergun, R. Tauler and M. Pons, Low-resolution structures of transient protein-protein complexes using small-angle X-ray scattering, *J. Am. Chem. Soc.* **131**, 4378–4386, (2009).

[24] X. Xing, Q. Liu, W. Wang, K. Zhang, T. Li, Q. Cai, G. Mo, W. Cheng, D. Wang, Y. Gong, Z. Chen, D. Qiu and Z. Wu, Shape evolution with temperature of a thermotolerant protein (*PeaT*1) in solution detected by small angle X-ray scattering, *Proteins* **81**, 53–62, (2013).

[25] P. Dedecker, F. C. De and J. Schryver, Hofkens, Fluorescent proteins: Shine on, you crazy diamond, *J. Am. Chem. Soc.* **135**, 2387–2402, (2013).

[26] D. M. Francis, B. Rozycki, A. Tortajada, G. Hummer, W. Peti and R. Page, Resting and active states of the ERK2:HePTP complex, *J. Am. Chem. Soc.* **133**, 17138–17141 (2011).

[27] J. M. Christie, A. S. Arvai, K. J. Baxter, M. Heilmann, A. J. Pratt, O'A. Hara, S. M. Kelly, M. Hothorn, B. O. Smith, K. Hitomi, G. I. Jenkins and E. D. Getzoff, Plant UVR8 photoreceptor senses UV-B by tryptophan-mediated disruption of cross-dimer salt bridges, *Science* **335**, 1492–1496, (2012).

[28] G. Tresset, Le C. Coeur, J. F. Bryche, M. Tatou, M. Zeghal, A. Charpilienne, D. Poncet, D. Constantin and S. Bressanelli, Norovirus capsid proteins self-assemble through biphasic kinetics via long-lived stave-like intermediates, *J. Am. Chem. Soc.* **135**, 15373–15381, (2013).
[29] S. Kler, R. Asor, C. Li, A. Ginsburg, D. Harries, A. Oppenheim, A. Zlotnick and U. Raviv, RNA encapsidation by SV40-derived nanoparticles follows a rapid two-state mechanism, *J. Am. Chem. Soc.* **134**, 8823–8830, (2012).
[30] P. V. Konarev, V. V. Volkov, A. V. Sokolova, M. H. J. Koch and D. I. Svergun, PRIMUS: A windows PC-based system for small-angle scattering data analysis, *J. Appl. Cryst.* **36**, 1277–1282, (2003).
[31] A. V. Semenyuk and D. I. Svergun, GNOM-a program package for small-angle scattering data processing, *J. Appl. Cryst.* **24**, 537–540, (1991).
[32] D. I. Svergun, Restoring low resolution structure of biological macromolecules from solution scattering using simulated annealing, *Biophys. J.* **76**, 2879–2886, (1999).
[33] D. I. Svergun and V. V. Volkov, Uniqueness of ab initio shape determination in small-angle scattering, *J. Appl. Cryst.* **36**, 860–864, (2003).
[34] D. I. Svergun, C. Barberato, M. H. J. Koch, L. Fetler and P. Vachette, Large differences are observed between the crystal and solution quaternary structures of allosteric aspartate transcarbamylase in the R state, *Proteins* **27**, 110–117, (1997).
[35] B. Rozycki, Y. C. Kim and G. Hummer, SAXS ensemble refinement of ESCRT-III CHMP3 conformational transitions, *Structure*, **19**, 109–116, (2011).
[36] C. D. Putnam, M. Hammel, G. L. Hura and J. A. Tainer, X-ray solution scattering (SAXS) combined with crystallography and computation: Defining accurate macromolecular structures, conformations and assemblies in solution, *Q. Rev. Biophys.* **40**, 191–285, (2007).
[37] D. Svergun, C. Barberato and M. H. J. Koch, CRYSOL-a program to evaluate X-ray solution scattering of biological macromolecules from atomic coordinates, *J. Appl. Cryst.* **28**, 768–773, (1995).
[38] A. Yaghmur, L. Paasonen, M. Yliperttula, A. Urtti and M. Rappolt, Structural elucidation of light activated vesicles, *J. Phys. Chem. Lett.* **1**, 962–966, (2010).
[39] R. Sun, J. Zhang and Y. Wang, Liquid crystal conformation of chemical substances and biological membrane interactions, *Sci. China (B)* **27**, 261–270, (1997).
[40] B. Angelov, A. Angelova, R. Mutafchieva, S. Lesieur, U. Vainio, V. M. Garamus, G. V. Jensen and J. S. Pedersen, SAXS investigation of a cubic to a sponge (L3) phase transition in self-assembled lipid nanocarriers, *Phys. Chem. Chem. Phys.* **13**, 3073–3081, (2011).
[41] G. Tresset, The multiple faces of self-assembled lipidic systems, *PMC Biophys.* **2**, 3, (2009).

[42] L. H. Yang, V. D. Gordon, A. Mishra, A. Som, K. R. Purdy, M. A. Davis, G. N. Tew and G. C. Wong, Synthetic antimicrobial oligomers induce a composition-dependent topological transition in membranes, *J. Am. Chem. Soc.* **129**, 12141–12147, (2007).
[43] D. C. Turner and S. M. Gruner, X-ray-diffraction reconstruction of the inverted hexagonal (Hii) phase in lipid water-systems, *Biochem.-US* **31**, 1340–1355, (1992).
[44] P. E. Harper, D. A. Mannock, R. N. A. H. Lewis, R. N. McElhaney and S. M. Grune, X-ray diffraction structures of some phosphatidylethanolamine lamellar and inverted hexagonal phases, *Biophys. J.* **81**, 2693–2706, (2001).
[45] O. Francescangeli, M. Pisani, V. Stanic, P. Bruni and T. M. Weiss, Evidence of an inverted hexagonal phase in self-assembled phospholipid-DNA-metal complexes, *Europhys. Lett.* **67**, 669–675, (2004).
[46] D. Wang and J. Hao, Self-assembly fibrillar network gels of simple surfactants in organic solvents, *Langmuir* **27**, 1713–1717, (2011).
[47] D. Wang, G. Wei, R. Dong and J. Hao, Multiresponsive viscoelastic vesicle gels of nonionic C(1)(2)EO(4) and anionic AzoNa, *Chemistry* **19**, 8253–8260, (2013).
[48] F. Ma, X. Chen, Y. Zhao, X. Wang, Q. Li, C. Lv and X. Yue, A nonaqueous lyotropic liquid crystal fabricated by a polyoxyethylene amphiphile in protic ionic liquid, *Langmuir* **26**, 7802–7807, (2010).
[49] I. Martiel, L. Sagalowicz and R. Mezzenga, A reverse micellar mesophase of face-centered cubic Fm3m symmetry in phosphatidylcholine/water/organic solvent ternary systems, *Langmuir* **29**, 15805–15812, (2013).
[50] H. Maeda, J. Wu, T. Sawa, Y. Matsumura and K. Hori, Tumor vascular permeability and the EPR effect in macromolecular therapeutics: A review, *J. Controlled Release* **65**, 271–284, (2000).
[51] T. Sato, H. Sakai, K. Sou, R. Buchner and E. Tsuchida, Poly(ethylene glycol)-conjugated phospholipids in aqueous micellar solutions: Hydration, static structure, and interparticle interactions, *J. Phys. Chem. B* **111**, 1393–1401, (2007).
[52] J. Wang, X. Xing, X. Fang, C. Zhou, F. Huang, Z. Wu, J. Lou and W. Liang, Cationic amphiphilic drugs self-assemble to the core-shell interface of PEGylated phospholipid micelles and stabilize micellar structure. Philos, *Trans. R. Soc. London, Ser. A* **371**, 20120309, (2013).
[53] N. Tang, G. Du, N. Wang, C. Liu, H. Hang and W. Liang, Improving penetration in tumors with nanoassemblies of phospholipids and doxorubicin, *J. Natl. Cancer Inst.* **99**, 1004–1015, (2007).
[54] L. Arleth, B. Ashok, H. Onyuksel, P. Thiyagarajan, J. Jacob and R. P. Hjelm, Detailed structure of hairy mixed micelles formed by phosphatidylcholine and PEGylated phospholipids in aqueous media, *Langmuir* **21**, 3279–3290, (2005).

[55] D. Löf, M. Tomšič, O. Glatter, Fritz-G. Popovski and K. Schillén, Structural characterization of nonionic mixed micelles formed by $C_{12}EO_6$ surfactant and P123 triblock copolymer, *J. Phys. Chem. B* **113**, 5478–5486, (2009).
[56] O. Glatter, Scattering studies on colloids of biological interest (amphiphilic systems), *Prog. Colloid Polym. Sci.* **84**, 46–54, (1991).
[57] O. Glatter, Evaluation of small-angle scattering data from lamellar and cylindrical particles by the indirect transformation method, *J. Appl. Cryst.* **13**, 577–584, (1980).
[58] L. K. Shrestha, T. Sato and K. Aramaki, Shape, size, and structural control of reverse micelles in diglycerol monomyristate nonionic surfactant system, *J. Phys. Chem. B* **111**, 1664–1671, (2007).
[59] J. Jansson, K. Schillén, M. Nilsson, O. Söderman, G. Fritz, A. Bergmann and O. Glatter, Small-angle X-ray scattering, light scattering, and NMR study of PEO-PPO-PEO triblock copolymer/cationic surfactant complexes in aqueous solution, *J. Phys. Chem. B* **109**, 7073–7083, (2005).
[60] J. Jansson, K. Schillén, G. Olofsson, R. C. Silva and W. Loh, The interaction between PEO-PPO-PEO triblock copolymers and ionic surfactants in aqueous solution studied using light scattering and calorimetry, *J. Phys. Chem. B* **108**, 82–92, (2004).
[61] L. K. Shrestha, S. C. Sharma, T. Sato, O. Glatter and K. Aramaki, Small-angle X-ray scattering (SAXS) study on nonionic fluorinated micelles in aqueous system, *J. Colloid interface Sci.* **316**, 815–824, (2007).
[62] S. C. Sharma, C. Rodríguez-Abreu, L. K. Shrestha and K. Aramaki, Short haired wormlike micelles in mixed nonionic fluorocarbon surfactants, *J. Colloid Interface Sci.* **314**, 223–229, (2007).
[63] L. K. Shrestha and K. Aramaki, Structural investigation of diglycerol monolaurate reverse micelles in nonpolar oils cyclohexane and octane, *J. Oleo Sci.* **58**, 235–242, (2009).
[64] L. K. Shrestha, O. Glatter and K. Aramaki, Structure of nonionic surfactant (Glycerol α- monomyristate) micelles in organic solvents: A SAXS study, *J. Phys. Chem. B* **113**, 6290–6298, (2009).
[65] M. V. Petoukhov and D. I. Svergun, Applications of small-angle X-ray scattering to biomacromolecular solutions *Int. J. Biochem. Cell Biol.* **45**, 429–437, (2013).
[66] A. Grishaev, J. Wu, J. Trewhella and A. Bax, Refinement of multidomain protein structures by combination of solution small-angle X-ray scattering and NMR data, *J. Am. Chem. Soc.* **127**, 16621–16628, (2005).
[67] A. Grishaev, L. Guo, T. Irving and A. Bax, Improved fitting of solution X-ray scattering data to macromolecular structures and structural ensembles by explicit water modeling, *J. Am. Chem. Soc.* **132**, 15484–15486, (2010).
[68] M. Ballauff, SAXS and SANS studies of polymer colloids, *Curr. Opin. Colloid Interface Sci.* **6**, 132–139, (2001).

[69] T. Kamal, T. J. Shin and S. Y. Park, Uniaxial tensile deformation of poly (ε-caprolactone) studied with SAXS and WAXS techniques using synchrotron radiation, *Macromolecules* **45**, 8752–8759, (2012).

[70] L. Yang, R. H. Somani, I. Sics, B. S. Hsiao, R. Kolb, H. Fruitwala and C. Ong, Shear-induced crystallization precursor studies in model polyethylene blends by in-situ rheo-SAXS and rheo-WAXD, *Macromolecules* **37**, 4845–4859, (2004).

[71] J. Schuster, R. Köhn, M. Döblinger, A. Keilbach, H. Amenitsch and T. Bein, *in situ* SAXS study on a new mechanism for mesostructure formation of ordered mesoporous carbons: Thermally induced self-assembly, *J. Am. Chem. Soc.* **134**, 11136–11145, (2012).

[72] P. Xiao, H. Li, S. Huang, H. Wen, D. Yu, Y. Shang, J. Li, Z. Wu, L. An and S. Jiang, Shear effects on crystalline structures of poly (L-lactide), *CrystEngComm* **15**, 7914–7925, (2013).

[73] Z. Jiang, Y. Wang, L. Fu, B. Whiteside, J. Wyborn, K. Norris, Z. Wu, P. Coates and Y. Men, Tensile deformation of oriented poly(ε-caprolactone) and its miscible blends with poly(vinylmethylether), *Macromolecules* **46**, 6981–6990, (2013).

[74] N. Hosono, M. A. Gillissen, Y. Li, S. S. Sheiko, A. R. Palmans and E. W. Meijer, Orthogonal Self-Assembly in folding block copolymers, *J. Am. Chem. Soc.* **135**, 501–510, (2013).

[75] A. Blanazs, R. Verber, O. O. Mykhaylyk, A. J. Ryan, J. Z. Heath, C. W. Douglas and S. P. Armes, Sterilizable gels from thermoresponsive block copolymer worms, *J. Am. Chem. Soc.* **134**, 9741–9748, (2012).

[76] K. Johnson, Y. S. Huang, S. Huettner, M. Sommer, M. Brinkmann, R. Mulherin, D. Niedzialek, D. Beljonne, J. Clark, W. T. Huck and R. H. Friend, Control of intrachain charge transfer in model systems for block copolymer photovoltaic materials, *J. Am. Chem. Soc.* **135**, 5074–5083, (2013).

[77] V. Ho, B. W. Boudouris, B. L. McCulloch, C. G. Shuttle, M. Burkhardt, M. L. Chabinyc and R. A. Segalman, Poly(3-alkylthiophene) diblock copolymers with ordered microstructures and continuous semiconducting pathways, *J. Am. Chem. Soc.* **133**, 9270–9273, (2011).

[78] Z. Mingjiang, E. K. Kim, J. P. McGann, S.-E. Chun, J. F. Whitacre, M. Jaroniec, K. Matyjaszewski and T. Kowalewski, Electrochemically active nitrogen-enriched nanocarbons with well-defined morphology synthesized by pyrolysis of self-assembled block copolymer, *J. Am. Chem. Soc.* **134**, 14846–14857, (2012).

[79] S. Lee, M. J. Bluemle and F. S. Bates, Discovery of a Frank-Kasper theta phase in sphere-forming block copolymer melts, *Science* **330**, 349–353, (2010).

[80] T. Lunkenbein, M. Kamperman, Z. Li, C. Bojer, M. Drechsler, S. Forster, U. Wiesner, A. H. Muller and J. Breu, Direct synthesis of inverse hexagonally ordered diblock copolymer/polyoxometalate nanocomposite films, *J. Am. Chem. Soc.* **134**, 12685–12692, (2012).

[81] T. Wang, J. Zhuang, J. Lynch, O. Chen, Z. Wang, X. Wang, D. LaMontagne, H. Wu, Z. Wang and Y. C. Cao, Self-assembled colloidal superparticles from nanorods, *Science* **338**, 358–363, (2012).
[82] T. Wang, X. Wang, D. LaMontagne, Z. Wang and Y. C. Cao, Macroscale lateral alignment of semiconductor nanorods into freestanding thin films, *J. Am. Chem. Soc.* **135**, 6022–6025, (2013).
[83] C. Lu, A. J. Akey, C. J. Dahlman, D. Zhang and I. P. Herman, Resolving the growth of 3D colloidal nanoparticle superlattices by real-time small-angle X-ray scattering, *J. Am. Chem. Soc.* **134**, 18732–18738, (2012).
[84] Y. Nagaoka, O. Chen, Z. Wang and Y. C. Cao, Structural control of nanocrystal superlattices using organic guest molecules, *J. Am. Chem. Soc.* **134**, 2868–2871, (2012).
[85] T. Li, X. Zan, R. E. Winans, Q. Wang and B. Lee, Biomolecular assembly of thermoresponsive superlattices of the tobacco mosaic virus with large tunable interparticle distances, *Angew. Chem. Int. Ed.* **52**, 6638–6642, (2013).
[86] R. J. Macfarlane, B. Lee, M. R. Jones, N. Harris, G. C. Schatz and C. A. Mirkin, Nanoparticle superlattice engineering with DNA, *Science* **334**, 204–208, (2011).
[87] M. R. Jones and C. A. Mirkin, Bypassing the limitations of classical chemical purification with DNA-programmable nanoparticle recrystallization, *Angew. Chem. Int. Ed.* **52**, 2886–2891, (2013).
[88] W. Wang, X. Chen, Q. Cai, G. Mo, L. S. Jiang, K. H. Zhang, Z. J. Chen, Wu, Z. H. and W. Pan, *In situ* SAXS study on size changes of platinum nanoparticles with temperature, *Eur. Phys. J. B* **65**, 57–64, (2008).
[89] W. Wang, K. Zhang, Q. Cai, G. Mo, X. Q. Xing, W. D. Cheng, Z. J. Chen and Z. H. Wu, Real-time SAXS and ultraviolet-visible spectral studies on size and shape evolution of gold nanoparticles in aqueous solution, *Eur. Phys. J. B* **76**, 301–307, (2010).
[90] A. Plech, V. Kotaidis, A. Siems and M. Sztucki, Kinetics of the X-ray induced gold nanoparticle synthesis, *Phys. Chem. Chem. Phys.* **10**, 3888–3894, (2008).
[91] B. Abécassis, F. Testard, O. Spalla and P. Barboux, Probing *in situ* the nucleation and growth of gold nanoparticles by small-angle X-ray scattering, *Nano. Lett.* **7**, 1723–1727, (2007).
[92] J. Polte, T. T. Ahner, F. Delissen, S. Sokolov, F. Emmerling, A. F. Thünemann and R. Kraehnert, Mechanism of gold nanoparticle formation in the classical citrate synthesis method derived from coupled in situ XANES and SAXS evaluation, *J. Am. Chem. Soc.* **132**, 1296–1301, (2010).
[93] J. Polte, R. Erler, A. F. Thünemann, S. Sokolov, T. T. Ahner, K. Rademann, F. Emmerling and R. Kraehnert, Nucleation and growth of gold nanoparticles studied via in situ small angle X-ray scattering at millisecond time resolution, *ACS Nano* **4**, 1076–1082, (2010).
[94] J. Polte, A. F. Thünemann, R. Erler, F. Emmerling and R. Kraehnert, SAXS in combination with a free liquid jet for improved time-resolved *in-situ* studies of the nucleation and growth of nanoparticles, *Chem. Commun.* **46**, 9209–9211, (2010).

[95] H. C. Lichtenegger, T. Schoberl, M. H. Bartl, H. Waite and G. D. Stucky, High abrasion resistance with sparse mineralization: Copper biomineral in worm jaws, *Science* **298**, 389–392, (2002).
[96] X.-Q. Xing, Y. Gong, Q. Cai, G. Mo, R. Du, Z.-J. Chen and Z.-H. Wu, Hierarchical structure and biomineralization in cricket teeth, *Chin. Phys. C* **37**, 028001, (2013).
[97] V. James, J. Kearsley, T. Irving, Y. Amemiya and D. Cookson, Using hair to screen for breast cancer, *Nature* **398**, 33–34, (1999).
[98] P. Dumas and L. Miller, The use of synchrotron infrared microspectroscopy in biological and biomedical investigations, *Vib. Spectrosc* **32**, 3–21, (2003).
[99] N. Ohta, T. Oka, K. Inoue, N. Yagi, S. Kato and I. Hatta, Structural analysis of cell membrane complex of a hair fibre by micro-beam X-ray diffraction, *J. Appl. Crystallogr.* **38**, 274–279, (2005).
[100] P. Saengkaew, W. Ussawawongaraya, S. Khaweerat, S. Rugmai, S. Ouajai, J. Luengviriya, S. Sanorpim, M. Tirarattanasompot and S. Rhianphumikarakit, A preliminary X-ray study on human-hair microstructures for a health-state indicator, *World Acad. Sci., Eng. Technol.* **59**, 2333–2337, (2011).
[101] A. A. Vazina, W. Bras, I. P. Dolbnya, V. N. Korneev, N. F. Lanina, A. M. Matyushin, P. M. Sergienko and A. V. Zabelin, Peculiarities of human hair structural dynamics, *Nucl. Instrum. Methods Phys. Res., Sect. A* **543**, 153–157, (2005).
[102] L. Kreplak, A. Franbourg, F. Briki, F. Leroy, D. Dalle and J. Doucet, A new deformation model of hard alpha-keratin fibers at the nanometer scale: Implications for hard alpha-keratin intermediate filament mechanical properties, *Biophys. J.* **82**, 2265–2274, (2002).
[103] A. M. Aksirov, V. S. Gerasimov, V. I. Kondratyev, V. N. Korneev, G. N. Kulipanov, N. F. Lanina, V. P. Letyagin, N. A. Mezentsev, P. M. Sergienko, B. P. Tolochko, V. A. Trounova and A. A. Vazina, Biological and medical application of SR from the storage rings of VEPP-3 and "Siberia-2". The origin of specific changes of small-angle X-ray diffraction pattern of hair and their correlation with the elemental content, *Nucl. Instrum. Methods. Phys. Res., Sect. A* **470**, 380–387, (2001).
[104] F. Briki, B. Busson, B. Salicru, F. Esteve and J. Doucet, Breast-cancer diagnosis using hair, *Nature* **400**, 226–226, (1999).
[105] V. A. Trounova, A. A. Vazina, N. F. Lanina, P. M. Sergienko and V. I. Kondratyev, Correlation between element concentrations and X-ray diffraction patterns in hair, *X-Ray Spectrom.* **31**, 314–318, (2002).
[106] G. L. Corino and P. W. French, Diagnosis of breast cancer by X-ray diffraction of hair, *Int. J. Cancer* **122**, 847–856, (2008).
[107] F. Briki, B. Busson, L. Kreplak, P. Dumas and J. Doucet, Exploring a biological tissue from atomic to macroscopic scale using synchrotron radiation: Example of hair, *Cell Mol. Biol.* **46**, 1005–1016, (2000).

[108] M. E. Rafik, F. Briki, M. Burghammer and J. Doucet, *In vivo* formation steps of the hard alpha-keratin intermediate filament along a hair follicle: Evidence for structural polymorphism, *J. Struct. Biol.* **154**, 79–88, (2006).
[109] M. E. Rafik, J. Doucet and F. Briki, The intermediate filament architecture as determined by X-ray diffraction modeling of hard alpha-keratin, *Biophys. J.* **86**, 3893–3904, (2004).
[110] L. Kreplak, C. Merigoux, F. Briki, D. Flot and J. Doucet, Investigation of human hair cuticle structure by microdiffraction: Direct observation of cell membrane complex swelling, *Biochim Biophys Acta* **1547**, 268–274, (2001).
[111] X. Xing, R. Du, Y. Li, B. Li, Q. Cai, G. Mo, Y. Gong, Z. Chen and Z. H. Wu, Structural change of human hair induced by mercury exposure, *Environ. Sci. Technol.* **47**, 11214–11220, (2013).
[112] W. Cheng, X. Xing, D. Wang, K. Zhang, Q. Cai, G. Mo, Z. Chen and Z. Wu, Small-angle X-ray scattering study on nanostructural changes with water content in red pine, American pine, and white ash, *J. Wood Sci.* **57**, 470–478, (2011).
[113] K. Liang, Y. Gong, J. Fu, S. Yan, Y. Tan, R. Du, X. Xing, G. Mo, Z. Chen, Q. Cai, D. Sun and Z. Wu, Microstructural change of degummed Bombyx mori silk: An *in situ* stretching wide-angle X-ray scattering study, *Int. J. Bio. Macromol.* **57**, 99–104, (2013).
[114] B. Lee, C.-T. Lo, S. Seifert, N. L. D. Rago, R. E. Winans and P. Thiyagarajan, Anomalous small-angle X-ray scattering characterization of bulk block copolymer/nanoparticle composites, *Macromolecules* **40**, 4235–4243, (2007).
[115] S. A. Pabit, X. Qiu, J. S. Lamb, L. Li, S. P. Meisburger and L. Pollack, Both helix topology and counterion distribution contribute to the more effective charge screening in dsRNA compared with dsDNA, *Nucleic Acids Res.* **37**, 3887–3896, (2009).
[116] S. A. Pabit, S. P. Meisburger, L. Li, J. M. Blose, C. D. Jones and L. Pollack, Counting ions around DNA with anomalous small-angle X-ray scattering, *J. Am. Chem. Soc.* **132**, 16334–16336, (2010).
[117] Y. Sanada, I. Akiba, K. Sakurai, K. Shiraishi, M. Yokoyama, E. Mylonas, N. Ohta, N. Yagi, Y. Shinohara and Y. Amemiya, Hydrophobic molecules infiltrating into the poly(ethylene glycol) domain of the core/shell interface of a polymeric micelle: Evidence obtained with anomalous small-angle X-ray scattering, *J. Am. Chem. Soc.* **135**, 2574–2582, (2013).
[118] J. R. Levine, J. B. Cohen, Y. W. Chung and P. Georgopoulos, Grazing-incidence small-angle X-ray scattering: New tool for studying thin film growth, *J. Appl. Cryst.* **22**, 528–532, (1989).
[119] S. K. Sinha, E. B. Sirota, S. Garoff and H. B. Stanley, X-ray and neutron scattering from rough surfaces, *Phys. Rev. B* **38**, 2297–2311, (1988).
[120] M. Rauscher, T. Salditt and H. Spohn, Small-angle X-ray scattering under grazing incidence: The cross section in the distorted-wave Born approximation, *Phys. Rev. B* **52**, 16855–16863, (1995).

[121] R. Lazzari, F. Leroy and G. Renaud, Grazing-incidence small-angle X-ray scattering from dense packing of islands on surfaces: Development of distorted wave Born approximation and correlation between particle sizes and spacing, *Phys. Rev. B* **76**, 125411, (2007).
[122] W. D. Cheng, G. Mo, X. Q. Xing, D. H. Wang, Y. Gong, Q. Cai, Z. J. Chen and Z. H. Wu, GISAXS and SAXS studies on the spatial structures of Co nanowire arrays, *Chin. Phys. C* **35**, 875–879, (2011).
[123] S. Bernstorff, P. Dubček, I. Kovačević, N. Radić and B. Pivac, Si nanocrystals in SiO_2 films analyzed by small angle X-ray scattering, *Thin Solid Films* **515**, 5637–5640, (2007).
[124] G. Renaud, R. Lazzari, C. Revenant, A. Barbier, M. Noblet, O. Ulrich, F. Leroy, J. Jupille, Y. Borensztein, C. R. Henry, J.-P. Deville, F. Scheurer, J. Mane-Mane and O. Fruchart, Real-time monitoring of growing nanoparticles, *Science* **300**, 1416–1419, (2003).

Chapter 8

Crystal-based X-ray Medical Imaging Using Synchrotron Radiation and Its Future Prospect

Masami Ando

Research Institute of Science and Technology, Tokyo University of Science, Yamasaki 2641, Noda, Chiba 278-8510 Japan

Naoki Sunaguchi

Graduate School of Medicine, Nagoya University, Nagoya, Aichi 461-8673, Japan

Yongjin Sung

Department of Mechanical Engineering and Biomedical Engineering, University of Wisconsin-Milwaukee, 3200 N. Cramer St., Milwaukee, WI 53211, USA

Daisuke Shimao

Department of Radiological Technology, Hokkaido University of Science, 7-Jo 15-4-1 Maeda, Teine, Sapporo, Hokkaido 006-8585, Japan

Jong-Ki Kim

Biomedical Engineering and Radiology, School of Medicine, Catholic University of Daegu, Daegu 705-718, Korea

Gang Li

Beijing Synchrotron Radiation Facility,
Institute of High Energy Physics, Chinese Academy of Sciences,
PO Box 918, 2-7, 19B Yuquan Road, Beijing 100049, P. R. China

Yoshifumi Suzuki

Graduate School of Engineering, Kyushu Institute of Technology,
Kitakyushu 804-8550, Japan

Tetsuya Yuasa

Graduate School of Engineering and Science, Yamagata University,
Jonan 4-3-16, Yonezawa, Yamagata 992-8510, Japan

Kensaku Mori

Department of Media Science, Graduate School of Information Science,
Nagoya University, Furo-cho, Chikusa-ku, Nagoya 464-8601, Japan

Shu Ichihara

Clinical Research Center, Department of Pathology,
Nagoya Medical Center, Sanno-maru 4-1-1,
Nakaku, Nagoya 460-0001, Japan

Rajiv Gupta

Department of Radiology, Massachusetts General Hospital
and Harvard Medical School, Boston, MA 02114, USA

Abstract

Synchrotron produces high-energy photons with a large photon flux, which is extremely useful in diverse research fields including physics, chemistry, biology, and medicine. The large beam flux of synchrotron radiation, in tandem with crystal optics, allows generating X-rays with very high spatial and temporal coherence. Such X-rays remove complications due to the irregular wave front or broad energy spectrum of conventional X-ray source; thus, a variety of novel X-ray imaging techniques have been developed and demonstrated using synchrotron radiation. Recently, X-ray phase-contrast imaging (XPCI) has drawn people's attention as a next-generation X-ray imaging technique owing to its capacity to generate high-contrast images of low-atomic-number materials such as soft tissues. In this article, we provide detailed explanation on the fundamental physics of X-ray optics for visualizing soft tissue and its medical applications using synchrotron radiation. In the last part we would like to provide a future aspect of clinical and pathological application.

Keywords: crystal-analyzer based imaging (CABI); DEI; X-ray dark-field imaging (XDFI); X-ray phase-contrast imaging (XPCI).

8.1. Introduction

Clinical soft-tissue imaging using traditional X-ray fluoroscopy, radiography or computed tomography (CT) uses absorption or attenuation difference between tissues as the primary contrast mechanism [138]. X-ray absorption, arising primarily from photoelectric absorption and Compton scattering, is markedly different in bones, soft tissues and air. However, among different soft tissue types, the attenuation coefficient is very similar; thus, conventional X-ray imaging cannot adequately differentiate between them. For example, while a calcified atherosclerotic plaque can be differentiated from a non-calcified plaque using traditional CT, the soft tissue components of the plaque such as the fibrous cap and atheroma are essentially indistinguishable. Similarly, one cannot distinguish the attenuation of a small tumour from surrounding healthy tissue using conventional CT, unless it is calcified, causes architectural distortion, or has dense cell packing. For these purposes, we must administer iodinated contrast agents [109], or use alternative imaging techniques such as MRI [23, 40].

Focused on medical imaging, clinical tests of human intravenous coronary angiography (IVCAG) were performed at five synchrotron laboratories in the world already in the late 20th century. More than 400 patients underwent IVCAG in three US and European synchrotron laboratories [34, 37, 107, 132] and another 55 patients at Photon Factory in Japan [1, 2, 43, 92, 93].

Multi-slice X-ray CT (MSCT) that can provide information on 3D location of plaques in coronary arteries, was also developed almost at a same time using a hospital X-ray tube.

On the other hand X-ray phase-contrast imaging (XPCI) can dramatically enhances the image contrast for soft tissues [10, 19–21, 29, 30, 38, 45, 46, 51, 53, 57, 142] by utilizing an alternative contrast mechanism, namely, the coherent scattering of X-rays by the electron clouds of materials. The phase-contrast signal manifests as angular deviation of incident X-rays or distortion of wave fronts [141]. X-rays are highly energetic photons, and propagate like bullets through a specimen. As a result, the phase-contrast signal, which originates from the wave nature of X-rays, is often dominated by the attenuation-contrast, and it is difficult to directly record the phase-contrast signal in a conventional X-ray imaging set-up. To overcome this difficulty, a variety of XPCI methods have been proposed. Here, we briefly summarize the principles of various XPCI techniques. Recent review articles [9, 14, 15–17, 52–54, 65, 66, 125] provide a complete list and detailed description of the existing XPCI techniques. Also the authors would like very much appreciate a great contribution in this field by Russian groups [101–104, 117].

In Section 8.2, we briefly explain the working principle of a variety of XPCI systems for biomedical specimens. In Section 8.3, we explain the theory of X-ray dynamical diffraction [11, 47, 49, 50, 55, 99, 136, 146], which is essential for explaining the image formation in crystal-analyzer based imaging (CABI). In Section 8.4, we provide detailed explanation of experimental setup and sample preparation. In Section 8.5, we explain numerical algorithms to extract the phase-contrast signals from raw intensity images. In addition, we describe tomographic reconstruction algorithms specifically designed for CABI, which increases the accuracy and speed of reconstruction. In Section 8.6, we show the images of a variety of biological specimens acquired with our CABI system; the images clearly show the broad range of medical applications of XPCI. Finally in Section 8.7, we will discuss future aspects of medical applications.

8.2. A Variety of X-ray Imaging for Biomedical Specimens

8.2.1. *X-ray interferometer imaging*

X-ray interferometer was invented by Bonse and Hart in 1965 [13]. In principle, X-rays are split by a crystal into two beams, one of which passes through a sample and the other through a free space.

The former is called a sample beam, and the latter called a reference beam. The wave front of the sample beam is distorted after passing through a sample. The coherent interference of this distorted sample beam and the reference beam generates a modulated fringe pattern at the detector plane. Analyzing the fringe pattern, one can acquire the information of wave front distortion reflecting the projected electron density of sample. The first demonstration of X-ray interferometry with biological samples was first reported by Ando and Hosoya in 1972 [3]. X-ray interferometer combined with computed tomography (CT) was also demonstrated to provide the 3-D structure of biological specimens [77–81, 145]. It is noteworthy that the interferometry system is very sensitive to small temperature variation in the air or vibration from the surrounding setup [28, 77–81, 145]. In addition, a specimen whose optical thickness is larger than 2π makes the interpretation of fringe pattern difficult. Due to these limitations, X-ray interferometry has been applied only to a small specimen in a vibration-isolated setup.

8.2.2. *Propagation-based XPCI*

Propagation-based XPCI has evolved from the experimental observation that a local disturbance to the phase of a wave generates amplitude variation of the wave while it progresses in the free space after the point of disturbance. Therefore, by recording the intensity distribution at multiple locations after a specimen, one can calculate the phase profile introduced by the specimen. This propagation-based XPCI does not rely on any optical elements for implementation, and thus this can provide high signal-to-noise ratio [94, 116, 148, 149]. However, in order to apply this technique to clinical imaging, a couple of limitations have to be addressed, which include a source with high spatial coherence and a detector with a small pixel size and high dynamic range [17]. A computational algorithm has been developed to relax these stringent requirements.

8.2.3. *Grating-based XPCI*

Grating-based XPCI originated from Talbot [129] is typically built using two or three gratings. A pair of gratings blocks the X-rays that transmit a specimen without bending, while it allows the refracted X-rays to pass through the pair. The refraction of X-rays is due to the electron density difference within the specimen. To remove the ambiguity in determining the phase, multiple images are typically recorded for different relative positions of the two gratings. When an X-ray source with small spatial coherence is used, the third grating can be added to the setup at the cost of reduced beam flux. Grating-based imaging has been demonstrated first using synchrotron and later with a lab-scale X-ray source. See the details of the grating interferometry in the above nice papers [12, 32, 70, 75, 82, 96, 97, 139, 140]. Grating-based XPCI provides high accuracy in phase measurement, but it is known that its signal-to-noise ratio is relatively low compared to other techniques.

8.2.4. *Crystal analyser-based XPCI*

The reflection or transmission efficiency of a crystal dramatically changes when the angle of incidence is changed close to the Bragg angle [118, 119]. Crystal analyser-based XPCI uses this high angular sensitivity of a crystal as a method to pick up the small angular deviation caused by phase-contrast signal. There are two types of X-ray optics; one uses a reflection (Bragg) type of analyser crystal which is called diffraction-enhanced imaging (DEI) (Fig. 8.1); this is in principle a double-crystal arrangement in a parallel mode where both two diffractions take place under Bragg reflection (Fig. 8.2). Both monochromator-collimator (MC) and Bragg reflection angle analyzer (BAA) in Fig. 8.1 diffract X-rays in the mode of Bragg reflection. The other optics is called X-ray dark-field imaging (XDFI) (Fig. 8.3) where Bragg reflection type MC, same one as in DEI, and Laue (transmission) type (Fig. 8.4) angle analyzer (LAA) are used. Chapman first demonstrated DEI [21, 95, 100]. He was not only a pioneer in Bragg case but also even in Laue case which he proposed two years earlier [19]. We have developed optics by adopting a thin and a fixed thickness of crystal in the transmission Laue mode named XDFI [4–9] where two images having opposite refraction components are simultaneously obtained in a single shot. Using Takagi-Taupin

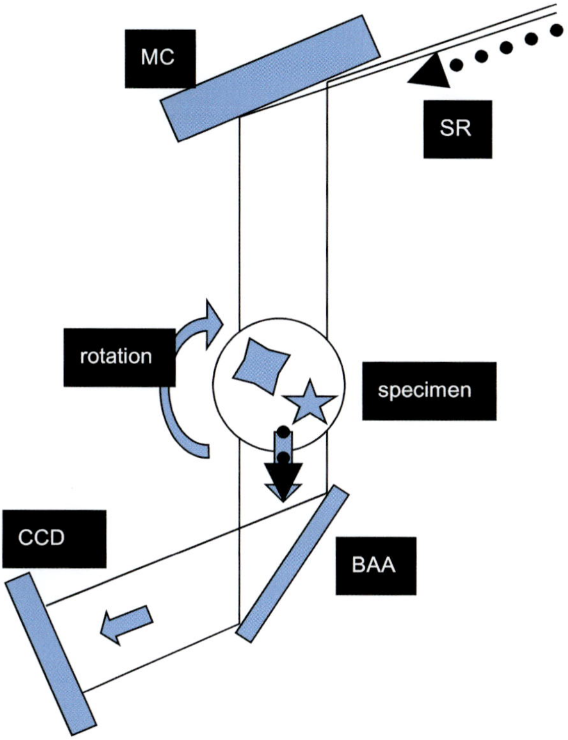

Fig. 8.1. Optics of DEI that comprises MC: monochromator-collimator, sample on a sample rotation stage, BAA: Bragg angle analyzer and CCD camera. MC receives monochromatic synchrotron radiation that was made by a pre-monochromator system upstream. The MC shown in this drawing adopts asymmetric diffraction because of producing narrower beam divergence and wider beam width of outgoing beam.

equation [127, 128, 131] one can estimate dependence of the spatial resolution on LAA thickness [126]. For the remaining part of this article, we will provide detailed description of its technique and applications to medical imaging. As an extension of DEI, Wu [143] has recently demonstrated a method to further increase the angular sensitivity by adopting multiple diffractions in the Bragg analyser. In the following work most of X-ray pictures have been taken at X-ray energy of 35 keV under 440 diffraction. Sometimes one needs lower X-ray energy such as 17.5 keV that is half of 35 keV; this case can be

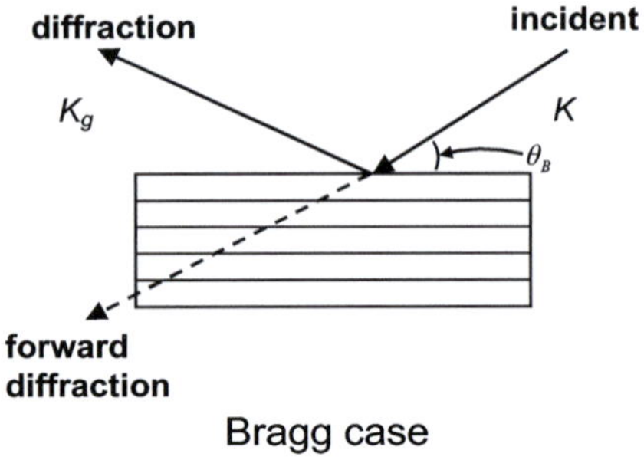

Fig. 8.2. Bragg case geometry, where the incident beam K becomes diffracted beam K_g that comes out from a same surface as the incident beam comes in. Θ_B is the Bragg angle. Only in case of a thin crystal the forward diffracted beam is produced.

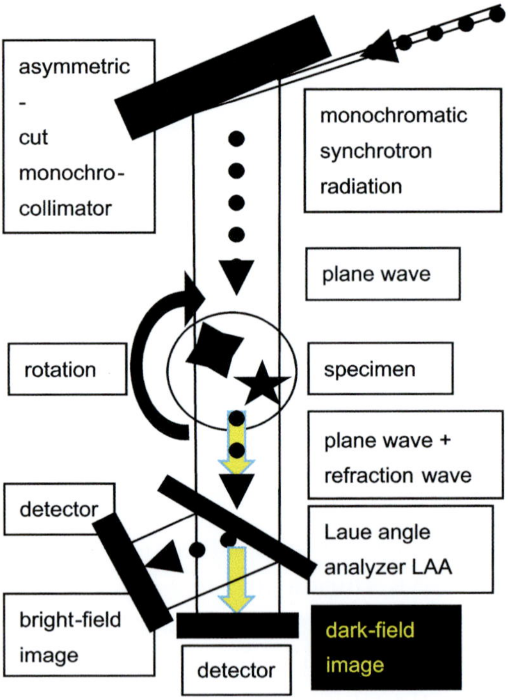

Fig. 8.3. Optics of X-ray dark-field imaging (XDFI). Plane parallel X-ray that is made by asymmetric diffraction of MC reaches a specimen from top produces refraction X-ray; this refracted X-ray and simply transmitted X-ray is angle analysed by a Laue angle analyser (LAA). Under a special thickness condition one can attain the condition of XDFI.

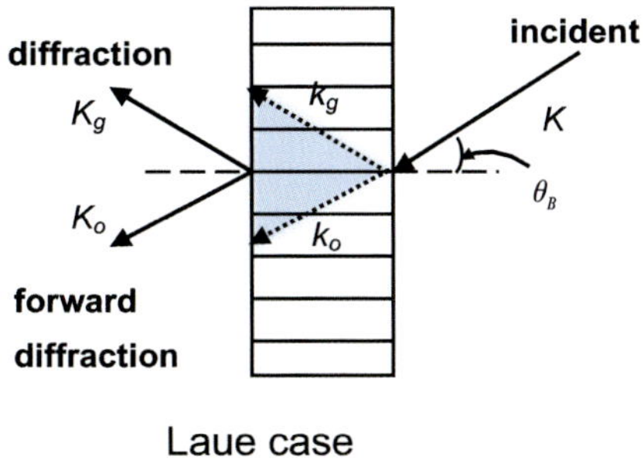

Fig. 8.4. Laue geometry. The X-ray beam K produces two beams in a crystal one k_o along the same direction of the incident beam and the other diffraction k_g as shown in the figure. Thus the outgoing beam splits into two beams in vacuum, one K_o called the forward diffraction and K_g the diffraction, as shown in the figure, at the backside of the crystal. Θ_B is the Bragg angle. The triangle drawn with gray which is formed by k_o, k_g and the base at crystal backside where the beam comes out is called the Borrmann fan. In the figure K_o and K_g are only drawn as representative of the beams outgoing from the other side of the crystal.

realized by using 220 diffraction index under the same Bragg angle. 35 keV has been chosen as a highest X-ray energy under the current synchrotron instrumentation condition and 17.5 keV was chosen as a lowest X-ray energy. In Table 8.1 are listed basic diffraction parameters such as d-spacing, Bragg angle, intrinsic Darwin width, divergence of outgoing beam after asymmetric diffraction for 111, 220 and 440 diffraction indices and the above two X-ray energies.

8.3. Theory of X-ray Dynamical Diffraction in Crystal

The X-ray dynamical diffraction theory accurately describes interaction of X-ray plane waves with perfect crystal lattice in both geometry, Bragg case and Laue case. When the electric field of X-ray plane waves in a crystal is weak, the electric polarizability P of a crystal can be related to the electric field E as

$$P = -\frac{e^2}{m\omega^2}\rho(\vec{r})E, \tag{8.1}$$

where e is the electron charge, m the mass of electron, ω the frequency of the electric field, and $\rho(\vec{r})$ the three-dimensional distribution of

Table 8.1. Basic parameters of diffraction for different diffraction indices and for different X-ray energies 35 keV and 17.5 keV. If choosing 440 for 35 keV X-rays and 220 for 17.5 keV X-rays one can switch the X-ray energy from one to the other without changing a crystal arrangement. This is one reason why the 440 optics has been chosen for almost all works described in this text.

		35 keV, 0.354 Å					17.5 keV, 0.709 Å				
Diffraction Index	d angstrom	Θ degree	ω_0 arcsec	Instrumental function arcsec	α degree	$b^{1/2}$	Θ degree	ω_0 arcsec	Instrumental function arcsec	α degree	$b^{1/2}$
440	0.96	10.6	0.36	0.05	10.2	0.14	21.7	0.77	0.08	21.3	0.1
220	1.92	5.29	1.13	0.23	4.89	0.2	10.6	2.31	0.32	10.2	0.14
111	3.14	3.23	1.64	0.43	2.83	0.26	6.48	3.29	0.84	6.08	0.18

electrical charge inside crystal. Let us define the electric susceptibility $\chi(\vec{r})$ as follows:

$$\chi(r) = -r_e \frac{\lambda^2}{\pi} \rho(\vec{r}), \tag{8.2}$$

where r_e is the classical radius of electron and λ the wavelength. Suppose $\rho(\vec{r})$ can be represented by a sum of harmonic components:

$$\rho(\vec{r}) = \frac{1}{\mathrm{v}_c} \sum_g F_g \exp(i\vec{g} \cdot \vec{r}), \tag{8.3}$$

where v_c represents the volume of a crystal unit cell, F_g the crystal structure factor, and g the diffraction index. We can similarly expand $\chi(\vec{r})$ into terms of harmonic components.

$$\chi(\vec{r}) = \sum_g \chi_g \exp(i\vec{g} \cdot \vec{r}). \tag{8.4}$$

Substituting Eqs. (8.3) and (8.4) for $\rho(\vec{r})$ and $\chi(\vec{r})$ in Eq. (8.2), χ_g can be related to F_g as

$$\chi_g = \frac{r_e \lambda^2}{\pi \mathrm{v}_c} F_g. \tag{8.5}$$

The electric field E inside a crystal can be represented by a superposition of many plane wave components that propagate along the forward diffraction $\vec{k}_o$ and the diffraction $\vec{k}_g$.

Thus $E(\vec{r})$ can be represented by a sum of Fourier components as follows:

$$\begin{aligned} E(\vec{r}) &= \exp\{(i\vec{k}_o \cdot \vec{r} - i\omega t)\} \sum_g E_g \exp(i\vec{g} \cdot \vec{r}) \\ &= \exp(-i\omega t) \sum_g E_g \exp(i\vec{k}_g \cdot \vec{r}), \end{aligned} \tag{8.6}$$

where the wave vector $\vec{k}_0$ for the incident wave and $\vec{k}_g$, the wave vector of the gth plane wave component has a relation through the reciprocal vector $\vec{g}$ of crystal such that

$$\vec{k}_g = \vec{k}_0 + \vec{g}, \tag{8.7}$$

as shown in (Fig. 8.5).

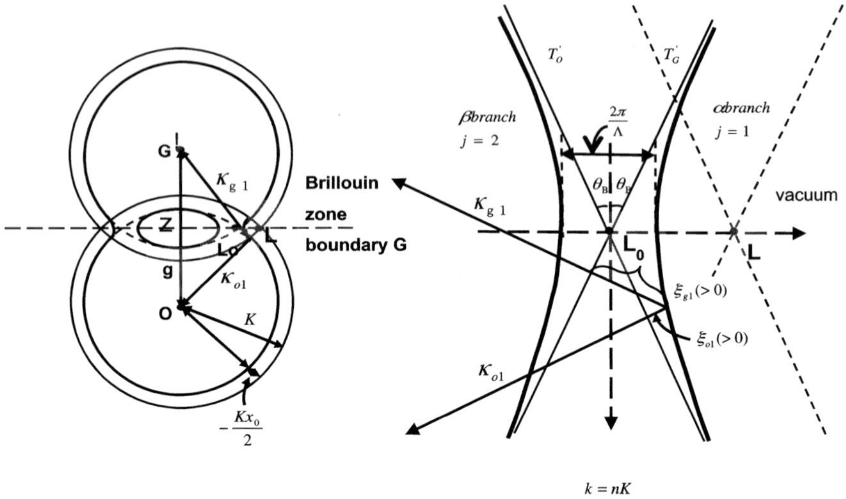

Fig. 8.5. Left shows relation of dispersion surface comprising two diffraction waves k_o and k_g and the reciprocal lattice g in crystal. k_o is produced from K in vacuum through refraction. k_o is smaller than K by $\frac{K\chi_o}{2}$ due to refraction. k_g is produced by k_o and g through diffraction. L and Lo are the Laue point in vacuum and in crystal, respectively. Right shows details around L and L_o. Since k is large surface of sphere O and G can be plane as shown with thin straight lines in the right. Set of k satisfying equation (8.9) becomes hyperbolas as shown with thick curves.

Using the Maxwell equations $\mathrm{rot}E = -\mu_0 \partial H/dt$ and $\mathrm{rot}H = \partial D/dt = -i\omega D$, where μ_o is the magnetic permeability, H the magnetic field, D the dielectric flux density, the following equation of E

$$\mathrm{rot}(\mathrm{rot}E) = K^2(1+\chi)E \tag{8.8}$$

can be obtained, where $K(= \omega/c = 2\pi/\lambda)$ is the wave number of X-rays in vacuum and the relation $E = \varepsilon_0^{-1}(1+\chi)^{-1}D \sim \varepsilon_0^{-1}(1-\chi)D$ has been used.

By use of Eqs. (8.4) and (8.6) one can obtain

$$\chi(\vec{r})E(\vec{r}) = \sum_g \sum_h \chi_{g-h} E_h \exp(i\vec{k}_g \cdot \vec{r}). \tag{8.9}$$

Put (8.9) into (8.8), then (8.9) is reduced to

$$\frac{{k_g}^2 - K^2}{K^2} E_g = \sum_h \chi_{g-h} E_h, \tag{8.10}$$

Equation (8.10) shows a relation between K the wave number in vacuum, k_g that in crystal, χ_g and $\chi_{\bar{g}}$ interaction parameters between E_g gth component of diffracted plane waves and crystal atoms. By replacing K by $k = K\left(1 + \frac{\chi_o}{2}\right)$ (Fig. 8.5) because the difference between K and k is negligibly small, Eq. (8.10) can be rewritten as

$$\left({k_0}^2 - k^2\right)\left({k_g}^2 - k^2\right) = K^4 P \chi_g \chi_{\bar{g}}, \tag{8.11}$$

where P the polarization component is 1 for the σ polarization and $\cos 2\Theta_B$ for the π polarization. This is the basic equation of X-ray dynamical diffraction.

If gap of k_0 and k_g from k (Fig. 8.6) is set such as $\xi_o = k_0 - k, \xi_g = k_g - k$, respectively, where ξ_o and ξ_g are called resonance error or excitation error thus

$$\xi_o \xi_g = \frac{1}{4} K P \chi_g \chi_{\bar{g}} = \frac{\pi^2 \cos^2 \Theta_B}{\Lambda^2}, \tag{8.12}$$

$$\text{where} \quad \Lambda = \frac{2\cos\Theta_B}{K|P||\chi_g|}. \tag{8.13}$$

This corresponds to a minimum distance between α branch and β branch in Fig. 8.6.

$$\text{Also if } W = \frac{(\Theta_B - \Theta - \Delta\Theta_o)\sin 2\Theta_B}{|P||\chi_g|} \quad \text{is defined} \tag{8.14}$$

This corresponds to fine angle rotation of crystal at order of 1 arc second or less.

$$\text{where} \quad \Delta\Theta_o = \frac{-\chi_o}{\sin\Theta} \tag{8.15}$$

In crystal diffraction there are two types from the view point of boundary conditions, one is Bragg case where incident X-rays produce only diffracted beam coming from the same surface where the

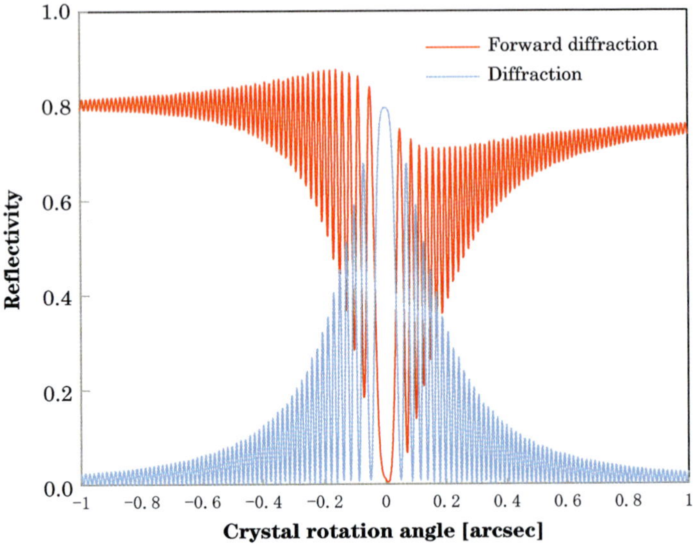

Fig. 8.6. Theoretical rocking curve of forward diffraction (FD) and diffraction (D) of a transmission Laue case under the condition of X-ray energy at 35 keV and the crystal thickness of 1088 μm. Sum of the intensity of FD and D together with photo-electron based absorption is ~0.8 for all the diffraction angle. 20% of gap from 100% comes from the photoelectric absorption of X-rays by crystal.

incident X-rays come in and the other case is Laue case where X-rays are incident onto a crystal surface to produce four Bloch waves, two towards the incident direction and the other two towards the diffraction direction; each two comprises α branch and β branch corresponding to two dispersion surfaces in a reciprocal lattice (Fig. 8.5).

In the Bragg case the boundary condition for the waves at incidence surface of crystal is as follows: towards the incidence direction

$$E_0{}^a = E_{01} \quad \text{for } W \leq -1 \quad \text{and} \quad E_{02} \quad \text{for } W \geq 1 \tag{8.16}$$

$$E_g{}^a = E_{g1} \quad \text{for } W \leq -1 \quad \text{and} \quad E_{g2} \quad \text{for } W \geq 1 \tag{8.17}$$

and there produce no Bloch waves in crystal under the angular condition $|W| \leq 1$; this corresponds to a total reflection of X-rays at surface.

As a result the amplitude of diffraction waves is obtained as follows:

$$E_g{}^a = (|P|)/P \exp(i\alpha_g)(-W \mp \sqrt{W^2-1})) \; E_o{}^a. \tag{8.18}$$

As was done in the Laue case the intensity profile of the Bragg condition can be described as follows:

$$\frac{I_g^W}{I_0} = \frac{|E_g^a|^2}{|E_0^a|^2}(|W|-\sqrt{W^2-1})^2 \quad \text{for } |W| \geq 1 \quad \text{and} \quad I \quad \text{for } |W| < 1. \tag{8.19}$$

In the Laue case the boundary condition at incidence surface of crystal for the waves is as follows:

$$E_0{}^a = E_{01} + E_{02}, \tag{8.20}$$

where $E_0{}^a$ corresponds to the amplitude of the incidence beam into crystal $E_0{}^a \exp(iK_0 \cdot r)$ and suffix 0 means incidence beam direction and *1* or 2 corresponds to α branch and *2* to β branch of the Bloch waves formed in crystal towards the incidence direction, respectively and the other boundary condition towards the diffraction direction

$$0 = E_{g1} + E_{g2}, \tag{8.21}$$

where g means the Bloch waves towards the diffraction direction and *1* and *2* corresponds to α branch and 2 to β branch (Fig. 8.5), respectively.

The amplitude of X-ray waves in crystal can be described as follows:

$$E_{0j} = \frac{1}{2}\left(1 \pm \frac{\mathrm{W}}{\sqrt{W^2+1}}\right) E_0{}^a \tag{8.22}$$

$$E_{gj} = \frac{1}{2}\frac{|P|}{P}\exp(i\alpha_g)\frac{\pm 1}{\sqrt{W^2+1}} E_o{}^a \tag{8.23}$$

These have dependence of W parameter defined in (8.14), where suffix j means *1* or *2*, while 1 for α branch and 2 for β branch. In the Laue case one needs one more boundary condition at the other side of crystal where the X-ray beam goes out after passing through the crystal such that along the incident direction

$$E_{01}\exp(ik_{01z}H) + E_{02}\exp(ik_{02z}H) = E_d{}^a\exp(iK_{0z}H), \qquad (8.24)$$

and along the diffraction direction

$$E_{g1}\exp(ik_{g1z}H) + E_{g2}\exp(ik_{g2z}H) = E_g{}^a\exp(iK_{gz}H). \qquad (8.25)$$

Using the conditions (8.18) and (8.19) one can obtain the intensity profile of diffraction waves and forward diffraction waves as follows:

$$\frac{I_g^W}{I_0} = \frac{|E_g^a|^2}{|E_0^a|^2} = \frac{\sin^2(\pi H\sqrt{w^2+1})/\Lambda}{w^2+1}, \qquad (8.26)$$

$$\frac{I_d^W}{I_0} = \frac{|E_d^a|^2}{|E_0^a|^2} = \frac{W^2+\cos^2(\pi H\sqrt{W^2+1})/\Lambda}{W^2+1}, \qquad (8.27)$$

where these neglect photo electron process of absorption under which condition the relation $I_g^w/I_0 + I_d^w/I_0 = 1$ is obtained for all W. Furthermore I_g^w/I_0 has maximum 1 and I_d^w/I_0 has minimum 0 under the condition

$$H = 1/2(\Lambda + N), \qquad (8.28)$$

where N is integer.

The Bragg-case angle analyzer BAA in Fig. 8.1 creates an image on a single CCD camera so that in order to deduce refraction contrast image one needs two exposures at both flanks of the Bragg peak. The rocking curve obtained by combination of two Bragg peaks is a result of convolution of the two Bragg peaks where the first Bragg peak usually adopts asymmetric one; the degree of asymmetry can be found in the Table 8.1 where the basic parameters involved in two X-ray energies are listed.

Figure 8.6 shows the calculated forward diffraction and diffraction profiles of a 1088 μm thick LAA that is the Laue case (Fig. 8.4) used in XDFI (Fig. 8.3). I_{FD} corresponding to the convex curve downward and I_D to the convex curve upward may change their transmission when the incident angle of X-rays into LAA is varied. The central intensity of I_{FD} becomes minimum at $W = 0$, where W

indicates the angle parameter defined in Eq. (8.14) and I_D becomes maximum at $W = 0$ at a fixed thickness of LAA as predicted in Eq. (8.28). That the left hand side of the rocking curve is higher than the right hand side comes from the difference of photoelectric process absorption between branches α and β. Λ in Eq. (8.13) is the Pendellösung fringe periodicity [49, 50]. $2\pi/\Lambda$ which is the shortest distance between two apexes of the dispersion surfaces in the right figure of Fig. 8.5 is given, for instance, as 1088 μm for 440 diffraction of silicon single crystal at the X-ray energy of $E = 35\,\text{keV}$. The rocking curve is typically acquired by rocking a LAA crystal for the fixed angle of incident radiation, and the recorded value represents the transmissivity or reflectivity of the crystal as a function of the rocking angle. The sum of the percent intensity of the forward diffraction I_d^w/I_0 and the diffraction I_g^w/I_0 becomes not 1 but approximately 0.8 because of photo-electric absorption in practice.

8.4. Experiment

8.4.1. *Experimental set-up*

CABI can be classified into two X-ray optics DEI and XDFI; DEI comprises an asymmetrically-cut Bragg-type monochromator-collimator (MC), a Bragg-case angle analyzer (BAA) and a CCD camera (Fig. 8.1). A specimen is placed in between MC and BAA. The MC receives monochromatic X-ray beams from a double-crystal monochromator (DCM) upstream. The DCM has a function to provide monochromatic X-ray beams and thus to reduce a high power white synchrotron radiation with $\Delta E/E|_{E=35\text{keV}} \sim 10$ down to the energy bandwidth of $\Delta E/E \sim 2.9 \times 10^{-4}$. This MC improves greatly angular-divergence if it adopts asymmetry diffraction that means its crystal surface has a certain angle α against the diffracting planes. This asymmetric cut crystal makes the beam divergence of an outgoing beam smaller by a factor $b = \sin(\Theta_\text{B} - \alpha)/\sin(\Theta_\text{B} + \alpha)$ where Θ_B is the Bragg angle. The beam size is expanded by a factor of $1/b$. For example, when $\Theta_\text{B} = 12.0°$ (440 diffraction at 31 keV), and $\alpha = 11.1°$, $b = 0.04$; the beam after the MC is $60 \times 70\,\text{mm}^2$.

The Pendellösung effect is not utilized but the analyzer is tuned to the tail of the intrinsic RC [59]. The non-scattered beam is not reflected but transmitted in the forward direction, while the scattered beam is reflected in the 2θ direction. The two beams are recorded simultaneously just as XDFI. An algorithm has been developed to separate the contributions of absorption, refraction and scattering [59, 60].

The experimental setup for XDFI comprises an X-ray source, an asymmetrically-cut Bragg-type MC, a Laue-case angle analyzer (LAA) and two CCD cameras (Fig. 8.4). The specimen is placed in between MC and the LAA. XDFI records refraction and scattering separately using two cameras. The reflected beam from the first crystal, expanded by the asymmetry of the crystal, has a very small divergence so that the Pendellösung fringe is observed in the analyzer crystal that has uniform thickness with unevenness smaller than 2 μm. When the crystal thickness is chosen properly, and the analyzer is tuned to the peak position the non-deviated beam is diffracted by angle $2\Theta_B$, while the refracted and scattered part passes through the crystal in the forward direction [6–8, 121]. The two beams are recorded simultaneously.

For phase contrast computed tomography, a specimen is mounted on a rotational stage with smallest precession and record 360 projection images while rotating the specimen around a vertical axis by 180°. Its angular interval is 0.5°. It takes 2–7 seconds per each projection image for data acquisition, approximately 1 s for data transfer from a CCD camera to a PC, and 1 s for saving in the PC and approximately 5 s for recording background by removing specimen from the beam incident every 10 frames. Thus the total data acquisition time for one tomogram is approximately 1.5 hours. Notably, the time for data transfer and storage can significantly be reduced by equipping a camera with large memory and transferring all the images after completing the data acquisition.

8.4.2. *Sample preparation*

All human specimens imaged in this study are obtained either from autopsies or excisional biopsies and mastectomies, with the entire experimental process cleared by the Institutional Review Boards of the institutes that provided specimens such as Nagoya Medical Center (NMC), Catholic University of Daegu and Massachusetts General Hospital (MGH). All specimens are immersed in a formalin bath with 10% concentration. The same specimens in case of MGH are also imaged with multi-detector CT (Definition Flash, Siemens Medical Solutions) at 80 kV acceleration energy and with the beam current of 154 mA. The 3-D volume reconstructed with 0.6-mm slice thickness is compared with XPCI tomographic images. The tissue specimens from excisional breast biopsies or mastectomies have been embedded in paraffin before histological sectioning at NMC. The surface of the paraffin block is polished using a fine sand paper to suppress surface unevenness during experiment at KEK. For imaging, all the specimens are put in a square water bath to let the X-ray beam in and out vertically to the water bath wall. Before recording images, air bubbles are carefully removed from the field of view. After imaging, we stain all the specimens with hematoxylin and eosin for histology images that are manually correlated with the phase and attenuation images.

Things to remind us as users are (1) content of 10% formalin or 4% formaldehyde should be precisely described; (2) specify human tissue; (3) specify organ; (4) describe how to remove infectious possibility; (5) describe that the experimenter will take care for possible touch to formalin and distribution of formalin vapor into air in the experimental hall; (6) in case of using tissue from patient description about whether patient may or may not receive disadvantages, whether or not privacy of patient is defended, whether or not which study is used for, and in case of being able to produce intelligent properties whether abandon the patent or return to patient, whether patient is satisfied with description in relation to all these items, whether form or procedure has been approved at IRB. (7) In the IRB should not describe such that 'fully paying attention to personal information' but clearly how to defend patient's each information. In case of bringing samples out of the institute each institute to which specimens will be in use for study or with whom specimens will be handled.

8.5. Algorithm and Data Processing

8.5.1. *Algorithm for computation of the angle of refraction*

Both in DEI and XDFI the rocking curves describe the amount of angular deviation in the incident beam due to the refraction in the object before the crystal. If the object being imaged is a pure "phase object," i.e. it has no absorption, the refraction angle can be directly calculated from the rocking curve in DEI and the FD or D image in XDFI.

In order to obtain correct value of refraction from a rocking curve of both angle analyzers, BAA in DEI and LAA in XDFI curve fitting is crucial. A simpler and linear approximation using a maximum slope at the full-width at half maximum of the rocking curve has been first proposed by Chapman [20] in DEI. Later Rigon, Chou [22, 105, 106] developed a technique on the base of the first order Taylor expansion at the mid-slope of the low angle side and of the high angle side of the rocking curves. Furthermore a variety of algorithms for approximating rocking curve have been proposed by many [14, 33, 39, 42, 56, 62, 66, 76, 150, 151].

Later another fitting algorithms have been developed such as extended DEI, multi-image radiography and Gaussian curve fitting; a single-valued inverse function introduced by Maksimeko *et al.* [73, 74], the Voigtian function introduced by Candes *et al.* [18] and the Pearson VII function by Kitchen [58, 59] can clearly differentiate soft tissue with a large dynamic range.

In case of LAA in XDFI rocking curve fitting with a polynomial function was proposed by Sunaguchi *et al.* [121]

$$I^{FD}(\theta) = \sum_{k=0}^{n} a_k^F \theta^k \quad \text{and} \quad I^{D}(\theta) = \sum_{k=0}^{n} a_k^D \theta^k, \tag{8.29}$$

where I^{FD} and I^D represent the forward diffraction and the diffraction components of the measured intensities, respectively as shown in Fig. 8.7. In this equation, θ is the rocking angle of the LAA crystal, while a_k^F and a_k^D are the, fitting coefficient for the forward diffraction and the diffraction rocking curves, respectively. In practice, human

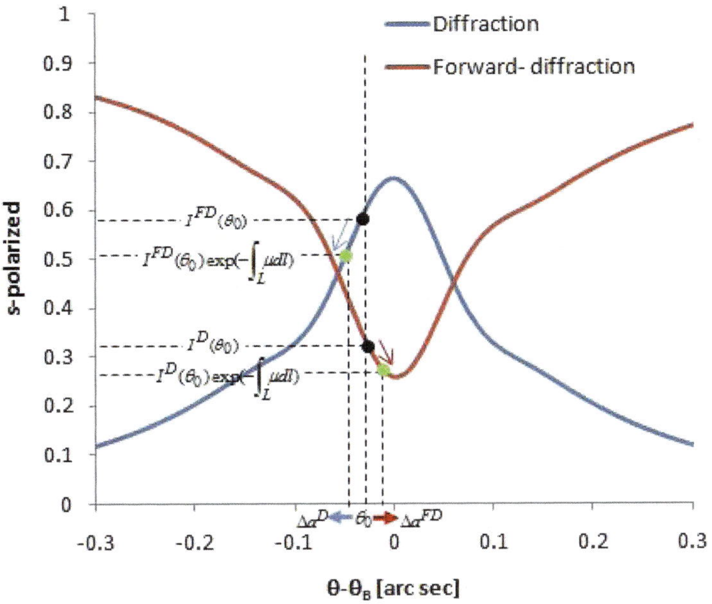

Fig. 8.7. Rocking curve of LAA; the central angular position Θ_o is intrinsic. However images may cause angular displacement. Genuine angular position can be recovered by considering the tissue absorption and by iteration of the curve fitting. Smoothness of the FD and D peaks come from convolution result between the theoretical curve and the incident beam that has 0.05 arc sec.

tissue has non-negligible absorption [98] and the observed intensity is due to both the refraction and the absorption. Because both FD and D beams are subject to the same amount of absorption, we may rewrite ${I_{obs}}^{FD}$ and ${I_{obs}}^{D}$ as follows:

$$ {I_{obs}}^{FD} = I^{FD}(\theta_0) \exp\left(-\int_L \mu dl\right), \tag{8.30} $$

$$ {I_{obs}}^{D} = I^{D}(\theta_0) \exp\left(-\int_L \mu dl\right). \tag{8.31} $$

From the Eqs. (8.17) and (8.18), we obtain

$$ \sum_{k=0}^{n} \left(\frac{a_k^{FD}}{I_{obs}^{FD}} - \frac{a_k^{D}}{I_{obs}^{D}} \right) {\theta_0}^k = 0. \tag{8.32} $$

We numerically solve this algebraic equation using the Newton–Raphson method [133] to obtain the refraction angle θ_0, with the absorption effect corrected, for each pixel in the projection images. Typically, the optimal degree n for the polynomial is selected 6 following Akaike's information criterion.

When the LAA is placed in the linear segment to the left of the minimum in the dark-field rocking curve, there is differential enhancement of the edges: refraction to the left of this placement angle causes an increase in the observed intensity while that to the other side decreases the observed intensity. In order to avoid ambiguity of sense of refraction that happens when the angular position of LAA is placed at the minimum of the rocking curve (Fig. 8.6 and 8.7) use of the left wing of FD and D curves (Fig. 8.6) is recommended. Also this way of use of the rocking curves will give us a maximum dynamic range of grey scale in contrast of image. We note that bones and calcifications impart large angles of refraction to the incident beam, so the measurement points do not fall on this linear portion; this may introduce CT reconstruction artefacts.

8.5.2. *Tomographic reconstruction*

The principle of the crystal-based X-ray optics imaging was already described in Section 4. Here, tomographic reconstruction techniques that have already been proposed for the crystal-based X-ray optics imaging, is described. Based on the previous works [33], a direct reconstruction algorithm for DEI [74] and for XDFI [122, 123] has been derived. The XDFI optics collects a set of projection images or forward-diffraction and diffraction images, and thereby abstracts a set of projections with respect to angular deviation as described by Section 4. If one regards the quantitative relationship in measurement process as important, one should select FBP method with signum function filter (SGN-FBP). In order to apply SGN-FBP method, one must make clear the quantitative relationship between physical quantities observed and to be estimated in a cross section of interest. The measurement process of angular deviation is represented as

$$\Delta\alpha(x) = \int_{-\infty}^{\infty} \frac{\partial}{\partial x}\delta(x,z)dz, \tag{8.33}$$

where the incident beam propagates toward the positive z-axis. $\delta(x,z)$ and $\Delta\alpha(x)$ are a distribution of phase shift and an angular deviation at the beam position x, respectively. Equation (8.33) means that the angular deviation is equal to the Radon transform of the derivative of phase shift distribution with respect to x. Taking the Fourier transform of Eq. (8.33) with respect to x, we get the following equation after some simple manipulation:

$$\int_{-\infty}^{\infty} \delta(x,z)dz = \Delta\alpha(x) \otimes F^{-1}\left(\frac{1}{2\pi i\rho}\right), \tag{8.34}$$

where x is the pixel coordinate, δ is the real part of refractive index of a specimen, $\Delta\alpha$ is the angular deviation, $\otimes$ is the convolution operation and F^{-1} is the inverse Fourier transform where ρ is the spatial frequency. The inverse Fourier transform term on the right hand side of Eq. (8.33) is

$$F^{-1}\left(\frac{1}{2\pi i\rho}\right) = \mathrm{sgn}(x) = \begin{cases} 1 & (x > 0) \\ 0 & (x = 0) \\ -1 & (x < 0) \end{cases} \tag{8.35}$$

Using this relation, we can interpret that Eq. (8.34) simply tells that the Radon transform of the refractive index can be connected to a convolution of the function sgn(x) with the angular deviation recorded with XDFI. Therefore, the refractive index map can be obtained by applying the standard filtered back projection to the right hand side of Eq. (8.34). More detailed explanation of this algorithm can be found in [121, 122]. There are also very good review articles of synchrotron radiation-based medical imaging [14, 125].

We note that when a sample includes hard tissues such as calcification or bone, crystal-based XPCI can generate artifacts in a refraction-contrast CT image, whereas an absorption-contrast CT does not. This is because the angle of refraction from those hard tissue regions does not fall on linear portion of the rocking curve.

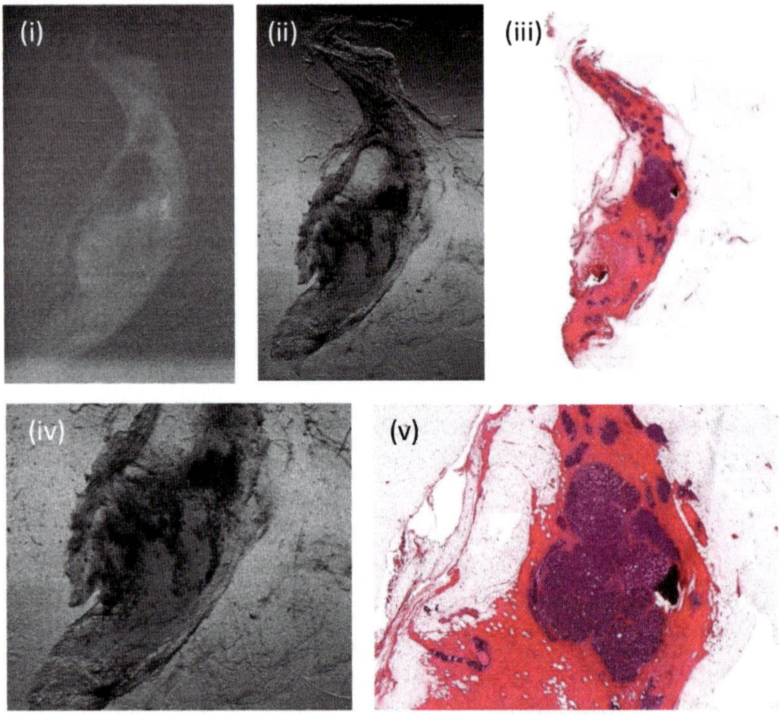

(a)

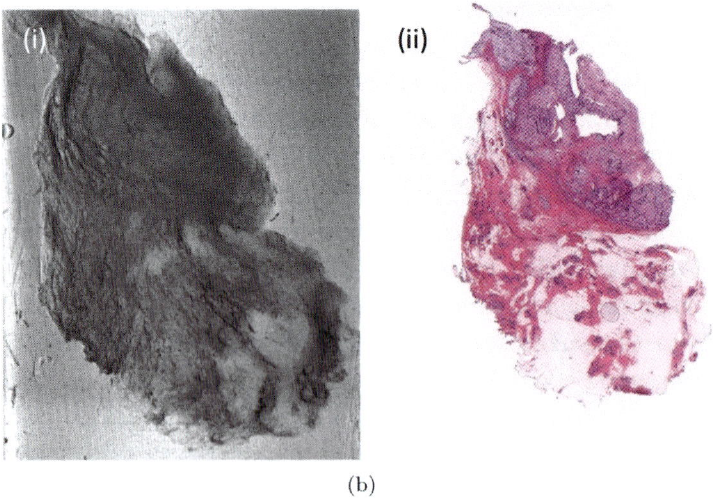

(b)

Fig. 8.9. (Figure on facing page) Breast tissue excisional biopsy specimens with benign tumours. (a) Specimen from a 33-year-old woman with a small breast mass and bloody nipple discharge: (i) absorption; (ii) phase; and (iii) a matched H&E-stained slide. Panels (iv) and (v) are zoomed-in views of (ii) and (iii), respectively. (b) Tissue specimen from a 42-year-old woman with a long-term follow up of a slowly enlarging mass that was surgically removed: (i) phase image and (ii) H&E-stained histopathology slide.

on histopathology as shown in Fig. 8.9(Aiv,v), in which a diagnosis of benign nodular adenosis was made. Figure 8.9(B) shows another benign breast tumour in a 42-year-old woman whose slowly enlarging mass was surgically removed. The phase image (Fig. 8.9(Bi)) shows well-demarcated, homogeneous tissue compressing the normal ducts and the TDLU. The histopathology, shown in Fig. 8.9(Bii), confirmed the diagnosis of a phyllodes tumour, a circumscribed fibroepithelial neoplasm composed of a double-layered epithelial component and a hypercellular stromal component.

Figures 8.10(A, B) show examples of invasive carcinoma in two different patients. The first patient (Fig. 8.10(A)), a 38-year-old woman with left breast stiffness, demonstrated an infiltrating mass on mammography and MRI. The phase contrast image (Fig. 8.10(Ai)) shows an expansile lesion that permeates the breast parenchyma, with preserved normal TDLU and multiple cysts, an appearance confirmed by histopathology (Fig. 8.10(Aii)) where a diagnosis of invasive carcinoma of no special type was made. Figure 8.10(B) shows a phase image and the corresponding histopathology images from a 45-year-old woman with right nipple haemorrhagic discharge and multifocal breast tumour with infiltration of the nipple. The infiltrative nature of the tumour, with adjacent skin invasion is well demonstrated on both the phase image Fig. 8.10(Bi) as well as the histopathology slide (Fig. 8.10(Bii)).

Figure 8.11(A) shows the case of a 54-year-old woman with discharge from the left nipple. Mammography revealed heterogeneous dense breast parenchyma with suspicious microcalcifications. The absorption image (Fig. 8.11(Ai)) of the excisional biopsy specimen is inconclusive because of poor soft tissue contrast. The phase

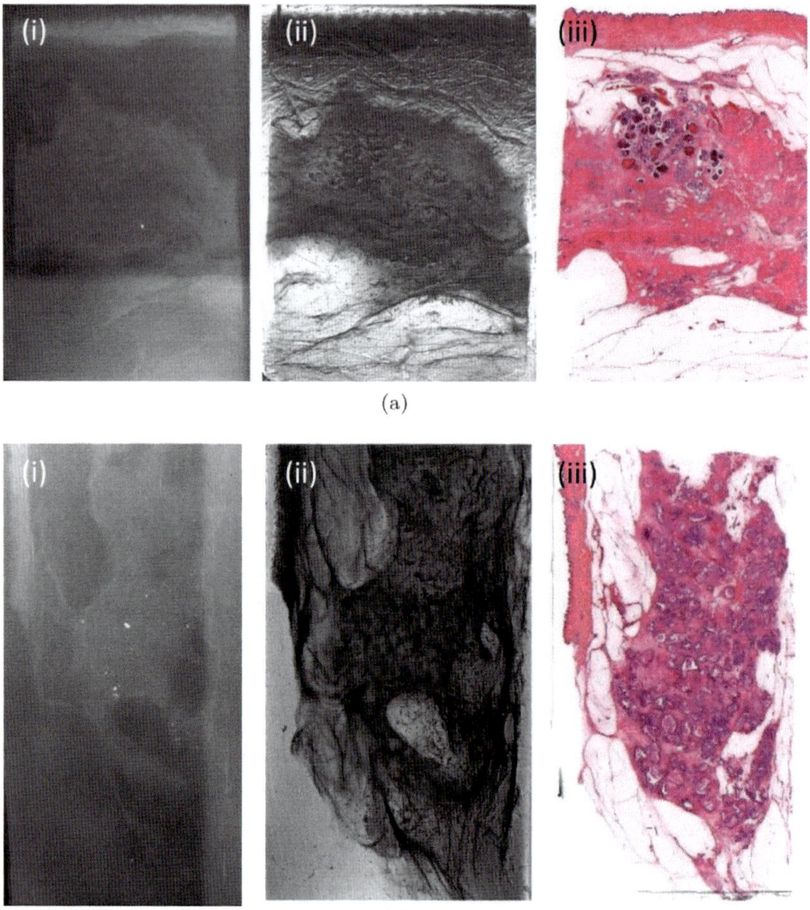

Fig. 8.11. Breast tissue specimens with ductal carcinoma in situ (DCIS). (a) Specimen from a 54-year-old woman with discharge from the left nipple: (i) absorption; (ii) phase images; and (iii) a matched H&E-stained slide. (b) A second specimen from a different patient with the same pathological condition: a 37-year-old woman with a palpable stiffness in the left breast; (i) absorption; (ii) phase images; and (iii) a matched H&E-stained slide.

8.6.1.2. *Articular cartilage and bone*

Articular cartilage is another promising organ to find out early rheumatoid arthritis. Many such works have been done [6, 7, 24–27, 31, 63, 67–69, 72, 85–90, 111–115, 137]. In Fig. 8.12 is shown

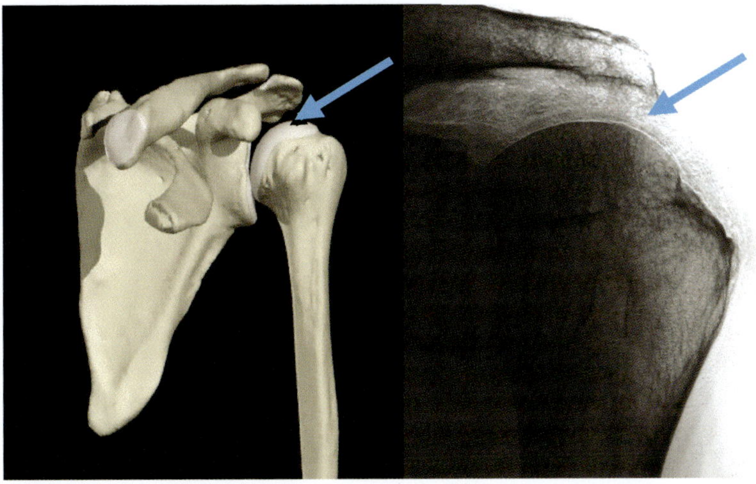

Fig. 8.12. XDFI image of shoulder in the right from a cadaver. MRI image is shown in the left. Articular cartilage is clearly seen. FOV was 9 cm × 9 cm. X-ray energy was 35 keV. This x-ray image was taken at BL20B at SPring-8. Courtesy Shimao: allowed use of this image.

an example of articular cartilage at human shoulder excised from a cadaver [5]. This was taken at BL20B at SPring-8. Also lung imaging using animal has been performed and attempted at visualizing human stomach by [130].

8.6.2. *3D and 2.5D tomosynthesis imaging*

8.6.2.1. *DCIS and virtual endoscopy*

In clinical treatment it is believed that medical doctors will request 3D information of sick part, especially in case of surgery. In Fig. 8.13 is shown 3D image of milk duct (ductas lecteferi) of DCIS (ductal carcinoma *in situ*). The data was acquired through an X-ray CCD camera (X-FDI6 Photonics Science, East Sussex UK) that has a view size of 8.7 mm (h) × 6.9 mm (w) with 16-bit and 1392 × 1040 pixels. The size of each pixel is 6.3 μm × 6.6 μm. The upper limit of the specimen size was determined by the detector (CCD camera) size rather than the beam size. The beam size at Photon Factory in Tsukuba is 30 mm × 30 mm. The DCIS sample was rotated by step of 0.2 degree

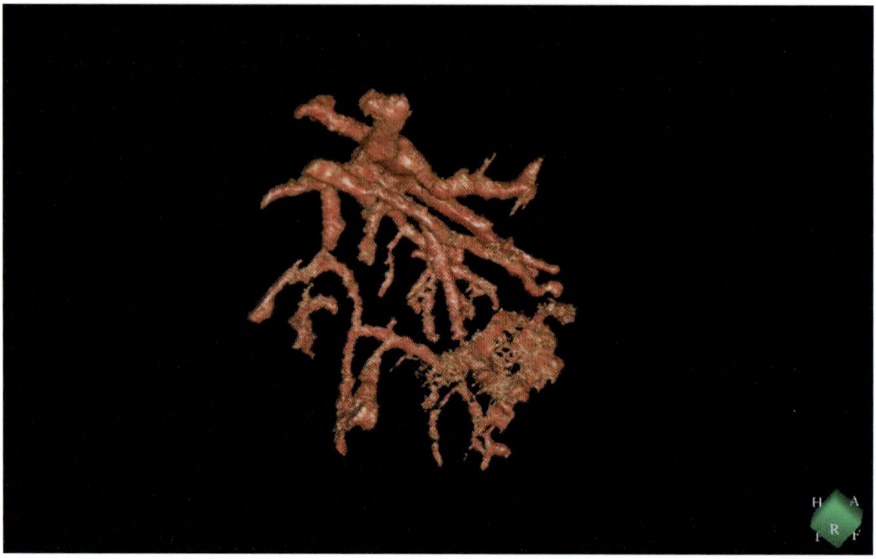

Fig. 8.13. 3D reconstructed image of DCIS. The view size is approximately 3.8 mm × 4 mm. This was used for endoscopy shown in Fig. 8.16.

so that total number frames of X-ray image was 900. Furthermore by helping of the algorithm developed by Mori [83, 84] inside milk duct is visible as shown in Fig. 8.14. This is a result of multiple calcifications occurring in the necrotic zone of the continuous DCIS of Fig. 8.13, resulting in the linear and branching microcalcifications. Dicotomatous branching of the ducts was also confirmed with virtual endoscopic observation. In virtual endoscopy, comedo-type necrosis appeared as a rope-like material (Fig. 8.14) occupying the center of the ducts [44].

8.6.2.2. *Eye specimen*

Using XDFI, we have imaged the anterior chamber of an enucleated eye. The sagittal slices of the reconstructed differential phase map (Figs. 8.15(A) (i)–(iv)) show markedly increased detail and contrast, especially at the interfaces. These images clearly demonstrate the margins of the cornea, iris, pupil, ciliary body, angle, lens, and the suspensory ligament of the lens. These differential phase images

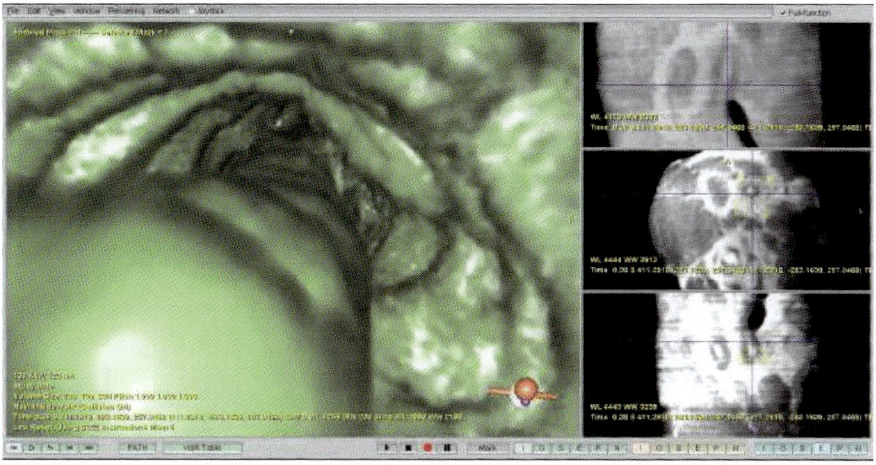

Fig. 8.14. 3D reconstructed image of mammary DCIS that is virtual endoscopy. The original 3 way of reconstructed images are shown in the right figure. The cross in the middle shows the entrance with which endoscopy starts. The comedotype necrosis appeared as a continuous rope-like structure within the duct. The size of the duct is around 0.6 mm.

were converted into the refractive index images in the coronal and sagittal (Figs. 8.15(B) (i) and (ii), respectively) planes where the same structures can be visualised. For comparison, the anterior chamber of the same specimen, imaged using standard multidetector CT (MDCT), and H&E (haematoxylin and eosin) staining are shown in Figs. 8.15(B) (iii) and (iv), respectively. We have also imaged the posterior globe and the site of optic nerve insertion of the same specimen using XDFI. Figure 8.15(C) clearly shows the tri-laminar structure of the globe consisting of the retina, the choroid and the sclera. The insertion of the optic nerve into the globe is clearly shown along with the central retinal artery, the nerve fascicles and the optic nerve sheath which is contiguous with the sclera.

The dark-field CT slice through the globe also shows separation of the retinal and choroidal layers from the underlying sclera at multiple places. This may be due to the post-mortem condition, prolonged formalin-fixation, and contraction of the vitreous humour. Irrespective of the aetiology, the retinal and choroidal detachments are well demonstrated by XDFI. In fact, the phase contrast image can show

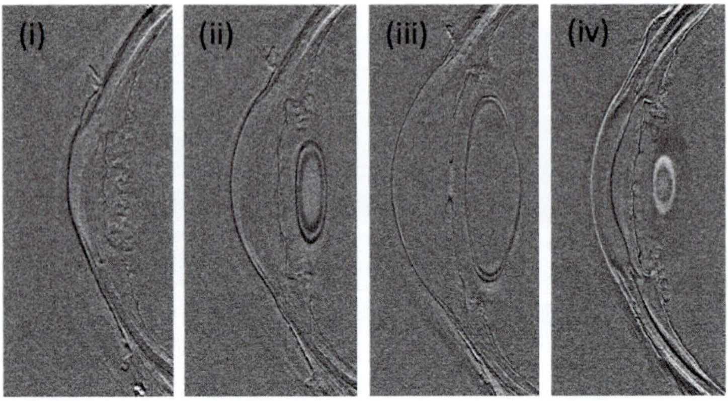

(a)

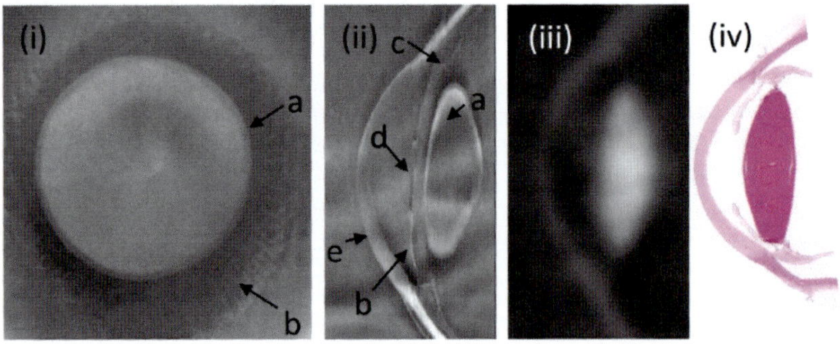

(b)

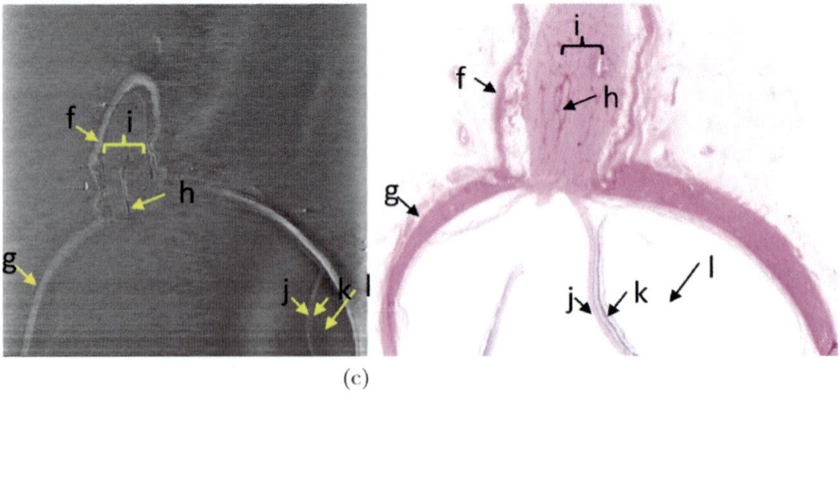

(c)

Fig. 8.15. (Figure on facing page) Anterior and posterior chambers of an eyeball imaged with XDFI. (a) Tomographic slices (sagittal plane) of the differential phase map of an anterior chamber ((i)–(iv)). (b) Refractive index images in the coronal (i) and sagittal (ii) planes, a matched absorption image from conventional MDCT (iii), and a histological H&E slide showing this specimen (iv); the difference in the imaging and histological appearance are due to tissue retraction during histological processing and sample preparation. (c) Globe and optic nerve insertion of the same specimen. a: lens; b: iris muscle; c: ciliary body; d. pupil; e. cornea; f. optic nerve sheath; g. sclera; h. central retinal artery; i. nerve fascicles; j. retina; k. choroid; l. sub-choroidal space.

whether only one or both layers are involved. The histology images confirm the presence of retinal and choroidal detachment. The histological processing of the specimen likely worsened the detachment of the retina and the choroid, as can be seen in Fig. 8.15(C), where it extends up to the optic nerve insertion.

8.6.2.3. *Atherosclerotic plaque*

We also applied XDFI and MDCT to a heavily diseased iliac artery containing atherosclerotic plaque. Figure 8.16 shows an absorption slice (Fig. 8.16(i)) and a phase tomographic slice from XDFI (Fig. 8.16(ii)), an approximately matched CT image from an MDCT system (Fig. 8.16(iii)), and the corresponding histopathology slide (Fig. 8.16(iv)). Figure 8.16(i), the absorption image, shows only calcifications in this atherosclerotic plaque specimen. The phase image (Fig. 8.16(ii)), on the other hand, shows a lot of details of the atherosclerotic plaque components. For example, the tri-laminar structure of the vessel wall consisting of the adventitia, the media and the intima can be appreciated. In addition, the intima, which is contiguous with the fibrous cap over the atheroma within the plaque, is clearly visualised. These structures are distinct from the calcification which shows artefacts from a phase saturation effect. The phase shift value of the round plastic rod placed adjacent to the specimen is highly uniform, which shows the fidelity of our method. The details seen on the phase image of plaque morphology are well matched with the corresponding histology slide (Fig. 8.16(iv)).

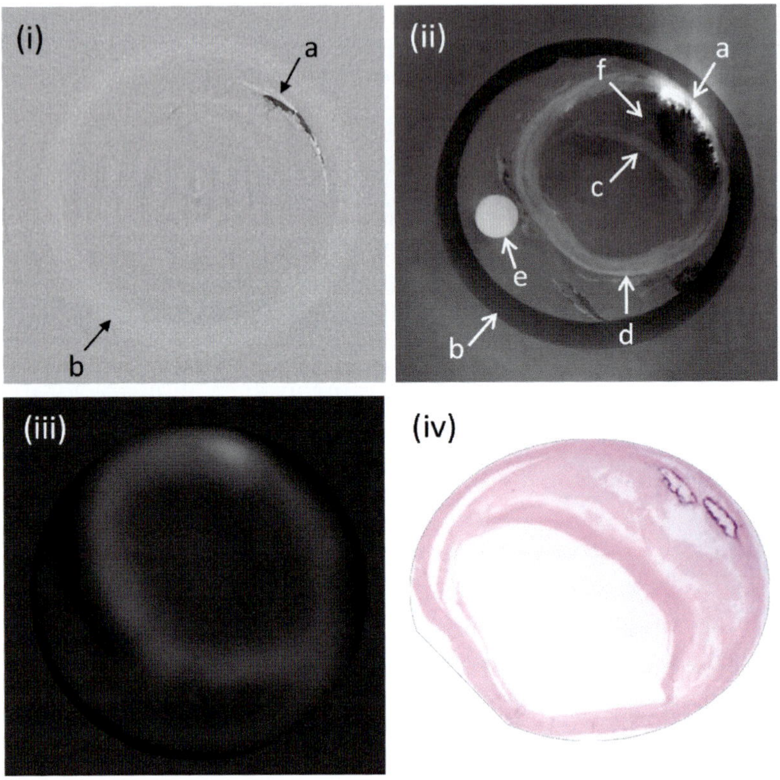

Fig. 8.16. An iliac artery specimen with atherosclerotic plaque. Absorption (i) and phase (ii) tomographic slices from XDFI, a matched CT slice from MDCT (iii), and the corresponding H&E-stained histopathology slide (iv). a: calcification; b: plastic container; c: fibrous cap; d: three-layer arterial wall; e: plastic rod; and f: soft plaque.

In this chapter, we presented an X-ray dark-field imaging method and demonstrated its capability by imaging an eye, components of an atherosclerotic plaque specimen, and breast tissue samples with benign and malignant tumours. The results obtained with XDFI clearly demonstrate the superior capability of phase contrast imaging for soft-tissue discrimination in both 2D projection and 3D tomographic domains. In the eye specimen, XDFI was able to discriminate soft-tissue structures, such as the iris, separated by aqueous humour on both sides, which have nearly equal absorption. Superiority of

XDFI in imaging soft-tissue was further demonstrated with a diseased iliac artery containing atherosclerotic plaque and breast samples with benign and malignant tumours. XDFI on breast tumours discriminated between the normal and diseased TDLU, which is significant because of the importance of knowing whether the tumour margin is displacing or infiltrating the breast parenchyma. In the past, such information could only be gathered by histopathological examination. While the small sample size precludes us from deriving definite phase contrast signatures for different types of breast cancers, the fact that the increased soft tissue contrast enables better visualisation of the invasiveness of cancer and destruction of native architecture is unmistakable. Phase contrast imaging can also distinguish *in situ* disease from invasive cancer by differences in X-ray refractivity and localisation in the normal structures.

8.6.2.4. *Brain tumor*

It has been most difficult to see brain inside skull so that seeing brain by removing skull was done by Connor *et al.* [26] and Seo [110] (see Fig. 8.17). Tumor infiltration in surrounding normal tissue of malignant glioma, crucial for survival rate and to determine the margin of surgical resection, often accompanies with microvascular proliferation (MVP) that was invisible with current clinical imaging methods. Figure 8.17, obtained from a DEI-CT study of rat glioma, shows histologic assay-compatible contrast among edema, ring-shaped viable tumor, central necrosis, and multiple hypointense speckle-like structures, identified as microvessel proliferation (MVP) structures with typical diameter of 30–160 μm. Potential α-smooth muscle actin in hypertrophic cells may contribute partly to this contrast. Figure 8.18, a 3D image reconstructed from serial 2D DEI-CT data, delineate three compartmental volumes, edema (green), viable tumor (yellow) and central necrosis (blue) in different colors. The architectures of the semi-cylindrical morphology of MVP (red colored) were more clearly visualized with partial sprouting features in peritumoral edema where tumor infiltration occurred. This MVP presents circular lumen and hyperplasia endothelial cell layer at mild

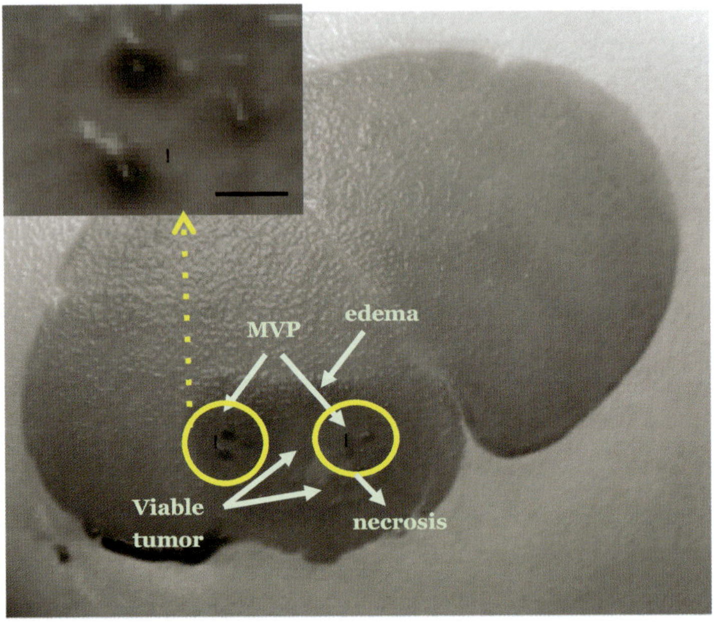

Fig. 8.17. A selected tumor slice of 2D reconstructed DEI images shows contrast among edema, ring-shaped viable tumor and central necrosis, which was compatible with histology assay. In addition, multiple hypointense speckle-like structures, identified as microvessel proliferation (MVP) structures with typical diameter of 30–160 μm, appeared in peritumoral edema. Lumen and part of hyperplasia cell layer appeared with contrast in an expanded area of MVP.

stage, but grows in a more sophisticated manner to a glomerulus-like structure containing three or more vessel lumina in more advanced stage of tumor infiltration. Therefore it is important to realize high-resolution skull-based brain imaging with diffraction-based SR technique in the near future to determine the tumor infiltration for surgical or radiation treatment of malignant glioma.

8.6.2.5. *Tomosynthesis*

Since the radiation dose will have to be taken into consideration it will be better to introduce an idea of tomosynthesis [35, 147] because by using the algorithm 'shift and add' or 'sgn shepp filter with Fourier back projection' one could expect dramatic decrease of

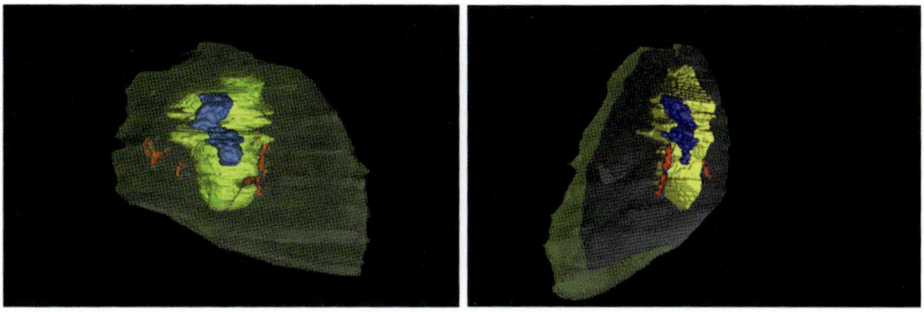

Fig. 8.18. Two different angle views of a 3D image reconstructed from serial 2D DEI-CT data delineate three compartmental volumes, edema (green), viable tumor (yellow) and central necrosis (blue) in different colors. The architectures of the semi-cylindrical morphology of MVP (red colored) were visualized with partial sprouting features in peritumoral edema where tumor infiltration occurred.

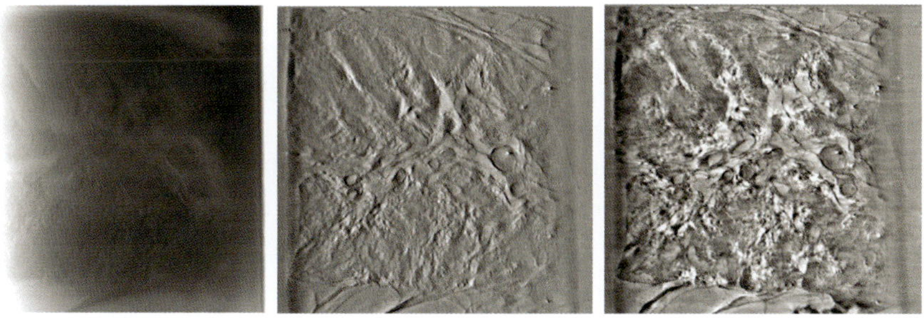

(a) (b) (c)

Fig. 8.19. Tomosynthesis using (a) shift and add, (b) sgn shepp filter with Fourier back projection of algorithm, (c) CT image. (a) and (b) were reconstructed by 55 projections while (c) was by 900.

absorption radiation dose to reconstruct pseudo 3D or 2.5D image [114, 115, 123]. Figure 8.19 shows two kinds of tomosynthesis using algorithm (a) shift and add (b) sgn shepp filter with Fourier back projection and other (c) CT image. (a) and (b) were reconstructed by 55 projections while (c) was by 900. Sample was DCIS. Apparent differences of image quality between (b) and (c) in the view plane, in other words, contrast and the spatial resolution in the plane vertical to the X-ray propagation are not seen so that the tomosynthesis can be useful and promising for clinical survey. The absorption radiation dose in case of tomosynthesis can be only 1/16 of that for CT.

8.7. Future Prospect

A goal of DEI and XDFI optics will be a clinical use and a pathological use. In case of clinical use one needs a large field of view (FOV), while in case of a pathological use one appreciates a high performance of a detector with a reasonable FOV ∼ say 25 mm in square and with a fine pixel equivalent to or smaller than 2–3 microns. A high performance X-ray source other than synchrotron radiation would be a serious factor for both purposes to achieve. That a variety of compact radiation sources are under development [41, 64, 71, 134, 135, 144] will support this direction.

Asymmetric diffraction shown at MC in Figs. 8.1 and 8.3 is a key technology not only to make the divergence of X-ray outgoing beam small but also the size of the outgoing beam larger. FOV for clinical purpose and a medium FOV for pathological purpose will be achievable by adopting is one solution. The current largest FOV achieved is 80 mm × 80 mm for visualizing articular cartilage of a shoulder excised from a cadaver ([6]). When designing clinical system one may use LAA cut out from a silicon boule. Largest diameter of floating zone (FZ) silicon single crystal is 200 mm while that of Czochralski (Cz) grown silicon crystal is 300 mm. Furthermore authors have been informed that already development of 450 mm diameter has been completed. Thus the maximum field of view (FOV) with 140 mm × 140 mm will be available from FZ crystal and 210 mm × 210 mm will be available from a Czochralski (Cz) crystal and possibly FOV with 318 mm × 318 mm will be available in a nearest future if using Cz crystal. Furthermore one could look forward to using a Cz-FZ crystal newly developed. A way to improve crystal quality is as follows: first Cz crystal has been grown due to lower cost and this crystal was put into FZ furnace to improve its crystal quality towards FZ crystal. Figure 8.20 shows test X-ray topography. A wafer for tesing was cut out from a boule and chemically etched by ourselves in a laboratory so that uneveness may remain. The contrast of images

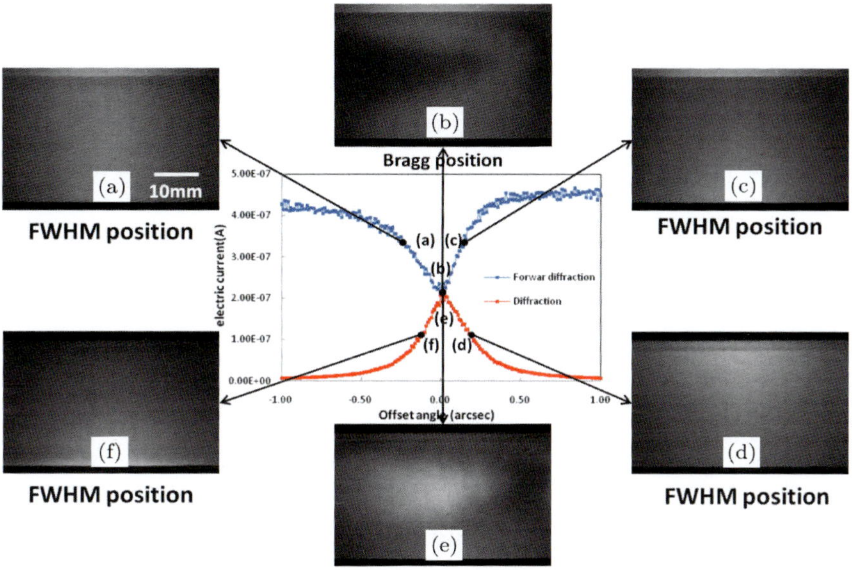

Fig. 8.20. X-ray topographic study of a Cz-FZ silicon wafer. Diffraction index is 440, X-ray energy is 35 keV. FWHM of the forward diffraction and diffraction shows 0.40 arc sec and 0.33 arc sec, respectively which is quite close to a theoretical value of 0.385 arc sec. There seems almost no diformation remaining in a crystal. FOV of the images shown is 33 mm vertical and 49 mm horizontal.

can be explained such that mechno-chemical polishing can lead to homogeneous contrast. The images shown are smaller as a first step. As a next step towards a maximum size 400 mm and even 450 mm. They can provide us enough size for clinical purpose of FOV with size of 280 mm $\times$ 280 mm and 315 mm $\times$ 315 mm. The radiation dose at 35 keV is approximately 1/7 of that at 17.2 keV which is believed to be the center of the spectrum in use for mammography and the absorption radiation at 35 keV will be 1/17 of that at 17.2 keV. The further the X-ray goes up the smaller the skin radiation dose and absorption radiation dose becomes. Naturally the most appropriate X-ray energy should be fixed by considering the synchrotron radiation spectrum.

At present, a CCD camera with pixel size of 7.4 μm is in use mostly for pathological image, while in clinical diagnosis the pixel size of 50 μm can be used. This will additionally reduce the exposure time to 4–20 msecs. The CCD camera (X-FDI 1.00:1, Photonic Science Ltd.) used for the recording has a 7.4 μm pixel size and a reasonably large field of view (36.1 mm (h) $\times$ 24.0 mm (v)). The CCD chips have 16-bit dynamic range and are air-cooled.

Large size of 2D detector has been already commercially available from Hamamatsu Photonics, Shimadzu, Hologic, Anrad, Fuji, GE, Canon, Photonic Science, Rigaku, Toshiba and Konica-Minolta.

In a real clinical mode separation of imaging signal from the very strong signals from bones is crucial. Already a way to remove too large refraction component from a mixed image has been proposed [124].

Among the above three key components the X-ray source will be a key element so that one could say that such an X-ray source will be able to open a door to a tomorrow's clinical tool using refraction imaging with which the system can save a lot of lives.

Additionally absorption radiation dose has been estimated (NIST database [91]) and measured [108] because in case of applying synchrotron radiation to clinical survey skin radiation dose and absorption radiation dose will be crucial. An example was taken in a case of routine mammography at hospital where X-ray energy representative is 17.5 keV and an expecting novel system using refraction contrast by DEI or by XDFI at 35 keV. Approximate calculation says that skin radiation dose at 35 keV may decrease by a factor 7 compared to 17.5 keV and absorption dose may decrease by a factor 17 compared to 17.5 keV. This means refraction-based imaging seems quite promising from a radiation dose point of view.

Last but not the least, just a few words about a perspective for a sort of X-ray pathology will be expressed. Figure 8.21 explains significant improvement of contrast and spatial resolution of soft tissue by thinning LAA from 2124–290 μm. Specimen is a papilloma breast cancer tissue. Computational prediction of spatial resolution [126] using Takagi-Taupin equation [127, 128, 131] can lead to how much improvement can be expected; 2124 μm will give us the spatial resolution $\sim$50 μm and 290 μm will give 10 μm. This could expect us being able to open X-ray pathology if the X-ray technique can provide the spatial resolution better than 5 μm at

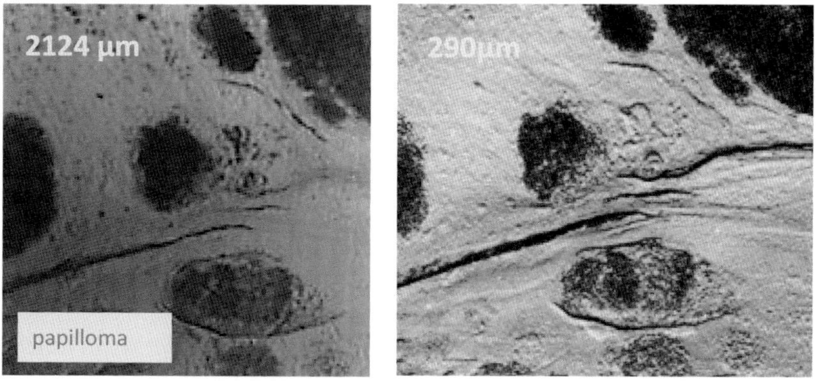

Fig. 8.21. How spatial resolution will be improved by thinning LAA down from 2124 μm to 290 μm. Specimen was used papilloma breast cancer tissue. Takagi-Taupin equation tells us significant improvement from 50 μm to 10 μm. Further improvement of the spatial resolution by even thinning of LAA can be expected.

a first stage and hopefully around 1 μm at a second stage. Needless to say that FOV should coincide typical pathology specimen size approximately 30 mm $\times$ 30 mm. In this paper, we describe the most recent advancements in XDFI and apply it to a variety of human tissue specimens. Specifically, the thickness of an analyser crystal has been reduced, which is installed in transmission Laue mode, to 150 μm. This provides the spatial resolution of 1.3 μm and 4 μm in the vertical and horizontal directions, respectively. In addition, we adopt an asymmetric-cut monochromator-collimator, which enables full-field imaging of a large specimen without scanning the beam across it. The phase images obtained with the proposed XDFI system clearly demonstrate soft tissue discrimination capability in various specimens. The beam size, which may be estimated from the FWHM along each direction, is 72 μm and 1160 μm along the vertical and horizontal directions, respectively. The sample plane is located 38 m away from the source, and the CCD camera is at 0.7 m from the sample. The spatial resolution given for this imaging geometry is 72 μm $\times$ 0.7 m/38 m = 1.3 μm in the vertical direction and 1160 μm $\times$ 0.7 m/38 m = 21.4 μm in the horizontal direction. MC collimates the beam, and this reduces the effective source size, and thus improves the spatial resolution. However, the beam diffraction within the LAA lowers the spatial resolution. From a numerical integration of the Takagi and Taupin equation [126, 128, 131], the degradation in spatial resolution can be calculated as a function of the crystal thickness. The final spatial resolution of the current XDFI system is theoretically given as 1.3 μm in the vertical direction and 4 μm in the horizontal direction. We note that these values can be further improved by narrowing the wavelength spread and by thinning the LAA down to 100 μm or less. XDFI can be performed in a high-speed mode using two cameras, which record forward diffraction (FD) and diffraction (D) images simultaneously. Alternatively, in this study we used a single camera to record the FD and D images sequentially.

8.8. Conclusion

If one will set a goal of refraction-based imaging at clinical use DEI and XDFI can be a candidate if a compact, less expensive and powerful enough light source in X-ray domain will be available. Survey starting from Intravenous Coronary Angiography [1, 2, 34, 37, 43, 48, 61, 92, 93, 107, 120, 132], describing X-ray dynamical diffraction theory, details of crystal-based optics of DEI and XDFI and their applications have been reviewed precisely. Since a number of good reviews on DEI using a Bragg angle analyser have been published a bit emphasis was set on XDFI using a Laue angle analyser.

Acknowledgement

Authors would like to express their sincere thanks to the Photon Factory, KEK and the SPring-8 where all data described in this chapter have been achieved at either facility. They also appreciate Professor Seishi Kikuta for his kind permission of partial use of his equations in the chapter of X-ray dynamical diffraction theory in his book entitled X-ray Scattering and Synchrotron Radiation Science – X-ray Basic (published by University of Tokyo Press, in Japanese). Also encouragement has continuously been given to the authors on XDFI work from Professor Yoshinori Chikaura at Kyushu University to whom we would like to deeply express their thanks. Kazuyuki Hyodo was especially thanked for his great contribution to many photos used in this chapter that were taken at his station at Photon Factory. Last but not the least some of the X-ray images have been taken by Anton Maksimenko, Eiko Hashimoto, Huo Qinkai, Masaki Sakai, Yuki Nakao and Sho Nakagawa to whom the authors would like to thank. Anton Maksimenko is especially thanked for his great contribution to creation of refraction based CT algorithm.

Bibliography

[1] M. Akisada, M. Ando, K. Hyodo, S. Hasegawa, K. Konishi, K. Nishimura, A. Maruhashi, F. Toyofuku, A. Suwa and K. Kohra, An attempt at coronary angiography with a large size monochromatic SR beam, *Nucl. Instr. & Meth. A* **264-1**, 713–718, (1986a).

[2] M. Akisada, K. Hyodo, M. Ando, A. Maruhashi, K. Konishi, F. Toyofuku, K. Nishimura, S. Hasegawa, A. Suwa and E. Takenaka, Synchrotron radiation at the photon factory for non-invasive coronary angiography-experimental studies, *J. Cardiography* **16**, 527, (1986b).

[3] M. Ando and S. Hosoya, Proc. of the 6th Int'l Conference on X-Ray Optics and Microanalysis, University of Tokyo Press, ed. G. Shinoda, K. Kohra and T. Ichinokawa 63–68, (1972).

[4] M. Ando, A. Maksimenko, H. Sugiyama, W. Pattanasiriwisawa, K. Hyodo and C. Uyama, Simple X-ray dark-and bright-field imaging using achromatic Laue optics., *Jpn. J. Appl. Phys.* **41**, L1016–L1018, (2002).

[5] M. Ando, H. Sugiyama, T. Kunisada, D. Shimao, K. Takeda, H. Hashizume and H. Inoue, Construction of X-ray dark-field imaging with a view size of 80 mm square and first visualization of human articular cartilage at femoral head under a nearly clinical condition, *Jpn. J. Appl. Phys.* **43**, Part 2, No.9A/B, L1175–L1177, (2004).

[6] M. Ando, H. Sugiyama, T. Kunisada, D. Shimao, K. Takeda, H. Hashizume and H. Inoue, Construction of X-ray dark-field imaging with a view size of 80 mm square and first visualization of human articular cartilage at femoral head under a nearly clinical condition, *Jpn. J. Appl. Phys.* **43**, L1175–L1177, (2004a).

[7] M. Ando, H. Sugiyama, A. Maksimenko, E. Rubenstein, J. Roberson, D. Shimao, E. Hashimoto and K. Mori, X-ray dark-field imaging and its application to medicine Radiat. *Phys. Chem.* **71**, 899–904, (2004b).

[8] M. Ando, H. Bando, T. Endo, S. Ichihara, E. Hashimoto, K. Hyodo, T. Kunisada, G. Li, A. Maksimenko, D. Shimao, H. Sugiyama, T. Yuasa and E. Ueno, Refraction-based 2D, 2.5D and 3D medical imaging: stepping forward to a clinical trial European J Radiology, **68S**, 532–536,

(2008). (Proceedings of 5th Medical Application for Synchrotron Radiation, MASR2007, Saskatoon, Canada, 2007/8/26–30).

[9] M. Ando, N. Sunaguchi, Y. Wu, S. Do, Y. Sung, A. Louissaint, T. Yuasa, S. Ichihara and R. Gupta, X-ray phase contrast imaging in the dark field: implementation and evaluation using excised tissue specimens, *European Radiology* **23**, 3021–3029, (2013).

[10] F. Arfelli, V. Bonvicini, A. Bravin, G. Cantatore, E. Castelli, L. D. Palma, M. Di Michiel, M. Fabrizioli, R. Longo, R. H. Menk, A. Olivo, S. Pani, D. Pontoni, P. Poropat, M. Prest, A. Rashevsky, M. Ratti, L. Rigon, G. Tromba, A. Vacchi, E. Vallazza and F. Zanconati, Mammography with synchrotron radiation: phase-detection techniques, *Radiology* **215**, 286–293, (2000).

[11] A. Authier, Dynamical theory of X-ray diffraction International Union of Crystallography Monographs on Crystallography Oxford University Press, (2003).

[12] M. Bech, T. H. Jensen, O. Bunk, T. Donath, C. David, T. Weitkamp, G. Le Duc, A. Bravin, P. Cloetens and F. Pfeiffer, Advanced contrast modalities for X-ray radiology: phase-contrast and dark-field imaging using a grating interferometer, *Z. Med. Phys.* **20**, 7–16, (2010).

[13] U. Bonse and M. Hart, An X-ray interferometer, *Appl. Phys. Lett.* **6**(8), 155–156, (1965).

[14] A. Bravin, Exploiting the X-ray refraction contrast with an analyzer: the state of the art, *J. Phys. D: Appl. Phys.* **36**, A24–A29 PII: S0022-3727 (03) 60103-8, (2003).

[15] A. Bravin, J. Keyriläinen, M. Fernández, S. Fiedler, C. Nemoz, M.-L. Karjalainen-Lindsberg, M. Tenhunen, P. Virkkunen, M. Leidenius, K. von Smitten, P. Sipilä and P. Suortti, High-resolution CT by diffraction-enhanced X-ray imaging: mapping of breast tissue samples and comparison with their histo-pathology, *Phys. Med. Biol.* **52**, 2197–2211, (2007a).

[16] A. Bravin, V. Mocella, P. Coan, A. Astolfo and C. Ferrero, A numerical wave-optical approach for the simulation of analyzer-based X-ray imaging, *Opt. Express* **15**, 5641–5648, (2007b).

[17] A. Bravin, P. Coan and P. Suortti, X-ray phase-contrast imaging: from pre-clinical applications towards clinics, *Physics in medicine and biology* **58**(1), R1–35, (2013).

[18] E. J. Candes, J. K. Romberg and T. Tao, Stable signal recovery from incomplete and inaccurate measurements, *Communications on Pure and Applied Mathematics* **59**(8), 1207–1223, (2006).

[19] D. Chapman, W. Thomlinson, F. Arfelli, N. Gmür, Z. Zhong, R. Menk, R. E. Johnston, D. Washburn, E. Pisano and D. Sayers, Mammography imaging studies using a Laue crystal analyzer, *Rev. Sci. Instrum.* **67**. 1–5, (1996).

[20] D. Chapman, W. Thomlinson, R. E. Johnston, D. Washburn, E. Pisano, N. Gmür, Z. Zhong, R. Menk, F. Arfelli and D. Sayers, Diffraction enhanced X-ray imaging, *Phys. Med. Biol.* **42**, 2015–2025, (1997).

[21] D. Chapman, E. Pisano, W. Thomlinson, Z. Zhong, R. E. Johnston, D. Washburn, D. Sayers and K. Malinowska, Medical applications of diffraction enhanced imaging, *Breast Disease* **10**, 197–207, (1998).
[22] C. Y. Chou, M. A. Anastasio, J. G. Brankov, M. N. Wernick, E. M. Brey, D. M. Connor and Z. Zhong, An extended diffraction-enhanced imaging method for implementing multiple-image radiography, *Phys. Med. Biol.* **52**, 1923–1945, (2007).
[23] B. Chu, A. Kampschulte, M. S. Ferguson *et al.*, Hemorrhage in the atherosclerotic carotid plaque: a high-resolution MRI study, *Stroke* **35**(5), 1079–1084, (2004).
[24] P. Coan, Development and application of the analyzer-based imaging technique with hard synchrotron radiation PhD Thesis Université Joseph Fourier, Grenoble, France, (2006a).
[25] P. Coan, F. Bamberg, P. C. Diemoz, A. Bravin, K. Timpert, E. Mützel, J. G. Raya, S. Adam-Neumair, M. F. Reiser and C. Glaser, Characterization of osteoarthritic and normal human patella cartilage by computed tomography X-ray phase-contrast imaging: a feasibility study, *Invest. Radiol.* **45**, 437–444, (2010).
[26] D. M. Connor, H. Benveniste, F. A. Dilmanian, M. F. Kritzer, L. M. Miller and Z. Zhong, Computed tomography of amyloid plaques in a mouse model of Alzheimer's disease using diffraction enhanced imaging, *Neuroimage* **46**, 908–914, (2009a).
[27] D. M. Connor, H. D. Hallen, D. S. Lalush, D. R. Sumner and Z. Zhong, Comparison of diffraction-enhanced computed tomography and monochromatic synchrotron radiation computed tomography of human trabecular bone, *Phys. Med. Biol.* **54**, 6123–6133, (2009b).
[28] C. David, B. Nohammer, H. H. Solak and E. Ziegler, Differential X-ray phase contrast imaging using a shearing interferometer, *Applied Physics Letters* **81**(17), 3287–3289, (2002).
[29] T. J. Davis, T. E. Gureyev, D. Gao, A. W. Stevenson and S. W. Wilkins, X-ray image contrast from a simple phase object, *Phys. Rev. Lett.* **74**, 3173–3176, (1995a).
[30] T. J. Davis, D. Gao, T. E. Gureyev, A. W. Stevenson and S. W. Wilkins, Phase-contrast imaging of weakly absorbing materials using hard X-rays, *Nature* **373**, 595–598, (1995b).
[31] P. C. Diemoz, A. Bravin, C. Glaser and P. Coan, Comparison of analyzer-based imaging computed tomography extraction algorithms and application to bone–cartilage imaging, *Phys. Med. Biol.* **55**, 7663–7679, (2010).
[32] P. C. Diemoz, P. Coan, I. Zanette, A. Bravin, S. Lang, C. Glaser and T. Weitkamp, A simplified approach for computed tomography with an X-ray grating interferometer, *Opt. Express* **19**, 1691–1698, (2011).
[33] F. A. Dilmanian, Z. Zhong, B. Ren, X. Y. Wu, L. D. Chapman, I. Orion and W. C. Thomlinson, Computed tomography of X-ray index of refraction using the diffraction enhanced imaging method, *Phys. Med. Biol.* **45**, 933–946, (2000).

[34] W.-R. Dix, Intravenous coronary angiography with synchrotron radiation, *Prog. Biophys. Molec. Biol.* **63**, 159–191, (1995).
[35] J. T. Dobbins III and D. J. Godfrey, Digital X-ray tomosynthesis: current state of the art and clinical potential, *Phys. Med. Biol.* **48**, R65, (2003).
[36] D. L. Donoho, Compressed sensing. Information Theory, IEEE Transactions on **52**(4), 1289–1306, (2006).
[37] F. Esteve, H. Elleaume, B. Bertrand, A. M. Charvet, S. Fiedler, G. Le Duc, S. Corde, C. Nemotz, M. Rennier, J. L. Lafaix, J. J. Leplat, P. Suortti, W. Thomlinson and J. F. Le Bas, Coronary angiography with synchrotron X-ray source on pigs after iodine or gadolinium intravenous injection, *Acad Radiol* **9**(suppl 1), S92–S97, (2002).
[38] S. Fiedler, A. Bravin, J. J. Keyriläinen, M. Fernández, P. Suortti, W. Thomlinson, M. Tenhunen, P. Virkkunen and M. L. Karjalainen-Lindsberg, Imaging lobular breast carcinoma: comparison of synchrotron radiation DEI–CT technique with clinical CT, mammography and histology, *Phys. Med. Biol.* **49**, 175–188, (2004).
[39] X. Gao, S. Luo, H. Yin, B. Liu, M. Xu, Q. Yuan, X. Gao and P. Zhu, A micro-tomography method based on X-ray diffraction enhanced imaging for the visualization of micro-organs and soft tissues, *Comput. Med. Imaging Graph.* **30**, 339–347, (2006).
[40] S. C. Goehde, P. Hunold, F. M. Vogt, S. C. Goehde, P. Hunold, F. M. Vogt, W. Ajaj, M. Goyen, C. U. Herborn, M. Forsting, J. F. Debatin and S. G. Ruehm, Full-body cardiovascular and tumor MRI for early detection of disease: feasibility and initial experience in 298 subjects, *American Journal of Roentgenology* **184**(2), 598–611, (2005).
[41] T. Hirai, H. Yamada, M. Sasaki, D. Hasegawa, M. Morita, Y. Oda, J. Takaku, T. Hanashima, N. Nitta, M. Tahakashi and K. Murata, Refraction contrast 11×-magnified X-ray imaging of large objects by MIRRORCLE-type table-top synchrotron, *J. Synchrotron Radiat.* **13**, 397–402, (2006).
[42] Z. F. Huang, K. J. Kang, Z. Li, P. P. Zhu, Q. X. Yuan, W. X. Huang, J. Y. Wang, D. Zhang and A. M. Yu, Direct computed tomographic reconstruction for directional-derivative projections of computed tomography of diffraction enhanced imaging, *Appl. Phys. Lett.* **89**(4):041124-041124-041123, (2006).
[43] K. Hyodo, K. Nishimura and M. Ando, Coronary angiography project at the photon factory using a large monochromatic beam. Handbook on synchrotron radiation IV, ed. by S. Ebashi *et al.*, *Elsevier Sci. Pub.*, Amsterda, (1991), 55.
[44] S. Ichihara, M. Ando, A. Maksimenko, T. Yuasa, H. Sugiyama, E. Hashimoto, K. Yamasaki, K. Mori, Y. Arai and T. Endo, 3-D reconstruction and virtual ductoscopy of high-grade ductal carcinoma in situ of the breast with casting type calcifications using refraction-based X-ray CT, *Virchows Archiv* **452**, 41–47, (2008).

[45] V. N. Ingal and E. A. Beliaevskaya, X-ray plane-wave topography observation of the phase-contrast from a non-crystalline object, *J. Phys. D: Appl. Phys.* **28**, 2314–2317, (1995).

[46] V. N. Ingal, E. A. Beliaevskaya, A. P. Brianskaya and R. D. Merkurieva, Phase mammography—a new technique for breast investigation, *Phys. Med. Biol.* **43**, 2555–2567, (1998).

[47] R. W. James, The optical principles of the diffraction of X-rays (London: G. Bell and Sons Ltd), (1962).

[48] X. M. Jiang, Private communication: International meeting on shanghai synchrotron radiation source, (1997).

[49] N. Kato, A theoretical study of pendellosung fringes. I. General considerations, *Acta Crystallographica* **14**(5), 526–532, (1961a).

[50] N. Kato, A theoretical study of pendellosung fringes. II. Detailed discussion based upon a spherical wave theory, *Acta Crystallographica* **14**(6), 627–636, (1961b).

[51] J. Keyriläinen, M. Fernández and P. Suortti, Refraction contrast in X-ray imaging, *Nucl. Instrum. Methods Phys. Res. A* **488**, 419–427, (2002).

[52] J. Keyriläinen, M. Fernandez, M.-L. Karjalainen-Lindsberg, P. Virkkuneng, M. Leidenius, K. von Smitten, P. Sipilä, S. Fiedler, H. Suhonen, P. Suortti and A. Bravin, Toward high-contrast breast CT at low radiation dose, *Radiology*, **249**, 321–332, (2008).

[53] J. Keyriläinen, A. Bravin, M. Fernández, M. Tenhunen, P. Virkkunen and P. Suortti, Phase-contrast X-ray imaging of breast, *Acta Radiologica* **51**(8), 866–884, (2010).

[54] J. Keyriläinen, M. Fernańdez, A. Bravin, M.-L. Karjalainen-Lindsberg, M. Leidenius, K. von Smitten, M. Tenhunen, A. Kangasmaki, P. Sipilä, C. Nemoz, P. Virkkuneng and P. Suortti, Comparison of in vitro breast cancer visibility in analyzer-based computed tomography with histopathology, mammography, computed tomography and magnetic resonance imaging, *J. Synchrotron Radiat.* **18**, 689–696, (2011).

[55] S. Kikuta, X-ray Scattering and Synchrotron Radiation Science X-ray Basic (in Japanese) ISBN 978-4-13-062831-0, (2011).

[56] M. Z. Kiss, D. E. Sayers and Z. Zhong, Measurement of image contrast using diffraction enhanced imaging, *Phys. Med. Biol.* **48**, 325–40, (2003).

[57] M. J. Kitchen, D. Paganin, R. A. Lewis, N. Yagi, K. Uesugi and S. T. Mudie, On the origin of speckle in X-ray phase-contrast images of lung tissue, *Phys. Med. Biol.* **49**, 4335–4348, (2004).

[58] M. J. Kitchen, K. M. Pavlov, K. K. Siu, R. H. Menk, G. Tromba and R. A. Lewis, Analyser-based phase-contrast image reconstruction using geometrical optics, *Phys. Med. Biol.* **52**, 4171–4187, (2007).

[59] M. J. Kitchen, D. M. Paganin, K. Uesugi, B. J. Allison, R. A. Lewis, S. B. Hooper and K. M. Pavlov, X-ray phase, absorption and scatter retrieval using two or more phase contrast images, *Opt. Express* **18**, 19994–20012, (2010).

[60] M. J. Kitchen, D. M. Paganin, K. Uesugi, B. J. Allison, R. A. Lewis, S. B. Hooper and K. M. Pavlov, Phase contrast image segmentation using a Laue analyzer crystal, *Phys. Med. Biol.* **56**, 515–534, (2011).
[61] V. I. Kondratyev, G. N. Kulipanov, M. V. Kuzin, N. A. Mezentsev, S. I. Nesterov and V. F. Pinduyrin, First test on subtraction bronchography study at the angiography station of the VEPP-3 storage ring Medical Applications of Synchrotron Radiation ed., Ando M and Uyama C Springer ISBN 4-431-70229-6 29-32, (1997).
[62] I. Koyama, A. Momose, J. Wu, T. T. Lwin and T. Takeda, Biological imaging by X-ray phase tomography using diffraction-enhanced imaging Japan, *J. Appl. Phys.* **44**, 8219–21, (2005).
[63] T. Kunisada, D. Shimao, H. Sugiyama, K. Takeda, T. Ozaki and M. Ando, X-ray dark field imaging of human articular cartilage: Possible clinical application to orthopedic surgery, *European Journal of Radiology* **68S**, S18–S21, (2008).
[64] D. H. Larsson, P. A. C. Takman, U. Lundström, A. Burvall and H. M. Hertz, A 24 keV liquid-metal-jet X-ray source for biomedical applications, *Rev. Sci. Instrum.* **82**, 123701, (2011).
[65] R. Lewis, Medical applications of synchrotron radiation X-rays, *Phys. Med. Biol.* **42**, 1213–1243, (1997).
[66] R. A. Lewis, C. J., A. P. Hufton, S. Evans, R. H. Menk, F. Arfelli, L. Rigon, G. Tromba, D. R. Dance, I. O. Ellis, A. Evan, E. Jacobs, S. E. Pinder and K. D. Rogers, X-ray refraction effects: application to the imaging of biological tissues, *Br. J. Radiol.* **76**, 301–8, (2003). J. Li, Z. Zhong, R. Lidtke, K. E. Kuettner, C. Peterfy, E. Aliyeva and C. Muehleman, Radiology of soft tissue of the foot and ankle with diffraction enhanced imaging, *J. Anat.* **202**, 463–470, (2003).
[67] J. Li, J. M. Williams, Z. Zhong, K. E. Kuettner, M. Aurich, J. Mollenhauer and C. Muehleman, Reliability of diffraction enhanced imaging for assessment of cartilage lesions, *Ex Vivo Osteoarthritis Cart.* **13** 187–197, (2005).
[68] J. Li, Z. Zhong, D. Connor, J. Mollenhauer and C. Muehleman, Phase-sensitive X-ray imaging of synovial joints, *Osteoarthritis Cartilage* **17**, 1193–1196, (2009).
[69] J. Li, N. Wilson, I. A. Zelazny, J. Meyer, Z. Zhong and C. Muehleman, Assessment of diffraction-enhanced synchrotron imaging for cartilage degeneration of the human knee joint, *Clin. Anat.* doi:10.1002/ca.22106, (2012). AQ5.
[70] Ke Li, Y. Ge, J. Garrett, N. Bevins, J. Zambelli and G.-H. Chen, Grating-based phase contrast tomosynthesis imaging: Proof-of-concept experimental studies, *Medical Physics* **41**, 011903-1-011903-11, (2014).
[71] Lyncean Technologies, 2012. The Compact Light Source (Palo Alto, CA: Lyncean Technologies, Inc) www.lynceantech.com/.
[72] S. Majumdar, A. S. Issever, A. Burghardt, J. Lotz, F. Arfelli, L. Rigon, G. Heitner and R. H. Menk, Diffraction enhanced imaging of articular

cartilage and comparison with micro-computed tomography of the underlying bone structure, *Eur. Radiol.* **14**, 1440–1448, (2004).

[73] A. Maksimenko, Appl. Nonlinear extension of the X-ray diffraction enhanced imaging, *Phys. Lett.* 90, 154106-1-154106-3, (1997).

[74] A. Maksimenko, T. Yuasa, M. Ando and E. Hashimoto, *Appl. Phys. Lett.* **91**, 234108, (2007).

[75] S. A. McDonald, F. Marone, C. Hintermüller, G. Mikuljan, C. David, F. Pfeiffer and M. Stampanoni, Advanced phase-contrast imaging using a grating interferometer, *J. Synchrotron Radiat.* **16**, 562–572, (2009).

[76] R. Menk, L. Rigon and F. Arfelli, Diffraction-enhanced X-ray medical imaging at the elettra synchrotron light source, *Nucl. Intrum. Methods A* **548**, 213–20, (2005).

[77] A. Momose, T. Takeda, Y. Itai and K. Hirano, *Nature Medicine* **2**, 473–475, (1986).

[78] A. Momose, Demonstration of phase-contrast X-ray computed tomography using an X-ray interferometer. Nuclear Instruments and Methods in Physics Research Section A: Accelerators, Spectrometers, Detectors and Associated Equipment, **352**(3), 622–628, (1995a).

[79] A. Momose, Demonstration of phase-contrast X-ray computed-tomography using an X-ray interferometer Nucl. Instrum. Methods, *Phys. Res. A* **352**, 622–628, (1995b).

[80] A. Momose, T. Takeda, Y. Itai, A. Yoneyama and K. Hirano, Perspectives for medical applications of phase-contrast X-ray imaging Medical Applications of Synchrotron Radiation AQ6 ed M Ando M and C Uyama (Tokyo: Springer) pp 54–61, (1998).

[81] A. Momose, T. Takeda, A. Yoneyama, I. Koyama and Y. Itai, Wide-area phase-contrast X-ray imaging using large X-ray interferometers, *Nucl. Instrum. Methods Phys. Res. A* **467**, 917–920, (2001).

[82] A. Momose, S. Kawamoto, I. Koyama, Y. Hamaishi, K. Takai and Y. Suzuki, Demonstration of X-ray Talbot interferometry, *Japanese Journal of Applied Physics* **42**(7B), L866–L868, (2003).

[83] K. Mori, A. Urano, J. Haseawa, J. Toriwaki, H. Anno and K. Katada, Virtualized Endoscope System - An application of virtual reality technology to Diagnostic Aid -IEICE TRANS. INF. & SYST. 79-D: 809–819, (1996).

[84] K. S. Mori, Y. Suenaga and Toriwaki, Fast software-based volume rendering using multimeda instructions on PC platforms and its application to virtual endoscopy. In: Anne V. Clough AAA (ed) Medical Imaging 2003: Physiology and Function: Methods,Systems and Applications. The International Society for Optical Engineering, San Diego, California, USA, pp. 111–122, (2003).

[85] C. Muehleman, D. R. Sumner and Z. Zhong, Refraction effects of diffraction-enhanced radiographic imaging: a new look at bone, *J. Am. Podiatric Med. Assoc.* **94**, 453–455, (2004a).

[86] C. Muehleman, S. Majumdar, A. S. Issever, F. Arfelli, R. H. Menk, L. Rigon, G. Heitner, B. Reime, J. Metge, A. Wagner, K. E. Kuettner

and J. Mollenhauer, X-ray detection of structural orientation in human articular cartilage, *Osteoarthritis Cart.* **12** 97–105, (2004b).
[87] C. Muehleman, J. Li, Z. Zhong, J. G. Brankov and M. N. Wernick, Multiple-image radiography for human soft tissue, *J. Anat.* **208**, 115–124, (2006).
[88] C. Muehleman, J. Li, D. Connor, C. Parham, E. Pisano and Z. Zhong, Diffraction-enhanced imaging of musculoskeletal tissues using a conventional X-ray tube, *Acad. Radiol.* **16**, 918–923, (2009a).
[89] C. Muehleman, J. Li, A. Schiff and Z. Zhong, Diffraction-enhanced imaging for achilles tendon lesions: a preliminary study, *J. Am. Podiatric Med. Assoc.* **99**, 95–99, (2009b).
[90] C. Muehleman, D. Fogarty, B. Reinhart, T. Tzvetkov, J. Li and I. Nesch, In-laboratory diffraction-enhanced X-ray imaging for articular cartilage, *Clin. Anat.* **23**, 530–538, (2010).
[91] NIST (National Institute of Standards and Technology) database for X-Ray Mass Attenuation Coefficients and Mass Energy Absorption Coefficients http://www.nist.gov/pml/data/xraycoef/
[92] S. Ohtsuka, Y. Sugishita, T. Takeda, Y. Itai, K. Hyodo and M. Ando, Dynamic intravenous coronary arteriography using synchrotron radiation and its application to the measurement of coronary blood flow. *Jpn. Circ. J.* **61-5**, 432–440, (1997).
[93] S. Ohtsuka, Y. Sugishita, T. Takeda, Y. Itai, J. Tada, K. Hyodo and M. Ando, Dynamic intravenous coronary angiography using 2D monochromatic synchrotron radiation, *The British Journal of Radiology*, **72**, 24–28, (1999).
[94] A. Olivo, F. Arfelli, G. Cantatore *et al.*, An innovative digital imaging set-up allowing a low-dose approach to phase contrast applications in the medical field, *Medical physics* **28** 1610, (2001).
[95] O. Oltulu, Z. Zhong, M. Hasnah, M. N. Wernick and D. Chapman, Extraction of extinction, refraction and absorption properties in diffraction enhanced imaging, *J. Phys. D: Appl. Phys.* **36**, 2152–2156, (2003).
[96] F. Pfeiffer, T. Weitkamp, O. Bunk and C. David, Phase retrieval and differential phase-contrast imaging with low-brilliance X-ray sources. *Nature Physics* **2**(4), 258–261, (2006).
[97] F. Pfeiffer, M. Bech, O. Bunk, P. Kraft, E. F. Eikenberry, C. H. Broñnimann, C. Gruñzweig and C. David, Hard-X-ray dark-field imaging using a grating interferometer, *Nature Materials* **7**(2), 134–137, (2008).
[98] E. Photon, Proton and Neutron Interaction Data for Body Tissues. ICRU Report 46, (1992).
[99] Z. Pinsker, Dynamical scattering of X-rays in crystals Springer, (1977).
[100] E. Pisano, R. E. Johnston, D. Chapman, J. Geradts, M. V. Iacocca, C. A. Livasy, D. B. Washburn, D. E. Sayers, Z. Zhong, M. Z. Kiss and W. C. Thom;inson, Human breast cancer specimens: diffraction-enhanced imaging with histologic correlation—improved conspicuity of lesion detail compared with digital radiography, *Radiology* **214**, 895–901, (2000).

[101] K. M. Podurets, V. A. Somenkov and S. S. h. Shil'shtein, Refraction-contrast radiography Zh. Tekh. Fiz. **59** 115, Sov. Phys. Tech. Phys. **34-6** 654–657, (1989a).

[102] K. M. Podurets, V. A. Somenkov and S. S. h. Shil'shtein, Neutron radiography with refraction contrast, *Physica B* **156 & 157**, 691, (1989b).

[103] K. M. Podurets, V. A. Somenkov, R. R. Chistyakov and S. S. h. Shil'shtein, Visualization of internal domain structure of silicon iron crystals by using neutron radiography with refraction contrast, *Physica B* **156 & 157**, 694, (1989c).

[104] V. V. Protopopov and J. Sobota, X-ray dark-field refraction-contrast imaging of micro-objects, *Opt. Commun.* **213**, 267–279, (2002).

[105] L. Rigon *et al.*, A new DEI algorithm capable of investigating sub-pixel structures, *J. Phys. D: Appl. Phys.* **36**, A107–A112, (2003).

[106] L. Rigon, F. Arfelli and R. H. Menk, Generalized diffraction enhanced imaging to retrieve absorption, refraction and scattering effects, *J. Phys. D: Appl. Phys.* **40**, 3077–3089, (2007).

[107] E. Rubenstein, R. Hofstadter, H. D. Zeman, A. C. Thomson, J. N. Otis, G. S. Brown, J. Giacomini, H. J. Gordon, R. S. Kernoff, D. C. Harrison and W. Thomlinson, Transvenous coronary angiography in humans using synchrotron radiation, *Proc. Nat. Acad. Sci.* USA. **83**, 9724–9728, (1986).

[108] H. Sato, D. Shimao and M. Ando, Investigation of absorbed radiation dose in refraction-enhanced breast tomosynthesis by a Laue case analyzer, *Radiation Protection Dosimetry* 145, 1–3, (2011).

[109] S. Schroeder, A. F. Kopp, A. Baumbach *et al.*, Noninvasive detection and evaluation of atherosclerotic coronary plaques with multislice computed tomography1, *Journal of the American College of Cardiology* **37**(5), 1430–1435, (2001).

[110] S. Seo, N. Sunaguchi, M. Ando, K.-H. Choi, H.-T. Kim, W.-S. Chang, K.-H. Kim and J.-K. Kim, Visualization of microvascular proliferation as a tumor infiltration structure in rat glioma specimens using the diffraction-enhanced imaging in-plane CT technique, *Phys. Med. Biol.* **57**, 1251–1262, (2012).

[111] D. Shimao, K. Mori, H. Sugiyama and K. Hyodo, Refraction-contrast articular cartilage image: comparison of depiction abilities between in-line holographic method and a laue type analyzer method, *Japanese Journal of Applied Physics* Vol. 44, No. 1A, 2005, pp. 450–451, (2005a).

[112] D. Shimao, H. Sugiyama, K. Hyodo, T. Kunisada and M. Ando, Evaluation of X-ray dark-field imaging in visualization of nearly clinical articular cartilage, *Nucl. Instrum. Methods Phys. Res. A* **548**, 129–134, (2005b).

[113] D. Shimao, H. Sugiyama, K. Hyodo, T. Kunisada and M. Ando, Articular cartilage depicted at optimized angular position of Laue angular analyzer by X-ray dark-field imaging, *Appl. Radiat. Isot.* **64**, 868–874, (2006).

[114] D. Shimao, T. Kunisada, H. Sugiyama and M. Ando, Refraction enhanced tomosynthesis of finger joint by X-Ray dark-field, *Imaging Jpn. J. Appl. Phys.* **46**, L608–L610, (2007).

[115] D. Shimao, T. Kunisada, H. Sugiyama and M. Ando, Shift-and-add tomosynthesis of a finger joint by X-ray dark-field imaging: Difference due to tomographic angle European, *Journal of Radiology* **68S**, S27–S31, (2008).

[116] A. Snigirev, I. Snigireva, V. Kohn, S. Kuznetsov and Schelokov, On the possibilities of X-ray phase-contrast microimaging by coherent high-energy synchrotron radiation, *Rev. Sci. Instrum* **66**, 5486–5492, (1995).

[117] V. A. Somenkov, A. K. Tkalich and S. S. h. Shil'stein, Zh. Tech. Fiz. **61** 197, *Sov. Phys. Tech. Phys.* **36-11** 1309, (1991).

[118] H. Suhonen, M. Fernández, A. Bravin, J. Keyriläinen and P. Suortti, Refraction and scattering of X-rays in analyzer-based imaging, *J. Synchrotron Radiat.* **14**, 512–521, (2007).

[119] H. Suhonen, Scattering and refraction as contrast mechanisms in X-ray imaging PhD Thesis, Report Series in Physics HU-P-D155, University of Helsinki, Finland, (2008).

[120] Z. Sun, Diagnostic accuracy of multislice CT angiography in Peripheral arterial disease, *J. of Vascular and Interventional Radiology* 17, 1915–1921, (2006).

[121] N. Sunaguchi, T. Yuasa, Q. Huo, S. Ichihara and M. Ando, X-ray refraction-contrast computed tomography images using dark-field imaging optics, *Appl. Phys. Lett.* **97**, 153701–153703, (2010).

[122] N. Sunaguchi, T. Yuasa, Q. Huo and M. Ando, Convolution reconstruction algorithm for refraction-contrast computed tomography using a Laue-case analyzer for dark-field imaging, *Optics letters* **36**(3), 391–393, (2011a).

[123] N. Sunaguchi, T. Yuasa, Q. Huo, S. Ichihara and M. Ando, Refraction-contrast tomosynthesis imaging using dark-field imaging optics, *Appl. Phys. Lett.* **99**, 103704, (2011b).

[124] N. Sunaguchi, T. Yuasa and M. Ando, Iterative reconstruction algorithm for analyzer-based phase-contrast CT of hard and soft tissue, *Appl. Phys. Letters* **103**, 143702-1-143702-4, (2013).

[125] P. Suortti, J. Keyriläinen and W. Thomlinson, Analyser-based X-ray imaging for biomedical research, *J. Phys. D: Appl. Phys.* **46**, 494002-1-494002-25, (2013).

[126] Y. Suzuki, Y. Chikaura and M. Ando, Computer simulation on spatial resolution of X-ray bright-field Imaging by dynamical diffraction theory for a Laue-case crystal analyzer, *J. Appl. Phys.* **110**, 084902-1-084902-4, (2011).

[127] S. Takagi, Dynamical theory of diffraction applicable to crystals with any kind of small distortion, *Acta Crystallographica* **15**(12), 1311–1312, (1962).

[128] S. Takagi, A dynamical theory of diffraction for a distorted crystal, *J. Phys. Soc.* Japan. **26**(5), 1239–1253, (1969).

[129] H. F. Talbot, Facts relating to optical science No. IV Phil. Mag. **9**, 419–427, (1836).

[130] L. Tang, G. Li, Y. S. Sun, J. Li and X. P. Zhang, Synchrotron-radiation phase-contrast imaging of human stomach and gastric cancer: in vitro studies, *J. Synchrotron Radiat.* **19**, 319–22, (2012).
[131] D. Taupin, Dynamic theory of X-ray diffraction in crystals, *Bull. Soc. Fr. Mineral. Crystallogr.* **87**, 469, (1964).
[132] W. Thomlinson, Transvenous angiography in humans Proc. of the Intern. school of E. Fermi — CXXVIII Course — biomedical applications of synchrotron radiation eds E. Brattini and A. Balerna IOS Press Amsterdam pp. 127–153, (1995).
[133] J. Tjalling, Ypma, Historical development of the Newton–Raphson method, *SIAM Rev.* **37**(4), 531–551, (1995).
[134] T. Tuohimaa, M. Otendal and H. M. Hertz, Phase-contrast X-ray imaging with a liquid-metal-jet-anode microfocus source, *Appl. Phys. Lett.* **91**, 074104, (2007).
[135] J. van Heekeren, A. Kostenko, T. Hanashima, H. Yamada, S. Stallinga, S. E. Offerman and L. J. van Vliet, Characterization of an X-ray phase-contrast imaging system based on the miniature synchrotron MIRRORCLE-6X, *Med. Phys.* **38**, 5136–5145, (2011).
[136] M. von Laue, Roentgenstrahlinterferenzen akademische verlag, (1960).
[137] A. Wagner, M. Aurich, N. Sieber, M. Stoessel, W. D. Wetzel, K. Schmuck, M. Lohmann, B. Reime, J. Metge, P. Coan, A. Bravin, F. Arfelli, L. Rigon, R. H. Menk, G. Heitner, T. Irving, Z. Zhong, C. Muehleman and J. A. Mollenhauer, Options and limitations of joint cartilage imaging: DEI in comparison to MRI and sonography, *Nucl. Instrum. Meth. A.* **548**, 47–53, (2005).
[138] S. Webb, M. A. Flower, Webb's physics of medical imaging. CRC Press, (2012).
[139] T. Weitkamp, A. Diaz, C. David, F. Pfeiffer, M. Stampanoni, P. Cloetens and E. Ziegler, X-ray phase imaging with a grating interferometer, *Opt. Express* **13**, 6296–6304, (2005).
[140] T. Weitkamp, C. David, O. Bunk, J. Bruder, P. Cloetens, F. Pfeiffer, X-ray phase radiography and tomography of soft tissue using grating interferometry. *Eur. J. Radiol.* **68**, S13–17, (2008).
[141] M. Wernick, O. Wirjadi, D. Chapman, Z. Zhong, N. P. Galatsanos, Y. Yang, J. G. Brankov, O. Oltulu, M. A. Anastasio and C. Muehleman, Multiple-image radiography, *Phys. Med. Biol.* **48**, 3875–3895, (2004).
[142] S. Wilkins, T. Gureyev, D. Gao, A. Pogany and A. Stevenson, Phase-contrast imaging using polychromatic hard X-rays, *Nature* **384**(6607), 335–338, (1996).
[143] Y. Wu, K. Hyodo, N. Sunaguchi, T. Yuasa and M. Ando, Development of high sensitivity X-ray multiple-times-diffraction enhanced imaging (M-DEI) optics Journal of Physics: Conference Series, *IOP Publishing*, pp 192008, (2013).

[144] K. Yamada *et al.*, A trial for fine and low-dose imaging of biological specimens using quasi-monochromatic laser-Compton X-rays, *Nucl. Instrum. Methods Phys. Res. A* **608**, S7–S10, (2009).
[145] A. Yoneyama, A. Momose, I. Koyama, E. Seya, T. Takeda, Y. Itai, K. Hirano and K. Hyodo, *J. Synchrotron Rad.* **9**, 277–281, (2002).
[146] W. H. Zachariasen, Theory of X-Ray Diffraction in Crystals (New York: Wiley), (1945).
[147] B. G. Ziedses des Plantes, Eine neue methode zur differenzierung in der Roentgenographie (Planigraphie), *Acta Radiol.* **13**, 182–192, (1932).
[148] J. Zhang, Y. Chen, G. Li and X. Jiang, Imaging characters of the lung cancer phantoms under the simulative clinical condition performed with hard X-ray in-line holography, *Journal of Instrumentation* **8**, 1–9, (2013).
[149] Y. Zhao *et al.*, High-resolution, low-dose phase-contrast X-ray tomography for 3D diagnosis of human breast cancers, *Proc. Natl. Acad. Sci.* USA. **109**, 18290–18294, (2012).
[150] Z. W. Zhong, Thomlinson, D. Chapman and D. Sayers, Implementation of diffraction-enhanced imaging experiments: at the NSLS and APS, *Nucl. Instrum. Methods Phys. Res. A* **450**, 556–567, (2000).
[151] P. P. Zhu, J. Y. Wang, Q. X. Yuan, W. X. Huang, H. Shu, B. Gao, T. D. Hu and Z. Y. Wu, Computed tomography algorithm based on diffraction-enhanced imaging setup, *Appl. Phys. Lett.* **87**, 264101, (2005).

Chapter 9

X-ray Imaging and Its Applications

Tiqiao Xiao and Honglan Xie

Shanghai Synchrotron Radiation Facility (SSRF), Shanghai Institute of Applied Physics, Chinese Academy of Sciences, China

Abstract

Micro-CT and dynamic imaging are dominating methods at X-ray imaging beamline of Shanghai Synchrotron Radiation Facility (SSRF). Driven by user requirements, quantitative CT imaging techniques were developed, especially the propagation-based phase contrast X-ray imaging. Two kinds of low-dose and fast micro-CT imaging methods based on EST (equally sloped tomography) and CS (compressed sensing theory) reconstruction algorithms respectively were developed, in which one set of micro-CT projections could be collected in 1s or less time. Additionally, X-ray imaging methods such as local micro-CT, grating-based X-ray phase-contrast imaging also were developed. For future development of X-ray imaging at SSRF, there are three trends. The first trend is to improve spatial resolution, in which spatial resolution of 20nm with full field X-ray imaging applied to materials science, soft condensed matter, biology, and cultural heritage, will be achieved. The second trend is to improve temporal resolution and to offer new scientific opportunities to follow a system evolving within a short time range (ps-μs-ms). Equipped with adequate sample environments (furnaces, fields, traction or compression machines, etc.), *in-situ* investigation on structural evolution will be practicable. Finally, quantitative analysis to static structures and evolution of three-dimensional structures will be especially emphasized. Multi-model CT imaging techniques, including X-ray fluorescence tomography, X-ray diffraction tomography and small-angle X-ray scattering (SAXS) tomography are also under developing at SSRF.

Keywords: X-ray imaging; X-ray in-line phase-contrast imaging; X-ray micro-CT; CT reconstruction; X-ray fluorescence CT.

9.1. Introduction

Conventional X-ray imaging techniques like radiography or computed tomography (CT) rely on differences in X-ray absorption of the sample. The image contrast, originated by a variation of density, composition or thickness of the sample, is based exclusively on the detection of amplitude variation of the transmitted X-rays. The main limitation of this technique is the poor enhancement of weakly absorbing details in soft tissues. Therefore, the conventional X-rays can only be used to observe hard tissues (such as bones or teeth) and cannot obtain good images of soft tissues (such as tumor or cartilage).

X-rays produced by a third-generation synchrotron are 10^8 times brighter than the X-rays from a conventional X-ray machine in a hospital. Moreover, the X-rays produced by a third-generation synchrotron can be used to obtain the phase contrast images of high quality for soft tissue structures with low density. Phase sensitive techniques are based on the observation of the phase-shifts produced by the sample on the incoming X-ray wave. In general, the phase part becomes important if the source has a high spatial coherence, as it occurs in synchrotron radiation (SR) sources. In the X-ray energy range of $15 \sim 25\,\text{keV}$, the phase shift is up to 1000 times more sensitive to variation of the structure and composition of soft biological tissue or light material which consists of atoms with low atomic number Z when compared to absorption [1–3]. Therefore it is possible to reveal phase effects even if absorption is very low or negligible. Synchrotron X-ray images have a much higher resolution than conventional X-rays which means they have the advantage of also being able to reveal fine details of soft tissues. The phase-contrast X-ray imaging can be combined with tomographic techniques to obtain the three-dimensional distribution of the real part of the refractive index of the sample. Further, with the continuous energy tunability of synchrotron radiation X-rays we can select the best imaging energy to image fine structures of tissues that are previously unseen. Moreover, the narrow pulse width time structure of the third-generation synchrotron radiation X-rays is very suitable for real-time imaging of *in vivo* micro-imaging. Up to now, the synchrotron radiation X-ray phase-contrast imaging (XPCI) has revealed a great application potential on early diagnosis and treatment research of tumor and cerebral vascular disease. For example, the synchrotron radiation X-ray phase-contrast imaging can discover micro-calcification lesions that the traditional mammography was unable to confirm; the synchrotron radiation X-ray micro-angiography imaging can display the blood vessel in the brain with 20 to 30 microns in diameter, however the traditional existing medical imaging equipment cannot detect the blood vessel with the diameter below 200 micron. Over the recent years, SR studies have demonstrated that X-ray phase-contrast imaging could substantially change the face of X-ray imaging, as it would provide dramatic image improvements in fields as diverse as biomedicine, material science, industry and others. Early diagnosis of tumours, imaging of blood vessels, detection of microfractures and delamination in composite materials are realized through synchrotron radiation X-ray imaging which are currently not accessible by means of conventional X-ray imaging techniques.

With the electron beam energy of 3.5 GeV, just lower than APS, ESRF and SPring-8, storage ring emittance of 3.9 nm·rad, Shanghai Synchrotron Radiation Facility (SSRF) will be the fourth energy in the world. SSRF consists of a 150 MeV LINAC, a booster that can increase the electron energy from 150 MeV to 3.5 GeV in 0.5 second, and a 3.5 GeV electron storage ring. The 20 cell, 4 fold structure is used to store an electron beam (average current 300 mA) with low emittance (minimum 3.9 nm·rad) and long lifetime (top-up mode). There are 40 bending magnets, 16 standard straights (6.5 meters) and 4 long straights (12 meters) along the ring, so more than 60 beamlines could be installed in the ring. SSRF has 7 initial beamlines and experimental stations. These 7 beamlines are used for macromolecular crystallography, XAFS, hard X-ray microfocus, X-ray imaging and biomedical application, soft X-ray spectromicroscopy, diffraction and small angle X-ray scattering, respectively.

The X-ray imaging and biomedical applications beamline (BL13W1) of SSRF devotes itself to developing and applications of dynamic in-line X-ray phase-contrast imaging (IL-PCI), X-ray microscopic computerized-tomography (Micro-CT) and other new techniques of X-ray imaging. These are two scientific cases, one is low dose, nondestructive, high resolution, dynamic and three-dimensional X-ray imaging for the inner microstructure of soft tissues and low Z materials; the other is nondestructive, high resolution, three-dimensional X-ray imaging for the inner microstructure of those sample of paleontology, archaeology and geology. As shown in Fig. 9.1, the BL13W1 beamline is composed of, a 2.0 Tesla wiggler, 6 set of filters for high heat load reducing, double crystal monochromator cooled with liquid nitrogen. Three sets of digital X-ray detector are employed with pixel size of 0.37 μm, 0.74 μm, 1.85 μm, 3.7 μm, 5.9 μm, 9 μm and 13 μm, respectively. The monochromator can work well at the photon energy range of 8–72.5 keV. Up to now, the filters and the monochromator could work well at 200 mA, 3.5 GeV

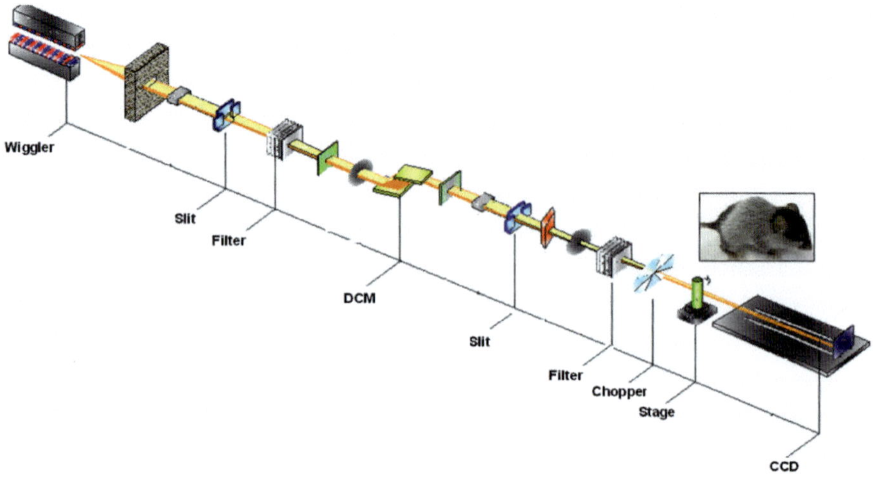

Fig. 9.1. Beamline BL13W1 layout.

Fig. 9.2. Photo of endstation.

and 17 mm wiggler gap. Figure 9.2 is the photo of the experimental station. Imaging methods of micro-CT, in-line phase contrast imaging (including dynamic imaging) and X-ray fluorescence CT could be provided for scientific researches on high Z and low Z samples including biomedical, material and paleontology samples. Spatial resolution of 0.8 μm and time resolution of 1 ms, could be achieved. Since 6 May, 2009, SSRF X-ray imaging beamline, which mainly focused on biomedical applications, has been put into operation and formally opened to users. By the end of 2016, 34014 hours were provided for 4233 users and 1023 user projects were performed. More than 307 user's papers have been published. These user research fields cover biomedicine (30%), material science (45%), paleontology, botany, geology, archaeology, environment science and so on. In this chapter some typical research results obtained mainly by using BL13W1 at SSRF will be introduced.

9.2. SR X-ray Imaging Methods

9.2.1. *Quantitative X-ray imaging*

The X-ray propagation-based phase-contrast imaging (PPCI) has a very simple experimental setup and requires no additional optics in the imaging geometry which is identical to the conventional absorption imaging geometry. PPCI only requires that the beam is sufficiently spatially coherent and the sample-to-detector distance (SDD) can be long enough. In conventionally absorption imaging, the SDD should be as short as possible in order to avoid blurring. However, with spatially coherent illumination, the Fresnel diffraction is observed at a certain SDD, thus the edge-enhancement information, which is proportional to the Laplacian of refractive index distribution of the sample, is observed even for a phase object. The edge-enhanced images produced by PPCI approach are often very useful in themselves, and the results allow visualizing the boundaries of regions with different refraction properties. However, quantitative information about the sample is always important in the analysis. Moreover, great attention was attracted to exploit the greater contrast available in the PPCI. Fortunately, the phase information contained in PPCI radiographies fringes could be extracted by means of phase-retrieval that enables the phase shift imposed on the wave by the sample to be recovered from the diffracted intensity distribution in the image plane or planes [4]. Several phase retrieval algorithms have been developed, including analytical methods and iterative methods. In general, the phase retrieval requires at least two intensity measurements, taken at two different SDDs. When PPCI combines with CT, i.e. the X-ray propagation-based phase-contrast CT (PPCT), typically hundreds or thousands of projections will be taken. Taking PPCT data at two different SDDs will deliver a higher radiation dose to the samples, which could hinder its biomedical applications. Definitely, the phase retrieval utilizing only one SDD PPCT data is preferable and more feasible, especially when dose is an essential issue in the experiment.

The X-ray imaging group of SSRF has developed three kinds of methods for quantitative in-line phase-contrast micro-CT. They are phase retrieval with a single SDD PPCT data, phase retrieval with two SDDs PPCT data and phase retrieval based on phase-attenuation duality(PAD), respectively [5–7]. The performance of different single SDD phase retrieval algorithms was studied [8].

9.2.1.1. *Single SDD phase retrieval*

The Modified Bronnikov algorithm (MBA) [9], which is revised based on Bronnikov algorithm [10], is a method to reconstruct directly three-dimensional (3D) refractive index from single SDD X-ray PPCT data. However, the absorption correction factor (ACF) value in MBA was determined using a semi-empirical approach, which may be cumbersome and time consuming since the ACF is varying with samples, and thus inconvenient in general applications. An approach to determine the ACF value simply and accurately for MBA algorithm was proposed by the X-ray imaging group of SSRF [5]. This approach is based on phase-attenuation duality that is valid for quasi-homogeneous and weakly absorbing samples. The proposed ACF takes the wavelength, SDD and phase-attenuation duality of PPCT into account. Moreover, a practical method for calculating the optimal phase-attenuation duality value directly is proposed based on the first Born-type approximation phase-retrieval algorithm [11]. The method was evaluated with experimental PPCT data collected at BL13W1 at SSRF. A sample with wires of nylon, polystyrene and PMMA, which were bought from GoodFellow company (GoodFellow, Huntingdon, UK), was investigated at 18 keV with effective pixel size of 3.7 μm and SDD = 20 cm. The reconstructed result is shown in Fig. 9.3(a), while Fig. 9.3(b) shows the

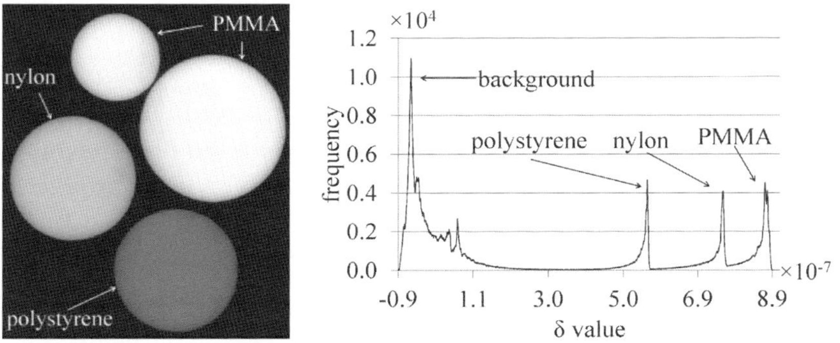

(a) (b)

Fig. 9.3. Experimental results: (a) Reconstructed result of the wire sample; (b) Histogram of (a), the labels indicate the peak for each material [5].

histogram of Fig. 9.3(a). As shown in both images, three materials can be well distinguished and there is a qualitative agreement with ideal values, which are polystyrene ($\delta = 7.24 \times 10^{-7}$) < nylon($\delta = 8.01 \times 10^{-7}$) < PMMA($\delta = 8.23 \times 10^{-7}$).

9.2.1.2. *Two SDDs phase retrieval*

Compared with phase retrieval with a single SDD and multiple SDDs, the phase retrieval with two SDDs images could improve the precision apparently in quantitative X-ray in-line phase contrast imaging. However, the critical point that determines the retrieval efficiency of this method is the optimization of SDD z_1 and z_2. The problem has been investigated systematically based on computer simulation and related experiments [7]. The results show that the highest retrieving precision could be obtained while $z_2 = 3z_1$, as shown in Fig. 9.4. This indicates a quick and easy way to get a relatively higher phase retrieval precision in many fields, such as biomedicine. Combined with micro-tomography, higher precision quantitative information and density resolution should be achieved, such as volume and thickness measurements.

9.2.1.3. *PAD phase retrieval*

Weak absorption is the main needed condition in phase retrieval of in-line X-ray phase-contrast CT. The absorption of most samples, however, cannot be neglected in practice. It is needed to find new algorithms of attaining quantitative information. The phase retrieval algorithm of PAD is used to realize the quantitative imaging of different density parts in a mixed contrast sample. Extending the validity of the PAD method, the PPCT of a sample with hybrid compositions of both the light and dense components with 60 keV is investigated. The results show that the PAD-based PPCT is effective in imaging both the weakly and strongly absorbing components simultaneously. Compared with the direct PPCT technique, the PAD-based PPCT technique demonstrates its excellent capability in material discrimination and characterization, as shown in Fig. 9.5 [8].

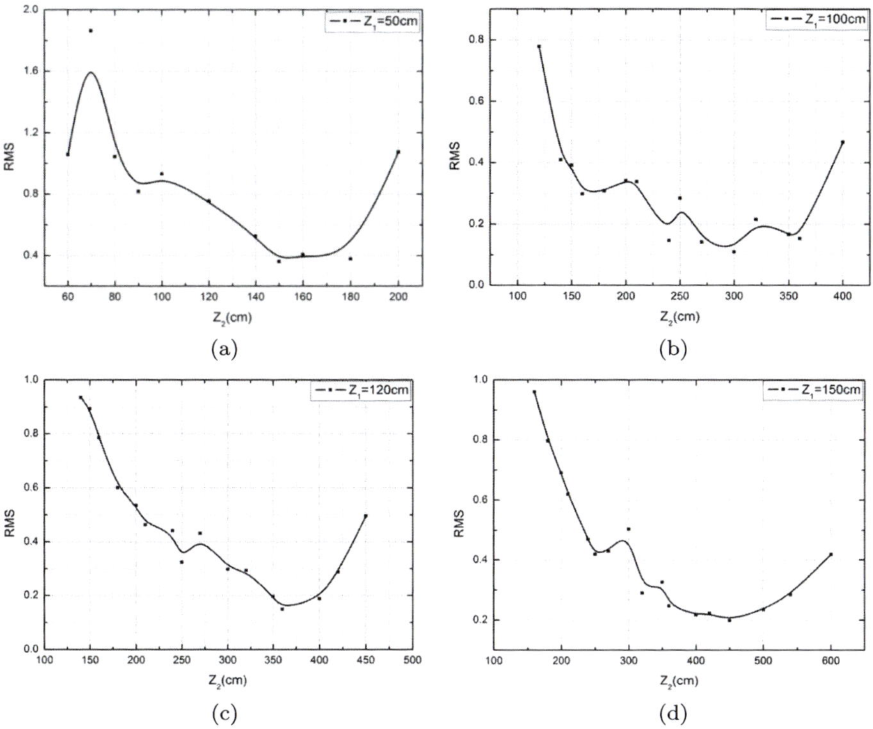

Fig. 9.4. Retrieving error versus z_2, while z_1 was set to (a) 50 cm, (b) 100 cm, (c) 120 cm, and (d) 150 cm, respectively [7].

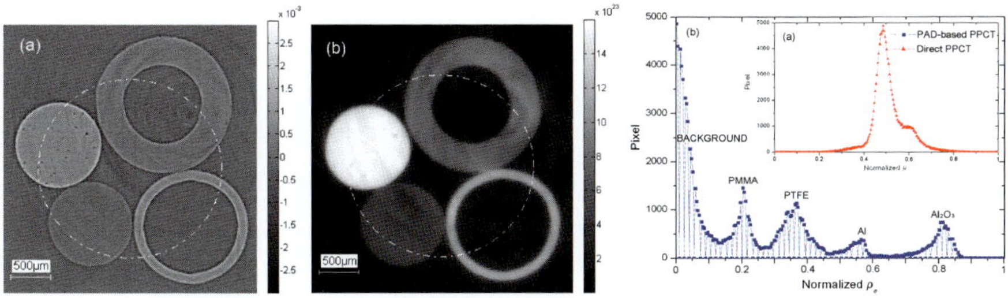

Fig. 9.5. Reconstructions and histograms of the electron density distribution of the samples with (a) direct and (b) PAD-based PPCTs at 60 keV [8].

9.2.1.4. *Grating-based Differential Phase Contrast Imaging (GDPC)*

X-ray grating-based differential phase contrast (GDPC) imaging technique has been widely applied in various fields in the past few years, especially in biomedical diagnostics [12–14]. For low-Z materials, GDPC could provide higher phase sensitivity than conventional absorption-based X-ray imaging and propagation-based X-ray phase contrast imaging. Combined with CT, quantitative information, such as absorption coefficient (absorption contrast), refractive index decrement (phase contrast) and scattering factor (dark-field), could be retrieved from one measurement simultaneously [14].

A GDPC imaging system has been built at BL13W1 of SSRF [15, 16]. The system is based on the grating interferometer, which includes a phase grating (G_1) and a transmission grating (G_2), as shown in Fig. 9.6. The incident beam is firstly diffracted by G_1 to generate a beam array, then the sub-beams in the array interfere with each other, and lastly form periodic fringe patterns after propagation for a certain distance. Grating G_2 acts as a transmission mask for the detector and helps to record the intensity variation of the fringe pattern. The distance between G_1 and G_2 can be adjusted depending on the X-ray energy. The phase-stepping technique [17]

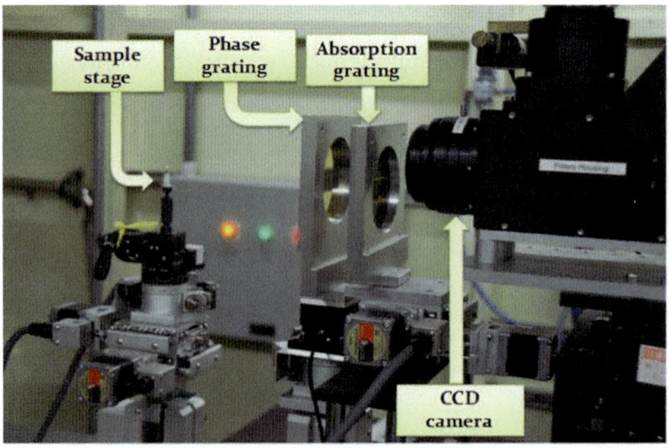

Fig. 9.6. Photo of experimental setup of GDPC [15].

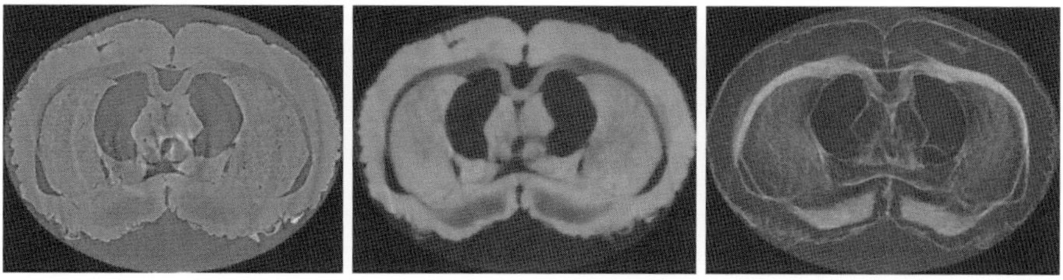

Fig. 9.7. Reconstructed slices of absorption, phase and dark-field contrast of mouse brain [15].

was applied for data acquisition, in which the scanning of the grating G_1 over one fringe period (typical 3–8 steps) by a nanopositioning stage.

Slices of absorption contrast, phase contrast and dark-field of mouse brain were reconstructed as well, as shown in Fig. 9.7. It is clearly depicted various tissues without any staining, such as caudate putamen (CPu), lateral ventricle (LV) and corpus callosum (cc). It is demonstrated that the GDPC system could resolve materials of minute density difference, and the three types of images could be complementary to each other for biomedical applications.

9.2.1.5. *Local CT*

In some cases, what people care about is not the whole sample, but only a part of the sample (e.g., in medical, the status of the tumor patient's tumor area). And in other cases, due to many reasons, some samples could not be broken to meet the X-ray beam or detector size (e.g., the internal structures of some rare ancient jade artifacts and fossils). Also to achieve high spatial resolution imaging to interested structure inside a large sample, the tiny field of view (FOV) of the detector usually cannot cover the whole sample. The traditional filtered back projection (FBP) algorithm for CT reconstruction can not work well in all these cases mentioned above. Thus, searching for a means to reduce exposure, which can perfectly reconstruct the region of interesting (ROI), at the same time has been of great interest in recent decades.

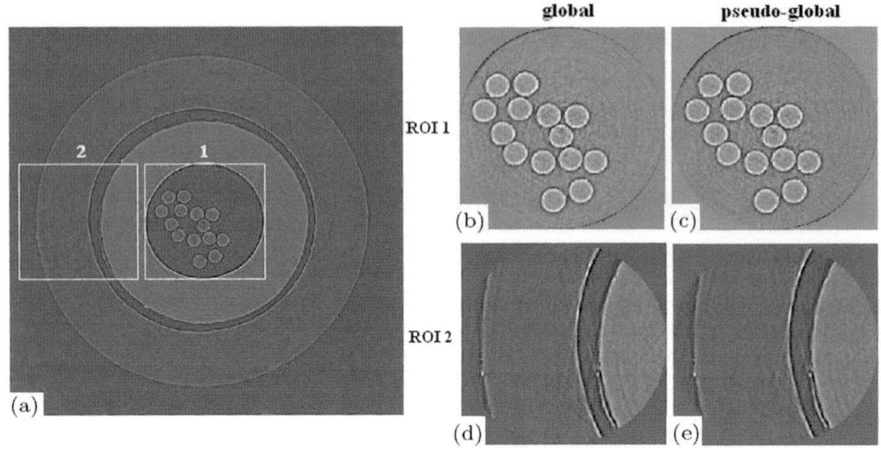

Fig. 9.8. Whole sample slice of the standard sample from a full-view scan (888 × 888 pixels) and the reconstructed ROIs (258 × 258 pixels). (a) is the whole sample slice of the standard sample in which the ROIs are indicated; (b), (d) are the ROIs taken from (a); (c), (e) are the ROIs reconstructed by pseudo-global tomography using local projections [19].

W. R. Madych shows that different parts of a projection have different contributions to the recovery of a fixed point. When the part of a projection is sufficiently far away from the point, it contains mainly low-frequency information which can often be ignored [18]. This means that we can use the projections on lines passing the region of exposure (ROI plus an extra margin) to reconstruct the ROI. The lost low-frequency information can be compensated by the projection extension. According to these, pseudo-global tomography was developed, which works well in local region reconstruction no matter where the region is. Figure 9.8 gives an example of the local reconstruction of a standard sample [19]. Conventional wisdom suggests that samples having a very asymmetric distribution of attenuation coefficient that passes outside the FOV will present the greatest challenge to ROI tomography [20–22]. Consequently we chose 2 regions as the ROIs. The difference between the ROIs taken from the whole sample slice and reconstructed by pseudo-global tomography using local projections cannot be distinguished by eyes.

9.2.2. *Dynamic X-ray imaging*

With the rapid development of detector techniques, the spatial resolution of X-ray phase contrast micro-tomography (XPCMT) increased gradually. However, the low temporal resolution of XPCMT limits its application fields, such as investigating the dynamic phenomena. The FBP algorithm is a typical XPCMT reconstruction algorithm. However, FBP requires high completeness of projection data. In order to reduce the radiation dose and speed up the data collecting time for micro-CT, two fast micro-CT imaging systems based on compressed sensing (CS) theory and equally sloped tomography (EST) are being developed at BL13W1 of SSRF [23–24].

9.2.2.1. *Compressed Sensing CT*

In 2006, Candes *et al.* proposed the CS theory [25], which stated that a sparse image could be reconstructed accurately when sampling frequency was much less than the Shannon/Nyquist sampling frequency. The CS theory could greatly lessen the amount of sampled data, and reduce the difficulty of subsequent data transmission, processing and storage. Combining CS and XPCMT, the projection number used for

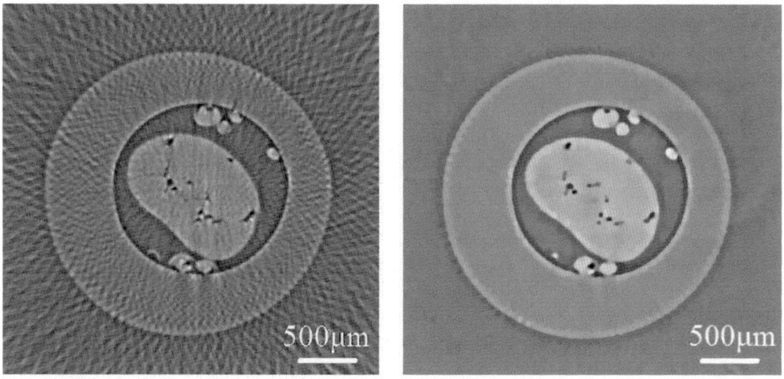

(a) (b)

Fig. 9.9. Reconstruction slices of the ant sample with (a) the FBP algorithm and (b) the CS-XPCMT algorithm respectively, both algorithms using 52 projections. (Provided by Chen Rong-Chang) image reconstruction would declines sharply, and moreover an incomplete data also could be employed to reconstruct an image with high quality results. Figure 9.9 shows the CT slices of the live ant sample [23]. With the CS-XPCMT algorithm, the number of projections acquired was reduced to 52 over the 180° range, which was about one tenth of the FBP ones. The images shows severe streaking artifacts appearing in the undersampled FBP image, however the streaking artifacts are alleviated and good reconstruction of sample details is obtained in the undersampled CS-XPCMT image.

9.2.2.2. *Equally sloped tomography*

BL13W1 has built a fast and low dose imaging system based on EST. EST was first proposed by Miao and collaborators in 2005 [26]. EST-based micro-CT may be the right solution for all the above mentioned

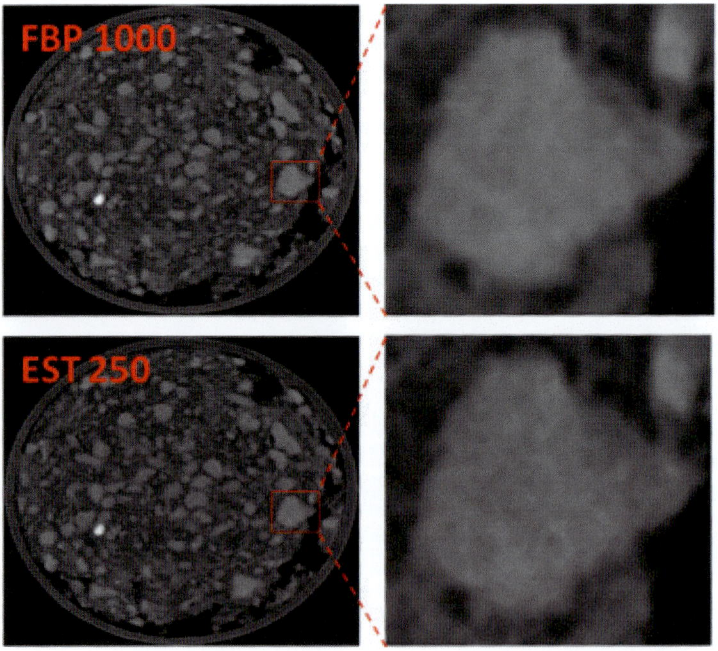

Fig. 9.10. The reconstruction results comparison between FBP and EST (Provided by Zhou Guang-Zhao).

problems in which the projections are equally sloped angled, not the traditional equally spatial angled. The reconstruction results of a cement material by FBP from 1000 projections and EST from 250 projections are shown in Fig. 9.10. It is obvious that comparable image quality could be achieved by EST with only one-quarter of the projections by FBP. Based on EST, *in vivo* micro-CT with 75% dose reduction and high pressure micro-CT with incomplete angle range of projections (samples in the diamond-anvil cell) has been realized on BL13W1 [24]. An *in vivo* the EST-based micro-CT imaging system will open to all users of BL13W1.

9.2.3. *X-ray fluorescence imaging*

X-ray fluorescence (XRF) is the emission of characteristic "secondary" (or fluorescent) X-rays from a material that has been excited by bombarding with high-energy X-rays or gamma rays, and is widely used for elemental analysis and chemical analysis [27]. High sensitivity and sub-micrometer resolution make hard X-ray fluorescence microanalysis with synchrotron radiation very attractive to many fields of science and technology. By combining X-ray fluorescence microanalysis with tomographic techniques, the element distribution on an arbitrary virtual section through the sample can be obtained without the need for destructive sample preparation. X-ray fluorescence computed tomography (XFCT) with synchrotron radiation was first proposed in 1986 by Boisseau [28] and has been applied in the fields of botany, biomedical, medicine, material, mineralogy [29–33] and so on. Many synchrotron radiation facilities have XFCT imaging system, for example, SPring-8, APS, ESRF, NSLS, etc.

In 2010, the μ-XRF mapping and XFCT system was established in BL13W1 at SSRF. The test sample and biomedical sample reconstructions proved the stability and reliability of this system [34–36]. Recently, ordered subsets expectation maximization (OSEM) algorithm has been introduced into XFCT to speed up the data acquisition process and improve the imaging accuracy. OSEM is based on the statistical theory and is widely used in PET (positron emission tomography) and SPECT (single photon emission computed tomography) reconstructions. It has been proven to be effective to the XFCT reconstruction and be able to decrease the dose by using a larger angular sampling interval [37].

In XFCT, the absorption correction and dose reduction are usually the main problems to be confronted. Especially for soft tissues, most elements of interest have lower atomic numbers whose characteristic X-rays have the poor penetrating power. Approaches have been proposed to overcome the absorption of the incident X-ray and X-ray fluorescence [38, 39]. However, the attenuation maps at the energies of the incident X-ray and X-ray fluorescence have to be obtained, and an extra dose was introduced. Absorption-corrected ordered subsets expectation maximization (AC-OSEM) algorithm has been developed at SSRF for the application of XFCT to soft tissues [40]. Experiments on pathological soft tissue were carried out to demonstrate the effectiveness of the algorithm. Cirrhotic and normal livers of Sprague-Dawley (SD) rats are selected as the samples, and the AC-OSEM algorithm is employed to reconstruct the Fe and Zn distributions inside the livers. The reconstructed element distributions with and without absorption correction are shown in Fig. 9.11, which were reconstructed with 10 iterations and one subset. It is obvious that the lost information is partially recovered after the absorption correction. From the corrected images, we can find that the content of Fe in the cirrhotic liver is significantly higher than that in the normal one, whereas the content of Zn does not change apparently.

The fundamental radiation covers the energy range from 8.0 to 72.5 keV in BL13W1 beamline. The basic technique and experimental set-up for XRF mapping and XFCT at SSRF beamlines can be used in low ($E < 22\,\mathrm{keV}$) and high ($E \approx 70\,\mathrm{keV}$) energy excitation mode for more elemental imaging compare with other sources. The system in BL13W beamline is fit for non-destructive elemental imaging with resolution levels in the 20~500 μm range, attracting a wide and ever-growing user community issued from biology, life science, earth and environmental science, archaeology and materials science.

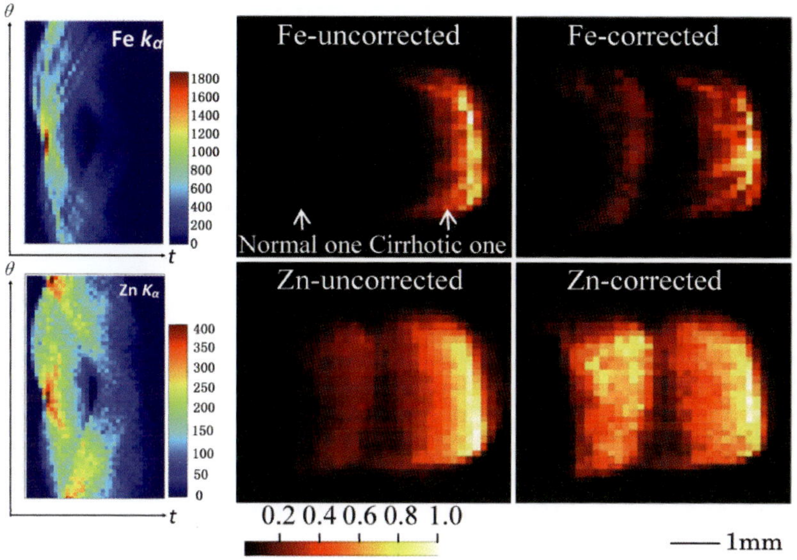

Fig. 9.11. The XFCT experimental sinogram and reconstructed Fe and Zn distributions in the livers of SD rats [40].

9.2.4. *Coherent X-ray Diffraction Imaging (CDI)*

Coherent X-ray Diffractive Imaging (CDI) is a lensless microscopic imaging technique. CDI is also called "coherent diffraction microscopy" which is firstly suggested by Sayre [41, 42] and demonstrated experimentally by J. Miao in 1999 [43]. Since it was first demonstrated in 1999, CDI has been rapidly developing into a materials imaging technique with resolutions approaching a few nanometers [44, 45]. CDI technique depends on phase retrieval algorithms which iterate back and forth between real and reciprocal space to recover the phase information of the diffraction patterns. Some iterative phase retrieval algorithms have been developed to solve the phase problem from the oversampled diffraction patterns, including the error reduction (ER), hybrid input-output (HIO), guided HIO (GHIO), difference map, shrinkwrap, Hamiltonian, oversampling smoothness (OSS) and so on.

According to the difference of imaging geometry, CDI is divided into different types, including planewave CDI [43, 46], scanning CDI or ptychographic CDI [47, 48], Bragg CDI [49, 50], Fresnel CDI [51]. To date, some new CDI methods come out, for example, surface CDI [52], poly CDI [53]. CDI can be applied to image thick objects at high resolution in 3D, so the future of CDI is bright for biology and materials/nanoscience.

9.2.5. *Instrumentation and software*

9.2.5.1. *Instrumentation*

The experimental station consists of two very flexible experimental tables. One is 2.5 m long and the other is 5.0 m long. The rail on the tables is used to adjust the distance between the sample and the X-ray CCD detector. The sample stage allows translation along all the three space direction with a resolution better than 1 μm and rotation in all three axes with the max speed of θ_z to be 90°/s. Ionization Chambers are used to monitor incident X-ray intensity. Four kinds of X-ray CCD detectors with different spatial resolution in the range from 0.19 μm/pixel to 24 μm/pixel for different samples are supplied. The fields of view of these X-ray CCD detectors are limited by the maximal beam size on samples which is 45 mm (H) $\times$ 5 mm (V). Table 9.1 shows the pixel size and field of view of X-ray CCD detectors on the BL13W1 endstation. Dynamic imaging experiments with the spatial-temporal resolution of 6.5 μm/pixel and 100 frame/s are carried out. Micro-CT experiments with the spatial resolution of 0.37 μm/pixel have been realized. An X-ray fluorescence detector also is supplied for developing XFCT method and X-ray fluorescence mapping method with the spatial resolution of 50 μm. The endstation can provide users several different X-ray imaging methods such as dynamic absorption or phase-contrast imaging, absorption or phase-contrast micro-CT, XFCT.

9.2.5.2. *Software*

The synchrotron based X-ray micro-CT (SR-μCT) has evolved as an increasingly accepted and utilized technique for characterizing the 3D internal structure of samples in the fields of material science and

Table 9.1. The parameters of X-ray CCD detectors on the BL13W1 endstation.

	X-ray CCD detectors on the BL13W1	objective lens	Pixel size	Field of view	scintillator (thickness)
1		1×	7.4 μm	15 mm × 5 mm	Yag:Ce, 100 μm
		1.25×	5.92 μm	12 mm × 5 mm	Yag:Ce, 100 μm
		2×	3.7 μm	7.6 mm × 5 mm	Yag:Ce, 100 μm
		4×	1.85 μm	3.8 mm × 3.8 mm	Yag:Ce, 50 μm
		10×	0.74 μm	1.5 mm × 1.5 mm	Yag:Ce, 50 μm
		20×	0.37 μm	0.8 mm × 0.8 mm	Yag:Ce, 20 μm
	Optique Peter, PCO2000	40×	0.19 μm	0.4 mm × 0.4 mm	Yag:Ce, 20 μm
2		AA50	0.65 μm	1.33 mm × 1.33 mm	GdOS:Tb, 10 μm
			2.25 μm	2.3 mm × 2.3 mm	GdOS:Tb, 10 μm
	HAMAMATSU Flash4.0	AA40	6.5 μm	13 mm × 5 mm	LSO:Ce 10 μm
3			9 μm	36 mm × 5 mm	GdOS:Tb
			13 μm	45 mm × 5 mm	GdOS:Tb
	Photonic Science, VHR				
4			24 μm	45 mm × 5 mm	GdOS:Tb
	Princeton Instruments, Quad-RO 4320				

life sciences. Some recent progress in detector development tends to decrease the pixel size while increasing the pixel number. As a consequence, a large number of SR-μCT projections (in the order of 1000) are required for existing CT algorithms to generate reconstructions without significant artifacts. Subsequently the CT data amount has increased accordingly, sometimes exceeding several Gigabytes. Therefore, tomographic data processing is a heavy computing task in CT experiments.

The X-TRACT is a Windows program for processing and reconstructing X-ray tomography data [54]. It is GPU-enabled for speed and has been parallelized so it can run on many CPU & GPU processors. To run X-TRACT you need a Windows machine connected to the internet. X-TRACT can either run in local mode, using the resources of the local machine, or in remote mode, in which case the local version of X-TRACT acts as a client for a server version of X-TRACT running in parallel mode on a computer cluster with large numbers of CPUs and/or GPUs. For large datasets, remote operation is usually much faster.

Under the close cooperation between imaging group at SSRF and INFN-Trieste section (Italy), the PITRE (Phase-sensitive X-ray Image processing and Tomography Reconstruction) software has been developed. It is a freeware that supports phase retrieval for propagation-based phase contrast imaging/tomography, extracting apparent absorption, refractive and scattering information from diffraction enhanced imaging (DEI), and tomographic reconstruction for conventional absorption CT data and for PPCT phase retrieval and DEI-CT extracted information [55]. The software can be obtained from http://webint.ts.infn.it/en/research/exp/beats2/ pitre.html or https://sites.google.com/site/rongchangchen/pitre-download.

9.3. Applications

9.3.1. *Biomedicine*

9.3.1.1. *Vascular*

Combining with angiography, the SR imaging has made the vascular structures possible to be visualized in experimental and clinical studies. The synchrotron radiation angiography may provide a novel solution to directly monitor the success of middle cerebral artery occlusion. Group of GuoYuan Yang randomly prepared twenty adult SD rats for middle cerebral artery occlusion models with different suture head silicone coatings, and then performed the *in vivo* imaging at beamline BL13W1 of SSRF [56]. Silicone-coated suture was superior to uncoated suture for producing consistent brain infarction. Additionally, silicone coating length was an important variable controlling the extent of the ischemic lesion: infarcts affected predominantly the caudate–putamen with large variability (<2 mm), both the cortex and caudate–putamen (2–3.3 mm), and most of the

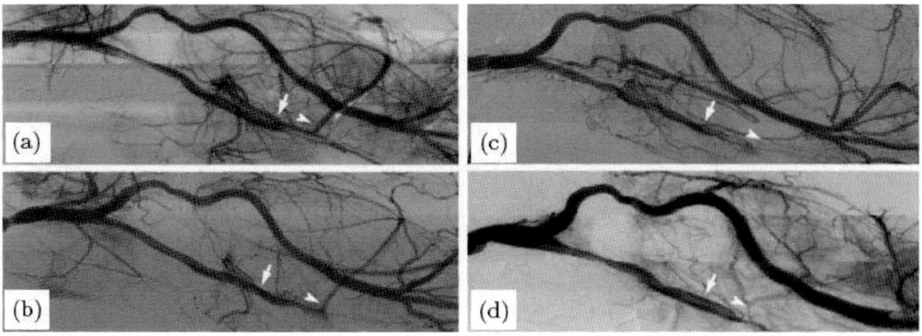

Fig. 9.12. Synchrotron radiation angiography images of middle cerebral artery (MCA) occlusion using silicone coated sutures with different coating lengths. The MCA was only partially occluded when the suture head was (a) not coated or (b) the coating length was <2 mm. The MCA was successful occluded when the coating length was 2.0 to 3.3 mm without affecting the hypothalamic artery (HTA), posterior cerebral artery (PCA), and anterior choroidal artery (AChA; c). The MCA was occluded when the coating length >3.3 mm together with AChA/PCA, and HTA (d) [56].

hemisphere, including the hypothalamus (>3.3 mm), as shown in Fig. 9.12 (see also Chapter 10 in this book).

TIEG1 (TGFβ inducible early gene 1) can induce apoptosis of cancer cells, but its role in inhibiting invasion and metastasis has not been reported and is unclear. Wei Jin *et al.* found that decreased TIEG1 expression is associated with increased EGFR (Epidermal Growth Factor Receptor) expression in breast cancer tissues and cell lines. TIEG1 significantly inhibits breast cancer cell invasion, suppresses mammary tumorigenesis in xenografts in mice, and decreases lung metastasis by inhibition of EGFR gene transcription and the EGFR signaling pathway. The microangiography experiment for blood vessels showed that the amount of blood vessels of tumors was decreased significantly in tumors expressing high levels of TIEG1 relative to control groups [57].

Using the SR angiography technique, the cerebral vasospasm in subarachnoid hemorrhage rats was detected [58]. 3D structures of acupuncture points were investigated by using in-line X-ray phase-contrast CT (IL-XPCT) and an accumulation of micro-vessels at each acupuncture point region was found [59] (see Chapter 10 in this book). The micro-angioarchitecture of rat spinal cord was investigated by using IL-XPCT and the resolution limit of the spatial blood supply network in the normal rat thoracic cord appeared to be as small as ~10 μm [60]. IL-XPCT could help to evaluate the validity of the pro- or anti-angiogenesis therapeutic strategies on microvasculature repair or regeneration.

9.3.1.2. *Lung tissues*

IL-XPCT at SSRF was used to observe the lungs for defining quantitatively the micro-structures for the purpose to realize early detection of lung diseases to improve the chances of successful treatment [61]. The precise view of lung airways and individual alveolus were achieved (see Fig. 9.13). The morphological parameters, such as diameter and alveolar surface area were measured. These parameters were of great importance in the diagnosis of diseases related to alveolus and alveolar scar. This offered a new perspective on the diagnosis of respiratory disease and may guide future work in the study of respiratory mechanism on the alveoli level.

9.3.1.3. *Biomedical materials for bone tissue regeneration*

The bone tissue is the largest tissue of human body, also it is the most likely to lead to defects. Every year millions of patients

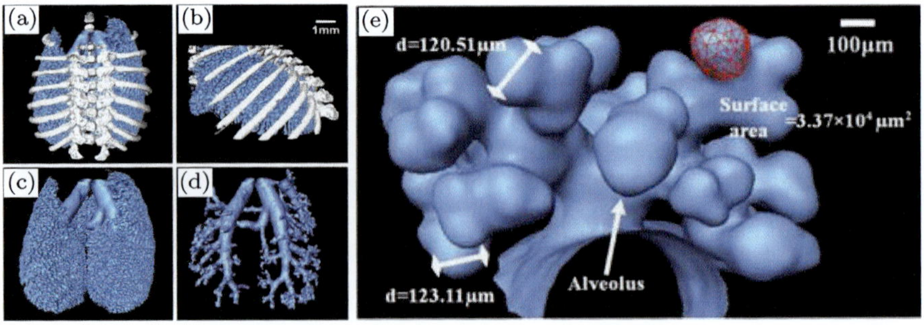

Fig. 9.13. 3D lung images. (a) Front view of the lung; (b) Right side view of the lung; (c) The bronchi and alveoli; (d) The bronchial tree; (e) Morphology parameters [61].

need surgery because of wounds, traffic accidents, aging population and natural disasters. Biomedical materials suitable for bone tissue regeneration should be endowed with biologically instructive properties such as bioactivity, biocompatibility, mechanical strength and controllable biodegradability. Synthetic three-dimensional scaffolds with an interconnective porous structure have been developed for osteogenesis and received a significant amount of attention in recent years. In order to quantitatively investigate the osteogenic effects on bone microstructure as well as to quantify the scaffold degradation within biopsy specimens sampled several months after implantation, the histomorphometric quantification using SR X-ray imaging was performed. The osteoinductive growth factor, recombinant human bone morphogenetic protein-2 (rhBMP-2) was incorporated into calcium/magnesium-doped silica-based scaffolds [62], or disulfide-containing PEG-based scaffolds [63], or calcium silicate/calcium phosphate cement scaffolds [64], or gelatin sponges based implant loaded with nanoparticle [65] for bioactivity and biodegradation promotion.

Calcium/magnesium-doped silica-based hierarchically mesoporous scaffolds (CMMS) impregnated with recombinant human bone morphogenetic protein-2 (rhBMP-2) were synthesized and assessed the bioactivity and biocompatibility of the scaffolds *in vitro*. The ectopic bone formation in mouse hindlimb muscle pocket and the bone regeneration in a rabbit femur cavity defect model were investigated by using SR X-ray imaging and histological analysis. The implantation of CMMS/rhBMP-2 resulted in an obvious bone formation in the thigh muscle pouches of mice, which indicated obvious osteoinduction of rhBMP-2. CMMS/rhBMP-2 scaffold significantly promoted bone regeneration, and both CMMS scaffolds gradually degraded and replaced by bone tissue after implanted in rabbit femur (see Fig. 9.14). Both degradation rate and growth factor liberation would depend on cell-secreted proteases during local tissue remodeling, thus the degradation could adapt to the reconstruction and lead to better regeneration.

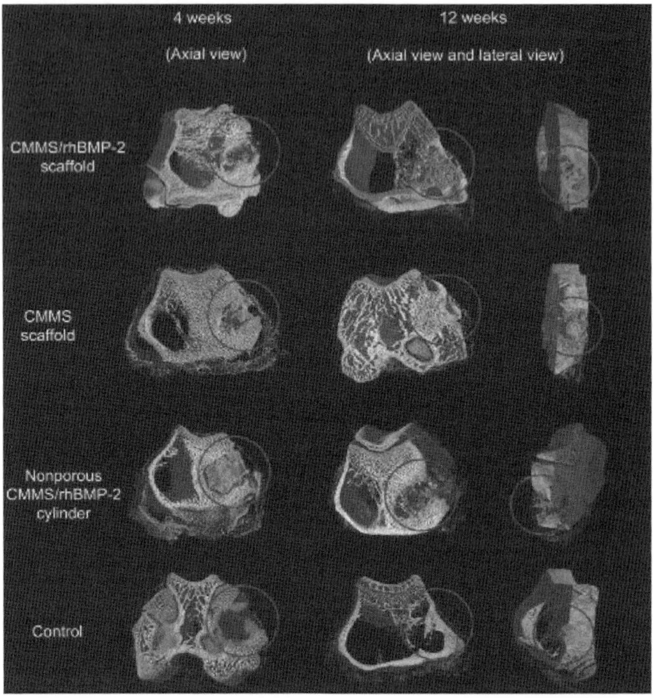

Fig. 9.14. 3D reconstruction of cross-section images by SR-μCT at 4 and 12 weeks after scaffolds were implanted in the rabbit femur (Provided by Han Guo).

9.3.2. *Material science*

BL13W1 could provide the high-energy and high-flux hard X-rays to realize the nondestructive dynamic or 3D imaging for materials. Users have carried out the research work in the fields of dendrite growth behavior in the solidification process of metallic alloy [66, 67], hydrogen evolution in the electrochemistry reaction [68], intermetallic compound (IMC) growth behavior and evolution rule of the bubbles in soldering process at the brazing interface [69, 70], 3D dendrite structures of alloy [71], the influence of foaming technology on the porous structure of polymer [72] and so on. The multi-energy synchrotron X-ray CT with a data-constrained modeling (DCM) approach has been used to characterize the physical structure of sample such as polymer, rock, and coals [73–76].

(a) t=1.5min (b) t=15min (c) t=57min

Fig. 9.15. 3D structure images of SiC–Al in microwave sintering [77].

9.3.2.1. *Metal–ceramic material*

The microstructure evolution of metal–ceramic materials Al-SiC was observed (see Fig. 9.15) and related sintering kinetics parameters were obtained. The different mechanisms of ceramic and metallic materials with microwave-dielectric loss and eddy current losses were analyzed, and the further analysis of the specific mechanism of interfacial polarization and micro-focus were carried out [77]. These results will help to reveal the mixed-interaction mechanisms of microwave sintering of metal–ceramic materials and offer some significant references for actual production.

9.3.2.2. *Fiber*

Using the special tensile testing device, high-resolution in-situ observations were obtained showing the fracture process for randomly oriented short carbon fiber/epoxy (SCF/EP) composites. Based on this method, the response to mechanical load of two different samples containing untreated SCFs and SCFs treated with oxidation were compared (see Fig. 9.16). The proportion of broken fibers to other types of damaged fiber in the sample with oxidation-treated fibers increased by about 6%. It was also found that oxidation treatment reduced the ineffective lengths of the composite fibers by about 20%, thereby improving the mechanical properties of the composites [78].

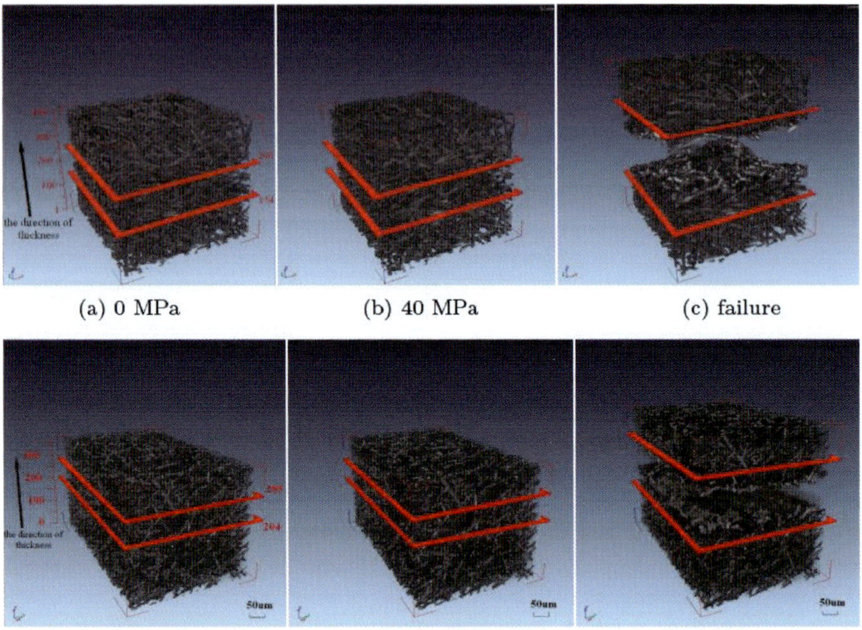

(a) 0 MPa (b) 40 MPa (c) failure

(d) 0 MPa (e) 40 MPa (f) failure

Fig. 9.16. SR-CT 3D images of Sample 1 with 15% SCFs (untreated) under (a) 0 MPa load, (b) 40 MPa load, and (c) at sample failure. Sample 2 with 15% SCFs (oxidation treatment) under (d) 0 MPa load, (e) 40 MPa load, and (f) at sample failure [78].

9.3.2.3. *Alloy*

The dendrite growth of a solidifying Sn-Bi binary alloy under a direct current (dc) electric field was observed *in situ* by dynamic X-ray imaging at BL13W1 (see Fig. 9.17) [79]. With increasing dc density, the dendrite morphology was changed from columnar dendritic to equiaxed cellular to equiaxed dendritic. This study opens a new window to in situ investigate the microstructure evolution of an optically opaque alloy during solidification under imposed external fields, such as electric field, magnetic field and supersonic field, etc. Using a combination of electron backscatter diffraction and X-ray synchrotron tomography, it has been demonstrated that α-Mg (Zn) dendrites grow with a six secondary branched structure behind the

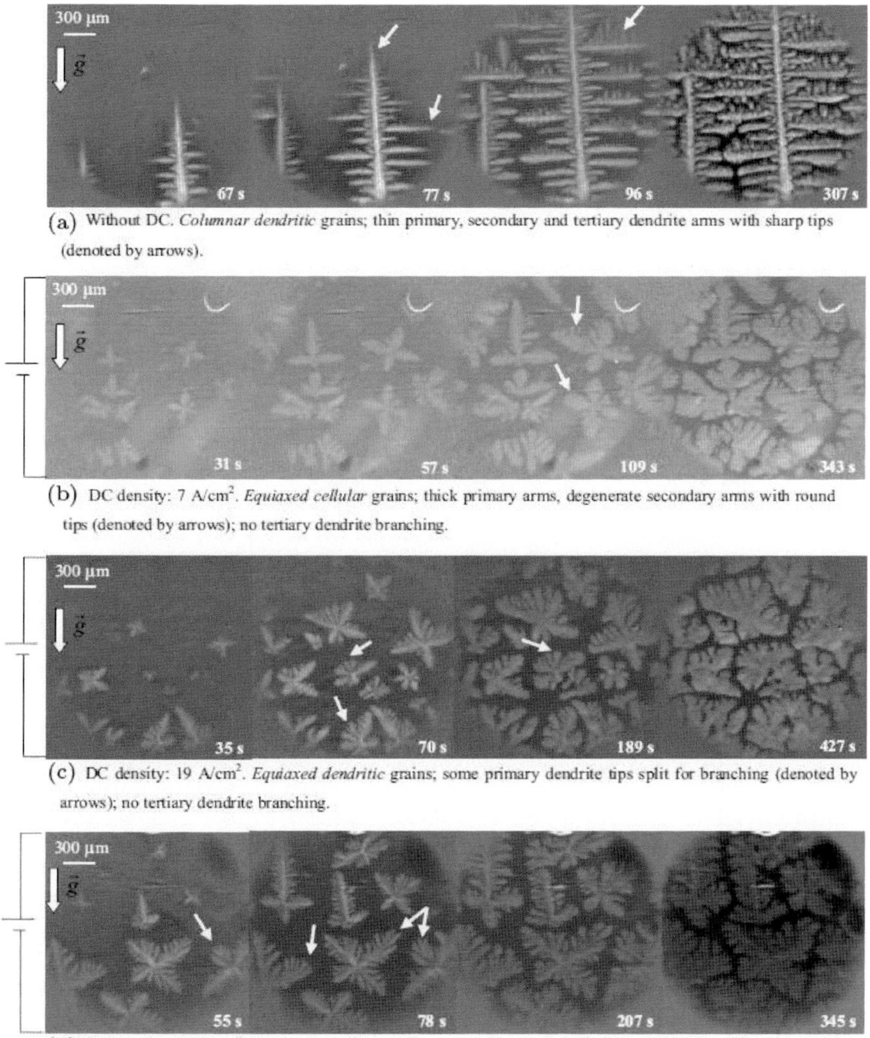

(a) Without DC. *Columnar dendritic* grains; thin primary, secondary and tertiary dendrite arms with sharp tips (denoted by arrows).

(b) DC density: 7 A/cm^2. *Equiaxed cellular* grains; thick primary arms, degenerate secondary arms with round tips (denoted by arrows); no tertiary dendrite branching.

(c) DC density: 19 A/cm^2. *Equiaxed dendritic* grains; some primary dendrite tips split for branching (denoted by arrows); no tertiary dendrite branching.

(d) DC density: 32 A/cm^2. *Equiaxed dendritic* grains; some primary dendrite tips split for branching (denoted by arrows); no tertiary dendrite branching.

Fig. 9.17. *In situ* observations on the dendritic growth of a solidifying Sn-12 wt%Bi alloy under DC field. Cooling rate is 1.5 k/min [79].

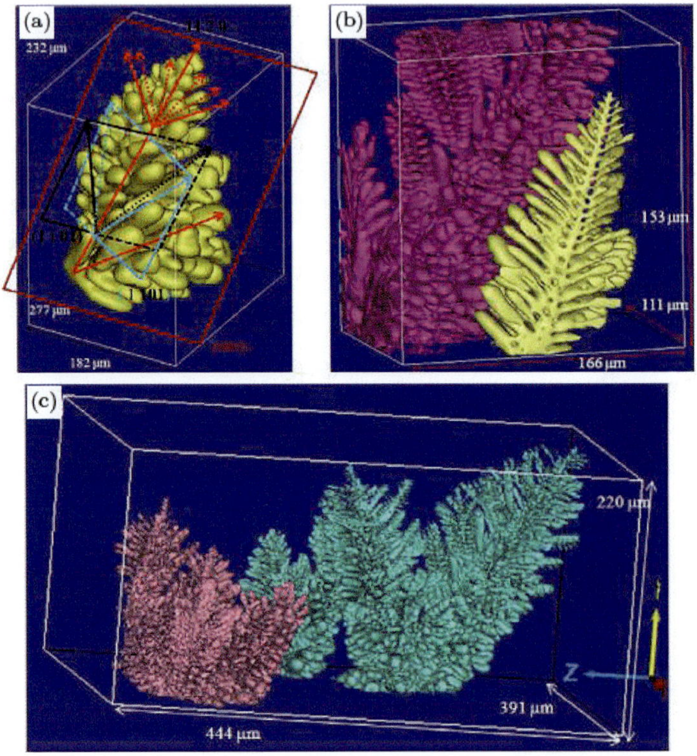

Fig. 9.18. (a) A reconstructed morphological image of α-Mg dendrite and its growth directions; (b, c) 3-D reconstructed dendrites exactly validate branched structure analysis of (a) (Provided by Guanyun Peng).

advancing dendrite tip (see Fig. 9.18) [71]. The result can be used for 3D finite element modeling for studying mechanical behavior of materials.

9.3.2.4. *Polymer*

The SR-μCT and scanning electron microscopy-ultramicrotomy were combined to study the distribution of micron-sized inorganic particles within a polymer matrix. It is demonstrated that the inorganic particles distributed in a polymeric matrix can form independent, but interpenetrating clusters of varying size and fractal dimension

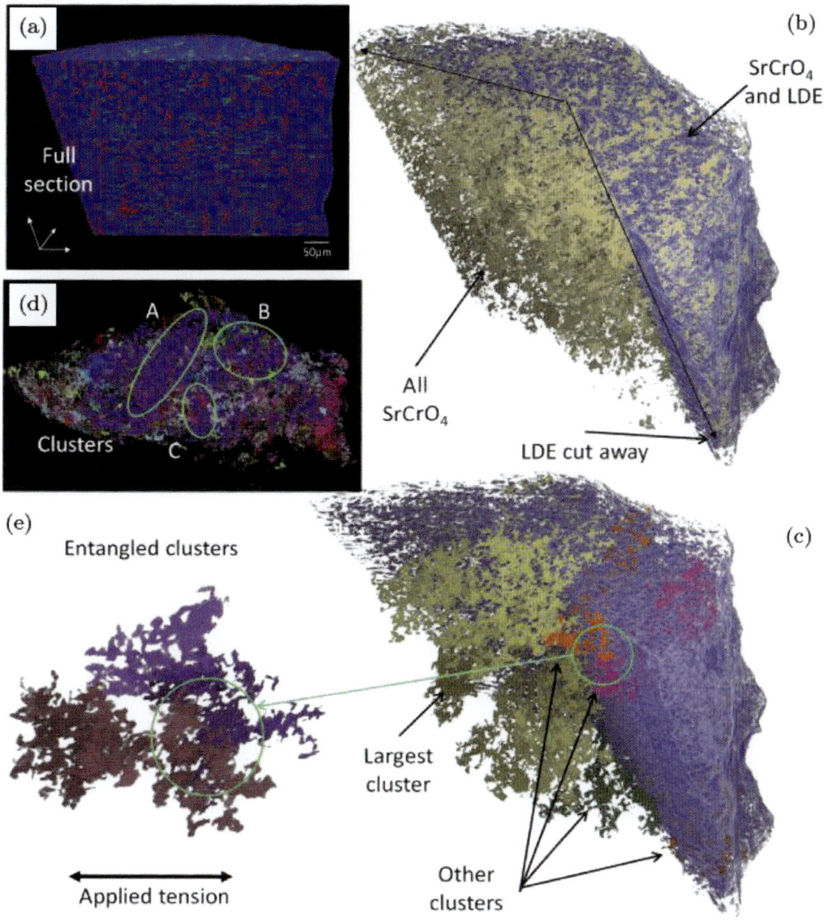

Fig. 9.19. Embedded fractal clusters. (a) Section through composite containing $SrCrO_4$ (red) embedded in an epoxy matrix with two densities nominated as low density or LDE (green) and normal (blue). (b) 3D reconstruction of LDE and all the $SrCrO_4$ particles. (c) 3D reconstruction of LDE and the five largest $SrCrO_4$ clusters. (d) View of all clusters (rendered in different colours) with the largest cluster in blue. (e) Entanglement of clusters [73].

with the largest having a fractal dimension of 2.36 (see Fig. 9.19). Because the strontium chromate is soluble, it can act as a template for the formation of void clusters with the same characteristics as the original strontium chromate ($SrCrO_4$) clusters. These clusters

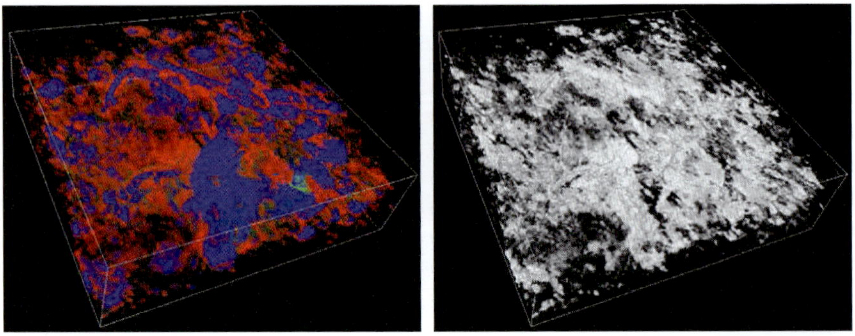

(a) (b)

Fig. 9.20. (a) The 3D view of distributions of void, minerals in the sample sub-volume. Void is displayed as red, blue and green means different minerals. (b) Distribution of void in a sub-volume of the coal sample [74].

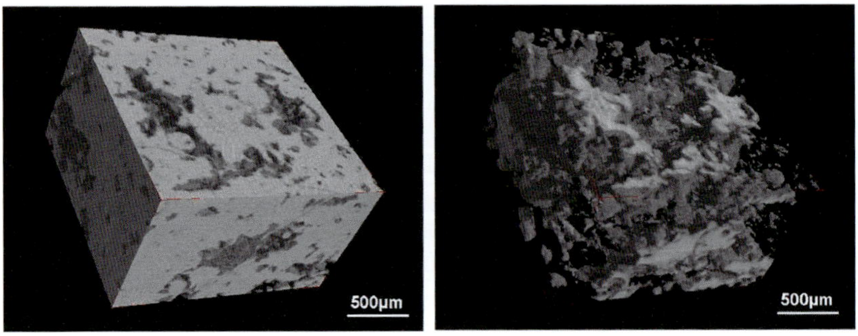

(a) (b)

Fig. 9.21. The three-dimensional compositional distribution of part of limestone sample. (a) dolomite and (b) calcite [76].

(void or inorganic phase) can act as either reservoirs for drug delivery, inhibitors in self-healing coatings or mechanical reinforcement in polymers [73].

9.3.2.5. *Oil shale*

3D physical structure of tight oil and gas reservoirs samples such as coal and limestone have been characterized with a DCM technique using multi-energy X-ray CT data (see Fig. 9.20 and Fig. 9.21)

[74–76]. The DCM approach is used to derive compositional volume fractions on each voxel, by minimizing the differences between the expected and CT reconstructed absorption coefficients. Voxel based 3D compositional maps are generated with colors and intensities representing compositions and volume fractions respectively. The quantitative results demonstrated that the approach can reveal significantly finer details of compositional distributions in the sample region of interest. The elevated spatial resolution is crucial for coal-bed methane reservoir evaluation and understanding the transformation of the minerals during coal processing and the physical properties of oil and gas reservoirs.

9.3.3. *Agriculture*

9.3.3.1. *Soil*

Fertilization has long been an important agronomic practice to enhance soil quality and productivity. Compared to the widely used medical CT and industrial CT, the SR-μCT has higher resolution, stronger contrast, and faster scanning speed, and is therefore an excellent tool for studying microstructures of soil aggregate (~1–10 mm in diameter). A detailed study of the internal microstructure of aggregates which conducted at BL13W1 can provide information regarding soil aggregation processes and thus soil quality [80, 81]. The treatments were no fertilizer (CK), chemical fertilizer (NPK), and chemical fertilizer plus organic manure (NPK+OM). The scanned aggregates were composed of small aggregates as shown in the 2D images (Fig. 9.22(a)). An overall perspective of the pore structure can be inferred from the 3D images of the pore systems (Fig. 9.22(b)). With high resolution tomography and image analysis, the measurement of aggregate microstructure can provide very detailed characterization of soil aggregation.

9.3.3.2. *Bamboo*

The bamboo is one of the world's fastest growing plants. The full height is reached by intercalary growth of each node. However, it is

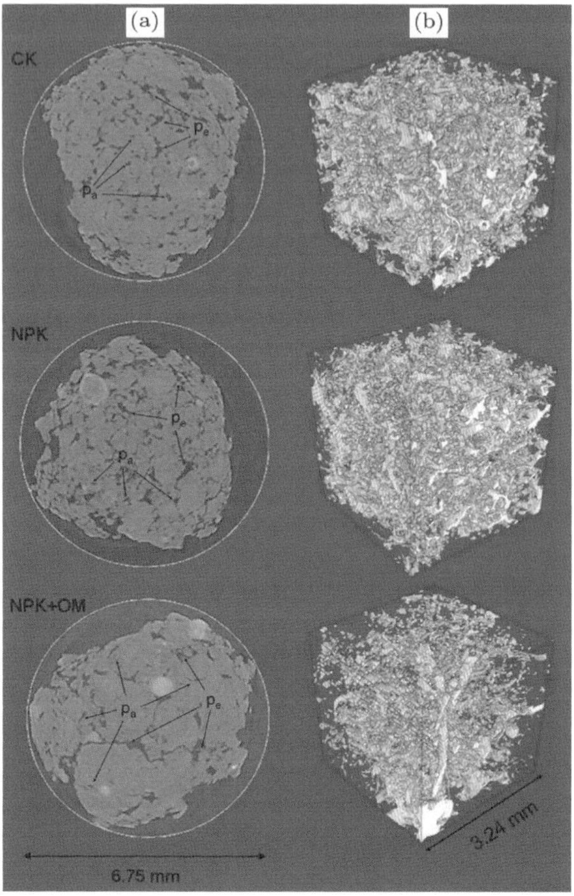

Fig. 9.22. Representative 2D and 3D visualizations of soil aggregate structures from the different fertilization treatments: NPK+OM, chemical fertilizer and organic manure; NPK, chemical fertilizer; CK, no fertilizer. p_e, inter-sub-aggregate pores; p_a, intra-sub-aggregate pores [80].

very difficult to detect by traditional methods the complex vascular system in a bamboo node. The complex structure of bamoo nodes has been revealed by SR-μCT. Two kinds of bamboo node samples were detected (see Fig. 9.23) [82]. The complex 3D network of vascular bundles has been directly obtained for the first time and some starch together surrounding the bud base was founded which will provide energy for later stages of development. The characteristic

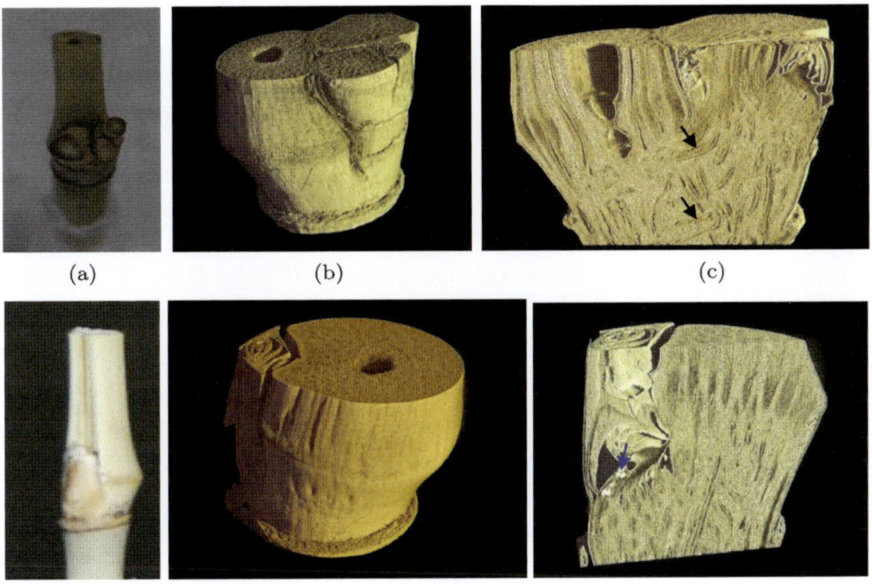

(a) (b) (c)

(d) (e) (f)

Fig. 9.23. (a) The photo, (b) 3D image and (c) longitudinal section of a node sample including the stem, two branches and one bud; (d) The photo, (e) 3D image and (f) longitudinal section of a node sample including the stem and one bud [82].

microstructures of adhensive at the interface and their penetration in wood/bamboo composite material were investigated systemically by μCT at SSRF [83].

9.3.4. *Pharmacy*

9.3.4.1. *Pharmacy*

In pharmaceutical industry, pharmaceutical structures play an important role in the pharmaceutical function and effective. Osmotic drug delivery systems (ODDS), delivering active agents by constant inner osmotic pressure, provide desired drug concentrations at the absorption site, maintaining the plasma drug level within the therapeutic range for a period of time, and thus can reduce the dosing frequency. SR-μCT provides quantitative structural information such as the shape and size of any dosage form. The 2D and 3D images for a number of solid dosage forms are acquired and used to learn their functions and structures, which are exemplified by the following examples. The internal structure of the tablet core for a monolith osmotic drug delivery system, felodipine sustained-release tablet, is visualized via SR-μCT during the drug release process [84, 85]. That the three-dimensional fractal values of surface of the tablet core correlates well with the drug release rate, indicates that it is a key quantitative indicator reflecting the drug release performance and can be used in the quality control of oral controlled drug delivery systems. The mixing and handling of granular materials are fundamental process operations in pharmaceutical industry. Using SR-μCT, the blend homogeneity of a model two-component mixture system is investigated [86]. The results indicate that the mixture homogeneity is improved with increasing rotations time, whereas the segregation increased with longer vibration time. Figure 9.24 shows the 3D structures of (a) controlled release kinetics of monolithic osmotic pump tablets and (b) mixing and segregation of granules.

9.3.4.2. *Traditional Chinese Medicines (TCMs)*

Microscopic structures investigation of traditional Chinese medicines (TCMs) is important to its quality evaluation. Compared with

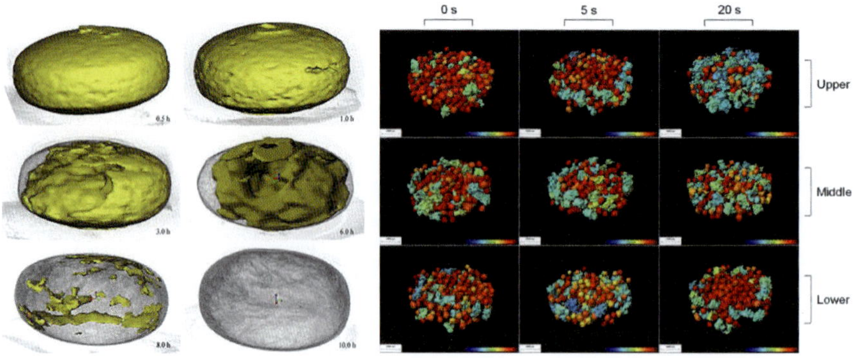

(a) (b)

Fig. 9.24. (a) The 3D structures of controlled release kinetics of monolithic osmotic pump tablets and (b) mixing and segregation of granules [84, 86].

(a) (b) (c)

Fig. 9.25. (a) 3D structures of a part of the main root of the wild ginseng, (b) calcium oxalate cluster crystals and (c) one calcium oxalate cluster crystal [88, 89].

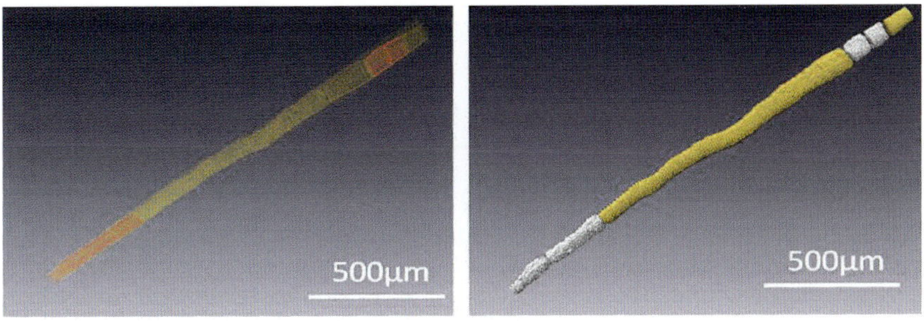

(a) (b)

Fig. 9.26. The 3D reconstruction of the vittae of Fructus Foeniculi. (a) 3-dimensional rendering of the overall vittae wall; (b) 3-dimensional rendering with the oil-filled chamber rendered in yellow and the empty oil chamber rendered in grey-white [88, 89].

conventional microscopic identification methods, XPCMT can realize clearer and finer microstructures of TCMs with better spatial and temporal resolutions nondestructively. Furthermore, for the nondestructive character, it provides the feasibility of microstructures quantitative research of TCMs. Many investigations on different kinds of TCMs have been done in recent years [87–89], including root and rhizome, fruit, flower, leaf and so on. Figure 9.25 shows a 3D structure of a part of the main root of the wild ginseng and the calcium oxalate cluster crystals' volume distribution. Figure 9.26 shows a 3D image of the vittae wall and internal volatile oil filling status. Meanwhile, we could also obtain the quantitative information about the volume of the oil and the vittae. The results demonstrate that the XPCMT is a promising new method for the identification of TCMs, especially for the valuable and difficult-slicing samples. It could realize the quantitative analysis for any inner structures interested of TCMs nondestructively.

9.3.5. *Others*

9.3.5.1. *Archaeology*

In the field of archaeology, on account of the protection of important culture heritage, the noninvasive characterization of 3D CT at BL13W1 can acquire the inner high-resolution structure of the sample. Eye bead is the article with western Asia style, and it is the evidence of East-West culture communication during the Warring State Period (475–221 BC). But there was still no scientific evidence of the production technology of eye bead. Group of Changsui Wang utilized SR-μCT with high revolution and relatively bigger field of view to perform the 3D imaging for eye bead samples (see Fig. 9.27) [90]. They firstly established nondestructive identification method for faience, glazed pottery and glass, found the ablate air bubble that would not appear in the normal glass phase, and testified the inlay production technology of eye bead according to micro analysis.

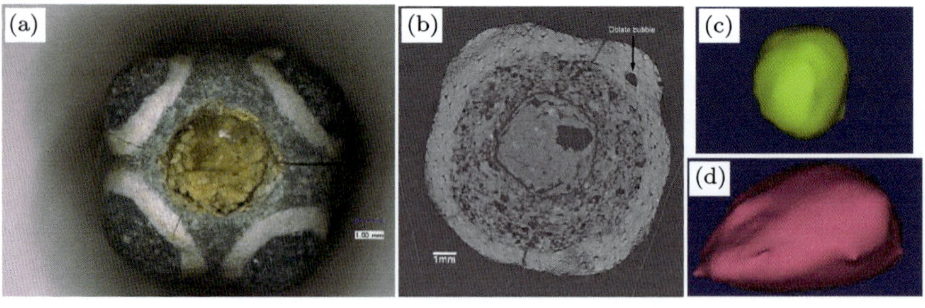

Fig. 9.27. (a) View of sample from the perforation, (b) one transverse section with outlined white layers, (c) the 3D reconstruction of a spherical air bubble distributed in the glass phase, and (d) the 3D reconstruction of an oblate air bubble located at the boundary of a horned eye and glaze [90].

9.3.5.2. *Paleontology*

Synchrotron radiation X-ray phase contrast CT technology has been considered to be the best way for imaging research of paleontologic fossil in the field of paleontology. With the application of Scanning Electron Microscope (SEM) and SR-μCT of SSRF, the research group headed by Xiguang Zhang successfully examined some crustacean fossils mainly focusing on a few phosphatocopines with nearly complete preservation of stalked eyes, antennula and limbs [91]. SEM has been well applied for examining details of the microstructure in the outer surface of these fossils during the investigation, and SR-μCT of BL13W1 has also provided precise data concerning some internal structures that are unable to be determined by SEM technology alone.

9.3.5.3. *Industry*

(1) Microstructure of diesel injector

In order to meet more stringent emission and fuel consumption regulations of diesel engines, the key issues of engine development are thermal efficiency improvement and emission reduction. The fuel, at high pressure (from 150 to 250 MPa), is injected into cylinders through a nozzle and broken up into tiny droplets, called spray atomization. The sprays have significant effects on the fuel combustion process and the formation of emissions [92–93]. With increasing injection pressure, the spray becomes more sensitive to the internal geometries of the nozzle such as the inlet chamfer radius, the diameter, and the conicity level.

Since 2009, Group of Zhijun Wu have been using SR-μCT of BL13W1 beamline at SSRF to analyze the internal microstructure of different diesel injectors [94]. Figure 9.28 shows the 3D structure of diesel injectors using different manufacturing methods: computerized numerical control drilling (CNCD) and electrical discharge machining (EDM) drilling. The simulation of internal flow in the diesel nozzle is carried out based on the 3D geometrical from SR-μCT. The results show that an uneven distribution of orifice inlet radii has significant influence on the flow characteristics [95].

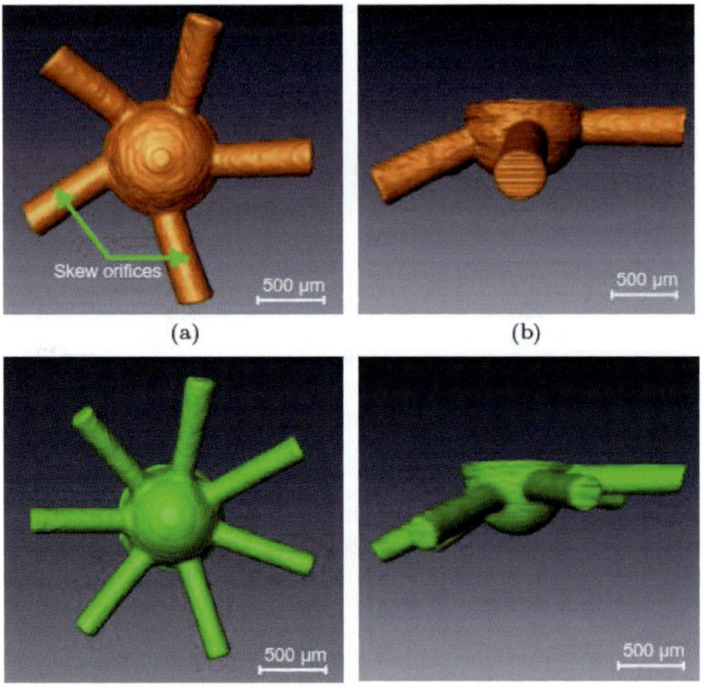

Fig. 9.28. 3D digital models of diesel nozzle orifices manufactured by the CNCD and EDM [94] (Reprinted with the permission of the Global Rights Department of Journal of Zhejiang University-SCIENCE A (Applied Physics & Engineering)).

(2) Microstructure of combustion chamber deposits

Combustion chamber deposits (CCDs) are generally found on the top of pistons or on the cylinder head in diesel engines due to incomplete combustion and physical or chemical interactions between fuel or oil components and the hot chamber surface under the high-temperature, high-pressure heterogeneous mixing conditions. Consequently, the engine performance, fuel economy and emissions may be affected [96, 97]. In order to reveal the formation mechanism of CCDs, physical structures have been investigated except the chemical content. Zhang *et al.* introduce SR-μCT to nondestructively analyze the whole CCD sample and discuss the formation mechanism of CCDs [98, 99]. Figure 9.29 shows the 3D and 2D structures of CCD sample on the piston surface from a large-scale two-stroke diesel

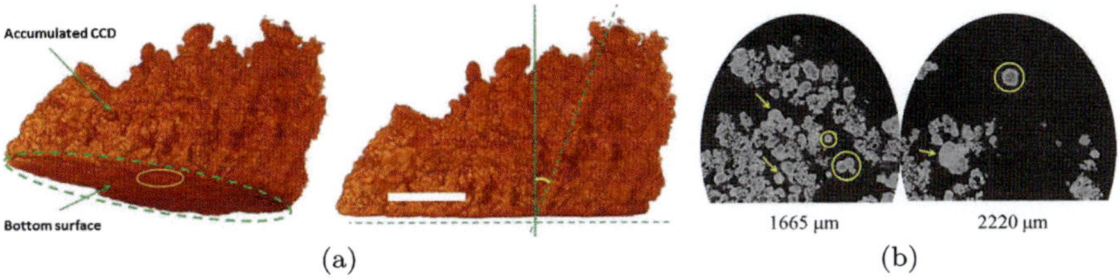

Fig. 9.29. (a) 3D and (b) 2D structure of the CCD sample [99].

engine. The white bar represents 1 mm. The growth direction of the PCD is not perpendicular to the bottom surface and the inclined angle is approximately 18 degrees (green line). The largest thickness of the PCD is approximately 2.6 mm. The symmetric distribution and smooth boundary (yellow arrow) on the top surface of the PCD indicated that the main formation mechanism could be due to the condensation of an incomplete combust gaseous mixture.

9.4. Perspective

After seven years' of user operation, the user community of the beamline has been built with two-thirds of them being biomedicine and material science users. Also its biomedical and material science applications have had some remarkable achievements. Micro-CT and dynamic imaging are the dominating methods. Driven by user requirements, quantitative CT imaging techniques have been developed, especially for the phase retrieval in the propagation based phase contrast imaging. Two kinds of low-dose and fast micro-CT imaging methods based on EST or CS reconstruction algorithm respectively have been developed which realized to complete a micro-CT data collection in 0.5 s. Other X-ray methods such as local micro-CT, and X-ray phase-contrast imaging with grating interferometer also are being developed.

For future development of X-ray imaging at SSRF, there are two trends. The first obvious trend is the improvement of spatial resolution. Higher spatial resolution X-ray imaging towards fulfilling the nanoscale challenge is a clear requirement originating from many different scientific communities, which include materials science, soft condensed matter, biology, and cultural heritage. The second obvious trend is the improvement of temporal resolution, made possible by the specific development of X-ray detectors and undulator sources, which will offer new scientific opportunities to follow a system evolving within a short time (μs–ms range). The X-ray imaging coupled with adequate sample environment (furnaces, fields, traction or compression machines, etc.) can produce *in situ* results. Two new X-ray imaging beamlines named as "fast X-ray imaging beamline" and "X-ray nano-CT beamline" of SSRF phase-II project have being proposed to be constructed in future. The temporal resolution of X-ray imaging at SSRF will be improved to 100 ps and the spatial resolution of X-ray imaging at SSRF will be improved to 20 nm. Fast tomography to follow 3D dynamic structural evolution and nano tomography to obtain 3D nano structures will be developed. They are very important for material sciences. The combination of techiniques will be considered to be developed, such as diffraction tomography, small-angle X-ray scattering (SAXS) tomography in future at SSRF.

Bibliography

[1] T. Davis, D. Gao and T. Gureyev *et al.*, Phase contrast imaging of weakly absorbing materials using hard X-rays, *Nature* **373**, 595–598, (1995).

[2] S. W. Wilkins, T. E. Gureyev and D. Gao *et al.*, Phase contrast imaging using polychromatic hard X-rays, *Nature* **384**, 335–338, (1996).

[3] Richard Fitzgerald, Phase-sensitive X-ray imaging, *Physics Today* **53**(7), 23–26, (2000).

[4] P. Cloetens *et al.*, Holotomography: Quantitative phase tomography with micrometer resolution using hard synchrotron radiation X-rays, *Applied Physics Letters* **75**(19), 2912–2914, (1999).

[5] R. C. Chen *et al.*, Phase retrieval in quantitative X-ray microtomography with a single sample-to-detector distance, *Opt. Lett.* **36**, 1719, (2011).

[6] R. C. Chen *et al.*, Comparison of single distance phase retrieval algorithms by considering different object composition and the effect of statistical and structural noise, *Opt. Express* **21**(6), 7384, (2013).

[7] Y. Q. Ren *et al.*, Optimization of image recording distances for quantitative X-ray in-line phase contrast imaging, *Opt. Express* **19**, 4170, (2011).

[8] H. Q. Liu *et al.*, Phase Retrieval for Hard X-ray Computed Tomography of Samples with Hybrid Compositions, *Chinese Optics Letters* **10**, 121101, (2012).

[9] A. Groso, R. Abela and M. Stampanoni, Implementation of a fast method for high resolution phase contrast tomography, *Optics Express* **14**(18), 8103–8110, (2006).

[10] A. V. Bronnikov, Theory of quantitative phase-contrast computed tomography. *Journal Of The Optical Society Of America A-optics Image Science and Vision* **19**(3), 472–480, (2002).

[11] T. E. Gureyev *et al.*, Optical phase retrieval by use of first Born-and Rytov-type approximations. *Applied Optics* **43**(12), 2418–2430 (2004).

[12] M. Willner, J. Herzen, S. Grandl *et al.*, Quantitative breast tissue characterization using grating-based X-ray phase-contrast imaging, *Phys. Med. Biol.* **59**, 1557–1571, (2014).

[13] A. Momose, W. Yashiro, K. Kido *et al.*, X-ray phase imaging: from synchrotron to hospital, *Phil. Trans. R. Soc. A* **372**, 20130023, (2014).

[14] N. Bevins, J. Zambelli, K. Li, Z. Qi and G. H. Chen, Multicontrast X-ray computed tomography imaging using Talbot-Lau interferometry without phase stepping, *Med. Phys.* **39**(1), 424, (2012).

[15] J. C. Qi, Y. Q. Ren, G. H. Du, R. C. Chen, Y. D. Wang, Y. He and T. Q. Xiao, Multiple Contrast Micro-Computed Tomography System Based on X-ray Grating Imaging, *Acta Optica Sinica*, **33**(10), 1034001, (2013).

[16] Y. Xi, B. Q. Kou, H. H. Sun, J. C. Qi, J. Q. Sun, J. Mohr, M. Borner, J. Zhao, L. X. Xu, T. Q. Xiao and Y. J. Wang, *J. Synchrotron Radiat.* **19**, 821, (2012).

[17] T. Weitkamp, A. Diaz, C. David, F. Pfeiffer, M. Stampanoni, P. Cloetens and E. Ziegler, X-ray phase imaging with a grating interferometer, *Opt. Express* **13**(16), 6296–6304, (2005).

[18] W. R. Madych, Tomography, approximate reconstruction, and continuous wavelet transforms, *Applied and Computational Harmonic Analysis* **7**(1), 54–100, (1999).

[19] W. H. Chen *et al.*, Pseudo-global tomography for local micro-computed tomography with high-brightness synchrotron X-rays, *Chinese Optics Letters*, **12**(2), 023401, (2014).

[20] J. C. Gore and S. Leeman, The reconstruction of objects from incomplete projections, *Physics in Medicine and Biology* **25**(1), 129, (1980).

[21] M. Muller and G. R. Arce, Truncation artifacts in tomographic reconstructions from projections, *Applied Optics*, **35**(20), 3902–3914 (1996).

[22] A. Katsumata *et al.*, Effects of image artifacts on gray-value density in limited-volume cone-beam computerized tomography. *Oral Surgery Oral Medicine Oral Pathology Oral Radiology and Endodontology* **104**(6), 829–836, (2007).

[23] L. Xu *et al.*, Fast X-ray micro-tomography for low Z materials, *Journal of Instrumentation* p. submitted.

[24] Guangzhao Zhou *et al.*, Development of a low dose and fast imaging system for phase contrast micro-CT at SSRF, *Journal of Synchrotron Radiation* submitted.

[25] E. J. Candes, J. Romberg and T. Tao, Robust uncertainty principles: Exact signal reconstruction from highly incomplete frequency information, *Ieee Transactions on Information Theory* **52**(2), 489–509, (2006).
[26] J. Miao, F. Foster and O. Levi, Equally sloped tomography with oversampling reconstruction, *Phys Rev B* **72**(5), 052103, (2005).
[27] B. Beckhoff, B. Kanngießer, N. Langhoff, R. Wedell and H. Wolff, Handbook of Practical X-Ray Fluorescence Analysis, *Springer* ISBN 3-540-28603-9, (2006).
[28] P. Boisseau, Determination of Three-Dimensional Trace Element Distributions by the Use of Monochromatic X-ray Microbeams, Ph.D Dissertation, M.I.T. Cambridge, 1986.
[29] S. Kim, T. Punshon, A. Lanzirotti *et al.* Localization of iron in Arabidopsis seed requires the vacuolar membrane transporter VIT1, *Science* **314**, 1295–1298 (2006).
[30] M. de Jonge, C. Holzner, S. Baines *et al.* Quantitative 3D elemental microtomography of Cyclotella meneghiniana at 400-nm resolution, *Proceedings of the National Academy of Sciences* **107**, 15676–15680, (2006).
[31] T. Takeda, J. Wu, Q. Huo *et al.*, X-ray fluorescent CT imaging of cerebral uptake of stable-iodine perfusion agent iodoamphetamine analog IMP in mice, *Journal of Synchrotron Radiation* **16**, 57, (2008).
[32] M. Naghedolfeizi, J. Chung, R. Morris *et al.*, X-ray fluorescence microtomography study of trace elements in a SiC nuclear fuel shell, *Journal of Nuclear Materials* **312**, 146–155, (2003).
[33] L. Lemelle, A. Simionovici, R. Truche *et al.*, A new nondestructive X-ray method for the determination of the 3D mineralogy at the micrometer scale, *American Mineralogist.* **89**, 547, (2004).
[34] Deng Biao, Yang Qun, Xie Honglan *et al.*, First X-ray fluorescence CT experimental results at SSRF X-ray imaging Beamline, *Chinese Physics C* **35**, 402–404, (2011).
[35] Jinke Qiu, Biao Deng *et al.*, Internal elemental imaging by scanning X-ray fluorescence microtomography at the hard X-ray microprobe beamline of the SSRF: Preliminary experimental, *Nuclear Instruments and Methods in Physics Research B* **269**, 2662–2666, (2011).
[36] Deng Biao, Yang Qun, Du Guohao, Tong Yajun, Xie Honglan and Xiao Tiqiao, The progress of X-ray fluorescence computed tomography at SSRF, *Nuclear Instruments and Methods in Physics Research B* **305**, 5–8, (2013).
[37] Qun Yang, Biao Deng *et al.*, Fast and accurate X-ray fluorescence CT imaging by ordered-subsets expectation maximization algorithm, *Journal of Synchrotron Radiation* **19**, 210–215, (2012).
[38] T. Yuasa *et al.*, *IEEE Transactions on Nuclear Science*, **44**, 54, (1997).
[39] G. Rust and J. Weigelt, *IEEE Transactions on Nuclear Science*, **45**, 75, (1998).
[40] Qun Yang, Biao Deng, Guohao Du, Honglan Xie, Guangzhao Zhou, Tiqiao Xiao and Hongjie Xu, X-ray fluorescence CT with absorption correction for biomedical samples, *XRS* **43**, 278–285, (2014).

[41] D. Sayre, Imaging Processes and Coherence in Physics, eds. M. Schlenker, M. Fink, J. P. Goedgebuer, C. Malgrange, J. C. Vienot and R. H. Wade, vol. 112 (Springer Lecture Notes in Physics Berlin: Springer, 1980), pp. 229–235.
[42] D. Sayre, Direct Methods of Solving Crystal Structures, ed. H. Schenck (New York: Plenum, 1991), pp. 353–356.
[43] J. Miao, P. Charalambous, J. Kirz and D. Sayre, *Nature* **400**, 342, (1999).
[44] Y. Takahashi *et al.*, High-resolution diffraction microscopy using the plane-wave field of a nearly diffraction limited focused X-ray beam, *Phys. Rev. B Condens. Matter* **80**, 054103, (2009).
[45] Y. Takahashi *et al.*, High-resolution projection image reconstruction of thick objects by hard X-ray diffraction microscopy, *Phys. Rev. B* **82**, 214102, (2010).
[46] H. Jiang *et al.*, Quantitative 3D imaging of whole, unstained cells by using X-ray diffraction microscopy, *Proc. Nat. Acad. Sci.* **107**, 11234–11239, (2010).
[47] J. M. Rodenburg *et al.*, Hard-X-ray lensless imaging of extended objects, *Phys. Rev. Lett.* **98**, 034801, (2007).
[48] P. Thibault *et al.*, High-resolution scanning X-ray diffraction microscopy, *Science* **321**, 379–382, (2008).
[49] I. K. Robinson *et al.*, Reconstruction of the shapes of gold nanocrystals using coherent X-ray diffraction, *Phys. Rev. Lett.* **87**, 195505, (2001).
[50] G. J. Williams *et al.*, Three-dimensional imaging of microstructure in Au nanocrystals, *Phys. Rev. Lett.*, **90**, 175501, (2003).
[51] G. J. Williams *et al.*, Fresnel coherent diffractive imaging, *Phys. Rev. Lett.* **97**, 025506, (2006).
[52] T. Sun *et al.* Three-dimensional coherent X-ray surface scattering imaging near total external reflection, *Nat Photon* **6**(9), 588–592, (2012).
[53] B. Abbey, L. W. Whitehead, H. M. Quiney, D. J. Vine, G. A. Cadenazzi, C. A. Henderson, K. A. Nugent, E. Balaur, C. T. Putkunz, A. G. Peele, G. J. Williams and I. McNulty, *Nat. Photonics* **5**, 420, (2011).
[54] T. E. Gureyev *et al.*, Toolbox for advanced X-ray image processing, *Proc. SPIE* **8141**, 81410B-14, (2011).
[55] R. C. Chen *et al.*, PITRE: software for phase-sensitive X-ray image processing and tomography reconstruction, *Journal of Synchrotron Radiation* **19**, 836–845, (2012).
[56] Yongjing Guan, Yongting Wang, Falei Yuan, Haiyan Lu, Yuqi Ren, Tiqiao Xiao, Kemin Chen, David A. Greenberg, Kunlin Jin and GuoYuan Yang, Effect of suture properties on Stability of middle cerebral artery occlusion evaluated by synchrotron radiation angiography, *Stroke* **43**, 888, (2012).
[57] Wei Jin, Bobin Chen, Jiyu Li, Hua Zhu, Mark Huang, Shengmei Gu, Qiaoqiao Wang, Jiaying Chen, Sanjian Yu, Jiong Wu and Zhiming Shao, TIEG1 inhibits breast cancer invasion and metastasis by inhibition of EGFR transcription and the EGFR signaling pathway, *Molecular and Cellular Biology* **32**, 50, (2012).

[58] Jun Cai, Yuhao Sun, Falei Yuan, Lujia Chen, Chuan He, Yuhai Bao, Zuoquan Chen, Meiqing Lou, Weiliang Xia, Guo-Yuan Yang and Feng Ling, A Novel Intravital Method to Evaluate Cerebral Vasospasm in Rat Models of Subarachnoid Hemorrhage: A Study with Synchrotron Radiation Angiography, *plos one* **7**, e33366, (2012).

[59] Dongming Zhang, Xiaohui Yan, Xinyi Zhang, Chenglin Liu, Ruishan Dang, Tiqiao Xiao and Peiping Zhu, Synchrotron radiation phase-contrast X-ray CT imaging of acupuncture points, *Analytical and Bioanalytical Chemistry* **401**, 803, (2011).

[60] Jianzhong Hu *et al.*, Visualization of microvasculature by X-ray in-line phase contrast imaging in rat spinal cord, *Physics in Medicine & Biology* **57**, N55, (2012).

[61] L. Zhang, D. Y. Li and S. Q. Luo, Non-invasive microstructure and morphology investigation of the mouse lung: Qualitative description and quantitative measurement, *Plos One*, **6**(2), e17400, (2011).

[62] Chenglong Dai *et al.*, Osteogenic evaluation of calcium/magnesium-doped mesoporous silica scaffold with incorporation of rhBMP-2 by synchrotron radiation-based mCT, *Biomaterials* **32**, 8506, (2011).

[63] Fan Yang *et al.*, Bone regeneration using cell-mediated responsive degradable PEG-based scaffolds incorporating with rhBMP-2, *Biomaterials* **34**, 1514, (2013).

[64] Jing Zhang *et al.*, RhBMP-2-loaded calcium silicate/calcium phosphate cement scaffold with hierarchically porous structure for enhanced bone tissue regeneration, *Biomaterials* **34**, 9381, (2013).

[65] Lingyan Cao *et al.*, Vascularization and bone regeneration in a critical sized defect using 2-N,6-O-sulfated chitosan nanoparticles incorporating BMP-2, *Biomaterials* **35**, 684, (2014).

[66] Q. Dong *et al.*, In situ observation of columnar-to-equiaxed transition in directional solidification using synchrotron X-radiation imaging technique, *Materials Science and Engineering a-Structural Materials Properties Microstructure and Processing* **530**, 271–276, (2011).

[67] Q. Dong *et al.*, Anaxial columnar dendrites in directional solidification of an Al-15 wt.% Cu alloy, *Materials Letters* **65**(21–22), 3295–3297, (2011).

[68] J. F. Wu, M. Q. Xu and G. C. Zhao, Graphene-based modified electrode for the direct electron transfer of Cytochrome c and biosensing, *Electrochemistry Communications*, **12**(1), 175–177, (2010).

[69] H. T. Ma *et al.*, In-situ study on growth behavior of Ag3Sn in Sn-3.5Ag/Cu soldering reaction by synchrotron radiation real-time imaging technology, *Journal of Alloys and Compounds* **537**, 286–290, (2012).

[70] L. Qu *et al.*, In situ study of the real-time growth behavior of Cu6Sn5 at the Sn/Cu interface during the soldering reaction, *Scripta Materialia* 72–73, 43–46, (2014).

[71] M. Y. Wang *et al.*, Growth orientations and morphologies of alpha-Mg dendrites in Mg-Zn alloys, *Scripta Materialia* **67**(7–8), 629–632 (2012).

[72] Z. Xing *et al.*, Preparation of microcellular polystyrene/polyethylene alloy foams by supercritical CO2 foaming and analysis by X-ray microtomography, *Journal of Supercritical Fluids* **82**, 50–55, (2013).
[73] A. E. Hughes *et al.*, Revelation of intertwining organic and inorganic fractal structures in polymer coatings, *Advanced Materials* (2014) p. DOI:10.1002/adma.201400561.
[74] H. Wang *et al.*, Data-constrained modelling of an anthracite coal physical structure with multi-spectrum synchrotron X-ray CT, *Fuel* **106**, 219–225, (2013).
[75] W. H. Chen *et al.*, A synchrotron-based local computed tomography combined with data-constrained modelling approach for quantitative analysis of anthracite coal microstructure, *Journal of Synchrotron Radiation*, **21**(3), 586–593, (2014).
[76] Y. D. Wang *et al.*, Synchrotron-based data-constrained modeling analysis of microscopic mineral distributions in limestone international, *Journal of Geosciences* **4**(2), 344–351, (2013). A. Groso, R. Abela and M. Stampanoni, Implementation of a fast method for high resolution phase contrast tomography, *Optics Express* **14**(18), 8103–8110, (2006).
[77] Y. C. Li *et al.*, In situ investigation on the mixed-interaction mechanisms in the metal-ceramic system's microwave sintering, *Acta Materialia* **66**, 293–301, (2014).
[78] X. F. Hu *et al.*, In situ observations of fractures in short carbon fiber/epoxy composites, *Carbon* **67**, 368–376, (2014).
[79] T. M. Wang *et al.*, Evolution of dendrite morphology of a binary alloy under an applied electric current: An in situ observation, *Physical Review E* **81**(4), (2010).
[80] H. Zhou *et al.* Effects of organic and inorganic fertilization on soil aggregation in an Ultisol as characterized by synchrotron based X-ray micro-computed tomography, *Geoderma* **195–196**(0), 23–30, (2013).
[81] H. Zhou *et al.* Effects of vegetation restoration on soil aggregate microstructure quantified with synchrotron-based micro-computed tomography, *Soil and Tillage Research* **124**(0), 17–23, (2012).
[82] Guanyun Peng, *Holzforschung*, **67**, 1–5. doi:10.1515/hf-2013-0080, (2013).
[83] G. Y. Peng, *Spectroscopy and Spectral Analysis* **33**(3), 829–833, (2013).
[84] X. Z. Yin *et al.*, *Journal of Pharmacy and Pharmacology* **65**, 953–959, (2013).
[85] X. Z. Yin *et al.*, *The AAPS Journal* **15**, 1025–1031, (2013).
[86] R. H. Liu *et al.*, *International Journal of Pharmaceutics* **445**, 125–133, (2013).
[87] Y. L. Xue *et al.*, Investigation of characteristic microstructures of wild ginseng by X-ray phase contrast microscopy, *Acta Physica Sinica* **59**, 5496–5507, (2010).
[88] L. L. Ye *et al.*, Application of X-ray phase contrast micro-tomography to the identification of traditional Chinese medicines, *Journal of Instruments* **8**(8), C07006, (2013).

[89] L. L. Ye *et al.*, X-ray phase contrast micro-tomography and its application in quantitative 3D imaging study of wild ginseng characteristic microstructures, *Acta Optica Sinica* **33**, 1234002, (2013).
[90] Yimin Yang, Lihua Wang, Shuya Wei, Guoding Song, Jonathan Mark Kenoyer, Tiqiao Xiao, Jian Zhu and Changsui Wang, Nondestructive analysis of dragonfly eye beads from the warring states period, excavated from a Chu Tomb at the Shenmingpu Site, Henan Province, China, *Microscopy and Microanalysis*, **19**, 335, (2013).
[91] Xiguang Zhang and Brian R. Pratt, The first stalk-eyed phosphatocopine crustacean from the lower cambrian of China, *Current Biology* **22**, 2149, (2012).
[92] R. Payri, F. J. Salvador, J. Gimeno *et al.*, Effects of nozzle geometry on direct injection diesel engine combustion process, *Applied Thermal Engineering*, **29**(10), 2051–2060, (2009).
[93] S. Som, A. I. Ramirez, D. E. Longman *et al.*, Effect of nozzle orifice geometry on spray, combustion, and emission characteristics under diesel engine conditions [J], *Fuel*, **90**(3), 1267–1276, (2011).
[94] Z.-j. Wu, Z.-l. Li, W.-d. Huang *et al.*, Comparisons of nozzle orifice processing methods using synchrotron X-ray micro-tomography. *Journal of Zhejiang University-SCIENCE A (Applied Physics & Engineering)* **13**(3), 1–7, (2012).
[95] W. Huang, Y. Gao, Z. Li *et al.*, Three-dimensional investigations of flow characteristics in a diesel nozzle, *Atomization and Sprays* **23**(4), 343–361, (2013).
[96] G. T. Kalghatgi, Combustion Chamber Deposits in Spark-Ignition Engines: A Literature Review[A]. In SAE paper 952443[C], (1995).
[97] G. Shu, L. Dong, X. Liang, A review of experimental studies on deposits in the combustion chambers of internal combustion engines[J], *International Journal of Engine Research* **13**(4), 357–369, (2012).
[98] Z. Xusheng, P. Guanyun, D. Guohao *et al.* Investigation on the microstructure and pore characteristics of carbon deposit on the piston top land of marine diesel engine using synchrotron X-ray microtomography, *Scientia Sinica Technologica (in Chinese)* **44**(9), 953–960, (2014).
[99] X. Zhang, G. Peng, G. Du *et al.*, Investigating the microstructures of piston carbon deposits in a large-scale marine diesel engine using synchrotron X-ray microtomography, *Fuel*, **142**(15), 173–179, (2015).

Chapter 10

Synchrotron Radiation Applications in Medicine

Yifeng Peng and Liangqi Wang

Radiology Department of Putuo Hospital,
Shanghai University of Traditional Chinese Medicine,
164 Lanxi Road, Shanghai 200062, China

Chenglin Liu

School of Physics and Electronics,
Yancheng Teachers University,
2 South Xiwang Avenue, Yancheng 224007, China

Alberto Bravin

European Synchrotron Radiation Facility,
71, Avenue des Martyrs,
38043 Grenoble Cedex, France

Gang Li

Beijing Synchrotron Radiation Facility,
IHEP, Chinese Academy of Sciences,
19 Yuquan Road, Beijing 100039, China

Shaoliang Chen

Zhongshan Hospital, Fudan University,
180 Fenglin Road, Shanghai 200032, China

Yongting Wang and Guo-Yuan Yang

Neuroscience and Neuroengineering Research Center,
Med-X Research Institute and School of Biomedical Engineering,
Shanghai Jiao Tong University,
800 Dongchuan Road, Shanghai 200240, China

Xinyi Zhang

Department of Physics,
National Laboratory for Surface Physics and
Synchrotron Radiation Research Center, Fudan University,
220 Handan Road, Shanghai 200433, China

Abstract

Phase-contrast imaging (PCI) on synchrotron radiation (SR) light source has become a paradigm in X-ray biomedical imaging. *In vitro* and *in vivo* biomedical research has attracted high attention thanks to its capability of providing high diagnostic significant images in a wide range of organs or viscera, including the breast, joints, cartilage, lung, eyes, vasculature, brain and central nervous system. In this chapter, the basic principle, main phase-contrast imaging methods and their bio-medical applications are introduced. Various imaging methods are described and evaluated. The prospect and limitations of synchrotron-based medical imaging are briefly discussed as well.

Keywords: Acupuncture points; analyzer-based imaging; angiography; breast cancer; computed tomography; diffraction enhancement imaging; grating interferometry (GI); in-line X-ray phase contrast micro-computed tomography (IL-XPμCT); interferometry; lung cancer; medical imaging; phase contrast; propagation-based phase-contrast imaging (PBI); vasculature; X-ray dark field image; X-ray fluorescence.

10.1. Introduction

Since their discovery at the end of the 20th century, X-rays have become a powerful and irreplaceable diagnostic tool. In the course of 21st century magnetic resonance imaging (MRI), single-photon emission computed tomography (SPECT), positron emission tomography (PET) and ultrasound (US) were also established; nowadays they form with X-rays a powerful portfolio of diagnostic techniques routinely used in hospitals. Furthermore, synchrotron radiation sources offer additional possibilities for exploring new medical imaging methods.

Images can be generated according to X-ray absorption difference (known as absorption contrast) among different parts inside an object. X-ray images show us the internal structure invisible with naked eyes. Over a hundred years, people have made great progress in X-ray imaging technology to acquire clearer and higher-resolution images with advanced science and technology from different fields [1]. With view to over a century of history, the imaging technology has developed in three major areas, including angiography, photofluorography (sometimes called just fluorography or radiography) and computed tomography (CT). Imaging has perfectly developed in these three areas, and imaging quality has much improved. However, all imaging technologies always depend on absorption contrast that refers to different X-ray absorption degrees due to density and composition differences of the internal structure of living samples or human being. Then what advantages does synchrotron radiation have? Why was it said that synchrotron radiation can do something that other light sources cannot do? High brightness and coherence of synchrotron radiation are usually credited. Brightness of the synchrotron radiation is several of orders of magnitude higher than conventional light sources. In addition, synchrotron radiation is partially coherent, which is better than others' as compared to other laboratory light sources. Synchrotron radiation made breakthrough in image research in 1980s, marked by the first human coronary artery angiogram obtained by the Stanford Synchrotron Radiation Lightsource (SSRL) at the United States in 1986 [1]. High brightness of synchrotron radiation can greatly reduce the dose of contrast agent for angiography or digital subtraction angiography, and as a result, a safer intravenous contrast agent can be used. However, the synchrotron radiation angiography technology as well as the other X-ray imaging technology also takes images depending on absorption contrast. It is well known that amplitude and phase of X-ray change as passing the object to obtain absorption contrast and phase contrast respectively. The other important breakthrough in the synchrotron radiation X-ray imaging therefore the application of phase contrast. In the last few years of the 20th century, phase contrast imaging is realized initially by using diffraction enhancement imaging (DEI), which is another milestone in history. Around 2000, many contrast imaging methods appeared, and image resolution was far higher, to the micron level from the millimeter level of conventional X-ray imaging. Similar to conventional X-ray CT technology, the synchrotron radiation imaging technology goes into the third development direction, a combination of contrast imaging and CT, which is known as X-ray phase contrast micro-computed tomography (XPμCT). Coherent diffraction imaging (CDI) developed in recent years even helped to increase resolution to the nanometer level. (See Sections 10.10 and 10.15). At the time when resolution and definition are greatly improved, the advanced synchrotron radiation imaging technology can meet the need of doctors who want to observe *in vivo* organs of living beings or people from three dimensions. We were surprised that the synchrotron radiation imaging technology goes in almost the same track though it appeared a century later than Roentgen's X-ray imaging technology. However, it makes a higher-level progress such that space resolution is increased to the micro-nanometer level from the millimeter-submilimeter level. Phase contrast is used to take images of details of small density difference for absorption contrast which is only used to reflect parts having large density difference. Owing to X-ray light source of super high brightness and complete coherence provided by FEL, CDI helps people realize their centurial dream in dynamic imaging single cell, single particle and single molecule. In the following sections, the basic principle, main phase-contrast imaging methods and their bio-medical applications are introduced.

10.2. Basic Principle

When X-rays travel through an object, both the amplitude and phase of the wave front are modified. In conventional X-ray transmission imaging, the contrast is generated from the change in the amplitude.

The amplitude contrast (absorption contrast) is usually insufficient for investigating soft tissue structures in a body. This is because all soft tissues have very similar absorption and scattering properties. On the other hand, the phase-contrast X-ray imaging can solve some of the basic problems of conventional radiographic X-ray imaging, which encounters severe difficulties with objects containing details having small differences in absorption. Because the phase-shift cross section is much larger than the absorption cross section for light elements, high sensitivity to soft tissue is expected even when a moderate dose is deposited. X-ray phase information is usually missing in a transmission image. Therefore special techniques are necessary to generate phase contrast.

The properties of synchrotron radiation X-ray beams which make them interesting to medical research consist in their extremely high intensity and broad-band energy spectrum. Synchrotron generated X-rays with sufficient spatial coherence can make it possible to detect the effects of the phase perturbations caused by the objects, even when absorption is negligible. Synchrotron radiation sources provide several orders of magnitude more flux than a conventional X-ray tube and show a smooth and continuous spectrum as compared to the sharply peaked characteristic emission lines from a tube. Basically, the high intensity allows monochromatic beams to be generated in a very broad range of energies. The standard problem of beam hardening in both medical imaging and therapy is eliminated by the monochromatic beams since the energy spectrum does not change with passage through tissue (only the intensity changes). The tunable spectrum allows enhancement of images and the delivery of therapeutic doses by selection of the most effective energy for a given procedure. An extended description of all these characteristics is reported in [2]. We invite all readers who are interested in this subject to address that review.

10.3. Phase-Contrast Imaging Methods

There are multiple techniques developed to exploit the phase-contrast imaging (PCI). Generally five main categories are classified: the propagation-based imaging (PBI) methods [3], the analyzer-based imaging (ABI) methods [4], the interferometric methods based on the use of crystals [5], the grating interferometric (GI) [6] and the non interferometric methods [7]. These PCI modalities have been extensively investigated to access their potential for biomedical imaging.

10.3.1. *Propagation-based phase-contrast imaging (In-line holography)*

Propagation-based X-ray phase-contrast imaging (Figs. 10.1 and 10.2) permits the visualization of tissues that have very similar X-ray absorption properties and may benefit a variety of biomedical imaging applications. The propagation-based phase-contrast imaging (PBI) is the easiest way to observe the phase-contrast and is similar to in-line holography. When a coherent wave-front traverses the sample and the distorted wave-front propagates sufficiently far from the sample, the tiny differences in phase propagation lead to interference,

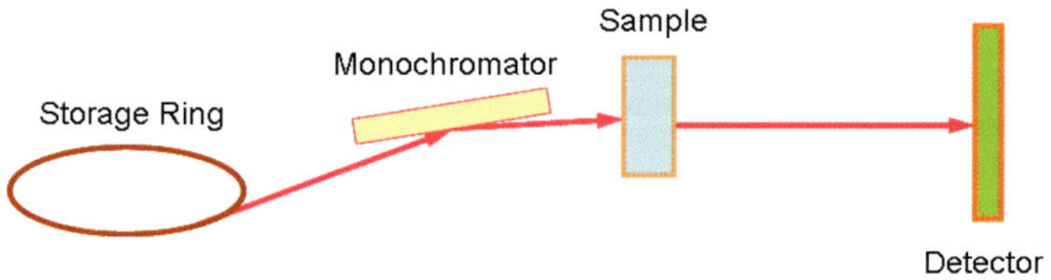

Fig. 10.1. Schematic setup for in-line phase contrast imaging (radiography).

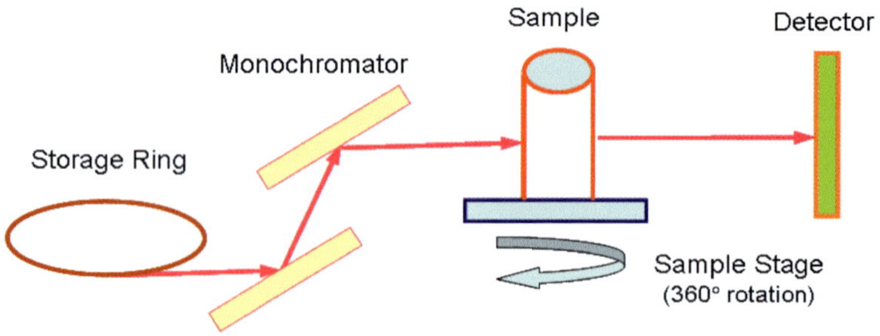

Fig. 10.2. Schematic setup for in-line phase contrast imaging (computed tomography-CT).

and differentiations of intensity can be observed in the image plane. Under quite general cases, the Laplacian of phase $d^2\varphi/\mathrm{d}x^2 + d^2\varphi/\mathrm{dy}^2$ (with the X-ray beam propagating along the z direction) is proportional to the longitudinal gradient of intensity [8]. This differential transport-of-intensity equation (TIE) can be settled to yield φ when the incident intensity is given and the transmitted intensity is recorded at an appropriate distance, or at two or more distances in the linear regime. We will discuss in the following that phase retrieval from TIE is available under some conditions. Phase CT imaging can be used because the total phase change which is determined by the object is proportional to the projected electron density. PBI visualizes the object structure without further data processing because it creates edge-enhancement of the sample, so phase retrieval may not be essential to determine useful results. Already pioneering researches in the late 1990s demonstrated that small objects utilizing polychromatic radiation from a micro-focus X-ray tube could be imaged at high spatial and contrast resolution [9], giving an imaging tool of potential clinical interest [10]. The phase retrieval methods and their approximations have been studied extensively, and similarities but also key differences can be found between the seven algorithms commonly used [11, 96].

10.3.2. *Analyzer-based imaging (ABI)*

Monochromatic X-ray beams and perfect crystal analyzers are utilized in the ABI method (Fig. 10.3). In first approximation, the analyzer reflects only the rays that make the correct angle (Bragg angle) with the atomic (Bragg) planes of the crystal, which works as an extremely narrow angular slit. In practice the reflection is allowed for a narrow range of angles, called reflectivity curve.

In a typical case the full width at half maximum (FWHM) of the reflectivity curve is a few microradians, and its shape is approximately triangular. The analyzer is tuned to reflect at the angle that corresponds to the peak of the intrinsic (without an object) reflectivity curve (top), or at the mid-slope on either side of the reflectivity curve (low or high). At the top, the so-called apparent

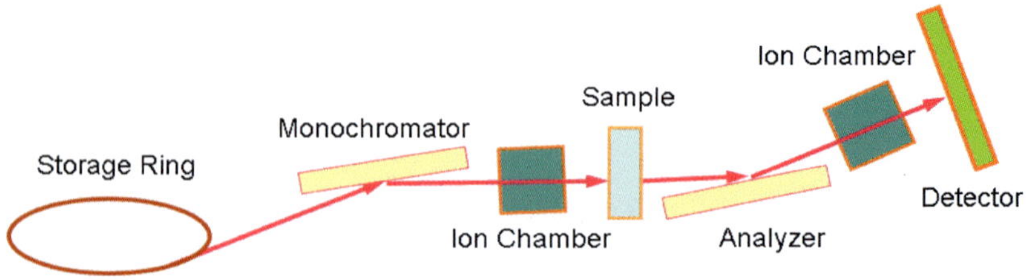

Fig. 10.3. Schematic setup for analyzer-based imaging (ABI).

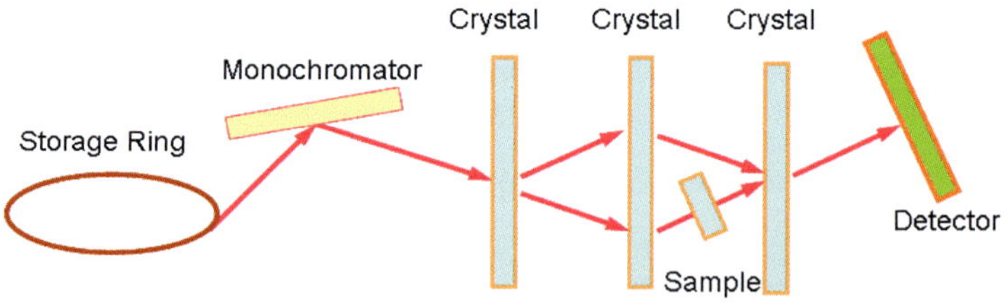

Fig. 10.4. Schematic setup for crystal interferometry (CI).

absorption (or extinction) contrast is obtained because the scattering and refracted rays are rejected. At the low and high positions, refraction-contrast is showed through the changes in propagation direction that are converted to intensity changes [4].

As mentioned before, a single image is often enough for diagnosis when ABI is utilized for tissue or organ imaging. However, more images are required when phase retrieval or separation of refraction from absorption-contrast is wanted. Algorithms based on adequate curve fitting images at two or more positions along the rocking curve (RC) are necessary.

10.3.3. *Interferometry*

The interferometry (Fig. 10.4) is a direct experimental method used for mapping the scattering properties of the object by measuring the phase changes occurred within the object. The interferometer was introduced by [12], and a device appropriate for bio-medical imaging was constructed in 1990s [13, 97, 99]. More recent advancements and applications have been reviewed by [14, 15]. Three parallel silicon crystal wafers, which stick out from a monolith perfect crystal ingot, constitute the crystal interferometer (CI) in its simplest form. In the first wafer (S) the incident X-ray beam is diffracted and split in two identical monochromatic wave-fronts which are split again in the second wafer (M). The inner beams traverse the third wafer (A), and an area detector records the interference pattern. A wedge-shaped phase shifter placed in one of the beams gives straight interference fringes with regular intervals (carrier fringes), and when the object is put in the other beam the fringes bend. The displacement of the carrier fringes is in accord with the phase shift caused by the object. When a Fourier method is used, the phase map in a single projection can be constructed faster [16].

The interferometric method is most suitable for small and smooth phase gradients. The limitation of disseminating the technique has been contributed by its intrinsic constraints resulting from small field of view (a few centimeters), and from the extreme requirements of stability and alignment of the interferometer.

10.3.4. *Grating interferometry (GI)*

X-ray GI provides substantially increased contrast over conventional absorption-based imaging methods, and therefore can provide new and complementary information [100]. GI is based on an optical phenomenon found out in the early 19th century [17], and explained as Fresnel diffraction in a periodic grating [18]. At regular distances behind the grating, the image of the grating is repeated, $d_T = 2p^2/\lambda$, where p is the period of the grating. At fractional distances the same patterns appear shifted and/or with fractional period. A phase grating is utilized in X-ray applications, where the grooves generate a periodic variation in the phase of the wave-front behind the grating. There is no phase shift when no object is placed in the X-ray beam. When an object is installed in front or behind the phase grating in the beam, the wave-front is deviated by refraction, which causes the shift of intensity pattern. Obviously, transverse coherence over a phase grating period is necessary. When compared with patterns without the object, three images can be obtained after application of a suitable algorithm: they are respectively the map of absorption, differential phase shift and dark field.

10.3.5. *Non interferometric methods*

The most known non-interferometric method is the coded-aperture: the detector is divided by a mask into a pattern of sensitive and insensitive regions between adjacent pixels, and a pre-sample mask creates the same pattern of beams that impinges on the boundaries of sensitive and insensitive regions. The original of Olivo *et al.* [19] was further developed in the recent years [7, 20]. The beams are deviated by refraction in the sample, resulting in intensity variation at the detector. The set-up is very simple, and polychromatic radiation from an X-ray tube can be used.

The implementation of the technique at X-ray energies close to 100 keV has been already demonstrated, opening promising perspectives in view of potential clinical applications [21].

10.3.6. *Comparison of the different PCI techniques*

A detailed comparison among the PCI methods is not easy because it is not possible to evaluate by extrapolating data because the different techniques are applied at various experimental conditions. Several experimental and set-up parameters may apply in each PCI method. In theory, these parameters should be optimized to maximize the sensitivity of the technique that depends on the features of both the sample to be studied, and the source/X-ray beam (e.g. source size and divergence, spatial and longitudinal coherence, divergence, energy, etc). In practice, mechanical and/or laboratory limitations determine the real experimental configurations. For all these reasons, there are only very few meritorious trials to theoretically and/or experimentally compare the sensitivity and performance of the methods. *Ad hoc* experiments are in fact necessary for detailed experimental comparisons that are performed with same physical parameters: source characteristics, energy, divergence, etc.

One of the few papers comparing the techniques is [22]. Three techniques are compared: propagation-based imaging, analyzer-based imaging and grating interferometry. The signal-to-noise ratio and the figure of merit are quantitatively compared for the three techniques on the same phantoms and using the same X-ray source and detector. Principal dependencies of the signal upon the numerous acquisition parameters, the spatial resolution and X-ray energy are discussed in detail. Among the three techniques, theoretically the most sensitive technique resulted to be the ABI; these results were experimentally confirmed using the same experimental parameters.

Only very few laboratories in the world can install and utilize the different PCI set-ups from the same X-ray source. The restricted number of suitable X-ray sources has limited the utilization of PCI techniques in the imaging of tissues and organs. Actually, transverse coherence (a small source) of the beam (e.g. GI, non-interferometric grating imaging and PBI), or longitudinal (high monochromaticity) coherence (e.g. ABI (Fig. 10.3) and CI (Fig. 10.4)) is highly required in some of the methods.

10.4. Biomedical Imaging Applications of SR

10.4.1. *Breast*

An early diagnosis has been demonstrated to be one of the key factors in the breast cancer management. The breast is a highly radiosensitive organ and therefore the dose is an important constraint in X-ray diagnosis. Infiltrating breast cancer shows extremely weak absorption contrast with respect to glandular tissues and therefore only subtle gray scale differences may appear on mammograms. For these reasons, their beast has been one of the early applications of phase contrast imaging, with the aim of improving the significance of diagnostic images. A clinical program for X-ray phase contrast mammography with SR has been started in March 2006 at the SYRMEP (Synchrotron Radiation for Medical Physics) beamline of Elettra, the SR facility in Trieste, Italy [23–25]. Scientists choose the lowest energy such that the mean glandular dose (MGD) is less or equal to the one delivered in the hospital using the conventional radiography, GE Senographe DS clinical unit. Two mammograms of the same breast

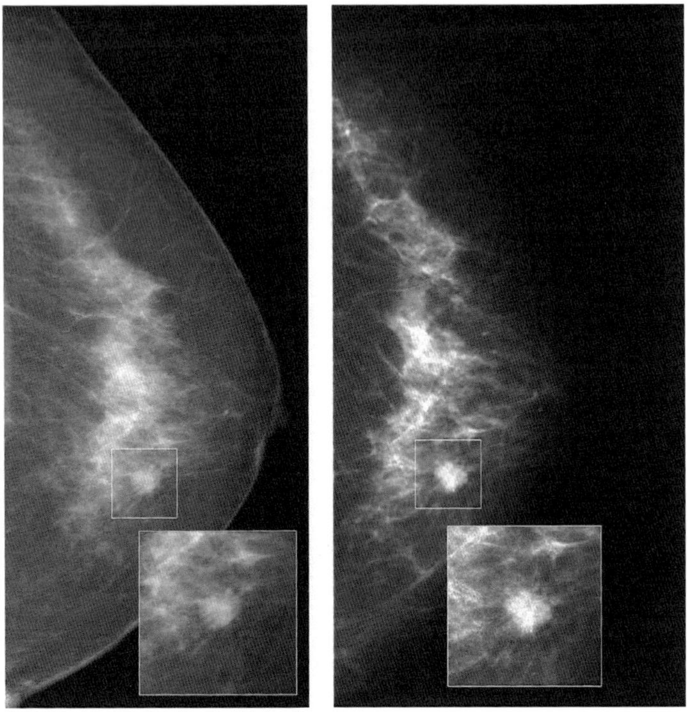

Fig. 10.6. (a) The breast imaging with mammogram with GE Senographe DS and (b) PPI, the tumor with the PPI is revealed more clearly at fewer radiation dose (adapted from [24] with permission).

are shown in Fig. 10.6: mammogram with GE Senographe DS on the left (MGD = 1.28 mGy) and phase propagation X-ray imaging (PPI) mammography on the right (MGD = 0.86 mGy). The conventional radiography suggests a probably benign lesion, but the PPI mammogram suggests a suspected lesion even with lower radiation dose. Some details are discussed in [26]. Furthermore, researchers demonstrated that a distribution of micrometer-sized calcifications in the human breast that were not visible in clinical X-ray mammography at diagnostic dose levels can produce a significant dark-field signal in a grating-based X-ray phase-contrast imaging [27]. So far, several studies have been performed using the PBI, ABI or GI. Pioneering works on excised breast tissue samples showed the improved contrast [23, 28]. These researches have led to the first clinical trials at the ELETTRA SR facility in Trieste, Italy [29, 30]. Further development in CT imaging has led to demonstrate that computed tomography PCI can be performed at clinically compatible X-ray doses [31], therefore allowing the 3D visualization of tumors in full breasts [32].

10.4.2. *Musculoskeletal phase-contrast imaging*

Diagnosis of osteoarthritis (OA) and cartilage lesions implies the analysis of both soft tissue (i.e., cartilage) and subchondral bony details. Current monitoring methods include MRI and CT imaging; nevertheless both show very severe limitations in terms of image quality, contrast or resolution, in particular in either visualizing the subcondral bone and the intra-cartilage damage (MRI) or the cartilage itself (CT). Complementary information can be provided by histological analysis which is invasive and requires complex preparation procedures. X-ray PCI has been playing an important role in the capability of early and accurate visualization of OA and of cartilage and tendons (Fig. 10.7). The reconstructed thin sections extracted from the PBI-CT volumes depict the borders and structure of muscle tissue, ligaments, vessels, and nerves with high contrast and detail, as shown in Fig. 10.7. Phase-contrast enhanced computed tomography (PC-CT) presents soft tissue contrast surpassing that of conventional CT with a clear discrimination of ligamentous, muscular, neural, and vascular structures. In addition, phase-contrast imaging images show cartilage and meniscal calcifications that are not perceptible on conventional CT or on MRI. Researchers have also demonstrated that coded-aperture-based X-ray phase contrast imaging implemented with conventional, uncollimated, non-microfocal laboratory sources can correctly visualize the cartilage layer in slices of an excised rat tibia and detect possible damages to this layer [33]. Z. Izadifar *et al.* indentified X-ray imaging ABI as a promising technique for visualization and characterization of TE (tissue engineering) scaffold in *in situ* cartilage, and the structural features of the scaffold can be characterized in CT-DE

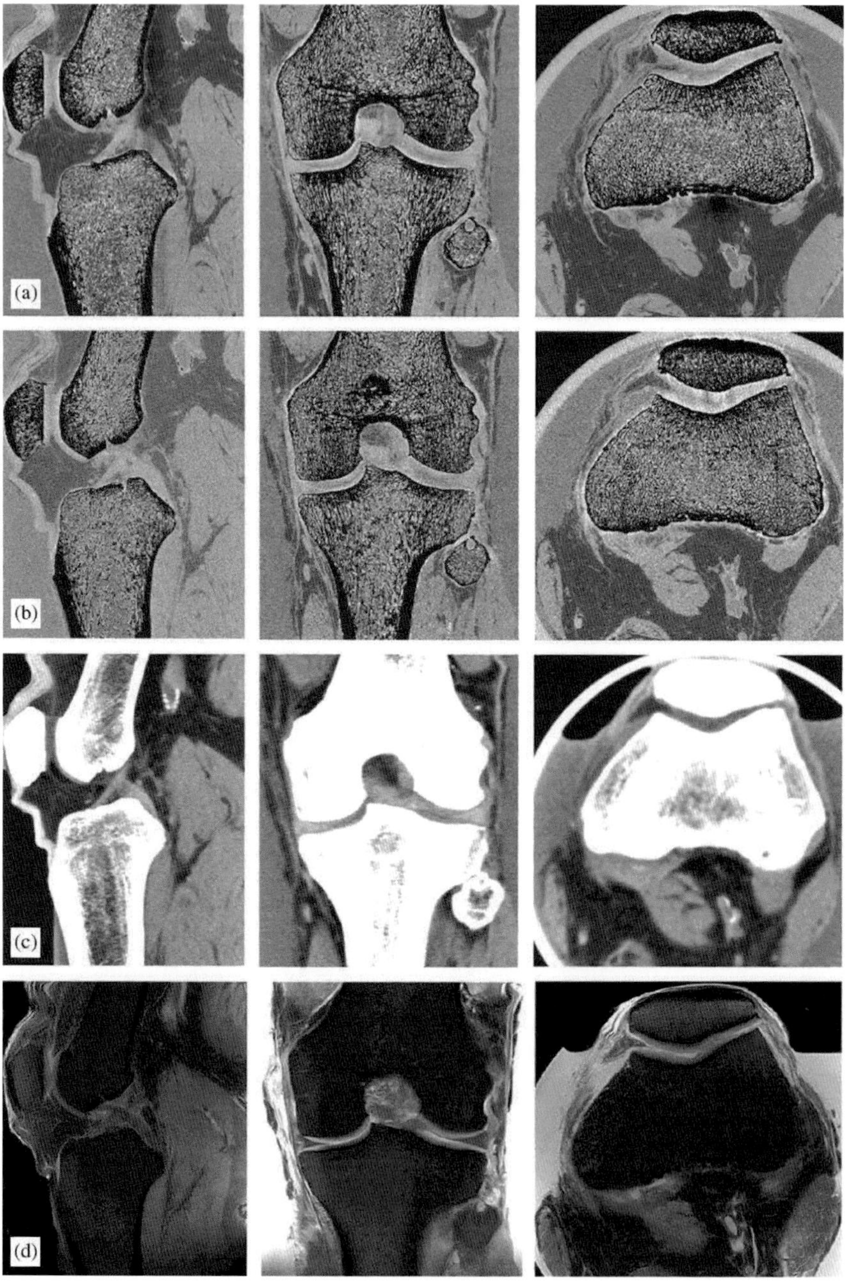

Fig. 10.7.

Fig. 10.7. (figure on facing page) (a) Corresponding sagittal, axial, and coronal sections acquired using PBI, (b) SR-CT imaging, (c) clinical CT, and (d) a 3-T MRI. The PBI images show markedly superior soft tissue contrast and delineation as well as cartilage depiction as compared with the PCI absorption and the clinical CT with details approximating the quality of the MRI. (adapted from [34] with permission).

images at 40 μm resolution and 40 keV photon energy (an optimum lower dose) [35].

10.4.3. *Airway and lung*

10.4.3.1. *Airway and lung*

In conventional radiography, the airway and lung are weakly absorbing organs, and lesions are seen as opaque masses. The group of K. K. W. Siu directly examined the airway surfaces in the nasal airways and tracheas of anaesthetized mice using phase contrast X-ray imaging (PCXI). They also achieved high resolution three-dimensional (3D) imaging of the small airways in mice using PC-CT to elucidate the structure-function relationships produced by airway disease. With this method they established non-invasive imaging able to detect and measure aspects of the biology and physiology of surface fluids present on airways in order to develop novel measures able to validate the success of proposed genetic or pharmaceutical therapies for cystic fibrosis (CF) airway disease [36]. Bacterial clearance depends on a layer of airway surface liquid (ASL) consisting of both a mucus layer that traps, kills, and inactivates bacteria and a periciliary liquid layer that keeps the mucus at an optimum distance from the underlying epithelia, to maximize ciliary motility and clearance of bacteria. The airways in CF patients and animal models of CF demonstrate abnormal ASL secretion and reduced antimicrobial properties. X. Luan *et al.* developed a synchrotron-based imaging technique to visualize the ASL layer and measure the effect of bacteria on ASL secretion. (Fig. 10.8) Intense scattering contrast from the lung was shown in the early DEI studies [37]. Due to scatter rejection, when the analyzer was changed into the peak position, the

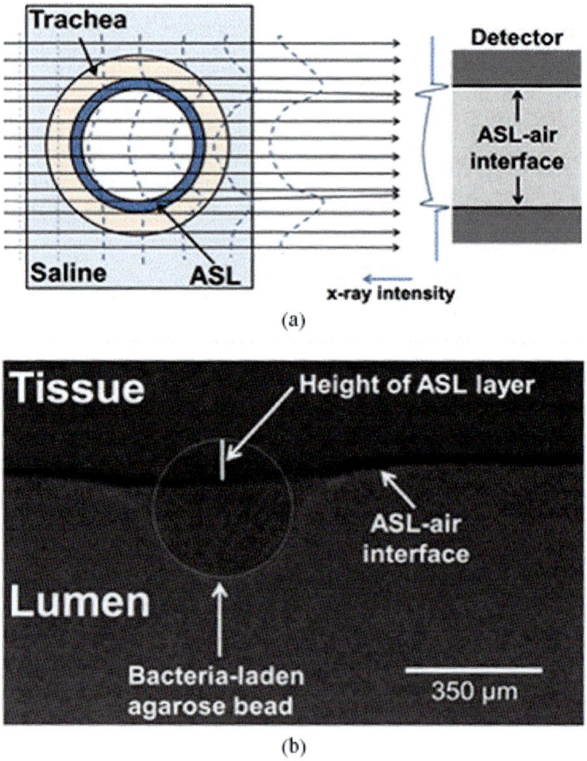

Fig. 10.8. The process of the bacteria laden agarose bead in the airway is described by the PCI. (a) Schematic showing the detection of ASL–air interface in the lumen of an intact trachea using PCI. (b) Measurement of ASL height in an intact trachea as the distance between the ASL–air interface (arrow) and the edge of the agarose bead touching the surface epithelium by a researcher blinded to the experimental conditions (adapted from [38] with permission).

apparent absorption in mouse lung was about four times larger than normal absorption. Both DEI and PBI techniques have been used in the research of lung structure and function (e.g., see [39, 40]). One research direction is focused on the newborn lung clearance imaging which is taken dynamically [41–43, 93, 95] (Fig. 10.9). Rapid succession of PBI images of rabbits at birth locates airway liquid and gives us the overview of lung aeration. This offers new and essential information on the transition to air-breathing of the newborn rabbit

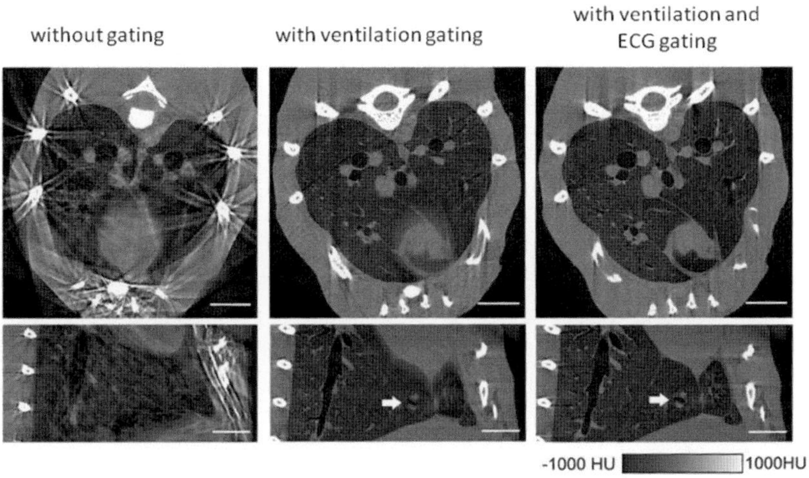

Fig. 10.9. The lung PCI imaging without/with ventilation gating, the imaging are clearer with ventilation and ECG (Electrocardiograph) gating (adapted from [44] with permission).

pups. In an *in vivo* mice model researchers have obtained the map of fluid movement in airways and of the liquid reduction on airway surface associated with cystic fibrosis [36, 45]. The speckle pattern could probably be used for observing pathological changes in lung structure because it is originated from, and proportional to, the number of alveoli traversed by the beam [43]. In excised rat and mouse tissues, researchers obtained the image of vessel stenosis and even of individual alveoli [46]. Atelectasis in an impaired lung has been detected by DEI [47, 92], and this imaging has been used as an instance in development of a variant of DEI technique. Iterative algorithm is used to retrieve absorption, refraction and scattering images [39, 48, 49].

10.4.3.2. *Lung cancer*

Lung cancer has the highest incidence and is the leading cause of death according to the 2012 statistics published by the World Health Organization (WHO). Early detection is crucial for a successful management of lung cancer. Chest radiography and CT are the principal diagnostic methods, but unfortunately they still lack sensitivity in the early detection. Synchrotron radiation IL-XPμCT method is a new potential candidate as a high sensitive diagnostic tool. J Zhang *et al.* investigated the feasibility of the clinical utilization of this method for the early lung cancer detection [50].

The lung cancer samples used in this study were composed of three parts. A formalin fixed pig lung was used to simulate the human lung tissue. A little lump of the formalin fixed rat liver embedded in the pig lung was used as the human lung cancer. The artificial materials used were the skin and muscle of the human chest. The pig lung and the rat liver were enveloped by such artificial materials. The rat liver and the human lung cancer have approximately the same Hounsfield unit that indicates the rat liver tissue is comparative to human lung cancer in the attenuation of X-rays. The artificial materials were about 9 cm thick designed as skin-tissue-equivalent materials with the same proportion in elements and the same properties in scattering and absorption to X-rays compared to the human chest. The rat liver (representing human lung cancer) was cut into different sizes, and embedded in two areas of pig lung (representing human lung tissues), whose thicknesses were 1 cm (named as thin lung) and 4 cm (named as thick lung) to simulate thin or thick area of human lung tissues, respectively.

The imaging experiments were performed at 13W1 of SSRF. The X-ray energy is 35 keV. The distance between sample and detector (DSD) is from 11 cm to 176 cm. The pixel size of the detector is 9 μm. Results are shown in Fig. 10.10: Fig. 10.10(a) is the thin lung recorded at a DSD of 40 cm and Fig. 10.10(b) corresponds to thick lung acquired at DSD of 60 cm. This work demonstrates that the lung cancer sample in thin lung with millimeter scale can be well recognized using IL-XPμCT under the above simulated conditions.

10.4.4. *Solid organs*

Both DEI and PBI techniques were used for imaging gastric cancer samples [51]. G Li *et al.* tried to image the rat liver in PCI in-line, and treated the sample with dehydration via the natural volatilization. The 3D structures of the hepatic lobules can be revealed clearly,

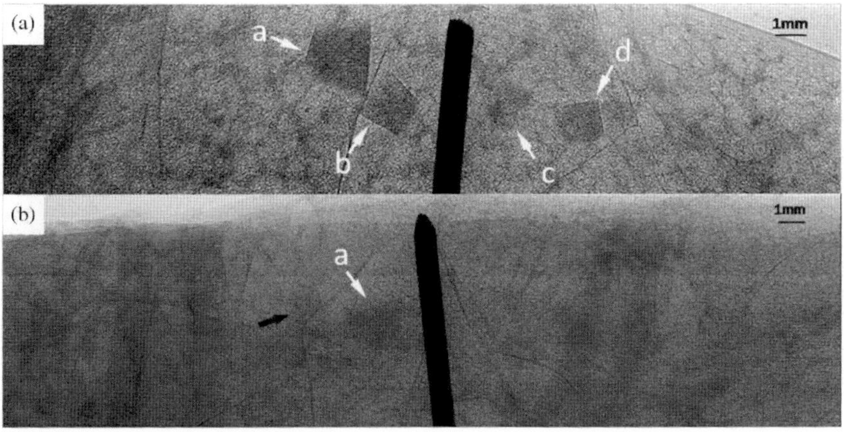

Fig. 10.10. (a) For thin lung, four rat livers (representing human lung cancer) (a, b, c and d) with different sizes ($2.2 \times 2 \times 1.5\,\text{mm}^3$, $1.4 \times 1.1 \times 1\,\text{mm}^3$, $1.3 \times 1 \times 1\,\text{mm}^3$ and $0.8 \times 1 \times 1\,\text{mm}^3$, approximately) can be discerned clearly (pointed by white arrows). (b) For thick lung, only the biggest one ($2.2 \times 2 \times 1.5\,\text{mm}^3$) shows adequate contrast to the normal lung tissues (pointed by a white arrow) while other smaller ones have too low contrast to be clearly discerned (pointed by a black arrow) (reprinted from [50]).

including the interlobular structures and the sinusoidal vessels [52] (Fig. 10.11). It indicates that IL-XPμCT is an effective way to non-invasively reveal the liver micro anatomy structures with high spatial resolution (sub-micrometer) and good contrast (using the phase contrast) that are invisible by the conventional CT and MRI.

PBI was recently used in the detection and characterization of early changes after ischemia-reperfusion in a standardized rat liver transplantation model [53]. PBI has also been used for imaging glomeruli and tubular structures of kidney in hamsters, which had developed sclerosis [54] and for the aging kidney evaluation [55] (Figs. 10.12 and 10.13). In the mouse kidney at the age of ten months, the vessels are smooth; the distribution of vessels and capillary loops are regular and homogeneous; and the contour of the kidney is round. Then at the age of 18 months, the kidney is shown with slight atrophy; the renal cortex is less regular with the vessels slightly disorder running; and the capillary loops are less homogeneous. Moreover at

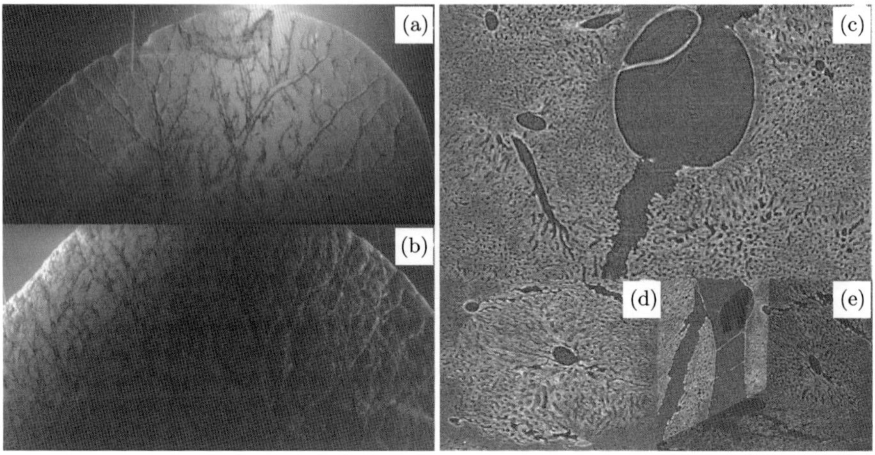

Fig. 10.11. (a) DEI project images of the normal rat liver (b) and the fibrosis rat liver of the 8 weeks model. (c, d and e) The 3D internal micro structures obtained using IL-XPμCT based on propagation. The micro vessels with different diameters, valves, hepatic lobules and sinusoids can be seen clearly (reprinted from [52]).

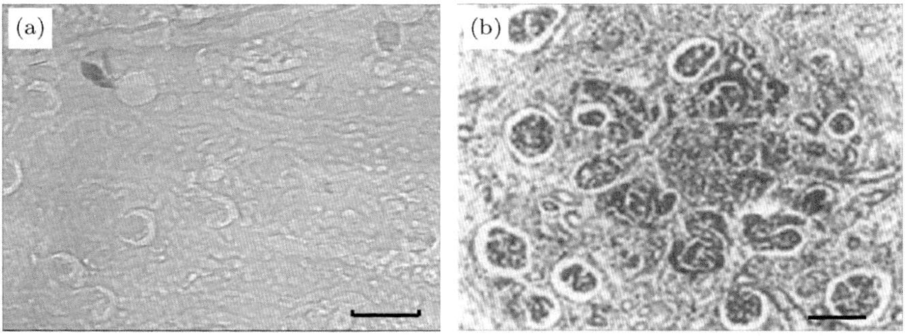

Fig. 10.12. (a) The section imaging of the propagation-based PCI with (b) histology imaging, the renal glomeruli, tubules, and collecting ducts are demonstrated on the PCI, there is high consistence between the PCI and the histology. Scale bars, (a) 50 μm, (b) 30 μm (reprinted from [55]).

the age of 24 months, it is revealed with evident atrophy and deformation, the vessels configuration is in complete disorder; the capillary loops are also disordered and reduced; and the cortex are thinning. At present, the clinical conventional advanced imaging technology

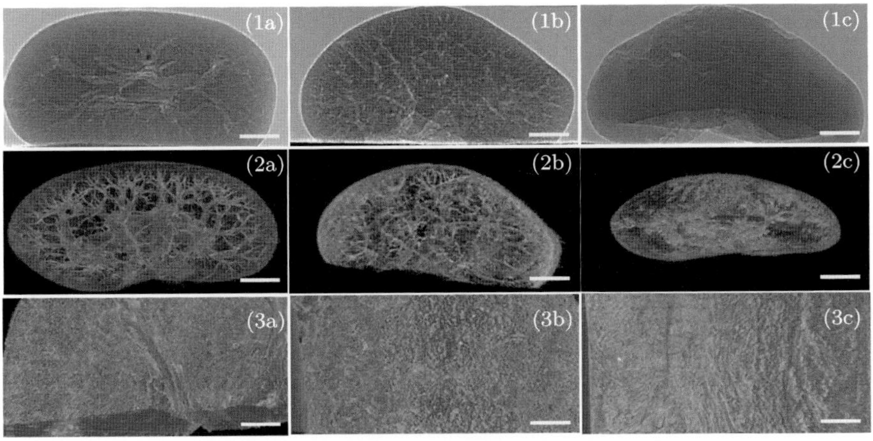

Fig. 10.13. (1a, 2a, 3a) The mouse kidneys imaging at different ages of 10 months, (1b, 2b, 3b) 18 months and (1c, 2c, 3c) 24 months respectively. (1a, 1b, 1c) are the radiographs, (2a, 2b, 2c) are the 3D reconstruction after the whole kidney CT scanning, while the (3a, 3b, 3c) are the 3D reconstruction local slices (mainly the cortex including part medulla) after high resolution scanning (1.85 mm CCD). Scale bars are 1 mm (rows 1 and 2) and 200 μm (row 3) (reprinted from [55]).

including the CT and MR cannot make the renal glomerulus visible directly.

10.4.5. *Vasculature, circulation and soft tissue*

10.4.5.1. *Imaging of rat micro-vessels in vivo*

It is interesting to study the use of physiological saline for contrast enhancement in DEI and interferometric imaging of small-caliber hepatic vessels of rats *in vivo* [56, 57]. Gas-filled micro-bubbles are a standard contrast agent in ultra-sonography. Arfelli *et al.* [58] proved that the same product could be used as a PCI contrast agent. PCI with a crystal interferometer has been applied in micro-CT of atherosclerotic plaque in mice that had taken high-cholesterol diet [59, 60]. The non-calcified plaque, not visible in high-resolution absorption CT images, was instead clearly seen, and different plaque components were also identified based on the phase maps. PBI was used to show micro-vessels of mouse *ex vivo* [61–63] (Fig. 10.14),

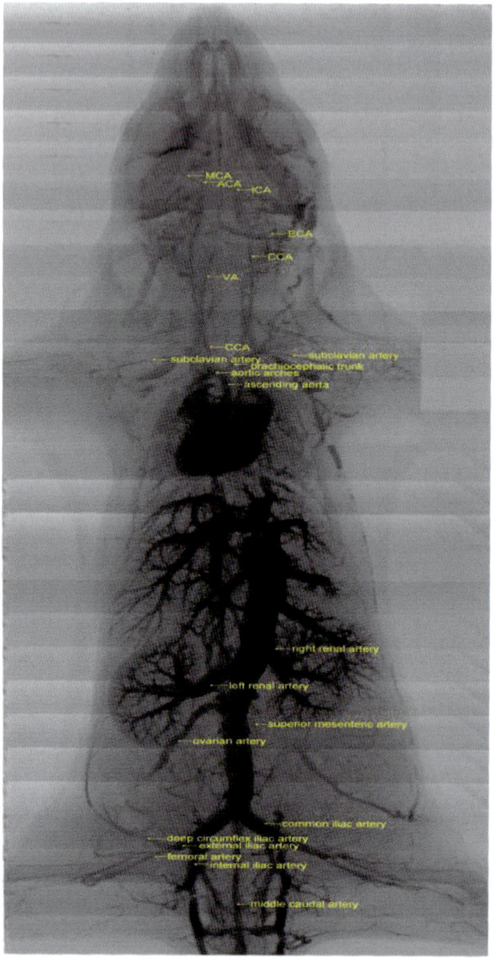

Fig. 10.14. Mouse whole body vessels imaging with PCI and $BaSO_4$ as contrast media (adapted from [61] with permission).

and to study angiogenesis in subcutaneous tumor *in vivo* [64]. Tang *et al.* studied the feasibility of microbubbles as PCI agents for angiography applications. Microbubbles highlighted the lumen of the renal microvessels, acting as a positive contrast medium in *ex vivo* imaging, and present a sharp contrast in SR phase contrast images of renal vasculature. Therefore, the microbubble-based PCI may become a new *in vivo* imaging tool in the study of tumor progression and therapeutic response [64]. Gaseous contrast agents have been applied in other PCI applications: ambient air in the high resolution CT imaging of the liver vascular tree in postmortem mice [65] and CO2 to image micro-vasculature in mouse kidneys [66]. Micro-vasculature of the spinal cord has been resoundingly imaged without a contrast agent in a mouse model by [67].

10.4.5.2. *Brain microvasculature imaging in living rodents*

Animal models are commonly used to study brain vascular diseases such as ischemic and hemorrhagic stroke. The ability to monitor the dynamic changes of brain vasculature in a living animal is of paramount importance to study the mechanisms of these diseases. Using synchrotron radiation based contrast enhanced imaging methods, it is possible to study the cerebral blood vessel perfusion and function in real time at high resolution.

Live rodent brain angiography was developed at the beam line 13W of the SSRF [68]. The contrast agent was administered by a remote controlled injection pump. The experimental set up is illustrated in Fig. 10.15. Several layers of live-imaging graphs were juxtaposed to yield an image of the full brain vasculature.

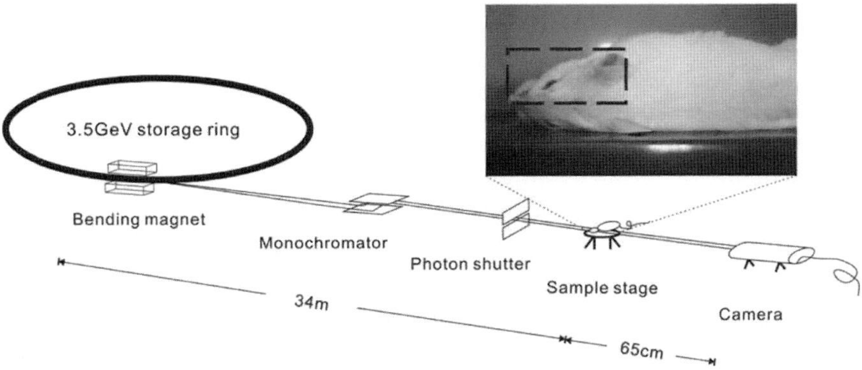

Fig. 10.15. Schematic illustration of synchrotron radiation angiography at the SSRF (reprinted from [61, 62]).

The X-ray beam produced in the storage ring was delivered to the animal brain after monochromatization by a set of silicon crystal monochromators. The source to animal distance was 34 meters and the animal to the CCD detector distance was 0.65 m. The dashed frame on the picture in the upper right corner shows the foot print of the beam on the mouse.

One important application of synchrotron radiation angiography (SRA) is to monitor brain vessel perfusion in living rodents before and after brain ischemia. The suture induced middle cerebral artery occlusion model is widely used in experimental studies of cerebral ischemia. In this model, a suture was created into the inner carotid artery to occlude blood supply to the middle cerebral artery. The suture can be withdrawn to allow reperfusion, which mimics reperfusion injury in clinical situation. However, several complications are associated with this model, including high animal mortality and variability of brain infarction. The SRA method was utilized to optimize the mouse and rat intraluminal suture middle cerebral artery occlusion (MCAO) model. It was found that there are variations of the patency of posterior communication arteries (PcomA) in C57BL/6 mice, which affects the blood supply in the posterior cerebral artery (PCA) area, resulting in high mortality and instability of intraluminal suture permanent MCAO model in these mice. SRA was used to measure the distance between the middle cerebral artery (MCA) and the PCA in C57BL/6 mice. The silicone coating length of suture tip was optimized based on this data, which reduced the model mortality to zero after permanent MCAO in C57BL/6 mice and produced stable brain infarct volume independent of the patency of PcomAs [69]. A similar method was used to optimize rat intraluminal MCAO model. The SR angiography results, together with histology data, proved that silicone coating length can affect the cerebral blood flow and brain infarct volume after MCAO in rats (Fig. 10.16) [70]. Suture-induced thrombosis is another potential complication of experimental stroke models. The formation of thrombosis can affect reperfusion rate after suture withdrawal. SRA data show that appropriate heparin management

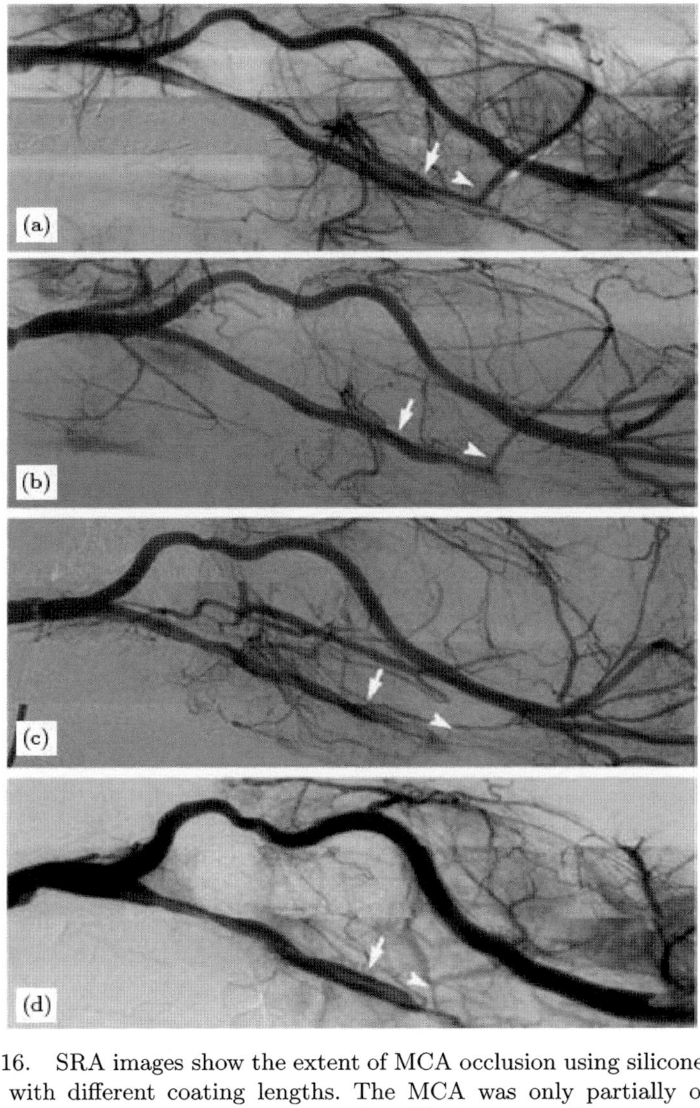

Fig. 10.16. SRA images show the extent of MCA occlusion using silicone-coated sutures with different coating lengths. The MCA was only partially occluded when (a) the suture head was not coated or (b) the coating length was <2 mm. (c) The MCA was successful occluded when the coating length was 2.0 to 3.3 mm without affecting the hypothalamic artery (HTA), PCA, and anterior choroidal artery (AChA). (d) The MCA was occluded when the coating length >3.3 mm together with AChA/PCA, and HTA (reprinted from Guan [70].

can be used to improve the reproducibility of reperfusion model in mice [71].

In addition to the ischemic brain, SR based biomedical imaging was also used to study the hemorrhagic brain, including brains with intracranial hemorrhage (ICH) and subarachnoid hemorrhage (SAH). ICH and SAH are two types of hemorrhagic stroke with the highest mortality rate in humans, which are mainly caused by spontaneous rupture of the cerebral small vessels with vascular abnormalities, such as arteriovenous malformations and brain aneurysms. Various approaches have been tested on animal models; however, most of the methods were histological analysis *in vitro* which one cannot review the vascular structure and function information under physiological status. Before the development of SR imaging methods, monitoring the hemodynamic and structural evaluation of cerebral vasculatures *in vivo* after ICH and SAH has remained critical yet formidable in small experimental animals. Using SR angiography, the vasospasm of anterior communicating artery (AcomA) and MCA was directly observed in two different ICH models of rat. The models were generated by autologous blood injection into the cisterna magna and pre-chiasmatic cistern. Detailed changes in the diameters of anterior cerebral artery (ACA), MCA, and AcomA were observed in the two different models [72]. The changes of blood flow in arteries that supply the cerebral cortex and striatum in mice after ICH were also characterized by combining SR angiography and Laser Speckle Contrast Imaging. The results showed that the blood flow in the cortical branches of MCA significantly decreased at 20 min after blood injection, while they recovered after three days. The blood flow in MCA, PCA and ACA were not changed three days after ICH according to SR angiographic results. Interestingly, the blood flow of lenticulostriate artery was significantly increased compared to the control group three days after ICH. These results show that SRA provided a unique tool, which could directly evaluate and monitor the structural and functional changes in cerebral vascular in small animal models of hemorrhagic stroke.

The third application of SRA is to monitor functional angiogenesis in experimental small animal models. Traditionally, histology and fluorescence immunohistochemistry were used to characterize microvessels and angiogenesis. However, these methods cannot provide information about the function of these newly grown vessels. No information regarding their perfusion status can be acquired. With the development of live synchrotron radiation angiography, the density of perfused cerebral vessels can be calculated, which correlates to vascular function more directly than histology results. With this method, it is now possible to evaluate the angiogenic efficacy of many interventions.

Gene therapy and stem cell transplantation are two areas that are pursued by many stroke researchers. It has been shown that various treatments can promote angiogenesis based on immunofluorescence staining against endothelial cell surface markers and DNA replication markers. Using SRA, it was shown that heterologous Netrin-1 overexpression in the ipsilateral hemisphere after cerebral ischemia in mice can improve functional angiogenesis and functional recovery. This was the first demonstration of functional angiogenesis induced by gene therapy in the ischemic brain *in vivo* (Fig. 10.17) [73]. It was subsequently demonstrated that human endothelial progenitor cell transplantation could also improve angiogenesis in the peri-infarct zone in mice brain [74]. These works suggest that synchrotron radiation angiography can be a valuable tool to assess functional angiogenesis after treatment. This method can be easily adapted to evaluate microvessel structures in other parts of the body in addition to the brain.

10.4.6. *Eyes and auditive system*

As it is known, the lens, a major optical component of the eye, has a gradient refractive index, which is required to provide sufficient refractive power and image quality. The refractive index variations across the lens are dependent on the distributions and concentrations of the varying protein classes. Hoshino *et al.* analyzed properties of eye lens by PCI imaging [75]. They measured the refractive index in the *in situ* eye lens from five species including pig, frog, mouse, newt and fish, using a specially constructed X-ray Talbot GI. Their

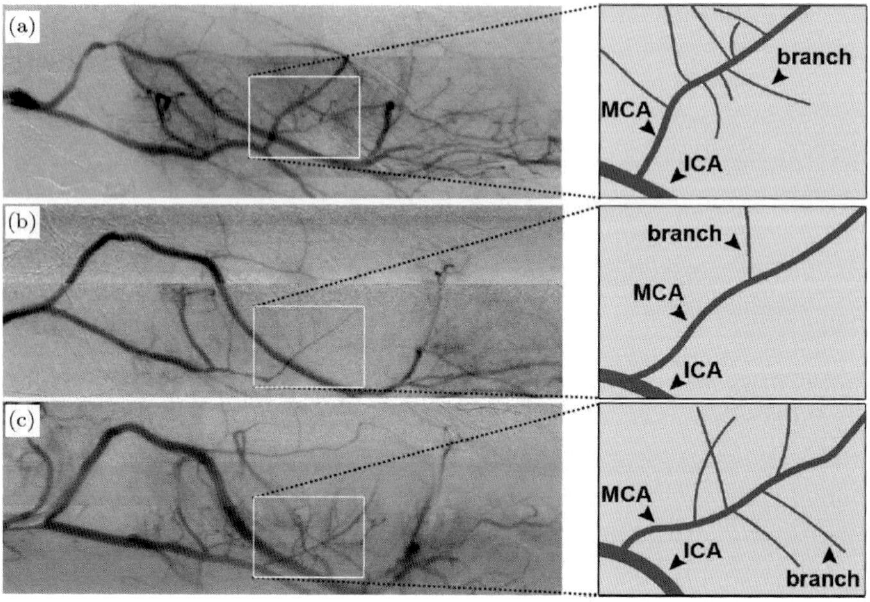

Fig. 10.17. Live synchrotron radiation angiography revealed newly formed functional vessels in transient middle cerebral artery occlusion (tMCAO) mice treated with adeno-associated viral vector–Netrin-1 (AAV–NT-1). Photomicrographs show the perfusion of the middle cerebral artery (MCA) territory in (a) normal mouse brain and in (b) saline-treated (c) or AAV–NT-1–treated mouse brain 28 days after tMCAO. MCA branches perfused by contrast media are illustrated on the right. ICA indicates internal carotid artery (reprinted from [73]).

measurements have been conducted in two planes: the one containing the optic axis (the sagittal plane) and the plane orthogonal to this (the equatorial plane) (Fig. 10.18). Refractive index profiles were measured in both sagittal and equatorial planes of each lens. This work shows that X-ray microtomography is able to detect subtle fluctuations in the index gradient that earlier methods have been unable to detect. This could provide very useful information about growth and development of the lens as well as insights into these processes for other organs.

Methods of GI and PBI have been also used to image cochlear tissue, which show the capacity of visualizing soft tissue structures around bone [76, 94]. It is demonstrated that cochlear soft tissue

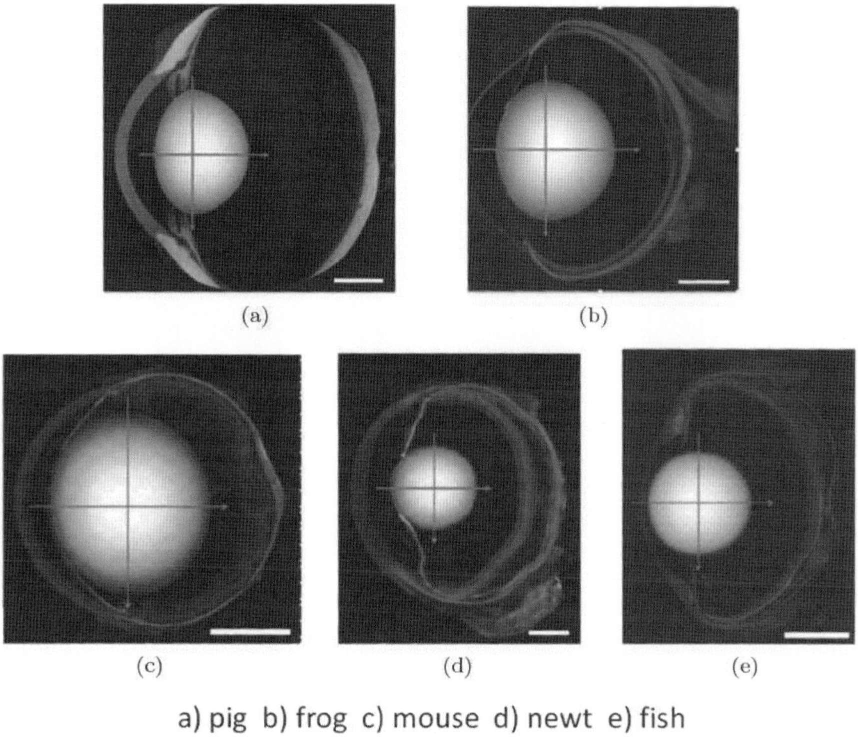

Fig. 10.18. Images of (a) porcine; (b) ranine; (c) murine; (d) newt; (e) piscine eyes in the sagittal plane. The position of the equatorial plane is marked with the blue arrow and the optic axis, along which the sagittal refractive index profiles were measured, is marked with a red arrow. The scale bars in the right hand lower corner are equal to (a) 4 mm (porcine); (b) 2 mm (ranine); (c) 1 mm (murine); (d) 0.5 mm (newt); (e) 1 mm (piscine) (adapted from [75] with permission).

structures in the micron range can be imaged with hard X-ray PCI; and it is likely that the PCI of the soft tissue structures is sufficient to image the soft tissue structures even if surrounded by bone.

10.4.7. *Brain and nerve imaging*

Despite the deep insights offered by the conventional (CT or MRI) imaging methods, their sensitivity and/or spatial resolution are insufficient to study the structures of brain and nerves at the cellular level.

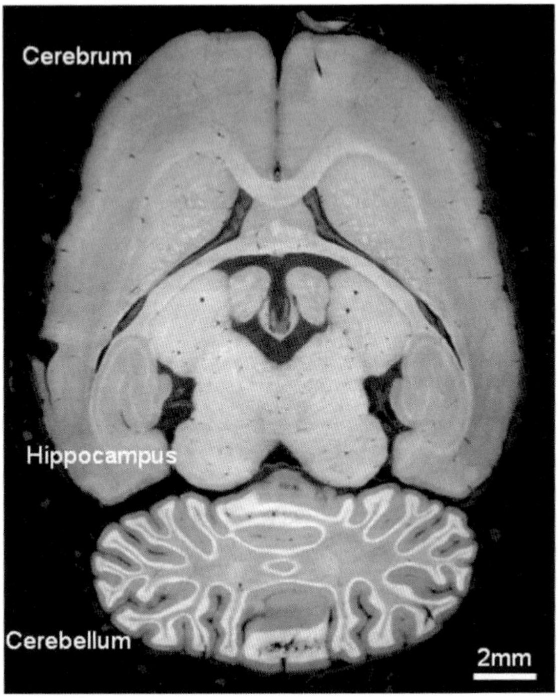

Fig. 10.19. The rat brain imaging with PCI-CT (adapted from [81]).

To exploit the potential of X-ray PCI for neuro imaging, a few studies have been performed on brain tissues. The application of the method allowed identifying portions of human brain tissue besides the blood vessels, the stratum moleculare, the stratum granulosum and the white matter (Fig. 10.19) [77]. By applying a phase retrieval method [78] that only requires a single PBI image per projection, PBI-CT images showed demarcated tissue borders at the gray/white matter boundaries of a mouse brain [39]; also detailed imaging of the brainstem, including the ventral cochlear nucleus, spinal tract of the trigeminal nerve and inferior cerebellar peduncle. Compared to other phase retrieval methods, this single image one has evident advantages in terms of dose and acquisition time saving. Another important application is the detection of main signatures of Alzheimer's disease (AD). Current attempts in brain imaging aim at visualizing amyloid plaques in living patients; these are distinguishing features of the AD. Their visualization might help in early diagnosis and assess the progression of the pathology. Amyloid plaques cannot be seen using conventional X-ray absorption-based imaging because the radiographic contrast is too low and the plaques are too small for the standard CT clinical resolution. Proof of principle micro-CT studies have been performed using either ABI [79] and GI [80] in mice; both techniques showed their ability in visualizing the plaques as small nodules in the cortex and hippocampus. These extraordinary results pave the way to the assessment of anti-amyloid compounds. Another considerable result of PCI in neuroimaging is the detection of fine brain features and cells without applying any stain or contrast agent.

10.4.8. *Obstetrics gynecology and andrology*

A phase-contrast X-ray microtomography system using the "Talbot" type grating-based interferometer has been built at the SPring-8 synchrotron radiation facility. It allows us to obtain three-dimensional structural information without sectioning, and it is particularly useful to examine anomalies that take place during development. Mouse fetus was dynamically traced by M. Hoshino *et al.* [82]. Figure 10.20 shows approximately sagittal tomographic sections of mouse fetuses at gestational days of 9, 10, 11, 13 and 15. At day 9, the fetus is strongly flexed in a dorsally convex direction. At the center of the embryo is the heart in the early development stage. Improved reproductive imaging can increase our understanding of the biologic mechanisms underlying the development of follicles, ovocytes, and corpora lutea.

PCI imaging provides insight into ovarian tumor biology and pathologic mechanisms underlying female infertility. PCI imaging of human prostatic tissue *ex situ* improved diagnostic and treatment tools for prostate cancer. Results to date indicate that phase contrast CT details structures within the prostatic tissue have far higher resolution than conventional imaging modalities. Synchrotron in-line X-ray phase contrast imaging computed tomography (XPCI-CT) was performed on the prostate *ex vivo* at the Canadian Light Source

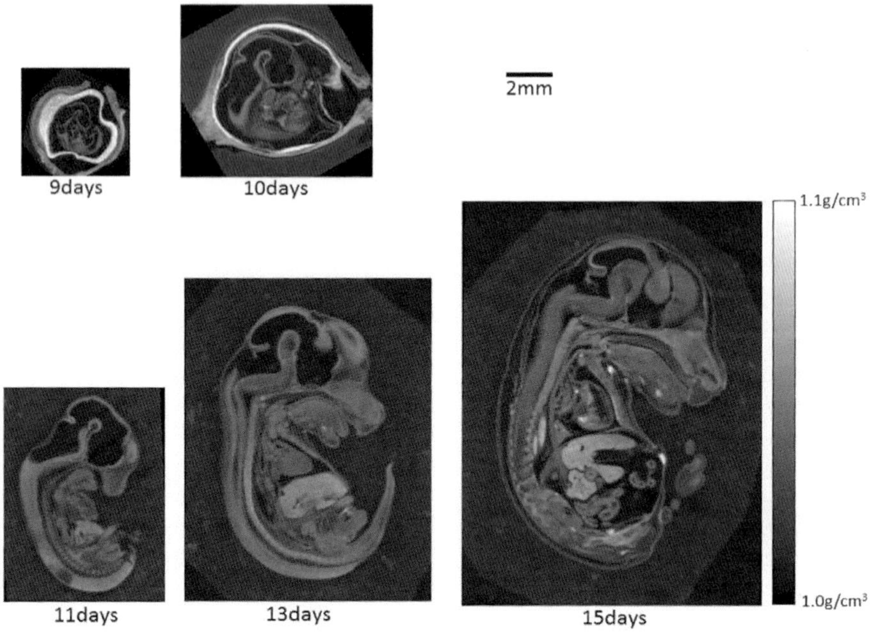

Fig. 10.20. Tomographic sections of mouse fetuses at gestational days of 9, 10, 11, 13 and 15 (adapted from [82] with permission).

and compared to histology. The XPCI-CT images of an invasive prostatic urothelial neoplasm in a dog were reported by [83]. The use of XPCI-CT represents an opportunity to image the prostate gland with greater detail than is currently possible with clinically available diagnostic tomographic imaging modalities such as CT and MRI [83].

10.4.9. *Acupuncture points*

To study the character of acupuncture points and their medical effects during acupuncture is a very important, but it is also a very hard subject related to traditional Chinese medicine (TCM). Xinyi Zhang and co-workers have been extensively studying the existence, or say the material basis of acupuncture points and what happens during acupuncture. Using synchrotron X-ray fluorescence (XRF), they determined that at three out of four acupuncture points they examined the contents of Ca, Fe, Cu and Zn were significantly higher than in surrounding tissues, with closely similar ratios of Cu to Fe at points Jianshi, Tiaokou and Xiajuxu, but not Ximen [84, 85]. These elements may be important in the function of acupuncture. Discussing this result, ESRF News reported that "no one currently knows what effect high metal concentrations have during acupuncture, but this study seems to support the idea that the practice has some physical aspects rather than just creating a placebo effect" [86].

Consequently, the local structure of these trace elements has been studied by the X-ray absorption fine structure (XAFS) spectroscopy [87]. They found that X-ray absorption near-edge structure (XANES) spectra suggest that the atomic structures around Fe in or next to the acupuncture points are different. The Fe K-edge EXAFS was investigated at the Sanyinjiao acupuncture area (SYJ in) and its surrounding tissues (SYJ out), as shown in Fig. 10.21(a,b), which demonstrates that the bond lengths of Fe–O and Fe–N coordination become longer on going from the surrounding tissues to the acupuncture point, resulting in an enhanced high-spin octahedral field in the acupuncture point area. The calculated electronic structures are shown in Fig. 10.21(c,d). It is clear that the electrons of the acupuncture point tend to a much higher spin arrangement than those of surrounding tissues, that could be favorable to strengthen the magnetotransport properties in the acupuncture point by the enhanced spin-coherence the high-spin configuration of Fe ions, and could mediate the O_2 adsorption in the heme as well. It is known that at a much high spin state heme has more activity in carrying oxygen than at lower spin states.

Thirdly, the medical imaging of acupuncture points has been studied by X. Zhang's group. They found an accumulation of micro-vessels in the acupuncture point area. Some details of this study will be shown bellow. The XRF and XAFS studies for acupuncture points are out of the main scope of this chapter and therefore they will not be discussed. Readers who are interested in deepening these aspects are invited to refer to reported references.

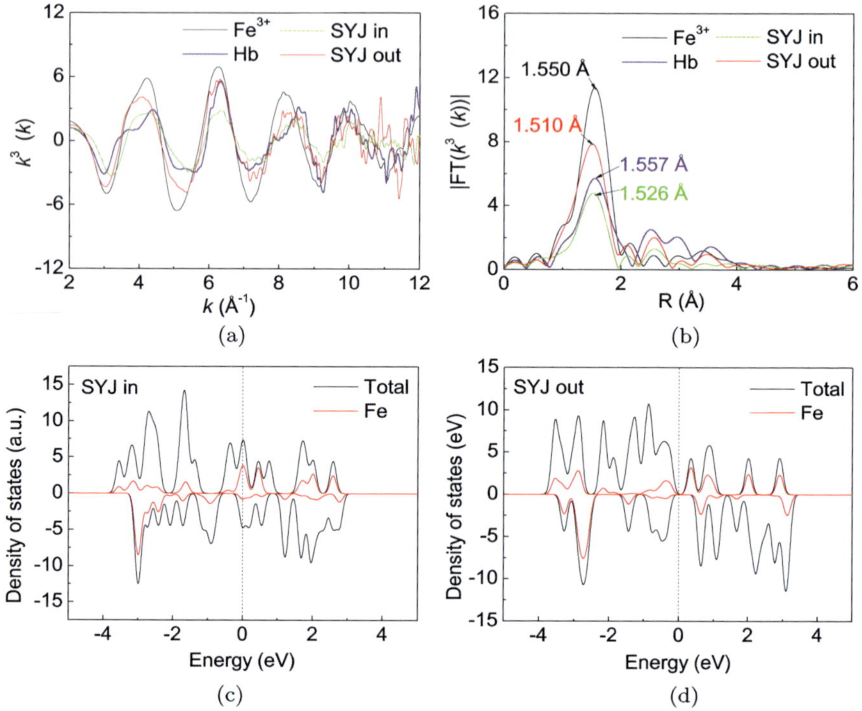

Fig. 10.21. (a) Fe K-edge EXAFS Oscillation functions and (b) Fourier transform (FT) spectra of samples of Sanyinjiao acupuncture area (SYJ in) and surrounding tissues (SYJ out). The EXAFS results for Fe(NO3)3 aqueous solution and haemoglobin (Hb) are also shown for comparison. In the FT curves, the strong peak at about 1.5 Å corresponds to the nearest coordination shell of Fe–N (and Fe–O) for all the samples. Figures 10.21(c,d) show the calculated densities of states (DOSs) for the models of SYJ in and SYJ out tissues, respectively. The DOSs for the acupuncture area in (c) are evidently spin-polarized, indicating a high-spin arrangement of the Fe 3*d* electrons. In contrast, the DOSs for the acupuncture point nearby tissues in (d) suggests a low-spin arrangement of the Fe 3*d* electrons (reprinted from [87].

10.4.9.1. *X-ray phase contrast computed tomography (XPCT)*

Three-dimensional topographic structures of acupuncture points were investigated by using the in-line X-ray phase contrast micro-computed tomography (ILXPμCT) method by [89]. Experiments were performed at the 16-pole wiggler X-ray imaging and biomedical beamline of SSRF. A monochromatic X-ray beam of 20 keV was used for the measurements. The beam cross-section was 45 (horizontal) $\times$ 5 (vertical) mm^2 and the photon flux on the sample was about 5×10^{10} photons/(mm^{2}*s). After traversing the sample, the X-ray beam was recorded by a CCD camera with an effective pixel size of 9 μm. The imaging distance was about 1 m and the exposure time of each image is 2 ms. For image reconstruction, a filtered back-projection algorithm was applied to the normalized projections using in-house developed software (BL13W scientists, SSRF). Two acupuncture points Zhongji, and Zusanli were studied. They are frequently used in acupuncture, and the acupuncture at these points shows distinct clinical effects. Zhongji is located in the abdomen at 1/13 of the distance from the pubic symphysis joint to the joint of the chest and abdomen, and Zusanli is located at about 0.3 cm below the tibial tuberosity, on the dorsolateral side of the hindlimb. The samples were air-dried and placed in a tube filled with air, of 4.5 mm in diameter, and then were fixed on an accurately rotatable sample stage. 3D images of acupuncture point were performed by the standard method by Zhang D. M. *et al.* [89]. Rotating the sample stage around the axis of the cylinder-shaped sample, they firstly obtained the projections of X-ray phase-contrast images shown in step 1 of Fig. 10.22, which were taken every 0.15°; in total 1,200 images were

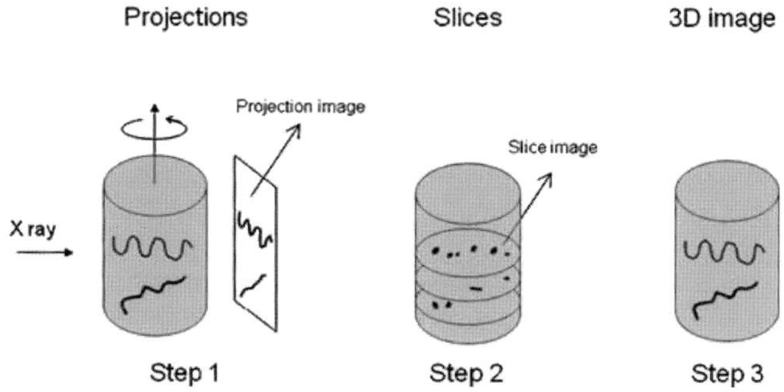

Fig. 10.22. The standard method to perform CT imaging consists of three steps: step 1, generation of projection images; step 2, after application of the filtered back-projection algorithm, slice images were obtained; and step 3, all slice images were added together to form a 3D image (reprinted from [89]).

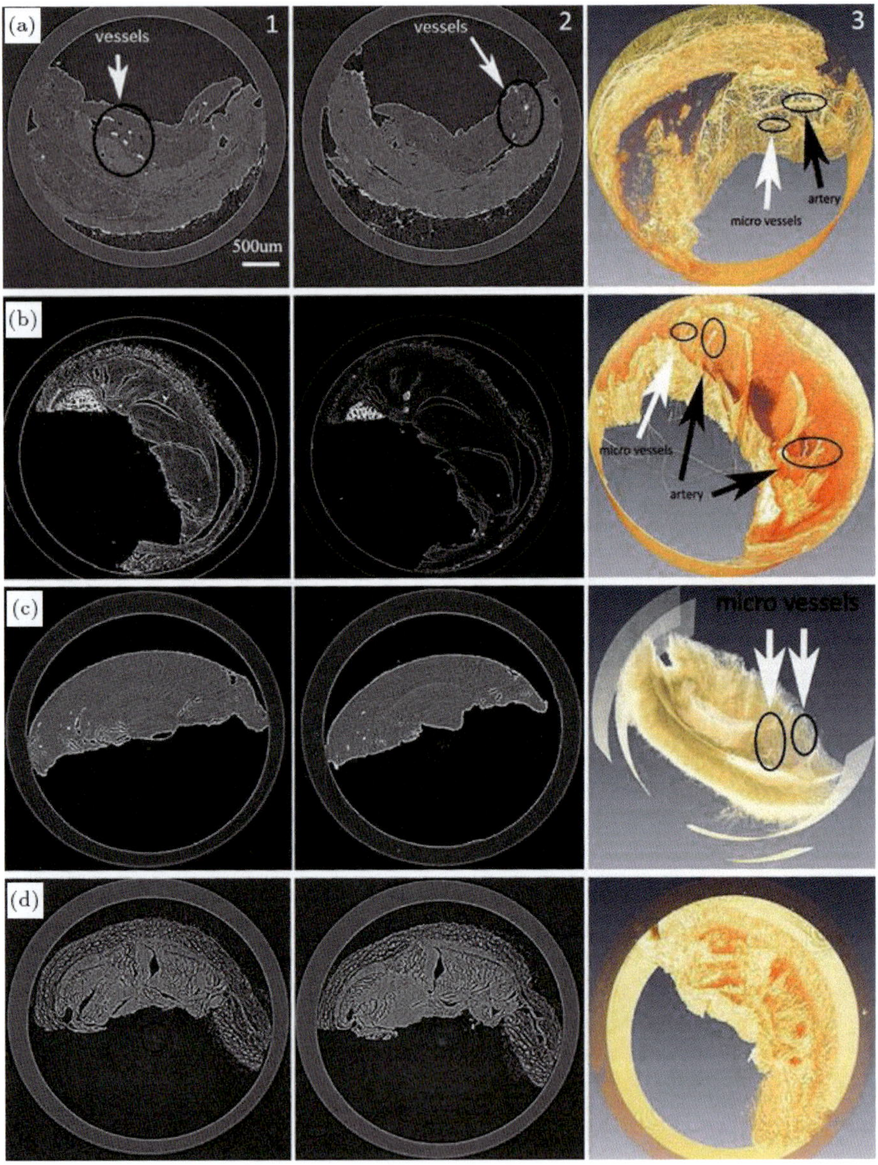
(a)
1
vessels
500um
2
vessels
3
micro vessels
artery
(b)
micro vessels
artery
(c)
micro vessels
(d)

Fig. 10.23.

Fig. 10.23. (figure on facing page) ILXPμCT results of acupuncture point regions and the surrounding tissues with spatial resolution of about $15\,\mu$ for (a) Zhongji acupuncture point region, (b) surrounding tissues away from Zhongji, (c) Zusanli acupuncture point region, and (d) surrounding tissues away from Zusanli. The first and second columns are reconstructed slice images of the 102nd slice and the 398th slice, respectively. The third column shows the 3D renderings of these samples. These 3D images were resliced along the axis perpendicular to the surface of a cylinder-like sample (reprinted from [89]).

collected between 0° and 180°. Then 400 images thick volume was resliced perpendicularly to the rotation axis (step 2). The distance between two neighboring slices was about $30\,\mu$m. Finally, all slice images were added together to form a 3D image (step 3).

An accumulation of micro-vessels at each acupuncture point region has been found, as shown in Fig. 10.23. A long blood vessel around with many fine branches is displayed clearly in the acupuncture points region (denoted by black circles in Fig. 10.23(a,c). In contrast, in the surrounding tissues, apart from some apparent thick vessels, no complicated vessels with the same branches as shown in Fig. 10.23(a,c) can be seen (Fig. 10.23(b,d)). This is the first time that 3D images have revealed the specific structures of acupuncture points. The accumulation of micro-vessels in acupuncture point regions is a clear characteristic, which cannot be seen in acupuncture point surrounding tissues.

Chenglin Liu *et al.* [88] did another experiment with samples of the muscle tissue selected from outside the hind limb of the rabbit, whose shape is a cylinder with a height of 39.69 mm and a diameter of 11.34 mm approximately, as shown in Fig. 10.24(a). There are two acupuncture points in the sample, Housanli (called Zusanli in the human body) at B and Shangjuxu at E. Letters A, C, D and F in Fig. 10.24 refer to the area of no acupuncture point. Figure 10.24(b) is one of the XPCI projection images, and Fig. 10.24(c) is the 3D reconstructed image of the entire sample. At the area of acupuncture point, the small micro-vascular and their bifurcation can be clearly seen around the thick blood vessels, but only few thick vases with no fine structures can be seen in non-acupuncture point areas.

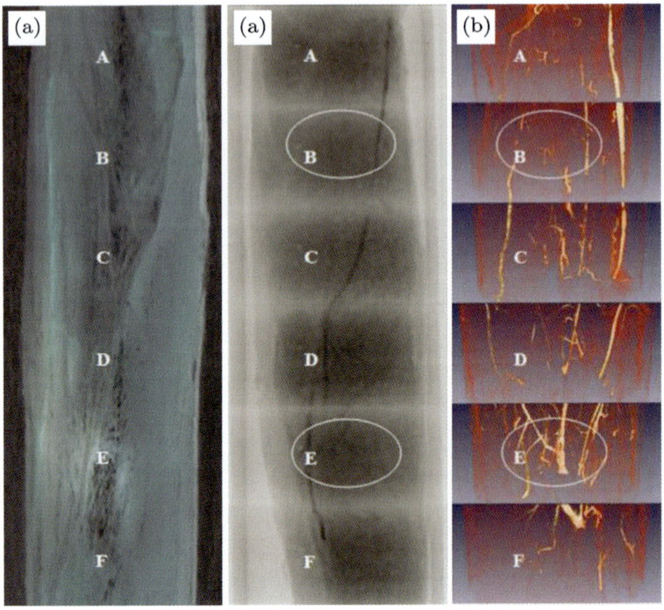

Fig. 10.24. (a) Photo of the rabbit sample, (b) its XPCI projection image, and (c) 3D reconstructed images (reprinted from [88]).

For quantitatively expressing the micro-vascular aggregation in acupuncture point regions, the relative distribution of vascular areas (Aa%) and the relative average gray degree (Gg%) of blood vessels were calculated using an image analysis software [90]. The relative vascular distribution Aa% is defined as the ratio of blood vessel area to the acupuncture point areas, and the relative average gray degree Gg% is the ratio of the average gray degree of the blood vessels region to the average gray degree of acupuncture point areas. The vascular density (AaGg%) is the product of Aa% and Gg% of the vascular [91]. Fig. 10.25 shows the calculation results. It can be clearly seen that the vascular density (AaGg%) is higher in the acupuncture point B and E.

10.4.9.2. *Dark field imaging*

The topographic structures of acupuncture points were also investigated by using the synchrotron based X-ray Dark Field

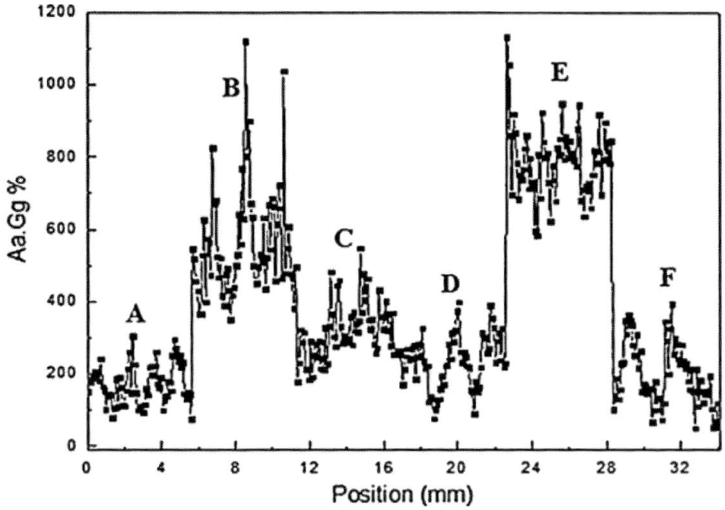

Fig. 10.25. The calculation results of the vascular density (reprinted from [88]).

Image (XDFI) method [84, 85]. Four acupuncture points were studied: Sanyinjiao, Neiguan, Zusanli and Tianshu. Xiaohui Yan *et al.* have found that at acupuncture point regions there exists the accumulation of micro-vessels. The images taken in the surrounding tissue out of the acupuncture points do not show such kind of structure. The XDFI experiment was performed at the BL14B endstation at the Photon Factory, KEK, where the synchrotron radiation X-rays were produced by a 5 Tesla wiggler. The photon energy was 36 keV, and the beam size was approximately 19.7 mm (H) $\times$ 33.5 mm (W). A silicon Si crystal in Laue geometry was used as an analyzer, whose diffraction plane was (440) and thickness was about 1 mm. The images were recorded by mammographic film of Kodak MinR 2000. The exposure time for each image was 2 minutes. Both the acupuncture points and the surrounding tissues far away from the acupuncture points were examined. As an example, Fig. 10.26(A) shows an image taken at Sanyinjiao acupuncture point. A long blood vessel around with some fine branches is clearly displayed. In the regions adjacent to the acupuncture points, there are no such micro-vessels accumulations (Fig. 10.26(b)).

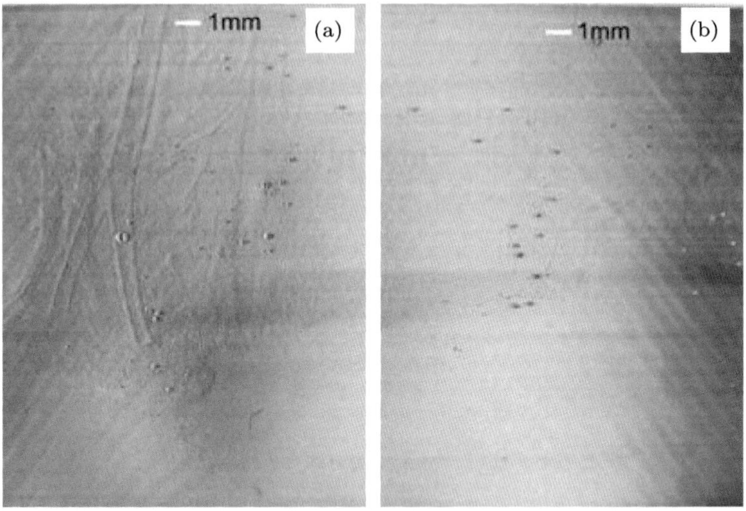

Fig. 10.26. (a) XDFI images at Sanyinjiao acupuncture point region and (b) surrounding tissues (reprinted from [84, 85]).

10.5. Conclusions

Phase-contrast imaging (PCI) on SR light source has become a paradigm in laboratory X-ray biomedical imaging. *In vitro* and *in vivo* biomedical research has attracted much attention thanks to its capability of providing high diagnostic significant images in a wide range of diseases, including for instance those of the breast, joints, cartilage, lung and central nervous system. The first clinical breast imaging research was performed at the ELETTRA SR source; results demonstrated that PCI mammography (using the PBI technique) produces exams of improved specificity and sensitivity compared to conventional absorption-based X-ray radiography. However, the results of its clinical use are not yet available. The pathway towards the clinical application is still long. How to extend the range of applications towards higher X-ray energies (and therefore thicker samples), how to diminish the imaging time, and how to increase the sensitivity of PCI should be considered in the design of new PCI set-ups with conventional sources. There are still some technical limitations preventing PCI to be applied in the clinical practice: the main factor is the fact that for different reasons, all techniques can only use part of the beam generated by a conventional X-ray source. From this point of view, various powerful compact X-ray source projects have been proposed and represent a promising solution to circumvent the present limitation preventing the clinical implementation [95, 98]. These new projects aim at generating quasi monochromatic X-ray beams with flux densities that are just between those available at the large scale SR sources and at the clinical X-ray generators.

Bibliography

[1] W. Otha, Linton, Medical Applications of X Rays, Beam Line, *Summer* **25**(2), 25–34, (1995).

[2] A. Bravin, P. Coan and P. Suortti, X-ray phase-contrast imaging: from pre-clinical applications towards clinics, *Phys, Med. Biol.* **58**, R1–R35, (2013).

[3] A. Snigirev, I. Snigireva, V. Kohn *et al.*, On the possibility of X-ray phase contrast microimaging by coherent high-energy synchrotron radiation, *Rev. Sci. Instrum.* **66**, 5486–92, (1995).

[4] D. Chapman, W. Thomlinson, Z. Zhong *et al.*, Diffraction enhanced X-ray imaging, *Phys. Med. Biol.* **42**, 2015–25, (1997).

[5] A. Momose, T. Takeda, Y. Itai *et al.*, Phase-contrast X-ray computed tomography for observing biological soft tissues, *Nat. Med.* **2**, 473–5, (1996).

[6] T. Weitkamp, A. Diaz, C. David *et al.*, X-ray phase imaging with a grating interferometer, *Opt. Express* **13**, 6296–304, (2005).

[7] A. Olivo and R. Speller, A coded-aperture technique allowing X-ray phase contrast imaging with conventional sources, *Applied Physics Letters* **91**, 074106, (2007).

[8] M. R. Teague, Deterministic phase retrieval: a Green's function solution, *J. Opt. Soc. Am.* **73**, 1434–41, (1983).

[9] S. W. Wilkins, T. E. Gureyev, D. Gao *et al.*, Phase-contrast imaging using polychromatic hard X-rays, *Nature* **384**, 335–8, (1996).

[10] R. Kincaid, A. Krol, S. Fourmaux *et al.*, Development of ultrafast laser-based X-ray *in vivo* phase-contrast micro-CT beamline for biomedical applications at advanced laser light source (ALLS) *Proc. Soc. Photo-Opt. Instrum. Eng* **7078**, 707818.1–707818.12, (2008).

[11] A. Burvall, U. Lundström, P. A. C. Takman *et al.*, Phase retrieval in X-ray phase-contrast imaging suitable for tomography, *Opt. Express* **19** 10359–76, (2011).

[12] U. Bonse and M. Hart, An X-ray interferometer, *Appl. Phys. Lett.* **6**, 155–6, (1965).

[13] A. Momose, Demonstration of phase-contrast X-ray computed-tomography using an X-ray interferometer, *Nucl. Instrum. Methods A* **352**, 622–8, (1995).
[14] A. Momose, Phase-sensitive imaging and phase tomography using X-ray interferometers, *Opt. Express*, **11**, 2303–14, (2003).
[15] A. Momose, Recent advances in X-ray phase imaging, *Japan J. Appl. Phys.* **44**, 6355–67, (2005).
[16] M. Takeda, H. Ina and S. Kobayashi, Fourier-transform method of fringe-pattern analysis for computer-based topography and interferometry, *J. Opt. Soc. Am.* **72**, 156–60, (1982).
[17] H. F. Talbot, Facts relating to optical science No. IV *Philos. Mag.* **9**, 419–27, (1836).
[18] L. Rayleigh, On copying diffraction-gratings, and on some phenomena connected therewith, *Philos. Mag.* **11**, 196–205, (1881).
[19] A. Olivo, F. Arfelli, G. Cantatore *et al.*, An innovative digital imaging set-up allowing a low-dose approach to phase contrast applications in the medical field *Med. Phys.* **28**, 1610–9, (2001).
[20] P. R. T. Munro, K. Ignatyev, R. D. Speller and A. Olivo, Phase and absorption retrieval using incoherent X-ray sources *Proc. Nat'l Acad. Sci. USA* **109**, 13922–7, (2012).
[21] P. C. Diemoz, M. Endrizzi, C. E. Zapata *et al.*, X-ray phase-contrast imaging with nanoradian angular resolution, *Phys. Rev. Lett.* **110**(13), 138105, (2013).
[22] P. C. Diemoz, A. Bravin, M. Langer *et al.*, Analytical and experimental determination of signal-to-noise ratio and figure of merit in three phase-contrast imaging techniques, *Opt. Express* **20**(25), 27670–90, (2012).
[23] F. Arfelli, V. Bonvicini, A. Bravin *et al.*, Mammography with synchrotron radiation: phase-detection techniques, *Radiology* **215**, 286–93, (2000).
[24] D. Dreossi, A. Abrami, F. Arfelli *et al.*, The mammography project at the SYRMEP beamline, *European Journal of Radiology* **68S**, S58–S62, (2008).
[25] E. D. Pisano, R. E. Johnston, D. Chapman *et al.*, Human breast cancer specimens: diffraction-enhanced imaging with histologic correlation improved conspicuity of lesion detail compared with digital radiography, *Radiology* **214**, 895–901, (2000).
[26] J. Keyriläinen, A. Bravin, M. Fernández *et al.*, Phase-contrast X-ray imaging of breast, *Acta Radiologica* **51**(8), 866–884, (2010).
[27] M. Thilo, R. Jens, A. Gisela *et al.*, On a dark-field signal generated by micrometer-sized calcifications in phase-contrast mammography, *Phys. Med. Biol.* **58**, 2713–32, (2013).
[28] F. Arfelli, M. Assante, V. Bonvicini *et al.*, Low-dose phase contrast X-ray medical imaging, *Phys. Med. Biol.* **43**, 2845–52, (1998).
[29] E. Castelli, F. Arfelli, D. Dreossi *et al.*, Clinical mammography at the SYRMEP beam line, *Nucl. Instrum. Methods A* **572**, 237–40, (2007).

[30] E. Castelli, M. Tonutti, F. Arfelli *et al.*, Mammography with synchrotron radiation: first clinical experience with phase-detection technique, *Radiology*, **259**, 684–94, (2011).
[31] J. Keyriläinen, M. Fernández, M. L. Karjalainen-Lindsberg *et al.*, Toward high-contrast breast CT at low radiation dose. *Radiology* **249**(1), 321–7, (2008).
[32] Y. Zhao, E. Brun, P. Coan *et al.*, *Proc Nat'l Acad Sci U S A* **109**(45), 18290–4, (2012).
[33] M. Marenzana, C. K. Hagen, P. Das Neves Borges *et al.*, Visualization of small lesions in rat cartilage by means of laboratory-based X-ray phase contrast imaging, *Phys. Med. Biol.* **57**, 8173–84, (2012).
[34] A. Horng, E. Brun, A. Mittone, S. Gasilov *et al.*, Cartilage and Soft Tissue Imaging Using X-rays Propagation-Based Phase-Contrast Computed Tomography of the Human Knee in Comparison With Clinical Imaging Techniques and Histology *Investigative Radiology,* **49**(9), 627–34, (2014).
[35] Z. Izadifar, L. D. Chapman and X. B. Chen, Computed tomography diffraction-enhanced imaging for *in situ* visualization of tissue scaffolds implanted in cartilage, *Tissue engineering. Part C. Methods* **20**, 140–8, (2014).
[36] K. K. W. Siu, K. S. Morgan, D. M. Paganin *et al.*, Phase contrast X-ray imaging for the non-invasive detection of airway surfaces and lumen characteristics in mouse models of airway disease, *European Journal of Radiology* **68**(3 Suppl), S22-6, (2008). K. Siu (eds), 6th International Conference on Medical Applications of Synchrotron Radiation *Am, Inst. of Phys (AIP)* Conf. Proc. p 1266, (2010).
[37] Z. Zhong, W. Thomlinson, D. Chapman *et al.*, Implementation of diffraction-enhanced imaging experiments: at the NSLS and APS, *Nucl. Instrum. Methods A* **450**, 556–67, (2000).
[38] X. J. Luan, A. V. Campanucci, M. Nair *et al.*, Pseudomonas aeruginosa triggers CFTR-mediated airway surface liquid secretion in swine trachea, *PNAS* **111**, 12930–5, (2014).
[39] M. A. Beltran *et al.*, Interface-specific X-ray phase retrieval tomography of complex biological organs, *Phys. Med. Biol.* **56**, 7353–69, (2011).
[40] D. W. Parsons, M. Kaye, D. Martin *et al.*, High-resolution visualization of airspace structures in intact mice via synchrotron phase contrast X-ray imaging (PCXI), *J. Anat.* **213**, 217–27, (2008).
[41] S. B. Hooper, M. J. Kitchen, M. L. Siew *et al.*, Imaging lung aeration and lung liquid clearance at birth using phase contrast X-ray imaging, *Clin. Exp. Pharmacol. Physiol.* **36**, 117–25, (2009).
[42] M. J. Kitchen, R. M. Lewis, M. Wallace *et al.*, Dynamic measures of regional lung air volume using phase contrast X-ray imaging, *Phys. Med. Biol.* **53**, 6065–77, (2008).
[43] R. A. Lewis, N. Yagi, M. J. Kitchen *et al.*, Dynamic imaging of the lungs using X-ray phase contrast, *Phys. Med. Biol.* **50**, 5031–40, (2005).

[44] T. Sera, H. Yokota, K. Fujisaki *et al.*, development of high-resolution 4d in vivo-ct for visualization of cardiac and respiratory deformations of small animals. *Phys Med Biol.* **53**, 4285–4301, (2008).
[45] M. Donnelley, K. K. W. Siu, R. A. Jamison *et al.*, Synchrotron phase-contrast X-ray imaging reveals fluid dosing dynamics for gene transfer into mouse airways, *Gene Ther.* **19**, 8–14, (2011).
[46] L. Zhang and S. Luo, Micro soft tissues visualization based on X-ray phase-contrast imaging, *Open Med. Inform. J* **5**(Suppl. 1), 19–25, (2011).
[47] D. M. Connor, Z. Zhong, H. D. Foda *et al.*, Diffraction enhanced imaging of a rat model of gastric acid aspiration pneumonitis, *Acad. Radiol.* **18**, 1515–21, (2011).
[48] M. J. Kitchen, D. M. Paganin, U. Kentaro *et al.*, X-ray phase, absorption and scatter retrieval using two or more phase contrast images, *Opt. Express* **18**, 19994–20012, (2010).
[49] M. J. Kitchen, D. M. Paganin, U. Kentaro *et al.*, Phase contrast image segmentation using a Laue analyzer crystal, *Phys. Med. Biol.* **56**, 515–34, (2011).
[50] J. Zhang, Y. Chen, G. Li *et al.*, *Journal of instrumentation* **8**, C07002, 2013. doi:10.1088/1748-0221/8/07/C07002.
[51] L. Tang, G. Li, Y. S. Sun *et al.*, Synchrotron-radiation phase-contrast imaging of human stomach and gastric cancer: *in vitro* studies, *Journal of Synchrotron Radiation* **19**, 319–22, (2012).
[52] G. Li, X. Y. Zhao, X. H. Xiao and Y. Chen, Investigating the liver micro structures using synchrotron hard X-ray phase contrast micro tomography, Presented in 7th International Symposium on Medical Applications of Synchrotron Radiation October 17–20, 2012, Shanghai, China, (2012).
[53] M. Brandlhuber, M. Armbruster, B. Zupanc *et al.*, A Novel and Sensitive Approach for the Evaluation of Liver Ischemia-Reperfusion Injury After Liver Transplantation, *Invest Radiol.* **51**(3), 170–6, (2016).
[54] J. Wu, T. Takeda, T. Lwin *et al.*, Imaging renal structures by X-ray phase-contrast microtomography, *Kidney Int.* **75**, 945–51, (2009).
[55] Y. F. Peng, Z. H. Zhao, G. H. Du *et al.*, Experimental investigation of mouse aging with SR PCI technology, *Journal of instrumentation* **8**(8), (2013). 6-910.1088/1748-0221/8/08/C08004.
[56] T. Takeda, A. Yoneyama, J. Wu *et al.*, In vivo physiological saline-infused hepatic vessel imaging using a two-crystal-interferometer based phase-contrast X-ray technique, *Journal of Synchrotron Radiation* **19**, 252–6, (2012).
[57] X. Zhang, Q. X. Yuan, X. R. Yang *et al.*, Medical application of diffraction enhanced imaging in mouse liver blood vessels, *Chinese Physics C* **33**(11), 986–990, (2009).
[58] F. Arfelli, L. Rigon and R. H. Menk, Microbubbles as X-ray scattering contrast agents using analyzer-based imaging, *Phys. Med. Biol.* **55**, 1643–58, (2010).

[59] M. Shinohara, T. Yamashita, H. Tawa *et al.*, Atherosclerotic plaque imaging using phase-contrast X-ray computed tomography, Am. J. Physiol. *Heart Circ. Physiol* **294**, H1094–100, (2008).
[60] M. Takeda, T. Yamashita, M. Shinohara *et al.*, Beneficial effect of antiplatelet therapies on atherosclerotic lesion formation assessed by phase-contrast X-ray CT imaging, *Int. J. Cardiovasc. Imaging* **28**, 1181–91, (2012).
[61] P. Liu, J. Q. Sun, J. Zhao *et al.*, Microvascular imaging using synchrotron radiation, *J. Synchrotron Rad.* **17**, 517–521, (2010).
[62] X. X. Liu, J. Zhao, J. Q. Sun *et al.*, Lung cancer and angiogenesis imaging using synchrotron radiation, *Phys. Med. Biol.* **55**, 2399–2409, (2010).
[63] Y. Xi, R. B. Tang, Y. J. Wang *et al.*, Microbubbles as contrast agent for in-line X-ray phase-contrast imaging, *Appl. Phys. Lett.* **99**, 011101, (2011).
[64] R. Tang, Y. Xi, W. M. Chai, Y. Wang *et al.*, Microbubble-based synchrotron radiation phase contrast imaging: basic study and angiography applications, *Phys. Med. Biol.* **56**, 3503–3512, (2011).
[65] C. M. Laperle *et al.*, Low density contrast agents for X-ray phase contrast imaging: the use of ambient air for X-ray angiography of excised murine liver tissue, *Phys. Med. Biol.* **53**, 6911–23, (2008).
[66] U. Lundstrom, D. H. Larsson, A. Burvall *et al.*, X-ray phase contrast for CO(2) microangiography, *Phys. Med. Biol.* **57**, 2603–17, (2012).
[67] J.-Z. Hu, T. D. Wu, L. Zeng *et al.*, Visualization of microvasculature by X-ray in-line phase contrast imaging in rat spinal cord, *Phys Med. Biol.* **57**, N55–63, (2012).
[68] F. Yuan, Y. Wang, Y. Guan *et al.*, Synchrotron radiation microangiography of mouse cerebrovasculature, *Frontier in Bioscience* (Elite Ed) **5**, 517–24, (2013).
[69] F. Yuan, Y. Tang, X. Lin *et al.*, Optimizing Suture Middle Cerebral Artery Occlusion Model in C57BL/6 Mice Circumvents Posterior Communicating Artery Dysplasia, *J Neurotrauma* **29**(7), 1499–505, (2012).
[70] Y. Guan, Y. Wang, F. Yuan *et al.*, Effect of suture properties on stability of middle cerebral artery occlusion evaluated by synchrotron radiation angiography, *Stroke* **43**, 888–891, (2012).
[71] X. Lin, P. Miao, J. Wang *et al.*, Surgery-related thrombosis critically affects the brain infarct volume in mice following transient middle cerebral artery occlusion, *PLoS One* **8**(9), e75561, (2013).
[72] J. Cai, Y. Sun, F. Yuan *et al.*, A Novel Intravital Method to Evaluate Cerebral Vasospasm in Rat Models of Subarachnoid Hemorrhage: A Study with Synchrotron Radiation Angiography, *PLoS One.* **7**(3), e33366, (2012).
[73] H. Lu, Y. Wang, X. He *et al.*, Netrin-1 hyper expression in mouse brain promotes angiogenesis and long-term neurological recovery after transient focal ischemia, *Stroke* **43**, 838–843, (2012).
[74] C. Chen, X. Lin, J. Wang *et al.*, Effect of HMGB1 on the paracrine action of EPC promotes post-ischemic neovascularization in mice, *Stem Cells.* **32**(10), 2679–89, (2014).

[75] M. Hoshino, K. Uesugi, N. Yagi *et al.*, Optical Properties of In Situ Eye Lenses Measured with Xray Talbot Interferometry: A Novel Measure of Growth Processes. *Plos One* **6**(9), e25140, (2011).

[76] C.-P. Richter, S. Shintani-Smith, A. Fishman *et al.*, Imaging of cochlear tissue with a grating interferometer and hard X-rays, *Micros. Res. Tech.* **72**, 902–7, (2009).

[77] G. Schulz, T. Weitkamp, I. Zanette *et al.*, High-resolution tomographic imaging of a human cerebellum: comparison of absorption and grating-based phase contrast, *J. R. Soc. Interface* **7**, 1665–76, (2010).

[78] D. Paganin, S. C. Mayo, T. E. Gureyev *et al.*, Simultaneous phase and amplitude extraction from a single defocused image of a homogeneous object *J. Microsc.* **206**, 33–40, (2002).

[79] D. M. Connor, H. Benveniste, F. A. Dilmanian *et al.*, Computed tomography of amyloid plaques in a mouse model of Alzheimer's disease using diffraction enhanced imaging, *Neuroimage* **46**, 908–14, (2009).

[80] B. R. Pinzer, M. Cacquevel, P. Modregger *et al.*, Imaging brain amyloid deposition using grating-based differential phase contrast tomography, *Neuroimage* **61**, 1336–46, (2012).

[81] Naoto Yagi 2016 private communication.

[82] M. Hoshino, K. Uesugi and N. Yagi, Phase-contrast X-ray microtomography of mouse fetus, *Biology Open*, **1**, 269–274, (2012). doi: 10.1242/bio.2012430.

[83] J. E. Montgomery, J. M. Wesolowski, B. Wolkowski *et al.*, Demonstration of synchrotron X-ray phase contrast imaging computed tomography of infiltrative transitional cell carcinoma of the prostatic urethra in a dog, *J. Med. Imag.* **3**(1), 015504, 2016. doi:10.1117/1. JMI.3.1.015504.

[84] X. H. Yan, X. Y. Zhang, C. L. Liu *et al.*, Do acupuncture points exist? *Phys. Med. Biol.* **54**, N143–N150, (2009).

[85] X. H. Yan, X. Y. Zhang, L. C. Liu *et al.*, Imaging Study on Acupuncture points, *Journal of Physics: Conference Series* **186**, 012100, (2009).

[86] ESRF News 2009 Do acupuncture points really exist? *ESRF News* Number 50 June P. 22.

[87] C. L. Liu, D. M. Zhang, Q. H. Liu *et al.*, not published

[88] C. L. Liu, X. H. Wang, H. Xu *et al.*, X-ray phase-contrast CT imaging of the acupoints based on synchrotron radiation, *J. Electron Spectrosc. Relat. Phenom* **196**, 80–84, (2014).

[89] D. M. Zhang, X. H. Yan, X. Y. Zhang *et al.*, Synchrotron radiation phase-contrast X-ray CT imaging of acupuncture points, *Anal. Bioanal. Chem.* **401**(3), 803–808, (2011).

[90] C. J. Gao, E. Y. Chen and G. J. Cai, X-anatomy study and computer image analysis of distribution of microvessel density in pulmonary carcinoma, *Anat and Clin.* **10**(2), 98–100, (2005).

[91] W. F. Lu, Z. H. Dong, Z. J. Liu *et al.*, Detection of microvasculature in rat hind limb using synchrotron radiation, *J. Surg. Res.* **164**(1), e193–e199, (2010).

[92] D. M. Connor, E. B. Cole, Z. Zhong *et al.*, Preliminary performance measurements from a second generation diffraction enhanced imaging system, *Proc. SPIE* **8313**, 83134G, (2012).
[93] M. J. Kitchen, R. A. Paganin, D. Lewis *et al.*, On the origin of speckle in X-ray phase contrast images of lung tissue, *Phys. Med. Biol.* **49**, 4335–48, (2004).
[94] M. J. Kitchen, K. M. Pavlov, K. K. W. Siu *et al.*, Analyzer-based phase contrast image reconstruction using geometrical optics, *Phys. Med. Biol.* **52**, 171–87, (2007).
[95] R. A. Lewis, Medical phase contrast X-ray imaging: current status and future prospects, *Phys. Med. Biol.* **49**, 3573–83, (2004).
[96] J. Li, P. Zhu and Y. Sun, An approximate reconstruction method for helical cone-beam differential phase-contrast computed tomography images, *Phys. Med. Biol.* **57**, 2347–56, (2012).
[97] A. Momose, S. Kawamoto, I. Koyama *et al.*, Demonstration of X-ray Talbot interferometry, *Japan J. Appl. Phys. Lett.* **42**, L866–8, (2003).
[98] I. Nesch, D. P. Fogarty, T. Tzvetkov *et al.*, The design and application of an in-laboratory diffraction-enhanced X-ray imaging, *Rev Sci Instrum.* **80**(9), 093702, (2009).
[99] Y. Takeda, W. Yashiro, T. Hattori *et al.*, Differential phase X-ray imaging microscopy with X-ray Talbot interferometer, *Appl. Phys. Express* **1**, 117002, (2008).
[100] T. Weitkamp, C. David, O. Bunk *et al.*, X-ray phase radiography and tomography of soft tissue using grating interferometry, *Eur. J. Radiol.* **68**, S13–7, (2008).

Chapter 11

Synchrotron Radiation Applications on High-Pressure Research

Bo Zou, Kai Wang, Shourui Li and Guangtian Zou

State Key Laboratory of Superhard Materials,
Jilin University, Qianjin Street 2699,
Changchun 130012, China

Abstract

As an important and matured experimental technology to probing materials' structure, synchrotron radiation attracts huge amount of users including condensed-matter physicists, physical chemists, and chemical biologists to investigate the properties of matter by its unparalleled capability. On the other hand, high pressure experiment technologies provide a powerful probe of our understanding of the physical properties of molecules and materials, but at the same time, high pressure experiments are heavily confined on obtaining accurate structural information due to the tiny sample volume (of order 100 μm) in the pressure-generator diamond-anvil cell, which can be perfectly remedied by the synchrotron radiation owing to its high energy and high brilliance. Great efforts have been made to take advantages of the unique characteristics of synchrotron radiation for high pressure research, including: a) high intensity in a broad energy spectral range, b) high brilliance (small angular divergence), c) tunable energy with the use of monochromators, d) pulsed time structure, e) coherence, and f) high polarization of the radiation. The combination of high pressure and synchrotron radiation provides a significant and effective method to explore the detail information of structure and physical properties under extreme conditions. Significant advances have occurred in high pressure synchrotron X-ray diffraction, extended absorption, inelastic scattering, imaging, and emission spectroscopy. With the steady development of high-pressure techniques combined with synchrotron radiation, new windows are opened for exploring the novel behaviors of materials under extreme conditions, including atomic coordination, structures, and bonding character.

Keywords: High pressure; Diamond anvil cell; Large volume press.

11.1. Introduction

Pressure is a fundamental thermodynamic variable that can be easily tuned in experiment, leading to significant advances in such extensive areas of science and technology as condensed matter physics, geophysics, astrophysics, chemistry, biology etc [1–5]. As mentioned by Hemley and Ashcroft, the pressure range observed in the universe varying over 60 orders of magnitude [6], which is larger than any other physical variable. High pressure investigations are useful for understanding the properties of molecules and materials. From the fascinating phenomena represented in Fig. 11.1, it is very easy to understand the origin of the increasing interest in high pressure studies. The significant effect of high pressure is the short of interatomic and intermolecular distances accompanied by the reduction of volume. On one hand, high pressure studies can acquire knowledge about the behavior of materials under extreme conditions which is necessary for understanding the physical and chemical evolutions occurring in the Earth and planetary interiors [7]. On the other hand, in many cases, scientists and engineers can use the high pressure technique to synthesize innovative materials with high elastic moduli and hardness [8, 9]. Moreover, these materials may have interesting optoelectronic, thermal, magnetic, semiconducting or superconducting properties which can be recovered to ambient conditions. Due to substantial technological advances during the past few decades, research in the field of high pressure chemistry has also progressed significantly [10]. Furthermore, the treatments of biological materials under high pressure are of considerable interest [11]. Because of its outstanding importance, high pressure science and technology are giving significant contributions to our deepest understanding of the ultimate properties of matter.

High pressures can be relatively easily applied to small samples with a controlled manner in laboratory, and the resulting changes in

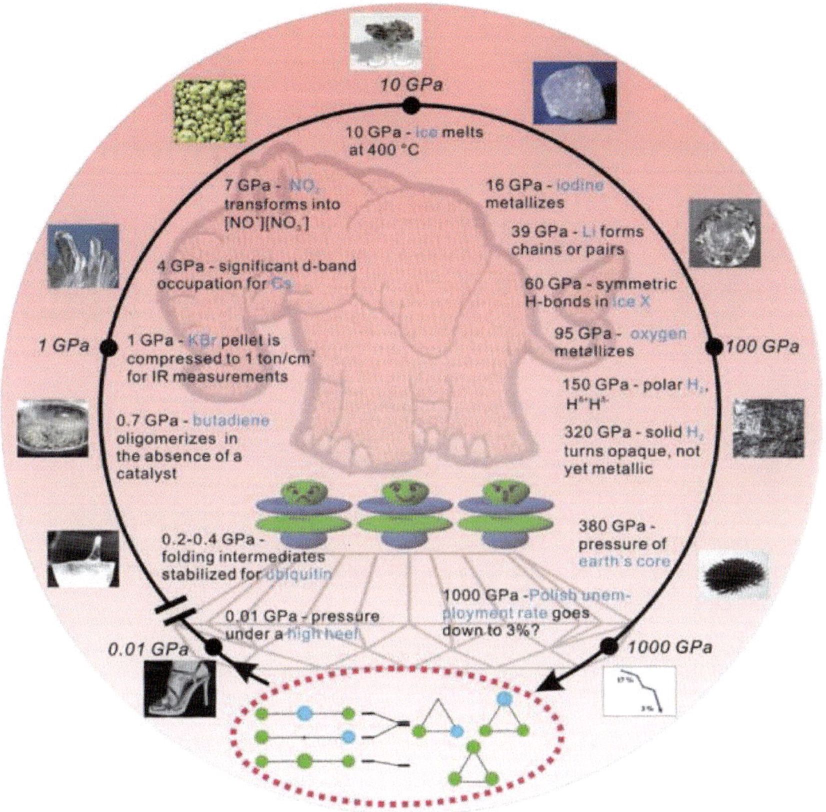

Fig. 11.1. The fascinating phenomena observed with selected examples under high pressure. Reproduced from [12] with permission from Wiley-VCH.

material properties can be explored systematically by means of *in situ* high pressure techniques. During the past decades, high pressure scientists and engineers have focused on pressure cell modifications, and the integration of analytical techniques with high pressure apparatus [13]. As a result of the developments in high pressure techniques, such as the diamond anvil cell (DAC) and the large volume press (LVP) [14, 15], it is possible to significantly expand the pressure-temperature range under static conditions. For example, we can use the laser-heated DAC technique [16] to simulate the high pressure-high temperature conditions of the Earth interior, and

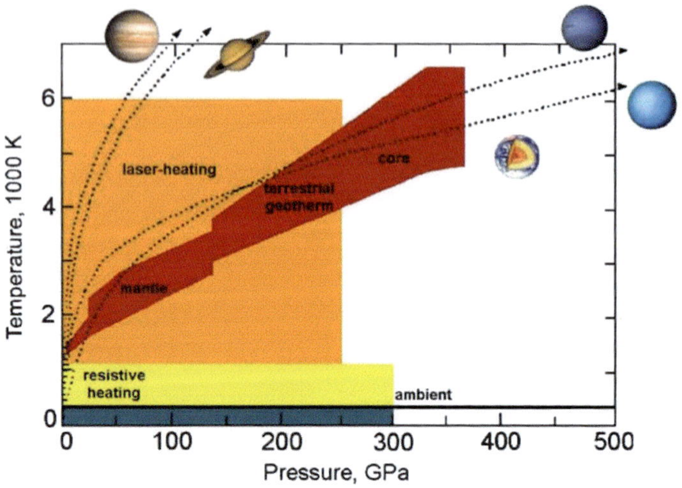

Fig. 11.2. The regions of pressure and temperature accessible by static compression techniques and the estimated conditions in some planets. Reproduced from [17] with permission from The Royal Society of Chemistry.

then investigate the crystal structure, phase diagrams, and thermal equations of state of materials under extreme conditions. Figure 11.2 shows the detailed view of static pressures and temperatures ranges available by laboratory equipments and compared to the regions of interest in Earth and planetary sciences.

The static compression experiments require *in situ* measurements under such extreme pressure and temperature conditions. Many kinds of *in situ* measurement methods based on high pressure apparatus [13], such as Raman scattering, X-ray diffraction (XRD), electrical measurements, magnetic studies, will enable detailed insight into the changes of atomic structures in materials. Due to the very small size of the samples (of order 100 μm), the brilliant and small beams of synchrotron radiation as an experimental tool had a profound effect on high pressure research. The advent of synchrotron radiation as light source allows micron-sized crystals to be used, because an intense beam can be focused down to a size smaller than that of samples in DAC. Additionally, using synchrotron radiation techniques, both the quality of signals and the measurement time have improved significantly, and the available pressure-temperature range is now much larger than before [18].

Synchrotron X-ray based techniques have had a significant impact on the development of material science. A great breakthrough in X-ray characterization of materials occurred when intense synchrotron X-ray sources became available for experiment use in the mid 1970s. Then in 1977, the combination of high pressure instrument with synchrotron radiation beam was pioneered by Buras *et al.* [19]. They opened new directions in investigating the properties of materials under high pressure. By using brilliant synchrotron radiation, the high pressure data are collected with markedly high speed, improved resolution and low noise [20]. In the past 35 years, many storage rings in different countries with strong radiation damping and with high intensity have beamlines devoted to high pressure research. Up to now, the applications of synchrotron radiation have been made in various high pressure techniques to investigate the high pressure crystal structure, electronic structure, and bonding information. For example, high pressure synchrotron XRD experiments can not only identify the high pressure phase but also measure intensity data for structural refinements; high pressure inelastic X-ray scattering gives information on the arrangement of the atoms through the observation of the electron distributions at high pressure [21], and high pressure X-ray absorption spectroscopy allow a unique insight into the local structure and bonding information for atoms in many types of materials.

Up to now, high pressure researches can be carried out at many synchrotron beamlines in the world. More accurate and complete descriptions of high pressure apparatus at synchrotron sources can be found in many literatures and available web pages. The most important high pressure experimental beamline stations are located in the US, Europe, and Asia, such as Advanced Light Source (APS), European Synchrotron Radiation Facility (ESRF), SPring-8 etc. Table 11.1 is the brief overviews of some beamlines and synchrotron radiation stations dedicated to high pressure investigations. In addition, many high pressure stations have corresponding support

Table 11.1. Selected stations and techniques for high pressure experiments at different synchrotron sources. This table includes many high pressure dedicated beamlines, and some beamlines which allow for high pressure studies as one of techniques. The information was collected from their websites and related references.

Laboratory	Beamline	High pressure techniques	Cell type	References
Beijing Synchrotron Radiation Facility	4W2	X-ray powder diffraction X-ray single crystal diffraction	DAC	[22]
Shanghai Synchrotron Radiation Facility	BL15U1	X-ray powder diffraction	DAC	[23]
National Synchrotron Radiation Research Center	BL01C2	X-ray powder diffraction	DAC	[24]
Spring-8	BL04B2	X-ray powder diffraction	DAC	[25–27]
	BL10XU	X-ray powder diffraction		
	BL12XU	Inelastic X-ray scattering X-ray absorption spectroscopy X-ray emission spectroscopy		
	BL39XU	X-ray magnetic circular dichroism spectroscopy X-ray absorption fine structure		
	BL12B2	Extended X-ray absorption fine structure		
	BL43IR	Infrared microspectroscopy		
	BL04B1	X-ray powder diffraction X-ray radiography	LVP	
	BL14B1	X-ray powder diffraction X-ray absorption fine structure Transmission radiography		
	BL22XU	X-ray powder diffraction X-ray single crystal diffraction Resonant X-ray diffraction	DAC LVP	
Photon Factory	BL-18C	X-ray powder diffraction	DAC	[28]
	AR-NE1A	X-ray powder diffraction Mössbauer Spectroscopy		
	AR-NE5C	X-ray powder diffraction	LVP	

(*Continued*)

Table 11.1. (*Continued*)

Laboratory	Beamline	High pressure techniques	Cell type	References
	AR-NE7A	X-ray powder diffraction X-ray imaging		
Indus-2	BL-08	X-ray absorption fine structure	DAC	[29]
	BL-11 BL-12	X-ray powder diffraction		
Pohang Light Source	5A XRS-MS	X-ray powder diffraction	DAC	[30]
European Synchrotron Radiation Facility	ID27	X-ray powder diffraction X-ray single crystal diffraction X-ray fluorescence X-ray Raman scattering	DAC LVP	[31–33]
	ID09A	X-ray powder diffraction X-ray single crystal diffraction Diffuse x-ray scattering	DAC	
	ID20	Inelastic X-ray Scattering		
	ID24	X-ray absorption spectroscopy		
	ID28	Inelastic X-ray scattering		
	BM01A	X-ray powder diffraction X-ray single crystal diffraction		
PETRA III	P02-ECB	X-ray powder diffraction X-ray single crystal diffraction	DAC LVP	[34]
Soleil	ODE	X-ray absorption spectroscopy X-ray magnetic circular dichroism	DAC	[35, 36]
	CRISTAL	Macromolecular crystallography		
	LUCIA	X-ray absorption spectroscopy		
	MARS	X-ray scattering		
	SWING	Small angle X-ray scattering		
Diamond Light Source	I15	X-ray powder diffraction X-ray single crystal diffraction	DAC	[37]
Swiss Light Source	MS-PD	X-ray powder diffraction	DAC	[38]
ELETTRA Synchrotron Light Laboratory	SISSI	Infrared spectroscopy	DAC	[39]
	XRD1 XRD2	X-ray diffraction		

(*Continued*)

Table 11.1. (*Continued*)

Laboratory	Beamline	High pressure techniques	Cell type	References
DAFNE-Light	SINBAD-IR	Infrared spectroscopy	DAC	[40]
Siberian Synchrotron and Terahertz Radiation Centre	4	X-ray diffraction	DAC	[41]
ALBA Synchrotron Light Source	BL04-MSPD	X-ray powder diffraction	DAC	[42]
Advanced Photon Source	3-ID-B,C,D	Nuclear resonant scattering Inelastic X-ray scattering	DAC	[16, 43, 44]
	4-ID-D	Magnetic reflectivity Magnetic X-ray scattering		
	13-BM-C	X-ray single crystal diffraction		
	15-ID-B,C,D	X-ray single crystal diffraction X-ray scattering		
	16-BM-B	White Laue single crystal-diffraction Energy dispersive X-ray diffraction		
	16-BM-D	X-ray powder diffraction X-ray single crystal diffraction X-ray absorption near-edge structure		
	16-ID-B	X-ray powder diffraction X-ray single crystal diffraction		
	16-ID-D	Nuclear resonant scattering Inelastic X-ray scattering X-ray Raman scattering X-ray emission spectroscopy		
	13-BM-D	X-ray absorption fine structure Tomography	DAC LVP	
	13-ID-C,D	Microdiffraction Inelastic X-ray scattering X-ray absorption fine structure		

(*Continued*)

Table 11.1. (*Continued*)

Laboratory	Beamline	High pressure techniques	Cell type	References
National Synchrotron Light Source	U2A	Infrared spectroscopy	DAC	[45, 46]
	X17B3	X-ray powder diffraction	DAC	
	X17C	X-ray powder diffraction	DAC	
		X-ray single crystal diffraction		
	X17B2	X-ray powder diffraction	LVP	
Cornell High Energy Synchrotron Source	B1	X-ray powder diffraction	DAC	[47]
Advanced Light Source	1.4.4	Infrared spectromicroscopy	DAC	[48]
	12.2.2	X-ray powder diffraction		
		X-ray imaging		
		X-ray absorption spectroscopy		
Canadian Light Source	HXMA	X-ray powder diffraction	DAC	[49]
Brazilian Synchrotron Light Laboratory	XPD	X-ray powder diffraction	DAC	[50]
	DXAS	X-ray absorption spectroscopy		
Australian Synchrotron	Powder Diffraction	X-ray powder diffraction	DAC	[51]

facilities for sample preparation and characterization in order to efficient use of beam time. Such as the cryogenic loading facility, laser heating systems, pressure control systems, and on-line pressure measurement systems.

11.2. High Pressure Experimental Techniques

In recent decades, high pressure experimental techniques played a critical role in the development process of high pressure science and a breakthrough of high pressure experimental techniques would bring a huge development space of high pressure science. High pressure experimental techniques have widespread application. It mainly includes the following three aspects: (1) the generation and measurement of high pressure, (2) high pressure *in situ* measurement techniques, (3) the combination of high pressure and other extreme

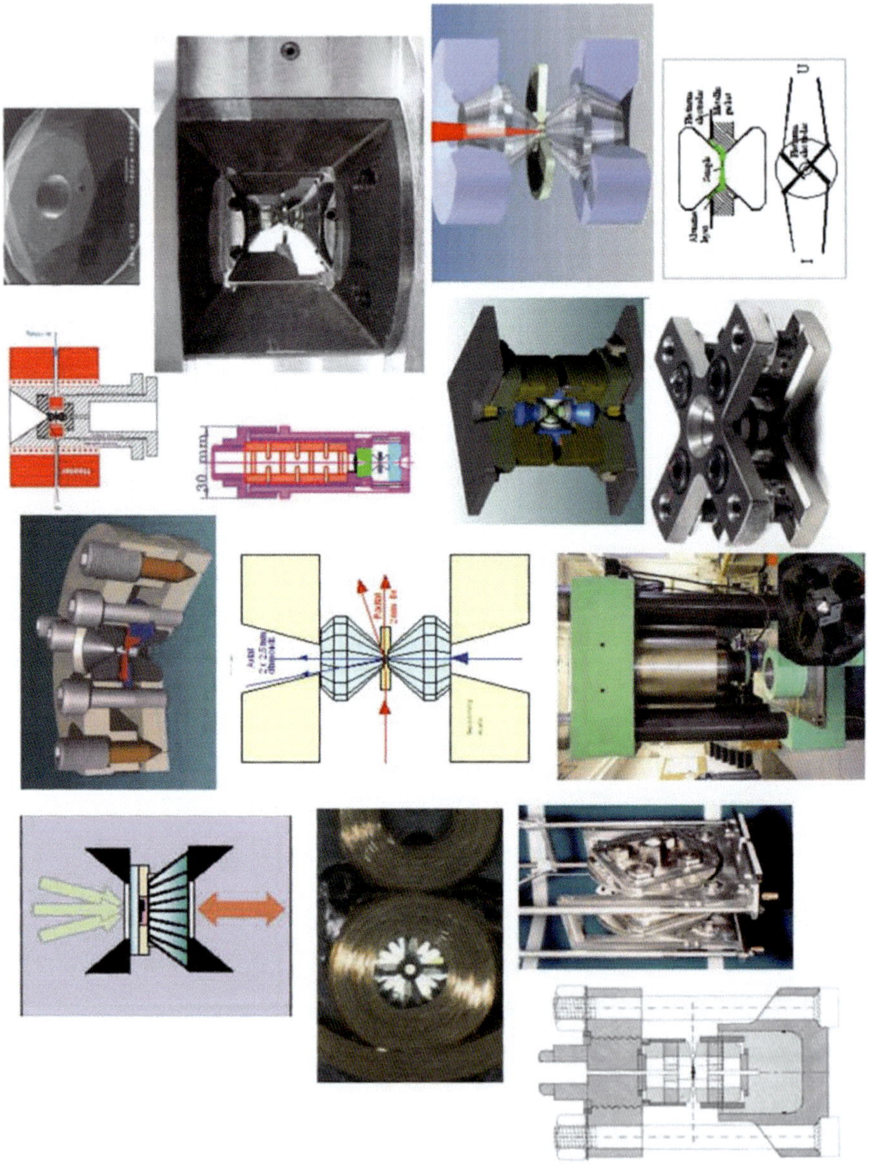

Fig. 11.3. Some examples of the static high pressure instruments now available in the laboratory. Reproduced from [52] with permission from Taylor & Francis Group.

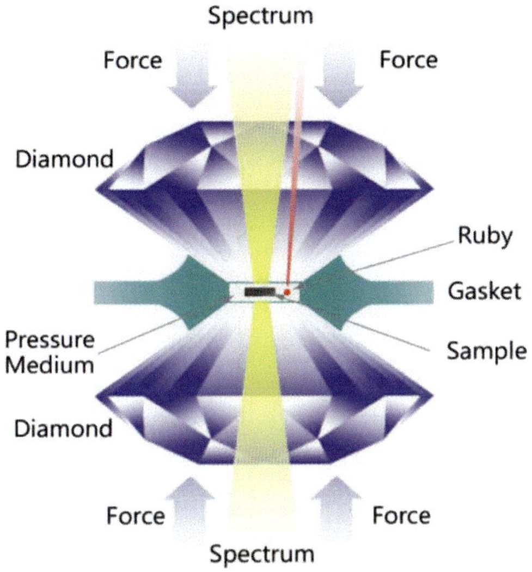

Fig. 11.4. The schematic of DAC for ultrahigh pressure generation.

conditions, such as low- or high-temperature combined with high pressure. In this section, we will mainly introduce the devices and techniques used to generate static high pressure. The following figure shows some selected high pressure apparatus used to generate static pressure in laboratory.

11.2.1. *The diamond anvil cell*

As early as 1950, the use of diamond for pressure generation was first reported [53] and was regarded as the start of a revolution in the high pressure techniques. The basic principle of the DAC is very simple (see Fig. 11.4). Samples were placed in the metal gasket hole between the flat and parallel culets of two diamond anvils. When the two opposed anvils are pushed together by external force, the sample is subjected to high pressure. During the last decades, many types of DAC have been developed and optimized for different high pressure experiments [54–56]. According to the different ways of external forces being applied, they can be divided into mechanical and pneumatic-driven systems. The mechanical DACs can generate pressure through clamps, threads and screws etc [57]. For the purpose of regulating the applied pressure in DACs, the cell needs to be taken off from the optical pathway of the equipment and readjusted after the operation. Such an operation is inconvenient and waster time when heating, cooling and vacuum chambers are dealt with. By using the membrane DAC, a remote control of pressure can be realized and no realignment is needed when adjusting the pressure.

The selection of diamonds for high pressure anvils are based on their mechanical and optical qualities. The diamond with negligible amounts of impurities and defect-free single crystal diamonds can be used as the high pressure anvils. For high pressure optical spectroscopy studies, low luminescence is also a critical consideration for diamond selection. The representative diamond anvils are 16-sided cut stones with the flat culet anvil set parallel to the (100) plane. The diameter ratio between the two diamond faces provides an indication of the highest pressure that can be achieved. Beveled and double-beveled cuts are often used to get higher pressure. The most common replacement of diamond is moissanite, cubic zirconia and sapphire [58]. A basic component of the diamond anvil system is the metallic gasket. The gasket is a metallic foil with an initial thickness of 100–250 μm, placed between the diamonds. Before use, the gasket was preindented to a typical thickness of 30–50 μm. Then we drill a cylindrical hole in its center to produce a sample chamber. Generally, high strength stainless steel can be used as gasket material in high pressure experiment. However, rhenium is more suitable for ultrahigh pressure and high pressure-high temperature experiment, owing to its higher yield strength. Boron and beryllium gaskets are frequently used to permit the X-ray beam to enter a diamond anvil cell from the side [59]. Diamond powder and cubic-boron nitride (c-BN) are also used as gasket in some special experiments [60]. Samples can be loaded in the gasket hole with pressure transmitting medium to ensure hydrostatic pressure conditions. Mixtures of methanol/ethanol/water (16:3:1) can be used up to 14.5 GPa at ambient temperature [61]. Commercial silicon oil is also available when pressures are higher than 15 GPa. Rare gases are the best hydrostatic pressure media, especially for the low temperature experiment and for the chemical inertia [14]. The ruby fluorescence method is a widely used technique to measure *in situ* the exact pressure in the sample chamber [62]. A ruby chip is put in the sample chamber and the pressure can be monitored by the pressure dependent R_1 ruby fluorescence lines shift. The resolution of the R_1 and R_2 lines are also used as a characterization of the quasi-hydrostatic condition. What's more, the knowledge of the equation of state (EOS) of some specific materials (Such as Au, Pt and NaCl) is frequently used to determine the pressure in XRD experiment [2, 63].

11.2.2. *High-pressure high-/low-temperature experiments*

The modern technique of high pressure and high temperature experiments can offer useful insight into the possible composition and state of the matter in earth's inner core. There are two main techniques used to perform experiments under this condition: the resistive heating and the laser heating. The resistive heating method can be divided into the external heating which the whole pressure chamber is heated with external high-temperature elements and internal heating which is when the heating element is placed inside a pressure chamber. In the method of the external heating, the wire windings are placed externally around the diamond and sample region. Temperature sensors are placed to the sample area as close as possible for reducing the errors in the temperature measurement [64]. There is a limit temperature not much greater than 1000°C for the reason of the diamond failure at high temperature [65, 66]. Because of these difficulties, physicists improved the method with many resistance furnace configurations, and so far the maximum temperatures can be get about 1300°C [67]. In the method of the internal heating, the heated area is only in the sample chamber which heated by a resistive wire bridged the conducting steel sheets in the hole. The gasket here is constructed from a thin layer of alumina sandwiched between two layers of conducting stainless steel, where alumina plays a role of the electrical insulation. Temperature is determined by measuring the emitted thermal radiation and the temperature is stable [68]. It has great potential even if there is a difficult experimental procedure to follow.

Laser heating is another main technique used to reach the high temperature conditions and the highest temperature in the diamond-anvil cell is achieved using this technique. Heating is obtained by high-power laser irradiation to the sample. Laser beams are tightly focused onto the sample with a 20–50 μm hot-spot, and the temperature can exceed 4000°C. As the small heated region in DAC, there is an extremely large temperature gradients ranging up to 400°C per micron. This is the main disadvantage of laser heating. So the double sided laser heating technique was developed to give a more uniform temperature distribution. Temperatures are measured by monitoring thermal radiation signals by spectroradiometry [69]. Because of its properties of convenience and controllable, laser heating system is widely applied by high pressure scientists.

Due to its simplicity and small size, DAC is suitable for low-temperature measurements, which are indispensable for a better understanding of many materials. Low temperature can be reached down to a few millikelvins by currently available techniques, and commercial cryostats device combined with DAC is the most widely used high pressure-low temperature apparatus. The temperature variations in a DAC is over the range of $0 < T < 300\,\mathrm{K}$ in high pressure experiment and the low temperature apparatus for optical and XRD measurements has been extensively reported [70, 71]. However, the focusing and collecting optics has a small working distance and it must fit to the cryostat windows. This is the main difficulty of this technique.

11.2.3. *The large volume multi-anvil press*

Large volume press (LVP), also called large anvil cells (LACs), came into use in the 1950s. Compared to the DAC, the maximum attainable temperature and pressure of LVP are limited. However, the sample volumes of LVP about 10^3–10^6 times those of the DAC and the P-T conditions in LVP sample chamber are more uniform. As an important complements of DAC, LVP is widely used in synthesis reactions and earth science research, especially in systematic studies on complex systems and accurate measurements of bulk physical properties. After decades of rapid development, LVP has become a necessary in high-P-T research using synchrotron and neutron sources.

There are various types of LVPs used at synchrotron sources and scientific applications. The cubic-anvil apparatus (DIA) is the earliest type which has a relative sampler device. A typical representative of LVP for synchrotron use is the Kawai-type apparatus. It mainly consists of the following components (See Fig. 11.5). (1) An outer anvil is consist of six hardened steel cylinder, connected to a hydraulic press that provides the force; (2) an inner anvil is consist of eight WC cubes, each cube has one corner truncated into a triangular face; (3) synchrotron radiation source for diffraction and imaging; (4) collimator and X-ray detectors; (5) an electrical heating system

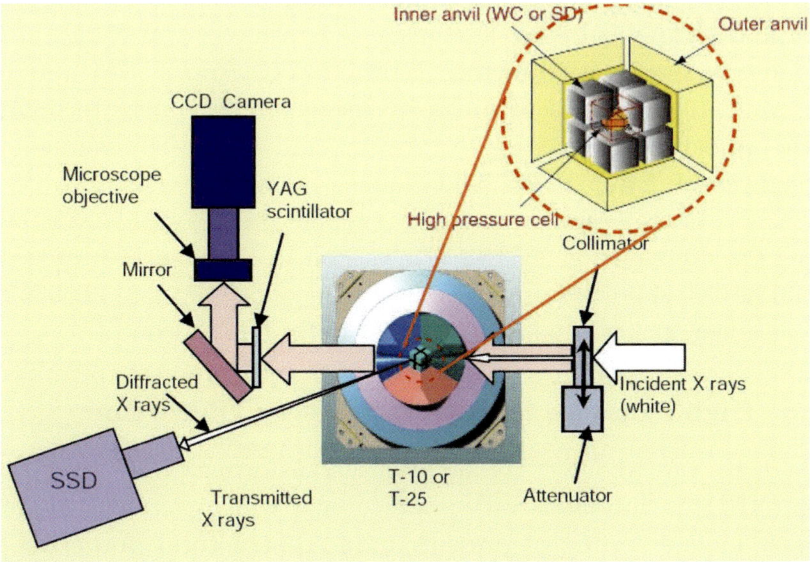

Fig. 11.5. Sketch of the LVP and the beamline setup for the *in situ* measurements of high pressure XRD and imaging. Reproduced from [44] with permission from Springer.

for temperature control. Other types of LVP used at synchrotron sources share almost the same parts [72, 73]. The Kawai-assembly which is a Kawai-type apparatus use a large DIA as outer anvil is a later type that can provide more maximum attainable temperature and pressure. In recent years, many new types of LVP with specific features have been developed to address certain particular needs in high pressure science. Moreover, several *in situ* experimental techniques (such as X-ray diffraction and imaging, ultrasonic interferometry, electrical conductivity, thermal conductivity and diffusivity etc.) have been used in the LVP to characterize the high pressure properties of materials.

To understand the dynamics of the Earth, kinetics of phase transformations provides an essential method. Combined with synchrotron source, LVP is the ideal tool with the advantages of homogeneous and stable temperature generation for studying phase relations. For examples, Fei [74] and Ono [75] studied the earth's rocky inner by making the phase diagrams of the rock's major component. Li [76] studied the elasticity of minerals and materials at high pressure and temperature by analyzing the result of the X-ray image and ultrasonic interferometer. Terasaki [77], Rutter [78] and Poe [79] studied viscosities of melts by using falling sphere techniques with radiographic imaging. Katayama and Tsuji [80] studied the structures of many melts and liquids by using techniques of X-ray absorption measurement and X-ray microtomography. Moreover, using the double-stage system (Kawai-cell in DIA) and sintered diamond anvils, simultaneous P-T conditions of 50 GPa and 2000 K can be obtained to study petrology under lower mantle conditions.

11.3. High Pressure Synchrotron X-ray Diffraction

As a direct way to detect the structural transformation, X-ray diffraction (XRD) technique is very powerful and widely accepted to get a deeper understanding of pressure-induced structural properties for a large variety of materials. With the development of new techniques, combination of synchrotron radiation and powder diffraction methods opened a new door for structural investigation of polycrystalline materials together with single crystals. Most notably, this combination is characterized by high resolution and permits precise refinement of the acquired data. Here we introduce the high pressure synchrotron XRD experiments on powders, single crystals, and nanomaterials in three sections. Some attractive achievements are also reviewed in each section.

11.3.1. *Polycrystalline samples*

There are two different powder diffraction method applied in high pressure research: angle-dispersive X-ray diffraction (ADXRD) employing monochromatic X-rays and energy-dispersive X-ray diffraction (EDXRD) employing fixed angle geometry. Compared with EDXRD technique, ADXRD technique is more promising with the benefit of advanced 2D detectors and with the improvement of synchrotron radiation sources. A wider field of application will exist for ADXRD, as this technique is characterized by high resolution and permits Rietveld refinement of the acquired data. In recent years, gorgeous HP research works have been achieved experimentally thanks to the feasible synchrotron X-ray techniques. The following examples provide some applications of high pressure ADXRD. Pressure-induced metal-to-semiconductor transitions were observed in Na by both XRD experiment and theoretical calculations [81]. The insulating Na has been determined to adopt the *c*-axis highly compressed double hexagonal close-packed structure, analogous to the Ni_2In-type structure where the Na ionic cores form the Ni sublattice and the interstitially localized electron maxima occupy the in sublattice. By compressing rhodium in fluid hydrogen at ambient temperature, Li *et al.* found that the fcc rhodium metal absorbs hydrogen and expands unit-cell volume by two discrete steps to form NaCl-typed fcc rhodium monohydride at 4 GPa and fluorite-typed fcc RhH_2 at 8 GPa. The low-temperature experiments show that RhH_2 can be recoverable after releasing pressure cryogenically to ambient pressure [82]. In the $Ce_{75}Al_{25}$ metallic glass, Zeng *et al.* discovered a long-range topological order corresponding to a single crystal of indefinite length (See Fig. 11.6) [83]. By using molecular dynamics

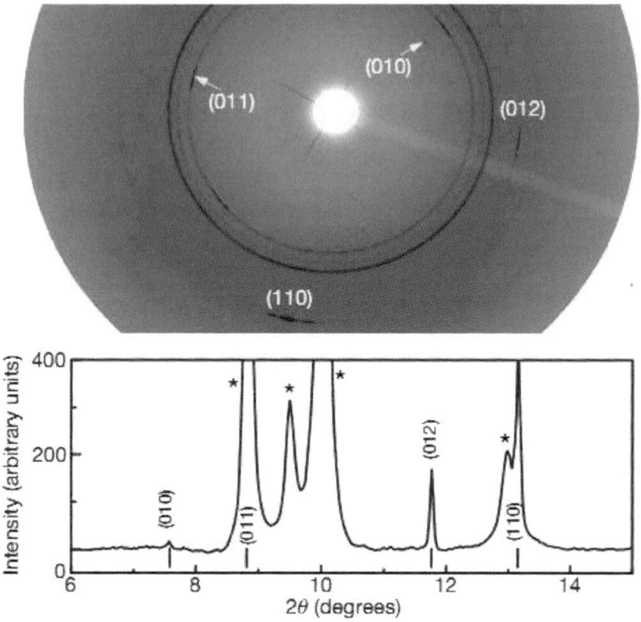

Fig. 11.6. X-ray CCD image and integrated diffraction pattern taken at 190 GPa from the transparent Na sample. Asterisks indicate reflections from the Re gasket. Reproduced from [81] with permission from Nature.

simulations and synchrotron X-ray techniques, they elucidate that the mismatch between the large Ce and small Al atoms frustrates the crystallization and causes amorphization, but a long-range fcc topological order still exists. Sun *et al.* reported that in the superconducting iron chalcogenides, a second superconducting phase suddenly reemerges above 11.5 GPa, after the T_c drops from the first maximum of 32 K at 1 GPa [84]. However, high-pressure polycrystalline ADXRD results of the two iron chalcogenides confirm that, to the first degree, the basic tetragonal crystal structure persists throughout the pressure range studied. Therefore, it is proposed that the disappearance of T_c in SC-I, the re-emergence of higher T_c in SC-II, and the final non-superconducting region reflect detailed structural variances within the basic tetragonal unit cell.

New frontiers that are being developed now lie in the high pressure supramolecular chemistry especially [85–91]. The hydrogen-bonded supramolecular motifs can exhibit rich phenomena under high-pressure conditions, such as reversible pressure-induced amorphization (PIA) [92] and phase transitions. Meanwhile, the cooperativity between hydrogen bonding and π-stacking interactions within ammonium squarate has been studied utilizing high-pressure techniques [93]. Guanidinium nitrate (GN) collapses to the three-dimensional (3D) hydrogen-bonded structure at ~1 GPa via rearrangements of hydrogen bonds, adopting a more efficient crystal packing [94]. As in most supramolecular systems, cyanuric acid–melamine 1:1 adduct (CA•M) has hydrogen-bonded networks in their molecular crystal structures [95]. By ADXRD experiments, Wang *et al.* investigated a pressure-induced irreversible phase transition at 4.4 GPa. A space group change from $C2/m$ to its subgroup $P2_1/m$ is proposed. It is proposed that this phase transition was due to supramolecular rearrangements brought about by changes in the hydrogen bonding networks.

11.3.2. *Single crystals*

Single-crystal XRD is the chief method for characterizing the atomic structure of new materials and in discerning materials that appear similar by other experiments. Compared with powder XRD, single-crystal XRD provides much more complete and accurate information on structure at high pressures [96]. Synchrotron applied in high-pressure single-crystal investigation calls for the improvement of the experimental instruments and techniques. The large angle is very important in single-crystal XRD to get the accurate structure information. Conical diamond and seat design in DAC and high-pressure single-crystal techniques were described by Reinhard Boehler and Ronald Miletich *et al.* [97]. This design allows for larger aperture making such diamond cells suitable to significantly increase the pressure range in single-crystal XRD, and it also reduces the anvil absorption due to reduced anvil thickness. Synchrotron radiation facility has been also improved to obtain better and higher-quality signal. In European Synchrotron Radiation Facility (ESRF), bright sources of X-rays are two in-vacuum undulators with a magnetic period of

23 mm in the short and ultrashort wavelength domains. In the high-pressure macromolecular crystallography mode, the wavelength is selected by a monolithic Si two-reflection monochromator. The DAC is settled on a three-axis goniometer which is mounted onto a stack of three orthogonal translations, and normal-beam geometry is used.

High-pressure single-crystal synchrotron has been widely implied in many areas. The single-crystal EDXD method was first used for H_2 to 26.5 GPa. Solid hydrogen remains in the hexagonal close-packed structure under the pressure conditions and exhibit increasing structural anisotropy with pressure [98]. Under ambient conditions, magnesiowüstite crystallizes in the NaCl (B1) structure. At high pressures, the wüstite end-member goes through two phase transitions from B1 through a rhombohedral distortion to the NiAs structure while the MgO end member remains in the B1 structure throughout the P-T range of the Earth [99]. This technique is also used to determine the structure of organic materials. L-cystine consists of molecules in the zeitterionic form, and crystallizes in the hexagonal space group $P6_122$. And the single-crystal structures have been determined at room temperature at pressures between 0.4 and 3.7 GPa [100]. The structural change of some metal-organic frameworks (MOFs) were investigated under high pressures using *in situ* single crystal X-ray diffraction. For example, single crystal ZAG-4 is highly compressible with reversibility possible within the pressure range of 10 GPa [101]. Moreover, it has important applications in high-pressure biophysics: macromolecular crystallography [102]. The A-form of d(GGTATACC) can withstand high pressures up to ~1.9 GPa. The gradual loss of long-range order between 1.6 and 1.9 GPa may be related closely to packing effects. The high-resolution structures show that geometry of base pairs is well preserved under compression up to at least 1.39 GPa [103]. Recently, Zhang *et al.* [104] reported a multigrain single-crystal X-ray diffraction analysis technique in a diamond anvil cell to study high pressure structure of $(Mg,Fe)SiO_3$ postperovskite. Crystallographic orientations of over 100 crystallites were simultaneously determined in a coarse-grained polycrystalline sample.

11.3.3. *Nanomaterials*

Nanomaterials as new structural and functional substances has many good characters such as surface effect, volume effect, quantum size effect, quantum tunneling effect, dielectric confinement effect. These characteristics make nanomaterials show different physical and chemical properties from common materials. Thus it is essential to investigate structural changes of nanomaterials under high pressure. High pressure is an effective physical parameter to vary distances between atoms, so that we can study the relationships between properties and structure of the nanomaterials. The study of nanoparticles under high pressure is considered an available path to expand the range of solid state materials. Although XRD is routinely carried out for the structural study, it may be indispensable to use synchrotron X-rays to obtain better sensitive and precise experimental dates for structural study of nanomaterials under high pressure. Recently, scientists applied a nano/submicron focused X-ray beam to study nanoscale single crystalline phases in DAC at high pressures [105]. Hence we can obtain sufficient Bragg reflections of nonmaterial from the DACs with small sample chamber and have a reliable quantitative analysis from the diffraction pattern.

Recently, scientists have obtained many achievements in the study of nanomaterials under high pressure using synchrotron XRD. Experimental evidence suggests that size-dependent properties in nanoparticles under pressure have been shown to influence compressibility and crystalline phase stability [106, 107]. For instance, in metal powders it is observed that the melting temperature decreases significantly with a decrease in the grain size [108]. Moreover, the amorphous phases can also be achieved through the high pressure treatment of nanocrystals. Wang *et al.* studied Y_2O_3 with particle sizes ranging from 5 nm to 1 μm at high pressure using synchrotron XRD and Raman spectroscopy techniques. They found that nanometer-sized Y_2O_3 particles are shown to be more stable than their bulk counterparts, and a grain size-dependent crystalline-amorphous transition was discovered in these materials [109]. The same size-depended structural changes of PbTe nanocrystals under high pressure was also investigated using synchrotron XRD [110]. The second family of materials to be considered has been grouped in the terms of semiconducting nanomaterials. For example, CdSe nanocrystal has been used as a model for high pressure optical and structural studies [111]. Zou *et al.* have studied the stability of metastable ZB- and WZ-MnS nanocrystals under high pressure using synchrotron XRD. They found that the ZB-nanoparticles and ZB/WZ-nanobipods are stable below their critical pressure, 5.3 and 2.9 GPa, respectively. When pressures exceed the critical point, all these metastable MnS NCs directly convert to the stable rock salt MnS [112]. It provides us a new insight and an opportunity to study structural stability of metastable phase NCs. The pressure induced structural phase transition has also been observed experimentally in other semiconducting nanoparticles such as ZnSe [113] and $CuGaS_2$ [114]. Carbon nanomaterials will constitute the last case to be discussed. High pressure behaviors of carbon nanomaterials have also been widely discussed by scientists and investigated by XRD, Raman scattering and other experimental methods [115]. *In situ* synchrotron XRD measurements of SWNT and C_{60}-peapod samples under high pressure were performed. It was found that the phase stabilities of SWNT and C_{60}-peapod pseudo-two-dimensional crystals in a wide pressure–temperature region and hexagonal diamond was synthesized by HPHT treatment of C_{60}-peapods [116,117]. Another type of carbon material, solvated C_{60}, show some striking features under high pressure. Wang *et al.* synthesized a long-range ordered material constructed from units of amorphous carbon clusters. It can be preserved back to ambient conditions and exhibit exceptional hardness [118].

11.4. High-Pressure Extended X-ray Absorption Fine Structure

The phenomenon of X-ray damping when passing through the material is defined as X-ray absorption. Generally, the intensity damping arises from photoelectric effect, Compton scattering and Rayleigh scattering. X-ray absorption spectroscopy (XAS) can be obtained by recording the incident intensity I_0 and the intensity I transmitted through the sample. By tuning the photon energy with a monochromator, the energy range of X-ray beam (typically 0.1–100 keV) can be realized. When the photon energy is adequate to excite the inner electron, the X-ray absorption coefficient (μ) increases rapidly. The discrete energies where μ increases rapidly are called absorption edges. Typically, μ increases sharply at the absorption edge, and then exhibits an oscillation feature versus photon energy. Such complicated fashion can occur in the energy range towards 1000 eV above the edge. There are two main regions above the absorption range in a typical XAS. The region from the edge to 50 eV is called Near-edge X-ray Absorption Fine Structure (NEXAFS), while the region over 50 eV to more than 1000 eV above the edge is termed as Extended X-ray Absorption Fine Structure (EXAFS).

The physical mechanism of EXAFS is the interference of the outgoing wave from the central atom and the backscattered part from the surrounding atoms. The definition of edges depends on the excited inner electrons. In other words, the K-, L- and M-edges means the principal quantum number n is 1, 2 and 3. Apart from the K-edge, the lower energy edges (L- and M-edges) are also expected to display the EXAFS effects. Because of the nature of short-range order, EXAFS can provide abundant structural information on the local environment surrounding the absorbing atom. For a given atom, the average interatomic distance, the disorder about the average, as well as the kind and number of surrounding atoms can be quantified by analyzing the EXAFS spectrum. Because the edges of different types of atoms are at the discrete and easily separable energies, the local environment surrounding each atom can be collected by tuning the photon energy to the specific range. This makes great contribution to understand the whole structure of the material. Furthermore, the sample can be in the crystalline or amorphous solids, solution or gas phase. This extensively extends the application of EXAFS and can compensate XRD where periodic structures are required.

The EXAFS effects have been discovered for over 80 years [119]. However, EXAFS had not attracted much attention of scientists due to poverty in experimental equipments and the lack of proper explanation. In 1971, Sayers Dale E. and the co-workers employed Fourier analysis of the experimental data to obtain the radial structure function. The crystal structures of amorphous and crystalline Ge were determined and analyzed. Their preliminary results imply that EXAFS is a powerful technique to explore material structures. Especially, the availability of synchrotron X-ray source opened new possibilities for applications of EXAFS in 1974 because the synchrotron X-ray offers high brightness and broad range than normal X-ray sources. Then almost various applications of EXAFS for structure determinations have been realized with synchrotron X-ray source. The applications comprise catalysts, biology, surface characterizations and structure characterizations of solutions and amorphous solids. The derailed applications of EXAFS have been reviewed elsewhere [120, 121]. Here, we are focused on the application of EXAFS in high-pressure studies with synchrotron X-ray source as the light source. In the latter section, we will present some examples of EXAFS and synchrotron application in high-pressure investigations.

Since the availability of synchrotron X-ray source in 1970s, EXAFS have been extensively applied in high-pressure studies. Especially, the sample for high-pressure experiments does not require the crystallinity of the material. This feature is useful when XRD technique becomes invalid. Lots of fruitful results of high-pressure exploring have been achieved by combining EXAFS and synchrotron techniques.

In 1978, R. Ingalls and the co-workers reported the high-pressure EXAFS using synchrotron as X-ray source for the first time [122]. This work has provided a new powerful tool for structure determination under high-pressure conditions. They examined the Fe-Fe, Fe-S and S-S distances versus pressure in pyrite FeS_2. Fe-S spacings proved to be less compressible than the lattice constants, while S-S bonds displayed high compressibility compared with the lattice constants. The atomic positions could be obtained by analyzing the EXAFS data. However, the two nearest-neighbor Fe-F distances had not been resolved in FeF_2 owing to restriction on experimental conditions. Then, a series of substances have been investigated using high-pressure EXAFS. For example, atomic spacings in Ge and NaBr were examined up to 5.2 and 2.1 GPa, respectively, using high-pressure EXAFS technique, and pointed out the high-pressure EXAFS data could be collected by carefully tuning the diamond anvil positions [123]. Analysis of EXAFS of CuBr indicated that no change in coordination occurred in the transformation from the zinc-blende to the tetragonal phase [124]. The evolutions of fine-structure information on α-quartz GeO_2, Cu, KBr and $KNbO_3$ versus pressure, up to 5.8, 10, 8 and 15.8 GPa, were traced [125, 126]. EXAFS of RbCl demonstrated that the mechanism of coexistence of phases was reasonable across B1 to B2 phase transition at 0.55 GPa [127].

Since the 1990s, high-pressure EXAFS of ZnTe, GeI_4, Ge_3N_4 and nanocrystalline GaSb have been performed to monitor the behaviors and stabilities [128–130]. For example, EXAFS of ZnTe at the Zn K-edge demonstrated two phase transitions at 9.5 and 12 GPa combined by XRD experiments. EXAFS of GeI_4 on the K-edge provided the local atomic scale description which indicated that GeI_4 became disordered at 12 GPa and had undergone a transition from the molecular crystal to a metallic glass at 25 GPa. High-pressure EXAFS of liquid krypton and gallium were conducted to probe the pressure dependence of local structural environments, such as the distance and asymmetry of the nearest-neighbor distribution, which helped one to understand the potential function [131]. Crystalline and liquid selenium and tellurium were investigated by high-pressure EXAFS measurements up to 2.5 and 6 GPa, respectively [132]. Large-volume presses were also applied for *in situ* structural measurements on liquids up to several gigapascals, combined with synchrotron X-ray radiation [80].

High-pressure EXAFS of amorphous GeO_2 were undertaken to investigate the local mechanism for the amorphous-amorphous transition [133]. It was proposed that amorphous GeO_2 experienced a transformation from a fourfold to fivefold coordinate structure between 13 and 30 GPa. The transition was continuous, evidenced by the gradual increase in the distance of Ge-O bond in the intermediate pressure range. Synchrotron X-ray absorption spectroscopy was performed to probe the response of metallic glass $Ce_{75}Al_{25}$ to external pressure [134]. The $4f$ delocalization from ambient localized state $4f^1$ to an itinerant state $4f^0$ at 5 GPa, which accounted for the volume collapse of ~8.6%. This work established a new type of electronic polymorphism. The amorphization of ZrW_2O_8 and $ZrMo_2O_8$ at high pressure was examined by tracing the XAS at the W L_1-edge and Mo K-edge [135].

Meanwhile, scientists are searching new anvil materials for high-pressure studies. Sintered boron carbide has been employed as the anvils to obtain the full XAFS spectrum [136]. Especially, nanocrystalline diamond anvils have been used to overcome the glitch problem, which is a serious problem from single crystal diamond anvils. The high-pressure glitch-free EXAFS can be realized over a wide energy range [137, 138]. Also, the use of nanocrystalline diamond anvils is a breakthrough in high-pressure EXAFS study.

11.5. High-Pressure Inelastic X-ray Scattering

Inelastic X-ray scattering (IXS) has been extensively applied in materials science, especially since the advent of third generation synchrotron sources. The extremely brilliant and high energy synchrotron light sources make measurements of IXS more easily and reliably. Simply speaking, the incident X-ray photon with energy E_1 and wavevector k_1 is scattered and evolved into another photon with energy E_2 and wavevector k_2 during the typical IXS process. Then the momentum transfer is $4\pi\sin(\theta/2)/\lambda$, where θ stands for the scattering angle. Through the analysis of IXS data, one can obtain the identical physical parameter, the dynamic structure factor $S(q, w)$, which contains abundant rich information about electron and photon dynamics. Especially, the dispersion of the whole Brillouin zone can be traced using the X-ray wavelength comparable to the wavevector. As early as 1923, the quantum theory of IXS has been proposed [139]. IXS has been considered as a new powerful tool to probe the dynamics of phonons and electrons combined with the third generation synchrotron sources.

The electronic structure and dynamics have been widely investigated using IXS. The $S(q, w)$ of electrons in Li was examined for various directions with the synchrotron storage ring [140]. The temperature dependence of $S(q, w)$ of electrons in Al was measured both in the solid and in the liquid phase. It was found that the shifts of the Plasmon energy for $q < q_c$ (q_c = Plasmon cutoff vector) arised from the alteration of the electron density at rising temperature, and the unique double-peak structure for $q > q_c$ reduced progressively with wakening of the ion-core potential [141]. The $S(q, w)$ of diamond and LiF were measured using IXS and *ab initio* calculations. It was proposed that the interaction of the excited electron in the conduction band with the hole in the valence band played a crucial role in the estimates of IXS [142]. The deviations from the dipole approximation in IXS about the fluorine K edge in LiF stemmed from an s-type exciton [143]. In Mott insulators $Ca_2CuO_2Cl_2$, the observed dispersive charge excitations across the Mott gap provided direct support evidence for the Hubbard model in interpretation of charge excitations across the Mott gap [144]. The charge-density excitation spectra of MgB_2 revealed a unique cosinelike and periodic energy dispersion versus momentum transfer along c-axis, which was due to the strong coupling of single-particle and collective degrees of freedom [145].

The collective dynamics in the liquid phase and phonon spectroscopy in solids have been probed using IXS techniques with the synchrotron sources. The measurement of the propagating excitation with a velocity in water implied the collective dynamics of liquids could also be probed using IXS techniques [146]. The collective dynamics of counterions between F-actin filaments in aqueous solution was observed using high-resolution IXS at the molecular length scale, and the unexpected dynamics of ions interacting with the cages of the nearest neighbors was observed [147]. Eberhard Burkel gave the detailed explanation for the theory and applications about the phonon dispersion curves in materials science. In the faced-centered cubic (fcc) ö-plutonium–0.6 weight % gallium alloy, the abnormal features of the phonon dispersion curves were observed and proved to be due to strong coupling between the lattice and the $5f$ valence instabilities and the phase transitions of plutonium [148].

In the last decade, resonant inelastic X-ray scattering (RIXS) has been used as a powerful probe of electronic excitations. During the RIXS process, the X-ray photon near to the absorption threshold excites the core electron, and then the excited state decays by the photon emitting process. Thus, the second-order process nature of RIXS makes the intensity of signal much weaker because of the low efficiency of X-ray emitting. The availability of third generation radiation sources overcomes this obstacle and promotes RIXS to the forefront of X-ray photon science. RIXS investigation of electron and elementary excitations has been summarized in the reference [149, 150].

In the past, many experimental challenges (typically the low-flux light source) have limited the observation of pressure effects to a narrow energy window. Since the advent of synchrotron sources, the pressure effects of high-energy electronic levels can be traced. The first directly measurements of electronic excitations in hcp ^4He at 61.5 MPa and 4.3 K was performed by N. Schell [151]. The high-brilliant source and improvements of experimental techniques (the focusing optics, gasket and cell design) promote the remarkable progress in high-pressure IXS investigations. Ho-kwang Mao has reviewed the improvements and progress in high-pressure IXS studies in the literature [152].

F. Occelli measured the phonon dispersion of the single crystal argon versus pressure up to 20 GPa, and demonstrated it was feasible to perform the single crystal IXS in a DAC to very high pressure. The results showed the photon energy could be strongly tuned with increasing pressure as a result of the nonadditive part of the interaction [153]. The IXS near the K-edge spectra of carbon implied that the π-bonds of bridging carbon atoms between graphite layers converted to ö-bonds, while the nonbridging carbon atoms remained as π-bonds at 17 GPa [154]. Especially, high-pressure IXS technique has been applied in geophysical materials. The lattice dynamics and the acoustic wave velocities of the corresponding systems have been surveyed [155, 156]. The valence change in $YbAl_3$ from 2.75 at ambient pressure to 2.93 at 30 GPa was monitored using IXS and theoretical methods [157]. The *in situ* synchrotron IXS spectra of Na-borate glasses to 25 GPa revealed that the boron coordination transformation was linear, which was in contrast to the multiple pathways in Li- and pure-borate glassed [158]. The electronic excitations in various materials at high pressure have been reviewed in the literature [21].

In the simplest metal Na, the plasmon excitations of body-centered cubic structure to 52 GPa and fcc structure from 64 to 97 GPa were examined using momentum dependent IXS [159]. The focusing systems for high-pressure IXS are improved in the synchrotron source stations, and enhancing the capability for high-pressure experiments [160, 161].

X-ray Raman scattering (XRS) is the nonresonant inelastic scattering of X-rays; it stems from the electronic excitations of the bound electrons in an atom. The high-brilliance of synchrotron sources can compensate the small scattering cross-section of XRS, and make XRS widely used in material science. The decomposition of water (H_2O) under high pressure and extensive X-radiation was examined using XRS, combined with XRD and optical Raman spectroscopy. This crystalline solid comprising of O-O and H-H bonds, which was stable over the large pressure and temperature range, was a new phase [162]. XRS spectra of benzene were measured up to 20 GPa, and the spectral changes were correlated with pressure induced phase transitions [163]. XRS spectra of the silicon L edge in SiO_2 were examined up to 74 GPa, which indicated the fivefold-coordinated silicon formation at high pressure [164]. N. Hiraoka pointed out how to minimize the influence of background in XRS [165].

High-pressure RIXS have been employed in investigation of transition metal and compounds. For example, RIXS of Gd indicated Kondo-like aspects during the delocalization of $4f$ electrons, which revealed a prolonged and continuous delocalization process with increasing pressure [166]. The band formation of $4f$ electrons in Ce was observed at high pressure using RIXS [167]. In TmTe, the pressure-induced valence change was monitored using RIXS at the Tm L3 edge. The experimental data could be well interpreted with the Anderson impurity model, whereas the plateau in Tm valence and hybridization pressure dependences suggested there was an n-channel Kondo (NCK) screening process [168].

11.6. High Pressure X-ray Imaging

The thorough knowledge of microstructure of materials under various conditions is crucial for understanding properties and structure-property relationships, and for generating new structures with target properties. The two-dimensional (2D) techniques comprising of transmission electron microscopy (TEM) and scanning electron microscopy (SEM) have been extensively used in materials science. Especially, the three-dimensional (3D) imaging has attracted much attention because it can provide direct 3D examination of corresponding materials. Nowadays, X-ray computed tomography (XCT), neutron computed tomography (NCT), and magnetic resonance imaging (MRI) are three of the most used 3D imaging techniques. The physical basis of XCT is the dependence of attenuation coefficient versus the sample position. In other words, the attenuation coefficients as a function of the position in a material are obtained. This can be realized by recording the radiographs with the sample rotating around a fixed axis between 0° and 180°. Then the algorithm of back-projection is used for reconstruction of the 3D visualization of the sample. The X-ray source can be an ordinary X-ray tube or the synchrotron radiation. Generally, there are two modes of X-ray microtomography. The first one is the conventional absorption tomography, which depends on the difference in the linear attenuation coefficients in the material. In this mode, the absorption contrast image of the sample can be obtained via recording the transmitted photons directly. And the sample must be placed very close to the detector. The second mode is the phase contrast mode. This mode can be performed when the X-ray beam is coherent and the distance between the sample and the detector is increased compared with the absorption imaging. The interference of the X-ray beam results in the spatial intensity distribution, which is recorded with the detector. The advantage of this mode is that it shows much more sensitivity compared with the absorption contrast imaging.

The holotomography modes requires complete images taken at several sample-detector distances, and is considered as the extension of the phase contrast mode [169]. This mode allows one to perform quantitative analysis in 3D mass density of the sample.

Because of its nondestructive nature, high resolution X-ray imaging is considered as an attractive scientific technique and has been used in many fields of materials science. Since the advent of synchrotron radiation source, synchrotron X-ray source has been used for X-ray microtomographic imaging as it delivers a very high flux and the optimum X-ray energy can be easily and continuously selected using single crystal monochromators [170]. The high-throughput X-ray microtomographic imaging system in Advance Photon Source (APS) enables one to perform the X-ray imaging experiments with 100 nm or better resolution in several minutes. The full 3D structural data of the dynamically evolving material on 10 min scale can be examined [171]. The X-ray microtomography and radiography imaging at BESSY II is installed for absorption, refraction and phase contrast investigations with the resolution up to submicrometer range. The set-ups for imaging investigation can offer moderate resolutions for the large field view to high resolution (down to $0.35\,\mu$m), and the energy of monochromatic synchrotron beam ranges from 6 to 80 keV [172]. The spatial resolution of detectors for synchrotron X-ray imaging have been improved to submicrometer range based on melioration in the scintillator, microscope and CCD camera. The spatial resolution down to 250 nm is available using the scintillator at shorter wavelength [173].

The synchrotron X-ray imaging with high resolution and high sensitivity has been used in many nondestructive evaluation applications including materials science and life science. For example, the first mechanical tests on metal matrix composites were performed at ESRF. The reinforcements in the matrix, the corresponding cracks and the mechanism for plastic deformation were detected and analyzed with *in situ* high resolution synchrotron X-ray phase contrast imaging [174]. L. Salvo *et al.* gave the detailed discussion about the progress in X-ray microtomography of metallic materials including the semi-solid and metallic foams, and proposed the external mechanical test machine or a furnace could be applied for *in situ* imaging experiments [175]. J. Baruchel *et al.* gave the recent advances in synchrotron X-ray imaging techniques and summarized the application of the imaging in many aspects such as characterizing the structure and defects in the material, the chemical composition, as well as density and morphology of the material [176].

Especially, the 3D nondestructive X-ray imaging is the novel scientific advance in a wide range of geological and clinical investigations. This technique has been performed to analyze the interior of fossils and meteorites, to examine the texture of igneous and metamorphic rocks, and to descript the porosity and permeability in soils, which is important in understanding the evolution of life, mantle metasomatism and planetary differentiation [177, 178]. Even the water flow processes in a sand can be visualized using synchrotron X-ray microtomography [179]. The X-ray imaging is usually termed as X-ray computed tomography (CT) in clinical practice. The CT imaging can be performed on the whole organ or the whole body with very short time and high resolution [180]. The X-ray imaging has become a crucial component in the toolkit for every investigator in geological and clinical fields.

Recently, the new imaging technique X-ray diffraction contrast tomography (DTC) has been developed for characterization polycrystalline materials [181]. This technique combines the principles of X-ray microtomography and 3D X-ray diffraction microscopy (3DXRD), and can be considered as an extension of 3DXRD. This advance opens new possibilities for mapping the grain shape and grain orientation in polycrystalline materials in a nondestructive way. Especially, the microtexture and local grain boundary characters comprising of orientation distribution function and grain boundary crystallographic planes can be obtained [182]. A modified strategy, including incorporation of the Friedel pair method and extension of scan range to 360°, is applied in DTC to improve the accuracy and the grain indexing routines [183]. Hence, this new tool is expected to be applied in the processes which are driven by grain boundary characters.

X-ray imaging is irreplaceable in compression science for two reasons. On one hand, the imaging technique is nondestructive and can characterize the initial state of the sample. On the other hand, the technique is suitable for materials in various states including liquids and amorphous materials where X-ray diffraction is invalid. Synchrotron X-ray imaging has become an effective tool for exploring materials under extreme conditions. For example, the density and viscosity of Fe-S liquids were measured under high pressure and temperature conditions using Stokes' method [184]. Additionally, the equation of state of liquids can be expected using X-ray imaging advance. Synchrotron X-ray radiography was used to image and measure stress and strain at diamond tip up to the pressure of 300 GPa. The 3D distribution of strain in gaskets and the diamond deformation were directly examined and the results revealed that gasket materials W and Fe was enhanced by nearly two orders of magnitude at ultrahigh pressure [185]. Russell J. Hemley *et al.* summarized the development of X-ray imaging until 2005.

Synchrotron X-ray imaging has been developed combined with the LVP apparatus at the GeoSoilEnviroCARS (GSECARS), the advanced photon source (APS) in 2005 [186]. This means the imaging of materials under high pressure and temperature conditions can be realized using X-ray microtomography. The relative light and thin Al alloy was applied to replace the containment ring of Drickamer anvil apparatus considering it was transparent to X-rays, and shows modest strength. The spatial resolution is scaled down to a few micrometers [73]. The detailed evolution of microstructure could be obtained in shear deformation experiments, implying the texture evolution and shear deformation could be expected with *in situ* high-pressure microtomography experiments. The LVP apparatus for *in situ* microtomography under high pressure and temperatures was developed in Spring-8. The time for collecting a data set for imaging was much shorter, which minimized the change of the field view during heating experiments. This new advance opens up new possibilities for investigating the equation of state, interfacial tension of noncrystalline materials including glassed and melts under various pressure and temperature conditions.

With the convenient LVP facility, *in situ* measurements of materials evolution and quenched samples can be traced. For example, the imaging measurements of refractory magnesium silicate glasses were performed [187]. The volume changes and isothermal bulk moduli were consistent with the results of ultrasonic experiments. The volume changes of amorphous $MgSiO_3$ were obtained at 2 GPa and the temperature up to 1000 °C. This work opens new possibilities for studies of glasses and liquids within magmas in planet interiors. The *in situ* microtomography can also be used to probe the interfacial tension between alloys and silicate melt with the seddile drop method under high pressure and temperatures. It showed the interfacial tension decreased significantly with sulphur content increasing [188]. Meanwhile, the 3D microtomography can be used to monitor the microstructure of materials quenched from various pressure and temperature conditions. For instance, the porosity of molten rocks in mantle peridotite was investigated and it was proposed that the porosity resulted from the balance between melting rate and viscous compaction, rather than the melt topology change [189].

The large shear deformation of multiphase composites under high pressure and high temperature could be examined with *in situ* X-ray microtomographic imaging. The detailed information on textural evolution from load-bearing framework to interconnected weaker layers were observed, which was of great importance in understanding the mechanical properties of materials in Earth's deep mantle [190]. The new facility based on X-ray imaging technique was developed for *in situ* macroscopic strain measurements under high pressure and temperature conditions at National Synchrotron Light Source (NSLS), Brookhaven National Laboratory. The experiments were performed with a DIA-type high-pressure apparatus. The modified vertical anvils are independent so as to allow superimposition of strain on the hydrostatic strain. The direct measurements on olivine rheological properties were performed using synchrotron-based Deformation-DIA (D-DIA) high-pressure apparatus. The reported experimental data including flow parameters were consistent with the flow laws, and the activation volume for olivine dislocation creep under high temperature was $0 \pm 5\,\text{cm}^3/\text{mol}$ [191]. The deformation properties of San Carlos olivine single crystals were examined at pressures from 3.5 to 8.5 GPa and temperatures from 1373 to 1673 K. The X-ray diffraction and imaging techniques were utilized to detect the constant applied stress and specimen strain rates, respectively. The results demonstrated a slip transition from a-slip to c-slip with increasing pressure, which might explain the discrepancy of seismic-velocity attenuation at low pressure and high pressure [192]. The examination of shear deformation of isolated melts of metal and metal sulphide in the olivine matrix indicated that melt could be efficient interconnected due to large strains in a dynamic process. This further suggested percolation was still a viable mechanism for melts migration and segregation in the mantle [193].

The highest pressure of LVP facility is limited due to the large sample volume. The feasible method to overcome this problem is to develop *in situ* X-ray microtomographic imaging technique based on DAC because DAC can generate much higher pressure. Liu *et al.* combined synchrotron X-ray imaging and DAC techniques and studies the abnormal behavior of amorphous selenium. The two step crystallization process was demonstrated in which the monoclinic form crystallized and then transformed to the trigonal phase through X-ray diffraction and imaging measurements. The equation of state of amorphous materials could be directly determined using X-ray microtomography. This advance extends the range of high pressure X-ray imaging using LVP facility in which the lower pressure is limited owing to large sample [194]. Xiao *et al.* presented the mass density measurement could be directly performed with X-ray microtomography in a DAC. An algorithm was developed to solve the missing angle and noise problems [195]. A new iterative method was introduced to improve the spatial resolution of X-ray tomography in a DAC and to overcome the limited imaging angular range problem [196]. Comparison of Sn volume data this work and the well-known EoS indicated the method was suitable for materials in various states.

11.7. High-Pressure X-ray Emission Spectroscopy

X-ray emission spectroscopy (XES) has been tremendously used in material science since the discovery of X-ray by Röntgen. The fields of XES application relate to correlated electron materials, magnetic materials and nanomaterials. The basic mechanism for XES is easily understood. An electron in the outer orbital of an atom fills the vacancy in the inner orbital, and thus the photon with the energy corresponding to the difference in ionization energies in the orbital is generated. The emission lines are determined by the atomic selection rules $\Delta l = \pm 1, \Delta j = \pm 1, 0$. The X-ray emission lines are classified based on the orbits in which the transition is generated. The Greek letters and Arabic numbers stand for the orbital of vacancy and the orbital of electron, respectively. Transition to a K shell produces the K emission line, and transition to the second shell generates L spectra. Importantly, there are usually two values of j for each l as a result of spin orbital coupling. So the transition can produce two possible emission lines except $l = 0$. As XES involves the electron energy distribution, it can be adopted to monitor the electron state including the spin state.

The advent of synchrotron paves the way for XES developments. The brilliance and the broad energy range of synchrotron sources can overcome some experimental problems. For example, the close-lying core hole or satellite can give rise to overlaying lines which makes the analysis of XES data difficult. This problem has been solved using monochromatized synchrotron radiation instead of the broadband photon sources [197].

Synchrotron-based XES has been applied to probe basic properties of many materials including liquid and solid materials. XES was used to examine the hydrogen bond in liquid water by comparing X-ray emission spectra of liquid water and the water molecular. The a_1-symmetry valence-bond proved to be crucial in the hydrogen bonding in liquid water [198]. For the compound $FeSi_{1-x}Ge_x$ ($x = 0$, 0.11, 0.19, and 0.44), the increase in the magnetic moment was observed with increasing x [199] by monitoring the Fe-K_β emission. The result also indicated there was a transition of magnetic order versus x. Actually, the systematic investigation on 12 Fe-K_β XES was performed to probe valence orbital composition. The results showed the spin state contribution was responsible for the K_β main line, while the valence to core region proved to be sensitive to the chemical environment around the Fe atom [200].

Compared with the mature synchrotron XRD technique, the development of high-pressure X-ray emission spectroscopy (XES) was limited mainly by the strong attenuation of X-ray source below 10 keV through diamond anvils. This energy range covers all the K absorption edges of $3d$ transition atoms and L absorption edge of $4f$ rare earth elements. Fortunately, this problem has been overcome by the use of new gasket materials (Be and Be-B) which exhibit much weaker absorption of X-ray photons as well as new cell designs. Meanwhile, XES can be interfaced with the laser heated DAC, which allows for exploring the electronic spin state across larger high pressure and temperature range [201]. Combination of synchrotron X-ray sources and developed experimental techniques enables one to examine the electron state of highly correlated transition metal compounds.

Nowadays, synchrotron-based XES is a well-established method for direct measurements of the electron state and local magnetic properties of atoms. Take the $3d$ transition atom Fe for example, the high spin Fe-K_β emission line consists of a main peak with an energy of 7058 eV and a satellite at lower energy side owing to the core hole exchange interaction. In the low spin of Fe^{2+}, the total magnetic moment is equal to zero. So the collapse of $3d$ magnetic moment can be characterized by the disappearance of the lower energy satellite because the exchange interaction vanishes.

The pressure dependence of electronic states and magnetic properties of simple metals and compounds of $3d$ transition metals have been extensively investigated using the *in situ* XES technique. The first high-pressure XES study of FeS was performed by J.-P. Rueff and C.-C. Kao [202]. The pressure-induced high spin to low spin transition in FeS was evidenced by disappearance of the low energy satellite in the Fe-K_β emission line. This study opens new possibilities for monitoring the electronic properties under high pressure. In FeO, a constant finite magnetic moment was observed, indicating no high spin to low spin transition and the high spin state was preserved up to 143 GPa [203]. The magnetic properties of alloy $Fe_{64}Ni_{36}$ was examined by monitoring high resolution XES. The two step reduction of local moment amplitude was raised from transitions from a high spin state to a low spin state at 5 GPs to a nonmagnetic state above 15 GPa [204]. In contrast, Fe_2O_3 underwent a high spin to low spin transition at 60 GPa, and a method was introduced to separate the crystallographic and electronic phase transitions [205]. The magnetic collapse in Fe_3C, the transformation from a magnetic state to nonmagnetic state in $FeCO_3$, the phase transition at 12 GPa and continuous reduction of spin moment in $Fe_{1.01}Se$ were probed by high resolution XES [206]. In addition, the spin state transition in ferromagnetic $La_{0.82}Sr_{0.18}CoO_3$ and the process for the spin magnetic moments reduction in multiferroic $BiMnO_3$ were traced by monitoring the corresponding K_β emission lines [207]. Meanwhile, the disappearance of the satellite peak at the α-phase transition in Fe implied the collapse of the magnetization in Fe. In γ-Fe the short range magnetic collapse was evidenced under high pressure and temperature conditions, up to 30 GPa and 1400 K [208]. In compressed Ge, the valence bandwidth was examined from XES. The width remained nearly unchanged in the semiconducting phase below 10 GPa, while the metallic phase above 10 GPa showed positive dependence with increasing pressure [209]. Importantly, the difference of valence band width between experimental data and calculated results within DFT framework in the semiconducting phase gave the challenge for the improvement in DFT methods. It was proposed that the neglected electronic correlation effects were responsible for this difference.

As one of the most important applications, XES has been applied to examine the electronic state of deep Earth minerals. (Mg,Fe)O, $(Mg,Fe)SiO_3$ and $(Mg,Fe)(Si,Al)O_3$, which is expected to dominate the mineral properties of Earth's lower mantle [210]. There was a transition from high spin to low spin in ferropericlase $(Mg_{0.83}Fe_{0.17})O$ in the pressure range 60–70 GPa, which corresponded to depths of 2000 km in the lower mantle. The further analysis indicated the observed high spin to low spin transition in magnesiowüstite could give rise to an abnormal behavior between the two states. The low-spin, high-pressure phase displayed a much higher bulk modulus and sound velocity than the high-spin, low-pressure phase [211]. The iron-iron exchange interaction proved to be the important factor responsible for the stability of high spin state in the (Mg,Fe)O system, and the relating high-pressure phase diagram had been reexamined [212]. In high pressure and temperatures conditions, the spin state of iron in the lower mantle ferropericlase was examined. The results showed there was a progressive spin transition over a large pressure-temperature range [213]. Iron-bearing silicate perovskite is the most abundant material in the lower mantle. In $(Mg_{0.9}Fe_{0.1})\ SiO_3$, two electronic transitions were observed at 70 GPa and 120 GPa, respectively. These two transitions were attributed to partial and full electron pairing in Fe, and it was suggested that nonconvecting layers might exist in the lowermost mantle [214]. In $(Mg_{0.6}Fe_{0.4})\ SiO_3$, the intermediate spin Fe^{2+} is preserved up to ~125 GPa, and then transformed to the low spin state at ~135 GPa. This transition accounted for the 1% volume decrease deduced from XRD measurements [215]. Moreover, the spin-pairing transition in $(Mg,Fe)(Si,Al)O_3$ was demonstrated, and the observed residual magnetic moment was significantly higher than that in the aluminum free one [216]. The pressure-induced spin crossover in the range 20–100 GPa was observed and analyzed in $(Mg_{0.88}Fe_{0.09})(Si_{0.94},Al_{0.10})O_3$ [217]. The thorough knowledge of spin states in these minerals has profound implications for compositional layering in the Earth's lower mantle.

Bibliography

[1] B. Sundqvist, Fullerenes under high pressures, *Adv. Phys.* **48**, 1–134, (1999).

[2] C. S. Zha, H. K. Mao and R. J. Hemley, Elasticity of MgO and a primary pressure scale to 55 GPa, *Proc. Natl. Acad. Sci. U. S. A.* **97**, 13494–13499, (2000).

[3] P. F. McMillan, Chemistry at high pressure, *Chem. Soc. Rev.* **35**, 855–857, (2006).

[4] M. Choukroun and O. Grasset, Thermodynamic data and modeling of the water and ammonia-water phase diagrams up to 2.2 GPa for planetary geophysics, *J. Chem. Phys.* **133**, (2010).

[5] K. Kurpiewska and K. Lewinski, High pressure macromolecular crystallography for structural biology: a review, *Cent. Eur. J. Biol.* **5**, 531–542, (2010).

[6] R. J. Hemley and N. W. Ashcroft, The revealing role of pressure in the condensed matter sciences, *Phys. Today* **51**, 26–32, (1998).

[7] S.-H. Shim, A. Bengtson, D. Morgan, W. Sturhahn, K. Catalli, J. Zhao, M. Lerche and V. Prakapenka, Electronic and magnetic structures of the postperovskite-type Fe_2O_3 and implications for planetary magnetic records and deep interiors, *Proc. Natl. Acad. Sci. U. S. A.* **106**, 5508–5512, (2009).

[8] V. L. Solozhenko, O. O. Kurakevych, D. Andrault, Y. Le Godec and M. Mezouar, Ultimate metastable solubility of boron in diamond: synthesis of superhard diamondlike BC_5, *Phys. Rev. Lett.* **102**, 015506, (2009).

[9] E. Y. Zarechnaya, L. Dubrovinsky, N. Dubrovinskaia, Y. Filinchuk, D. Chernyshov, V. Dmitriev, N. Miyajima, A. El Goresy, H. F. Braun, S. Van Smaalen, I. Kantor, A. Kantor, V. Prakapenka, M. Hanfland, A. S. Mikhaylushkin, I. A. Abrikosov and S. I. Simak, Superhard semiconducting optically transparent high pressure phase of boron, *Phys. Rev. Lett.* **102**, 185501, (2009).

[10] M. Somayazulu, P. Dera, A. F. Goncharov, S. A. Gramsch, P. Liermann, W. Yang, Z. Liu, H.-k. Mao and R. J. Hemley, Pressure-induced bonding and compound formation in xenon–hydrogen solids, *Nat. Chem.* **2**, 50–53, (2010).

[11] D. D. Horikawa, K. I. Iwata, K. Kawai, S. Koseki, T. Okuda and K. Yamamoto, High hydrostatic pressure tolerance of four different anhydrobiotic animal species, *Zool. Sci.* **26**, 238–242, (2009).

[12] W. Grochala, R. Hoffmann, J. Feng and N. W. Ashcroft, The chemical imagination at work in very tight places, *Angew. Chem.-Int. Edit.* **46**, 3620–3642, (2007).

[13] W. A. Bassett, Diamond anvil cell, 50th birthday, *High Pressure Res.* **29**, CP5-186, (2009).

[14] A. Jayaraman, Diamond anvil cell and high-pressure physical investigations, *Rev. Mod. Phys.* **55**, 65–108, (1983).

[15] G. Y. Bussod, T. Katsura and D. C. Rubie, The large volume multi-anvil press as a high P-T deformation apparatus, *Pure Appl. Geophys.* **141**, 579–599, (1993).

[16] G. Shen, P. Chow, Y. Xiao, S. Sinogeikin, Y. Meng, W. Yang, H.-P. Liermann, O. Shebanova, E. Rod, A. Bommannavar and H.-K. Mao, HPCAT: an integrated high-pressure synchrotron facility at the Advanced Photon Source, *High Pressure Res.* **28**, 145–162, (2008).

[17] A. F. Goncharov and R. J. Hemley, Probing hydrogen-rich molecular systems at high pressures and temperatures, *Chem. Soc. Rev.* **35**, 899–907, (2006).

[18] W. Paszkowicz, High-pressure powder X-ray diffraction at the turn of the century, *Nucl. Instrum. Methods Phys. Res., Sect. B* **198**, 142–182, (2002).

[19] B. Buras, J. S. Olsen, L. Gerward, G. Will and E. Hinze, X-ray energy-dispersive diffractometry using synchrotron radiation, *J. Appl. Crystallogr.* **10**, 431–438, (1977).

[20] S. K. Lee, J.-F. Lin, Y. Q. Cai, N. Hiraoka, P. J. Eng, T. Okuchi, H.-k. Mao, Y. Meng, M. Y. Hu, P. Chow, J. Shu, B. Li, H. Fukui, B. H. Lee, H. N. Kim and C.-S. Yoo, X-ray Raman scattering study of $MgSiO_3$ glass at high pressure: Implication for triclustered $MgSiO_3$ melt in Earth's mantle, *Proc. Natl. Acad. Sci. U. S. A.* **105**, 7925–7929, (2008).

[21] J.-P. Rueff and A. Shukla, Inelastic X-ray scattering by electronic excitations under high pressure, *Rev. Mod. Phys.* **82**, 847–896, (2010).

[22] J. Liu, X. D. Li and Y. C. Li, The present status of high-pressure research at Beijing Synchrotron Radiation Facility, *J. Phys.: Condens. Matter* **14**, 10505, (2002).

[23] J. Guo, X.-J. Chen, J. Dai, C. Zhang, J. Guo, X. Chen, Q. Wu, D. Gu, P. Gao, L. Yang, K. Yang, X. Dai, H.-k. Mao, L. Sun and Z. Zhao, Pressure-driven quantum criticality in iron-selenide superconductors, *Phys. Rev. Lett.* **108**, 197001, (2012).

[24] C.-M. Lin, K.-T. Wu, T.-L. Hung, H.-S. Sheu, M.-H. Tsai, J.-F. Lee and J.-J. Lee, Phase transitions in Lu_2O_3 under high pressure, *Solid State Commun.* **150**, 1564–1569, (2010).

[25] Y. Ohishi, N. Hirao, N. Sata, K. Hirose and M. Takata, Highly intense monochromatic X-ray diffraction facility for high-pressure research at SPring-8, *High Pressure Res.* **28**, 163–173, (2008).

[26] O. Shimomura, K. Tsuji and N. Hamaya, The Spring-8 project and the high pressure group in Japan, *High Pressure Res.* **8**, 703–710, (1992).

[27] W. Utsumi, K. Funakoshi, Y. Katayama, M. Yamakata, T. Okada and O. Shimomura, High-pressure science with a multi-anvil apparatus at SPring-8, *J. Phys.: Condens. Matter* **14**, 10497–10504, (2002).

[28] T. Kikegawa, High pressure studies at the photon factory, *High Pressure Res.* **8**, 631–637, (1992).

[29] K. K. Pandey, G. Nandini, A. K. Mishra and M. S. Surinder, High pressure phase transition in Nd_2O_3, *J. Phys. Conf.* **377**, 012006, (2012).

[30] K. H. Cho, C. E. Seo, Y. S. Choi, Y. H. Ko and K. J. Kim, Effect of pressure on electric generation of PZT(30/70) and PZT(52/48) ceramics near phase transition pressure, *J. Eur. Ceram. Soc.* **32**, 457–463, (2012).

[31] D. Häuserma and M. Hanfland, Optics and beamlines for high-pressure research at the european synchrotron radiation facility, *High Pressure Res.* **14**, 223–234, (1996).

[32] A. Kvick, High-pressure facilities at the ESRF, *High Pressure Res.* **8**, 717–722, (1992).

[33] E. Schultz, M. Mezouar, W. Crichton, S. Bauchau, G. Blattmann, D. Andrault, G. Fiquet, R. Boehler, N. Rambert, B. Sitaud and P. Loubeyre, Double-sided laser heating system for in situ high pressure–high temperature monochromatic x-ray diffraction at the ESRF, *High Pressure Res.* **25**, 71–83, (2005).

[34] A. Rothkirch, G. D. Gatta, M. Meyer, S. Merkel, M. Merlini and H. P. Liermann, Single-crystal diffraction at the Extreme Conditions beamline P02.2: procedure for collecting and analyzing high-pressure single-crystal data, *J. Synchrotron Radiat.* **20**, 711–720, (2013).
[35] A. Dewaele, P. Loubeyre, P. Dumas and M. Mezouar, Oxygen impurities reduce the metallization pressure of xenon, *Phys. Rev. B* **86**, 014103, (2012).
[36] F. Baudelet, Q. Kong, L. Nataf, J. D. Cafun, A. Congeduti, A. Monza, S. Chagnot and J. P. Itié, ODE: a new beam line for high-pressure XAS and XMCD studies at SOLEIL, *High Pressure Res.* **31**, 136–139, (2011).
[37] M. I. McMahon, E. Gregoryanz, L. F. Lundegaard, I. Loa, C. Guillaume, R. J. Nelmes, A. K. Kleppe, M. Amboage, H. Wilhelm and A. P. Jephcoat, Structure of sodium above 100 GPa by single-crystal X-ray diffraction, *Proc. Natl. Acad. Sci. U. S. A.* **104**, 17297–17299, (2007).
[38] T. D. Bennett, P. Simoncic, S. A. Moggach, F. Gozzo, P. Macchi, D. A. Keen, J.-C. Tan and A. K. Cheetham, Reversible pressure-induced amorphization of a zeolitic imidazolate framework (ZIF-4), *Chem. Commun.* **47**, 7983–7985, (2011).
[39] L. Baldassarre, A. Perucchi, P. Postorino and S. Lupi, Infrared studies under pressure at the infrared beamline SISSI at ELETTRA, *High Pressure Res.* **29**, 639–643, (2009).
[40] M. Piccinini, M. Cestelli Guidi, A. Marcelli, P. Calvani, E. Burattini, A. Nucara, P. Postorino, A. Sacchetti, E. Arcangeletti, E. Sheregii, J. Polit and A. Kisiel, Far-infrared synchrotron radiation spectroscopy of solids in normal and extreme conditions, *Phys. Status Solidi C* **2**, 236–239, (2005).
[41] S. V. Rashchenko, A. Y. Likhacheva, A. D. Chanyshev and A. I. Ancharov, The application of in situ X-ray diffraction for the study of mineral reactions: The formation of lawsonite at 400°C and 25 kbar, *J. Struct. Chem.* **53**, 43–46, (2012).
[42] J. A. Sans, F. J. Manjón, C. Popescu, A. Muñoz, P. Rodríguez-Hernández, J. L. Jordá and F. Rey, Arsenolite: a quasi-hydrostacic solid pressure-transmitting medium, *J. Phys. Condens. Matter* **28**, 475403, (2016).
[43] G. Shen, Y. Wang, V. Prakapenka, C. J. Benmore, E. E. Alp, Y. Ding and W. Yang, High-pressure research at the Advanced Photon Source, *Sync. Rad. News* **23**, 32–38, (2010).
[44] Y. Wang, In High-Pressure Crystallography, Boldyreva, E., Dera, P., Eds.; Springer Netherlands: 2010; Vol. 0, p 81–96.
[45] L. Ehm, M. Vaughan, T. Duffy, Z. Liu, L. Wang, B. Li, D. Weidner, Z. Chen, S. Ghose and Z. Zhong, High-pressure research at the National Synchrotron Light Source, *Sync. Rad. News* **23**, 24–30, (2010).
[46] R. J. Hemley, A. F. Goncharov, R. Lu, V. V. Struzhkin, M. Li and H. K. Mao, High-pressure synchrotron infrared spectroscopy at the national synchrotron light source, *Nouv Cim D* **20**, 539–551, (1998).
[47] K. Brister, High pressure research at CHESS, *Nucl. Instrum. Methods Phys. Res., Sect. A* **319**, 333–339, (1992).

[48] M. Kunz, A. A. MacDowell, W. A. Caldwell, D. Cambie, R. S. Celestre, E. E. Domning, R. M. Duarte, A. E. Gleason, J. M. Glossinger, N. Kelez, D. W. Plate, T. Yu, J. M. Zaug, H. A. Padmore, R. Jeanloz, A. P. Alivisatos and S. M. Clark, A beamline for high-pressure studies at the Advanced Light Source with a superconducting bending magnet as the source, *J. Synchrotron Radiat.* **12**, 650–658, (2005).

[49] J. S. Tse, D. D. Klug, S. Desgreniers, J. S. Smith, R. Flacau, Z. Liu, J. Hu, N. Chen and D. T. Jiang, Structural phase transition in CaH_2 at high pressures, *Phys. Rev. B* **75**, 134108, (2007).

[50] F. F. Ferreira, H. P. S. Correa, M. T. D. Orlando, J. L. Passamai Jr, C. G. P. Orlando, I. P. Cavalcante, F. Garcia, E. Tamura, L. G. Martinez and J. L. Rossi, Pressure study of monoclinic ReO_2 up to 1.2 GPa using X-ray absorption spectroscopy and X-ray diffraction, *J. Synchrotron Radiat.* **16**, 48–56, (2008).

[51] K. S. Wallwork, B. J. Kennedy and D. Wang, The high resolution powder diffraction beamline for the Australian synchrotron, *AIP Conf. Proc.* **879**, 879–882, (2007).

[52] R. J. Hemley, Percy W. Bridgman's second century, *High Pressure Res.* **30**, 581–619, (2010).

[53] A. W. Lawson and T.-Y. Tang, A diamond bomb for obtaining powder pictures at high pressures, *Rev. Sci. Instrum.* **21**, 815–815, (1950).

[54] G. J. Piermarini and S. Block, Ultrahigh pressure diamond-anvil cell and several semiconductor phase transition pressures in relation to the fixed point pressure scale, *Rev. Sci. Instrum.* **46**, 973–979, (1975).

[55] H. K. Mao and P. M. Bell, Equations of state of MgO and ε Fe under static pressure conditions, *J. Geophys. Res.* **84**, 4533–4536, (1979).

[56] L. Merrill and W. A. Bassett, Miniature diamond anvil pressure cell for single crystal X-ray diffraction studies, *Rev. Sci. Instrum.* **45**, 290–294, (1974).

[57] M. I. Eremets and Y. A. Timofeev, Miniature diamond anvil cell: Incorporating a new design for anvil alignment, *Rev. Sci. Instrum.* **63**, 3123–3126, (1992).

[58] X. Ji-an, M. Ho-kwang and J. H. Russell, The gem anvil cell: high-pressure behaviour of diamond and related materials, *J. Phys.: Condens. Matter* **14**, 11549, (2002).

[59] H.-K. Mao and R. J. Hemley, Energy dispersive X-ray diffraction of micro-crystals at ultrahigh pressures, *High Pressure Res.* **14**, 257–267, (1996).

[60] L. Wang, W. Yang, Y. Xiao, B. Liu, P. Chow, G. Shen, W. L. Mao and H.-k. Mao, Application of a new composite cubic-boron nitride gasket assembly for high pressure inelastic X-ray scattering studies of carbon related materials, *Rev. Sci. Instrum.* **82**, 073902-4, (2011).

[61] G. J. Piermarini, R. A. Forman and S. Block, Viscosity measurements in the diamond anvil pressure cell, *Rev. Sci. Instrum.* **49**, 1061–1066, (1978).

[62] J. H. Eggert, K. A. Goettel and I. F. Silvera, Ruby at high pressure. I. Optical line shifts to 156 GPa, *Phys. Rev. B* **40**, 5724–5732, (1989).

[63] C.-S. Zha, W. A. Bassett and S.-H. Shim, Rhenium, an *in situ* pressure calibrant for internally heated diamond anvil cells, *Rev. Sci. Instrum.* **75**, 2409–2418, (2004).
[64] L. Ming, M. Manghnani, J. Balogh, S. Qadri, E. Skelton and J. Jamieson, Gold as a reliable internal pressure calibrant at high temperatures, *J. Appl. Phys.* **54**, 4390–4397, (1983).
[65] D. M. Adams, A. G. Christy, J. Haines and S. M. Clark, Second-order phase transition in PbO and SnO at high pressure: Implications for the litharge-massicot phase transformation, *Phys. Rev. B* **46**, 11358, (1992).
[66] A. Zinn, D. Schiferl and M. Nicol, Raman spectroscopy and melting of nitrogen between 290 and 900 K and 2.3 and 18 GPa, *J. Chem. Phys.* **87**, 1267, (1987).
[67] R. Miletich, D. R. Allan and W. F. Kuhs, High-pressure single-crystal techniques, *Rev. Mineral. Geochem.* **41**, 445–519, (2000).
[68] H. Mao, P. Bell and C. Hadidiacos, Experimental phase relations of iron to 360 kbar, 1400°C, determined in an internally heated diamond-anvil apparatus, (1987).
[69] S. Guoyin, W. Lin, F. Richard, M. Ho-kwang and J. H. Russell, A portable laser heating microscope for high pressure research, *J. Phys. Conf.* **215**, 012191, (2010).
[70] J. C. Chervin, B. Canny, M. Gauthier and P. Pruzan, Micro-Raman at variable low-temperature and very high pressure, *Rev. Sci. Instrum.* **64**, 203–206, (1993).
[71] J. D. Thompson, Low-temperature pressure variations in a self-clamping pressure cell, *Rev. Sci. Instrum.* **55**, 231–234, (1984).
[72] W. Utsumi, K. Funakoshi, S. Urakawa, M. Yamakata, K. Tsuji, H. Konishi and O. Shimomura, SPring-8 beamlines for high pressure science with multi-anvil apparatus, *Rev. High Pressure Sci. Technol.* **7**, 1484–1486, (1998).
[73] T. Katsura, K.-i. Funakoshi, A. Kubo, N. Nishiyama, Y. Tange, Y.-i. Sueda, T. Kubo and W. Utsumi, A large-volume high-pressure and high-temperature apparatus for in situ X-ray observation, 'SPEED-Mk.II', *Phys. Earth Planet. Inter.* **143–144**, 497–506, (2004).
[74] Y. Fei, J. Van Orman, J. Li, W. van Westrenen, C. Sanloup, W. Minarik, K. Hirose, T. Komabayashi, M. Walter and K. Funakoshi, Experimentally determined postspinel transformation boundary in Mg_2SiO_4 using MgO as an internal pressure standard and its geophysical implications, *J. Geophys. Res.* **109**, B02305, (2004).
[75] S. Ono, T. Katsura, E. Ito, M. Kanzaki, A. Yoneda, M. J. Walter, S. Urakawa, W. Utsumi and K. Funakoshi, In situ observation of ilmenite-perovskite phase transition in $MgSiO_3$ using synchrotron radiation, *Geophys. Res. Lett.* **28**, 835–838, (2001).
[76] B. Li, J. Kung, T. Uchida and Y. Wang, Pressure calibration to 20 GPa by simultaneous use of ultrasonic and X-ray techniques, *J. Appl. Phys.* **98**, 013521-5, (2005).

[77] H. Terasaki, S. Urakawa, K.-I. Funakoshi, Y. Wang, Y. Shibazaki, T. Sanehira, Y. Ueda and E. Ohtani, Interfacial tension measurement of Ni-S liquid using high-pressure X-ray micro-tomography, *High Pressure Res.* **28**, 327–334, (2008).

[78] M. D. Rutter, R. A. Secco, H. Liu, T. Uchida, M. L. Rivers, S. R. Sutton and Y. Wang, Viscosity of liquid Fe at high pressure, *Phys. Rev. B* **66**, 060102, (2002).

[79] B. T. Poe, C. Romano, C. Liebske, D. C. Rubie, H. Terasaki, A. Suzuki and K. Funakoshi, High-temperature viscosity measurements of hydrous albite liquid using *in situ* falling-sphere viscometry at 2.5 GPa, *Chem. Geol.* **229**, 2–9, (2006).

[80] Y. Katayama and K. Tsuji, X-ray structural studies on elemental liquids under high pressures, *J. Phys.: Condens. Matter* **15**, 6085, (2003).

[81] Y. Ma, M. Eremets, A. R. Oganov, Y. Xie, I. Trojan, S. Medvedev, A. O. Lyakhov, M. Valle and V. Prakapenka, Transparent dense sodium, *Nature* **458**, 182–185, (2009).

[82] B. Li, Y. Ding, D. Y. Kim, R. Ahuja, G. Zou and H. K. Mao, Rhodium dihydride (RhH_2) with high volumetric hydrogen density, *Proc. Natl. Acad. Sci. U. S. A.* **108**, 18618–18621, (2011).

[83] Q. Zeng, H. Sheng, Y. Ding, L. Wang, W. Yang, J. Z. Jiang, W. L. Mao and H. K. Mao, Long-range topological order in metallic glass, *Science* **332**, 1404, (2011).

[84] L. Sun, X. J. Chen, J. Guo, P. Gao, Q. Z. Huang, H. Wang, M. Fang, X. Chen, G. Chen and Q. Wu, Re-emerging superconductivity at 48 kelvin in iron chalcogenides, *Nature* (2012).

[85] S. R. Li, Q. Li, K. Wang, X. Tan, M. Zhou, B. Li, B. Liu, G. Zou, and B. Zou, Pressure-induced phase transition in guanidinium perchlorate: A supramolecular structure directed by hydrogen bonding and electrostatic interactions, *J. Phys. Chem. B* **115**, 11816–11822, (2011).

[86] Q. Li, S. R. Li, K. Wang, W. Li, J. Liu, B. Liu, G. Zou and B. Zou, Compression studies of face-to-face p-stacking interaction in sodium squarate salts: $Na_2C_4O_4$ and $Na_2C_4O_4 \cdot 3H_2O$, *J. Chem. Phys.* **137**, 184905–184910, (2012).

[87] S. R. Li, Q. Li, J. Zhou, R. Wang, Z. Jiang, K. Wang, D. Xu, J. Liu, B. Liu, G. Zou and B. Zou, Effect of high pressure on the typical supramolecular structure of guanidinium methanesulfonate, *J. Phys. Chem. B* **116**, 3092–3098, (2012).

[88] X. Tan, K. Wang, S. R. Li, H. Yuan, T. Yan, J. Liu, K. Yang, B. Liu, G. Zou and B. Zou, Exploration of the pyrazinamide polymorphism at high pressure, *J. Phys. Chem. B* **116**, 14441–14450, (2012).

[89] T. T. Yan, S. R. Li, K. Wang, X. Tan, Z. Jiang, K. Yang, B. Liu, G. Zou and B. Zou, Pressure-induced phase transition in N-H...O hydrogen-bonded molecular crystal oxamide, *J. Phys. Chem. B* **116**, 9796–9802, (2012).

[90] Q. Li, S. R. Li, K. Wang, X. Li, J. Liu, B. Liu, G. Zou and B. Zou, Pressure-induced isosymmetric phase transition in sulfamic acid: A combined Raman and X-ray diffraction study, *J. Chem. Phys.* **138**, 214505, (2013).
[91] S. R. Li, Q. Li, K. Wang, M. Zhou, X. Huang, J. Liu, K. Yang, B. Liu, T. Cui, G. Zou and B. Zou, Pressure-induced irreversible phase transition in the energetic material urea nitrate: Combined Raman scattering and X-ray diffraction study, *J. Phys. Chem. B* **117**, 152–159, (2013).
[92] K. Wang, D. F. Duan, R. Wang, A. Lin, Q. Cui, B. Liu, T. Cui, B. Zou, X. Zhang, J. Hu, G. Zou and H.-K. Mao, Stability of hydrogen-bonded supramolecular architecture under high pressure conditions: Pressure-induced amorphization in melamine-boric acid adduct, *Langmuir* **25**, 4787–4791, (2009).
[93] S. R. Li, K. Wang, M. Zhou, Q. Li, B. Liu, G. Zou and B. Zou, Pressure-induced phase transitions in ammonium squarate: A supramolecular structure based on hydrogen-bonding and pi-stacking interactions, *J. Phys. Chem. B* **115**, 8981–8988, (2011).
[94] R. Wang, S. R. Li, K. Wang, D. Duan, L. Tang, T. Cui, B. Liu, Q. Cui, J. Liu, B. Zou and G. Zou, Pressure-induced phase transition in hydrogen-bonded supramolecular structure: Guanidinium nitrate, *J. Phys. Chem. B* **114**, 6765–6769, (2010).
[95] K. Wang, D. F. Duan, R. Wang, D. Liu, L. Tang, T. Cui, B. Liu, Q. Cui, J. Liu, B. Zou and G. Zou, Pressure-induced phase transition in hydrogen-bonded supramolecular adduct formed by cyanuric acid and melamine, *J. Phys. Chem. B* **113**, 14719–14724, (2009).
[96] R. J. Hemley, H. K. Mao and V. V. Struzhkin, Synchrotron radiation and high pressure: new light on materials under extreme conditions, *J. Synchrotron Radiat.* **12**, 135–154, (2005).
[97] R. Boehler, K. De Hantsetters, New anvil designs in diamond-cells, *High Pressure Res.* **24**, 391–396, (2004).
[98] H. Mao, A. Jephcoat, R. Hemley, L. Finger, C. Zha, R. Hazen and D. Cox, Synchrotron X-ray diffraction measurements of single-crystal hydrogen to 26.5 gigapascals, *Science* **239**, 1131–1134, (1988).
[99] H. Mao, J. Shu, Y. Fei, J. Hu and R. J. Hemley, The wüstite enigma, *Phys. Earth Planet. Inter.* **96**, 135–145, (1996).
[100] S. A. Moggach, D. R. Allan, S. Parsons, L. Sawyer and J. E. Warren, The effect of pressure on the crystal structure of hexagonal L-cystine, *J. Synchrotron Radiat.* **12**, 598–607, (2005).
[101] K. J. Gagnon, C. M. Beavers and A. Clearfield, MOFs under pressure: the reversible compression of a single crystal, *J. Am. Chem. Soc.* **135**, 1252–1255, (2013).
[102] R. Fourme, E. Girard, R. Kahn, A. C. Dhaussy, M. Mezouar, N. Colloc'h and I. Ascone, High-pressure macromolecular crystallography (HPMX): status and prospects, *Biochim. Biophys. Acta* **1764**, 384–390, (2006).
[103] E. Girard, T. Prangé, A. C. Dhaussy, E. Migianu-Griffoni, M. Lecouvey, J. C. Chervin, M. Mezouar, R. Kahn and R. Fourme, Adaptation of

the base-paired double-helix molecular architecture to extreme pressure, *Nucleic Acids Res.* **35**, 4800–4808, (2007).

[104] L. Zhang, Y. Meng, P. Dera, W. Yang, W. L. Mao and H.-k. Mao, Single-crystal structure determination of (Mg,Fe)SiO_3 postperovskite, *Proc. Natl. Acad. Sci. U. S. A.* **110**, 6292–6295, (2013).

[105] L. Wang, Y. Ding, U. Patel, W. Yang, Z. Xiao, Z. Cai, W. L. Mao, H.-k. Mao, Studying single nanocrystals under high pressure using an X-ray nanoprobe, *Rev. Sci. Instrum.* **82**, -, (2011).

[106] H. S. Yuan, K. Wang, S. R. Li, X. Tan, Q. Li, T. T. Yan, B. Y. Cheng, K. Yang, B. B. Liu, G. T. Zou and B. Zou, Direct zircon-to-scheelite structural transformation in YPO4 and YPO4:$Eu3^{+}$ nanoparticles under high pressure, *J. Phys. Chem. C* **116**, 24837–24844, (2012).

[107] H. Yuan, K. Wang, S. Li, X. Tan, Q. Li, T. Yan, K. Yang, J. Liu, B. Liu, G. Zou and B. Zou, High-pressure stability and compressibility of zircon-type $YV_{1-x}P_xO_4$:Eu^{3+} solid-solution nanoparticles: An X-ray diffraction and Raman spectroscopy study, *J. Phys. Chem. C* **117**, 18603–18612, (2013).

[108] P. R. Couchman and W. A. Jesser, Thermodynamic theory of size dependence of melting temperature in metals, *Nature* **269**, 481–483, (1977).

[109] L. Wang, W. Yang, Y. Ding, Y. Ren, S. Xiao, B. Liu, S. V. Sinogeikin, Y. Meng, D. J. Gosztola, G. Shen, R. J. Hemley, W. L. Mao and H.-k. Mao, Size-Dependent Amorphization of Nanoscale Y_2O_3 at High Pressure, *Phys. Rev. Lett.* **105**, 095701, (2010).

[110] Z. Quan, Y. Wang, I.-T. Bae, W. S. Loc, C. Wang, Z. Wang and J. Fang, Reversal of hall-petch effect in structural stability of PbTe nanocrystals and associated variation of phase transformation, *Nano Lett.* **11**, 5531–5536, (2011).

[111] S. H. Tolbert and A. P. Alivisatos, Size dependence of a first order solid-solid phase transition: The wurtzite to rock salt transformation in CdSe nanocrystals, *Science* **265**, 373–376, (1994).

[112] X. Yang, Y. Wang, K. Wang, Y. Sui, M. Zhang, B. Li, Y. Ma, B. Liu, G. Zou and B. Zou, Polymorphism and formation mechanism of nanobipods in manganese sulfide nanocrystals induced by temperature or pressure, *J. Phys. Chem. C* **116**, 3292–3297, (2012).

[113] H. Karzel, W. Potzel, M. Köfferlein, W. Schiessl, M. Steiner, U. Hiller, G. M. Kalvius, D. W. Mitchell, T. P. Das, P. Blaha, K. Schwarz and M. P. Pasternak, Lattice dynamics and hyperfine interactions in ZnO and ZnSe at high external pressures, *Phys. Rev. B* **53**, 11425–11438, (1996).

[114] N. Xiao, L. Zhu, K. Wang, Q. Dai, Y. Wang, S. Li, Y. Sui, Y. Ma, J. Liu, B. Liu, G. Zou and B. Zou, Synthesis and high pressure transformation of metastable wurtzite structured $CuGaS_2$ nanocrystals, *Nanoscale* **4**, 7443–7447, (2012).

[115] A. San-Miguel, Nanomaterials under high-pressure, *Chem. Soc. Rev.* **35**, 876–889, (2006).

[116] S. Kawasaki, Y. Matsuoka, T. Yokomae, Y. Nojima, F. Okino, H. Touhara and H. Kataura, Effect of a liquid pressure-transmitting medium on the high pressure behavior of open- and closed-end single-walled carbon nanotubes and of C_{60}-peapods, *Phys. Status Solidi B* **241**, 3512–3516, (2004).

[117] Y. Zou, B. Liu, L. Wang, D. Liu, S. Yu, P. Wang, T. Wang, M. Yao, Q. Li, B. Zou, T. Cui, G. Zou, T. Wågberg, B. Sundqvist and H.-K. Mao, Rotational dynamics of confined C_{60} from near-infrared Raman studies under high pressure, *Proc. Natl. Acad. Sci. U. S. A.* **106**, 22135–22138, (2009).

[118] L. Wang, B. Liu, H. Li, W. Yang, Y. Ding, S. V. Sinogeikin, Y. Meng, Z. Liu, X. C. Zeng and W. L. Mao, Long-range ordered carbon clusters: A crystalline material with amorphous building blocks, *Science* **337**, 825–828, (2012).

[119] L. V. Azároff, Theory of extended fine structure of X-ray absorption edges, *Rev. Mod. Phys.* **35**, 1012–1021, (1963).

[120] B. K. Teo, Chemical applications of extended X-ray absorption fine structure (EXAFS) spectroscopy, *Acc. Chem. Res.* **13**, 412–419, (1980).

[121] E. Stern, Structure determination by X-ray absorption, *Contemp. Phys.* **19**, 289–310, (1978).

[122] R. Ingalls, G. A. Garcia and E. A. Stern, X-ray absorption at high pressure, *Phys. Rev. Lett.* **40**, 334–336, (1978).

[123] R. Ingalls, E. Crozier, J. Whitmore, A. Seary and J. Tranquada, Extended X-ray absorption fine structure of NaBr and Ge at high pressure, *J. Appl. Phys.* **51**, 3158–3163, (1980).

[124] J. M. Tranquada and R. Ingalls, X-ray-absorption study of CuBr at high pressure, *Phys. Rev. B* **34**, 4267–4277, (1986).

[125] J. Freund, R. Ingalls and E. Crozier, Extended X-ray absorption fine structure study of copper under high pressure, *Phys. Rev. B* **39**, 12537, (1989).

[126] A. I. Frenkel, F. M. Wang, S. Kelly, R. Ingalls, D. Haskel, E. A. Stern and Y. Yacoby, Local structural changes in $KNbO_3$ under high pressure, *Phys. Rev. B* **56**, 10869–10877, (1997).

[127] S. Kelly, R. Ingalls, F. Wang, B. Ravel and D. Haskel, X-ray-absorption fine-structure study of the B1-to-B2 phase transition in RbCl, *Phys. Rev. B* **57**, 7543–7550, (1998).

[128] M. P. Pasternak, R. D. Taylor, M. B. Kruger, R. Jeanloz, J. P. Itie and A. Polian, Pressure-induced amorphization of GeI_4 molecular-crystals, *Phys. Rev. Lett.* **72**, 2733–2736, (1994).

[129] C. L. Bull, P. F. McMillan, J. P. Itie and A. Polian, X-ray absorption and EXAFS studies of Ge coordination and bonding in high-pressure nitrides: beta-Ge_3N_4(phenacite) and gamma-Ge_3N_4 (spinel), *Phys. Status Solidi A* **201**, 909–916, (2004).

[130] C. E. M. Campos, J. C. de Lima, T. A. Grandi, J. P. Itie, A. Polian, J. C. Chervin, P. S. Pizani and E. B. Saitovich, The pressure-induced phase transition of mechanically alloyed nanocrystalline GaSb, *J. Phys.: Condens. Matter* **20**, -, (2008).

[131] L. Comez, A. Di Cicco, J. P. Itie and A. Polian, High-pressure and high-temperature X-ray absorption study of liquid and solid gallium, *Phys. Rev. B* **65**, -, (2002).
[132] Y. Katayama, O. Shimomura and K. Tsuji, EXAFS study on liquid selenium and liquid tellurium under high pressure, *J. Non-Cryst. Solids* **250**, 537–541, (1999).
[133] M. Baldini, G. Aquilanti, H. k. Mao, W. Yang, G. Shen, S. Pascarelli and W. L. Mao, High-pressure EXAFS study of vitreous GeO_2 up to 44 GPa, *Phys. Rev. B* **81**, 024201, (2010).
[134] Q.-s. Zeng, Y. Ding, W. L. Mao, W. Yang, S. V. Sinogeikin, J. Shu, H.-k. Mao and J. Z. Jiang, Origin of pressure-induced polyamorphism in $Ce_{75}Al_{25}$ metallic glass, *Phys. Rev. Lett.* **104**, 105702, (2010).
[135] T. Varga, A. P. Wilkinson, A. C. Jupe, C. Lind, W. A. Bassett and C.-S. Zha, Pressure-induced amorphization of cubic ZrW_2O_8 studied *in situ* and ex situ by synchrotron X-ray diffraction and absorption, *Phys. Rev. B* **72**, 024117, (2005).
[136] L.-R. Zheng, R.-Z. Che, J. Liu, Y.-H. Du, Y.-L. Zhou and T.-D. Hu, A new cell for X-ray absorption spectroscopy study under high pressure, *Chinese Phys. C* **33**, 701, (2009).
[137] M. Baldini, W. Yang, G. Aquilanti, L. Zhang, Y. Ding, S. Pascarelli and W. L. Mao, High-pressure EXAFS measurements of crystalline Ge using nanocrystalline diamond anvils, *Phys. Rev. B* **84**, 014111, (2011).
[138] N. Ishimatsu, K. Matsumoto, H. Maruyama, N. Kawamura, M. Mizumaki, H. Sumiya and T. Irifune, Glitch-free X-ray absorption spectrum under high pressure obtained using nano-polycrystalline diamond anvils, *J. Synchrotron Radiat.* **19**, (2012).
[139] A. H. Compton, A quantum theory of the scattering of X-rays by light elements, *Phys. Rev.* **21**, 483, (1923).
[140] W. Schülke, H. Nagasawa and S. Mourikis, Dynamic structure factor of electrons in Li by inelastic synchrotron X-ray scattering, *Phys. Rev. Lett.* **52**, 2065–2068, (1984).
[141] C. Sternemann, A. Kaprolat and W. Schülke, Effect of thermal vibration and the solid-liquid phase transition on electron dynamics: An inelastic X-ray scattering study on Al, *Phys. Rev. B* **57**, 622–626, (1998).
[142] W. A. Caliebe, J. A. Soininen, E. L. Shirley, C. C. Kao, K. Hämäläinen, Dynamic structure factor of diamond and LiF measured using inelastic X-ray scattering, *Phys. Rev. Lett.* **84**, 3907–3910, (2000).
[143] K. Hämäläinen, S. Galambosi, J. A. Soininen, E. L. Shirley, J. P. Rueff and A. Shukla, Momentum dependence of fluorine K-edge core exciton in LiF, *Phys. Rev. B* **65**, 155111, (2002).
[144] M. Hasan, E. Isaacs, Z. X. Shen, L. Miller, K. Tsutsui, T. Tohyama, and S. Maekawa, Electronic structure of Mott insulators studied by inelastic X-ray scattering, *Science* **288**, 1811–1814, (2000).
[145] Y. Q. Cai, P. C. Chow, O. D. Restrepo, Y. Takano, K. Togano, H. Kito, H. Ishii, C. C. Chen, K. S. Liang, C. T. Chen, S. Tsuda, S. Shin, C. C. Kao,

W. Ku and A. G. Eguiluz, Low-energy charge-density excitations in MgB2: Striking interplay between single-particle and collective behavior for large momenta, *Phys. Rev. Lett.* **97**, 176402, (2006).

[146] F. Sette, G. Ruocco, M. Krisch, U. Bergmann, C. Masciovecchio, V. Mazzacurati, G. Signorelli, and R. Verbeni, Collective dynamics in water by high energy resolution inelastic X-ray scattering, *Phys. Rev. Lett.* **75**, 850–853, (1995).

[147] T. E. Angelini, R. Golestanian, R. H. Coridan, J. C. Butler, A. Beraud, M. Krisch, H. Sinn, K. S. Schweizer and G. C. L. Wong, Counterions between charged polymers exhibit liquid-like organization and dynamics, *Proc. Natl. Acad. Sci. U. S. A.* **103**, 7962–7967, (2006).

[148] J. Wong, M. Krisch, D. L. Farber, F. Occelli, A. J. Schwartz, T. C. Chiang, M. Wall, C. Boro and R. Xu, Phonon dispersions of fcc δ-plutonium-gallium by inelastic X-ray scattering, *Science* **301**, 1078–1080, (2003).

[149] A. Kotani, Resonant inelastic X-ray scattering in d and f electron systems, *Eur. Phys. J. B* **47**, 3–27, (2005).

[150] L. J. P. Ament, M. van Veenendaal, T. P. Devereaux, J. P. Hill, J. van den Brink, Resonant inelastic X-ray scattering studies of elementary excitations, *Rev. Mod. Phys.* **83**, 705–767, (2011).

[151] N. Schell, R. O. Simmons, A. Kaprolat, W. Schülke and E. Burkel, Electronic excitations in hcp ^{4}He at 61.5 MPa and 4.3 K studied by inelastic X-ray scattering spectroscopy, *Phys. Rev. Lett.* **74**, 2535–2538, (1995).

[152] H. Mao, C. Kao and R. J. Hemley, Inelastic X-ray scattering at ultrahigh pressures, *J. Phys.: Condens. Matter*, **13**, 7847, (2001).

[153] F. Occelli, M. Krisch, P. Loubeyre, F. Sette, R. Le Toullec, C. Masciovecchio and J. P. Rueff, Phonon dispersion curves in an argon single crystal at high pressure by inelastic X-ray scattering, *Phys. Rev. B* **63**, 224306, (2001).

[154] W. L. Mao, H. Mao, P. J. Eng, T. P. Trainor, M. Newville, C. Kao, D. L. Heinz, J. Shu, Y. Meng and R. J. Hemley, Bonding changes in compressed superhard graphite, *Science* **302**, 425–427, (2003).

[155] G. Fiquet, J. Badro, F. Guyot, C. Bellin, M. Krisch, D. Antonangeli, H. Requardt, A. Mermet, D. Farber, C. Aracne-Ruddle, Application of inelastic X-ray scattering to the measurements of acoustic wave velocities in geophysical materials at very high pressure, *Phys. Earth Planet. Inter.* **143**, 5–18, (2004).

[156] T. S. Duffy, Synchrotron facilities and the study of the Earth's deep interior, *Rep. Prog. Phys.* **68**, 1811, (2005).

[157] R. S. Kumar, A. Svane, G. Vaitheeswaran, V. Kanchana, E. D. Bauer, M. Hu, M. F. Nicol and A. L. Cornelius, Pressure-induced valence change in $YbAl_3$: A combined high-pressure inelastic X-ray scattering and theoretical investigation, *Phys. Rev. B* **78**, 075117, (2008).

[158] S. K. Lee, P. J. Eng, H. Mao and J. Shu, Probing and modeling of pressure-induced coordination transformation in borate glasses: Inelastic X-ray scattering study at high pressure, *Phys. Rev. B* **78**, 214203, (2008).

[159] H. K. Mao, Y. Ding, Y. Xiao, P. Chow, J. Shu, S. Lebègue, A. Lazicki and R. Ahuja, Electronic dynamics and plasmons of sodium under compression, *Proc. Natl. Acad. Sci. U. S. A.* **108**, 20434–20437, (2011).
[160] C. Y. Huang, Y. Q. Cai, N. Hiraoka, C. C. Chen, S. C. Chung, Y. F. Song and K. L. Tsang, Cost-effective upgrade of a focusing system for inelastic X-ray scattering experiments under high pressure, *J. Synchrotron Radiat.* **15**, 50–54, (2007).
[161] A. Alatas, B. Leu, J. Zhao, H. Yavaş, T. Toellner and E. Alp, Improved focusing capability for inelastic X-ray spectrometer at 3-ID of the APS: A combination of toroidal and Kirkpatrick-Baez (KB) mirrors, *Nucl. Instrum. Methods Phys. Res., Sect. A* **649**, 166–168, (2011).
[162] W. L. Mao, H. Mao, Y. Meng, P. J. Eng, M. Y. Hu, P. Chow, Y. Q. Cai, J. Shu and R. J. Hemley, X-ray–induced dissociation of H_2O and formation of an O_2–H_2 alloy at high pressure, *Science* **314**, 636–638, (2006).
[163] M. Pravica, O. Grubor-Urosevic, M. Hu, P. Chow, B. Yulga and P. Liermann, X-ray raman spectroscopic study of benzene at high pressure, *J. Phys. Chem. B* **111**, 11635–11637, (2007).
[164] H. Fukui, M. Kanzaki, N. Hiraoka and Y. Q. Cai, Coordination environment of silicon in silica glass up to 74 GPa: An X-ray Raman scattering study at the silicon L edge, *Phys. Rev. B* **78**, 012203, (2008).
[165] N. Hiraoka and Y. Cai, High-pressure studies by X-ray Raman scattering, *Sync. Rad. News* **23**, 26–31, (2010).
[166] B. Maddox, A. Lazicki, C. Yoo, V. Iota, M. Chen, A. McMahan, M. Hu, P. Chow, R. Scalettar and W. Pickett, $4f$ delocalization in Gd: Inelastic X-ray scattering at ultrahigh pressure, *Phys. Rev. Lett.* **96**, 215701, (2006).
[167] J. P. Rueff, J. P. Itié, M. Taguchi, C. Hague, J. M. Mariot, R. Delaunay, J. P. Kappler and N. Jaouen, Probing the γ-α transition in bulk Ce under pressure: A direct investigation by resonant inelastic X-ray scattering, *Phys. Rev. Lett.* **96**, 237403, (2006).
[168] I. Jarrige, J. P. Rueff, S. Shieh, M. Taguchi, Y. Ohishi, T. Matsumura, C. P. Wang, H. Ishii, N. Hiraoka and Y. Cai, Pressure-induced valence anomaly in TmTe probed by resonant inelastic X-ray scattering, *Phys. Rev. Lett.* **101**, 127401, (2008).
[169] P. Cloetens, W. Ludwig, J. Baruchel, D. Van Dyck, J. Van Landuyt, J. Guigay and M. Schlenker, Holotomography: Quantitative phase tomography with micrometer resolution using hard synchrotron radiation x rays, *Appl. Phys. Lett.* **75**, 2912–2914, (1999).
[170] J. H. Kinney and M. C. Nichols, X-ray tomographic microscopy (XTM) using synchrotron radiation, *Annu. Rev. Mater. Sci.* **22**, 121–152, (1992).
[171] Y. Wang, F. De Carlo, D. C. Mancini, I. McNulty, B. Tieman, J. Bresnahan, I. Foster, J. Insley, P. Lane, G. von Laszewski, C. Kesselman, M.-H. Su and M. Thiebaux, A high-throughput X-ray microtomography system at the Advanced Photon Source, *Rev. Sci. Instrum.* **72**, 2062–2068, (2001).

[172] A. Rack, S. Zabler, B. R. Müller, H. Riesemeier, G. Weidemann, A. Lange, J. Goebbels, M. Hentschel and W. Görner, High resolution synchrotron-based radiography and tomography using hard X-rays at the BAMline (BESSY II), *Nucl. Instrum. Methods Phys. Res., Sect. A* **586**, 327–344, (2008).

[173] T. Martin and A. Koch, Recent developments in X-ray imaging with micrometer spatial resolution, *J. Synchrotron Radiat.* **13**, 180–194, (2006).

[174] J. Y. Buffière, E. Maire, P. Cloetens, G. Lormand and R. Fougères, Characterization of internal damage in a MMCp using X-ray synchrotron phase contrast microtomography, *Acta Mater.* **47**, 1613–1625, (1999).

[175] L. Salvo, P. Cloetens, E. Maire, S. Zabler, J. J. Blandin, J. Y. Buffière, W. Ludwig, E. Boller, D. Bellet and C. Josserond, X-ray micro-tomography an attractive characterisation technique in materials science, *Nucl. Instrum. Methods Phys. Res., Sect. A* **200**, 273–286, (2003).

[176] J. Baruchel, P. Bleuet, A. Bravin, P. Coan, E. Lima, A. Madsen, W. Ludwig, P. Pernot and J. Susini, Advances in synchrotron hard X-ray based imaging, *CR Phys.* **9**, 624–641, (2008).

[177] R. A. Ketcham and W. D. Carlson, Acquisition, optimization and interpretation of X-ray computed tomographic imagery: applications to the geosciences, *Comput Geosci-UK* **27**, 381–400, (2001).

[178] W. D. Carlson, Three-dimensional imaging of earth and planetary materials, *Earth. Planet. Sci. Lett.* **249**, 133–147, (2006).

[179] D. Wildenschild, J. Hopmans, M. Rivers and A. Kent, Quantitative analysis of flow processes in a sand using synchrotron-based X-ray microtomography, *Vadose Zone J.* **4**, 112–126, (2005).

[180] W. A. Kalender, X-ray computed tomography, *Phys. Med. Biol.* **51**, R29, (2006).

[181] W. Ludwig, A. King, P. Reischig, M. Herbig, E. Lauridsen, S. Schmidt, H. Proudhon, S. Forest, P. Cloetens and S. Roscoat, New opportunities for 3D materials science of polycrystalline materials at the micrometre lengthscale by combined use of X-ray diffraction and X-ray imaging, *Mater. Sci. Eng., A* **524**, 69–76, (2009).

[182] A. King, M. Herbig, W. Ludwig, P. Reischig, E. Lauridsen, T. Marrow, J. Buffière, Non-destructive analysis of micro texture and grain boundary character from X-ray diffraction contrast tomography, *Nucl. Instrum. Methods Phys. Res., Sect. A* **268**, 291–296, (2010).

[183] W. Ludwig, P. Reischig, A. King, M. Herbig, E. M. Lauridsen, G. Johnson, T. J. Marrow and J. Y. Buffiere, Three-dimensional grain mapping by X-ray diffraction contrast tomography and the use of Friedel pairs in diffraction data analysis, *Rev. Sci. Instrum.* **80**, 033905-9, (2009).

[184] R. Secco, M. Rutter, S. Balog, H. Liu, D. Rubie, T. Uchida, D. Frost, Y. Wang, M. Rivers and S. Sutton, Viscosity and density of Fe–S liquids at high pressures, *J. Phys.: Condens. Matter* **14**, 11325, (2002).

[185] R. J. Hemley, H. Mao, G. Shen, J. Badro, P. Gillet, M. Hanfland and D. Häusermann, X-ray imaging of stress and strain of diamond, iron, and tungsten at megabar pressures, *Science* **276**, 1242, (1997).

[186] Y. Wang, T. Uchida, F. Westferro, M. L. Rivers, N. Nishiyama, J. Gebhardt, C. E. Lesher and S. R. Sutton, High-pressure X-ray tomography microscope: Synchrotron computed microtomography at high pressure and temperature, *Rev. Sci. Instrum.* **76**, 073709-7, (2005).

[187] C. E. Lesher, Y. Wang, S. Gaudio, A. Clark, N. Nishiyama and M. Rivers, Volumetric properties of magnesium silicate glasses and supercooled liquid at high pressure by X-ray microtomography, *Phys. Earth Planet. Inter.* **174**, 292–301, (2009).

[188] H. Terasaki, S. Urakawa, K. Funakoshi, N. Nishiyama, Y. Wang, K. Nishida, T. Sakamaki, A. Suzuki and E. Ohtani, *In situ* measurement of interfacial tension of Fe-S and Fe-P liquids under high pressure using X-ray radiography and tomography techniques, *Phys. Earth Planet. Inter.* **174**, 220–226, (2009).

[189] W. Zhu, G. A. Gaetani, F. Fusseis, L. G. J. Montési, F. De Carlo, Microtomography of partially molten rocks: three-dimensional melt distribution in mantle peridotite, *Science* **332**, 88, (2011).

[190] Y. Wang, C. Lesher, G. Fiquet, M. L. Rivers, N. Nishiyama, J. Siebert, J. Roberts, G. Morard, S. Gaudio and A. Clark, *In situ* high-pressure and high-temperature X-ray microtomographic imaging during large deformation: A new technique for studying mechanical behavior of multiphase composites, *Geosphere* **7**, 40–53, (2011).

[191] L. Li, D. Weidner, P. Raterron, C. Jiuhua, M. Vaughan, M. Shenghua and B. Durham, Deformation of olivine at mantle pressure using the D-DIA, *Eur. J. Mineral.* **18**, 7–19, (2006).

[192] P. Raterron, E. Amiguet, J. Chen, L. Li and P. Cordier, Experimental deformation of olivine single crystals at mantle pressures and temperatures, *Phys. Earth Planet. Inter.* **172**, 74–83, (2009).

[193] D. Bruhn, N. Groebner and D. Kohlstedt, An interconnected network of core-forming melts produced by shear deformation, *Nature* **403**, 883–886, (2000).

[194] H. Liu, L. Wang, X. Xiao, F. De Carlo, J. Feng, H. Mao and R. J. Hemley, Anomalous high-pressure behavior of amorphous selenium from synchrotron X-ray diffraction and microtomography, *Proc. Natl. Acad. Sci. U. S. A.* **105**, 13229, (2008).

[195] X. Xiao, H. Liu, L. Wang and F. De Carlo, Density measurement of samples under high pressure using synchrotron microtomography and diamond anvil cell techniques, *J. Synchrotron Radiat.* **17**, 360–366, (2010).

[196] J. Wang, W. Yang, S. Wang, X. Xiao, F. De Carlo, Y. Liu and W. L. Mao, High pressure nano-tomography using an iterative method, *J. Appl. Phys.* **111**, 112626-5, (2012).

[197] J. Nordgren, G. Bray, S. Cramm, R. Nyholm, J. E. Rubensson and N. Wassdahl, Soft X-ray emission spectroscopy using monochromatized synchrotron radiation, *Rev. Sci. Instrum.* **60**, 1690–1696, (1989).

[198] J. H. Guo, Y. Luo, A. Augustsson, J. E. Rubensson, C. Såthe, H. Ågren, H. Siegbahn and J. Nordgren, X-ray emission spectroscopy of hydrogen bonding and electronic structure of liquid water, *Phys. Rev. Lett.* **89**, 137402, (2002).

[199] H. Yamaoka, N. Tsujii, H. Oohashi, D. Nomoto, I. Jarrige, K. Takahiro, K. Ozaki, K. Kawatsura and Y. Takahashi, Bulk electronic properties of $FeSi_{1-x}Ge_x$ investigated by high-resolution X-ray spectroscopies, *Phys. Rev. B* **77**, 115201, (2008).

[200] N. Lee, T. Petrenko, U. Bergmann, F. Neese and S. DeBeer, Probing valence orbital composition with iron Kβ X-ray emission spectroscopy, *J. Am. Chem. Soc.* **132**, 9715–9727, (2010).

[201] J. F. Lin, V. V. Struzhkin, S. D. Jacobsen, G. Shen, V. B. Prakapenka, H. K. Mao and R. J. Hemley, X-ray emission spectroscopy with a laser-heated diamond anvil cell: a new experimental probe of the spin state of iron in the Earth's interior, *J. Synchrotron Radiat.* **12**, 637–641, (2005).

[202] J. P. Rueff, C. C. Kao, V. V. Struzhkin, J. Badro, J. Shu, R. J. Hemley and H. K. Mao, Pressure-induced high-spin to low-spin transition in FeS evidenced by X-ray emission spectroscopy, *Phys. Rev. Lett.* **82**, 3284–3287, (1999).

[203] J. Badro, V. V. Struzhkin, J. Shu, R. J. Hemley, H. Mao, C. Kao, J. P. Rueff and G. Shen, Magnetism in FeO at megabar pressures from X-ray emission spectroscopy, *Phys. Rev. Lett.* **83**, 4101–4104, (1999).

[204] J. P. Rueff, A. Shukla, A. Kaprolat, M. Krisch, M. Lorenzen, F. Sette and R. Verbeni, Magnetism of Invar alloys under pressure examined by inelastic X-ray scattering, *Phys. Rev. B* **63**, 132409, (2001).

[205] J. Badro, G. Fiquet, V. V. Struzhkin, M. Somayazulu, H.-k. Mao, G. Shen and T. Le Bihan, Nature of the High-Pressure Transition in Fe_2O_3 Hematite, *Phys. Rev. Lett.* **89**, 205504, (2002).

[206] J. Chen, S. Haw, J. Lee, T. Chou, S. Chen, K. Lu, Y. Liang, Y. Lee, N. Hiraoka and H. Ishii, Pressure dependence of the electronic structure and spin state in $Fe_{1.01}$ Se superconductors probed by X-ray absorption and X-ray emission spectroscopy, *Phys. Rev. B* **84**, 125117, (2011).

[207] J. Chen, S. Haw, J. Lee, S. Chen, K. Lu, S. Chen, M. Deng, Y. F. Liao, J. Lin and B. Chen, Pressure-dependent electronic structures and orbital hybridization of Mn 3d states in multiferroic $BiMnO_3$: A combined X-ray absorption, X-ray emission, and resonant X-ray emission study, *Phys. Rev. B* **86**, 045103, (2012).

[208] J. P. Rueff, M. Mezouar and M. Acet, Short-range magnetic collapse of Fe under high pressure at high temperatures observed using X-ray emission spectroscopy, *Phys. Rev. B* **78**, 100405, (2008).

[209] V. V. Struzhkin, H. Mao, J. F. Lin, R. J. Hemley, J. S. Tse, Y. Ma, M. Y. Hu, P. Chow and C. C. Kao, Valence band X-ray emission spectra of compressed germanium, *Phys. Rev. Lett.* **96**, 137402, (2006).

[210] J. Badro, G. Fiquet, F. Guyot, J. P. Rueff, V. V. Struzhkin, G. Vankó, G. Monaco, Iron partitioning in Earth's mantle: toward a deep lower mantle discontinuity, *Science* **300**, 789–791, (2003).

[211] J. F. Lin, V. V. Struzhkin, S. D. Jacobsen, M. Y. Hu, P. Chow, J. Kung, H. Liu, H. Mao and R. J. Hemley, Spin transition of iron in magnesiowüstite in the Earth's lower mantle, *Nature* **436**, 377–380, (2005).

[212] J.-F. Lin, A. G. Gavriliuk, V. V. Struzhkin, S. D. Jacobsen, W. Sturhahn, M. Y. Hu, P. Chow and C.-S. Yoo, Pressure-induced electronic spin transition of iron in magnesiowustite-(Mg,Fe)O, *Phys. Rev. B* **73**, 113107, (2006).

[213] J. F. Lin, G. Vankó, S. D. Jacobsen, V. Iota, V. V. Struzhkin, V. B. Prakapenka, A. Kuznetsov and C. S. Yoo, Spin transition zone in Earth's lower mantle, *Science* **317**, 1740–1743, (2007).

[214] J. Badro, J. P. Rueff, G. Vankó, G. Monaco, G. Fiquet and F. Guyot, Electronic transitions in perovskite: Possible nonconvecting layers in the lower mantle, *Science* **305**, 383–386, (2004).

[215] J. F. Lin, S. Speziale, V. B. Prakapenka, P. Dera, B. Lavina and H. C. Watson, High-pressure X-ray diffraction and X-ray emission studies on iron-bearing silicate perovskite under high pressures, *High Pressure Res.* **30**, 230–237, (2010).

[216] J. Li, V. V. Struzhkin, H.-k. Mao, J. Shu, R. J. Hemley, Y. Fei, B. Mysen, P. Dera, V. Prakapenka and G. Shen, Electronic spin state of iron in lower mantle perovskite, *Proc. Natl. Acad. Sci. U. S. A.* **101**, 14027–14030, (2004).

[217] J. Li, W. Sturhahn, J. Jackson, V. Struzhkin, J. Lin, J. Zhao, H. Mao and G. Shen, Pressure effect on the electronic structure of iron in (Mg, Fe)(Si, Al) O_3 perovskite: a combined synchrotron Mössbauer and X-ray emission spectroscopy study up to 100 GPa, *Phys. Chem. Miner.* **33**, 575–585, (2006).

Chapter 12

X-ray Excited Optical Luminescence and Its Applications

Lijia Liu and Xuhui Sun

Soochow University-Western University Centre for Synchrotron Radiation Research, Institute of Functional Nano & Soft Materials (FUNSOM) and Collaborative Innovation Center of Suzhou Nano Science & Technology, Soochow University, Suzhou, Jiangsu, 215123, China

Abstract

The chapter introduces the spectroscopic technique which is used to study the luminescence mechanism of light-emitting materials: X-ray excited optical luminescence (XEOL). In combination with X-ray absorption fine structure, XEOL tracks the luminescence from a material under the excitation of a tunable synchrotron radiation source, providing element- and site-specific optical features. This chapter first introduces the basic concept of the XEOL process, followed by its experimental setup (including steady-state XEOL and time-resolved XEOL), and how to conduct data acquisition and analysis. Numerous examples are also provided to demonstrate the capability of such technique in analyzing practical material systems.

Keywords: X-ray excited optical luminescence (XEOL); time-resolved XEOL; nanomaterials; optical property.

12.1. XEOL and TR-XEOL: The Basics

12.1.1. *Introduction*

X-ray excited optical luminescence (XEOL) is a spectroscopic technique that monitors light emission from a substance upon X-ray excitation. Although the concept of "XEOL" was first introduced as early as when X-ray was first discovered, since some materials are found light-emitting upon X-ray excitation [1], more recent development of XEOL as an optical property probe is associated with the development of synchrotron radiation technology. Synchrotron radiation provides bright and highly collimated X-ray source, which allows the detection of weak light emitters and diluted samples. Its pulsed time structure facilitates the study of the luminescence dynamics in terms of luminescence decay lifetime. The most important reason that XEOL is regarded as a unique and valuable technique is its capability of being combined with X-ray absorption fine structure (XAFS) which probes the local chemical environment of a material with element and site specificity.

So far, XEOL has proved to be an effective analytical technique in studying luminescence mechanism of various material systems, such as semiconductors [2–5], rare earth compounds [6–9], multi-component materials [10, 11]. Early stage XEOL research is mainly on rare earth-based materials, since those materials are known as intense lumiphores [12]. An excellent book chapter written by Rogalev *et al.* in 2001, has addressed the theory basis of XEOL phenomenon and provided valuable insight to this analytical technique [13]. Time-resolved XEOL (TR-XEOL) has also been reviewed in several journal articles by Sham and co-workers [14, 15]. The main purpose of this chapter is to demonstrate how XEOL can be used to better understand the luminescence properties of materials. Due to increasing demand of energy-related devices and fast development of nanotechnology, the examples selected in this chapter emphasizes XEOL application on semiconductor nanomaterials and organic materials for organic light-emitting diode applications. We are first going to give a general introduction on synchrotron-based XEOL and TR-XEOL, addressing the issues that commonly occur during measurements, followed by some recent developments of XEOL and TR-XEOL instrumentation. Finally, several representative material systems are selected to demonstrate the XEOL analysis in practical systems.

12.1.2. *Mechanism of X-ray excited luminescence: emission from core-electron decay*

The key question that XEOL deals with is, of course, the luminescence behavior of a material in response to X-ray excitation. Luminescence from solids is a complicated process [16]. We shall start with an illustration of how the luminescence is produced from a material upon absorbing energy. A semiconductor has a band gap between its conduction band and valence band. In the case of a molecule, there is also a gap between its highest occupied molecular orbital (HOMO) and lowest unoccupied molecular orbital (LUMO). When the substance at ground state is stimulated by a photon containing energy equal to or greater than the gap, a valence electron (or electron at HOMO) is populated to the conduction band (or LUMO), the system is at its excited state. The excited electron and the hole left behind will find ways to bind together and form an exciton. Immediately afterward, the system relaxes to the ground state, and the exciton (electron-hole pair) recombines. There are several paths that the energy can be dissipated, either as photons or phonons. The former is called radiative recombination, which leads to luminescence, while the latter is non-radiative recombination, and will produce heat. The radiative recombination may undergo different channels. Shown in Fig. 12.1 as in the case of a semiconductor, the recombination takes place between an electron at the bottom of the conduction band and a hole at the top of the valance band. The optical photon emitted has energy close to the value of the band gap (process b1). Often in semiconductors, there are various defects in the form of lattice vacancies or impurity atoms that could act as electron or hole traps; an optical photon could also be generated during the recombination, but its energy is much less than the band gap of the material (process b2 and b3).

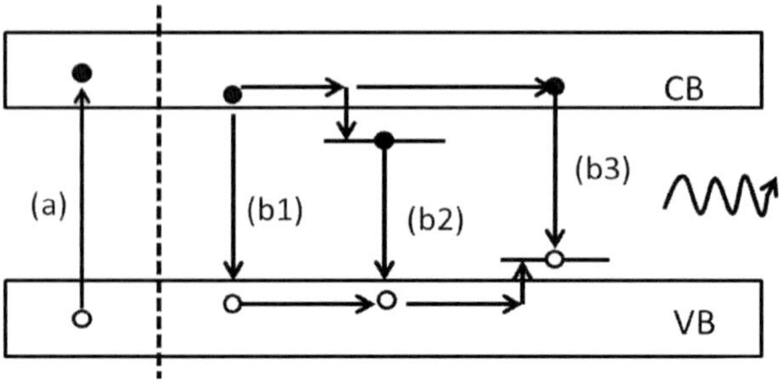

Fig. 12.1. Schematic illustration of exciation and radiative de-excitation processes: (a) Excitation: creation of hole-electron pair. (b1) near-band gap recombination, (b2) recombination involving an electron trap, (b3) recombination involving a hole trap.

In principle, as long as the excitation source has energy greater than the band gap (or HOMO-LUMO gap) of the material, luminescence could be generated. Conventional characterization techniques for luminescent property is the use of a xenon arc lamp, a laser, or an electron beam; these are referred to as fluorescence spectroscopy, photoluminescence (PL) spectroscopy, and cathodoluminescence (CL) spectroscopy, respectively. Xenon arc lamp is able to provide continuous broad spectrum that covers from ultraviolet to infrared, and the wavelength is tunable using a monochromator. The intensity, however, is relatively low. Laser source is able to provide light of high power density, but with single energy. On the other hand, synchrotron radiation source produces X-ray that could achieve both high intensity and energy tunability. Unlike PL or CL which monitors band-to-band transition, XEOL creates excitation events of core-to-bound, core-to-quasi-bound, core-to-continuum, and the luminescence is produced during multi-de-excitation paths of the core electron.

12.1.2.1. *XEOL*

XEOL usually requires the use of an X-ray source of energy within 10^2–10^4 eV, which is in the range of the core-level electron binding energy of various elements. Upon absorption of an X-ray photon, an electron at a core-level is excited to a previously unoccupied

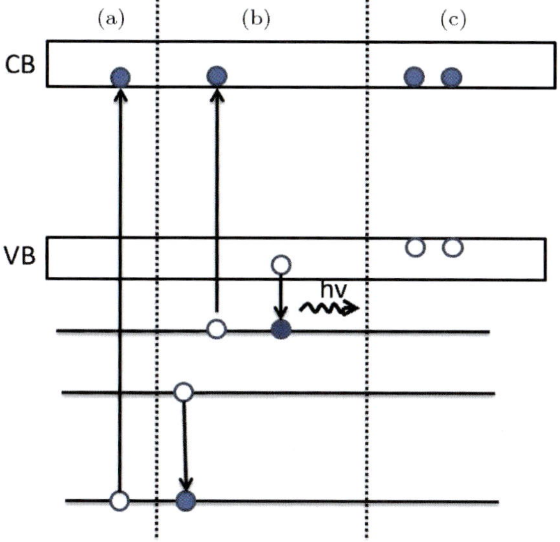

Fig. 12.2. Schematic illustration of the thermalization process. (a) Core-electron excitation; (b) filling the core-hole via Auger and X-ray fluorescence processes; (c) after the thermalization, holes are settled at the top of valance band (VB) and electrons are at the bottom of conduction band (CB).

electronic state and a core-hole is left behind. This process takes place within femtoseconds [14]. The system immediately responds by filling the core-hole with electrons at shallower levels, shown in Fig. 12.2. An auger electron is emitted, and a shallow-level hole is thus created, and will be further filled by still shallower electrons or valence electrons. The core-hole could also be filled with electrons at shallower levels while emitting an X-ray photon (fluorescence). The creation of core-hole leads to a cascade process, and as in the case of a semiconductor, the hole will eventually reach the top of the valence band. In the meantime, photoelectrons and auger electrons create more electrons and holes as they travel in the medium and lose their energy through inelastic scattering. The process is called "thermalization", which will eventually lead to holes reaching the top of the valence band and electrons to the bottom of the conduction band. The entire thermalization process is relatively fast, which is usually in the time scale of sub-picoseconds. After that, like the regular luminescence process, the electron and hole will recombine and the ones which go through the radiative channel will produce luminescence. The emitted photon has the energy close to the band gap of the material, and the lifetime is in the order of nanoseconds to sub-nanoseconds. In the meantime, defects, surface states, impurities could act as traps to electrons or holes, so that the luminescence observed usually lies in lower energy with a slightly longer lifetime (several tens of nanoseconds or longer).

As mentioned above, the biggest advantage that XEOL has is when it is combined with core-level spectroscopy, such as XAFS. Monitoring the optical yield for structural analysis has been proposed by researchers in the early days [17]. However, due to the low signal to noise ratio, as well as sample damage due to prolonged beam exposure, XEOL yield has been proved not an optimal method compared to electron yield or fluorescence yield spectra in terms of structural analysis [18]. On the other hand, as a near-edge probe, XEOL yield can be obtained with high quality which represents the optical decay relevant to the selected core-electron. X-ray absorption near-edge structure (XANES) is able to provide rich information about unoccupied density of states, oxidation state, and local chemical environment. XANES is part of the X-ray absorption fine structure, but it mainly focuses on the region from a few eV below up to around 50 eV above the absorption edge of the element of interest. In this region, the modulation is caused by transition from core-level to previously unoccupied bound and quasi-bound states, governed by the dipole selection rule, and hence the produced spectrum is regarded as the "fingerprint" of the material. Since XEOL is produced during the de-excitation of the core-electron, one can track the yield of optical photon as a function of excitation energy, and be able to obtain luminescence specific to a particular decay channel, or in another words, the optical fingerprint of the material.

If XEOL is obtained simultaneously with XANES, the data recorded will be a contour map containing information in two dimensions. Figure 12.3 shows a typical 2D XEOL-XANES map obtained from oxidized porous Si nanowires [19]. The horizontal axis is the

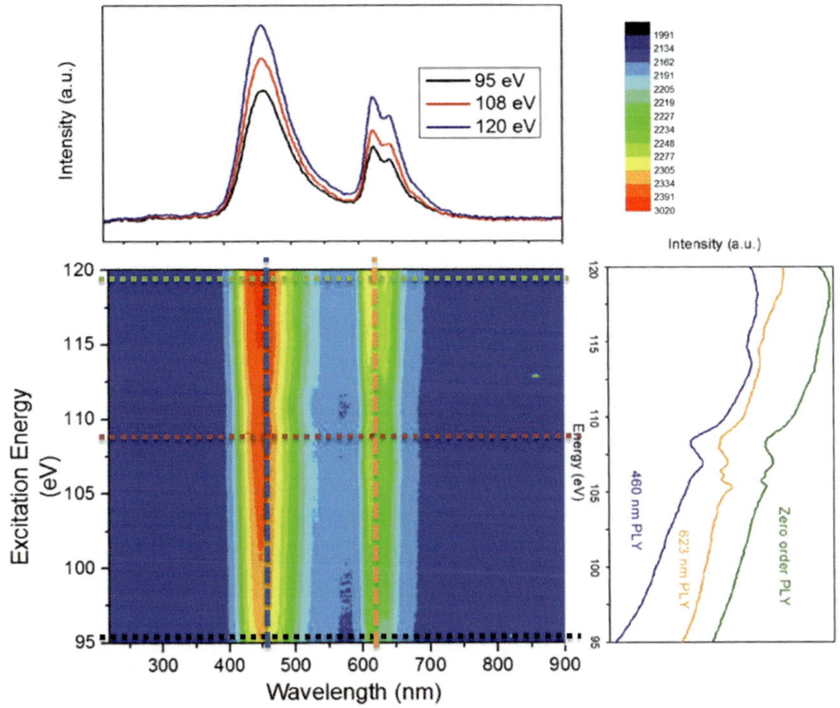

Fig. 12.3. Main graph: 2D XEOL-XANES map from oxidized porous Si nanowires. Color coding represents the luminescence intensity. Upper panel: XEOL spectra at selected excitation energy indicated as horizontal dotted lines in the map. Right panel: luminescence yield at selected wavelengths (indicated as vertical dashed lines in the map) and overall luminescence (zero order) plotted against excitation energy.

wavelength, the vertical axis is the excitation energy, and the color-coded z-axis is the luminescence intensity. In this map, a luminescence spectrum at any energy could be obtained by doing horizontal cuts from the map, which yields a line spectrum plotted as intensity against wavelength; this is essentially very similar to the spectrum commonly seen in conventional luminescence detection techniques. In the meantime, by doing vertical cuts at a selected wavelength or integrating the intensity along the entire horizontal axis, we are able to plot the evolution of luminescence intensity (partial or total) as a function of excitation energy.

12.1.2.2. *TR-XEOL*

Time-resolved XEOL takes advantage of the pulsed time structure of X-rays from a synchrotron radiation. Various synchrotron radiation facilities around the world could provide stable, pulsing sources with desired windows between pulses. For example, the Advanced Photon Source operates at a top-up mode, producing X-ray pulses of 80 ps and 153 ns dark gaps [14]. Such time structure could create an excitation event to monitor the time of light of photons emitted from a light-emitting material. The fundamental principle of TR-XEOL is the same as other time-resolved luminescence detection techniques. It measures the time-dependent decay of the chosen emission band following absorption of an X-ray pulse. From the decay curve, the lifetime of the excited state can be obtained. TR-XEOL have been successfully used to reveal the temporal behavior of band gap and defect-related luminescence bands from ZnS and other nanomaterials [20–22]. As in typical semiconducting materials, the luminescence from near band gap exciton recombination is usually short-lived, within the first 20 ns, and exhibits a narrow peak. Luminescence from various defects are usually of longer lifetime, and have a broad peak. By setting an appropriate time window, one could separate the luminescence profile of an emission band with particular lifetime from the ungated luminescence spectrum.

12.2. Experimental Setup and Data Interpretation

12.2.1. *Experimental techniques*

12.2.1.1. *Combined analysis of XEOL and XANES*

To perform a simple XEOL measurement, one needs to have an X-ray source, a photodetector, an optical spectrometer, and samples to be measured. The X-ray source can be provided by conventional X-ray tube, however, in order to fully utilize the advantage that XEOL possesses, a third-generation synchrotron radiation source is preferred.

The experiment is usually set up as follows: the sample is placed on a manipulator and facing toward the incident X-ray beam.

The incident angle depends on specific experiment requirement. The manipulator could be equipped with heating and/or cooling capability to allow temperature-dependent measurements. An optical lens is placed in front of the sample (backscatter mode) or behind the sample (transmission mode) to collect the luminescence signals, and analyzed by a spectrometer. For XEOL using hard X-ray, there is almost no restriction on the form of samples: they can be gas [23], liquid [24], and solids of any forms. At the soft X-ray region, the sample has to be ultrahigh vacuum compatible, so solids (powders, bulk or thin films) are the easiest and most straightforward samples to measure. Measurements of liquid samples are possible, but the liquids need to be sealed in a X-ray and optical transparent cell, which usually involves the use of a Si_3N_4, Be or polymer window. Luminescence is collected by an optical lens and fed to the spectrometer through an optical fiber.

At the Spherical Grating Monochromator (SGM) beamline at the Canadian Light Source, a motor controlled manipulator enables the precise control of sample position. A thermal couple is installed on the manipulator, as well as a liquid nitrogen inlet and outlet, so XEOL measurements can be done under various temperatures. The XEOL is collected by an optical lens placed in front of the sample, and the optical signal is transferred through an optical fiber to an Ocean Optics QE65000 portable spectrometer. In theory, for XEOL measurements, any type of spectrometer can be used, but due to the limited space at the endstation, a compact one is always preferred. Charge-coupled device (CCD) detector is used to achieve the optimum efficiency, since it captures the luminescence over the entire wavelength in one motion instead of scanning through the wavelength range at a fixed step.

A XEOL spectrum is obtained by collecting the luminescence from a sample at chosen excitation energy. More often, XEOL is measured simultaneously with the XANES scan across the absorption edge of the selected element. That is, at each energy step, one XEOL spectrum will be collected. When the scan is completed, it generates a matrix file, and can be plotted as a color coded contour map, as previously shown in Fig. 12.3. Meanwhile, the total counts (or partial counts from selected window) of luminescence are recorded at each excitation energy, creating a spectrum which records luminescence intensity as a function of excitation photon energy spectrum. This is usually referred to as total PLY (photoluminescence yield) in zero order (whole wavelength) or partial PLY at selected optical window.

The normal XANES scan usually produces two spectra, depending on the detection modes. The total electron yield (TEY) measures the current, and the X-ray fluorescence yield (FY) collects the emitted X-ray fluorescence photons (can be done either as total fluorescence or partial fluorescence). With the XEOL attachment, one energy scan could generate three types of spectra: TEY, FY (total and/or partial), and PLY (total and/or partial). Optical spectrometer also produces a 2D matrix file, which contain the full picture of luminescence evolution along the excitation energy.

12.2.1.2. *TR-XEOL*

The basic setup for TR-XEOL is similar to regular XEOL. For time-dependent experiment, the optical spectrometer should have time-resolved capability. A photomultiplier tube with sub-nanosecond time response is preferred. There are several synchrotron facilities with the capability of doing TR-XEOL, such as APS [25], CLS [26], and recently Diamond [27]. The set-ups are similar, and Fig. 12.4 shows a schematic TR-XEOL layout [14]. The optical luminescence is collected by objective lens and focused onto the entrance slit of a monochromator. After being dispersed, the monochromatic light is detected by a fast photomultipler tube (PMT). The output signal is then amplified and discriminated to be used as a trigger to the "start" channel of a time-to amplitude converter (TAC). A bunch clock signal, originating from the storage ring, is used to provide the "stop" signal to the TAC. The output signal from TAC is then directed into a multichannel analyzer (MCA) operating in the pulse-height analysis mode.

Recently, beamlines at CLS, APS and CHESS have employed streak cameras for TR-XEOL measurements [26, 27]. Streak cameras

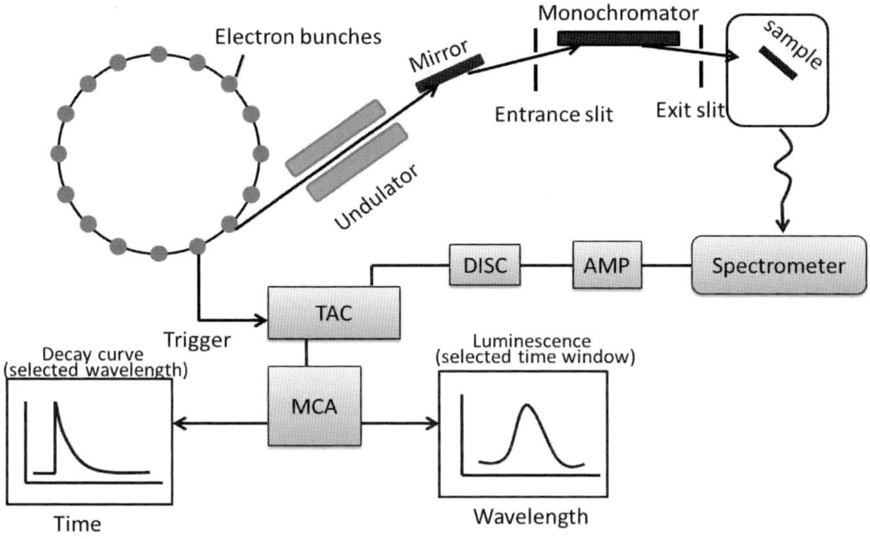

Fig. 12.4. Schematic diagram of a TR-XEOL experimental setup.

operate by converting a temporal event into a spatial profile, and are able to achieve sub-picosecond time resolution. It allows one to obtain the lifetime decay simultaneously with luminescence data, hence creating a 2D image of emission profile with x-axis being wavelength, y-axis being time, and z-axis being emission intensity. It is achieved by passing the XEOL through a spectrograph that disperses the light across the photocathode of the streak camera. The photoelectrons at the front of the streak tube are generated with a horizontal profile determined by the XEOL spectrum. The streak tube only applies a vertical deflection to the transiting photoelectrons so that horizontal profile is retained at the detector. The great advantage of employing a streak camera is that it is able to monitor multiple decays of different lifetimes within one shot.

Figure 12.5 shows examples of TR-XEOL image obtained using a streak camera installed at the spherical grating monochromator (SGM) beamline at the Canadian Light Source [26]. The samples tested were ZnO and stishovite (a high pressure polymorph of SiO_2). The equipment includes a Hamamatsu C5680 streak camera with a

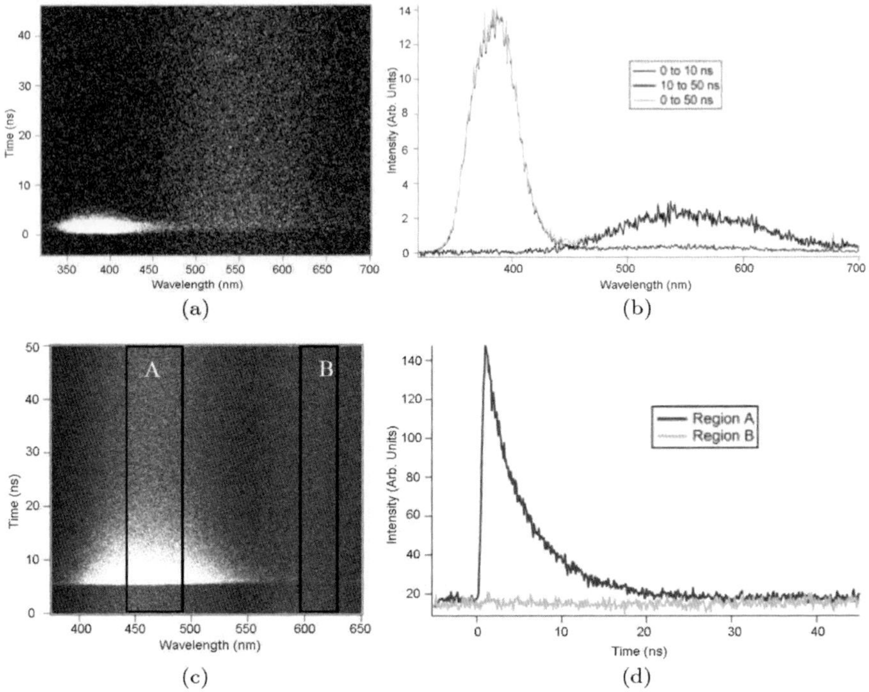

Fig. 12.5. 2D TR-XEOL images of (a) ZnO and (c) stishovite obtained using streak camera. (b) Horizontal line profiles showing emissions at different time windows. (d) Vertical line profiles showing decay curves of different luminescence region. Reprinted with permission from [26].

M5677 slow single sweep unit, a Bruker IS200 spectrograph, and a 30 mm WD lens system composed of 1 inch lenses and a sapphire UHV window. A more detailed description can be found in work done by Regier *et al.* [26] similar setup has been used to study the dynamic optical property of GaN-ZnO solid solution [28]. The streak camera produces a 2D image containing both temporal and spectral information. By doing horizontal cuts, the emission wavelength at selected time window can be revealed (Fig. 12.5(b)). Similarly, by doing vertical cuts, one could monitor the decay lifetime of any emission band of interest (Fig. 12.5(d)).

12.2.1.3. *Other techniques*

The recent progress in synchrotron technology enables more versatile applications of XEOL. For example, Hayashi *et al.* have applied XEOL in combination with X-ray standing wave and atom-resolved X-ray holography, in which they demonstrated that hard X-ray techniques can be used to measure light elements with the help of XEOL [29]. Villanova *et al.* were able to integrate XEOL with microanalysis, and have shown that μ-XRF and μ-XEOL can be used to study photovoltaic cells, providing a direct path for comprehensive understanding of the relationship of material, structural and optical properties [30].

12.2.2. *Data interpretation*

12.2.2.1. *Data collection and analysis*

Similar to all optical measurements, a "dark" spectrum should be measured for background subtraction. Any light from the environment causing unnecessary interference on the luminescence detection of the sample should be prevented from entering the sample stage. It is of particular important for weak light-emitting samples. The spectra quality could be optimized by adjusting the focus of the lenses and mirror positions. Increasing the integration time and the X-ray beam intensity could also enhance the signal, but it should to be done with extra caution. Increasing the integration time will increase the exposure time under X-ray. For samples that tend to suffer beam damage, such an action might cause the deformation of their surface structures. Increasing the beam intensity is often done by opening up the exit beam slit, which will inevitably cause the loss of the high resolution of XANES.

Depending on the specific type of spectrometers, XEOL data can be exported in either multi-column ASCII files, or a matrix file containing 2D information. The ASCII file contains wavelength, and total counts of optical photon at zero order and at selected wavelength windows. Most of the spectrometer control software contain

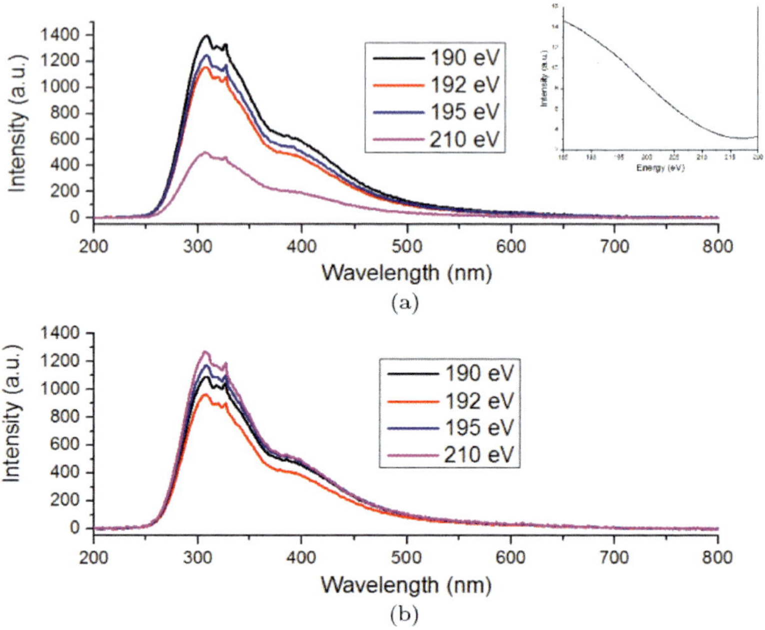

Fig. 12.6. XEOL of hexagonal BN powder with selected excitation energies across the B K-edge. (a) Raw spectra, (b) spectra normalized to the incident flux. The inset of (a) shows the incident photon flux across the B K-edge.

background subtraction function, but for the ones which do not, manual background subtraction should be done to (1) get the best signal to noise ratio and (2) to eliminate the possible light coming from the environment. One should also keep in mind that the total beam current is not constant throughout the experiment, so it is desirable to normalize each XEOL spectrum to the corresponding beam current. Figure 12.6 compares the XEOL data of BN powder of hexagonal phase plotted with and without normalizing to the beam current. It can be seen that the luminescence intensity varies along the photon energy scanned from below to above the B K-edge, but results with and without normalization yield very different trends. The incident photon flux exhibit a monotonic decrease across the absorption edge, so we have to take into account the optical response of the material under different excitation energies.

12.2.2.2. *Precautions and limitations*

In most cases, PLY provides the most valuable information that contains the site- and element-specificity of the luminescence properties. However, continuous collection of a PLY spectrum with a XANES scan is not always applicable, especially for weak light-emitting substances, or samples that tend to suffer beam damage. One option is to select several excitation energies to obtain a set of single XEOL spectra. The trend of luminescence intensity variation could be extrapolated by integrating each XEOL spectra. The excitation energy is usually selected where characteristic XANES feature appears or disappears, i.e., the peaks and valleys of a XANES spectrum, before the edge, and far beyond the edge. A peak in XANES usually corresponds to a newly-opened excitation channel, at which a different decay channel is likely to be observed.

If the sample is stable and bright enough to make a PLY measurement, in some cases, especially in the soft X-ray region, the PLY spectrum exhibits features that are similar to TEY or FY, but could be either completely inverted, or distorted (partial inversion). Figure 12.7 shows the PLY of BN crystals measured across the B K-edge [31]. Compared to the TEY and FY of BN, the PLY shows all the features from BN but completely inverted.

The inversion of PLY arises from different mechanisms. The most straightforward explanation is that the new excitation channel does not favor the optical decay. Since below the edge, all electrons at shallow levels contribute to the luminescence, as soon as the incident energy is sufficient to excite the core electron, a majority of the photons will be absorbed by that core electron. If the decay channel is preferentially non-radiative, a decrease in the luminescence will be observed.

However, other factors should also be taken into account. For example, one should keep in mind that there is an abrupt change in the X-ray probing depth below and above the absorption edge. For example, in the case of BN, at normal incidence, X-ray penetration depth below the B K-edge is as much as 620 nm [32]. Once past the edge, the penetration depth abruptly decreases to ~100 nm. The probing depth at this stage is very limited to the surface. Since

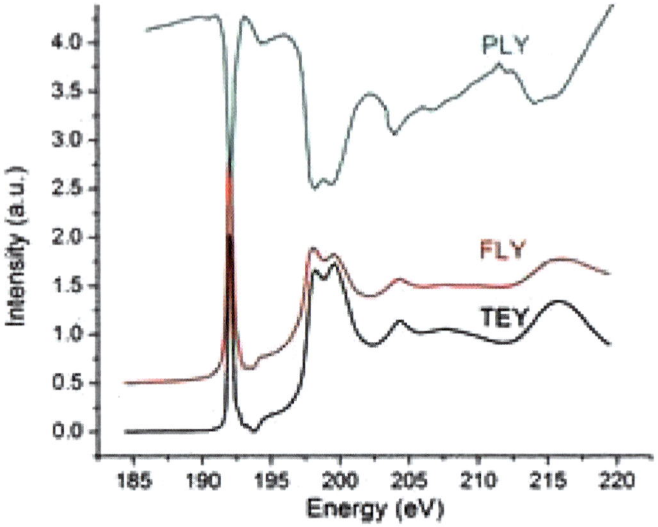

Fig. 12.7. XANES and PLY of hexagonal phase BN powder at the B K-edge. Adapted with permission from [31].

XEOL is a secondary process, during which Auger electrons produce secondary electron–hole pairs to undergo radiative recombination. When the penetration depth is deceased, a main fraction of Auger electrons can escape the solid without transferring the energy to the optical channel. If it is the bulk defect that contributes to the luminescence, decrease in probing depth would thus result in a decrease in luminescence intensity.

Overall, precautions should be taken before making an assumption of the luminescence mechanism. Since XEOL is a rather complicated process, the variation of luminescence intensity is dependent on various factors such as sampling depth, sample dimension, morphology and quality.

12.3. XEOL Applications in Material Analysis

12.3.1. *Multi-component nanomaterials*

The element and site specificity that XEOL processes can be used to examine materials that have multiple light-emitting components.

Below are some examples on how XEOL can be used to analyze the origin of the luminescence from a multi-component sample.

12.3.1.1. *Si-SiO_2: luminescence from complicated interfaces*

The biggest impact that XEOL has caused is that it successfully unravels the luminescence of porous Si [33]. Luminescence from nanostructured Si and its oxide has been investigated over the years by various techniques [34–38]. Elemental silicon is a semiconductor with an indirect band gap of ~1.3 eV. At common situation, when the valence electrons in bulk Si are excited, the near-band gap recombination is forbidden, and the energy is released by phonon emission, hence no luminescence is observed. However, when the size of silicon is reduced to below 5 nm, direct exciton recombination may occur near band gap due to quantum confinement effect [39, 40]. On the other hand, SiO_2 is a wide band gap insulator, and it is common to see visible luminescence from glassy SiO_2 due to the presence of defects and vacancies [37, 38, 41]. It has been long time since people believe that elemental Si at solid state is not light-emitting until the discovery of intense orange-red luminescence from porous silicon [35]. Since the surface of Si is inevitably covered with oxide layer of different thickness, the origin of luminescence from solid state silicon is always under debate, on whether the origin of the luminescence is Si or Si oxide.

Si K-edge XANES probes the unoccupied electronic states silicon and its oxide from the Si perspective. It measures the electronic transition of the Si 1s electron to previously unoccupied 3*p* states. Typical XANES spectra of Si and SiO_2 obtained using FY mode are shown in Fig. 12.8. It can be seen that elemental Si exhibit an edge jump at 1839 eV, while in oxidized Si, the peak shifts to 1846 eV. Based on the feature, we are able to distinguish the oxidation states of Si in a material. Based on such feature, we could obtain PLY simultaneously with XANES, and by comparing the spectra of PLY and XANES, we are able to track which form of Si is responsible for the observed luminescence.

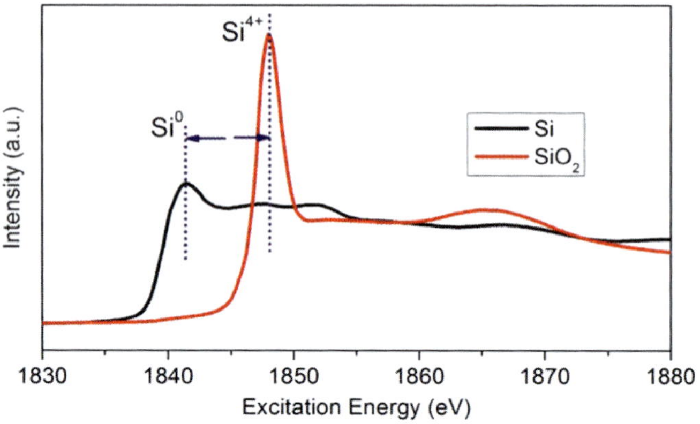

Fig. 12.8. XANES of Si and SiO_2 (obtained in FY mode) at the Si K-edge.

Figure 12.9 shows an early study of porous silicon luminescence in comparison with a bright light emitter siloxene using XEOL and XANES [33]. Siloxene is a Si-contained planar molecule in which Si is terminated with -H and -OH groups. From Fig. 12.8, we know that Si K-edge XANES can be used to distinguish the oxidation states of Si species easily. In Fig. 12.9(a), luminescence from siloxene is centered at 520 nm while porous silicon exhibits emission at 750 nm. The PLY spectra of porous Si and siloxene unambiguously show that the observed emission bands are of two different origins: luminescence of porous Si is solely from the elemental Si. Even for the sample containing trace amount of surface oxide, the oxide species does not contribute to the luminescence. The luminescence of siloxene however, comes from both Si and Si of higher oxidation states, and is hence attributed to Si-H and Si-OH.

A follow-up study on porous Si reveals that different treatment could modify the luminescence properties, and the shift in emission band can be tracked and assigned to the corresponding luminescence center [42]. Shown in Fig. 12.10, freshly prepared porous Si exhibit luminescence at 680 nm, and the luminescence yield (referred to as TLY in Fig. 12.10(a)) tracks the absorption coefficient similar to elemental Si. After air exposure, the luminescence shifts to 700 nm. The TLY shows features from both Si and SiO_x, especially the 700 nm

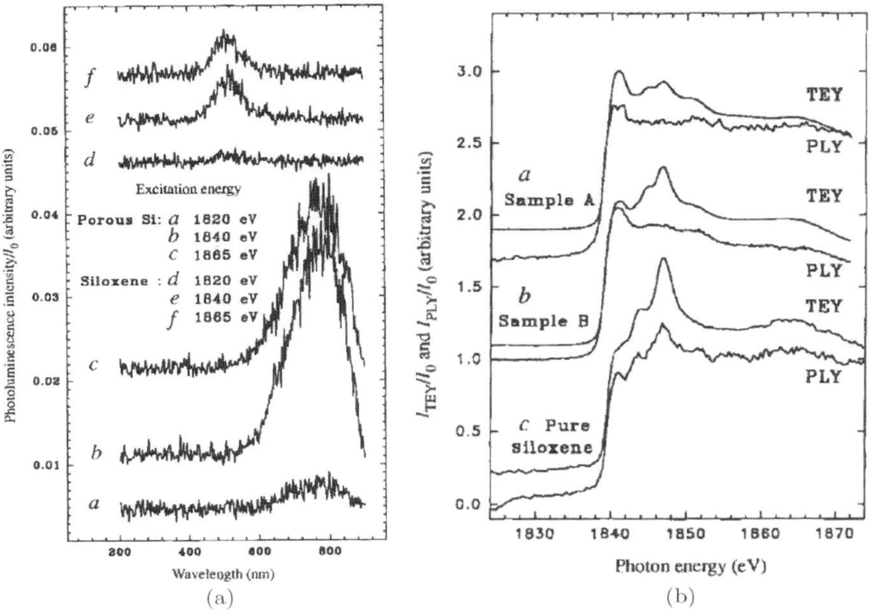

Fig. 12.9. (a) XEOL spectra from porous silicon and siloxene under selected excitation energies. (b) PLY-XANES of porous Si and siloxene across the Si K-edge. Sample a: freshly made porous Si, sample b: air-oxidized porous Si, sample c: pure siloxene. Adapted with permission from [33].

partial luminescence (PLY). The $AgNO_3$ treated porous Si results in further complicated luminescence, located at 450 nm, 580 nm, and a noticeable 340 nm band. The optical responses of luminescence at different regions behave differently across the Si K-edge, which provides direct evidence of the luminescence origins of each band.

Silicon nanowire (SiNW) is a one-dimensional nanostructure and has a more complicated Si/SiO_x interface due to crystal defects and native oxide. Luminescence from SiNW has been reported frequently. Depending on the synthesis strategy and sample treatment, luminescence from SiNW is found throughout the entire visible spectra [43–45]. Combing XEOL with Si K-edge XANES, one can attribute the observed luminescence to the correct origin. In our previous work, we have studied the luminescence properties of chain-like Si nanowires, which have structures of Si nanoparticles embedded in a

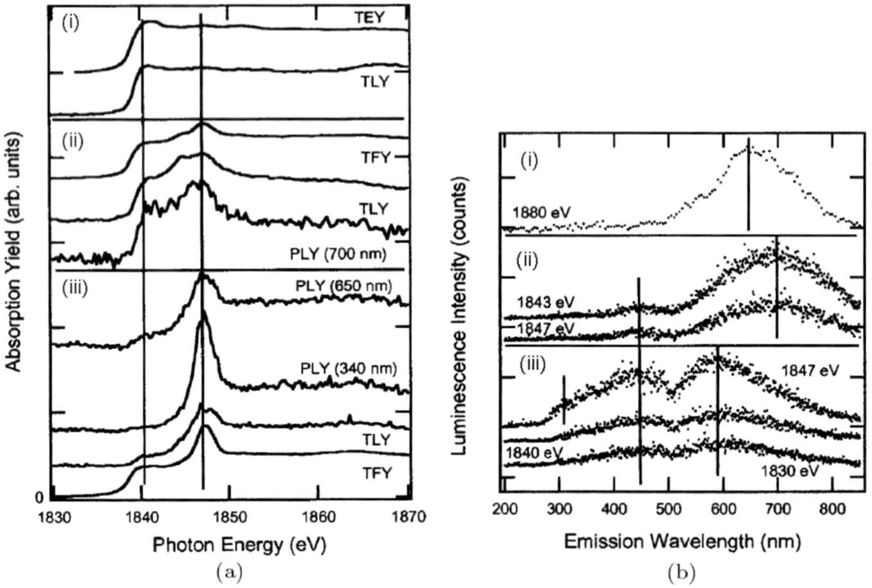

Fig. 12.10. (a) Si K-edge XANES and XEOL recorded in total luminescence yield (TLY)and partial luminescence yield (PLY); (b) Luminescence of porous silicon under different treatment excited at selected energies. Samples: (i) HF-refreshed porous Si, (ii) ambient-oxidized porous Si, (iii) $AgNO_3$-oxidized porous Si. Adapted with permission from Ref. 42. Copyright 2000, AIP Publishing LLC.

SiO_2 matrix [45]. The XEOL of the nanowires, shown in Fig. 12.11 exhibit an intense 460 nm emission with a 530 nm shoulder. From wavelength selected PLY, we can clearly see that the 460 nm emission is of pure SiO_2 feature, while the zero-order PLY contains resonance from elemental Si. As a result, we are able to conclude that the 460 nm emission is of oxide nature, while the 530 nm has a partial contribution from the nanosized Si and Si-SiO_2 interface.

A more careful study was carried out to examine thermal annealing induced luminescence from porous Si nanowires [19]. Figure 12.12 shows the XEOL and XANES at the Si K-edge of 800°C annealed porous Si nanowires. The 460 nm emission tracks the absorption profile of SiO_2, but a close look at the PLY reveals that the peak maximum exhibits a slight low-energy shift. Another emission band at 623 nm, however, follows the absorption profile of SiO_2. Although the

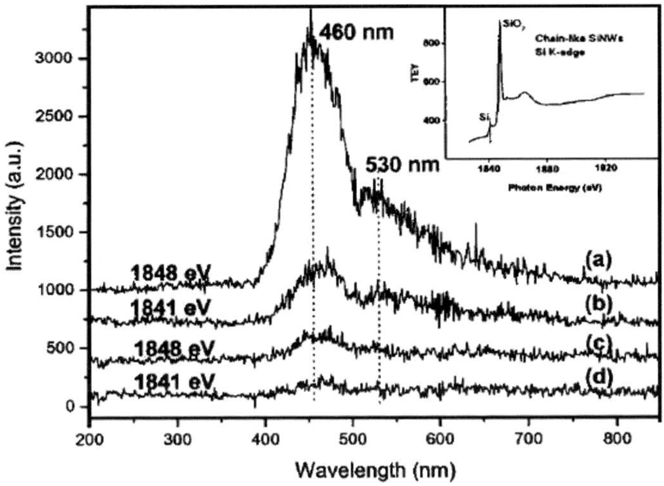

(a)

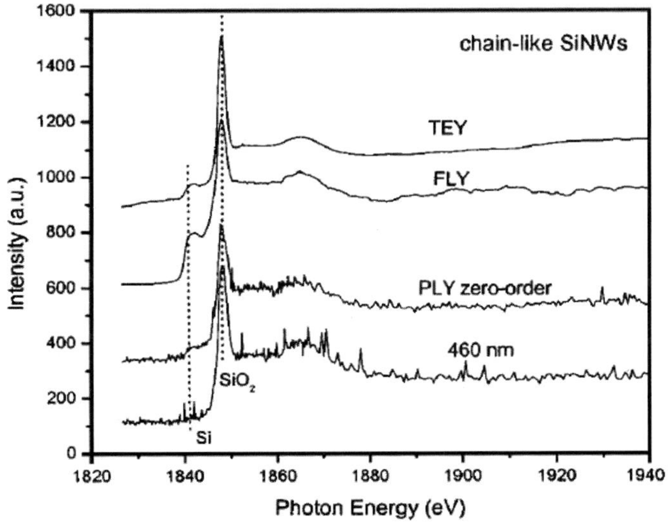

(b)

Fig. 12.11. (a) XEOL of chainlike Si nanowires (curves a and b) compared to normal Si nanowires (curves c and d) excited with energies at the Si K-edge and SiO_2 resonance position. (b) PLY and XANES of SiNW at the Si K-edge of chainlike Si nanowires. Adapted with permission from [45].

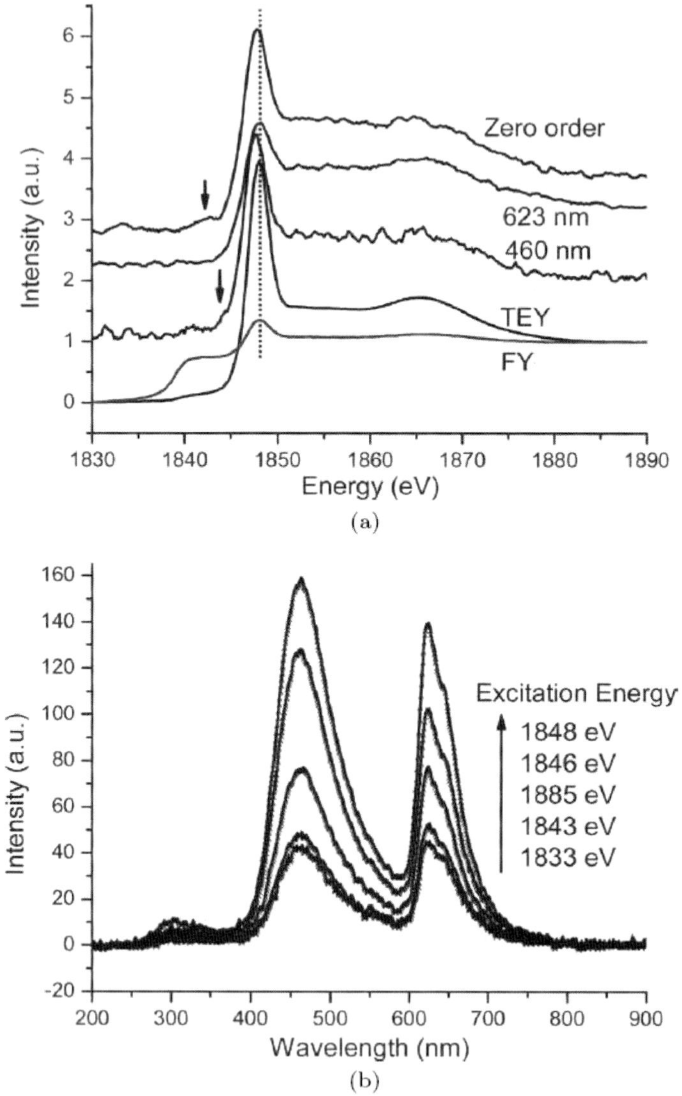

Fig. 12.12. (a) PLY-XANES and XEOL of oxidized porous Si nanowires at the Si K-edge, (b) selected XEOL spectra of excitation energy below to above the Si K-edge.

460 nm emission has widely accepted as an oxygen vacancy-related luminescence from SiO_2, PLY here provides direct evidence on the oxygen deficient nature (under coordinated Si) of such luminescence site. In contrast to the 623 nm, which is also a defect-related luminescence, is from fully coordinated SiO_2 site (i.e., oxygen rich).

12.3.1.2. *ZnO-ZnS*

Aside from nanostructured Si, ZnS and ZnO, nanostructures are also one of the representative systems that have been used in XEOL studies. ZnS is a wide band gap semiconductor, and it has a thermodynamically stable zinc blende phase (sphalerite) with band gap of 3.54 eV [46]. Bulk ZnS sphalerite is not light-emitting under room temperature. However, people have observed that ZnS of wurtzite structure exhibit interesting luminescence properties [20, 47, 48]. ZnS nanoribbons with dominantly wurtzite phase synthesized using Au-catalyzed high temperature method[48] are found having multiple emission bands. Shown in Fig. 12.13, ZnS nanoribbons show emission bands at 338 nm, 430 nm and 520 nm at low temperature, while only the 520 nm dominates at room temperature [47]. TEM image shows that a trace amount of sphalerite ZnS is present in the nanoribbons, especially at the interface between nanoribbons and the Au catalysts.

By monitoring the particular bands, and comparing to XANES of different phase ZnS as shown in Fig. 12.14, one can clearly see that the 338 nm emission tracks the absorption profile of wurtzite, while the 430 nm and 520 nm emission show more of the sphalerite feature. It is the first report where XEOL was used to distinguish the particular polytype involved in the luminescence.

Luminescence property of nanostructured ZnO has also been extensively studied [49–52]. ZnO has a band gap of 3.37 eV (wurtzite structure), and highly crystalline ZnO is a bright light emitter due to its direct band gap. As for nanostructured ZnO, their luminescence properties appear to be strongly morphology- and crystal quality-dependent [51, 53, 54]. Figure 12.15 shows the XEOL of ZnO of different morphologies [51]. The sharp narrow peak located at 380 nm is usually attributed to the near band gap emission. The broad green

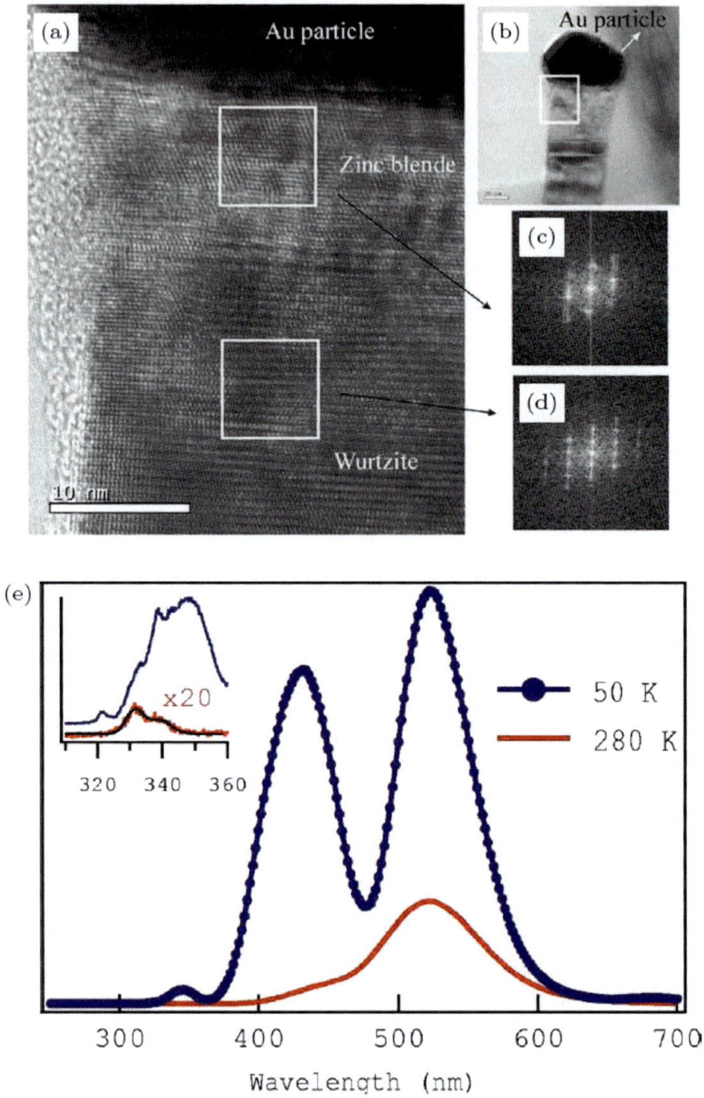

Fig. 12.13. (a) High-resolution TEM image of the boxed region of a ZnS nanoribbons shown in (b), (c) and (d) diffraction patterns of boxed regions in (a). (e) XEOL. Of ZnS nanoribbons excited using 1000 eV X-ray at 50 and 280 K. Adapted with permission from [47].

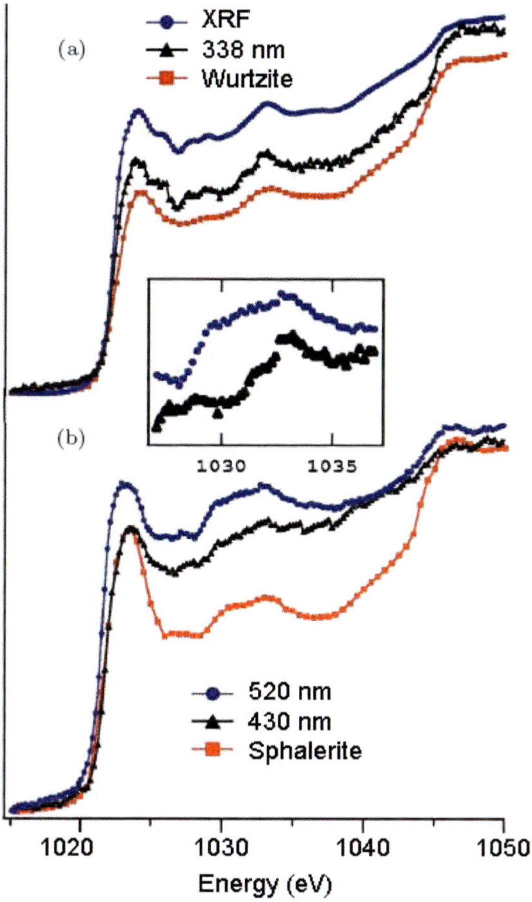

Fig. 12.14. XANES and PLY at selected wavelengths of ZnS nanoribbons at the Zn L_3-edge. Adapted with permission from [47].

emission (~480 nm) and a red emission (~770 nm) present in some samples are both defect-related. The luminescence from nanostructured ZnO will be discussed in more detail in the next section.

We then examine a heterostructure containing both ZnO and ZnS. Using a vapor-liquid-solid method, ZnO and ZnS form nanoribbons and are grown side by side, forming a biaxial structure. Such structure was found light-emitting, and the XEOL spectra are shown in Fig. 12.16 [10, 55]. It is found that the spectra obtained at selected

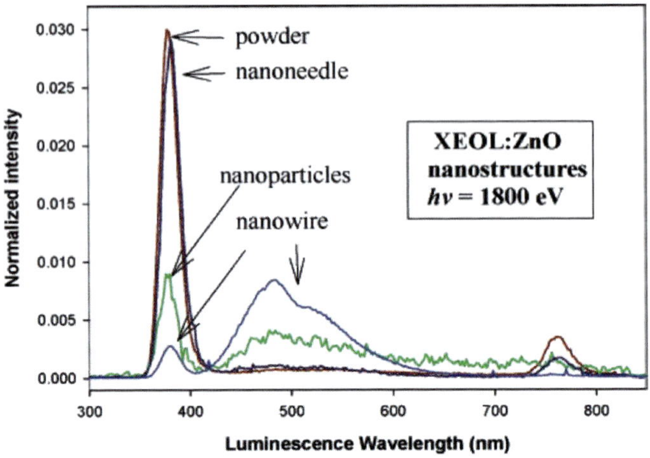

Fig. 12.15. XEOL of ZnO nanostructures of different morphologies. Reprinted with permission from [51].

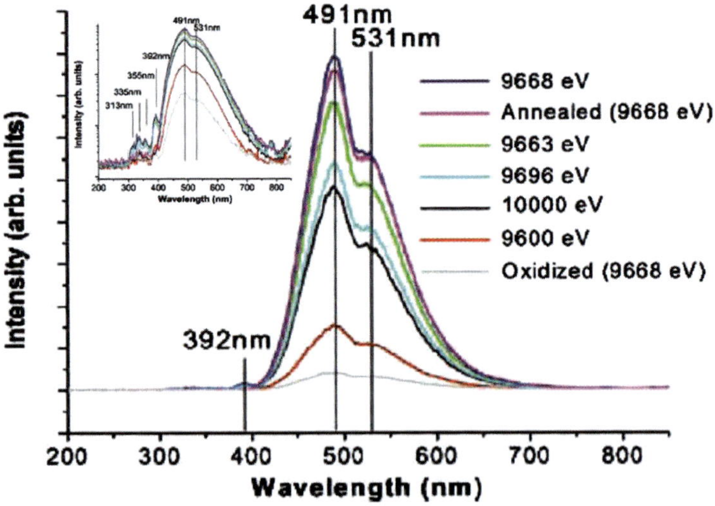

Fig. 12.16. XEOL spectra of ZnO-ZnS biaxial nanorods under various excitation energies. The inset shows the spectra plotted in natural log scale. Reprinted with permission from [10].

excitation energies across the Zn K-edge have similar profiles albeit intensities vary. XEOL shows an intense visible green emission at 491 nm with a 531 nm shoulder, which is similar to the XEOL of nanostructured ZnO in Fig. 12.15. A weak peak at 392 nm is noticeable, which is close to the ZnO near-band gap emission. However, all the characteristic ZnS luminescence is not observed. The near-band gap emission around 330 nm can only be identified when the spectra are plotted in natural log scale. The ZnS defect luminescence which is expected at 520 nm, is also missing.

However, simply based on XEOL result, it might be too soon to conclude that ZnO is the only emitter in this heterostructure. One might argue that the ZnS contribution could hide under the ZnO luminescence, especially their defected-related emission bands are close. This is also the limitation of laboratory luminescence measurements, since they just simply collect the total luminescence from all possible recombination channels between conduction band and valence band.

Here is when XEOL-XANES comes in handy. Based on the previous result of single component ZnS and ZnO nanostructure, we know that the PLY across the absorption edge tracks the XANES spectra. For this heterostructure sample, the S K-edge and Zn K-edge are chosen to examine the origin of the luminescence. As shown in Fig. 12.17(a), the PLY of single component ZnO and ZnS shows

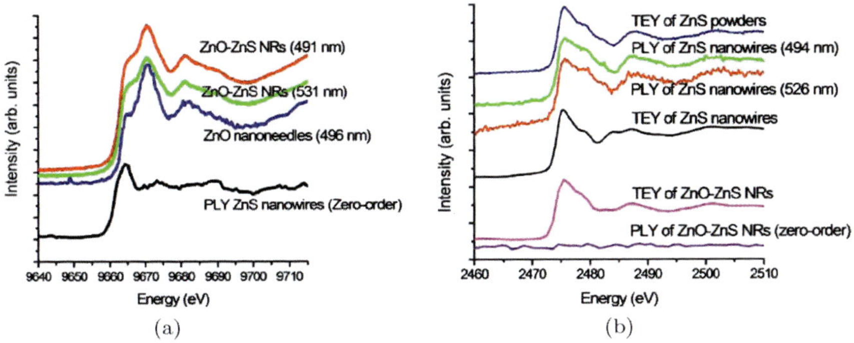

Fig. 12.17. (a) Zn K-edge and (b) S K-edge PLY and XANES of ZnO-ZnS biaxial nanorods. Reprinted with permission from [10].

distinct features above the edge. For biaxial ZnO-ZnS nanoribbons, the PLY spectra at both 491 nm and 531 nm wavelength track the emission of ZnO, while no ZnS feature is found. Another evidence is found when measuring the S K-edge PLY. Shown in Fig. 12.17(b), although the TEY of ZnO-ZnS nanorods shows ZnS features, but there is no variation of PLY across the S K-edge. XANES and XEOL provide clear evidence that although ZnO and ZnS are both light-emitters on their own, when combined, the ZnS luminescence contribution is negligible. The incoming X-ray photons absorbed by ZnS are released through a non-radiative decay.

12.3.2. *Surface states and defect in nanomaterials*

Nanomaterials often contain various defects, which strongly affect the luminescence properties of the materials. Using conventional luminescence detection techniques, it is difficult to correlate the type of defect with the observed luminescence band. A combination analysis of XEOL and TR-XEOL has been proved to be of particular useful to solve this problem. The exciton recombination is much faster than the recombination through defect centers. Upon applying gated time window, one is able to capture the event that takes place in a very short lifetime.

12.3.2.1. *SnO_2 nanomaterials*

Bulk SnO_2 is a semiconductor with a band gap of 3.6 eV (room temperature). It is not light-emitting unless it is made into a nanostructure. Visible luminescence has been observed from SnO_2 nanoparticles, nanowires, nanobelts [56, 57]. The luminescence spectra usually exhibits a broad peak centered at ~600 nm, and has been attributed to surface defects or oxygen vacancies. Zhou *et al.* have investigated a ribbon-like SnO_2 nanostructure using XEOL and TRXEOL [58]. The SnO_2 nanoribbons were grown by thermal evaporation using SnO as precursor. TR-XEOL using different time windows are shown in Fig. 12.18(a), with excitation energy at 491.1 eV. Although for ungated ones (0–570 ns, fully open), the luminescence

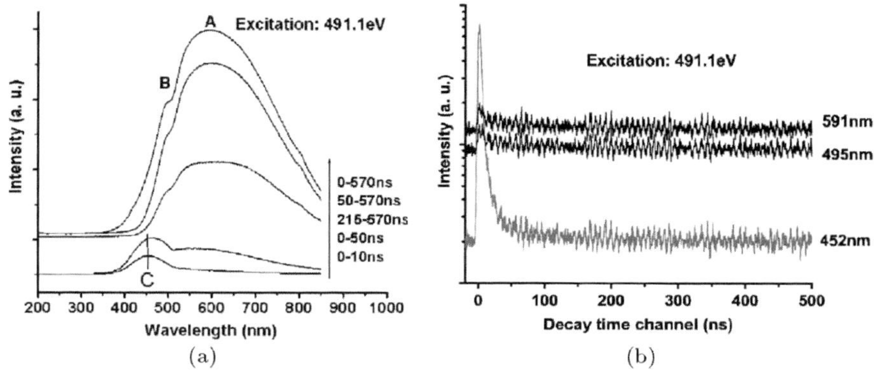

Fig. 12.18. (a) TR-XEOL spectra of SnO_2 excited at 491.1 eV photon energy and collected at selected time windows. (b) decay behavior of the emission bands indicated as A, B, and C in (a). Reprinted with permission from [58].

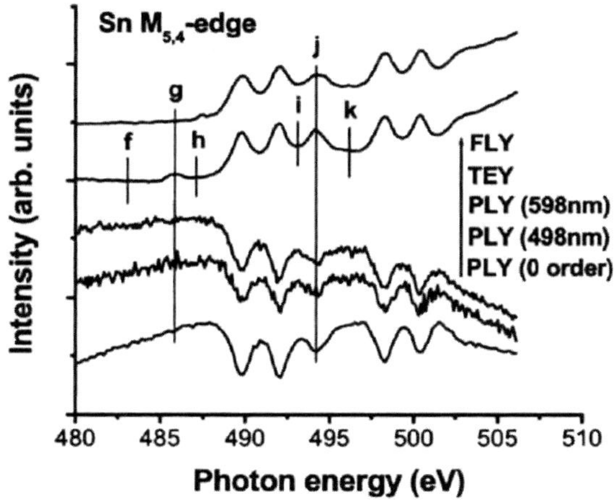

Fig. 12.19. Sn $M_{5,4}$-edge XANES and PLY of SnO_2 nanoribbons. Reprinted with permission from [59].

is dominated by visible luminescence, using fast windows (0–50 ns and 0–10 ns) clearly reveals an emission band at 452 nm. The decay behavior, shown in Fig. 12.18(b), shows that the 452 nm luminescence has a short lifetime.

When we examine the XANES of SnO_2 nanoribbons (Fig. 12.19), the two detection modes, TEY and FLY of surface and bulk sensitivity, respectively, show slight differences: an emerging of peak g and enhancing of peak j in TEY. The two features are associated with surface states, due to a large fraction of undercoordinated Sn atoms resulting from O vacancies. A careful examination of the energy difference between the surface state g and the conduction band minimum yields a value of 2.7 eV, which corresponds well with the fast decay band at 452 nm. So this band can be attributed to the recombination between electron at CB minimum and hole at the surface state. In a follow-up study, they measured the optical response of the two visible bands along the excitation energy across the Sn $M_{5,4}$-edge [59]. It can be seen that the PLY of the two visible bands tracks all the features in the XANES profiles but the defect states (feature g and i). PLY shows selective specificity, and its amplitude changes only when Sn at the normal lattice position is excited. As a result, it is reasonable to conclude that the observed visible luminescence is from the decay of trapped electrons in O vacancies to an intrinsic surface state. Excitation to a surface state in the lattice will not open up a new channel to the radiative decay, since the luminescence from charge carrier migration to defect sites still dominates. It is a good example showing that when the luminescence from a system is due to the energy transfer from absorbing ions to the luminescence center (e.g. surface defect), selectively opening the channel to surface state will not affect the overall luminescence intensity. In this case, PLY will lose specificity at that site.

12.3.3. *ZnO nanomaterials*

In Section 12.3.1.2, we have briefly mentioned that the luminescence from ZnO is highly dependent on the crystal quality and sample morphology. The commonly observed ZnO luminescence spectrum has a narrow near-band gap emission and a broad defect emission, the intensity ratio varies from case to case. It is important to understand the origin of the luminescence, especially the defect luminescence ZnO, in order to achieve better control in device design.

Anisotropic optical yield of ZnO nanostructure was first observed by Rosenberg *et al.* [60]. They examined ZnO nanowires and

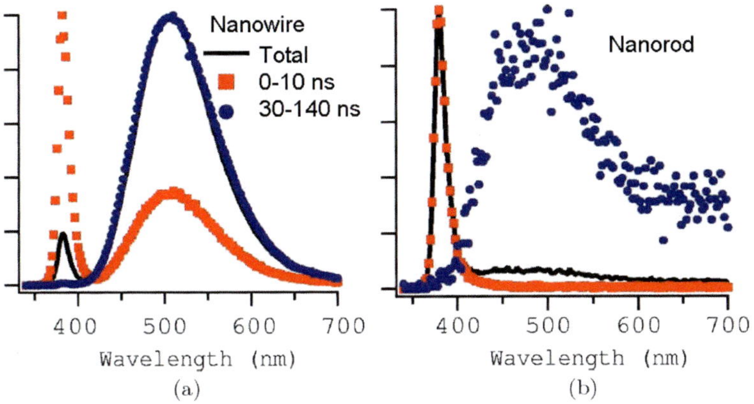

Fig. 12.20. TR-XEOL of (a) nanowires ZnO and (b) nanorods ZnO excited at 1050 eV. Reprinted with permission from [60].

nanorods of different luminescence behaviors. Figure 12.20 shows the TR-XEOL of the two ZnO samples excited at 1050 eV. In ungated spectra, the XEOL of nanowires is dominated by defect emission (broad band at ~500 nm), and near-band gap emission is relatively weak. The nanorods exhibit the opposite peak ratio, in which the luminescence is primarily from the band gap. The time-gated XEOL emphasizes the luminescence that of different lifetimes. It is clearly seen that the fast window (0–10 ns) captures the near-band gap luminescence (and some defect luminescence), while the slow window (30–140 ns) solely captures the defect emission. Close examination reveals that the slow emission from nanorod is slightly different from the one in nanowire.

While monitoring the two emission bands individually, PLY across the O K-edge show remarkably different features, shown in Fig. 12.21. The TEY spectra of the two ZnO, shown as reference, are similar, which represents the transition from O 1*s* electron to final states of *p* symmetry. The features (indicated by arrows) at 533 eV and 537 eV belong to the O p_z and p_{x+y} states, respectively. The p_z orbitals are along the *c*-axis of the nanowires, while the p_{x+y} orbitals are aligned toward the surface. X-ray is tuned to selectively excite O 1*s* electrons to a specific channel, the corresponding luminescence is

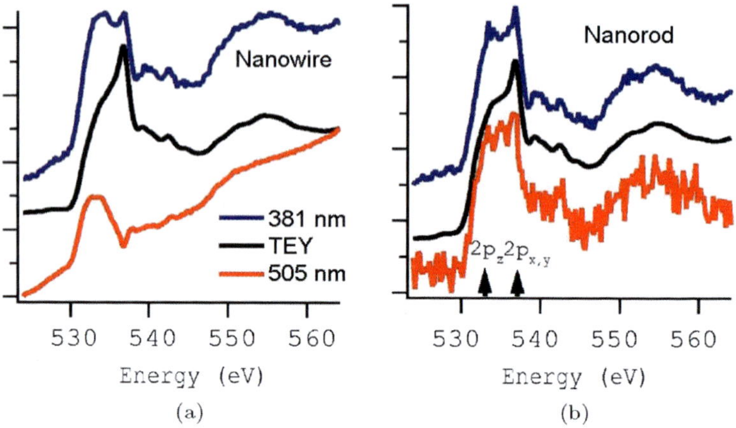

Fig. 12.21. PLY and TEY of (a) ZnO nanowires and (b) ZnO nanorods at the O K-edge. Reprinted with permission from [60].

produced. From the figure, we can see that the radiative de-excitation of the p_z states is much more efficient as the p_{x+y} states. Nanowires in particular, the defect emission is only from the p_z state decay. The experimental results prove the theoretical prediction which says c-axis is the unique axis for dipole formation, so that the luminescence is enhanced by a preferred radiating dipole. Luminescence from nanorods, on the other hand, does not show a significant anisotropy PLY. This suggests although the defect emission from the two samples are at similar position, PLY-XANES shows they are of different luminescence origin.

12.3.3.1. *GeO_2 nanomaterials*

GeO_2 nanomaterial is another class of semiconductors that recently has attracted much attention due to its unique electronic and optical properties. Its large band gap (5 eV) makes it a good candidate as a host material for developing luminescence devices. We investigated luminescence properties of GeO_2 nanowires in relate to their electronic structures [61]. Shown in Fig. 12.22, an intense visible luminescence is seen under X-ray excitation. The intensity varies as the incident photon energy tuned across the absorption edge of Ge and O.

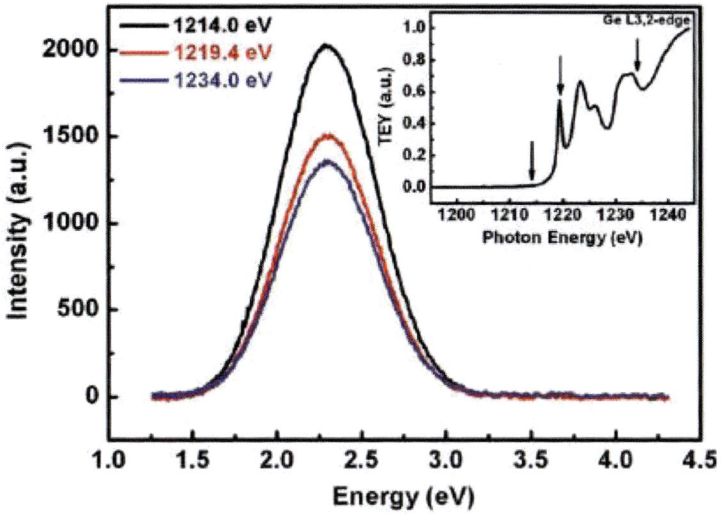

(a)

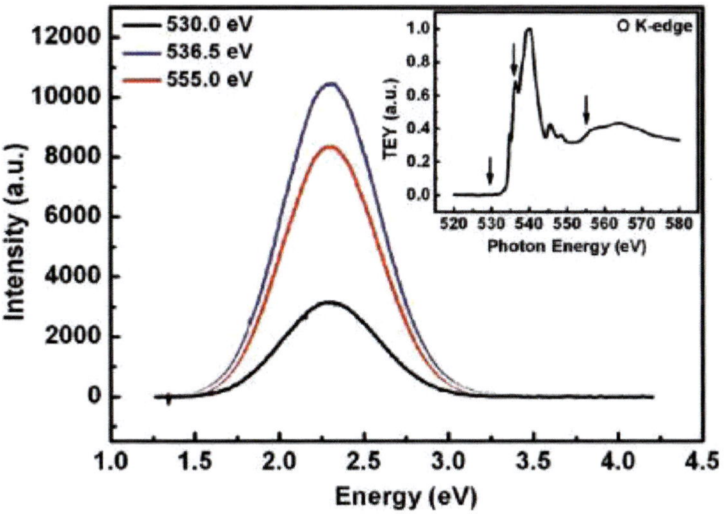

(b)

Fig. 12.22. XEOL of GeO_2 nanowires excited across (a) the Ge L-edge and (b) the O K-edge. The insets show the XANES at the corresponding edge and the arrows indicate selected excited energies for obtaining XEOL spectra. Reprinted with permission from [61].

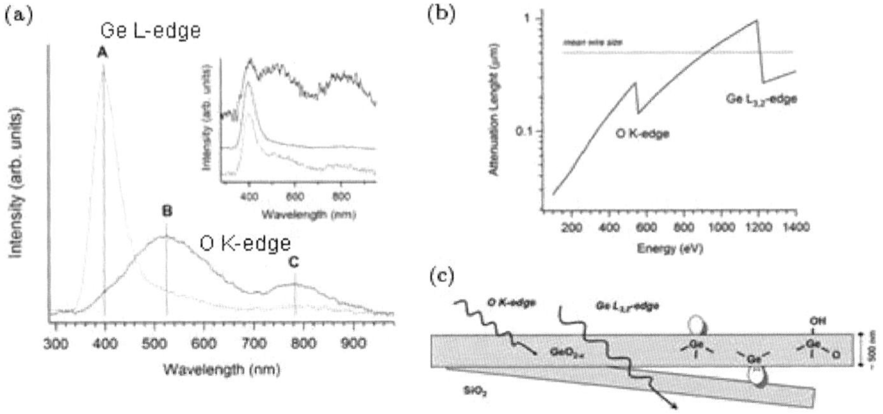

Fig. 12.23. (a) XEOL of GeO_2 nanowires at different excitation energies, (b) X-ray attenuation length of GeO_2 as a function of excitation energy, (c) shematic model of the GeO_2 nanowires on a SiO_2 substrate with surface, bulk, and interface defects. Adapted with permission from [62].

A more interesting phenomenon observed by Armelao *et al.* on the excitation energy-sensitivity of luminescence from GeO_2 nanowires [62]. Shown in Fig. 12.23, GeO_2 nanowires grown by a similar strategy as Peng *et al.*, exhibit very different emission spectra when excited at the O K- and Ge L_3-edge. At the O K-edge (530 eV), the GeO_2 nanowires exhibit two emission bands at 525 nm and 780 nm. Those two emissions, however, are replaced by an intense violet emission at 410 nm when the energy is tuned to the Ge L_3-edge (1220 eV). More interestingly, another abrupt transition took place when the energy is tuned from Ge L_3-edge resonance (1220 eV), to off-resonance 1230 eV, where the violet emission weakens significantly, and the other two emission bands appear.

Such phenomenon was attributed to the X-ray penetration-depth dependency of the observed luminescence. Shown in Fig. 12.23(b), the X-ray attenuation length gradually increases as the energy increases. When it reaches to the binding energy of certain core level of the element, due to the sudden increase in the absorption coefficient, the penetration depth of the X-ray abruptly drops. In GeO_2 nanowires configuration, the nanowires are grown in a random direction on a SiO_2 substrate, the defects present in the nanostructure are depth dependent: the surface is dominated by Ge-OH or dangling bonds, the bulk GeO_2 contains oxygen vacancies, and the GeO_2-SiO_2 contains two-folded Ge centers. Higher excitation energy (at the vicinity of Ge L_3-edge), it is likely to probe deep-lying defect, and the argument is further supported by doing angular dependent measurements [62]. In this manner, XEOL is successfully used as an effective analytical tool to trace the origin of the defect luminescence.

12.3.4. *Organic materials*

In organic molecules, the radiative recombination takes place at the HOMO-LUMO gap. Since the frontier orbitals can be calculated precisely, the site contribution to HOMO and LUMO bands is clear. By scanning across the absorption edge of the element (usually C, N, or O) and measuring the XEOL along the XANES, the origin of the luminescence is able to be assigned with element and site specificity. The pioneer work of XEOL study on OLED materials was conducted by Sham *et al.* in 2000 [63]. They compared the luminescence spectra of poly-vinyl-carbazole film using photoluminescence and X-ray, and proved the validity of investigating luminescence of organic molecule using XEOL. They pointed out the potential achieving site selectivity by doing XEOL-XANES. Later on, XEOL has been used to study lots of organic and organometallic compounds with aromatic ring structures containing C and N. The site specified luminescence properties of these molecules are revealed.

12.3.4.1. *Tris (8-hydroxyquinoline) aluminum (Alq_3)*

Tris (8-hydroxyquinoline) aluminum (Alq_3) is the foundation for small-molecule organic light-emitting device (OLED) research [64]. Extensive research has done to understand the electronic and optical properties of this molecule. Naftel *et al.* utilize XEOL to explore the origin of site specificity in Alq_3 luminescence [65]. Shown in Fig. 12.24, the C K-edge XANES of Alq_3 exhibit complex features. The first doublet at 286–288 eV is attributed to the transition of C

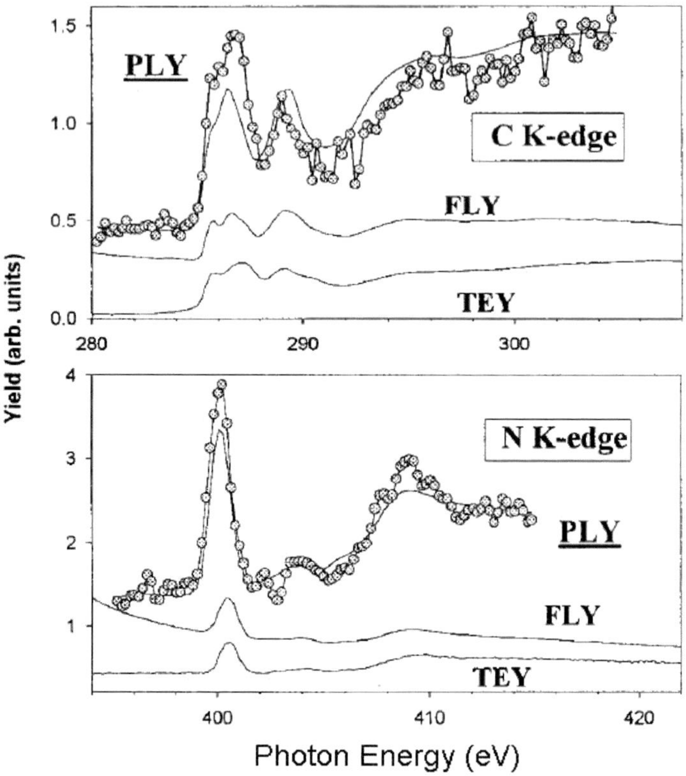

Fig. 12.24. TEY, FLY and total PLY XANES of Alq_3 at the C and N K edges. Reprinted with permission from [64].

$1s$ to π^* of aromatic carbons, while oscillations beyond 288 eV are transition to other C related features. It can be seen based on the peak ratio between the first doublet and the second peak at ~290 eV is greatly enhanced in PLY, compared to TEY or FLY. This indicates transitions to aromatic carbon π^* orbital facilitate the luminescence, which show some degree of site specificity. Similar behavior can be observed in the N K-edge XANES, too, in which the first N π^* peak is greatly enhanced in PLY. However, no enhancement was observed when exciting across the O K-edge and Al K-edge. It proves that the luminescence produced by Alq_3 is localized at aromatic C and N π^* region.

12.3.4.2. *Poly(N-vinylcarbazole)*

Poly(N-vinylcarbazole), denoted as PVK, is a hole conducting polymer, and has been explored as blue light emitting diodes. Photoluminescence measurements have shown that PVK emits light at 400 nm with a 380 nm shoulder which is from intrachain excimers. Naftel *et al.* used XEOL to examine the site-specific luminescence of this compound [66]. Under X-ray excitation, the luminescence spectrum of PVK is similar to PL results. Higher excitation energy reveals several new emission bands, which are attributed to the involvement of deeper lying LUMO orbitals. The most interesting features, however, is the optical response of the main emission across the absorption edge. Shown in Fig. 12.25, PLY of PVK at the C K-edge and the N K-edge are plotted together with XANES.

Similar to Alq3, at the C K-edge, three transitions within 285 eV and 292 eV arise from the C 1s to LUMO of p character. The first two peaks are attributed to transitions from C 1s to π^*. The splitting of π^* orbital is due to the lowering of symmetry upon the presence of N atoms, which is the similar case as the one in Alq3. PLY spectrum of PVK tracks the profiles of TEY and FY, suggesting the effective excimer recombination at the C site. An enhanced peak at 288 eV, at which the transition of C π^* of pyrrole carbon (C-N) character is directly populated. A similar situation is also seen in Alq_3, and the

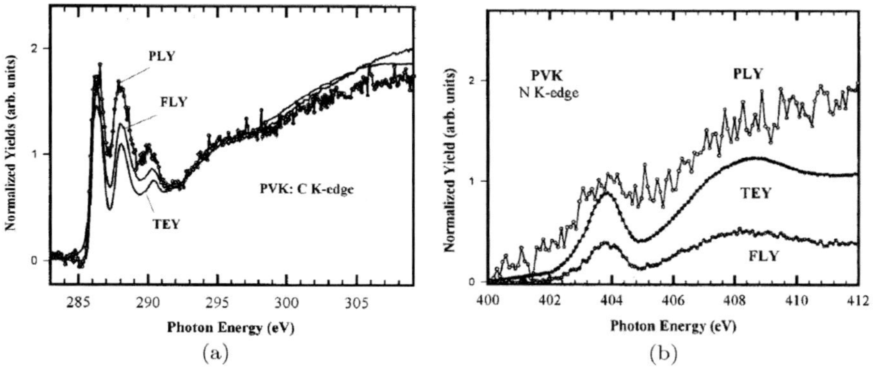

Fig. 12.25. TEY, FLY, and PLY XANES of PVK at (a) the C K-edge and (b) the N K-edge. Reprinted with permission from [66].

latter shows a bigger enhancement of the C-N π^* feature. As for the N K-edge, unlike Alq_3, in which a significant increase of the π^*-related luminescence is observed, the N K-edge PLY of PVK shows less sensitivity to the modulation of absorption coefficient. It indicates the N core excitation channel is not optically favorable. Although N is present in both Alq_3 and PVK, N is part of the aromatic ring in Alq3 but not in PVK. PLY XANES proves that the roles of N in participating the luminescence of the two compounds are entirely different.

12.3.4.3. *Other systems*

Due to the complexity of organic molecules and the limited space, we are not going to list all the studies done using XEOL and XANES. For example, P. S. G. Kim *et al.* have performed extensive studies on Au centered complexes with various ligands [67–69], J. Lv *et al.* have studied the morphology dependency of 2,5,8,11-tetra-tert-buylperylene crystals [70]. N. Peltekis *et al.* investigated the luminescence properties of magnesium phthalocyanine [71]. All research is established on the ultrahigh resolution of XANES at the C and N K-edge, so that selective excitation to a particular molecular level can be achieved. Taking advantage of the element and site specificity of XEOL, one can find the most effective luminescence channel in the molecule, and to establish a link between observed luminescence and its electronic structure.

12.4. Summary and Concluding Remarks

Unlike PL or CL, which is relatively straight forward, probes the recombination between excited valence electrons and holes left behind. The luminescence is either associated with band gap, traps, or luminescence centers. XEOL however, admittedly is a rather complicated process, especially when we are trying to utilize the element and site selectivity for optical property analysis. One of the reasons that XEOL is not yet widely used is that for luminescence characterizations, PL and CL techniques come rather handy, so people usually do not bother going to a synchrotron source to obtain a similar result. However, when choosing the right system, synchrotron-based XEOL and TR-XEOL in conjunction with XANES analysis could provide valuable information that would not otherwise be obtained by other means.

Bibliography

[1] B. H. Kelves, *Naked to the Bone: Medical Imaging in the Twentieth Century* (Rusgers University Press, New Brunswick, New Jersey, 1997).

[2] Y. F. Hu, K. H. Tan, P. S. Kim, P. Zhang, S. J. Naftel, T. K. Sham, I. Coulthard, and B. W. Yates, *Rev. of Scientific Instruments* **73**, 1379, (2002).

[3] P. Zhang, P. S. Kim and T. K. Sham, *Journal of Applied Physics* **91**, 6038, (2002).

[4] L. Liu, J. Chan and T.-K. Sham, *Journal of Physical Chemistry C* **114**, 21353, (2010).

[5] R. A. Rosenberg, Y. Choi, K. Vijayalakshmi, M. Kareev, J. Tchakhalian, S. Balaz and L. J. Brillson, *Applied Physics Letters* **102**, 192910, (2013).

[6] N. Subramanian, R. Kesavamoorthy, K. G. Rajan, M. Yousuf, S. Bera and S. V. Narasimhan, *Journal of Physics-Condensed Matter* **9**, 4769, (1997).

[7] M. Ishii, Y. Tanaka, S. Komuro, T. Morikawa, Y. Aoyagi and T. Ishikawa, *Journal of Electron Spectroscopy and Related Phenomena* **114**, 521, (2001).

[8] P. Tola, A. Retournard, J. Dexpertghys, M. Lemonnier, M. Pagel and J. Goulon, *Chemical Physics* **78**, 339, (1983).

[9] T. Paik, T. R. Gordon, A. M. Prantner, H. Yun and C. B. Murray, *Acs Nano* **7**, 2850, (2013).

[10] M. W. Murphy, X. T. Zhou, J. Y. P. Ko, J. G. Zhou, F. Heigl and T. K. Sham, *Journal of Chemical Physics* **130**, 084707, (2009).

[11] M. J. Ward, W.-Q. Han and T.-K. Sham, *Journal of Physical Chemistry C* **115**, 20507, (2011).

[12] A. P. D'Silva and V. A. Fassel, X-ray excited optical luminescence of the rare earths in *Handbook on the Physics and Chemistry of Rare Earths* (North-Holland, New York, 1979).

[13] A. Rogalev and J. Goulon, in *Chemical Applications of Synchrotron Radiation*, ed. T. K. Sham (World Scientific, 2002).

[14] T.-K. Sham and R. A. Rosenberg, *Chemphyschem* **8**, 2557, (2007).

[15] T. K. Sham, *International Journal of Nanotechnology* **5**, 1194, (2008).

[16] D. R. Viji, *Luminescence of Solids* (Plenum, New York, 1998).

[17] L. Soderholm, G. K. Liu, M. R. Antonio and F. W. Lytle, *Journal of Chemical Physics* **109**, 6745, (1998).

[18] N. Daldosso, F. Rocca, G. Dalba, P. Fornasini and R. Grisenti, *Journal of Porous Materials* **7**, 169, (2000).

[19] L. Liu and T.-K. Sham, *Small* **8**, 2371, (2012).

[20] R. A. Rosenberg, G. K. Shenoy, F. Heigl, S. T. Lee, P. S. G. Kim, X. T. Zhou and T. K. Sham, *Applied Physics Letters* **86**, 263115, (2005).
[21] R. A. Rosenberg, G. K. Shenoy, F. Heigl, S.-T. Lee, P. S. G. Kim, X. T. Zhou and T. K. Sham, *Appl. Phys. Lett.* **86**, 263115, (2005).
[22] L. Armelao, F. Heigl, S. Brunet, R. Sammynaiken, T. Regier, R. I. R. Blyth, L. Zuin, R. Sankari, J. Vogt and T.-K. Sham, *Chemphyschem* **11**, 3625, (2010).
[23] R. A. Rosenberg, C. R. Wen, K. Tan and J. M. Chen, *J. Chem. Phys.* **92**, 5196, (1990).
[24] T. K. Sham, P. Kristof and R. A. Holroyd, *Rev. Sci. Instrum.* **63**, 1198, (1992).
[25] F. Heigl, A. Jürgensen, X. T. Zhou, S. Lam, M. Murphy, J. Y. P. Ko, T. K. Sham, R. A. Rosenberg, R. Gordon, D. Brewe, T. Regier and L. Armelao, **879**, 1202, (2007).
[26] T. Z. Regier, J. M. Vogt, R. Sammynaiken, T. K. Sham, R. Garrett, I. Gentle, K. Nugent and S. Wilkins, 838, (2010).
[27] R. P. Taylor, A. A. Finch, J. F. W. Mosselmans and P. D. Quinn, *Journal of Luminescence* **134**, 49, (2013).
[28] M. J. Ward, T. Z. Regier, J. M. Vogt, R. A. Gordon, W. Q. Han and T. K. Sham, *Journal of Physics: Conference Series* **425**, 092006, (2013).
[29] K. Hayashi, T. Hayashi, T. Shishido, E. Matsubara, H. Makino, T. Yao and T. Matsushita, *Physical Review B* **76**, (2007).
[30] J. Villanova, J. Segura-Ruiz, T. Lafford and G. Martinez-Criado, *Journal of Synchrotron Radiation* **19**, 521, (2012).
[31] L. Liu, T.-K. Sham, W. Han, C. Zhi and Y. Bando, *Acs Nano* **5**, 631, (2011).
[32] X-ray calculator: http://henke.lbl.gov/optical_constants/
[33] T. K. Sham, D. T. Jiang, I. Coulthard, J. W. Lorimer, X. H. Feng, K. H. Tan, S. P. Frigo, R. A. Rosenberg, D. C. Houghton and B. Bryskiewicz, *Nature* **363**, 331, (1993).
[34] A. G. Cullis, L. T. Canham and P. D. Calcott, *Appl. Phys. Rev.* **82**, 909, (1997).
[35] L. T. Canham, *Appl. Phys. Lett.* **57**, 1046, (1990).
[36] D. S. English, L. E. Pell, Z. Yu, P. F. Barbara and B. A. Korgel, *Nano Lett.* **2**, 681, (2002).
[37] Y. D. Glinka, S. H. Lin, L. P. Hwang, Y. T. Chen and N. H. Tolk, *Phys. Rev. B* **64**, 085421, (2001).
[38] G. Pacchioni and G. Lerano, *Phys. Rev. Lett.* **79**, 753, (1997).
[39] C. Delerue, G. Allan and M. Lannoo, *Phys. Rev. B* **48**, 11024, (1993).
[40] T. Takagahara and K. Takeda, *Phys. Rev. B* **46**, 15578, (1992).
[41] J. H. Stathis and M. A. Kastner, *Phys. Rev. B* **35**, 2972, (1987).
[42] I. Coulthard, W. J. Antel Jr., J. W. Freeland, T. K. Sham, S. J. Naftel and P. Zhang, *Appl. Phys. Lett.* **77**, 498, (2000).
[43] T. K. Sham, S. J. Naftel, P. S. G. Kim, R. Sammynaiken, Y. H. Tang, I. Coulthard, A. Moewes, J. W. Freeland, Y. F. Hu and S. T. Lee, *Physical Review B* **70**, (2004).

[44] D. P. Yu, Q. L. Hang, Y. Ding, H. Z. Zhang, Z. G. Gai, J. J. Wang, Y. H. Zou, W. Qian, G. C. Xiong and S. Q. Feng, *Appl. Phys. Lett.* **73**, 3076, (1998).

[45] X. H. Sun, N. B. Wong, C. P. Li, S. T. Lee and T. K. Sham, *Journal of Applied Physics* **96**, 3447, (2004).

[46] U. Ozgur, Y. I. Alivov, C. Liu, A. Teke, M. A. Reshchikov, S. Dogan, V. Avrutin, S. J. Cho and H. Morkoc, *J. Appl. Phys.* **98**, 041301 (2005).

[47] R. A. Rosenberg, G. K. Shenoy, F. Heigl, S. T. Lee, P. S. G. Kim, X. T. Zhou and T. K. Sham, *Appl. Phys. Lett.* **87**, 253105, (2005).

[48] X. T. Zhou, P. S. G. Kim, T. K. Sham and S. T. Lee, *J. Appl. Phys.* **98**, 024312, (2005).

[49] S. Hayakawa, T. Hirose, L. Yan, M. Morishita, H. Kuwano and Y. Gohshi, *X-Ray Spectrometry* **28**, 515, (1999).

[50] N. E. Hsu, W. K. Hung and Y. F. Chen, *Journal of Applied Physics* **96**, 4671, (2004).

[51] X. H. Sun, S. Lam, T. K. Sham, F. Heigl, A. Jurgensen and N. B. Wong, *Journal of Physical Chemistry B* **109**, 3120, (2005).

[52] S. Larcheri, C. Armellini, F. Rocca, A. Kuzmin, R. Kalendarev, G. Dalba, R. Graziola, J. Purans, D. Pailharey and F. Jandard, *Superlattices and Microstructures* **39**, 267, (2006).

[53] O. Lobacheva, M. W. Murphy, J. Y. P. Ko and T.-K. Sham, *Canadian Journal of Chemistry-Revue Canadienne De Chimie* **87**, 1255, (2009).

[54] J. Smith, A. Akbari-Sharbaf, M. J. Ward, M. W. Murphy, G. Fanchini and T. K. Sham, *Journal of Applied Physics* **113**, (2013).

[55] M. Murphy, P. S. G. Kim, X. T. Zhou, J. G. Zhou, M. Coulliard, G. A. Botton and T. K. Sham, *J. Phys. Chem. C* **113**, 4755, (2009).

[56] J. Q. Hu, Y. Bando, Q. L. Liu and D. Golberg, *Adv. Funct. Mater.* **13**, 493, (2003).

[57] D. Calestani, L. Lazzarini, G. Salviati and M. Zha, *Cryst. Res. Technol.* **40**, 937, (2005).

[58] X. T. Zhou, F. Heigl, M. W. Murphy, T. K. Sham, T. Regier, I. Coulthard and R. I. R. Blyth, *Applied Physics Letters* **89**, (2006).

[59] X. T. Zhou, J. G. Zhou, M. W. Murphy, J. Y. P. Ko, F. Heigl, T. Regier, R. I. R. Blyth and T. K. Sham, *Journal of Chemical Physics* **128**, (2008).

[60] R. A. Rosenberg, G. K. Shenoy, L. C. Tien, D. Norton, S. Pearton, X. H. Sun and T. K. Sham, *Appl. Phys. Lett.* **89**, 093118, (2006).

[61] M. Peng, Y. Li, J. Gao, D. Zhang, Z. Jiang and X. Sun, *Journal of Physical Chemistry C* **115**, 11420, (2011).

[62] L. Armelao, F. Heigl, P.-S. G. Kim, R. A. Rosenberg, T. Z. Regier and T.-K. Sham, *Journal of Physical Chemistry C* **116**, 14163, (2012).

[63] T. K. Sham, R. Sammynaiken, Y. J. Zhu, P. Zhang, I. Coulthard and S. J. Naftel, *Thin Solid Films* **363**, 318, (2000).

[64] C. W. Tang and S. A. VanSlyke, *Appl. Phys. Lett.* **51**, 913, (1987).

[65] S. J. Naftel, P. Zhang, P. S. Kim, T. K. Sham, I. Coulthard, W. J. Antel, J. W. Freeland, S. P. Frigo, M. K. Fung, S. T. Lee, Y. F. Hu and B. W. Yates, *Applied Physics Letters* **78**, 1847, (2001).

[66] S. J. Naftel, P. S. G. Kim, T. K. Sham, R. Sammynaiken, B. W. Yates and Y. F. Hu, *Journal of Applied Physics* **93**, 5191, (2003).
[67] P. S. G. Kim, M. C. Brandys, Y. F. Hu, R. J. Puddephatt and T. K. Sham, *Journal of Luminescence* **105**, 21, (2003).
[68] P. S. G. Kim, Y. F. Hu, R. J. Puddephatt and T. K. Sham, *Physica Scripta* **T115**, 545, (2005).
[69] P. S. G. Kim, Y. F. Hu, Y. M. Yiu and T. K. Sham, *Journal of Electron Spectroscopy and Related Phenomena* **144**, 811, (2005).
[70] J. Lv, P. J. Y. Ko, Y. Zhang, L. Liu, X. Zhang, X. Zhang, X. Sun and T. K. Sham, *Applied Physics Letters* **98**, (2011).
[71] N. Peltekis, B. N. Holland, S. Krishnamurthy, I. T. McGovern, N. R. J. Poolton, S. Patel and C. McGuinness, *Journal of the American Chemical Society* **130**, 13008, (2008).

Chapter 13

A Compact X-ray Free-Electron Laser: SACLA

Hitoshi Tanaka, Takashi Tanaka, Kei Sawada, Makina Yabashi and Tetsuya Ishikawa

RIKEN SPring-8 Center Kouto 1-1-1, Sayo-cho, Hyogo 679-5148, Japan

Abstract

Japan constructed the first 'compact' X-ray free-electron laser (XFEL), and started user operation in 2012. This chapter describes the historical development of the compact XFEL concept, the underlying physical principles of XFEL, the details of the accelerator system, the details of the photon beam line, and the scientific applications, followed by a brief comment on the future trend of the XFEL facilities.

Keywords: Spontaneous radiation, undulator; shot noise; energy modulation; density modulation; microbunch; coherent radiation; FEL parameter; gain length; gain curve; laser saturation; exponential growth; optical guiding; electron source; thermionic-gun; low emittance injector; bunch compression; high gradient; C-band; short period; in-vacuum undulator; X-ray based alignment; reflective mirror; Be window; X-ray spectrometer; X-ray optics; diagnostic system; double-mirror system; double-crystal monochromator; elastic-emission machining; ultrahigh vacuum; beam monitor; focusing system; optical laser system; serial femtosecond crystallography (SFX); atom, molecular and optical science (AMO); coherent diffractive imaging (CDI); DAPHNIS; MPCCD detector; focusing; nonlinear; mutiphoton process; hollow atom; double-core hole; two-photon absorption; ultrafast phenomena; transient phenomena; chemical reaction; single shot; coherent diffraction imaging; structural analysis; protein; damage free.

13.1. Introduction

Coherent radiation in short-wavelength region was a long-awaited dream of scientists because it gives a powerful analytical tool for nano-world. The first practicable coherent X-ray source is Linac Coherent Light Source (LCLS) at the SLAC National Accelerator Laboratory in the U.S. which generated 8 keV coherent X-rays in 2009. Two years later, SPring-8 Angstrom Compact free-electron Laser (SACLA) started operation above 10 keV region. As of May 2014, only two coherent X-ray sources are in operation around the world, although a few sources will appear in Germany, Swiss, Korea and other countries within 5 years.

Extension of the classical laser invented in 1960 [1] to X-ray region was difficult due to the very short life time of the inner-shell energy levels. Instead, a free electron laser (FEL) was proposed by Madey in 1970 [2] as a method to generate coherent radiation regardless of the wavelength. Madey's FEL consists of an electron accelerator with an undulator which is placed in an optical resonant cavity. The undulator is usually composed of periodic alternating magnets which 'undulates' electron beam to generate photons. Inside the optical cavity, stimulated emission of the photons occurs in the undulator to produce coherent light. Although the Madey's scheme of FEL has no theoretical wavelength limit, the lack of high-reflectance mirrors for the optical cavity restricts the practical wavelength range. In X-ray region we do not have any high reflectance mirror at the normal incidence required for the optical cavities.

Realistic extension to the shorter wavelength became plausible when 'Self-Amplified Spontaneous Emission (SASE)' concept was discovered in mid 1980s [3]. When an extremely low-emittance electron bunches with fairly high peak current pass through a long undulator, interaction between emitted photons and electrons periodically modulates the electron density in the bunch with a period equal to the wavelength of photons that the undulator emit. This micro-bunched electrons wiggle coherently in the undulator to produce coherent radiation in the forward direction. This process of generating coherent radiation is called SASE. To prepare good quality electron bunches for SASE process to happen, we need linear accelerators which can easily produce low-emittance electron bunches, instead of ring ones in which the electron beam emittance is deteriorated by the synchrotron radiation process. The wavelength of the coherent radiation generated by the SASE process is that of the spontaneous radiation that the undulator generates. The wavelength of the on-axis undulator radiation is given by

$$\lambda_p = \frac{\lambda_m}{2\gamma^2}(1 + K^2) \tag{13.1}$$

where λ_p the photon wavelength, γ the electron energy in units of its rest energy, λ_m the period length of the undulator magnetic field, and K the undulator parameter [4]. Since the current technology can reduce the scale of λ_m down to dozens of mm, X-ray wavelength, say 0.1 nm, could be achievable with high energy electron beam with $\gamma \sim 10{,}000$ which corresponds to several GeV.

In 1990s Claudio Pellegrini and his coworkers proposed to build a coherent X-ray source based on SASE principle by using the existing linear electron accelerator at Stanford Linear Accelerator Center, which was later realized as LCLS. It uses 1/3 of the existing 2-mile linac to extract maximum 14.5 GeV electron beam followed by 130 m long undulator and photon beamlines [5]. The total length of the facility is as long as 2 km. Some Japanese scientists attended to the incubating meetings for LCLS without making any feedback to the Japanese community. Following to this US project, DESY group in late 1990s proposed TESLA project [6] which was a combination of an electron-positron linear collider for high energy physics and X-ray free electron laser (XFEL) for photon science. The linear collider was not approved for some reasons, but XFEL was decided to proceed as an EU project. This was the onset of the European X-ray Free Electron Laser project. It will construct a new superconducting linac inside an underground tunnel, followed by several undulator lines to produce coherent radiation with different wavelengths. The total length of the facility, 3.5 km, is even longer than that of LCLS. The European XFEL should have been completed in 2013, but is not yet completed as of May 2014. The hardware completion is announced to be within 2016, followed by the user operation in 2017.

One of the present authors (TI), and Dr. Hideo Kitamura who was in charge of undulator development in SPring-8, were frequently invited to the discussion meeting for the European XFEL. We were astonished by the wonderful performance of the SASE XFELs as coherent X-ray sources, but in the same time, felt disappointed to acknowledge the possible difficulty to construct many XFELs with the scale of LCLS or European XFEL because of the huge construction cost. Only a few parties could afford such expensive facilities. In addition, because SASE XFELs are the linear accelerator based light source capable of accommodating a smaller number of beamlines compared with the ring-based synchrotron radiation facilities, two or three XFEL facilities would never fulfill the demands of researchers. However, we realized that the adoption of the 'in-vacuum' undulators standardly utilized in SPring-8 storage ring, instead of 'out-of-vacuum' undulators used in both LCLS and European XFEL, could downsize the facility: smaller gap between magnets with the in-vacuum undulators could produce strong magnetic field for undulator operation along the electron beam path even when the magnetic period length is reduced. Since, in this case, we could make the electron energy smaller to generate a certain value of photon wavelength, the length of the linear accelerator would be shorter because it is more or less proportional to the electron energy. The idea of 'Compact SASE Source' was born in this way.

In 2001 we started an R&D program for the development of elementary technologies toward the compact SASE sources in RIKEN. We asked Dr. Tsumoru Shintake, who was then an associate professor in KEK, for helps in developing an injector and other accelerator components. He had been developing in KEK a C-band accelerator tube system which we proposed to use for the compact SASE source because its acceleration gradient was twice larger than the S-band system used in LCLS. He proposed a new type of injector system based on a classical thermionic electron gun followed by a velocity bunching system. SASE XFEL requires a low emittance, short bunched, and high density electron beam which, at the time, was believed to be generated only by laser-RF electron guns. A year later, Dr. Shintake joined RIKEN as a Chief Scientist. In the R&D program, he was in charge of accelerator development, Dr. Hideo Kitamura was in charge of in-vacuum undulator development, and Dr. Tetsuya Ishikawa was in charge of optics development. The R&D was successfully progressed until 2004, when we proposed construction of a prototype FEL working in the extreme ultra-violet (EUV) region.

In 2004 we were noticed that the ministry of education, culture, sports, science and technology in Japan (MEXT) would start a new projects to establish key technologies from 2006 fiscal year, to which we were invited to submit a proposal. We made a design study of a compact X-ray SASE source reaching 0.1 nm wavelength and published a conceptual design report (CDR) [7]. We held an International Review Meeting to access our proposal. The conclusion of the Review Meeting was generally positive, though the Review members admitted our project to be extremely 'ambitious'. The important recommendations that the Meeting gave us were (i) to change the linac energy from 6 GeV to 8 GeV to correspond to the full-energy injection to SPring-8 storage ring, and (ii) to prove the properness of our scheme by showing the lasing using the prototype machine. We invited Dr. Hitoshi Tanaka as a group leader of commissioning and operation. His group worked hard to lase at 49 nm wavelength the prototype machine, which was based on 250 MeV linac. The prototype machine, called SCSS (SPring-8 Compact SASE Source) was later used for both component development of SACLA and user experiments in EUV region.

We started construction of SACLA from 2006. Dr. Tetsuya Ishikawa was appointed as the project leader, and took responsibility for beamline/endstation construction as well. Dr. Noritaka Kumagai, who led the construction project of the SPring-8 storage ring, was invited to lead the accelerator construction group which made all the engineering tasks for constructions. In the accelerator construction group, Dr. Shintake led the accelerator R&D team, and Dr. Kitamura led the undulator R&D team. Construction progressed smoothly to find the completion in March 2011, on schedule and within the budget. At the time of accelerator completion, a FEL beamline with a double crystal monochromator as well as a MPCCD 2-D X-ray image detector were ready for operation.

Commissioning was carried out by a team led by Dr. Hitoshi Tanaka. Although the prepared electron beam based alignment described in the CDR [7] was not effective because of the difficulty of aligning the beam position monitors, but we found photon-based alignment was much more efficient. It was essential for the photon-based alignment to have made the monochromator and the image detector ready to use. It took only 2 months to observe the first lasing at 0.12 nm wavelength in the beginning of June 2011. Accelerator commissioning continued together with the beamline and end-station commissioning. In the late October 2011, we observed our shortest wavelength record of 0.063 nm. Finally in February 2012, we started user operation of SACLA with the same proposal review scheme as SPring-8.

Noteworthy features of SACLA, in addition to the first compact XFEL in the world, are: (i) it is the first, and the only one XFEL that collocate with third generation synchrotron light source, (ii) the low emittance electron beam of the SACLA linac could be injected to the SPring-8 storage ring in full energy, and (iii) there is an experimental hutch in which SACLA XFEL beam and SPring-8 X-ray undulator beam intersect. When we place a sample on the intersection, we can make such experiments that the SACLA XFEL beam is used as a pump pulse and SPring-8 undulator X-rays as probe beam. This chapter consists of seven sections including this introduction. The second section discusses the fundamentals of XFELs. The third and fourth sections describe SACLA hardware and beamlines, respectively. Applications of SACLA are discussed in the fifth section, followed by the future perspective (Section 13.6), and concluding summary (Section 13.7). The reader can find a relevant article from a bit different viewpoint in [8].

13.2. Fundamentals of XFELs

This section deals with fundamentals of SASE XFELs. The lasing mechanism is explained in a qualitative way and typical light source properties are given in universal forms with a couple of normalized parameters.

13.2.1. *Lasing mechanism*

As understood from the name containing the term "free electron", XFEL is another sort of laser source based on a principle different from conventional lasers and does not have any theoretical wavelength limit. XFEL is similar to ordinary synchrotron radiation (SR) in the point that an ultra-relativistic electron beam emit radiation. In both systems, the electron beam is accelerated by a high-energy accelerator and is injected to a device called an undulator generating a periodic magnetic field. Note, however, that two conditions should be satisfied to achieve lasing in XFELs: the quality of the electron beam is sufficiently high and the undulator is long enough. This is in contrast to conventional SR facilities, in which these two conditions may be desirable but not indispensable. Once the two conditions are satisfied, lasing takes place through four steps, which are explained in the following subsections in detail.

13.2.1.1. *Spontaneous radiation*

In common accelerators that make use of radio-frequency acceleration, the electron beam is bunched in the longitudinal direction. In XFEL accelerators, the bunch length l_p is typically of the order of several microns. When the electron beam is injected into the undulator as shown in Fig. 13.1, each electron moves along a sinusoidal trajectory with the magnetic period λ_u and emits radiation. The speed of the moving electron is always less than that of light and thus the electron is left behind from the radiation emitted by itself by a specific distance of λ_x while it travels the distance of λ_u. As a result, radiation at the wavelength of λ_x becomes in phase and thus its intensity is significantly enhanced. Detailed analysis shows that λ_x is proportional to λ_u/E^2, where E is the electron energy.

In most SR facilities, λ_x is in the range between nanometer and angstrom and is thus much shorter than l_b. Because the electrons in the electron beam distribute randomly over the whole bunch length,

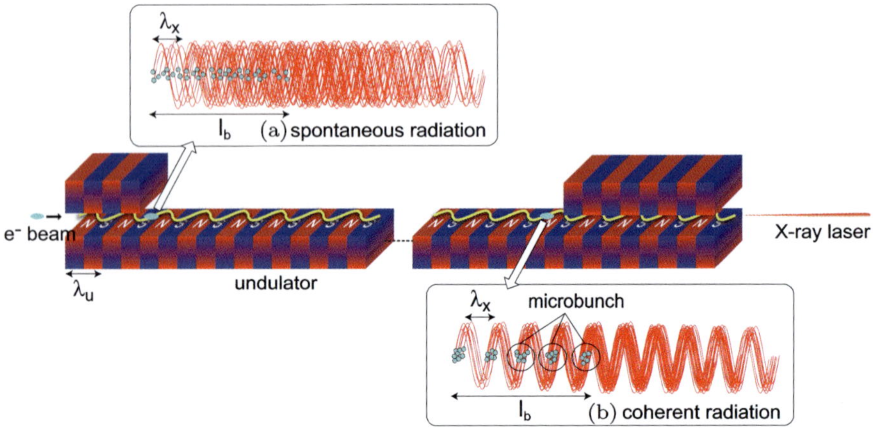

Fig. 13.1. Principle of operation of XFEL. A high-energy electron beam is injected to the periodic magnetic field generated by an undulator. (a) Spontaneous radiation emitted from an electron in which the electrons are distributed randomly. (b) Coherent radiation emitted from the microbunched electron beam.

radiation emitted from each electron also distributes randomly. Here, we note that the distribution of radiation is "random" but not "uniform". If the electron distribution is perfectly "uniform", electromagnetic waves emitted from individual electrons are cancelled out and thus no radiation is emitted from the electron beam. In practice, the electron beam always has some kind of density fluctuation: there exist a number of domains where the electron density is dense or sparse. Such a density fluctuation is referred to as "shot noise".

It is easy to understand that the total radiation power emitted from the electron beam is given as $N_e P_s$, where N_e denotes the number of electrons contained in the electron beam, while P_s denotes the radiation power emitted from a single electron. In other words, radiation emitted from each electron is integrated as a particle (photon). This is referred to as spontaneous radiation, which is the main radiation process utilized as light sources in conventional SR facilities, especially in short-wavelength regions. It is well known that the radiation power of spontaneous radiation linearly increases as the electron beam travels along the undulator.

In SASE XFELs, the spontaneous radiation emitted from an electron beam triggers the amplification process. In other words, it works as seeding light, or more specifically, it is amplified by the same electron beam. This is the origin of the word "Self-Amplified Spontaneous Emission".

13.2.1.2. *Energy modulation*

When the power of spontaneous radiation becomes sufficiently high, the electrons in the electron beam start to interact with it, and the laser amplification process is launched. In Fig. 13.2, an electron beam moving along the sinusoidal trajectory in the undulator is schematically illustrated together with radiation at the wavelength of λ_x propagating synchronously with the electron beam. Hereinafter, the propagation axis of the electron beam and radiation is defined as the Z axis, while the horizontal direction perpendicular to Z is defined as the X axis. Note that the electron trajectory is exaggerated in this figure, whose amplitude and period are usually of the order of 1 μm and 10 mm, respectively. The maximum deflection angle of the electron trajectory in the undulator is at most 0.1 mrad and the slightly deflected electrons interact, i.e., exchange energy with radiation and an energy modulation is induced in the electron beam, which is explained in more detail as follows.

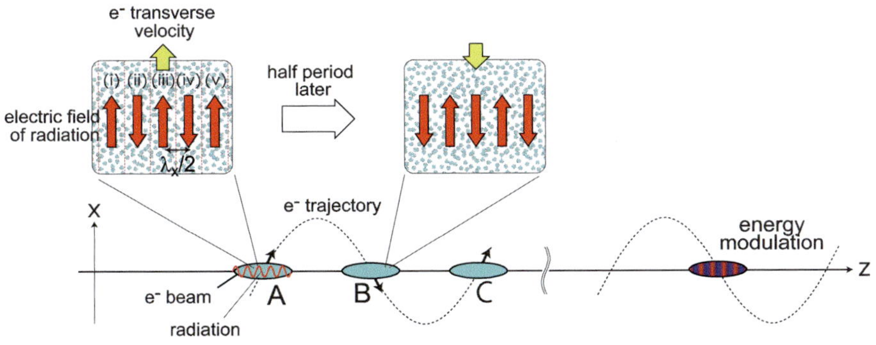

Fig. 13.2. Interaction between the electrons and photons (electromagnetic wave). The electrons are accelerated or decelerated by the electric field of radiation and the energy modulation is induced in the electron beam.

First, let us look to the relation between the electron motion and electric field of radiation when the electron beam is passing through the point A, as shown in the inset in the left side of Fig. 13.2. Because the radiation is an electromagnetic wave with the wavelength of λ_x, its electric vector flips alternately with the pitch of $\lambda_x/2$ along the longitudinal axis Z. The electrons, which have a finite and identical horizontal velocity component, are thus accelerated or decelerated according to their longitudinal positions with respect to the radiation. To be specific, the electrons positioned in the domains (i), (iii) or (v) are decelerated and lose energy, while those in (ii) and (iv) are accelerated and gain energy (note that an electron has a negative charge).

Next, let us look to the electrons passing through the point B away from the point A by half the period, where the horizontal velocity components of all the electrons are flipped. It is easy to understand that the electric field of radiation is also flipped, because the radiation goes ahead of the electrons by half the wavelength while the electrons travel half the period as already mentioned, and thus the relative phase advance of radiation is π. As a consequence, all the electrons are accelerated or decelerated in the same manner as at the point A, i.e., exchange energy with radiation. The situation is exactly the same at the point C, which is one period away from the point A. Summing up the above discussions, the electrons contained in the electron beam continuously exchange energy with radiation, and the energy of the electron beam is modulated along the propagation axis with the period equal to the wavelength of radiation λ_x. It should be noted that the average electron energy does not change in this process. Also note that this kind of energy exchange is not sustained with the radiation whose wavelength is different from λ_x, and thus no energy modulation is induced.

13.2.1.3. *Microbunch*

The energy modulation as explained above is eventually converted to density modulation, which is explained as follows.

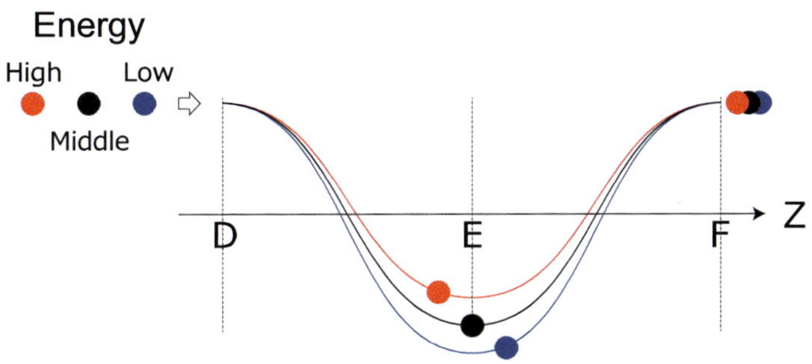

Fig. 13.3. Mechanism of bunching of the electron beam with the energy modulation, when it moves along the sinusoidal trajectory. The path length depends on the electron energy and thus the high-energy electron at the tail position catch up with the low-energy electron at the head position.

Let us consider the case when three different electrons having three different (high, middle and low) energies move along the sinusoidal trajectory in the undulator as shown in Fig. 13.3. We assume that they are arranged along the propagation axis in the order of lowest to highest energy.

Strictly speaking, all the electrons do not necessarily move along the same trajectory. Namely, the trajectory amplitude is inversely proportional to the electron energy, and thus the path length depends on it. As a result, the low-energy electron takes the longest time to travel from the point D to F among the three electrons, and thus they approaches to each other while they move along the sinusoidal trajectory. It is obvious that the electrons having the opposite energy order move apart from each other.

Now let us apply the above discussion to the electron beam having a periodic energy modulation. It is easy to understand that individual electrons are locally and periodically bunched while the electron beam moves along the sinusoidal trajectory, and thus the periodic energy modulation is converted to the periodic density modulation, which is also referred to as microbunch. The electron beam in which the microbunch is formed is hereinafter called the microbunched electron beam.

13.2.1.4. *Coherent radiation*

Once the microbunch is formed in the electron beam, the radiation process shifts from the spontaneous radiation as shown in Fig. 13.1(a) to that shown in Fig. 13.1(b). This radiation process is referred to as coherent radiation, in which radiation emitted from each electron is integrated coherently as a wave (electromagnetic wave). In this case, the field amplitude of radiation is N_e times that of radiation emitted from a single electron. Because the radiation power is proportional to the square of the field amplitude, the power of coherent radiation is N_e^2 times that of radiation emitted from a single electron, and thus N_e times that of spontaneous radiation. In general, N_e is of the order of 10^9–10^{10}, and thus the radiation emitted from a microbunched electron beam is much brighter than that from the normal electron beam without microbunch. Note that this discussions is valid only at the wavelength of λ_x, i.e., the pitch of the microbunch, and the spontaneous radiation is dominant for other wavelengths.

13.2.1.5. *Lasing mechanism*

In the former subsections, three processes necessary for lasing, i.e., induction of energy modulation, formation and growth of microbunch, and emission of coherent radiation, have been explained step by step. In practice, they occur simultaneously in the manner of positive feedback. To be specific, the microbunched electron beam, which is created by the interaction between the electrons and photons, amplifies the radiation through the emission of coherent radiation. Then the amplified radiation further facilitates the growth of microbunch. This is the lasing mechanism of XFELs.

13.2.1.6. *Comparison with conventional lasers*

Now let us compare the XFELs with conventional lasers in terms of the wavelength tunability. It is well known that the conventional lasers operate at discrete wavelengths, because they are based on the stimulated emission from bound electrons in atoms or molecules. In contrast, XFELs can in principle operate at any wavelengths. This is attributable to the fact that the high-energy "free electrons" work as the laser medium in XFELs. It should be noted, however, that the electrons in XFELs are not completely "free": the sinusoidal trajectory specified by the periodic magnetic field in the undulator imposes a sort of boundary condition on the electrons. In other words, by tuning the parameters that define the sinusoidal trajectory (magnetic period and strength, or electron energy), the lasing wavelength can be controlled. Furthermore, XFELs do not require the optical cavity indispensable in the conventional lasers. As a result, XFELs can operate even at short-wavelength regions in which we do not have any high reflectance mirror at the normal incidence.

Note that the above discussion ignores the historical background. The free electron laser, without the prefix of "X-ray", has been originally proposed based on the premise of the optical cavity, which has been the biggest obstacle toward shortening the wavelength. In order to overcome this difficulty and realize the X-ray laser, a new principle of FEL, i.e., SASE, has been proposed, in which the undulator is long enough to enhance the amplification gain and realize lasing even without the optical cavity. Most of the short-wavelength FELs including XFELs that are in operation or under construction are based on this principle.

13.2.2. *Light source properties*

As explained in the former sections, the principle of XFEL is different from that of two major light sources in photon science: SR and conventional laser. The characteristics and source performances of XFELs are thus different from the two sources, which are qualitatively explained in this section.

13.2.2.1. *Normalization factor*

Mathematically, the amplification process in XFELs is described by the so-called FEL equation. This is a series of basic equations that describe the motion of electrons in the electromagnetic field of radiation and growth of field amplitude by emission of coherent radiation from the microbunched electron beam. The FEL equation can be derived from classical electrodynamics and special theory of relativity. From this point of view, XFEL is completely different from the conventional laser in which quantum mechanics plays an essential role.

The FEL equation depends on a large number of parameters that specify the electron beam quality and undulator properties, and cannot be analytically solved in practical conditions. Thus, numerical simulations are usually carried out in order to evaluate quantitatively the performance of a specific XFEL facility, which give us the accurate light source specifications necessary for designing the optical components in the beamline and analyzing the experimental results. On the other hand, they do not lead to a clear understanding of characteristics of XFEL sources, because the results are quite parameter dependent and are not universal. Instead of the numerical method above, the light source properties can be characterized in a universal way by introducing a number of appropriate normalization factors. In this paper, two important normalization factors, "FEL parameter" and "gain length" are explained.

- FEL Parameter
  The FEL parameter is a dimensionless normalization factor given as a function of the current density of the electron beam, the undulator field strength, and the lasing wavelength. As explained later, it relates to the laser efficiency, gain length (refer to the next section), and spectral and temporal profiles. It is often denoted by the Greek letter ρ and is in the range of 10^{-2} to 10^{-4} in most XFEL facilities.
- Gain Length
  The gain length is defined as the undulator length for the laser to be amplified by the factor of $e \simeq 2.718$, the base of the natural logarithm. To be specific, the laser power scales as e^{L_u/L_g}, where L_u is the undulator length, and L_g denotes the gain length given as

$$L_g = \frac{\lambda_u}{4\pi\sqrt{3}\rho}, \tag{13.2}$$

with λ_u being the undulator period and ρ being the FEL parameter already introduced. Note that the above formulas have been derived by solving the FEL equation in the most simplified form in which the effects due to diffraction of radiation and finite energy spread and emittance of the electron beam are not taken into account. We note that even when these effects are not negligible, the exponential scaling of the laser power still holds, with the gain length being more or less longer than that given by Eq. (13.2).

13.2.2.2. *Gain curve*

Taking advantage of the two normalization factors introduced above, the laser amplification process in XFEL can be described in a universal manner.

Figure 13.4 explains qualitatively how the laser is amplified as the electron beam travels along the undulator, in which the laser power is plotted as a function of the undulator length. This is usually referred to as the gain curve of XFEL, and in Fig. 13.4, the undulator length is normalized by L_g, and the laser power is normalized by ρP_{beam}, where P_{beam} is the power of the electron beam. It should be stressed that the gain curve normalized in this manner is universal and does

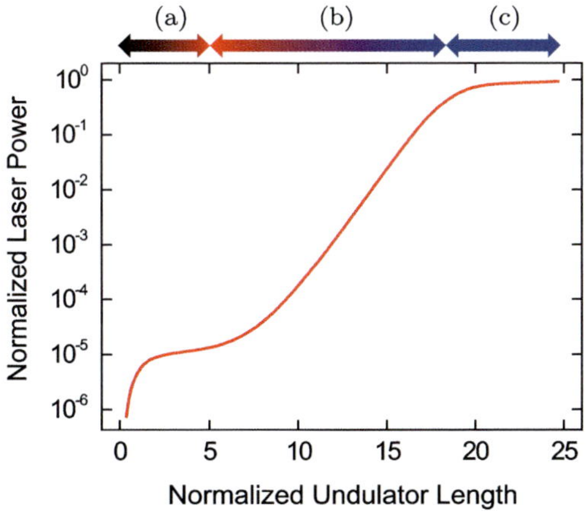

Fig. 13.4. Typical gain curve in XFEL. The undulator length is normalized by the gain length, while the laser power is normalized by the product of the FEL parameter and the electron beam power. Refer to the text for the meanings of the domains indicated by (a), (b) and (c).

not largely depend on the quality of the electron beam and undulator properties.

Now we find from Fig. 13.4 that the amplification process can be categorized into three stages according to the undulator length as follows.

- Emission of spontaneous radiation
  When the undulator is not long, e.g., $L_u < 5L_g$, the radiation power grows linearly as L_u (note that the ordinate is logarithmic). This indicates that the spontaneous radiation is dominant in this stage and the microbunch is not yet formed in the electron beam.
- Laser amplification
  The radiation power growth becomes nonlinear for $L_u \geq 5L_g$, suggesting that the radiation process is no more spontaneous and the laser amplification takes place. In particular, the radiation power grows exponentially as L_u in the range $10L_g \leq L_u \leq 15L_g$, which is thus referred to as the exponential growth regime.
- Saturation
  The power growth rate gradually drops for $L_u > 15L_g$ and the growth stops when L_u reaches $\sim 20L_g$. This phenomenon is known as the laser saturation.

Note that the normalized laser power at saturation is nearly equal to 1, which means that the maximum laser power that can be extracted in XFEL is roughly given by ρP_{beam}. Recalling that P_{beam} denotes the electron beam power, or more specifically, the input power to the laser medium, it is easy to understand that the FEL parameter ρ gives the energy conversion efficiency in XFEL. Also note that the undulator length of $20L_g$ to reach saturation can be rewritten as $\sim \lambda_u/\rho$ using Eq. (13.2), which means that the number of undulator periods to reach saturation is of the order of the inverse of the FEL parameter ρ.

13.2.2.3. *Spectral and temporal profiles*

The spontaneous radiation, which can be regarded as the seed light in SASE XFELs, is spatially and temporally incoherent. This is a consequence of the fact that the shot noise, or the density fluctuation in the electron beam, is the origin of spontaneous radiation as mentioned before. In other words, it contains a large number of transverse and longitudinal modes. This is also the case for conventional lasers in which the spontaneous emission works as the seed light.

In conventional lasers, the lasing mode can be actively selected by controlling the optical cavity or with other alternative methods, and thus the single-mode lasing is in principle possible. On the other hand, such an active mode selection cannot be applied in SASE XFELs, in which no optical cavities are available. As a result, the incoherence of seed light has a great impact on the characteristics of radiation in SASE XFELs, in contrast to the conventional lasers.

Figure 13.5 shows the typical profiles in the time and frequency domains of a single XFEL pulse, which has been retrieved from the result of a numerical simulation. It is found that each profile contains

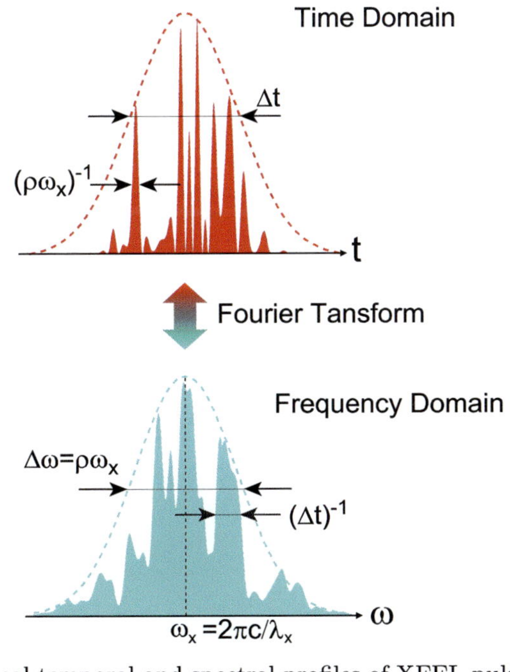

Fig. 13.5. Typical temporal and spectral profiles of XFEL pulse. A lot of spikes are contained within the envelopes indicated by dashed lines in both profiles.

a lot of spikes within an envelope indicated by a dashed line. The width of the envelope in the time domain, or the pulse width $(\Delta t)^{-1}$, is comparable to the bunch length of the electron beam. This is easily understood by recalling the fact that the laser medium in XFEL is the electron beam. The inverse of the pulse width, roughly determines the typical spike width in the frequency domain, which is easily understood from the theory of Fourier limit.

The width of the envelope in the frequency domain, or the bandwidth $\Delta\omega$, is defined by the laser gain bandwidth. Detailed analysis based on the FEL equation shows that gain bandwidth in XFEL is roughly given by the FEL parameter ρ. As a result, the bandwidth of the spectral envelope is of the order of $\rho\omega_x$, and the temporal width $(\rho\omega_x)^{-1}$, which is defined by the Fourier limit of the spectral bandwidth, roughly determines the typical spike width in the time domain.

Note that the spike profiles and positions of the XFEL pulse depend on the initial shot noise of the electron beam, and thus fluctuate from shot to shot, even if the electron beam parameters such as the emittance and current are kept constant. This is in contrast to the envelope profiles that do not change, as long as the beam parameters are the same.

13.2.2.4. *Spatial profile*

As mentioned in the previous section, the temporal and spectral profiles of XFEL contain a lot of spikes, or longitudinal modes. On the other hand, the spatial profiles of XFEL behave differently, and the transverse single-mode lasing is in principle possible, the reason for which is explained as follows.

In order to solve the FEL equation that takes into account the three-dimensional effects such as the finite emittance of the electron beam and diffraction of radiation, the amplified radiation is usually expanded into transverse eigenmodes, i.e., TEM_{nm} modes, where n and m are indices of the mode. Then detailed analysis shows that the gain of the lower spatial mode is usually higher than that of the higher mode, as long as the spatial profile of the electron beam is common and not too special. To be specific, TEM_{01} mode has the highest gain, TEM_{00} and TEM_{10} modes the second-highest gain, TEM_{11} the third-highest gain, and so on.

The above theorem can be applied to consider the spatial profile of the amplified radiation. First, we expand the seed light, which contains a lot of transverse modes, into a number of transverse eigenmodes. Then the radiation field of each transverse mode is amplified according to the gain calculated by solving the FEL equation relevant to each mode. After this operation, the intensity of higher modes become relatively low compared to the lowest mode TEM_{00} even if their initial intensities are comparable. Finally, all the transverse modes are combined together to give the spatial profile after amplification. It is easy to understand that the spatial profile of the amplified radiation becomes similar to that of TEM_{00} mode, which is in general the Gaussian beam, and thus the number of transverse mode significantly reduces as the amplification process advances. This effect is referred to as optical guiding, and plays an important role of improving the spatial coherence in XFEL.

An example of the optical guiding effect is shown in Fig. 13.6, in which the spatial profiles at different undulator lengths are plotted.

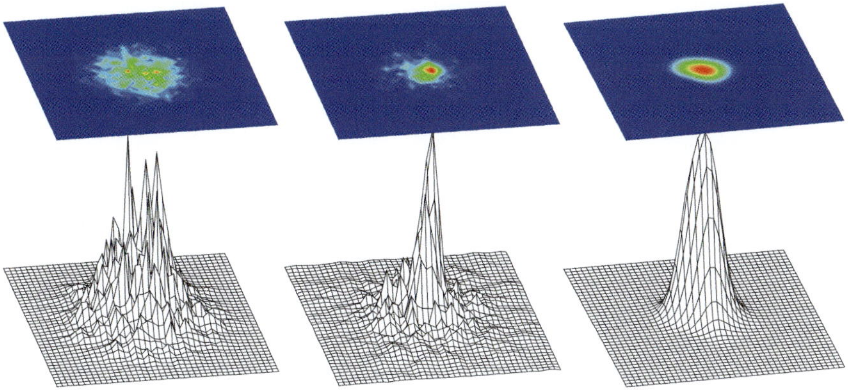

(a) L_u=4L_g (b) L_u=8L_g (c) L_u=16L_g

Fig. 13.6. Spatial profiles of XFEL pulse at different undulator lengths, which correspond to (a) spontaneous radiation, (b) early stage of amplification, and (c) before saturation.

Note that each profile is normalized by the maximum intensity. In the spontaneous radiation regime ($L_u = 4L_g$), the spatial profile is composed of a lot of spikes (modes). As the amplification goes on, the spikes are gradually eliminated and the profile reduces to a single Gaussian mode.

13.2.3. *Brief note for readers*

In this paper, explanations on the XFEL theory have been confined to qualitative aspects so that the readers can get the feeling of how it works. Those who have taken more interest in a quantitative description are recommended to refer to [9] and [10], and references therein.

13.3. Accelerator System

13.3.1. *Introduction*

An accelerator system for XFEL is presently a linear accelerator basis because a small-scale ring accelerator with a strong radiation excitation cannot provide required beam brilliance at high energy of around 10 GeV. This naturally leads to an accelerator system composed of four sub-systems. An electron source, which generates electron beams working as laser medium, locates at the top of the system. A following bunch compression system squeezes the generated electron beam into the running direction (time axis) up to a target peak current keeping quality of the electron beam. A main RF acceleration system accelerates the fully compressed electron beam with a high electron density up to final energy. A long undulator beamline develops a density modulation corresponding to a laser wavelength through the positive feedback interaction between the electron beam and spontaneous radiation and finally generates SASE XFEL by a single-pass stimulated radiation process.

SACLA had been designed and constructed aiming at compact and high-performance XFEL. Although SACLA takes the same system composition, several features are there, which are motivated from this design concept. The technology of short period in-vacuum undulators used at SACLA reduces the XFEL facility size by lowering

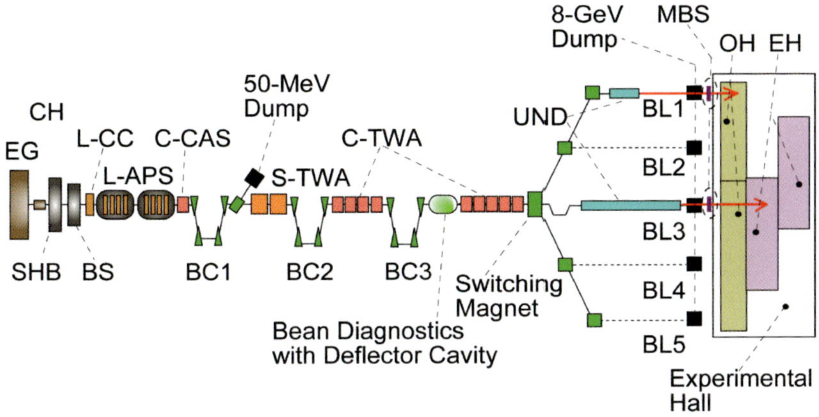

Fig. 13.7. Schematic configuration of SACLA [27]. EG, 500-kV electron gun; CH, chopper with collimator; SHB, 238-MHz sub-harmonic-buncher; BS, 476-MHz booster; L-APS, L-band alternating periodic structure (APS) typed standing-wave cavity; S(C)-TWA, S(C)-band travelling-wave acceleration tube; BC: bunch compressor; UND, undulator; BL, beam line; MBS, main beam shutter; OH, optical hutch; EH, experimental hutch.

the beam energy. In addition, adoption of a high gradient C-band acceleration system enables the further size reduction. On the other hand, lower beam energy severely raises a requirement on the beam brilliance. In order to clear this hurdle, a thermionic gun based low emittance injector was developed.

Another feature is a capability to install multi-SASE XFEL beamlines. SACLA can provide five XFEL beamlines at maximum. By introducing an AC switching magnet at the end of the main accelerator as shown in Fig. 13.7, five experimental groups at maximum can carry out different experiments simultaneously, which is quite effective for replying to various research demands increasing.

13.3.2. *Generation of high quality electron beam*

A highly brilliant electron beam, which has a small angular divergence and beam size (i.e., small beam emittance) and high peak current, is a key for a sufficiently high amplification gain [3] by fully developing the density modulation. Under the realistic condition that the beam energy, peak current and total charge of electrons are

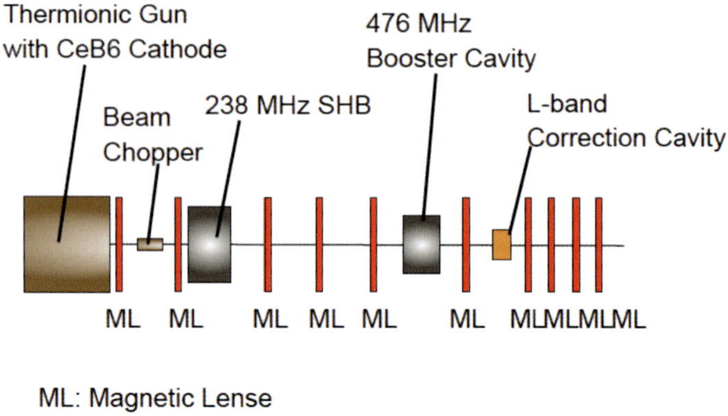

Fig. 13.8. Schematic view of the thermionic gun based low emittance injector.

around 10 GeV, 3 kA, and 1 nC, respectively, the required normalized emittance, which is extrapolated to the limit with kinematic energy of zero, is evaluated to be 1π mm.mrad. The world record of the lowest value was 25π mm.mrad in the 1980s [11]. Since then, the electron gun has been developed for twenty years in order to satisfy this target value of 1π mm.mrad. For this period, a photo-cathode RF gun has mainly been investigated. This system generates a highly intense and short electron beam by irradiating a short laser pulse with high intensity onto a photo-cathode. The RF cavity connected to the gun immediately accelerates the short electron beam pulse with high intensity before the space charge degrades the electron beam quality [12].

SACLA, however, solved this problem by revolving a conventional thermionic-gun based buncher system [13]. The thermionic-gun based low emittance injector at SACLA shown in Fig. 13.8 [8] adopts a CeB_6 suffix single crystal whose diameter is 3 mm as a thermionic cathode. A charging voltage of 500 kV extracts a homogeneous cylindrical electron beam with a peak current of 1 A and a duration of $\sim$2 μs from the cathode heated at 1800 K. Under these cathode parameters, the thermal theoretical emittance is 0.4 πmm.mrad when normalized, which surpasses the design target [30]. The cylindrical shape and beam homogeneity are critically important for keeping the beam emittance at low beam energy because space charge becomes linear at the central part of the beam under this condition. Initial beam energy of 500 keV enables to focus the beam with the localized magnetic lenses instead of a long solenoid coil, which can separate two functions, acceleration and transversal focusing with each other. A beam chopper combining a constant magnetic field with a pulsed electric field slices the beam into a 1 nsec pulse. A twin RF buncher composed of a 238 MHz sub-harmonic buncher (SHB) and a 476 MHz booster cavity gives an energy chirp on the beam for velocity bunching. The booster raises the beam energy up to about 700 MeV to suppress the emittance growth during the velocity bunching process [15]. Although the system is rather complicated compared with the photo-cathode RF gun, it enables almost maintenance-free and stable operation.

13.3.3. *Multi-stage bunch compression system*

Since SACLA requires a high peak current of 3.4 kA [8], an initial current of 1 A is ~3500 times increased by compressing the 1-nsec bunch into the time axis keeping the slice emittance. This compression factor is larger than that of LSLS [16] and EXFEL [17] by a magnitude of one order and hence, it is critically important to linearize the overall compression process. An original linearization scheme named "over-correction scheme" [18] was devised at SACLA to meet the high design target.

Figure 13.9 is a schematic view of the four-stage bunch compression system. The first stage is a bunch compressor based on a velocity bunching process in the injector, which compresses the bunch in a straight section of ~1 m by using dependence of the velocity on the beam energy. This scheme compresses a bunch of 1 nsec into that of 30 psec. The second to fourth compressors are the magnetic chicane type ones, which compress the bunch in a chicane composed of four identical bending magnets by using dependence of the pass-length on the beam energy. In order to efficiently capture the bunch and generate the energy chirp over the bunch, accelerators with three different RF frequencies, i.e., L-band (1.4 GHz) for the second, S-band

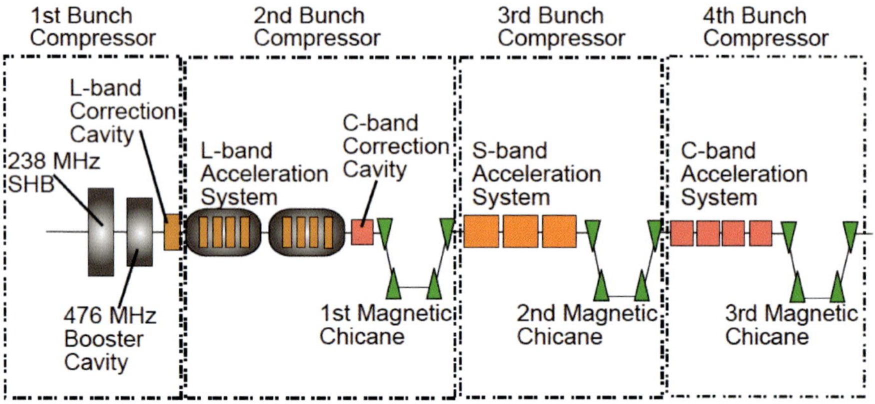

Fig. 13.9. Schematic view of the four-stage bunch compression system.

(2.8 GHz) for the third, and C-band (5.7 GHz) for the fourth are used according to the bunch length.

Two kinds of correction cavities, which locate at the upstream part of the system as shown in Fig. 13.9, suppress the non-linearity over the system to achieve a high compression factor of ~3500. The first L-band correction cavity linearizes the velocity bunching process. The second C-band correction cavity covers the linearization over the remaining bunching processes based on the magnetic chicanes. Since the longitudinal phase space is almost elect in the final (fourth) bunching process it requires the most severe linearization condition. The second and third processes don't need the rigorous linearization compared with the final one. The C-band correction cavity therefore gives an exceeded correction chirp for the second and third compressors in order to assure the sufficient linearization at the fourth compressor. The heart of this over-correction scheme is a dynamical enhancement of the correction chirp through the multi-stage compressor [18]. The initial slight chirp is gradually enhanced as the compression proceeds and hence, the correction-mismatch at the second and third compressors never becomes significant. Figure 13.10 shows variation of the energy chirp in the over-correction scheme.

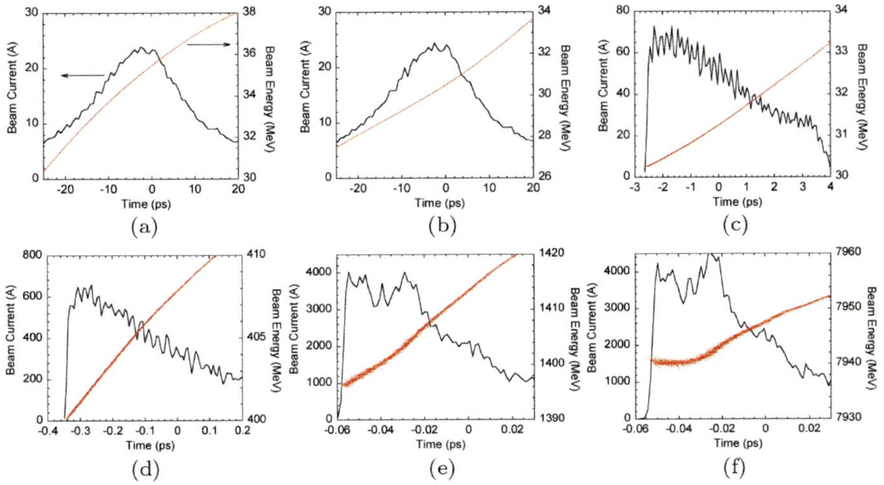

Fig. 13.10. Energy chirp distribution (red) and the current profile along the four-stage bunch compression system [18]. (a) Entrance of the C-band correction cavity. (b) Exit of the C-band correction cavity. (c) Exit of the 1st BC. (d) Exit of the 2nd BC. (e) Exit of the 3rd BC. (f) End of the C-band main linac.

13.3.4. *High-gradient and stable acceleration system*

Highly efficient and stable beam acceleration is important for a compact XFEL such as SACLA. Although it is a well-known fact that higher RF frequency can raise an acceleration gradient by increasing the density of an RF electric field in the cavity, the higher frequency imposes severe accuracy on manufacturing and finishing. At SACLA C-band of ~5.7 GHz was adopted as a RF frequency instead of X-band of ~11.4 GHz because the problem of an electric discharge at high gradient operation cannot be solved for the X-band system [19]. The maximum beam repetition rate was set to 60 Hz in consideration of difficulty in development of high power RF sources.

Figure 13.11 shows a schematic view of the C-band acceleration unit at SACLA [20]. In order to increase an acceleration gradient, a SLED-type [21] pulse compressor compresses the RF pulse from the 50 MW Klystron by about three times. A 3dB directional coupler equally divides this pulse into the two and each pulse is fed 1.8 m constant gradient acceleration structure. The high-power RF

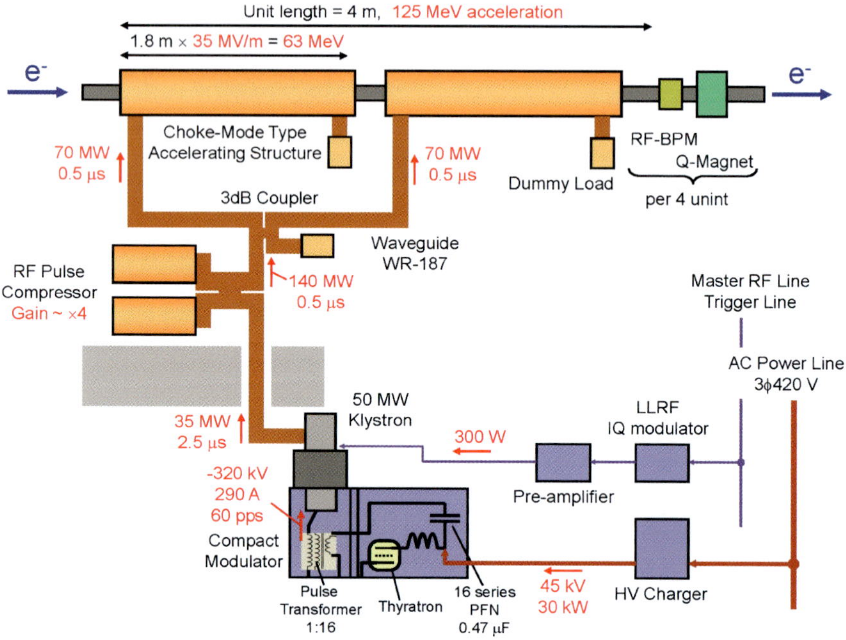

Fig. 13.11. Schematic view of the C-band acceleration unit [20].

equipment including a Klystron, modulator, charging power-supply, and control unit is also downsized to fit a limited space. The developed system has routinely provided a high gradient of 37MV/m [22]. Figure 13.12 shows a picture of the developed compact modulator where the Klystron and modulator are in a single insulation oil tank.

On the other hand, high laser performance requires the stable beam acceleration and precise accelerator control. This is because XFEL is an open loop system and stabilization using so-called feedback control is basically not applicable. Stability and precise control of all the trigger-driven equipment determines the laser performance. Tolerances of RF parameters are quite severe, 100 ppm for the RF voltage and 100 fsec for phase, respectively [23]. To meet these requirements, the precise timing system using an ultra-low noise twin oscillators and fiber with good temperature characteristics [24], the high precision inverter power-supply [25], high precision RF voltage and phase control system using IQ-modulator/demodulator [26],

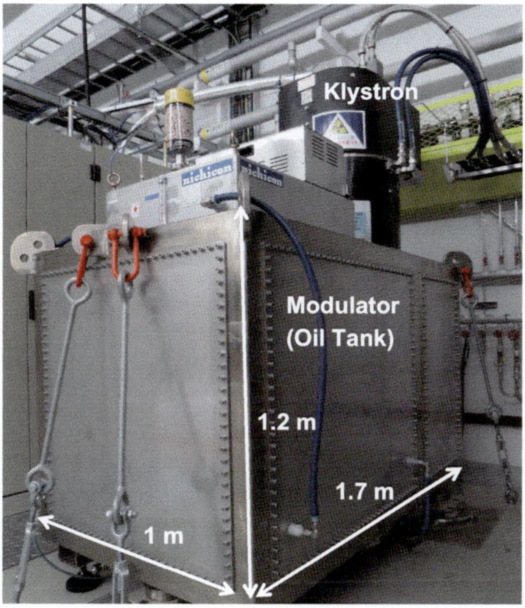

Fig. 13.12. Photo of the compact modulator where the Klystron and modulator are contained in a single insulation oil tank.

etc. were developed. Figure 13.13 shows achieved phase stability of 238 MHz SHB [27].

13.3.5. *High precision undulator system*

SACLA adopts a short-period in-vacuum undulator [28] to lower the beam energy keeping the target laser wavelength, which is essential for reducing the facility scale as much as possible. An undulator period of 18 mm with a maximum K-value of 2.2 enables us to generate SASE XFEL at 0.1 nm using an electron beam of 8 GeV. The adopted variable gap mechanism assures the sufficient beam aperture in the beam-tuning period and also enables experimental users to adjust the laser wavelength within ±15%. Although the short-period in-vacuum undulator with a variable gap mechanism has various advantages, how to achieve the precise undulator field distribution and beam orbit setting through the undulator beamline was a remaining issue to be solved.

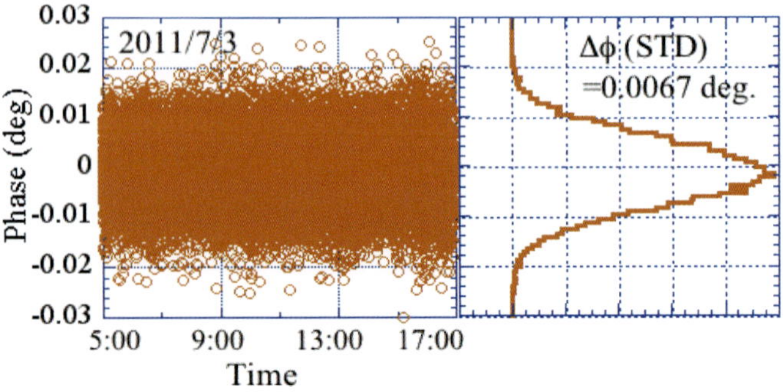

Fig. 13.13. Measured phase stability of 238 MHz SHB over 12 hrs [27].

The conventional field measurement method using precise scanning of magnetic sensors supported by massive granite [29] is not applicable to the assembled in-vacuum undulator of which magnet arrays are installed in the vacuum vessel. In order to assure the field quality of each undulator installed in the undulator hall, the *in situ* field measurement system called SAFALI (Self-Aligned Field Analyzer with Laser Instrumentation) [30] was developed. Instead of massive granite, the developed system introduces the laser-based 3D position-measurement, which enables the precise 3D position control together with the feedback control system. The sufficiently precision, 10 μm in rms has been achieved for the assembled in-vacuum undulator being 5 m long. Figure 13.14 shows a schematic view of the SAFALI system [30].

The X-ray based alignment and tuning method [31] was developed in order to precisely align the undulator segments composed of the undulator, quadrupole magnet, beam position monitors, etc. on the straight and to control the K-values of all the undulators with an error of less than 5×10^{-4}. This method aligns the undulator segments one-by-one from the upstream by guiding the undulator radiation onto the target in the optical hutch locating 100 m downstream from the end of the undulator. The K-values are also calibrated by analyzing the measured spectra of undulator radiation. Figure 13.15 shows the schematic illustration of the photon diagnostic system used for this method.

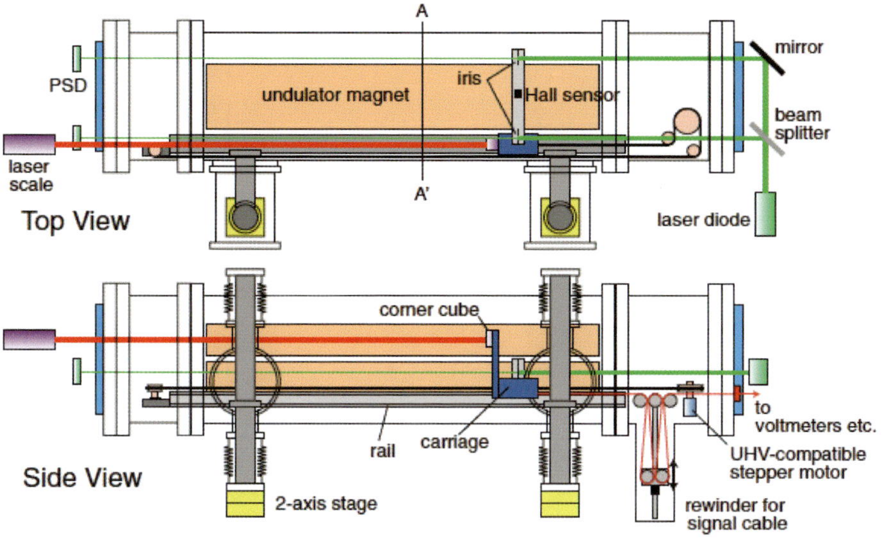

Fig. 13.14. Schematic view of the SAFALI system [30].

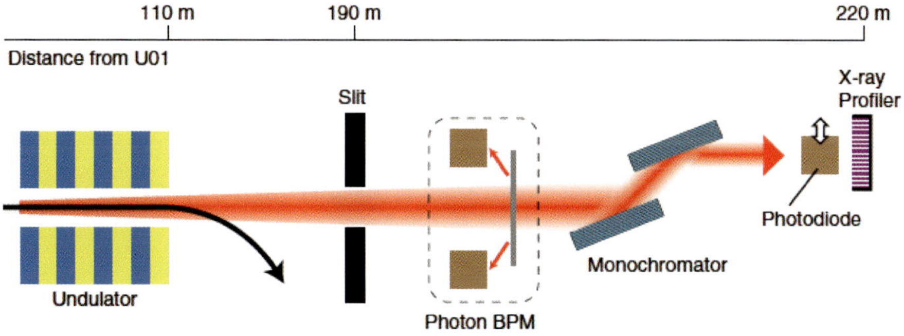

Fig. 13.15. Schematic illustration of the photon diagnostics system in SACLA. The top scale shows the distance from the entrance of the 1st undulator segment (U01) [31].

Owing to these developments, beam commissioning of the undulator beamline was smoothly completed at SACLA. A gain length of 2.3 m shown in Fig. 13.16 and pulse energy of ∼0.5 mJ have been routinely obtained at 10 keV.

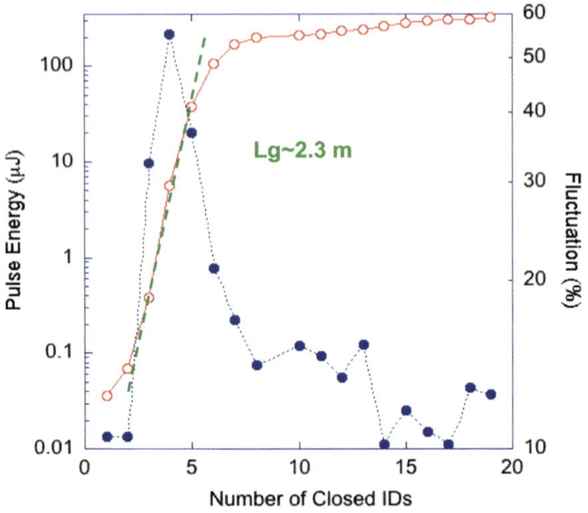

Fig. 13.16. Measured gain curve (red empty circles) at 10 keV together with the laser intensity fluctuation (blue filled circles).

13.4. Photon Beamline

13.4.1. *Introduction*

The advent of XFEL light sources has opened us exciting opportunities for direct observation on ultrafast dynamics of atoms and electrons in complex materials. The unique characteristics of XFEL reveal unknown, essential mechanisms of ultrafast reactions in proteins, catalyses, and quantum systems. Furthermore, ultraintense X-ray beam generated with a tight focusing system could be useful to verify the framework of quantum electrodynamics (QED) by probing extremely weak interactions between photon-vacuum or photon-photon.

To explore full potentials of XFEL sources, a photon beamline system, which delivers XFEL light from the undulator to endstations with manipulating and diagnosing the photon beam properties, plays a crucial role. In this chapter, we review the beamline system developed at SACLA [32]. Here we briefly summarize the history. To fully utilize coherent X-rays, we have started development of optical elements and methods in the late 1990s, just after the inauguration of

SPring-8. We constructed two unique beamlines at SPring-8: a 1-km beamline BL29XU and a 25 m undulator beamline 19LXU. In these beamlines, we successfully developed high-quality optics for coherent X-ray applications, for example, reflective mirrors and speckle-free Be windows. We also have developed a prototype of dispersive, single-shot X-ray spectrometer [33]. In parallel to these developments of X-ray optics, we built a photon beamline in the extreme-ultraviolet (EUV) region for the SCSS test accelerator (SCSS) [34], which is a prototype of SACLA with a 250-MeV beam energy in 2005. We have successfully observed a lasing at a wavelength of 49 nm in June 2006. Based on these experiences and knowledge, we started design work of the SACLA beamline system in 2006. Succeeding the completion of the Experimental Building, we constructed the beamline in 2010 and completed it before starting the machine operation. We first observed spontaneous radiation at the beamline in March 2011, and then started to perform fine tuning of the accelerator to achieve lasing. We quickly accomplished first lasing at a wavelength of 1 Å on June 7, only two months after the start of the tuning. The beamline has significantly contributed to this success by enabling precise alignment of the trajectory of the electron beam along the undulator segments [31]. We celebrated the inauguration of SACLA and started user operation in March 2012. Following the startup phase, we are currently shifting towards the steady operation phase. Initial scientific outcomes in the various fields, ranging from biology, chemistry, and quantum physics and optics, have been produced, which will be introduced in 13.5.

Next, we summarize a design concept of the SACLA beamline. Then we will introduce the basic optics and diagnostics, and describe end stations and some instruments.

13.4.2. *Design concept*

XFEL light has very unique characteristics such as ultrahigh brilliance, excellent transverse coherence, and femtosecond pulse duration. Since a jump of the XFEL source from the existing one is significant, we did not have any established guidelines to exploit full capabilities and potentials. For this purpose, we built a design concept of the beamline [35]. The essence introduced as follows.

First, we decided to utilize state-of-the-art X-ray optics in order to suppress unwanted speckles and wavefront distortions that are produced with non-ideal optics under coherent illumination. In particular, quality of reflective mirrors and Be windows are crucial. A large spatial acceptance of these optical elements are also important for utilizing full beam. Second, we installed these basic optical components into an optics hutch and use them as common systems for all end stations, instead of a distributed arrangement into each endstation. This centralized scheme improves the efficiency of construction and operation, and facilitates a quality control and upgrade. Third, we have developed a shot-to-shot non-destructive diagnostic system of XFEL properties, such as pulse energies, beam profiles with positions, and wavelength, and installed them also into the optics hutch. They are used for analyzing experimental data with high accuracies, as well as for providing useful feedback of accelerator operation. Fourth, we designed our end stations as much as flexible without any permanent experimental instruments targeting for specific applications. This large flexibility is particularly important for the initial startup phase, because it enables us to perform various trials with new ideas.

13.4.3. *Beamline design and performance*

Based on the design concept in the previous chapter, we built the first XFEL beamline BL3 [32], as schematically shown in Figs. 13.17 and 13.18. The optics hutch encloses common beamline optics and diagnostics. For the basic beamline optics, we have installed a couple of double-mirror systems for pink-beam ($\Delta E/E \sim 5 \times 10^{-3}$) utilization, and a silicon double-crystal monochromator (DCM) for monochromatic-beam ($\Delta E/E \sim 10^{-4}$) applications (Fig. 13.19). For the former systems, we employ 400-mm-long plane mirrors with vertical deflections. A combination of the two mirrors enables us to keep the exit beam axis to the endstation being parallel to the incident beam. We have three mirror chambers (M1, M2a, M2b). A pair of

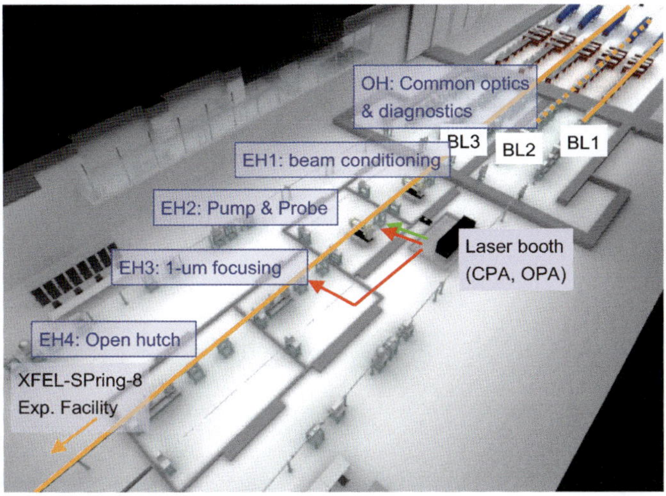

Fig. 13.17. Schematic view of SACLA experimental hall and beamlines.

M1 and M2a (M1 and M2b) is used with an incident angle of 4 mrad (2 mrad). With carbon coating, the cutoff energies are 7.5 keV and 15 keV respectively for these configurations. In order to avoid wavefront distortions, surface figures of these mirrors are precisely finished with an elastic-emission machining (EEM) technique. For the latter, we use silicon (111) monochromator crystals with mechanochemical polishing. To avoid contaminations on the crystal surfaces that could produce speckles, we designed the DCM chamber and mechanics with ultrahigh vacuum (UHV) compatibility. Also, we have reduced the number of degree of freedom for motion controls of crystals in order to increase stability. We did not apply crystal cooling, because the heatload is small, being much lower than 1 W. The DCM is used for producing monochromatic beam, while the intensity decreases to several percent of the incident SASE-XFEL light that has a larger spectral width of 5×10^{-4}.

We select these mirrors and the DCM with an exclusive manner. In each case, we could keep the position of XFEL light in the endstations with a height of 1420 mm from the floor level, which facilitate switching of the optics even during the experiments. Figure 13.19 shows the ranges of wavelengths covered with these optics.

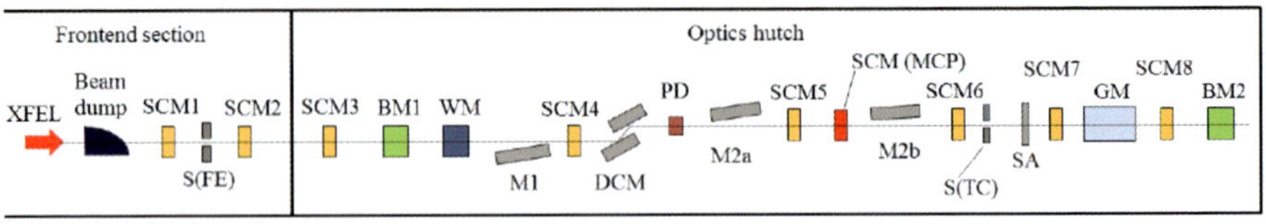

Fig. 13.18. Beamline configuration of BL3. S (FE): Frontend slit; SCM: screen monitor; WM: wavelength monitor; M1 (M2a, M2b): mirror; PD: photodiode; S (TC) transport-channel slit; SA: solid absorber; GM gas monitor [32].

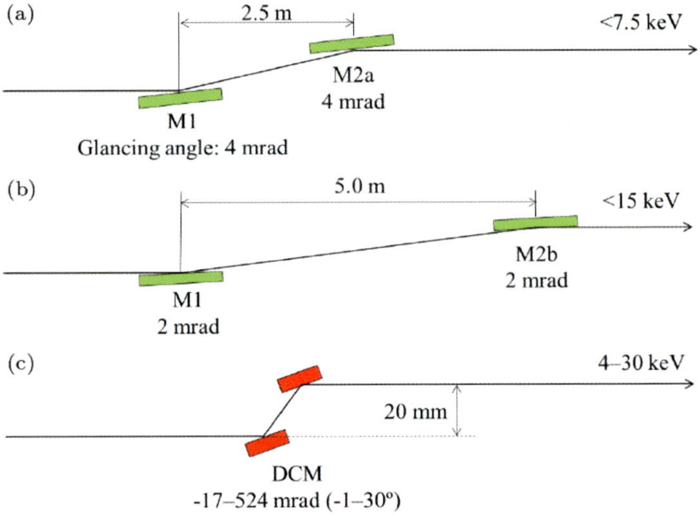

Fig. 13.19. Optical configuration and ranges of photon energy [32].

For controlling the beam intensity with orders of magnitudes, we installed silicon attenuators with different thicknesses (0.1 mm to 3 mm). Also, we have developed and installed a pulse selector that could pick up any designated pulses from 60-Hz pulse trains.

We introduce our diagnostic system for properties of XFEL light. We utilized different types of thin diamond foils for routine shot-to-shot monitoring of intensity, position, spatial profile, and wavelength with a non-destructive manner. For measuring intensity and position, we developed a beam monitor (BM) that combines a thin diamond foil that scatters a small portion of incident X-rays with quadrant photodiodes (PD) [36]. The sum of the PD outputs provides the beam intensity, while the subtracted signals give center-of-mass positions of the beam. For the thin foil, we utilized a CVD nano-crystalline diamond foil that gives a homogeneous scattering pattern without distorting the wavefront of the transmission beam. We found that the BM has a wide dynamic range over six orders of magnitudes. The CVD nano-crystalline diamond foil is also used for monitoring wavelength of XFEL light. A scattering angle of the Debye-Scherrer pattern is measured with a MPCCD detector attached on an arm of a diffractometer. For measuring a spatial profile of XFEL light, we developed a screen monitor (SCM) with a boron-doped thin diamond foil that has a high efficiency to produce fluorescence. These diagnostic devices are installed in the optics hutch. The shot-to-shot beam properties are saved on a database, and utilized for experimental analysis and for accelerator operation.

These beamline devices are utilized for precise alignment of the electron beam trajectory along undulators, which is required to achieve X-ray lasing. A central position of the monochromatic X-ray of spontaneous light is a good indication of the pointing of the electron beam traveling through an undulator segment. For this purpose, we combined the DCM and a dedicated screen monitor that uses a phosphor-coupled multi-channel plate (MCP) with high sensitivity. By measuring the center of mass, we could determine the beam trajectory with an accuracy of 1 μrad for each undulator segment, and compensated the deviations with steering magnets located between the segments [31].

13.4.4. *Endstations*

We have constructed four experimental hutches (EH) in the Experimental Hall of SACLA. EH1 (advanced optics) is used for R&D of advanced optics and diagnostics. In EH2 (Pump & Probe), one can perform pump and probe (PP) experiments with optical laser and unfocused XFEL light. EH3 (coherent focusing) encloses 1-μm focusing system [37], which is applicable to nano-crystal diffraction, coherent diffractive imaging (CDI), and non-linear X-ray optics with increase X-ray intensity. Optical laser is also available at EH3.

EH4 (versatile) is an open hutch for large instruments. In the downstream of the Experimental Hall, we built the SACLA-SPring-8 Experimental Facility for combinational experiments for SACLA and SPring-8. In this building, we constructed EH5 and installed a 50 nm focusing system [38]. High power laser systems are installed in this facility.

We introduce optical laser system used in EH2 and 3. The system consists of a Ti-Sapphire based mode-lock oscillator, a chirped pulse amplifier (CPA) and an optional optical parametric amplifier (OPA). CPA provides 30-fs laser pulses with a pulse energy of several mJ at a repetition rate of 1 kHz. Arrival timing jitters between optical and XFEL pulses can be measured with a timing monitoring tool at a resolution below 100 fs.

As a key topic, we introduce focusing systems developed and used at SACLA. For focusing X-ray beam, reflective optics (mirror), refractive optics (lens), and diffractive optics (zone plate) are widely utilized. For XFEL applications, mirrors are the most appropriate optics because of their high focusing efficiency (ideally close to 100%), an achromatic property, and high damage threshold due to small absorption of x-rays. However, there is a severe requirement on an accuracy of surface figure in order to avoid wavefront distortion. In addition, a sufficient mirror length is required for accepting the whole beam with a grazing incident condition. We have collaborated with a group of Prof. Kazuto Yamauchi (Osaka University) to develop state-of-the-art mirrors. In SPring-8, we achieved 7-nm focusing [38], which is the smallest spot of light. At SACLA, we have achieved 1-μm focusing at EH3, as mentioned, with a high intensity of 10^{18} W/cm^2. Figure 13.20 shows a damage profile of silicon substrate with single-shot irradiation of XFEL pulse. X-ray beam that has a large penetration depth ionize silicon atoms inside the substrate. Following collisions of fast electrons and particles, a needle-like crater was finally created. A high-temperature and high-pressure condition formed deeply inside the substrate energy density state is a very unique phenomenon viewing from high-energy density science. This 1-μm system is utilized for a variety of applications, such as serial femtosecond crystallography (SFX), atom, molecular, and optical science (AMO), coherent

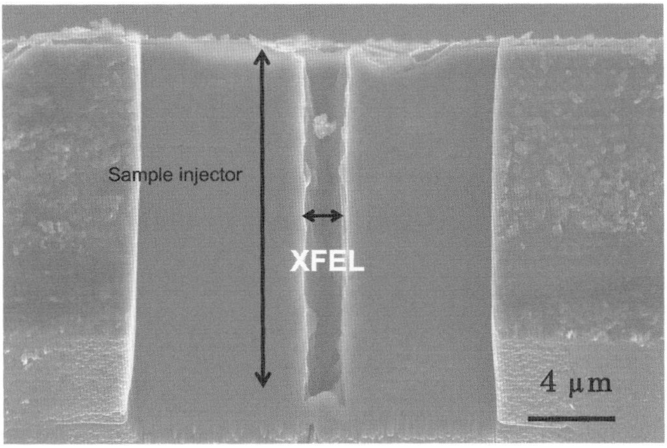

Fig. 13.20. Section profile of silicon substrate irradiated with single XFEL pulse. (Courtesy of Dr. Takahisa Koyama (JASRI))

diffractive imaging (CDI), and non-linear X-ray science, which are presented in Section 13.5.

As a typical experimental instrument, we introduce a platform developed for conducting XFEL diffraction and scattering experiments called "DAPHNIS" (Diverse Application Platform for Hard x-ray diffractioN In SACLA), which is capable of shot-to-shot measurement with fresh samples brought with different types of sample injectors mounted in a small sample chamber, shown in Fig. 13.21. Fast 2-dimensional detector called a short-work distance MPCCD detector [39] (SWD-MPCCD) could cover a wide diffraction angle for high-resolution structural analysis. The system can be utilized to analyze a broad range of samples such as small protein crystals and liquid. Furthermore, it is easily applicable to study of dynamics combining with optical lasers.

13.5. Applications

13.5.1. *Overview*

To study new science by XFEL, the first step is to clarify what are distinct features of XFEL compared with other light sources, for

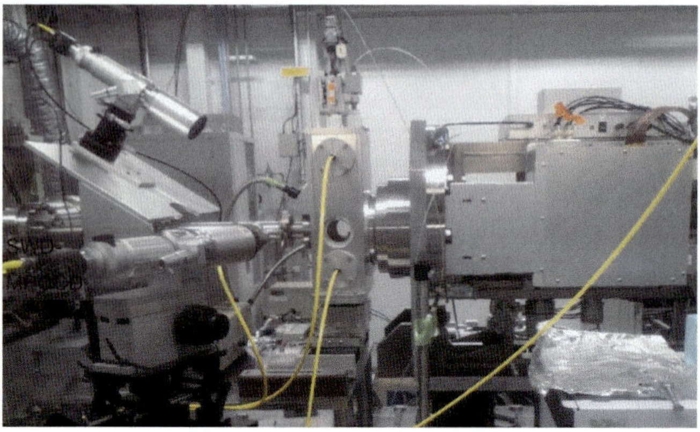

Fig. 13.21. Apparatus of DAPHNIS system.

example, X-rays by synchrotron radiation or visible lasers. The following is a list of important properties of XFEL:

1. High intensity: SACLA has achieved a pulse energy of 0.6 mJ/pulse. Such a high-energy pulse plays an important role for studying tiny phenomena such as nonlinear phenomena.
2. 100% coherence: SACLA has 100% spatial coherence. The coherence in the transverse direction is a key property for imaging. As for a time and frequency domain, namely the longitudinal direction, high coherence will be achieved by seeding.
3. Short pulse: Pulse width of a SACLA beam is up to now several femto seconds. Such short pulse enables us to reveal ultrafast phenomena such as chemical reaction and phase transition.
4. Short wavelength: X-rays have short wavelength of ∼1 Å that gives atomic resolution. The high resolution makes it possible to visualize atomic responses of various phenomena, which are known phenomenologically without proof of principle.
5. High frequency: Photon energy of X-rays is around 10 keV. The high frequencies of waves are useful to investigate deeper structures of an atom such as an electron in an inner shell and a nuclear.

Combination of the features makes best use of SACLA. Some of these features are not independent but related by the dispersion,

$\omega = ck$, where ω, c and k are the frequency, the speed of light in vacuum and the wavenumber, respectively. Since the open for public use in 2011, SACLA have delivered interesting research topics. In this chapter, we show selected experimental results using a beamline at SACLA.

13.5.2. *Focusing*

X-rays have very high frequency that is several orders of magnitude larger than the plasma frequency of resonance. In this region of frequency, an interaction between light and matter is very weak. It makes difficult to get desirable optical elements such as a cavity. Absence of a cavity in the X-ray region prevents from lasing of X-rays, leading to the construction of free electron lasers. Moreover, X-rays are sensitive to a surface of an optical element or a sample due to the very short wavelength. Thus it is important to realize a very flat surface of an element such as a mirror for manipulating an X-ray beam. A mirror is used not only for bending a beam but for focusing. For practical use of XFEL, a focusing mirror has several requirements: high efficiency for higher fluence, a spectral acceptance for the full bandwidth, high tolerance for the intense beam, elimination of speckles, and so on.

In 2012, we achieved focusing of XFEL to the area of $0.95 \times 1.30\ \mu m^2$ at the photon energy of 10 keV by reflective optics of the Kirkpatrick-Baez (KB) mirror geometry, as shown in Fig. 13.22 [37]. By the arrangement, the efficiency is near 100% that gives about 40000-fold increase in the fluence to the power density of 6×10^{17} W cm^{-2}. Realization of such a high power density paves the way for exploring new field of science such as high energy density science by means of optics.

Figure 13.23 shows the power density of SACLA beam achieved in 2014 [40]. The power density is 10^{20} Wcm^{-2}, which is 3 orders of magnitude larger than that recorded in 2012. This result is realized by two-stage reflective focusing system consisting of two KB geometries with to focal points. This value of 10^{20} Wcm^{-2} has important meaning. The relaxation time of the K-shell core hole of sub-femtosecond

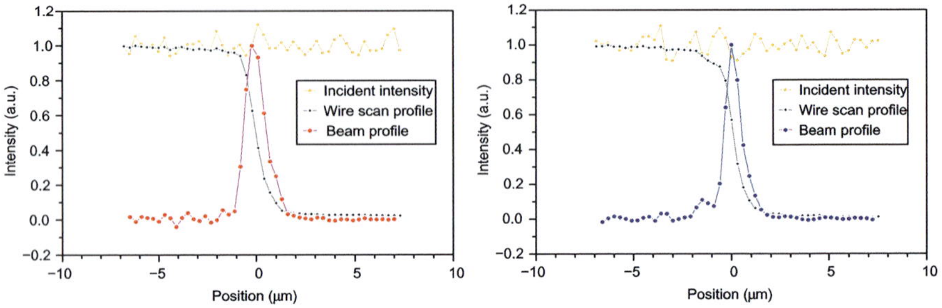

(a) (b)

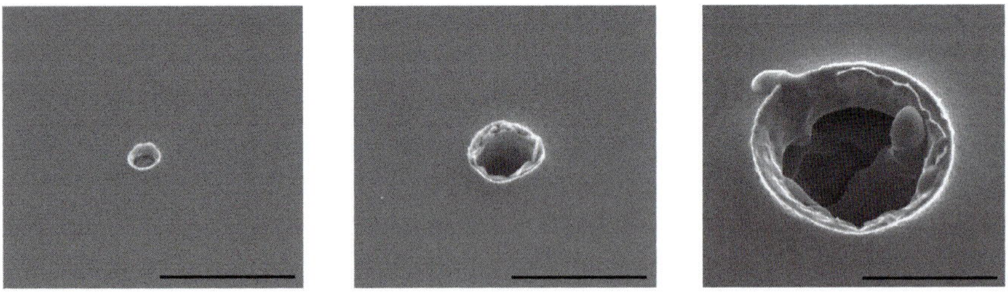

(c) (d) (e)

Fig. 13.22. Evaluation results for focusing beam performance [37]. (a)(b) Intensity distributions of the focused beam measured with the knife-edge scanning method. The beam size was 1.20 μm (FWHM) in the vertical direction (a) and 0.95 μm in the horizontal direction (b). (c)–(e) Irradiation imprints by focused beam with scale bars of 5 μm. A 1.4 μm layer of platinum coated onto a silicon substrate. The pulse energies of the beam were 0.09 μJ (c), 0.30 μJ (d) and 7.6 μJ (e), respectively.

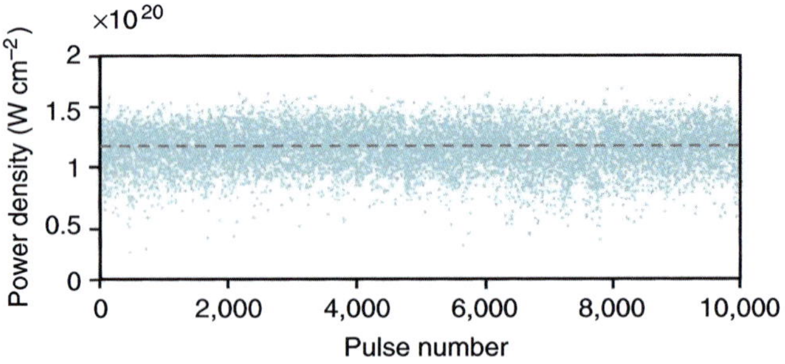

Fig. 13.23. Estimated power density of each XFEL pulse on focal plane [40]. The X-ray energy is 9.9 keV with the pulse duration, 7 fs. The average energy of the pulses is 42 μJ. The FWHM of the beam profile is 30 $\times$ 50 nm.

order disturbs atomic excitations by multi-photon processes. However, by applying a beam with the intensity of 10^{20} Wcm^{-2}, the number of excited states can dominate the decay rate. Thus the focused beam with the high power density is useful for realizing nonlinear phenomena including the amplification of X-rays, coherent scattering processes and ultrafast dynamics.

13.5.3. *Nonlinear phenomena*

Nonlinear optical phenomena, in general, are so small and needs strong intensity. In a visible region, an interaction between light and matter is relatively large. Nonlinear optical effects are observed by using conventional lasers. However, in the X-ray region, such an nonlinear interaction is extremely small. With a help of the intense beam, it now becomes possible to study nonlinear phenomena and higher order processes in the X-ray region.

13.5.3.1. *Multiphoton procecess*

A short wavelength and high frequency of X-rays have been used for atomic, molecular and optical physics (AMO). By using a beam at SACLA, we realized deep inner-shell multiphoton ionization of xenon atoms. Figure 13.24(a) shows the ion time-of flight of Xe [41]. We can see clear evidence of ions with a charge state of up to +26 and isotopes at different charges. In Fig. 13.24(b), a typical pathway of realizing Xe^{24+} is shown. At each photoionization process, the deep single-core vacancy is rapidly filled up within sub-femto seconds and the time scale of the corresponding Auger and Coster-Kronig decay cascade ranges of a few or tens of femto seconds. This study will play one of key roles for molecular structure determination by XFEL.

Another interesting phenomenon of multiphoton processes is realization of a hollow atom by creating a double core-hole state [42]. As shown in Fig. 13.25(a), an X-ray photon of 15 keV creates an single-core hole atom by photo exciting the K electron (P_1). The single-core hole state decays with a 170-as lifetime by emitting X-ray fluorescence (F_1) or the non-radiative Auger process (A_1), ejecting another

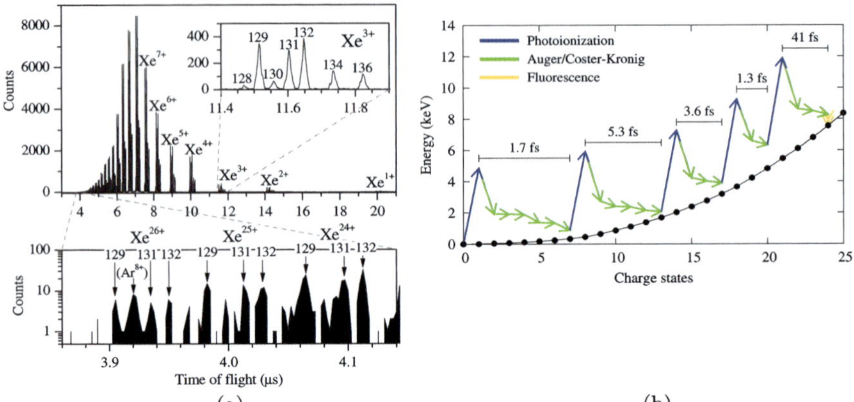

(a) (b)

Fig. 13.24. Evidence and behaviors of multiphoton ionization of Xe [41]. (a) The ion time-of-flight spectrum of Xe recorded at the photon energy of 5.5 keV at the full XFEL pulse energy. (b) An exemplary pathway of multiphoton multiple ionization of Xe. The black solid line with dots indicates the ground-configuration energy for given charge states. The energy of neutral Xe is set to zero.

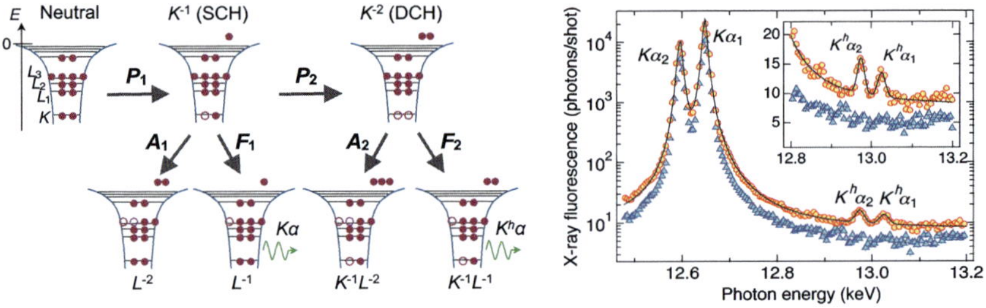

(a) (b)

Fig. 13.25. Schematic diagram of X-ray photoionization and decay processes in Kr (a), and X-ray fluorescence spectra of Kr (b) [42]. The circles and triangles were measured with high and low intensity X-rays, respectively. The solid line is the fitting with the Pearson VII function. The inset shows the magnified spectra around the hypersatellites. The vertical bar on each point indicates the standard error of the mean value.

electron. Intense X-rays can excite the remaining K electron before decay (P_2), producing a double-core hole atom with two K holes. The double-core hole state decays by F_2 or A_2. Figure 13.25(b) shows X-ray fluorescence spectra of krypton measured at two different intensities. The average pulse energies for high and low intensity pulses were 80.2 μJ and 49.9 μJ, respectively. Two prominent peaks correspond to $K\alpha$ lines that are the X-ray fluorescence from the single-core hole state. The higher peak is $K\alpha_1$, corresponding to transitions from the L_3 subshell to the K shell, while the other lower one is $K\alpha_2$. Another set of two peaks is observed on the tail of the highenergy side of the spectrum. These peaks are assigned to hyper-satellites (F_2 in Fig. 13.25(a)), corresponding to X-ray fluorescence from the double-core hole states. A single X-ray pulse can create the double-core hole of krypton because two X-ray photons ionize the same atom with about a 1Å diameter within the single-core hole lifetime of 170 as.

13.5.3.2. *Nonlinear X-ray optics*

Multiphoton ionizations we have discussed here are not explicitly nonlinear phenomena but sequential higher-order processes. It is interesting to study nonlinear X-ray optics by XFELs. However, non-linear X-ray effects often have competition with single and sequential multiphoton processes. Such competing processes make it difficult to observe higher-order nonlinear effects. To study nonlinear X-ray effects is not just an naive analogy of nonlinear effects in other regions of the spectrum. Especially in the hard X-ray region, an interaction between X-rays and an electron in a matter has different aspects compared with that in the visible region. For instance, in the frequency domain, X-rays can excite electrons in a core. Another example is the short wavelength of X-rays that is beyond the long wavelength approximation used in the visible region.

We report experimental evidence for second harmonic generation (SHG) at hard X-ray wavelengths [44]. Using a pumping beam of 7.3 keV generated by XFEL, we observed SHG in diamond. Figure 13.26 shows photon-energy histograms of the measured signal at the peak of the phase matching condition. We determine whether the measured second harmonic photons originate from the X-ray SHG in the crystal or from the residual second harmonic photons generated from the XFEL. A characteristic of the SHG is the nonlinearity with respect to the incident intensity. To differentiate the two

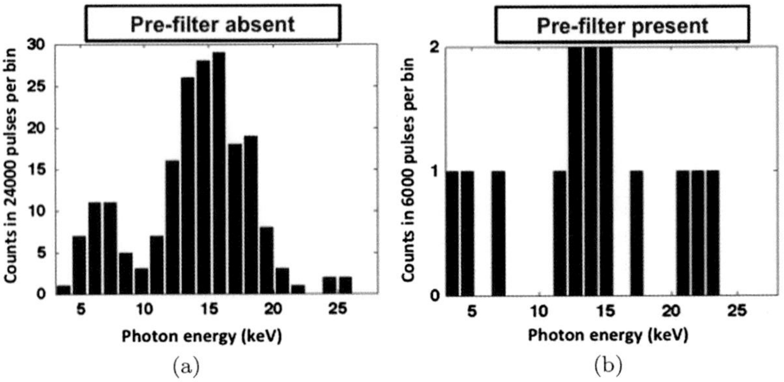

Fig. 13.26. Photon-energy histograms of the measured signal at the peak of the phase-matching condition [44]. The energy of a pump photon is 7.3 keV, and the SHG photon energy is 14.6 keV. The energy resolution of the detector is ∼30%. The data in part (a) are taken with no attenuation. The data in part (b) are taken with a 0.025 mm Al filter before the diamond crystal. The transmission through the filter is 0.43 and 0.9 at 7.3 and 14.6 keV, respectively.

possibilities, we insert a 0.025 mm Al prefilter before the diamond crystal to preferentially attenuate the fundamental. The resulting histogram is plotted in Fig. 13.26(b). In this case we count only 7 second harmonic photons in 6000 pulses. This count rate is approximately 4 times smaller than the count rate with no prefilter. Since the transmission of the undulator second harmonic through the prefilter is approximately 90%, the reduction of the SH signal by a factor of 4 indicates that most of the SH signal with no prefilter is generated in the crystal. The generated SHG is stronger than a factor of 10 than the background radiation. The SHG is generated over a narrow phase-matching condition. Our results indicate that it is possible to observe nonlinear X-ray processes in crystals at pump intensities exceeding 10^{16} Wcm^{-2}.

Higher order nonlinear effects are interesting topics. Here we show the first demonstration of third-order nonlinear effect of X-ray optics. We successfully observed two-photon absorption of a 5.6 keV X-rays by germanium. Figure 13.27 shows schematic diagram of two-photon absorption and experimental results [43]. In Fig. 13.27(a), the 10 keV peak is assigned to the germanium $K\alpha_1$ fluorescence by two-photon

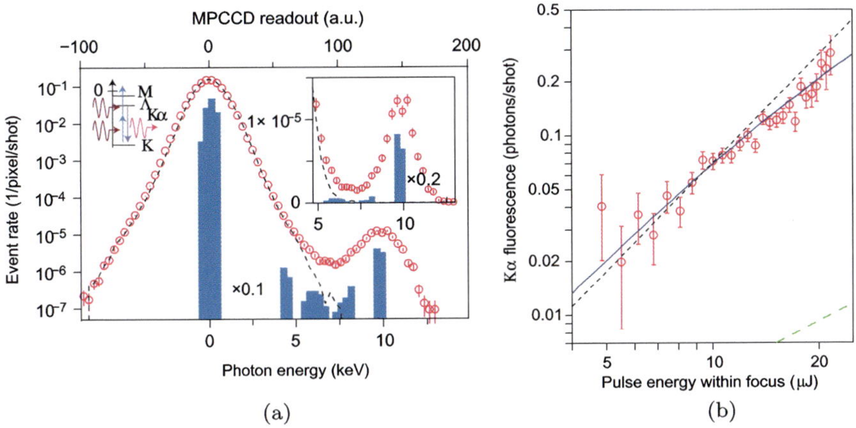

Fig. 13.27. Schematic of two-photon absorption and experimental results [43]. (a) Fluorescence spectrum of two-photon absorption. Circles and vertical lines indicate the raw spectrum and the statistical error, respectively. The insets shows magnified spectra around the fluorescence peak on a linear scale, and the energy diagram of two-photon absorption. (b) Pulse-energy dependence of germanium fluorescence by two-photon absorption. The slope of the dashed line is 2. The long-dashed line with a slope of 1 indicates one-photon absorption fluorescence by the second-harmonic radiation of XFEL.

absorption of X-rays. The dashed line is the noise spectrum of MPCCD that represents the photon-energy resolution. Vertical bars indicate the spectrum extracted by deconvolution of the noise spectrum. The residual contribution arises from the scattered pump X-rays Johansson monochromator suppresses it at very low levels. We estimate the $K\alpha_1$ fluorescence due to one-photon absorption of the higher-harmonic radiation by the XFEL, and ionization by thermal electrons, and confirm that these are much weaker than the signal from two-photon absorption. In Fig. 13.27(b), evidence of the nonlinearity is shown. The pulse-energy dependence of the $K\alpha_1$ fluorescence has a quadratic dependence, which is direct evidence of two-photon absorption.

13.5.3.3. *Elementary particle physics*

Strong intensity and short wavelength of a SACLA beam allows us to reveal structures of not only materials but vacuum. Quantum electrodynamics (QED) predicts existence of photon-photon (γ-γ) scattering processes as a nonlinear effect of vacuum. There are several experiments on γ-γ scattering in the visible region [46, 47]. However, these experiments were far above the prediction by QED, and no evidence of the event was observed. Even though no signals are observed, it is worthwhile to give a limit of the event. Such experiments are important not only for checking validity of QED but for searching new physics beyond the Standard Model. As for frequency dependence, the cross section of γ-γ scattering is proportional to ω^6. This means that shorter wavelength of light is more suitable for detection. X-rays have the cross section 10^{30} times larger than that of visible light.

We report the first demonstration of a direct search for γ-γ scattering in the X-ray region. Figure 13.28(a) shows a schematic of the experimental setup in the chamber. To hit two X-ray beams each other, we used a beam splitter and an interferometer technique. In Fig. 13.28(b), the energy spectrum after the timing cut is shown. The observed peaks are consistent with the null observation. Thus no events are observed in the signal region (18.1–19.9 keV). From the experimental results, we calculated the limit of the cross section of γ-γ scattering to be $<1.7 \times 10^{-24}$ m^2 at 95% confidence level. Comparison with other results is shown in Fig. 13.28(c). Our experiments are up to now far from the QED cross section. However, we can improve the monochromaticity of the SACLA beam by the seeding technique in the near future. Such an improvement of monochromaticity and the intensity will make a step further to the QED prediction.

13.5.4. *Ultrafast phenomena*

Ultrafast dynamics are of much interest over whole range of the spectrum. In the X-ray region, a short pulse can reveal ultrafast dynamics by visualizing what kind of reaction is going on inside an atom. To see ultrafast phenomena, realization of a short pulse is necessary. Moreover, determination the pulse duration is of importance because we can measure a profile of a pulse only in the frequency domain.

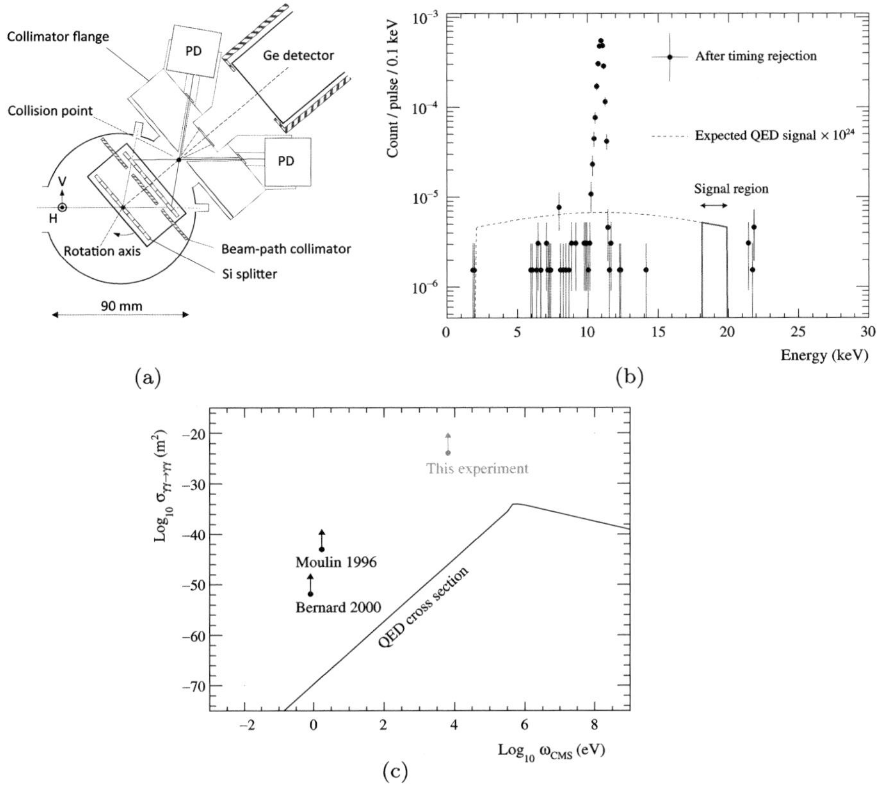

Fig. 13.28. Schematic of experimental setup and results [45]. (a) The beam-collision chamber. Four beams after the two blades of the silicon beam splitter. Horizontal (H) and vertical (V) directions are shown in a plane which is normal to the beam axis. (b) Energy spectrum obtained from the total injection of 6.5×105 pulses, after the timing rejection cut. The dotted line shows the expected QED signal shape int he full solid angle (signal strength is magnified by 1024). The solid line corresponds to the signal region from 18.1 keV ($\theta = 12.5°$) to 19.9 keV ($\theta = 0°$). (c) 95% confidence level upper limit on the QED cross-section of elastic γ-γ scattering obtained with SACLA. QED prediction and limits from other experiments in the visible region are also shown.

We realized the short pulse width of the SACLA beam, and determined the duration to be several femto seconds [48]. Figure 13.29 shows simulated profiles of XFEL in the time domain. The shortest pulse we have achieved has the duration of 4.5 fs.

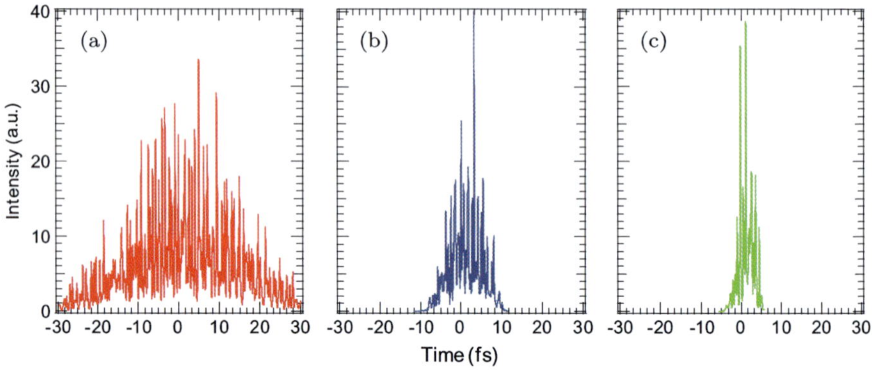

Fig. 13.29. Simulated temporal profiles of XFEL pulses, corresponding to the measured energy spectra [48]. The pulse durations were (a) $\Delta T_1 = 31$ fs, (b) $\Delta T_2 = 8.9$ fs and $\Delta T_3 = 4.5$ fs, respectively.

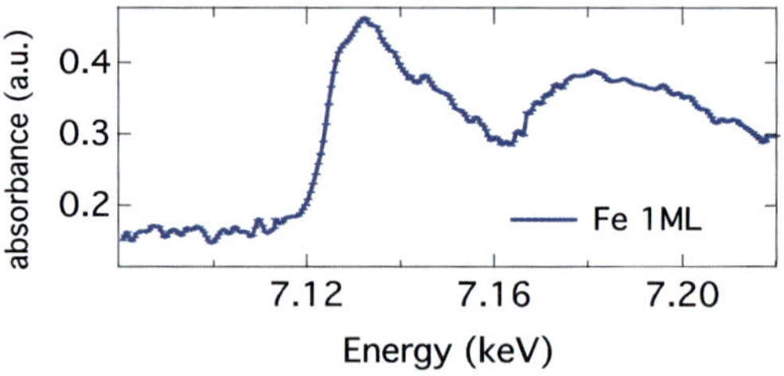

Fig. 13.30. Dispersive X-ray absorption spectroscopy (D-XAS) spectrum of Fe K-edge obtained from the liquid jet of 1M aqueous $Fe(III)(C_2O_4)_3^{3-}$ solution with 100 lm diameter [49].

Once a profile of a pulse is obtained, our next interest is to use the pulse for investigating ultrafast phenomena. To trace transient phenomena, we developed a method of dispersive X-ray absorption spectroscopy [49]. Usually XFEL has stochastic spike structures in the frequency domain. This behavior induces a severe problem in the spectral normalization process. We have introduced two primary split beams into a dispersive spectrometer for simultaneous measurement of signal and reference spectra. As shown in Fig. 13.30, absorption spectrum of Fe K-edge is obtained by a single-shot measurement using a relatively large bandwidth of $\Delta E/E \sim 5 \times 10^3$ of the beam.

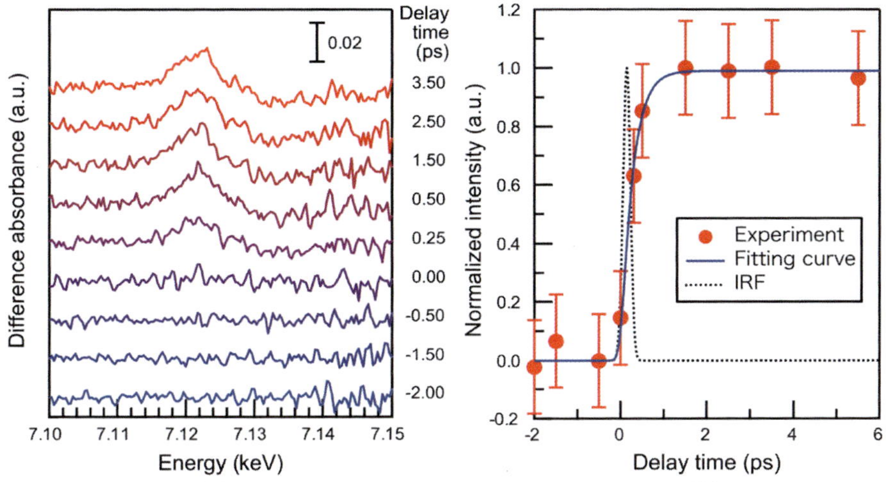

Fig. 13.31. Transient spectra of aqueous 0.5 M ammonium iron(III) oxalate trihydrate solutioncite [50]. (a) X-ray spectra of time-resolved difference absorption. (b) Temporal dependence of the absorbance change integrated over 7.120 ± 0.006 keV in the difference absorption spectra.

By using this method, we measured the difference X-ray absorption spectra of the $Fe(III)(C_2O_4)_3^{3-}$ complex around the K-edge of the Fe atom at several pump-probe time delays, as shown in Fig. 13.31 [50]. A positive absorbance change indicates an increased absorption by 400-nm photoexcitation. The peak position appearing at 7.12 keV is lower than the K-edge of the steady-state spectrum of the molecule. The shape of the difference spectra does not vary with the time delay. Figure 13.31 represents a transient phenomenon of the absorbance change. Assuming the instrumental response time of 200 fs expected from timing jitter of the X-ray pulse and the duration of 400-nm pulse, the least squares fitting of the experimental data provides the rise time of 260 ± 50 fs as shown in solid line. Our results open up a time-resolved X-ray spectroscopy for transient phenomena of the femtosecond time scale. In the future, it will be interesting to reveal transient dynamics of chemical reactions in an atomic scale with very short time scale of femto- and atto-seconds.

13.5.5. *Imaging and structural analysis*

An X-ray beam with high intensity is useful for analyzing structures of crystals, proteins and so on. One of the most powerful crystallographic methodology to non-crystalline material is the coherent diffraction imaging (CDI). However, under an intense beam, samples are always suffering from damages. To avoid this difficulty, short pulses provide a way of detecting diffraction signal from the sample before destroy.

13.5.5.1. *Coherent diffraction imaging*

We have developed an experimental instruments for imaging experiments at SACLA [51, 52]. By using the instruments, we demonstrated the CDI analysis of nano-particles to obtain the size distribution of particles as well as the electron density projection of individual particles. Figure 13.32 shows projection images of Au/Ag nano-boxes reconstructed from single-shot coherent X-ray diffraction patterns [53]. We reconstructed the two-dimensional electron density projection with sub-10 nm resolution from selected coherent

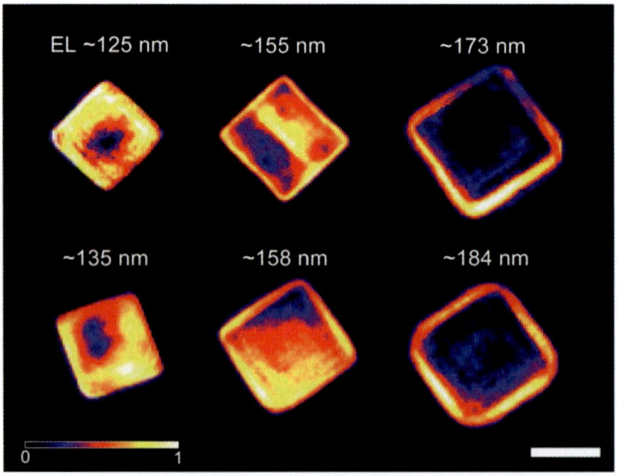

Fig. 13.32. Projection images of Au/Ag nanoboxes. The image of a nanobox with EL ~173 nm was reconstructed from the diffraction pattern. The scale bar is 100 nm.

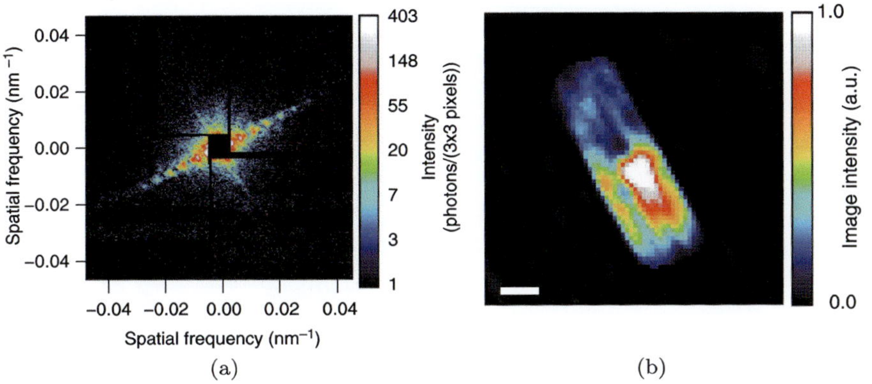

Fig. 13.33. Experimental coherent X-ray diffraction pattern and reconstruction. Measured CXD pattern (a) and reconstructed image (b) of a live M. lacticum cell, respectively [54]. The scale bar is 100 nm.

diffraction patterns. This method enables the simultaneous analysis of the size distribution of synthesized structures of particles at nanoscale resolution.

The short pulses at SACLA enable single-shot imaging almost free from damage. Thus a short pulse beam is useful for analyzing not only static materials such as metals but also a living matter. We first present a method for capturing snapshots of live cells in micro-liquid enclosure by XFEL [54]. The coherent X-ray diffraction (CXD) pattern in Fig. 13.33(a) contains a clear fringe extending in one direction. The fringe spacing and orientation indicates a sample width of 194nm and is tilted by 28°. The width is comparable to the diameter of an M. lacticum cell. We performed image reconstruction from the experimental CXD pattern. Figure 13.33(b) shows the reconstructed image of an M. lacticum cell. It is rod shaped with a width and length of ∼194 nm and ∼570 nm, respectively. This result agreed well with the known shape and size of M. lacticum cells.

Femtosecond X-ray pulses with high intensity from XFEL enable single-shot image allowing for instantaneous views of nano structures. On the other hand, coherent X-rays from synchrotron radiation are suitable for investigating steady-state features of a sample. Thus combination of these two X-rays sources can improve the quality of

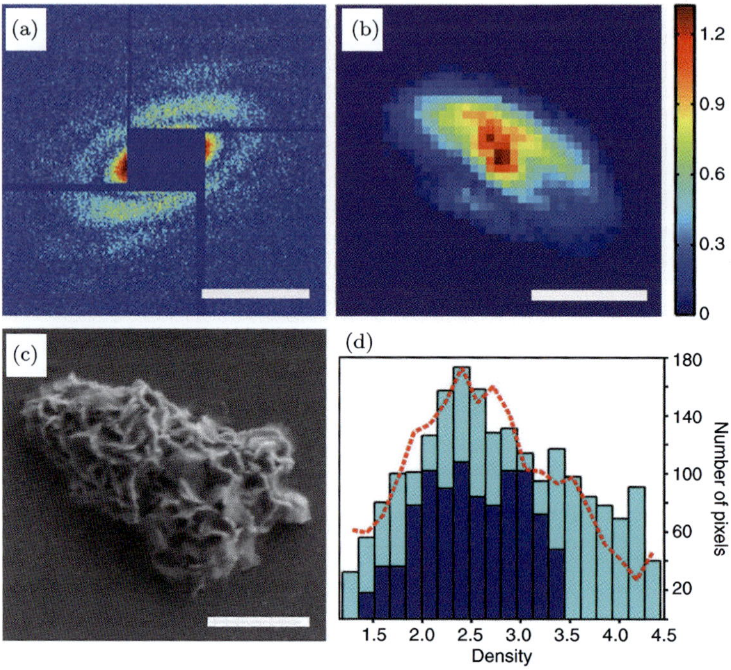

Fig. 13.34. Femtosecond single-shot XFEL imaging of an RNA microsponge [55]. (a) Single-shot femtosecond speckle pattern of an RNAi microsponge is obtained at SACLA. (b) Reconstructed image from (a) as the projected electron density. The colour bar displays the linear density scale in arbitrary unit. (c) SEM image of a similarly structured, RNAi microsponge. (d) Histogram of the electron density distribution of the small RNAi (blue) (b), compared with the density distribution of large RNAi (cyan). The broken red lines represent the average of histograms from large RNAi microsponges imaged at SPring-8. The scale bars are 10 μm^{-1}, 250 nm and 500 nm (a–c), respectively.

imaging. We successfully demonstrated a combination of XFEL and synchrotron X-rays to provide insights into nanostructure formation of a biological macromolecular complex of RNA interference micro sponges [55], as shown in Fig. 13.34. We propose that the combined applications of synchrotron X-rays and XFELs demonstrated here can establish a unique pathway to seamlessly perform *ab initio* structure analysis of non-resembling structures within cells, organelles and macromolecular complexes, from micron-scale architectures to nanoscale details.

13.5.5.2. *Structural analysis*

Analyzing structures of proteins are one of the most important missions of biomedical science. X-rays are fundamental tools to determine structures, because X-rays have short wavelength that brings high resolution. However, intense X-rays destroy protein samples. Thus radiation damage-free crystal analysis by using the femtosecond pulses has been expected but not been fully realized. Here we report a method of femtosecond crystallography for solving radiation damage-free crystal structures of large proteins at sub-angstrom spatial resolution [56]. As shown in Fig. 13.35, we demonstrated the performance of the method by determining a 1.9 Å radiation damage-free structure of bovine cytochrome c oxidase, a large (420-kDa), highly radiation-sensitive membrane protein. Our method of using large single crystals could not be obtained using the microbeam at SPring-8 at the same dose per image with a longer exposure time, or by a low-dose experiment). The ultrashort, femtosecond exposure time provides many advantages compared with lnger exposure time at a synchrotron beamline.

13.6. Future XFEL Sources

The splendid success of LCLS and SACLA has made the wide range of science community to recognize the enormous potential of the XFEL to expand the frontiers of the sciences in many fields as partly shown in the previous section. Demands for the XFEL sources are so high that much less than a half of the proposals can get beamtime at these facilities. Now that there are only two XFEL facilities in operation as of May 2014, the third one will come on stage in a couple of years. However, considering the rapid growth of the user demands, it is highly desirable to have far more numbers of facilities in operation. When we started designing the SACLA concept, there were only two precedent XFEL construction projects in the world, LCLS and European XFEL, the locations of which made a beautiful equilateral triangle with our SACLA when viewing from a particular direction. However, after we have proposed that XFELs

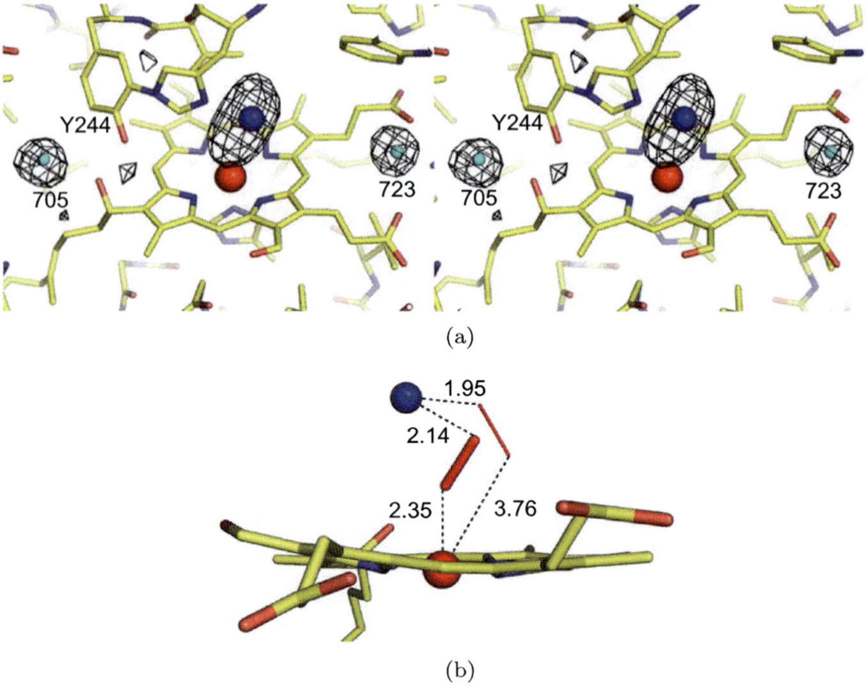

Fig. 13.35. The O_2-reduction site free of radiation damage [56]. (a) $F_o - F_c$ map in stereo at 1.9 Å resolution for the XFEL data. Electron density cages at the 3σ level (σ/0.041 e$^-$Å^3) are depicted along with the structural model of the O_2-reduction site, including the Fe(red) and Cu(blue) atoms and the oxygen atoms (cyan) of the two water molecules. The two water molecules 705 and 723 were not included in the structural refinements in order to compare the electron densities with the electron density between Fe_{a3} and Cu_B as well as the electron density near Tyr244. The peak height of both water molecules is 9.0σ, and the peak height near Tyr244 is 3.4σ. (b) Diagram of ligand binding. A stick model of the O_2-reduction site with a major component (95%; thick stick) and a minor component (5%; thin stick) of peroxide anions, both with an O–O distance of 1.55 Å. The colors used to designate the atom species are the same as in (a). The numerical values indicated between the metal atoms and the oxygen atoms of the peroxide anions represent the distance in angstroms.

could be constructed in a compact form with less cost, it becomes evident that most developed countries can afford XFEL. Actually, Switzerland and Korea have already started construction of their own XFEL facilities, and some other countries started planning to have their own facilities. Downsizing achieved by SACLA could be an efficacious way to increase the beamtime supply of XFELs.

Another way to increase the beamtime supply would be to enhance the repetition rate of a single linac so as to feed electron beams to many FEL lines. The operating two facilities are both using normal conducting linacs whose repetition rates are relatively low, being 120 Hz for LCLS and 60 Hz for SACLA. However, the European XFEL is promising much higher repetition rate with the use of a superconducting linac, to operate more than five independent FEL undulators simultaneously by switching the electron beams to different branches. At the current technical level, the normal conducting linacs with lower repetition rates can accelerate electron bunches with higher peak current and higher charges than the superconducting linacs with higher repetition rates, thus enabling to achieve higher peak brilliance and higher pulse energies. The construction and operation cost of the superconducting facility seem to be extremely high, but cost per a laser pulse would be comparable with, or, less than that of the normal conducting facilities if all the laser pulses could be utilized without waste. This might be one reason for the Department of Energy of the US to decide the LCLS upgrade with a superconducting linac which can increases the repetition rate in trade for the higher peak brilliance as well as pulse energy.

These two opposite trends of SASE XFEL development could polarize the future facilities. Even so, further downsizing of the SASE XFELs will continue in the future. We hope we can make such XFELs with less than 100 m length in some decades from now that dozens of universities, institutes and factories will be able to own their XFELs even inside of Japan. Two key technologies toward this direction would be shorter period undulators and higher gradient linacs. New acceleration technologies such as laser acceleration [57] should be developed seriously for this purpose.

Although SASE XFEL is a noteworthy light source, its laser pulses consist of quite a few number of modes. For some applications, it is desirable to use FEL pulses with a smaller number of modes, or ultimately, with a single mode. Such FEL would be generated by seeding technique which induces stimulus emission by introducing seed light. Since it is rather difficult in soft to hard X-ray regions to prepare a seed light by using an independent light source such as laser higher harmonics [58], self-seeding scheme was proposed in which a photon monochromator was inserted in a space prepared between long undulators. Electron beams detoured around the monochromator will run overlapped with the monochromatic seed light in the downstream long undulator to enhance stimulated emission [59]. In hard X-ray region, self-seeding scheme using a transmission beam of a diamond single crystal was proposed [60] and found to work beautifully at LCLS [61]. A similar scheme was adopted at SACLA. Soft X-ray self-seeding was tested recently at LCLS to show a beautiful result [62].

SACLA was designed to collocate with a third generation synchrotron X-ray source, SPring-8. SASE XFELs can deliver the X-ray pulses with extremely high peak brilliance and fs duration. The extreme high intensity frequently permits only 'diffract-before-destroy' mode, or more generally, 'measure-before-destroy' mode. Samples could be damaged in a single shot irradiation of the XFEL pulse. Meanwhile, SR can deliver the X-ray pulses with moderate peak intensity, but the averaged brilliance is high because of the higher repetition rate. Therefore, samples will not be damaged in a single shot so that the sample dynamics could be traced with a single sample. The non-destructive nature of the synchrotron observation will allow the tomographic reconstruction of 3D structure from a number of 2D projections from the various directions. This would be difficult for XFEL observation because sample could be damaged by the first shot. The different characteristics of these two sources could be well combined to create a new regime of the X-ray sciences. We hope we can open up a new regime by using our SACLA-SPring-8 interaction facility.

One of the future forms of the seeded XFEL could be the XFEL oscillator (XFELO) which uses an optical cavity composed of Bragg diffraction mirrors [63]. Although the original proposal of XFELO was with an energy recovery linac (ERL), any electron sources exceeding a certain critical performance would work. Current storage rings, however, can provide far less performance necessary for the XFELO operation. Within a decade, there will be upgraded storage rings for synchrotron radiation applications whose emittance is close to the diffraction limit of the emitted photons. Such upgrade are considered at ESRF, APS and SPring-8 as to make their electron beam emittance to be less than 100 pm rad. The possible smaller beam emittance in future would enable to construct XFELO with storage ring technology. We can allocate many beamlines operated simultaneously around this type of ring XFELO. The ring XFELO would be of extremely high repetition rate, but the peak brilliance wouldn't be so high. This characteristics is somewhat similar to that of SASE XFEL with superconducting linac. Therefore, SASE XFELs with superconducting linacs may be replaced by the ring XFELO in some distant future.

13.7. Concluding Summary

In this chapter, we intended to describe deep information of SACLA which may be difficult to touch in detail in usual journal papers and/or conference proceedings. Since XFELs and their applications are just emerging, the readers can find a lot of buds in these fields. The authors are more than happy if they can provide the readers with something important in this article. Last but not least, the authors wish to express their sincere gratitude to all who have contributed to the SACLA construction.

Bibliography

[1] T. Maiman, *Nature* **187**, 493, (1960).
[2] J. M. J. Madey, *J. Appl. Phys.* **42**, 1906, (1971).
[3] R. Bonifacio, C. Pellegrini and L. M. Narducci, *Opt. Comm.* **50**, 373, (1984).
[4] D. Attwood, *Soft X-Rays and Extreme Ultraviolet Radiation* (Cambridge University Press, New York, 1999).
[5] https://lcls.slac.stanford.edu.
[6] http://tesla.desy.de/new_pages/TDR_CD/start.html.
[7] http://xfel.riken.jp/research/before2010/SCSSCDR.pdf.
[8] T. Ishikawa, H. Aoyagi, T. Asaka, Y. Asano, N. Azumi, T. Bizen, H. Ego, K. Fukami, T. Fukui, Y. Furukawa, S. Goto, H. Hanaki, T. Hara,

T. Hasegawa, T. Hatsui, A. Higashiya, T. Hirono, N. Hosoda, M. Ishii, T. Inagaki, Y. Inubushi, T. Itoga, Y. Joti, M. Kago, T. Kameshima, H. Kimura, Y. Kirihara, A. Kiyomichi, T. Kobayashi, C. Kondo, T. Kudo, H. Maesaka, X. M. Marchal, T. Masuda, S. Matsubara, T. Matsumoto, T. Matsushita, S. Matsui, M. Nagasono, N. Nariyama, H. Ohashi, T. Ohata, T. Ohshima, S. Ono, Y. Otake, C. Saji, T. Sakurai, T. Sato, K. Sawada, T. Seike, K. Shirasawa, T. Sugimoto, S. Suzuki, S. Takahashi, H. Takebe, K. Takeshita, K. Tamasaku, H. Tanaka, R. Tanaka, T. Tanaka, T. Togashi, K. Togawa, A. Tokuhisa, H. Tomizawa, K. Tono, S. Wu, M. Yabashi, M. Yamaga, A. Yamashita, K. Yanagida, C. Zhang, T. Shintake, H. Kitamura and N. Kumaga, *Nature Photonics* **6**, 540, (2012).

[9] E. L. Saldin, E. A. Schneidmiller and M. V. Yurkov, *The Physics of Free Electron Lasers* (Springer, 1999).

[10] Z. Huang and K. J. Kim, *Phys. Rev. ST Accel. Beams* **10**, 034801, (2007).

[11] S. J. Russell, *NIM A* **507**, 304, (2003).

[12] D.T. Palmer, X.J. Wang, R.H. Miller, M. Babzien, I. Ben-Zvi, C. Pellegrini, J. Sheehan, J. Skaritka, H. Winick, M. Woodle and V. Yakimenko, in *Proc. of the Particle Accelerator Conference*, Vancouver, May, 2687, (1997).

[13] H. Tanaka *et al.*, in *Proc. of the 28th International Free Electron Laser Conference*, BESSY, Berlin, Germany, 27 Aug.-01 Sep., 769, (2006).

[14] K. Togawa *et al.*, *Phys. Rev. ST Accel. Beams* **10**, 020703, (2007).

[15] T. Shintake *et al.*, *Phys. Rev. ST Accel. Beams* **10**, 070701, (2009).

[16] Linac Coherent Light Source, (LCLS) Conceptual Design Report, SLAC-R-593, (Stanford, 2002).

[17] M. Altarelli, R. Brinkmann, M. Chergui, W. Decking, B. Dobson, S. Düsterer, G. Grübel, W. Graeff, H. Graafsma, J. Hajdu, J. Marangos, J. Pflüger, H. Redlin, D. Riley, I. Robinson, J. Rossbach, A. Schwarz, K. Tiedtke, T. Tschentscher, I. Vartaniants, H. Wabnitz, H. Weise, R. Wichmann, K. Witte, A. Wolf, M. Wulff and M. Yurkov, (eds): *XFEL: The European X-ray Free-Electron Laser, Technical design report*, DESY 2006-097, (DESY, Hamburg, 2006).

[18] K. Togawa, T. Hara and H. Tanaka, *Phys. Rev. ST Accel. Beams* **12**, 080706, (2009).

[19] J. Frisch *et al.*, *Proc. of the XXI Linear Accelerator Conference, Gyeongju*, Korea, 19–23 Aug., pp. 788–790, (2002).

[20] T. Inagaki, *Proc. of the XXIV Linear Accelerator Conference*, Victoria, British Columbia, Canada, 29 Sep.–3 Oct., pp. 1090–1094, (2008).

[21] Z. D. Farkas *et al.*, SLAC-PUB-1453, (Stanford 1974).

[22] T. Inagaki *et al.*, submitted.

[23] H. Tanaka *et al.*, *Proc. of the 31st International Free Electron Laser Conference*, Liverpool, UK, 23–28 Aug., 758, (2009).

[24] Y. Otake *et al.*, *Proc. of the 28th International Free Electron Laser Conference*, BESSY, Berlin, Germany, 27 Aug.–01 Sep., 645, (2006).

[25] T. Inagaki *et al.*, *Proc. of the 32nd International Free Electron Laser Conference*, Malmö, Sweden, 23–27 Aug., 525, (2010).

[26] Y. Otake *et al.*, *Proc. of the 4th Asian Particle Accelerator Conference*, RRCAT, Indore, India, 29 Jan.–02 Feb., 634, (2007).
[27] H. Tanaka, *Proc. of the 2nd International Particle Accelerator Conference*, San Sebastian, Spain, 4–9 Sep., 21, (2011).
[28] H. Kitamura, *J. Sync. Rad.* **5**, 184, (1998).
[29] T. Tanaka *et al.*, *J. Sync. Rad.* **14**, 416, (2007).
[30] T. Tanaka, T. Seike, A. Kagamihata, T. Schmidt, A. Anghel, M. Brügger, W. Bulgheroni, B. Jakob and H. Kitamura, *Phys. Rev. ST Accel. Beams* **12**, 120702, (2009).
[31] T. Tanaka, S. Goto, T. Hara, T. Hatsui, H. Ohashi, K. Togawa, M. Yabashi and H. Tanaka, *Phys. Rev. ST Accel. Beams* **15**, 110701, (2012).
[32] K. Tono, T. Togashi, Y. Inubushi, T. Sato, T. Katayama, K. Ogawa, H. Ohashi, H. Kimura, S. Takahashi, K. Takeshita, H. Tomizawa, S. Goto, T. Ishikawa and M. Yabashi, *New J. Phys.* **15**, 083035, (2013).
[33] M. Yabashi, J. B. Hastings, M. S. Zolotorev, H. Mimura, H. Yumoto, S. Matsuyama, K. Yamauchi and T. Ishikawa, *Phys. Rev. Lett.* **97**, 084802, (2006).
[34] M. Yabashi *et al.*, *Proc. FEL conf.* 785, (2006).
[35] M. Yabashi and T. Ishikawa, (eds.), XFEL/SPring-8 Beamline Technical Design Report, ver. 2.0, (RIKEN/JASRI, Hyogo, Japan, 2010).
[36] K. Tono *et al.*, *Rev. Sci. Instrum.* **82**, 023108, (2011).
[37] H. Yumoto, H. Mimura, T. Koyama, S. Matsuyama, K. Tono, T. Togashi, Y. Inubushi, T. Sato, T. Tanaka, T. Kimura, H. Yokoyama, J. Kim, Y. Sano, Y. Hachisu, M. Yabashi, H. Ohashi, H. Ohmori, T. Ishikawa and K. Yamauchi, *Nature Photonics* **7**, 43, (2012).
[38] H. Mimura, S. Handa, T. Kimura, H. Yumoto, D. Yamakawa, H. Yokoyama, S. Matsuyama, K. Inagaki, K. Yamamura, Y. Sano, K. Tamasaku, Y. Nishino, M. Yabashi, T. Ishikawa and Kazuto Yamauchi., *Nature Phys.* **6**, 122, (2010).
[39] T. Kameshima, S. Ono, T. Kudo, K. Ozaki, Y. Kirihara, K. Kobayashi, Y. Inubushi, M. Yabashi, T. Horigome, A. Holland, K. Holland, D. Burt, H. Murao and T. Hatsui, *Rev. Sci. Instrum.* **85**, 033110, (2014).
[40] H. Mimura, H. Yumoto, S. Matsuyama, T. Koyama, K. Tono, Y. Inubushi, T. Togashi, T. Sato, J. Kim, R. Fukui, Y. Sano, M. Yabashi, H. Ohashi, T. Ishikawa and K. Yamauchi, *Nature Commun.* **5**, 3539, (2014).
[41] H. Fukuzawa, S.-K. Son, K. Motomura, S. Mondal, K. Nagaya, S. Wada, X.-J. Liu, R. Feifel, T. Tachibana, Y. Ito, M. Kimura, T. Sakai, K. Matsunami, H. Hayashita, J. Kajikawa, P. Johnsson, M. Siano, E. Kukk, B. Rudek, B. Erk, L. Foucar, E. Robert, C. Miron, K. Tono, Y. Inubushi, T. Hatsui, M. Yabashi, M. Yao, R. Santra and K. Ueda, *Phys. Rev. Lett.* **110**, 173005, (2013).
[42] K. Tamasaku, M. Nagasono, H. Iwayama, E. Shigemasa, Y. Inubushi, T. Tanaka, K. Tono, T. Togashi, T. Sato, T. Katayama, T. Kameshima, T. Hatsui, M. Yabashi and T. Ishikawa, *Phys. Rev. Lett.* **111**, 043001, (2013).

[43] K. Tamasaku, E. Shigemasa, Y. Inubushi, T. Katayama, K. Sawada, H. Yumoto, H. Ohashi, H. Mimura, M. Yabashi, K. Yamauchi and T. Ishikawa, *Nature Photonics* **8**, 313, (2014).
[44] S. Shwartz, M. Fuchs, J. B. Hastings, Y. Inubushi, T. Ishikawa, T. Katayama, D. A. Reis, T. Sato, K. Tono, M. Yabashi, S. Yudovich and S. E. Harris, *Phys. Rev. Lett.* **112**, 163901, (2014).
[45] T. Inada, S. Adachi, T. Namba, S. Asai, T. Kobayashi, K. Tamasaku, Y. Tanaka, Y. Inubushi, K. Sawada, M. Yabashi and T. Ishikawa, *Phys. Lett. B* **732**, 356, (2014).
[46] F. Moulinm D. Bernard, and F. Amiranoff, *Z Phys. C* **72**, 607, (1996).
[47] D. Bernard, F. Moulin, F. Amiranoff, A. Braun, J. P. Chambaret, G. Darpentigny, G. Grillon, S. Ranc and F. Perrone, *Eur. Phys. J. D* **10**, 141, (2000).
[48] Y. Inubushi, K. Tono, T. Togashi, T. Sato, T. Hatsui, T. Kameshima, K. Togawa, T. Hara, T. Tanaka, H. Tanaka, T. Ishikawa and M. Yabashi, *Phys. Rev. Lett.* **109**, 144801, (2012).
[49] T. Katayama, Y. Inubushi, Y. Obara, T. Sato, T. Togashi, K. Tono, T. Hatsui, T. Kameshima, A. Bhattacharya, Y. Ogi, N. Kurahashi, K. Misawa, T. Suzuki and M. Yabashi, *Appl. Phys. Lett.* **103**, 131105, (2013).
[50] Y. Obara, T. Katayama, Y. Ogi, T. Suzuki, N. Kurahashi, S. Karashima, Y. Chiba, Y. Isokawa, T. Togashi, Y. Inubushi, M. Yabashi, T. Suzuki and K. Misawa, *Opt. Exp.* **22**, 1105, (2014).
[51] M. Nakasako, Y. Takayama, T. Oroguchi, Y. Sekiguchi, A. Kobayashi, K. Shirahama, M. Yamamoto, T. Hikima, K. Yonekura, S. Maki-Yonekura, Y. Kohmura, Y. Inubushi, Y. Takahashi, A. Suzuki, S. Matsunaga, Y. Inui, K. Tono, T. Kameshima, Y. Joti and T. Hoshi, *Rev. Sci. Instrum.* **84**, 093705, (2013).
[52] C. Song, K. Tono, J. Park, T. Ebisu, S.-N. Kim, H. Shimada, S.-S. Kim, M. G.-Jones, D. Nam, T. Sato, T. Togashi, K. Ogawa, Y. Joti, T. Kameshima, S. Ono, T. Hatsui, S. Iwata, M. Yabashi and T. Ishikawa, *J. Appl. Cryst.* **47**, 188, (2014).
[53] Y. Takahashi, A. Suzuki, N. Zettsu, T. Oroguchi, Y, Takayama, Y. Sekiguchi, A. Kobayashi, M. Yamamoto and M. Nakasako, *Nano Letters* **13**, 6028, (2013).
[54] T. Kimura, Y. Joti, A. Shibuya, C. Song, S. Kim, K. Tono, M. Yabashi, M. Tamakoshi, T. Moriya, T. Oshima, T. Ishikawa, Y. Bessho and Y. Nishino, *Nature Commun.* 5:3052 doi: 10.1038/ncomms4052, (2014).
[55] M. G.-Jones, Y. Bessho, S. Kim, J. Park, S. Kim, D. Nam, C. Kim, Y. Kim, D. Y. Noh, O. Miyashita, F. Tama, Y. Joti, T. Kameshima, T. Hatsui, K. Tono, Y. Kohmura, M. Yabashi, S. S. Hasnain, T. Ishikawa and C. Song, *Nature Commun.* **5**, 3798, (2014).
[56] K. Hirata, K. Shinzawa-Itoh, N. Yano, S. Takemura, K. Kato, M. Hatanaka, K. Muramoto, T. Kawahara, T. Tsukihara, E. Yamashita, K. Tono, G. Ueno, T. Hikima, H. Murakami, Y. Inubushi, M. Yabashi, T. Ishikawa,

M. Yamamoto, T. Ogura, H. Sugimoto, J.-R. Shen, S. Yoshikawa and H. Ago, *Nature Methods* 2014, doi:10.1038/NMETH.2962
[57] W. P. Leemans, B. Nagler, A. J. Gonsalves, Cs. Toth, K. Nakamura, G. R. Geddes, E. Esarey, C. Schroeder and S. M. Hooker, *Nature Physics* **2** 696, (2006).
[58] G. Lambert, T. Hara, D. Garzella, T. Tanikawa, M. Labat, B. Carre, H. Kitamura, T. Shintake, M. Bougeard, S. Inoue, Y. Tanaka, P. Salieres, H. Merdji, O. Chubar, O. Gobert, K. Tahara and M-E. Couprie, *Nature Physics* **4**, 296, (2008).
[59] J. Feldhaus, E. L. Saldin, J. R. Schneider, E. A. Schneidmiller and M. V. Yurkov, *Opt. Comm.* **140**, 341, (1997).
[60] G. Geloni, V. Kocharyan and E. Saldin, *J. Mod. Opt.* **58**, 1391, (2011).
[61] J. Amann, W. Berg, V. Blank, F.-J. Decker, Y. Ding, P. Emma, Y. Feng, J. Frisch, D. Fritz, J. Hastings, Z. Huang, J. Krzywinski, R. Lindberg, H. Loos, A. Lutman, H.-D. Nuhn, D. Ratner, J. Rzepiela, D. Shu, Y. Shvyd'ko, S. Spampinati, S. Stoupin, S. Terentyev, E. Trakhtenberg, D. Waltz, J. Welch, J. Wu, A. Zholents and D. Zhu, *Nature Photonics* **6**, 693, (2012).
[62] D. Ratner, J. W. Amann, D. K. Bohler, M. Boyes, D. Cocco, F. -J. Decker, Y. Ding, D. Fairley, Y. Feng, J. Hastings, P. Heimann, Z. Huang, J. Krzywinski, H. Loos, A. A. Lutman, G. Marcus, A. Marinelli, T. J. Maxwell, S. Moeller, P. Montanez, D. S. Morton, H.-D. Nuhn, D. Walz, J. Welch and J. Wu, *FEL2014 Abstracts*, 165, (2014).
[63] K.-J. Kim, Y. Shvyd'ko and S. Reiche, *Phys. Rev. Lett.* **100**, 244802, (2008).

Chapter 14

Femtosecond Imaging of Single Particles and Molecules Using X-ray Free-electron Lasers

Andrew V. Martin

ARC Centre for Excellence in Advanced Molecular Imaging, School of Physics, University of Melbourne, VIC, Australia

N. Duane Loh

Center for BioImaging Sciences, Department of Biological Sciences, National University of Singapore, Singapore. PULSE Institute at SLAC National Lab, 2575 Sand Hill Road, Menlo Park, CA 94025, USA

Abstract

X-ray free-electron lasers provide a greatly increased peak intensity on femtosecond time scales, opening up new opportunities to study the structure and dynamics of nanoparticles, viruses and biological macromolecules. These opportunities are being realized with new crystallography techniques to study nanoscale periodic samples and with coherent diffractive imaging techniques to study non-periodic samples. This chapter provides an introduction to coherent diffractive imaging techniques, a review of its applications and development at synchrotron sources, and finally a survey of the first imaging and crystallography results from X-ray laser experiments.

Keywords: Coherent X-ray diffraction; X-ray free-electron laser; coherent diffractive imaging; phase retrieval; serial femtosecond crystallography; ptychography; X-ray holography; tomography; single particle imaging.

14.1. Introduction

X-ray free-electron lasers (XFELs) produce pulses of coherent X-rays that are only tens of femtoseconds in duration and have up to a billion times higher peak brightness than synchrotron sources [1]. These properties are ideal for high resolution, time-resolved experiments on sub-picosecond time scales. Less intuitively, there is also great potential to use XFELs to study radiation sensitive samples like protein crystals. Although an XFEL pulse destroys the sample, atomic motion is slow compared to the time it takes for the pulse to traverse the sample [2] and XFEL diffraction is highly sensitive to the undamaged structure. This principle is commonly referred to as "diffraction-before-destruction" and it opens up the possibility of high resolution structural studies of sub-micron protein crystals [3], individual viruses [4], cells [5], transient magnetic phenomena [6] and nanoparticles [7]. Femtosecond pulses outrun the translational and rotational times of particles in solution or as an aerosol and XFELs have a unique capability to image samples that reside natively in these environments [7].

XFEL imaging relies on the high coherence of the beam and measures structural information indirectly in the form of a diffraction pattern. An image of the object's structure is then recovered via computational analysis. This technique is known as coherent diffractive imaging (CDI) and its application with XFELs has branched off from the substantial development of these methods at third-generation synchrotrons. CDI can be thought of as an extension of crystallography methods to non-periodic samples.

The XFEL implementations of CDI and crystallography are different in many ways to their synchrotron counterparts. One of the biggest differences is sample delivery. XFELs operate at high repetition rates and each measurement is destructive, so the best use of an XFEL source is to continuously inject new samples into the interaction region with a liquid jet or as an aerosol. For crystals, this technique has become known as serial femtosecond crystallography (SFX) [8]. For non-periodic samples, it is called single particle imaging.

In this chapter we review the theory of CDI (Section 14.2) and the experimental development of CDI with synchrotron sources (Section 14.3), which provided the context from which XFEL imaging emerged. Finally we focus in more detail on the development of XFEL imaging techniques (Sections 14.4, 14.5 and 14.6).

14.2. Theory of Coherent Diffractive Imaging

14.2.1. *Coherent scattering theory*

Since the theory of X-ray diffraction is well established and comprehensively covered elsewhere [9–11], here we summarise some key concepts that feature in coherent X-ray imaging techniques. The duration of an X-ray diffraction measurement is much longer than the oscillation times of the electromagnetic fields, so that a monochromatic coherent X-ray beam can be modeled by a complex scalar function [9]. The wave propagates according to a wave equation and because X-ray interactions are weak, the first Born approximation (kinematic scattering) is valid in a large number of cases [11]. Assuming a uniform (plane-wave) incident illumination, the scattered far-field intensity from an object can be modeled by

$$I(\mathbf{q}) = J_0 r_e^2 P |F(\mathbf{q})|^2 d\Omega, \tag{14.1}$$

where J_0 is the number of incident photons per unit area, r_e is the classical electron radius, P is a polarization factor, $d\Omega$ is the solid angle subtended by a pixel and $F(\mathbf{q})$ is the scattering factor. The scattering factor is related to the sample's electron density $\rho(\mathbf{r})$ by a Fourier transform, as follows

$$F(\mathbf{q}) = \int \rho(\mathbf{r}) e^{2\pi i \mathbf{q}\cdot\mathbf{r}} d\mathbf{r}. \tag{14.2}$$

For a non-periodic object of finite size, the diffraction pattern has a diffuse speckled structure as shown in Fig. 14.1(a). The smallest features of such a pattern are known as "speckles" and their typical size is inversely proportional to the object's size.

The structure factor is sampled on the points $\mathbf{q}$ that lie on the surface of the Ewald sphere [10, 11] as shown in Fig. 14.1(c). The Ewald

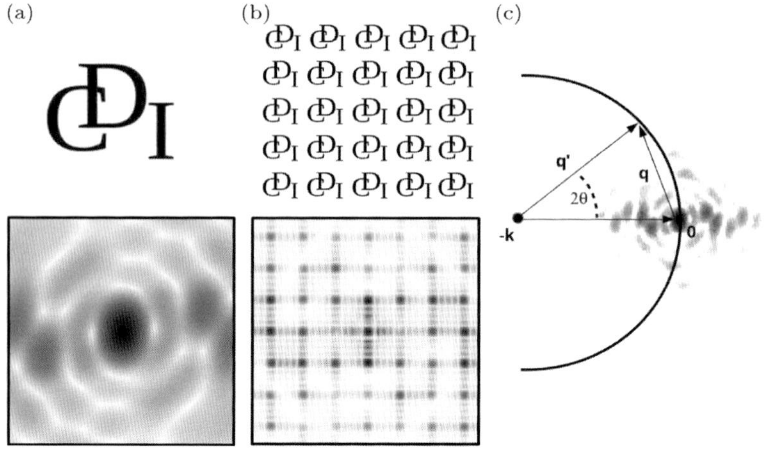

Fig. 14.1. (a) A non-periodic object and its diffraction pattern. (b) The same object tiled as a 5 × 5 periodic array and its Bragg diffraction. (c) A diagram of the Ewald sphere applied to non-periodic diffraction.

sphere is the set of all scattering vectors $\mathbf{q}'$ that have a magnitude equal to the wave vector of the incident illumination $\mathbf{k}$, such that $\mathbf{q} = \mathbf{q}' - \mathbf{k}$. Each scattering vector $\mathbf{q}'$ is associated with a unique scattering angle θ, which can be used to predict where on the detector that part of the diffracted X-ray wave will be measured. The Ewald sphere plays an important role in high resolution experiments such as crystallography and single molecule imaging.

For a perfectly ordered crystal of finite size, the square of the scattering factor can be written as

$$|F(\mathbf{q})|^2 = |F_u(\mathbf{q})|^2 S(\mathbf{q}), \tag{14.3}$$

where $F_u(\mathbf{q})$ is the scattering factor of a single unit cell and $S(\mathbf{q})$ is the shape transform of the crystal. For a perfect crystal with an integer number of unit cells, $S(\mathbf{q})$ is given by [10]

$$S(\mathbf{q}) = \prod_{i=1}^{3} \frac{\sin^2\left[\frac{\pi}{\lambda}(\mathbf{q} - \mathbf{q}_0) \cdot N_i \mathbf{a}_i\right]}{\sin^2\left[\frac{\pi}{\lambda}(\mathbf{q} - \mathbf{q}_0) \cdot \mathbf{a}_i\right]}, \tag{14.4}$$

where $\mathbf{a}_i$ is the ith lattice vector and N_i the number of unit cell along in this direction. The X-ray wavelength is denoted by λ and

$\mathbf{q}_0$ is the wavevector of the incident illumination. An example of the scattered intensity from a small crystal is shown in Fig. 14.1(b). The maximum peak intensity corresponds to that of $|F_u(\mathbf{q})|^2$, shown in Fig. 14.1(b), but at all other points the shape transform modulates the intensity. Counting fringes between peaks, inclusive of the peaks themselves, gives the number of unit cells along an axis of the crystal. Since XFEL beams result in very thin Ewald spheres, associated with low beam convergence and a small spread of wavelengths, the Bragg intensity measured in an XFEL crystallography experiment is only a cross-section of a broad structured peak.

Strain or disorder in the crystal lattice can be accounted for by modifying the definition of $S(\mathbf{q})$ [12]. When there are multiple molecules per unit cell, the ordering of molecules on the crystal surface is not necessarily consistent with a particular unit cell definition, and it is necessary to average $|F_u(\mathbf{q})|^2$ over the different definitions of the unit cell [13].

At low scattering angles the Ewald sphere is almost flat, such that $q_z \approx 0$ (where z is the axis of propagation), and one can see from Eq. (14.2) that the diffraction depends on the projected structure

$$\rho_{\text{proj}}(\mathbf{r}_\perp) = \int \rho(\mathbf{r})dz, \tag{14.5}$$

where $\mathbf{r}_\perp$ is a two-dimensional vector in the plane perpendicular to z. The X-ray wave at the exit surface plane of the sample $\psi(\mathbf{r}_\perp)$ can be modeled by multiplying the incident wave $\psi_{\text{Illum}}(\mathbf{r}_\perp)$ by a transmission function

$$\begin{aligned}\psi(\mathbf{r}_\perp) &= T(\mathbf{r}_\perp)\psi_{\text{Illum}}(\mathbf{r}_\perp)\\ &\equiv e^{-ir_e\lambda\rho_{\text{proj}}(\mathbf{r}_\perp)}\psi_{\text{Illum}}(\mathbf{r}_\perp),\end{aligned} \tag{14.6}$$

where λ is the x-ray wavelength and $T(\mathbf{r}_\perp)$ is called the transmission function. Many x-ray imaging experiments operate in the Fraunhofer diffraction regime, such that the propagation to the detector can be modeled by a Fourier transform

$$\psi(\mathbf{q}) = \int \psi(\mathbf{r}_\perp)e^{-2\pi i\mathbf{q}\cdot\mathbf{r}_\perp}d\mathbf{r}_\perp. \tag{14.7}$$

14.2.2. *Theory of coherent diffractive imaging algorithms*

Coherent diffractive imaging (CDI) techniques recover an image of the sample from a measured diffraction pattern. Measurement records the intensity of the scattered illumination and the phase information carried by the scattered wave is lost. The goal of CDI techniques is to recover the lost phase information, so that the scattered wave can be constructed, from which an image of the sample can be recovered.

Recovering the phase of the scattered wave requires a large system of non-linear equations to be solved for which there are no algorithms that are guaranteed to converge [14]. However, many algorithms have been developed that demonstrate rapid convergence in practice [14, 15]. Here we briefly summarize the key features of the problem in the context of uniform illumination (plane-wave CDI) which is pertinent to XFEL imaging. It should be noted though that many applications use structured illumination formed by focusing the beam prior to specimen (Fresnel CDI) in order to improve the reliability of image recovery as discussed further in Section 14.3. The geometry of each case is shown in Fig. 14.2.

The phase of the scattered wave can be recovered using *a priori* knowledge about the sample, which constrains the search for the solution. A powerful formalism for expressing constraints is based on projection operators [16]. A projection makes the minimal change to the input wave function required to satisfy the constraint. The modulus constraint specifies the intensity at the detector plane and is defined by the following projection operator

$$\psi_{\mathrm{mod}}(\mathbf{r}) \equiv P_{\mathrm{mod}}\psi(\mathbf{r}) = \mathcal{F}^{-1}[\widetilde{\psi}_{\mathrm{mod}}(\mathbf{q})], \tag{14.8}$$

where $\mathcal{F}^{-1}$ is the inverse Fourier transform and $\widetilde{\psi}_{\mathrm{mod}}(\mathbf{q})$ is given by

$$\widetilde{\psi}_{\mathrm{mod}}(\mathbf{q}) = \sqrt{I(\mathbf{q})}e^{i\arg(\widetilde{\psi}(\mathbf{q}))}, \tag{14.9}$$

where $I(\mathbf{q})$ is the measured diffraction pattern. The sampling required for the diffraction measurement ideally corresponds to a field-of-view in the object plane that is at least twice as wide as the

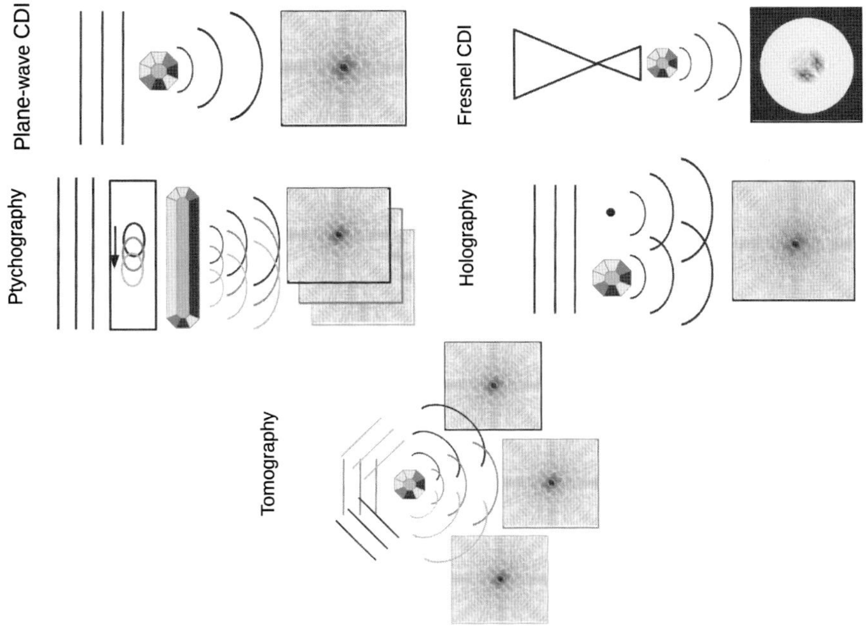

Fig. 14.2. A pictorial overview of some different coherent diffractive imaging techniques. Vertical lines represent uniform incident X-rays and semi-circular lines represent diffracted X-rays. The tomography diagram is illustrative of the essential idea only, as in practice the sample is rotated.

object, in order that the entire unaliased autocorrelation function can be resolved.

The size and shape of the object is constrained by using the support function $S(\mathbf{r})$ and is given by

$$P_{\text{sup}}\psi(\mathbf{r}) = \begin{cases} \psi(\mathbf{r}), & \mathbf{r} \in S(\mathbf{r}) \\ 0, & \mathbf{r} \notin S(\mathbf{r}). \end{cases} \tag{14.10}$$

A real object can be constrained by setting the imaginary part of $\psi(\mathbf{r})$ to be zero everywhere and positivity enforced by setting any negative values of $\psi(\mathbf{r})$ to be zero.

The simplest algorithm is known as the error-reduction algorithm (ER) and consists of alternating between the support constraint and the intensity constraint (see Table 14.1). The error-reduction algorithm is guaranteed not to increase the normalised Euclidean error,

Table 14.1. Some popular CDI algorithms.

ER	$\psi^{(n+1)} = P_{\text{sup}} P_{\text{mod}} \psi^{(n)}$
HIO[a]	$\psi^{(n+1)} = \psi^{(n)}_{\text{mod}} \in S$ $\psi^{(n)} - \beta \psi^{(n)}_{\text{mod}} \notin S$
RAAR[a]	$\psi^{(n+1)} = \beta \psi^{(n)} + 2\beta P_{\text{sup}} P_{\text{mod}} \psi^{(n)} + (1 - 2\beta) P_{\text{mod}} \psi^{(n)} - \beta P_{\text{sup}} \psi^{(n)}$

[a]The parameter β is a feedback parameter, which is typically set between 0.5−1.

defined by [14]

$$\epsilon_m = \frac{\sum_i |P_{\text{mod}} \psi(\mathbf{r}_i) - \psi(\mathbf{r}_i)|^2}{\sum_i |P_{\text{mod}} \psi(\mathbf{r}_i)|^2}, \tag{14.11}$$

where the index i is summed over all detector pixels. The ER algorithm always moves the iterate in the direction of the steepest gradient of the error metric, and converges to the closest minimum. The problem is that in general the error metric has many minima, and only the global minimum is an accurate solution. ER is not adequate for most phase retrieval applications. Since its behavior is stable in the presence of noise, it is good for the refinement of a close solution and in combination with other algorithms.

There are better algorithms for finding the global minimum like the hybrid input-output method (HIO) [15]. These algorithms employ a sophisticated feedback of the errors from the last iteration to avoid stagnation in local minima. HIO can be more sensitive than ER to measurement noise, leading the search away from the correct solution. Another way to understand this is that errors in the measured data cause the constraints not to be mutually compatible, and the phase retrieval problem is not well-posed. Alongside HIO, the difference map [16] defines a flexible and general class of algorithms with similar search behaviour and performance to HIO.

An algorithm that can compromise between local-minimum refinement and global minimum searching is the relaxed-alternating-averaged-reflections algorithm (RAAR), whose behavior varies according to the feedback parameter β. A β value close to 0 produces behavior close to the ER algorithm, while a β value close to 1 produces a global-minima search behaviour. Values of β around 0.8–0.9 represent a good compromise between these two behaviors.

Such tunablility of RAAR is also possible with a modified version of the difference map [17].

The motivation for further development of phase retrieval algorithms is to improve robustness and reliability in applications to experimental data. For example, experimental noise has been treated by modifying the feedback of CDI algorithms with a noise threshold [18] or by enforcing smoothness outside the support [19]. Another approach is to change the optimization method, like guiding the solution by performing multiple reconstructions [20]. Improvements in the robustness and reliability of phase retrieval methods have also been achieved by modifying the experiment to involve multiple measurements, partial coherence, holographic references or structured illumination, as discussed in Sections 14.2.3–14.2.5 and 14.3.

There is no general way to verify that the recovered image is correct. Instead, the reproducibility of the reconstruction is checked by performing multiple reconstructions from different random starting points. Each reconstruction must be corrected for any global phase offsets or complex conjugation that is present (because these operations also produce valid solutions and make no change to the physical interpretation). Then all the reconstructions can be averaged and any phases that fluctuate between reconstructions will average to zero. The information carried by the average can be measured with the Phase Retrieval Transfer Function (PRTF) [21],

$$\mathrm{PRTF}(\mathbf{q}) = \frac{|\langle \psi(\mathbf{q}) \rangle|}{\sqrt{I(\mathbf{q})}}. \tag{14.12}$$

The PRTF is often presented as an angular average and used to estimate resolution. Although there is no consensus on the specific values of the PRTF to use to calculate resolution, a typical value is 0.5. The PRTF is easier to interpret if a Wiener filter is applied [22].

14.2.3. *Partially coherent diffractive imaging*

The theory of X-ray scattering described so far has assumed a monochromatic and fully coherent illumination. A reduction in coherence reduces the contrast of the diffraction pattern. If this loss of contrast is not too severe, coherent diffractive imaging experiments are still possible provided that phase retrieval algorithms are modified. Instead of modeling the beam as a single complex function, partially coherent methods employ the mutual coherence function $J(\mathbf{r}, \mathbf{r}')$. The mutual coherence function can be concisely partitioned into a series of modes $\psi_n(\mathbf{r})$, weighted by coefficients η_n, that incoherently contribute to the diffraction pattern. The degree of coherence is specified by the values of the weights, which in turn indicates the effective number of modes required. For a detailed acount of partial coherence methods see the review by Quiney [23].

An accurate account of partial coherence in CDI has been shown to significantly improve the quality of reconstructions [24, 25]. The same formalism can be used to apply CDI with multiple wavelengths, which is an important techinque for using X-ray sources based on high harmonic generation [26]. The multiple wavelength approach can be adapted to account for a continuous broadband spectrum of incident wavelengths, which enables CDI experiments at third-generation synchrotron sources with greater flux and shorter exposure time [27].

14.2.4. *Ptychography and keyhole CDI*

Ptychography is a coherent imaging technique that can be applied to extended objects [28–30]. Instead of a support constraint, the phase retrieval problem is overdetermined by translating the beam across the sample and taking multiple diffraction measurements. To ensure that each measurement provides new information, the beam is either of finite size or has a non-uniform profile. To ensure convergence, there needs to be significant overlap between probes in neighboring positions, such that each point on the sample features in multiple measurements. The convergence of the ptychographic methods improves by increasing the overlap fraction of the probes [31]. The method has been extended to simultaneously recover the probe and the object [32, 33]. Another source of error is inaccurate knowledge of the probe position for each measurement and improved reconstructions can be obtained by refining the probe positions during the reconstruction [34, 35]. Ptychography has matured into a robust form of microscopy that is offered or planed to be offered as a permanent capability of many synchrotron beamlines around the world, notably in conjunction with fluorescence microscopy [36, 37]. A developing application of ptychography is to dynamic samples such as vibrating systems [38] or fluctuating ensembles [39].

An alternative approach to imaging extended objects is keyhole diffractive imaging [40], in which the probe is scanned across a sample and a separate reconstruction is performed for each measurement. The robustness of the reconstruction procedure is due to the convergent incident beam and the results are comparable to ptychography experiments. The advantages of keyhole CDI and ptychography have also been combined in a single technique [41].

14.2.5. *Holography*

Holography uses the interference between a known reference wave and the unknown scattered wave to greatly simplify the analysis required to recover the phase. The advantage is that the analysis involves a single Fourier transform or deconvolution, avoiding the need for iterative algorithms and their associated issues. In Fourier holography [42], the reference is a pinhole at sufficient separation from the object. While a pinhole leads to the simplest analysis, it introduces an undesirable trade off between resolution and signal-to-noise. Another alternative is to choose a larger structured reference for which there is a known linear relationship between the hologram and the object. For example, references with sharp corners can be used with a linear filter [43, 44], and when the reference is a uniformly redundant array, the object is recovered by a convolution with the reference itself [45].

If the illumination and reference are sufficiently well characterized, holographic data can be inverted using numerical linear methods [46], which can be robust and fast using conjugate gradient methods [47]. This last option provides a great flexibility of experimental geometries [48], relaxing the requirement that the reference be a separate, isolated X-ray scatterer or aperture.

14.2.6. *3D imaging: Tomography*

Tomography produces three-dimensional images by measuring the sample from different orientations. It can be implemented with coherent diffraction, because each coherent diffraction measurement contains information about the sample's structure projected along the axis of the beam. Recalling the theory of Section 14.2.1, each measurement samples the intensity of the Fourier transform of the electron density on the Ewald sphere. The diffraction measurements can be represented on these spherical surfaces and then averaged to construct a three-dimensional intensity volume [21, 49]. The corresponding phases for this three-dimensional intensity volume can be recovered with the iterative algorithms described in Section 14.2.2.

Fresnel CDI has also been developed into a tomographic technique, but the presence of the structured illumination means that it is easier to reconstruct an image from each 2D measurement first, and then merge the ensemble of reconstructed projected images into a three-dimensional representation of the object [50]. The method of assembling projection images is also employed in ptychographic tomography [51].

14.2.7. *Bragg CDI*

As discussed in Section 14.2, the structure of a Bragg peak is sensitive to both shape and lattice disorder, like strain, and CDI methods can be used to extract real space maps of nanoparticle strain and shape [52]. By rotating the crystal, the three-dimensional structure of the peak can be measured and converted into 3D image [53, 54].

14.3. Synchrotron Experiments and Applications of Coherent Diffractive Imaging

14.3.1. *Milestone experiments and applications to material science*

Miao *et al.* [55] were the first to demonstrate that phase retrieval methods could be applied to the X-ray diffraction of a non-periodic sample. They produced images of patterns of 100 nm gold dots, demonstrating the potential for CDI to be a high-resolution technique capable of revealing finer structural details than optical imaging. Since then, CDI methods have developed along multiple parallel directions to provide the robustness and versatility required to handle many applications in both material science and biology.

A key development that greatly broadened the applicability of CDI techniques was the dynamical determination of the support during the reconstruction [56]. Commonly known as the Shrinkwrap method, it removed the requirement that the shape of the object be known in advance. Since the structure of the object is unknown in most applications, Shrinkwrap has been adopted widely.

It was shown that using structured illumination, by focusing the beam prior to the sample plane with a zone plate, significantly improved the robustness of image reconstruction [57–59]. This method is called Fresnel CDI. The resulting diffraction pattern is closer to a shadow projection image of the sample, and has a greatly reduced dynamic range compared to diffraction using uniform incident illumination. Experimental demonstrations of CDI with partial coherence established the improvements in image quality that are gained by accurate beam characterization [24]. The exposure time in CDI experiments can be greatly reduced by using the broadband spectrum of the synchrotron using the partial coherence formalism with multiple wavelengths [27]. The same formalism can be used with high-harmonic X-ray sources that produce a spectrum of discrete wavelengths [26].

The development of CDI into a tomographic technique capable of 3D imaging meant that the penetrative properties of x-rays could be used to explore the internal structure of materials and biological samples. A thorough exploration of CDI tomography was conducted on a pyramid-like structure made of gold balls [21] (see Fig. 14.3). This study detailed the relationship between 2D and 3D images, explored the effects of unmeasured regions of the diffracted volume and explained how to estimate resolution with the PRTF. The technique was applied to study a ceramic nanofoam [60], revealing the 3D morphology of the internal voids in the porous structure.

Bragg CDI was first applied to gold particles [52], demonstrating that the information about the shape and micro-structure of the

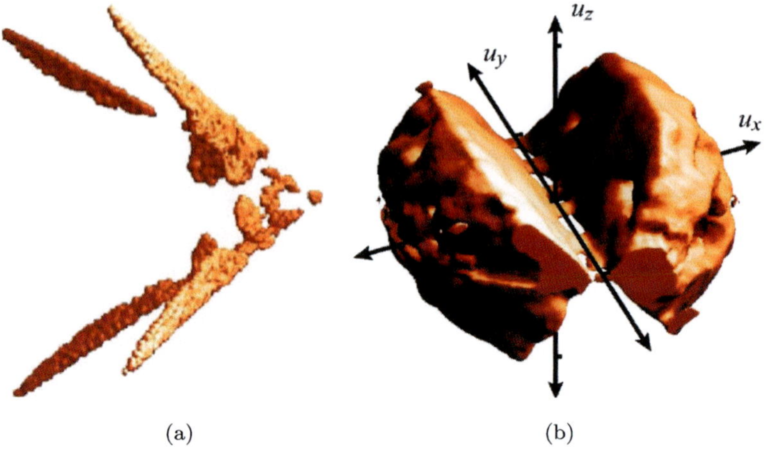

Fig. 14.3. Three-dimensional CDI image of a pyramid-like structure compose of gold balls and the phase retrieval transfer function of the reconstruction (adapted from [21] with permission).

gold contained in the diffraction near a Bragg peak was accessible to coherent imaging methods. It was applied in 3D to observe the interior structure of the particles like twinning [53] and to map the strain field [54]. Bragg CDI has diverse applications to study, for example, defects in crystals [61], zeolyte microcrystals [62] and nanoscale layered transistors [63].

High resolution ptychography has been used study materials for electrochemical storage [37], where notably the complex refractive index was shown to have chemical sensitivity. Other applications include imaging of nanoscale microchips [64] and high-resolution 3D imaging of nanoporous glass [65].

14.3.2. *Biological applications*

The potential for high-resolution biological imaging has been a driving motivation for the development of X-ray CDI [55, 66]. One major goal was to image a cell without staining or sectioning. The first 2D images of a biological cell were made of a Yeast cell [49] and the internal components could be resolved to around 30 nm, which competed well with alternative imaging techniques. As biological samples

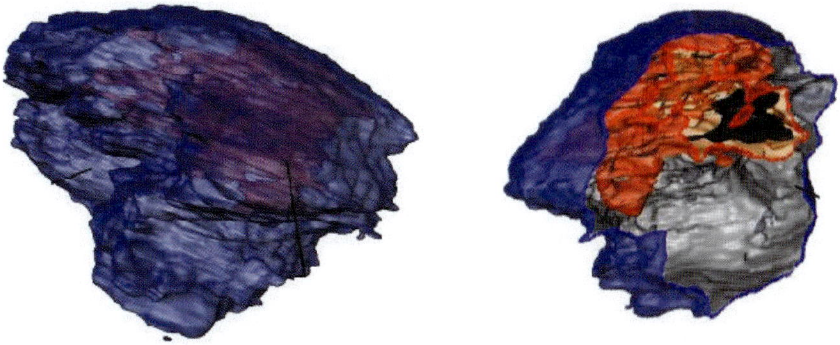

(a) (b)

Fig. 14.4. Three-dimensional CDI reconstruction of a red blood cell infected with the malaria parasite (reprinted from [67] with permission). The red region indicates the parasite and the blue and grey regions indicate the host red blood cell.

are radiation sensitive, it was necessary to freeze the cell. A yeast spore cell has been imaged in 3D with sufficient resolution to identify many internal components, like the nucleus, the mitochondria and vacuoles [66]. 3D CDI imaging has also been applied to study red blood cells infected with the malaria parasite [67] (see Fig. 14.4). This experiment was also notably a demonstration of Fresnel CDI tomography. Another notable biological application, and in fact the first 3D biological CDI image, was of a human chromosome [68] which resolved a high density axial structure internal to the chromosome. 3D ptychography has been developed and used to image a piece of bone [51]. The ptychographic approach was sufficiently robust to reveal fine details about variations in bone density and was also able to handle absorption in the sample.

14.4. Experimental Background for Coherent Imaging with X-ray Lasers

14.4.1. *X-ray Free-electron Laser Sources* (*XFELs*)

Madey first proposed in 1971 [69] to lase an electron bunch in a one-pass operation mode, which is suitable for linear particle accelerators.

This process begins with an electron gun, where a low-emittance packet of photoelectrons are ejected from cathodes after they have been illuminated by harmonic-upshifted optical laser pulses. This packet of electrons is then accelerated and shaped to prepare them for an oscillatory passage through ordered arrays of magnets called undulators. Unlike normal undulator sources where these oscillating electrons emit radiation in an uncorrelated fashion, extended undulators at XFELs micro-bunch pulsed packets of electrons into periodicities comparable to the wavelength of emitted radiation. When this condition is met, electrons in a packet interact 'constructively' with the photons produced in an earlier oscillation, which results in an exponential amplification in emitted radiation (i.e., lasing).

Source electron packets at XFELs are compressed to produce X-ray pulses that are in the femtosecond range, which serve as ultrafast shutters for sample examination. Pulse lengths down to 10 fs can be achieved with low-charge, hence low-emittance, electron packets. Depending on the XFEL, pulses are delivered at 10-27000 Hz in continuous pulsing mode, or "on-demand" at lower rates.

Four XFELs have been built and are currently operational, spanning soft to hard x-ray regimes [70–73] and several are planned for the future [74]. Typical photon pulse energies range from 10–100 μJ [70, 72, 73], up to several mJ [71]. For example, a millijoule pulse of 8keV photons could have up to approximately 10^{12} photons. When focused to a 100 nm spot, such a pulse will elastically scatter up to thousands of photons from single biomolecules that are only $\sim$10 nm across (see Fig. 14.5).

Radiation from XFELs tends to be linearly polarized due to the planar configuration of undulator magnets. To produce circularly polarized X-rays, which is important for imaging magnetic contrast [17, 42, 75], the main undulators could be re-configured into a helical arrangement [73], or a short 'afterburner' undulator segment could be appended. In the latter, the strong planar polarization would have to be suppressed [76].

For a more complete review on XFELs and an accessible explanation of the underlying lasing physics, refer to [74, 77].

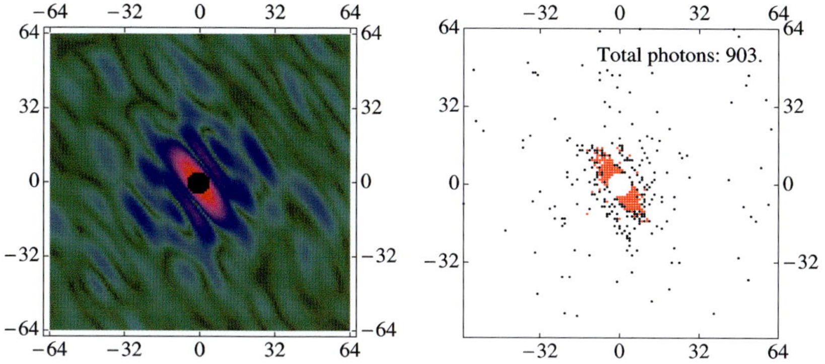

Fig. 14.5. Simulated damage-free elastic scattering from a biomolecule. Left: noiseless Ewald sphere section of Fourier intensities of an RNA Polymerase II molecule (PDB:POLR2D) which contains 9234 non-hydrogen atoms (2.5 Å photon wavelength; 53 mm sample-detector distance; $200 \times 200\,\mu\text{m}^2$ detector pixels). Missing low angle scattering data indicated by black, central disk. Right: simulated noisy diffraction pattern. Photon receiving pixels colored in black (1-photon) or red (>1 photon). Other scattering parameters: 10^{12} photons per pulse, with 20% transmission efficiency; 100 nm-diameter XFEL focus with top-hat intensity profile.

14.4.2. *Diffraction before destruction*

The high intensities of XFEL pulses provide excellent signal-to-noise for scattering, but also rapidly damage a sample's structure. Nevertheless, if these pulses are sufficiently short they can "self-shutter" [78] the exposure of the sample to large radiation doses. In this case, photon detectors should measure predominantly elastic scattering from the sample before damage, although inelastic photon scattering and fluorescence are unavoidable. This imaging principle is known as *diffraction before destruction*.

When an XFEL pulse impacts a sample, first a sample's inner shell electrons are ejected via the photoelectric effect. A cascade of Auger decay and secondary ionization events follow within femtoseconds, which produces a supply of mobile secondary electrons. The onset of catastrophic Coulombic explosion occurs around 5–10 femtoseconds when the inertia of the damaged sample's ions is overcome. For these reasons, the durations of typical X-ray pulses are designed to be in the femtosecond range.

Two approaches have been used to study XFEL radiation damage. The first uses molecular dynamics simulation to model atom-photon interactions, as well as Auger decay processes [2]. Secondary ionization by photoelectrons and how they re-interact with the sample have also been included [79]. The high computational complexity of simulating molecular dynamics typically limit these to samples with fewer atoms. A second approach uses continuum models to coarse-grain the molecular dynamics details [80], which permits the modeling of larger molecules.

Early experimental demonstrations show that the diffraction-before-destruction principle is valid for CDI at micron and sub-micron resolutions [81, 82]. High resolution demonstrations of the principle were first demonstrated with serial femtosecond crystallography (SFX). SFX damage studies were performed for lysozyme and photosystem I (PS1) [8, 83, 84] as a function of X-ray pulse duration, where it became evident that short pulses measure less damage. Remarkably, structure factors can still be collected at crystallographic reflections despite considerably longer X-ray pulse lengths [84]. Damage in SFX is different from those described by Debye-Waller factors in macromolecular crystallography: earlier parts of each XFEL pulse diffract Bragg reflections of an undamaged crystal to high resolution, while later parts of the pulse add only lower resolution reflections plus diffuse background due to an increasingly damaged crystal. Computational correction of these two damage effects was proposed [84]. Nevertheless, other studies [83] also suggest that our understanding of these damage processes is incomplete but progressing.

14.4.3. *Fast detectors for femtosecond imaging*

The high pulse repetition rate and brillance of XFELs enables high-throughput serial measurements of many individual weakly scattering samples. This serial diffractive imaging imposes unique requirements on data acquisition. To begin, X-ray detectors must record high-resolution diffraction patterns at the same rate as the arrival of X-ray pulses at XFELs ($>10 - 100\,\mathrm{Hz}$), while maintaining a low-noise linear response over a large dynamic range ($\sim$1–10^4 photons) on single diffraction patterns.

X-rays diffracted from samples are detected with photon integrators, since current electronics are not fast enough to discriminate between individual photons that arrive within femtosecond XFEL pulses. Both Charge-Coupled Device (CCD) [85, 86] and Complementary Metal-Oxide-Semiconductor (CMOS) [87] photon integrators have been used at XFELs. In both cases, they have silicon sensors for direct photon detection. Although CCDs tend to have lower electronic noise and have a longer history of development and characterization than CMOS detectors, CCDs' readout rates are lower than that of CMOS detectors [88].

Data loads from XFEL experiments are demanding. Consider a five-megapixel detector that records a diffraction pattern per pulse at pulse rates of 100 Hz. Several gigabytes of pixel values are read to disk per second, which is equivalent to 10–100 terabytes of data per hour of acquisition. Analysis methods are therefore essential to sort and discriminate these large datasets. The scale of data acquisition at XFELs is considerably different from that at synchrotrons.

Other types of detectors are used correlatively with X-ray detectors. For instance, ion mass spectrometers can measure ion debris from the sample after pulse illumination [85]. With these, sample detection and differentiation as well as compositional fingerprinting [7, 89] are possible. Energy dispersive X-ray spectrometers have also been employed in tandem with x-ray photon detection [90] to determine structural and emission states in optically pumped samples.

Although even more stringent requirements are needed for the upcoming European XFEL, where pulse repetition rates will reach 27000 Hz, exciting development in detector technology is expected to keep pace [88, 91].

14.4.4. *High-throughput sample delivery*

High-throughput sample delivery is necessary to fully utilize the rapid data acquisition rates at XFELs. Samples are commonly delivered in the liquid phase, gas phase, or adhered to solid substrates that are sequentially scanned by the X-ray pulses. Important considerations for sample delivery are ensuring a high likelihood that a sample is hit; maintaining the sample close to its native state (e.g., extended exposure to vacuum may freeze and cryo-damage solvated samples); and minimising background scattering from solvents or substrates. In most experiments, the orientation of the sample is also random and not known, so that determining sample orientation is a distinctive feature of SFX and imaging with XFELs.

Liquid phase delivery is used for Serial Femtosecond Crystallography (SFX), where micron-size crystals are confined in liquid jets of diameters greater than 0.5 μm. For thin jets, the Bragg reflections from micro-crystals are still well-detected against the diffuse background from water scattering. Liquid phase injectors are based on the design of the Rayleigh jet, where a back-pressured sample-laden liquid is introduced into the X-ray interaction chamber (Fig. 14.6(a)). This jet can be broken into a quasi-regular stream of droplets via the Rayleigh-Plateau instability [92] with piezo transducer [93]. Practically, the diameters of either the jet phase (>8 μm) or droplet phase (>16 μm) of bare Rayleigh jets are undesirably large for SFX.

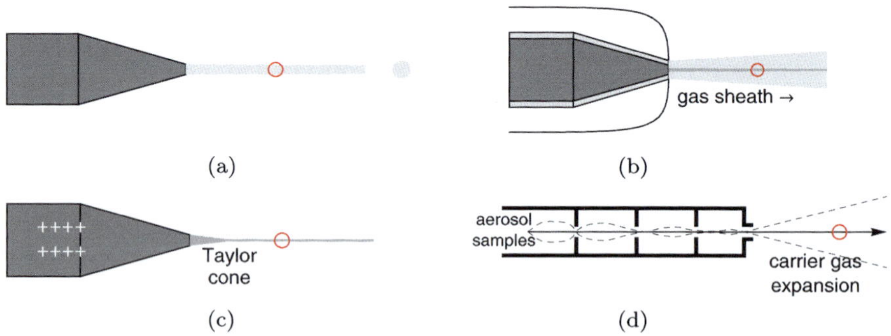

Fig. 14.6. Sample injectors. (a) Rayleigh jet, breaking up into droplets. (b) Gas dynamic virtual nozzle: co-axial gas flow (light gray) focuses liquid jet (dark gray). (c) Electrospray. (d) Aerodynamic focusing inlet: differentially-pumped chambers (high to low pressure, from left to right) with orifices that guide carrier streamlines (dashed), which in turn focus the trajectories of massive aerosols (solid arrow). Red circles indicate typical X-ray interaction regions (X-ray pulses going into the page).

Hence, Rayleigh jets are narrowed by a co-axial sheath gas flow in a Gas Dynamic Virtual Nozzle (GDVN) [Fig. 14.6(b)] to achieve submicron jets or droplets. Rayleigh jets can also be focused electrostatically [Fig. 14.6(c)] to produce a narrow jet (∼1–8 μm diameter) known as an electrospray [93]. Efforts to reduce the sample consumption of liquid injectors are directed at increasing the viscosity of the carrier liquid with polyethelene glycol (PEG) additives [94] or with crystals in the lipid cubic or sponge mesophases [95].

Gas phase sample injection is based on the aerodynamic focusing inlet [96–98], where aerosolized samples are manipulated by gas streams generated by orifices that separate differentially-pumped chambers [Fig. 14.6(d)]. Implementations typically differ in how samples are aerosolized upstream of the focusing inlet: electrospray or nebulization are popular options. The passage of aerosolized droplets through the inlet also dries them, so aerosol injection is commonly used for non-periodic imaging (e.g., single cells, viruses etc.), where minimal solvent scattering is crucial. The downside of gas phase injection is that typical diameters of particle sprays are several hundred microns. Typical hit rates are a few percent and sample consumption rates can be high.

Methods for substrate-bound sample delivery include adhering samples onto silicon nitride membranes [82] (flat or with fabricated wells), then cryo-protecting them against evaporative cooling in high-vacuum imaging chambers, or cryofreezing the sample onto grids [99] commonly used in cryo electron-microscopy. Fine-stepping motors then translate these samples across the fixed path of the x-ray pulses.

14.4.5. *Fluctuations in X-ray pulses*

To integrate and interpret many XFEL measurements requires knowledge about fluctuations in pulse parameters, and possibly methods to correct for them. SASE X-ray laser sources have good spatial coherence but poor temporal coherence, as compared to conventional visible light lasers. This is primarily due to pulse-to-pulse spatial and temporal inhomogeneities in the single-pass electron bunch (see Section 14.4.1). Further, the roughness of offset and focusing mirrors could introduce intensity and phase modulations on X-ray pulses [100]. For example, variations in key pulse parameters across soft and hard X-rays at the Linac Coherent Light Source (LCLS) from [71] are reproduced below.

- Electron bunch peak current, r.m.s., ∼5–12%.
- X-ray pulse power stability, r.m.s., ∼3–12%.
- Photon wavelength stability, r.m.s., ∼0.1–0.2%.
- Spectral stability, or bandwidth (FWHM), ∼0.2–1%. This can be suppressed by more than an order of magnitude with external seeding of electron bunch [73] or self-seeding [101].
- Pointing stability of centroids of the X-ray pulses, ∼10–20% of beam waist.

14.5. X-ray Laser Techniques and Applications — Crystallography

14.5.1. *Serial femtosecond crystallography data analysis*

Serial Femtosecond Crystallography (SFX) enables the collection of diffraction patterns from micron and sub-micron sized crystals. Due to radiation damage these crystals are generally unusable for macromolecular crystallography (MX) with synchrotron sources. Hence, SFX provides a new path to protein structure determination when large protein crystals are difficult to produce. Since each XFEL measurement is destructive, an SFX dataset comprises of single measurements of many crystals ($>10^4$) in random orientations. The goal of SFX data analysis is to assemble all of the Bragg intensities into a single three-dimensional (3D) dataset, which can then be passed on to the same structure determination routines that are used in MX.

SFX data processing begins with peak detection. A peak can be identified as a cluster of connected pixels above an intensity threshold which is set to exclude background scattering. A pattern is labeled a *hit* if it has sufficiently many of such clusters. The centroid of each cluster is used to assign each peak a spatial frequency, and then the pattern is passed to an auto-indexing program. For an example program see CrystFEL [102].

Any of the auto-indexing routines developed for MX can be used for SFX to assign each peak the correct *hkl* Miller indices to each Bragg peak. Indexed peaks collected from many diffraction hits are then merged into a single 3D dataset. The combination of extended shape transforms of nanocrystalline Bragg peaks and small bandwidth of XFEL pulses results in peaks of high partiality. A Monte Carlo approach is adopted to integrate each *hkl* reflection from many partial views measured in SFX [103]. For this to succeed, the user must specify wisely the radius of inclusion around each partial Bragg peak [103]. Further, sub-threshold peaks that are undetected prior to indexing but predicted by auto-indexing routines must also be integrated [102]. With sufficiently many collected peaks, the Monte Carlo integration here also averages away fluctuations in nanocrystal-independent variables (e.g., incident X-ray pulse intensity, uncertainty in photon wavelength and background subtraction, etc.). Schemes for correcting for pulse intensity fluctuations before merging are being investigated [102]. Estimation of the partiality of peaks using post-refinement has demonstrated significant improvement in data quality for simulated data [104]. Finally, the merged 3D structure factors are sent to structure determination algorithms used in MX (e.g., molecular replacement, etc.).

The differences of peak intensities in SFX are expectedly large, due to fluctuating illumination conditions or high peak partiality, which are absent in MX. To assess data quality the indexed hits are partitioned into two sets that are then compared against each other for similarity using the metric R_{split} [102]. Since typical values are higher than comparable metrics used in MX [8, 102, 105], there are ongoing efforts to understand the limitations of SFX data quality.

In the case of very small crystals there is the possibility for the coherent scattering in the neighborhood of each Bragg peak to provide extra information about the structure of the unit cell for *ab initio* reconstructions. It has been proposed that this information can be extracted from an SFX dataset after merging the data as a continuous intensity distribution, instead of integrating each peak [13, 106, 107]. Although this work is in its preliminary stages, it is a good example of the new opportunities that XFELs are providing for the field of crystallography.

14.5.2. *SFX experiments*

The first demonstration of SFX was on the molecule photosystem I, which produced scattering to around 8.5 Å limited by the detector geometry [3]. Data was collected on two detectors to capture the low and high angle scattering. Low angle scattering was recorded with higher angular resolution, so that the structure around individual Bragg peaks could be measured. Remarkably, interference fringes that spanned two Bragg peaks were observed, demonstrating the high coherence of the X-ray source and the finite size of the crystals. Bragg CDI reconstructions from these peaks recovered the shapes of the crystals and by counting fringes, it was shown that crystals as small as 250 nm had diffracted strongly in the XFEL beam. Using the Monte-Carlo integration approach, the Bragg intensities were obtained and an electron density was recovered using molecular replacement.

The first high resolution SFX experiment recovered the structure of lysozyme to better than 2 Å, a sufficient resolution for applications in structural biology. A significant conclusion from this study was that no effects of radiation damage could be observed, so that the damage-before-destruction principle was validated for high-resolution SFX. SFX has been applied to determine the structure of catepsin B, a molecule involved in sleeping sickness, and photosystem II [108].

Pump-probe SFX experiments can be implemented with the liquid injector by modifying the crystals upstream of the interaction region. For example, light activated molecules can be pumped with an optical laser. The time-delay between when the protein dynamics are initiated and the crystal is probed by the XFEL pulse is set by the speed of the crystals in the liquid jet. This technique has been used to study the dynamics of photosystem I-Ferredoxin cocrystals [109] and to observe light-activated states of photosystem II [90, 110]. The study of photosystem II is notable for combining diffraction with spectroscopy to pin down subtle conformations of the molecule. These experiments highlight the potential for XFELs to help uncover the molecular details underpinning photosynthesis.

14.6. X-ray Laser Techniques and Applications: Single-particle Imaging

14.6.1. *Hit detection and classification*

Crucial to single-particle imaging is first identifying which detector frames contain diffraction signal from a specific sample, *hits*. In early XFEL experiments with larger samples (>100 nm in diameter) a threshold criterion based on the overall photon count on each pattern, after known artifacts and background were suppressed, was effective for hit detection. Detected hits can be further categorized using particle size estimates from the autocorrelation support by clustering similar patterns together [111], or with machine learning methods [112]. Both hit detection and classification are more challenging for weak scatterers and/or with strong background scattering. For these situations noise-robust hit detection and classification are being developed [113].

14.6.2. *Diffraction pattern analysis*

Spatial correlations of the scattering densities within the sample can be directly interpreted from their diffraction intensities. For example, using suitable band-pass filters, a compact support of the scatterer can be recovered. This already sets upper limits on the size of the scatterer and is crucial to constrain attempts at phase retrieval. Further, the scaling of these spatial correlations are immediately evident from the diffraction intensities, which could reveal the fractal-like structure of individual disordered particles [98]. This method has been used to show differences in particle-to-particle disorder within aerosol soot [7] (Fig. 14.8).

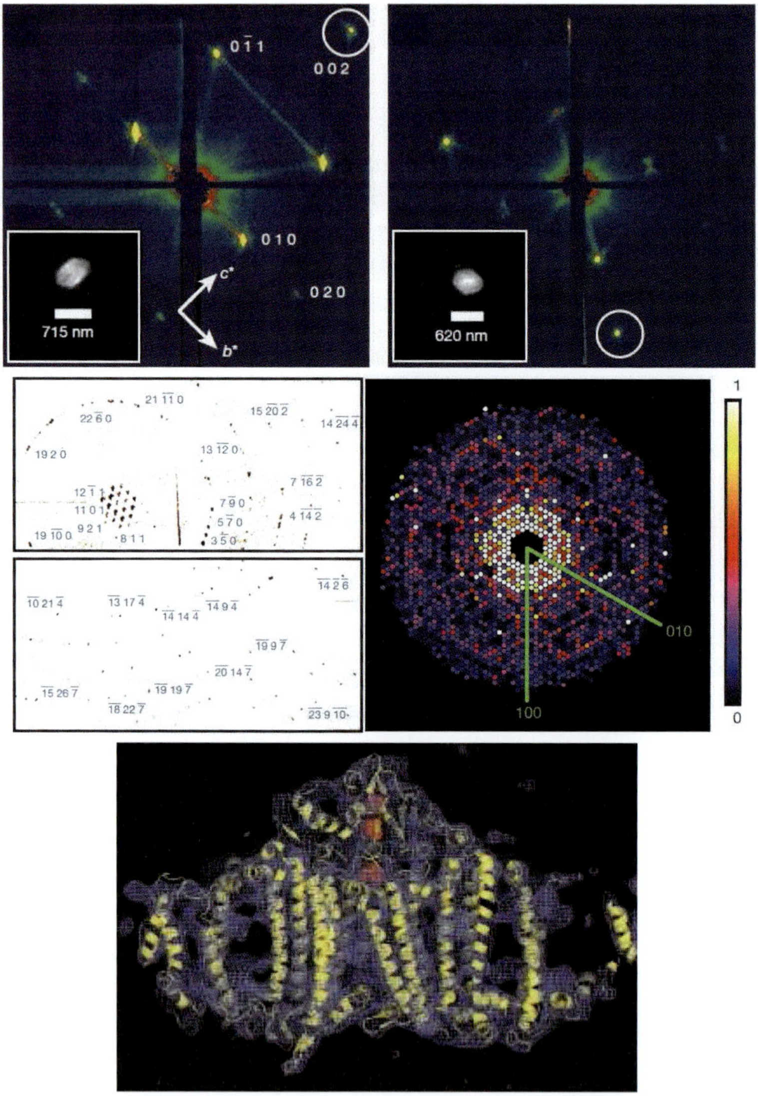

Fig. 14.7. Results from the first Serial Femtosecond Crystallography experiment on photosystem 1 (Adapted from [3] with permission). The top panel shows the diffraction to low order Bragg peaks with Bragg CDI images of the crystal shapes from the circled spots. The middle panel shows the high angle Bragg scattering on the left and the integrated intensities on the right. The bottom panel is an electron density map obtained from the data (see [3] for details).

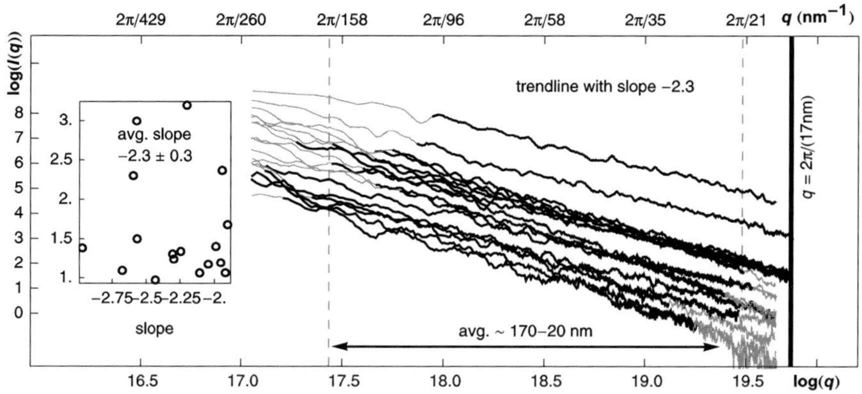

Fig. 14.8. Fractal morphology of aerosol soot. Morphological parameters such as fractal dimension can be measured from the slope of the intensity as a function of scattering vector, $I(q)$. Inset: particle size estimates from the autocorrelation support, adapted from [7].

The internal disorder of each particle can also be estimated, especially when prior knowledge of its constituents are imposed [114]. The data collected here is not unlike Small Angle X-ray Scattering (SAXS) for single particles. Here, particles are individually interrogated with XFELs, unlike SAXS with earlier generation x-ray sources that restrict us to averaging over many particles.

14.6.3. *Two-dimensional imaging*

CDI methods described in Section 14.2 have been used at XFELs to image inorganic and biological samples. When the beam is larger than the sample, the illumination is close to uniform across the sample and the assumptions of plane-wave CDI are valid. The recovered image is a projection of the sample as discussed in Section 14.2. Applications have been made to aerosolized fabricated nano-particles, viruses, nano-crystals and soot particles [7, 9, 115, 116]. To handle the abundance of data, automated phasing methods have been developed [116].

Unlike synchrotron CDI experiments, noise in the diffraction pattern cannot be reduced by increasing the exposure time. Hence, signal-to-noise is a limitation on the resolution that can be achieved.

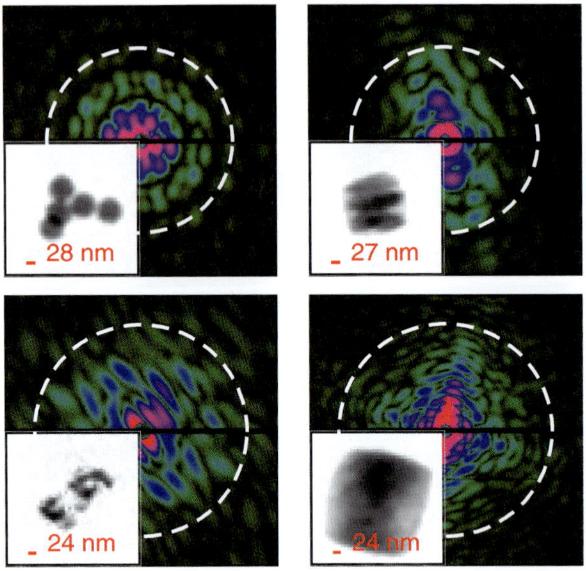

Fig. 14.9. Phase retrieval at LCLS. Diffraction intensities (in color), from which the inset 2D electron densities (grayscale) were reconstructed. Length of scale bars in insets equal the respective recovered resolutions (estimated from their 10-reconstruction phase-retrieval transfer function [Eq. 14.12]). Samples, injected by the aerodynamic focusing inlet, shown here are: (clockwise from top left) polystyrene spheres, iron oxide ellipsoids, soot, and salt nanocrystal coated with soot. Adapted from [7].

XFEL applications have increased the need for noise-robust phasing algorithms by using stronger constraints from prior knowledge [17] and/or by better managing the disagreement between noisy constraints [18].

In CDI, the diffraction data at very low scattering angles is never directly measured, because it overlaps with the intense, direct, unscattered X-ray beam. This missing data creates ambiguities in phase retrieval and hence image formation [117]. Fortunately, the low-resolution shape information of the scatterer is contained in this missing data, which is also redundantly represented in the measured intensities at higher scattering angles via the convolution theorem. Hence, a computational analogy to darkfield microscopy on the measured intensities can be used to recover a high-pass filtered image of the scatterer [118]. Direct replacement of the missing data is also possible if the scatterer's low resolution structure is known and can be suitably oriented using only high-resolution intensity measurements [4].

14.6.4. *Three-dimensional imaging*

The ability for X-rays to probe the interior of samples over a micron thick is the basis for many imaging applications. Tomography requires many views of the same object, which is not possible with destructive XFEL measurements. Hence, 3D XFEL imaging is only applicable to a sample that is reproducible, at least to some resolution. Smaller samples, like single molecules and viruses, tend to be more reproducible than larger samples, but they also scatter fewer photons. Hence, there has been a need to develop tomographic techniques with very low photon signals and where the particle orientation is not known. The goal in 3D single-particle XFEL imaging is to assemble the 3D Fourier intensities of a weak scatterer from very many of its noisy and incomplete (e.g., scatterer orientations are unknown) diffraction patterns. Once the 3D intensities are assembled, regular CDI phasing methods are used to recover the 3D image [21]. In this section, we give an overview of the algorithms developed to perform the 3D assembly.

The first intuitive approach was common arc methods [119–121] which were extended from angular reconstitution for single-particle electron microscopy [122]. They involve determining the relative orientations of triplets of randomly oriented diffraction patterns (i.e., Ewald sphere sections) by identifying pair-wise common arcs of intersection between them [Fig. 14.10]. Because such pair-wise matching relies only on the limited photon counts along these arcs, compared to methods that compare a 2D diffraction pattern against another, common arc methods require higher signal-to-noise ratios to succeed [120]. Diffraction pattern classification is essential for improving signal-to-noise prior to 3D assembly [119].

Very low signal-to-noise measurements can be assembled with graph-theoretic methods that represent the detector pixels as orthogonal directions of a high-dimensional Hilbert space. A noisy

Fig. 14.10. Common arc of intersection between two curved Ewald sphere sections of a biomolecule's Fourier intensities.

diffraction pattern is identified with a point in this high-dimensional space. A user-defined metric determines the distance between pairs of points (i.e., diffraction patterns) and a graph is constructed by connecting each point with its closest neighbors. By computing special trajectories on these graphs using optimal paths [123], the Laplacian eigenmap [124] or extremal geodesic paths [125], the relative orientations of each pattern can be assigned [125–127].

Another class of algorithm that performs well with very low signals are Bayesian methods. Statistical distributions of the underlying detection noise, known or assumed, can be used to identify patterns of like-orientations. A Bayesian reconstruction algorithm is the Expand-Maximize-Compress (EMC) [128, 129], which uses expectation maximization [130] to iteratively cluster datasets into a dense sampling of possible scatterer orientations [Fig. 14.11]. It has been shown to succeed at very low signal levels and modest additive background noise [131]. In EMC each pattern has a probability distribution across the possible orientations and is not restricted to any specific one. EMC's formalism for orientation discovery can correct for other unmeasured stochastic variables such as the fluctuations in incident photon intensity suffered by different copies of the target scatterer [132] or background photons [113]. EMC has also been

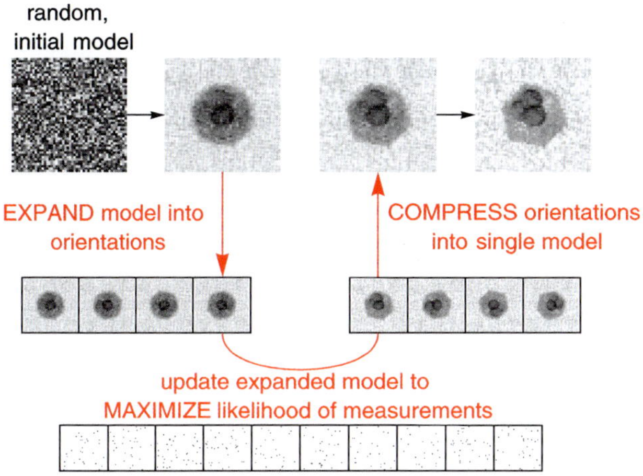

Fig. 14.11. Expand-Maximize-Compress (EMC) reconstruction algorithm, when applied to a 2D image reconstruction analog to 3D single-particle imaging [136]. Iterative reconstruction of the solution contrast (top right) from a random initial model (top left) is depicted in the top row. Steps of each EMC iteration label the red arrows. Whereas the noisy, unoriented input datasets (i.e., noisy images are in one of four possible $\pi/2$ in-plane rotations) in the bottom row describe the true image contrast in this 2D analog, similarly incomplete datasets describe a biomolecule's band-limited Fourier intensities in 3D single-particle imaging.

modified to use suitable Pearson correlations between patterns as an alternative measure for determining their orientation similarity [133]. Graph theoretic approaches can also be combined very successfully with Bayesian inference [134] using a framework known as Generative Topographic Mapping (GTM) [135], which was originally designed for visualizing a high-dimensional cloud of data points.

A promising alternative approach to single particle imaging is to use intensity correlations via a technique known as correlated X-ray scattering (CXS) [137]. Conveniently, the correlations due to a single scatterer's Fourier intensities can be computed from many random diffraction patterns despite not knowing the scatterers' orientation in each. A key advantage CXS has over other methods above is that multiple scatterers can diffract onto each pattern. Because spurious correlations from the relative orientations and positions of multiple scatterers in each pattern are different, they average away with sufficiently many random patterns leaving behind only the desired single-particle intensity correlations. CXS has been demonstrated in simulation when the particle has a known rotational symmetry [138, 139], and has been used to resolve the structure of azimuthally symmetric dimers from single-particle diffraction patterns collected at LCLS [140]. A major disadvantage of CXS is that it is not known how to obtain 3D Fourier intensities when there is no known particle symmetry [141].

We conclude by noting that the early success of assembly algorithms in simulated experiments are gradually being met with those from recent XFEL experiments [125, 132, 140].

Bibliography

[1] B. W. J. McNeil and N. R. Thompson, X-ray free-electron lasers, *Nature Photonics* **4**, 814–821, (2010).

[2] R. Neutze, R. Wouts, D. Van Der Spoel, E. Weckert and J. Hajdu, Potential for biomolecular imaging with femtosecond X-ray pulses, *Nature* **406**, 752–757, (2000).

[3] H. N. Chapman, P. Fromme, A. Barty, T. A. White, R. A. Kirian, A. Aquila, M. S. Hunter, J. Schulz, D. P. DePonte, U. Weierstall, R. B. Doak, F. R. N. C. Maia, A. V. Martin, I. Schlichting, L. Lomb, N. Coppola, R. L. Shoeman, S. W. Epp, R. Hartmann, D. Rolles, A. Rudenko, L. Foucar, N. Kimmel, G. Weidenspointner, P. Holl, M. Liang, M. Barthelmess, C. Caleman, S. Boutet, M. J. Bogan, J. Krzywinski, C. Bostedt, S. Bajt, L. Gumprecht, B. Rudek, B. Erk, C. Schmidt, A. Hömke, C. Reich, D. Pietschner, L. Strüder, G. Hauser, H. Gorke, J. Ullrich, S. Herrmann, G. Schaller, F. Schopper, H. Soltau, K.-U. Kühnel, M. Messerschmidt, J. D. Bozek, S. P. Hau-Riege, M. Frank, C. Y. Hampton, R. G. Sierra, D. Starodub, G. J. Williams, J. Hajdu, N. Timneanu, M. M. Seibert, J. Andreasson, A. Rocker, O. Jönsson, M. Svenda, S. Stern, K. Nass, R. Andritschke, C.-D. S. ter, F. Krasniqi, M. Bott, K. E. Schmidt, XiaoyuWang, I. Grotjohann, J. M. Holton, T. R. M. Barends, R. Neutze, S. Marchesini, R. Fromme, S. Schorb, D. Rupp, M. Adolph, T. Gorkhover, I. Andersson, H. Hirsemann, G. Potdevin, H. Graafsma, B. Nilsson and J. C. H. Spence, Femtosecond X-ray protein nanocrystallography, *Nature* **470**, 73–77, (2011).

[4] M. M. Seibert, T. Ekeberg, F. R. N. C. Maia, M. Svenda, J. Andreasson, O. Jonsson, D. Odic, B. Iwan, A. Rocker, D. Westphal, M. Hantke, D. P. DePonte, A. Barty, J. Schulz, L. Gumprecht, N. Coppola, A. Aquila, M. Liang, T. A. White, A. Martin, C. Caleman, S. Stern, C. Abergel, V. Seltzer, J.-M. Claverie, C. Bostedt, J. D. Bozek, S. Boutet, A. A. Miahnahri, M. Messerschmidt, J. Krzywinski, G. J. Williams, K. O. Hodgson,

M. J. Bogan, C. Y. Hampton, R. G. Sierra, D. Starodub, I. Andersson, S. Bajt, M. Barthelmess, J. C. H. Spence, P. Fromme, U. Weierstall, R. Kirian, M. Hunter, R. B. Doak, S. Marchesini, S. P. Hau-Riege, M. Frank, R. L. shoeman, L. Lomb, S. W. Epp, R. Hartmann, D. Rolles, A. Rudenko, C. Schmidt, L. Foucar, N. Kimmel, P. Holl, B. Rudek, B. Erk, A. Hömke, C. Reich, D. Pietschner, G. Weidenspointner, L. Strüder, G. Hauser, H. Gorke, J. Ullrich, I. Schlichting, S. Herrmann, G. Schaller, F. Schopper, H. Soltau, K.-U. Kühnel, R. Andritschke, C.-D. Schröter, F. Krasniqi, M. Bott, S. Schorb, D. Rupp, M. Adolph, T. Gorkhover, H. Hirsemann, G. Potdevin, H. Graafsma, B. Nilsson, H. N. Chapman and J. Hajdu, Single mimivirus particles intercepted and imaged with an X-ray laser, *Nature* **470**, 78–81, (2011).

[5] M. M. Seibert, S. Boutet, M. Svenda, T. Ekeberg, F. Maia, M. J. Bogan, N. Tmneanu, A. Barty, S. Hau-Riege, C. Caleman, M. Frank, W. H. Benner, J. Y. Lee, S. Marchesini, J. W. Shaevitz, D. A. Fletcher, S. Bajt, I. Andersson, H. N. Chapman, and J. Hajdu, Femtosecond diffractive imaging of biological cells, *Journal of Physics B* **43**, 194015, (2010).

[6] C. von Korff Schmising, B. Pfau, M. Schneider, C. M. Günther, M. Giovannella, J. Perron, B. Vodungbo, L. Müller, F. Capotondi, E. Pedersoli, N. Mahne, J. Lüning and S. Eisebitt, Imaging ultrafast demagnetization dynamics after a spatially localized optical excitation, *Phys. Rev. Lett.* **112**, 217203, (2014).

[7] N. D. Loh, C. Y. Hampton, A. V. Martin, D. Starodub, R. G. Sierra, A. Barty, A. Aquila, J. Schulz, L. Lomb, J. F. Steinbrener, R. L. shoeman, S. Kassemeyer, C. Bostedt, J. Bozek, S. W. Epp, B. Erk, R. Hartmann, D. Rolles, A. Rudenko, B. Rudek, L. Foucar, N. Kimmel, G. Weidenspointner, G. Hauser, P. Holl, E. Pedersoli, M. Liang, M. S. Hunter, L. Gumprecht, N. Coppola, C. Wunderer, H. Graafsma, F. R. N. C. Maia, T. Ekeberg, M. Hantke, H. Fleckenstein, H. Hirsemann, K. Nass, T. A. White, H. J. Tobias, G. R. Farquar, W. H. Benner, S. P. Hau-Riege, C. Reich, A. Hartmann, H. Soltau, S. Marchesini, S. Bajt, M. Barthelmess, P. H. Bucksbaum, K. O. Hodgson, L. Strüder, J. Ullrich, M. Frank, I. Schlichting, H. N. Chapman, and M. J. Bogan, Fractal morphology, imaging and mass spectrometry of single aerosol particles in flight, *Nature* **486**, 513–517, (2012).

[8] S. Boutet, L. Lomb, G. J. Williams, T. R. M. Barends, A. Aquila, R. B. Doak, U. Weierstall, D. P. DePonte, J. F. Steinbrener, R. L. shoeman, M. Messerschmidt, A. Barty, T. A. White, S. Kassemeyer, R. A. Kirian, M. M. Seibert, P. A. Montanez, C. Kenney, R. Herbst, P. Hart, J. Pines, G. Haller, S. M. Gruner, H. T. Philipp, M. W. Tate, M. Hromalik, L. J. Koerner, N. van Bakel, J. Morse, W. Ghonsalves, D. Arnlund, M. J. Bogan, C. Caleman, R. Fromme, C. Y. Hampton, M. S. Hunter, L. C. Johansson, G. Katona, C. Kupitz, M. Liang, A. V. Martin, K. Nass, L. Redecke, F. Stellato, N. Tîmneanu, D. Wang, N. A. Zatsepin, D. Schafer, J. Defever, R. Neutze, P. Fromme, J. C. H. Spence, H. N. Chapman and

I. Schlichting, High-resolution protein structure determination by serial femtosecond crystallography, *Science* **337**, 362–364, (2012).
[9] M. Born and E. Wolf, *Principles of Optics: Electromagnetic Theory of Propagation, Interference and Diffraction of Light.* (Cambridge University Press, 1999).
[10] B. E. Warren, *X-ray Diffraction.* (Addison-Wesley Publishing Company, 1969).
[11] D. M. Paganin, *Coherent X-ray Optics.* (Oxford University Press, 2006).
[12] R. A. Dilanian, V. A. Streltsov, H. M. Quiney and K. A. Nugent, Continuous X-ray diffractive field in protein nanocrystallography, *Acta Cryst. A* **69**, 108–118, (2013).
[13] R. A. Kirian, R. J. Bean, K. R. Beyerlein, O. M. Yefanov, T. A. White, A. Barty and H. N. Chapman, Phasing coherently illuminated nanocrystals bounded by partial unit cells, *Phil. Trans. R. Soc. Lon. B* **369**, 20130331, (2014).
[14] S. Marchesini, A unified evaluation of iterative projection algorithms for phase retrieval, *Review of Scientific Instruments* **78**, 011301, (2007).
[15] J. R. Fienup, Phase retrieval algorithms — a comparison, *Appl. Optics.* **21**, 2758–2769, (1982).
[16] V. Elser, Phase retrieval by iterated projections, *J. Opt. Soc. Am. A* **20**, 40–55, (2003).
[17] N.-T. D. Loh, S. Eisebitt, S. Flewett and V. Elser, Recovering magnetization distributions from their noisy diffraction data, *Phys. Rev. E* **82**, 061128, (2010).
[18] A. V. Martin, F. Wang, N. D. Loh, T. Ekeberg, F. Maia, M. Hantke, G. van der Schot, C. Y. Hampton, R. G. Sierra, A. Aquila, S. Bajt, M. Barthelmess, C. Bostedt, J. D. Bozek, N. Coppola, S. W. Epp, B. Erk, H. Fleckenstein, L. Foucar, M. Frank, H. Graafsma, L. Gumprecht, A. Hartmann, R. Hartmann, G. Hauser, H. Hirsemann, P. Holl, S. Kassemeyer, N. Kimmel, M. Liang, L. Lomb, S. Marchesini, K. Nass, E. Pedersoli, C. Reich, D. Rolles, B. Rudek, A. Rudenko, J. Schulz, R. L. Shoeman, H. Soltau, D. Starodub, J. F. Steinbrener, F. Stellato, L. Strüder, J. Ullrich, G. Weidenspointner, T. A. White, C. B. Wunderer, A. Barty, I. Schlichting, M. J. Bogan and H. N. Chapman, Noise-robust coherent diffractive imaging with a single diffraction pattern, *Optics Express* **20**, 16650–16661, (2012).
[19] J. A. Rodriguez, R. Xu, C.-C. Chen, Y. Zou and J. Miao, Oversampling smoothness: an effective algorithm for phase retrieval of noisy diffraction intensities, *J. Appl. Cryst.* **46**, 312, (2013).
[20] C.-C. Chen, J. Miao, C. W. Wang and T. K. Lee, Application of optimization technique to noncrystalline X-ray diffraction microscopy: Guided hybrid input-output method, *Phys. Rev. B* **76**, 064113, (2007).
[21] H. N. Chapman, A. Barty, S. Marchesini, A. Noy, S. Hau-Riege, C. Cui, M. R. Howells, R. Rosen, H. He, J. C. H. Spence, U. Weierstall, T. Beetz, C. Jacobsen and D. Shapiro, High-resolution *ab initio* three-dimensional X-ray diffraction microscopy, *J. Opt. Soc. Am. A* **23**, 1179–1200, (2006).

[22] J. Steinbrener, J. Nelson, X. Huang, S. Marchesini, D. Shapiro, J. J. Turner, and C. Jacobsen, Data preparation and evaluation techniques for X-ray diffraction microscopy, *Opt. Express* **18**, 18598, (2010).

[23] H. M. Quiney, Coherent diffractive imaging using short wavelength light sources, *Journal of Modern Optics* **57**, 1109–1149, (2010).

[24] G. J. Williams, H. Quiney *et al.*, Coherent diffractive imaging and partial coherence, *Physical Review B — Condensed Matter and Materials Physics* **75**, 104102, (2007).

[25] L. Whitehead, G. J. Williams, H. M. Quiney, D. J. Vine, R. A. Dilanian, S. Flewett and K. A. Nugent, Diffractive imaging using partially coherent X-rays, *Physical Review Letters* **103**, 243902, (2009).

[26] B. Chen, R. A. Dilanian, S. Teichmann, B. Abbey, A. G. Peele, G. J. Williams, P. Hannaford, L. V. Dao, H. M. Quiney and K. A. Nugent, Multiple wavelength diffractive imaging, *Phys. Rev. A* **79**, 023809, (2009).

[27] B. Abbey, L. W. Whitehead, H. M. Quiney, D. J. Vine, G. A. Cadenazzi, C. A. Henderson, K. A. Nugent, E. Balaur, C. T. Putkunz, A. G. Peele, G. J. Williams and I. McNulty, Lensless imaging using broadband X-ray sources, *Nat. Photonics* **5**, 420–424, (2011).

[28] H. M. L. Faulkner and J. M. Rodenburg, Movable aperture lensless transmission microscopy: A novel phase retrieval algorithm, *Phys. Rev. Lett.* **93**, 023903, (2004).

[29] J. M. Rodenburg, A. C. Hurst and A. G. Cullis, Transmission microscopy without lenses for objects of unlimited size, *Ultramicroscopy* **107**, 227–231, (2007).

[30] J. M. Rodenburg, A. C. Hurst, A. G. Cullis, B. R. Dobson, F. Pfeiffer, O. Bunk, C. David, K. Jefimovs and I. Johnson, Hard-X-ray lensless imaging of extended objects, *Physical Review Letters* **98**, 034801, (2007).

[31] O. Bunk, M. Dierolf, S. Kynde, I. Johnson, O. Marti and F. Pfeifer, Influence of the overlap parameter on the convergence of the ptychographical iterative engine, *Ultramic.* **108**, 481–487, (2008).

[32] A. Maiden and J. Rodenburg, An improved ptychographical phase retrieval algorithm for diffractive imaging, *Ultramicroscopy* **109**, 1256–1262, (2009).

[33] P. Thibault, M. Dierolf, O. Bunk, A. Menzel and F. Pfeiffer, Probe retrieval in ptychographic coherent diffractive imaging, *Ultramicroscopy* **109**, 338–343, (2009).

[34] M. Guizar-Sicairos and J. R. Fienup, Phase retrieval with transverse translation diversity: A nonlinear optimization approach, *Opt. Express* **16**, 7264, (2008).

[35] A. Maiden, M. Humphry, M. Sarahan, B. Kraus and J. Rodenburg, An annealing algorithm to correct positioning errors in ptychography, *Ultramicroscopy* **120**, 64–72, (2012).

[36] D. J. Vine, D. Pelliccia, C. Holzner, S. B. Baines, A. Berry, I. McNulty, S. Vogt, A. G. Peele and K. A. Nugent, Simultaneous X-ray fluorescence and ptychographic microscopy of cyclotella meneghiniana, *Optics Express.* **20**, 18287, (2012).

[37] D. A. Shapiro, Y.-S. Yu, T. Tyliszczak, J. Cabana, R. Celestre, W. Chao, K. Kaznatcheev, K. L. David, F. Maia, S. Marchesini, Y. S. Meng, T. Warwick, L. L. Yang and H. A. Padmore, Chemical composition mapping with nanometre resolution by soft X-ray microscopy, *Nat Photon* **8** (10), 765–769 (Oct., 2014). ISSN 1749-4885. URL http://dx.doi.org/10.1038/ nphoton.2014.207.

[38] J. N. Clark, X. Huang, R. J. Harder and I. K. Robinson, Dynamic imaging using ptychography, *Phyiscal Review Letters* **112**, 113901, (2014).

[39] P. Thibault and A. Menzel, Reconstructing state mixtures from diffraction measurements, *Nature* **494**, 68–71, (2013). ISSN 0028-0836.

[40] B. Abbey, K. A. Nugent, G. J. Williams, J. N. Clark, A. G. Peele, M. A. Pfeifer, M. D. Jonge and I. McNulty, Keyhole coherent diffractive imaging, *Nat. Phys.* **4**, 394–398, (2008).

[41] D. J. Vine, G. Williams *et al.*, Ptychographic fresnel coherent diffractive imaging, *Phys. Rev. A* **80**, 063823, (2009).

[42] S. Eisebitt, J. Lüning, W. F. Schlotter, M. Lörgen, O. Hellwig, W. Eberhardt and J. Stöhr, Lensless imaging of magnetic nanostructures by X-ray spectro-holography, *Nature* **432**, 885–888, (2004).

[43] S. G. Podorov, K. M. Pavlov and D. M. Paganin, A non-iterative reconstruction method for direct and unambiguous coherent diffractive imaging, *Optics Express* **15**, 9954–9962, (2007).

[44] M. Guizar-Sicairos and J. R. Fienup, Holography with extended reference by autocorrelation linear differential operation, *Opt. Express* **15**, 17592–17612, (2007).

[45] S. Marchesini, S. Boutet, A. Sakdinawat, M. J. Bogan, S. Bajt, A. Barty, H. N. Chapman, M. Frank, S. Hau-Riege and A. Szöke, Massively parallel X-ray holography, *Nature Photonics* **2**, 560–563, (2008).

[46] A. V. Martin and L. J. Allen, Direct retrieval of a complex wave from its diffraction pattern, *Optics Communications* **281**, 5114–5121, (2008).

[47] A. J. D'Alfonso, A. J. Morgan, A. V. Martin, H. M. Quiney and L. J. Allen, Fast deterministic approach to exit-wave reconstruction, *Phys. Rev. A.* **85**, 013816, (2012).

[48] A. V. Martin, A. J. DAlfonso, F. Wang, R. Bean, F. Capotondi, R. A. Kirian, E. Pedersoli, L. Raimondi, F. Stellato, C. H. Yoon and H. N. Chapman, X-ray holography with a customizable reference, *Nat Commun.* **5**, – (Aug., 2014). URL http://dx.doi.org/10.1038/ncomms5661.

[49] D. Shapiro, P. Thibault, T. Beetz, V. Elser, M. Howells, C. Jacobsen, J. Kirz, E. Lima, H. Miao, A. M. Neiman and D. Sayre, Biological imaging by soft X-ray diffraction microscopy, *Proceedings of the National Academy of Sciences of the United States of America* **102**, 15343–15346, (2005).

[50] C. T. Putkunz, M. A. Pfeifer, A. G. Peele, G. J. Williams, H. M. Quiney, B. Abbey, K. A. Nugent and I. McNulty, Fresnel coherent diffraction tomography, *Optics Express* **12**, 11746–11753, (2010).

[51] M. Dietrolf, A. Menzel, P. Thibault, P. Schneider, C. M. Kewish, R. Wepf, O. Bunk and F. Pfeiffer, Ptychographic X-ray computed tomography at the nanoscale, *Nature* **467**, 436–439, (2010).

[52] I. K. Robinson, I. A. Vartanyants, G. J. Williams, M. A. Pfeifer and J. A. Pitney, Reconstruction of the shapes of gold nanocrystals using coherent X-ray diffraction, *Physical Review Letters* **87**, 195505, (2001).
[53] G. J. Williams, M. A. Pfeifer, I. A. Vartanyants and I. K. Robinson, Three-dimensional imaging of microstructure in Au nanocrystals, *Physical Review Letters* **90**, 175501, (2003).
[54] M. A. Pfeifer, G. J. Williams, I. A. Vartanyants, R. Harder and I. K. Robinson, Three-dimensional mapping of a deformation field inside a nanocrystal, *Nature* **442**, 63–66, (2006).
[55] J. Miao, P. Charalambous, J. Kirz and D. Sayre, Extending the methodology of X-ray crystallography to allow imaging of micrometre-sized non-crystalline specimens, *Nature* **400**, 342–344, (1999).
[56] S. Marchesini, H. He, H. N. Chapman, S. P. Hau-Riege, A. Noy, M. R. Howells, U. Weierstall and J. C. H. Spence, X-ray image reconstruction from a diffraction pattern alone, *Phys. Rev. B* **68**, 140101, (2003).
[57] G. Williams *et al.*, Fresnel coherent diffractive imaging, *Physical Review Letters* **97**, 55061–55064, (2006).
[58] H. Quiney *et al.*, Diffractive imaging of highly focused X-ray fields, *Nature Physics* **2**, 101–104, (2006).
[59] L. Whitehead *et al.*, Fresnel diffractive imaging: An experimental study of coherence and curvature, *Phys Rev B* **77**, 104112, (2008).
[60] A. Barty, S. Marchesini, H. N. Chapman, C. Cui, M. R. Howells, D. A. Shapiro, A. M. Minor, J. C. H. Spence, U. Weierstall, J. Ilavsky, A. Noy, S. P. Hau-Riege, A. B. Artyukhin, T. Baumann, T. Willey, J. Stolken, T. van Buuren, and J. H. Kinney, Three-dimensional coherent X-ray diffraction imaging of a ceramic nanofoam: Determination of structural deformation mechanisms, *Phys. Rev. Lett.* **101**, 055501, (2008).
[61] Y. Takahashi, A. Suzuki, S. Furutaku, K. Yamauchi, Y. Kohmura and T. Ishikawa, Bragg X-ray ptychography of a silicon crystal: Visualization of the dislocation strain field and the production of a vortex beam, *Phys. Rev. B* **87**, 121201, (2013).
[62] W. Cha, N. C. Jeong, S. Song, H.-j. Park, T. C. Thanh Pham, R. Harder, B. Lim, G. Xiong, D. Ahn, I. McNulty, J. Kim, K. B. Yoon, I. K. Robinson and H. Kim, Core-shell strain structure of zeolite microcrystals, *Nat Mater* **12**(8), 729–734 (Aug., 2013). ISSN 1476-1122. http://dx.doi.org/10.1038/nmat3698.
[63] G. Xiong, O. Moutanabbir, M. Reiche, R. Harder and I. Robinson, Coherent X-ray diffraction imaging and characterization of strain in silicon-on-insulator nanostructures, *Advanced Materials*. p. n/a, (2014). ISSN 1521-4095. DOI:10.1002/adma.201304511. http://dx.doi.org/10.1002/adma.201304511.
[64] A. Schropp, P. Boye, A. Goldschmidt, S. Honig, R. Hoppy, J. Patommel, C. Rakete, D. Samberg, S. Stephan, S. Schroder, M. Burgmanner and C. Schroer, Non-destructive and quantitative imaging of a nano-structured microchip by ptychographic hard X-ray scanning microscopy, *J. Microsc.* **241**, 9, (2011).

[65] M. Holler, A. Diaz, M. Guizar-Sicairos, P. Karvinen, E. Frm, E. Hrknen, M. Ritala, A. Menzel, J. Raabe and O. Bunk, X-ray ptychographic computed tomography at 16nm isotropic 3d resolution, *Sci. Rep.* **4**, – (Jan., 2014). http://dx.doi.org/10.1038/srep03857.

[66] H. Jiang, C. Song, C.-C. Chen, R. Xu, K. S. Raines, B. P. Fahimian, C.-H. Lu, T.-K. Lee, A. Nakashima, J. Urano, T. Ishikawa, F. Tamanoi and J. Miao, Quantitative 3d imaging of whole, unstained cells by using X-ray diffraction microscopy, *Proc. Natl. Acad. Sci.* **107**, 11234–11239, (2010).

[67] M. Jones, G. van Rissen, B. Abbey, C. T. Putkunz, M. D. Junker, E. Balaur, D. J. Vine, I. McNulty, B. Chen, B. D. Arhatari, S. Frankland, K. A. Nugent, L. Tilley and A. G. Peele, Whole-cell phase contrast imaging and the nanocale using fresnel coherent diffractive imaging tomography, *Sciencific Reports* **3**, 2288, (2013).

[68] Y. Nishino, Y. Takahashi, N. Imamoto, T. Ishikawa and K. Maeshima, Three-dimensional visualization of a human chromosome using coherent X-ray diffraction, *Phys. Rev. Lett.* **102**, 018101, (2009).

[69] J. Madey, Stimulated emission of bremsstrahlung in a periodic magnetic field, *Journal of Applied Physics* **42**, 1906–1913, (1971).

[70] W. Ackermann, G. Asova, V. Ayvazyan, A. Azima, N. Baboi, J. Bähr, V. Balandin, B. Beutner, A. Brandt and A. Bolzmann, Operation of a free-electron laser from the extreme ultraviolet to the water window, *Nat. Photonics* **1**, 336–342, (2007).

[71] P. Emma, R. Akre, J. Arthur, R. M. Bionta, C. Bostedt, J. Bozek, A. Brachmann, P. H. Bucksbaum, R. Coffee, F. J. Decker, Y. Ding, D. Dowell, S. Edstrom, A. Fisher, J. Frisch, S. Gilevich, J. Hastings, G. Hays, P. Hering, Z. Huang, R. Iverson, H. Loos, M. Messerschmidt, A. Miahnahri, S. Moeller, H.-D. Nuhn, D. Pile, D. Ratner, J. Rzepiela, D. Schultz, T. Smith, P. Stefan, H. Tompkins, J. Turner, J. Welch, W. White, J. Wu, G. Yocky and J. N. Galayda, First lasing and operation of an angstrom-wavelength free-electron laser, *Nat. Photonics* **4**, 641–647, (2010).

[72] T. Ishikawa, H. Aoyagi, T. Asaka, Y. Asano, N. Azumi, T. Bizen, H. Ego, K. Fukami, T. Fukui, Y. Furukawa, S. Goto, H. Hanaki, T. Hara, T. Hasegawa, T. Hatsui, A. Higashiya, T. Hirono, N. Hosoda, M. Ishii, T. Inagaki, Y. Inubushi, T. Itoga, Y. Joti, M. Kago, T. Kameshima, H. Kimura, Y. Kirihara, A. Kiyomichi, T. Kobayashi, C. Kondo, T. Kudo, H. Maesaka, X. M. Maréchal, T. Masuda, S. Matsubara, T. Matsumoto, T. Matsushita, S. Matsui, M. Nagasono, N. Nariyama, H. Ohashi, T. Ohata, T. Ohshima, S. Ono, Y. Otake, C. Saji, T. Sakurai, T. Sato, K. Sawada, T. Seike, K. Shirasawa, T. Sugimoto, S. Suzuki, S. Takahashi, H. Takebe, K. Takeshita, K. Tamasaku, H. Tanaka, R. Tanaka, T. Tanaka, T. Togashi, K. Togawa, A. Tokuhisa, H. Tomizawa, K. Tono, S. Wu, yabashi, Makina, M. Yamaga, A. Yamashita, K. Yanagida, C. Zhang, T. Shintake, H. Kitamura and N. Kumagai, A compact X-ray free-electron laser emitting in the sub-ångström region, *Nat. Photonics* **6**, 540–544, (2012).

[73] E. Allaria, R. Appio, L. Badano, W. A. Barletta, S. Bassanese, S. G. Biedron, A. Borga, E. Busetto, D. Castronovo, P. Cinquegrana, S. Cleva, D. Cocco, M. Cornacchia, P. Craievich, I. Cudin, G. D'Auria, M. D. Forno, M. B. Danailov, R. De Monte, G. De Ninno, P. Delgiusto, A. Demidovich, S. Di Mitri, B. Diviacco, A. Fabris, R. Fabris, W. Fawley, M. Ferianis, E. Ferrari, S. Ferry, L. Froehlich, P. Furlan, G. Gaio, F. Gelmetti, L. Giannessi, M. Giannini, R. Gobessi, R. Ivanov, E. Karantzoulis, M. Lonza, A. Lutman, B. Mahieu, M. Milloch, S. V. Milton, M. Musardo, I. Nikolov, S. Noe, F. Parmigiani, G. Penco, M. Petronio, L. Pivetta, M. Predonzani, F. Rossi, L. Rumiz, A. Salom, C. Scafuri, C. Serpico, P. Sigalotti, S. Spampinati, C. Spezzani, M. Svandrlik, C. Svetina, S. Tazzari, M. Trovo, R. Umer, A. Vascotto, M. Veronese, R. Visintini, M. Zaccaria, D. Zangrando and M. Zangrando, Highly coherent and stable pulses from the FERMI seeded free-electron laser in the extreme ultraviolet, *Nat. Photonics* **6**, 699–704, (2012).

[74] J. N. Galayda, J. Arthur, D. F. Ratner and W. E. White, X-ray free-electron lasers — present and future capabilities [Invited], *J. Opt. Soc. Am. B* **27**, B106, (2010).

[75] T. Wang, D. Zhu, B. Wu, C. Graves, S. Schaffert, T. Rander, L. Müller, B. Vodungbo, C. Baumier, D. Bernstein, B. Bräuer, V. Cros, S. de Jong, R. Delaunay, A. Fognini, R. Kukreja, S. Lee, V. López-Flores, J. Mohanty, B. Pfau, H. Popescu, M. Sacchi, A. Sardinha, F. Sirotti, P. Zeitoun, M. Messerschmidt, J. Turner, W. Schlotter, O. Hellwig, R. Mattana, N. Jaouen, F. Fortuna, Y. Acremann, C. Gutt, H. Dürr, E. Beaurepaire, C. Boeglin, S. Eisebitt, G. Grübel, J. Lüning, J. Stöhr and A. Scherz, Femtosecond single-shot imaging of nanoscale ferromagnetic order in co/Pd multilayers using resonant X-ray holography, *Physical Review Letters* **108**, 267403, (2012).

[76] E. A. Schneidmiller and M. V. Yurkov, Obtaining high degree of circular polarization at X-ray free electron lasers via a reverse undulator taper, *Physical Review Special Topics-Accelerators and Beams* **16**, 110702, (2013).

[77] G. Margaritondo and P. Rebernik Ribic, A simplified description of X-ray free-electron lasers, *J Synchrotron Rad* **18**, 101–108, (2011).

[78] J. Solem, Imaging biological specimens with high-intensity soft X-rays, *Journal of the Optical Society of America B* **3**, 1551–1565, (1986).

[79] Z. Jurek, G. Faigel and M. Tegze, Dynamics in a cluster under the influence of intense femtosecond hard X-ray pulses, *Eur. Phys. J. D* **29**, 217–229, (2004).

[80] S. P. Hau-Riege, R. A. London and A. Szőke, Dynamics of biological molecules irradiated by short X-ray pulses, *Phys. Rev. E* **69**, 051906, (2004).

[81] H. N. Chapman, A. Barty, M. J. Bogan, S. Boutet, M. Frank, S. P. Hau-Riege, S. Marchesini, B. W. Woods, S. Bajt, W. H. Benner, R. A. London, E. Plönjes, M. Kuhlmann, R. Treusch, S. Dusterer, T. Tschentscher, J. R. Schneider, E. Spiller, T. Möller, C. Bostedt, M. Hoener, D. Shapiro,

K. O. Hodgson, D. Van Der Spoel, F. Burmeister, M. Bergh, C. Caleman, G. Huldt, M. M. Seibert, F. R. N. C. Maia, R. W. Lee, A. Szöke, N. Timneanu and J. Hajdu, Femtosecond diffractive imaging with a soft-X-ray free-electron laser, *Nat. Phys.* **2**, 839–843, (2006).
[82] H. N. Chapman, S. Hau-Riege, M. J. Bogan, S. Bajt, A. Barty, S. Boutet, S. Marchesini, M. Frank, B. W. Woods, W. H. Benner, R. A. London, U. Rohner, A. Szöke, E. Spiller, T. Möller, C. Bostedt, D. A. Shapiro, M. Kuhlmann, R. Treusch, E. Plönjes, F. Burmeister, M. Bergh, C. Caleman, G. Huldt, M. M. Seibert and J. Hajdu, Femtosecond time-delay X-ray holography, *Nature* **448**, 676–680, (2007).
[83] L. Lomb, T. R. M. Barends, S. Kassemeyer, A. Aquila, S. W. Epp, B. Erk, L. Foucar, R. Hartmann, B. Rudek, D. Rolles, A. Rudenko, R. L. shoeman, J. Andreasson, S. Bajt, M. Barthelmess, A. Barty, M. J. Bogan, C. Bostedt, J. D. Bozek, C. Caleman, R. Coffee, N. Coppola, D. P. DePonte, R. B. Doak, T. Ekeberg, H. Fleckenstein, P. Fromme, M. Gebhardt, H. Graafsma, L. Gumprecht, C. Y. Hampton, A. Hartmann, G. Hauser, H. Hirsemann, P. Holl, J. M. Holton, M. S. Hunter, W. Kabsch, N. Kimmel, R. A. Kirian, M. Liang, F. R. N. C. Maia, A. Meinhart, S. Marchesini, A. V. Martin, K. Nass, C. Reich, J. Schulz, M. M. Seibert, R. Sierra, H. Soltau, J. C. H. Spence, J. F. Steinbrener, F. Stellato, S. Stern, N. Timneanu, X. Wang, G. Weidenspointner, U. Weierstall, T. A. White, C. B. Wunderer, H. N. Chapman, J. Ullrich, L. Strüder and I. Schlichting, Radiation damage in protein serial femtosecond crystallography using an X-ray free-electron laser., *Phys. Rev. B Condens. Matter Mater. Phys.* **84**, 214111, (2011).
[84] A. Barty, C. Caleman, A. Aquila, N. Timneanu, L. Lomb, T. A. White, J. Andreasson, D. Arnlund, S. Bajt, T. R. M. Barends, M. Barthelmess, M. J. Bogan, C. Bostedt, J. D. Bozek, R. Coffee, N. Coppola, J. Davidsson, D. P. DePonte, R. B. Doak, T. Ekeberg, V. Elser, S. W. Epp, B. Erk, H. Fleckenstein, L. Foucar, P. Fromme, H. Graafsma, L. Gumprecht, J. Hajdu, C. Y. Hampton, R. Hartmann, A. Hartmann, G. Hauser, H. Hirsemann, P. Holl, M. S. Hunter, L. Johansson, S. Kassemeyer, N. Kimmel, R. A. Kirian, M. Liang, F. R. N. C. Maia, E. Malmerberg, S. Marchesini, A. V. Martin, K. Nass, R. Neutze, C. Reich, D. Rolles, B. Rudek, A. Rudenko, H. Scott, I. Schlichting, J. Schulz, M. M. Seibert, R. L. shoeman, R. G. Sierra, H. Soltau, J. C. H. Spence, F. Stellato, S. Stern, L. Strüder, J. Ullrich, X. Wang, G. Weidenspointner, U. Weierstall, C. B. Wunderer and H. N. Chapman, Self-terminating diffraction gates femtosecond X-ray nanocrystallography measurements, *Nat. Photonics* **6**, 35–40, (2012).
[85] L. Strüder, S. Epp, D. Rolles, R. Hartmann, P. Holl, G. Lutz, H. Soltau, R. Eckart, C. Reich, K. Heinzinger, R. Richter, L. Foucar, I. Schlichting, J. Ullrich, R. L. shoeman, *et al.*, Large-format, high-speed, X-ray pnCCDs combined with electron and ion imaging spectrometers in a multipurpose chamber for experiments at 4th generation light sources, *Nuclear Instruments and Methods in Physics Research Section A: Accelerators, Spectrometers, Detectors and Associated Equipment* **614**, 483–496, (2010).

[86] A. Kiyomichi, A. Amselem, T. Hirono, T. Ohata, R. Tanaka, M. Yamaga and T. Hatsui, Development of image data acquisition system for 2D detector at SACLA, *Proceedings of ICALEPCS2011, Grenoble, France* **947**, (2011).
[87] H. T. Philipp, M. Hromalik, M. Tate, L. Koerner and S. M. Gruner, Pixel array detector for X-ray free electron laser experiments, *Nuclear Instruments and Methods in Physics Research Section A: Accelerators, Spectrometers, Detectors and Associated Equipment* **649**, 67–69, (2011).
[88] A. Koch, M. Kuster, J. Sztuk-Dambietz and M. Turcato, Detector Development for the European XFEL: Requirements and Status, *Journal of Physics: Conference Series* **425**, 062013, (2013).
[89] B. Iwan, J. Andreasson, M. Bergh, S. Schorb, H. Thomas, D. Rupp, T. Gorkhover, M. Adolph, T. Möller, C. Bostedt, J. Hajdu and N. Tîmneanu, Explosion, ion acceleration and molecular fragmentation of methane clusters in the pulsed beam of a free-electron laser, *Phys. Rev. A* **86**, 033201, (2012).
[90] J. Kern, R. Alonso-Mori, R. Tran, J. Hattne, R. J. Gildea, N. Echols, C. Glockner, J. Hellmich, H. Laksmono, R. G. Sierra, B. Lassalle-Kaiser, S. Koroidov, A. Lampe, G. Han, S. Gul, D. DiFiore, D. Milathianaki, A. R. Fry, A. Miahnahri, D. W. Schafer, M. Messerschmidt, M. M. Seibert, J. E. Koglin, D. Sokaras, T. C. Weng, J. Sellberg, M. J. Latimer, R. W. Grosse-Kunstleve, P. H. Zwart, W. E. White, P. Glatzel, P. D. Adams, M. J. Bogan, G. J. Williams, S. Boutet, J. Messinger, A. Zouni, N. K. Sauter, V. K. Yachandra, U. Bergmann and J. Yano, Simultaneous femtosecond X-ray spectroscopy and diffraction of photosystem II at room temperature, *Science* **340**, 491–495, (2013).
[91] M. W. Tate, D. Chamberlain, K. S. Green, H. T. Philipp, P. Purohit, C. Strohman and S. M. Gruner, A medium-format, mixed-mode pixel array detector for kilohertz X-ray imaging, *Journal of Physics: Conference Series* **425**, 062004, (2013).
[92] J. W. Strutt and L. Rayleigh, On the instability of jets, *Proceedings of the London Mathematical Society.* pp. 4–13, (1879).
[93] U. Weierstall, R. B. Doak, J. C. H. Spence, D. Starodub, D. Shapiro, P. Kennedy, J. Warner, G. G. Hembree, P. Fromme and H. N. Chapman, Droplet streams for serial crystallography of proteins, *Experiments in Fluids* **44**, 675–689, (2007).
[94] R. G. Sierra, H. Laksmono, J. Kern, R. Tran, J. Hattne, R. Alonso-Mori, B. Lassalle-Kaiser, C. Glockner, J. Hellmich, D. W. Schafer, N. Echols, R. J. Gildea, R. W. Grosse-Kunstleve, J. Sellberg, T. A. McQueen, A. R. Fry, M. M. Messerschmidt, A. Miahnahri, M. M. Seibert, C. Y. Hampton, D. Starodub, N. D. Loh, D. Sokaras, T. C. Weng, P. H. Zwart, P. Glatzel, D. Milathianaki, W. E. White, P. D. Adams, G. J. Williams, S. Boutet, A. Zouni, J. Messinger, N. K. Sauter, U. Bergmann, J. Yano, V. K. Yachandra and M. J. Bogan, Nanoflow electrospinning serial femtosecond crystallography, *Acta Crystallographica Section D* **68**, 1584–1587, (2012).

[95] L. C. Johansson, D. Arnlund, T. A. White, G. Katona, D. P. DePonte, U. Weierstall, R. B. Doak, R. L. shoeman, L. Lomb, E. Malmerberg, J. Davidsson, K. Nass, M. Liang, J. Andreasson, A. Aquila, S. Bajt, M. Barthelmess, A. Barty, M. J. Bogan, C. Bostedt, J. D. Bozek, C. Caleman, R. Coffee, N. Coppola, T. Ekeberg, S. W. Epp, B. Erk, H. Fleckenstein, L. Foucar, H. Graafsma, L. Gumprecht, J. Hajdu, C. Y. Hampton, R. Hartmann, A. Hartmann, G. Hauser, H. Hirsemann, P. Holl, M. S. Hunter, S. Kassemeyer, N. Kimmel, R. A. Kirian, F. R. N. C. Maia, S. Marchesini, A. V. Martin, C. Reich, D. Rolles, B. Rudek, A. Rudenko, I. Schlichting, J. Schulz, M. M. Seibert, R. G. Sierra, H. Soltau, D. Starodub, F. Stellato, S. Stern, L. Strüder, N. Timneanu, J. Ullrich, W. Y. Wahlgren, x. Wang, G. Weidenspointner, C. B. Wunderer, P. Fromme, H. N. Chapman, J. C. H. Spence and R. Neutze, Lipidic phase membrane protein serial femtosecond crystallography, *Nat. Methods* **9**, 263–265, (2012).

[96] P. Liu, P. J. Ziemann, D. B. Kittelson and P. H. McMurry, Generating particle beams of controlled dimensions and divergence: II. Experimental evaluation of particle motion in aerodynamic lenses and nozzle expansions, *Aerosol Sc. & Tech.* **22**, 314–324, (1995).

[97] M. J. Bogan, W. H. Benner, S. Boutet, U. Rohner, M. Frank, A. Barty, M. M. Seibert, F. R. N. C. Maia, S. Marchesini, S. Bajt, B. W. Woods, V. Riot, S. P. Hau-Riege, M. Svenda, E. Marklund, E. Spiller, J. Hajdu and H. N. Chapman, Single particle X-ray diffractive imaging, *Nano Lett.* **8**, 310–316, (2008).

[98] M. J. Bogan, D. Starodub, C. Hampton and R. G. Sierra, Single-particle coherent diffractive imaging with a soft X-ray free electron laser: Towards soot aerosol morphology, *J. Phys. B-At. Mol. Op.* **43**, 194013, (2010).

[99] M. Nakasako, Y. Takayama, T. Oroguchi, Y. Sekiguchi, A. Kobayashi, K. Shirahama, M. Yamamoto, T. Hikima, K. Yonekura, S. Maki-Yonekura, Y. Kohmura, Y. Inubushi, Y. Takahashi, A. Suzuki, S. Matsunaga, Y. Inui, K. Tono, T. Kameshima, Y. Joti and T. Hoshi, KOTOBUKI-1 apparatus for cryogenic coherent X-ray diffraction imaging., *The Review of scientific instruments* **84**, 093705, (2013).

[100] A. Barty, R. Soufli, T. McCarville, S. L. Baker, M. J. Pivovaroff, P. Stefan, and R. M. Bionta, Predicting the coherent X-ray wavefront focal properties at the Linac Coherent Light Source (LCLS) X-ray free electron laser, *Opt. Express* **17**, 15508–15519, (2009).

[101] J. Amann, W. Berg, V. Blank, F. J. Decker, Y. Ding, P. Emma, Y. Feng, J. Frisch, D. Fritz, J. Hastings, Z. Huang, J. Krzywinski, R. Lindberg, H. Loos, A. Lutman, H. D. Nuhn, D. Ratner, J. Rzepiela, D. Shu, Y. V. Shvyd'ko, S. Spampinati, S. Terentyev, E. Trakhtenberg, D. Walz, J. Welch, A. Zholents and D. Zhu, Demonstration of self-seeding in a hard-X-ray free-electron laser, *Nat. Photonics* **6**, 693–698, (2012).

[102] T. A. White, R. A. Kirian, A. V. Martin, A. Aquila, K. Nass, A. Barty and H. N. Chapman, CrystFEL: a software suite for snapshot serial crystallography, *Journal of Applied Crystallography* **45**, 335–341, (2012).

[103] R. A. Kirian, X. Wang, U. Weierstall, K. E. Schmidt, J. C. H. Spence, M. Hunter, P. Fromme, T. A. White, H. N. Chapman and J. Holton, Femtosecond protein nanocrystallography — data analysis methods, *Opt. Express* **18**, 5713–5723, (2010).

[104] T. White, Post-refinement method for snapshot serial crystallography, *Phil. Trans. R. Soc. B* **369**, 20130330, (2014).

[105] R. A. Kirian, T. A. White, J. M. Holton, H. N. Chapman, P. Fromme, A. Barty, L. Lomb, A. Aquila, F. R. N. C. Maia, A. V. Martin, R. Fromme, X. Wang, M. S. Hunter, K. E. Schmidt and J. C. H. Spence, Structure-factor analysis of femtosecond microdiffraction patterns from protein nanocrystals, *Acta Crystallogr, A, Found Crystallogr* **67**, 131–140, (2011).

[106] J. C. H. Spence, R. A. Kirian, X. Wang, U. Weierstall, K. E. Schmidt, T. White, A. Barty, H. N. Chapman, S. Marchesini and J. Holton, Phasing of coherent femtosecond X-ray diffraction from size-varying nanocrsytals, *Optics Express* **19**, 2866, (2011).

[107] V. Elser, Direct phasing of nanocrystal diffraction, *Acta Crystallogr. A* **69**, 559–69, (2013). ISSN 1600-5724.

[108] J. Kern, R. Alonso-Mori, J. Hellmich, R. Tran, J. Hattne, H. Laksmono, C. Glöckner, N. Echols, R. G. Sierra, J. Sellberg, B. Lassalle-Kaiser, R. J. Gildea, P. Glatzel, R. W. Grosse-Kunstleve, M. J. Latimer, T. A. McQueen, D. DiFiore, A. R. Fry, M. Messerschmidt, A. Miahnahri, D. W. Schafer, M. M. Seibert, D. Sokaras, T.-C. Weng, P. H. Zwart, W. E. White, P. D. Adams, M. J. Bogan, S. Boutet, G. J. Williams, J. Messinger, N. K. Sauter, A. Zouni, U. Bergmann, J. Yano and V. K. Yachandra, Room temperature femtosecond X-ray diffraction of photosystem ii microcrystals, *Phys. Natl. Acad. Sci. USA* **109**, 9721–9726, (2012).

[109] A. Aquila, M. S. Hunter, R. B. Doak, R. A. Kirian, P. Fromme, T. A. White, J. Andreasson, D. Arnlund, S. Bajt, T. R. M. Barends, M. Barthelmess, M. J. Bogan, C. Bostedt, H. Bottin, J. D. Bozek, C. Caleman, N. Coppola, J. Davidsson, D. P. DePonte, V. Elser, S. W. Epp, B. Erk, H. Fleckenstein, L. Foucar, M. Frank, R. Fromme, H. Graafsma, I. Grotjohann, L. Gumprecht, J. Hajdu, C. Y. Hampton, A. Hartmann, R. Hartmann, S. Hau-Riege, G. Hauser, H. Hirsemann, P. Holl, J. M. Holton, A. Hömke, L. Johansson, N. Kimmel, S. Kassemeyer, F. Krasniqi, K. Kühnel, M. Liang, L. Lomb, E. Malmerberg, S. Marchesini, A. V. Martin, F. R. Maia, M. Messerschmidt, K. Nass, C. Reich, R. Neutze, D. Rolles, B. Rudek, A. Rudenko, I. Schlichting, C. Schmidt, K. E. Schmidt, J. Schulz, M. M. Seibert, R. L. Shoeman, R. Sierra, H. Soltau, D. Starodub, F. Stellato, S. Stern, L. Strüder, N. Timneanu, J. Ullrich, X. Wang, G. J. Williams, G. Weidenspointner, U. Weierstall, C. Wunderer, A. Barty, J. C. H. Spence and H. N. Chapman, Time-resolved protein nanocrystallography using an X-ray free-electron laser, *Opt. Express* **20**, 2706–2716, (2012).

[110] C. Kupitz, S. Basu, I. Grotjohann, R. Fromme, N. A. Zatsepin, K. N. Rendek, M. S. Hunter, R. L. Shoeman, T. A. White, D. Wang, D. James, J.-H. Yang, D. E. Cobb, B. Reeder, R. G. Sierra, H. Liu, A. Barty,

A. L. Aquila, D. Deponte, R. A. Kirian, S. Bari, J. J. Bergkamp, K. R. Beyerlein, M. J. Bogan, C. Caleman, T.-C. Chao, C. E. Conrad, K. M. Davis, H. Fleckenstein, L. Galli, S. P. Hau-Riege, S. Kassemeyer, H. Laksmono, M. Liang, L. Lomb, S. Marchesini, A. V. Martin, M. Messerschmidt, D. Milathianaki, K. Nass, A. Ros, S. Roy-Chowdhury, K. Schmidt, M. Seibert, J. Steinbrener, F. Stellato, L. Yan, C. Yoon, T. A. Moore, A. L. Moore, Y. Pushkar, G. J. Williams, S. Boutet, R. B. Doak, U. Weierstall, M. Frank, H. N. Chapman, J. C. H. Spence and P. Fromme, Serial time-resolved crystallography of photosystem ii using a femtosecond X-ray laser, *Nature* **513**(7517), 261–265 (Sept., 2014). ISSN 0028-0836. URL http://dx.doi.org/10.1038/nature13453.

[111] C. H. Yoon, P. Schwander, C. Abergel, I. Andersson, J. Andreasson, A. Aquila, S. Bajt, M. Barthelmess, A. Barty, M. J. Bogan, C. Bostedt, J. D. Bozek, H. N. Chapman, J.-M. Claverie, N. Coppola, D. P. DePonte, T. Ekeberg, S. W. Epp, B. Erk, H. Fleckenstein, L. Foucar, H. Graafsma, L. Gumprecht, J. Hajdu, C. Y. Hampton, A. Hartmann, E. Hartmann, R. Hartmann, G. Hauser, H. Hirsemann, P. Holl, S. Kassemeyer, N. Kimmel, M. Kiskinova, M. Liang, N.-T. D. Loh, L. Lomb, F. R. N. C. Maia, A. V. Martin, K. Nass, E. Pedersoli, C. Reich, D. Rolles, B. Rudek, A. Rudenko, I. Schlichting, J. Schulz, M. M. Seibert, V. Seltzer, R. L. shoeman, R. G. Sierra, H. Soltau, D. Starodub, J. F. Steinbrener, G. Stier, L. Strüder, M. Svenda, J. Ullrich, G. Weidenspointner, T. A. White, C. B. Wunderer and A. Ourmazd, Unsupervised classification of single-particle X-ray diffraction snapshots by spectral clustering, *Optics Express* **19**, 16542–16549, (2011).

[112] S. Kassemeyer, J. F. Steinbrener, L. Lomb, E. Hartmann, A. Aquila, A. Barty, A. V. Martin, C. Y. Hampton, S. Bajt, M. Barthelmess, T. R. M. Barends, C. Bostedt, M. Bott, J. D. Bozek, N. Coppola, M. Cryle, D. P. DePonte, R. B. Doak, S. W. Epp, B. Erk, H. Fleckenstein, L. Foucar, H. Graafsma, L. Gumprecht, A. Hartmann, R. Hartmann, G. Hauser, H. Hirsemann, A. Hömke, P. Holl, O. Jonsson, N. Kimmel, F. Krasniqi, M. Liang, F. R. N. C. Maia, S. Marchesini, K. Nass, C. Reich, D. Rolles, B. Rudek, A. Rudenko, C. Schmidt, J. Schulz, R. L. shoeman, R. G. Sierra, H. Soltau, J. C. H. Spence, D. Starodub, F. Stellato, S. Stern, G. Stier, M. Svenda, G. Weidenspointner, U. Weierstall, T. A. White, C. B. Wunderer, M. Frank, H. N. Chapman, J. Ullrich, L. Strüder, M. J. Bogan and I. Schlichting, Femtosecond free-electron laser X-ray diffraction data sets for algorithm development, *Opt. Express* **20**, 4149–4158, (2012).

[113] N. D. Loh, Effects of extraneous noise in Cryptotomography, *Proceedings of SPIE* **8500**, (2012).

[114] E. Pedersoli, N. D. Loh, F. Capotondi, C. Y. Hampton, R. G. Sierra, D. Starodub, C. Bostedt, J. Bozek, A. J. Nelson, M. Aslam, S. Li, V. P. Dravid, A. V. Martin, A. Aquila, A. Barty, H. Fleckenstein, L. Gumprecht, M. Liang, K. Nass, J. Schulz, T. A. White, N. Coppola, S. Bajt, M. Barthelmess, H. Graafsma, H. Hirsemann, C. Wunderer, S. W.

Epp, B. Erk, B. Rudek, A. Rudenko, L. Foucar, S. Kassemeyer, L. Lomb, D. Rolles, R. L. shoeman, J. F. Steinbrener, R. Hartmann, A. Hartmann, G. Hauser, P. Holl, N. Kimmel, C. Reich, H. Soltau, G. Weidenspointner, W. H. Benner, G. R. Farquar, S. P. Hau-Riege, M. S. Hunter, T. Ekeberg, M. Hantke, F. R. N. C. Maia, H. J. Tobias, S. Marchesini, M. Frank, L. Strüder, I. Schlichting, J. Ullrich, H. N. Chapman, P. H. Bucksbaum, M. Kiskinova and M. J. Bogan, Mesoscale morphology of airborne core–shell nanoparticle clusters: X-ray laser coherent diffraction imaging, *J. Phys. B: At. Mol. Opt. Phys.* **46**, 164033, (2013).

[115] A. V. Martin, J. Andreasson, A. Aquila, S. Bajt, T. R. M. Barends, M. Barthelmess, A. Barty, W. H. Benner, C. Bostedt, J. D. Bozek, P. H. Bucksbaum, C. Caleman, N. Coppola, D. P. DePonte, T. Ekeberg, S. W. Epp, B. Erk, G. R. Farquar, H. Fleckenstein, L. Foucar, M. Frank, L. Gumprecht, C. Y. Hampton, M. Hantke, A. Hartmann, E. Hartmann, R. Hartmann, S. Hau-Riege, G. Hauser, P. Holl, A. Hoemke, O. Jonsson, S. Kassemeyer, N. Kimmel, M. Kiskinova, F. Krasniqi, J. Krzywinski, M. Liang, N.-T. D. Loh, L. Lomb, F. R. N. C. Maia, S. Marchesini, M. Messerschmidt, K. Nass, D. Odic, E. Pedersoli, C. Reich, D. Rolles, B. Rudek, A. Rudenko, C. Schmidt, J. Schulz, M. M. Seibert, R. L. shoeman, R. G. Sierra, H. Soltau, D. Starodub, J. F. Steinbrener, F. Stellato, L. Strüder, M. Svenda, H. Tobias, J. Ullrich, G. Weidenspointner, D. Westphal, T. A. White, G. J. Williams, J. Hajdu, I. Schlichting, M. J. Bogan and H. N. Chapman, Single particle imaging with soft X-rays at the Linac Coherent Light Source, *Proceedings of SPIE* **8078**, 807809, (2011).

[116] H. J. Park, N. D. Loh, R. G. Sierra, C. Y. Hampton, D. Starodub, A. V. Martin, A. Barty, A. Aquila, J. Schulz, J. Steinbrener, R. L. Shoeman, L. Lomb, S. Kassemeyer, C. Bostedt, J. Bozek, S. W. Epp, B. Erk, R. Hartmann, D. Rolles, A. Rudenko, B. Rudek, L. Foucar, N. Kimmel, G. Weidenspointner, G. Hauser, P. Holl, E. Pedersoli, M. Liang, M. S. Hunter, L. Gumprecht, N. Coppola, C. Wunderer, H. Graafsma, F. R. Maia, T. Ekeberg, M. Hantke, H. Fleckenstein, H. Hirsemann, K. Nass, H. J. Tobias, G. R. Farquar, W. H. Benner, S. Hau-Riege, C. Reich, A. Hartmann, H. Soltau, S. Marchesini, S. Bajt, M. Barthelmess, L. Strueder, J. Ullrich, P. Bucksbaum, M. Frank, I. Schlichting, H. N. Chapman, M. J. Bogan and V. Elser, Toward unsupervised single-shot diffractive imaging of heterogeneous particles using X-ray free-electron lasers, *Optics Express* **21**, (2013).

[117] P. Thibault, V. Elser, C. Jacobsen and D. Shapiro, Reconstruction of a yeast cell from X-ray diffraction data, *Acta Crystallographica Section A* **62**, 248–261, (2006).

[118] A. V. Martin, N. D. Loh, C. Y. Hampton, R. G. Sierra, F. Wang, A. Aquila, S. Bajt, M. Barthelmess, C. Bostedt, J. D. Bozek, N. Coppola, S. W. Epp, B. Erk, H. Fleckenstein, L. Foucar, M. Frank, H. Graafsma, L. Gumprecht, A. Hartmann, R. Hartmann, G. Hauser, H. Hirsemann, P. Holl, S. Kassemeyer, N. Kimmel, M. Liang, L. Lomb, F. R. N. C. Maia, S. Marchesini, K. Nass, E. Pedersoli, C. Reich, D. Rolles, B. Rudek, A. Rudenko, J. Schulz,

R. L. shoeman, H. Soltau, D. Starodub, J. F. Steinbrener, F. Stellato, L. Strüder, J. Ullrich, G. Weidenspointner, T. A. White, C. B. Wunderer, A. Barty, I. Schlichting, M. J. Bogan and H. N. Chapman, Femtosecond dark-field imaging with an X-ray free electron laser, *Optics Express* **20**, 13501–13512, (2012).

[119] G. Huldt, A. Szőke and J. Hajdu, Diffraction imaging of single particles and biomolecules, *J. Struct. Biol.* **144**, 219–227, (2003).

[120] V. L. Shneerson, A. Ourmazd and D. K. Saldin, Crystallography without crystals. I. The common-line method for assembling a three-dimensional diffraction volume from single-particle scattering, *Acta Crystallographica Section A* **64**, 303–315, (2008).

[121] G. Bortel and M. Tegze, Common arc method for diffraction pattern orientation, *Acta Cryst. A* **A67**, 533–543, (2011).

[122] M. van Heel, Angular reconstitution: A posteriori assignment of projection directions for 3D reconstruction, *Ultramicroscopy* (1987).

[123] J. B. Tenenbaum, V. De Silva and J. C. Langford, A Global Geometric Framework for Nonlinear Dimensionality Reduction, *Science* (2000).

[124] R. Coifman, Diffusion maps, *Applied and Computational Harmonic Analysis* (2006).

[125] S. Kassemeyer, A. Jafarpour, L. Lomb, J. F. Steinbrener, A. V. Martin and I. Schlichting, Optimal mapping of X-ray laser diffraction patterns into three dimensions using routing algorithms, *Phys. Rev. E* (2013).

[126] D. Giannakis, P. Schwander and A. Ourmazd, The symmetries of image formation by scattering. I. Theoretical framework, *Opt. Express* **20**, 12799–12826, (2012).

[127] P. Schwander, D. Giannakis, C. H. Yoon and A. Ourmazd, The symmetries of image formation by scattering. II. Applications, *Optics Express* **20**, 12827–12849, (2012).

[128] V. Elser, Noise limits on reconstructing diffraction signals from random tomographs, *Information Theory, IEEE Transactions on* **55**, 4715–4722, (2009).

[129] N.-T. D. Loh and V. Elser, Reconstruction algorithm for single-particle diffraction imaging experiments, *Phys. Rev. E* **80**, 6705, (2009).

[130] A. Dempster, N. Laird and D. Rubin, Maximum likelihood from incomplete data via the EM algorithm, *J. R. Stat. Soc.* **39**(1), 1–38, (1977).

[131] H. T. Philipp, K. Ayyer, M. W. Tate, V. Elser and S. M. Gruner, Solving structure with sparse, randomly-oriented X-ray data, *Optics Express* **20**, 13129–13137, (2012).

[132] N. D. Loh, M. J. Bogan, V. Elser, A. Barty, S. Boutet, S. Bajt, J. Hajdu, T. Ekeberg, F. R. N. C. Maia, J. Schulz, M. M. Seibert, B. Iwan, N. Tîmneanu, S. Marchesini, I. Schlichting, R. L. shoeman, L. Lomb, M. Frank, M. Liang and H. N. Chapman, Cryptotomography: Reconstructing 3D fourier intensities from randomly oriented single-shot diffraction patterns, *Phys. Rev. Lett.* **104**, 25501, (2010).

[133] M. Tegze and G. Bortel, Atomic structure of a single large biomolecule from diffraction patterns of random orientations, *Journal of Structural Biology* **179**, 41–45, (2012).

[134] R. Fung, V. Shneerson, D. Saldin and A. Ourmazd, Structure from fleeting illumination of faint spinning objects in flight, *Nat Phys.* **5**, 64–67, (2009).

[135] C. M. Bishop, M. Svensén and C. K. I. Williams, GTM: The generative topographic mapping, *Neural Comput.* (1998).

[136] N. D. Loh, A minimal view of single-particle imaging with X-ray lasers, *Philos. Trans. R. Soc. B Biol. Sci.* **369**(1647), 20130328, (2014).

[137] Z. Kam, Determination of macromolecular structure in solution by spatial correlation of scattering fluctuations, *Macromolecules* **10**, 927–934, (1977).

[138] D. K. Saldin, H. C. Poon, V. L. Shneerson, M. Howells, H. N. Chapman, R. A. Kirian, K. E. Schmidt and J. Spence, Beyond small-angle X-ray scattering: Exploiting angular correlations, *Physical Review B* **81**, 174105, (2010).

[139] R. P. Kurta, M. Altarelli, E. Weckert and I. A. Vartanyants, X-ray cross-correlation analysis applied to disordered two-dimensional systems, *Phys. Rev. B* **85**, 184204, (2012).

[140] D. Starodub, A. Aquila, S. Bajt, M. Barthelmess, A. Barty, C. Bostedt, J. D. Bozek, N. Coppola, R. B. Doak, S. W. Epp, B. Erk, L. Foucar, L. Gumprecht, C. Y. Hampton, A. Hartmann, R. Hartmann, P. Holl, S. Kassemeyer, N. Kimmel, H. Laksmono, M. Liang, N. D. Loh, L. Lomb, A. V. Martin, K. Nass, C. Reich, D. Rolles, B. Rudek, A. Rudenko, J. Schulz, R. L. shoeman, R. G. Sierra, H. Soltau, J. F. Steinbrener, F. Stellato, S. Stern, G. Weidenspointner, M. Frank, J. Ullrich, L. Strüder, I. Schlichting, H. N. Chapman, J. C. H. Spence and M. J. Bogan, Single-particle structure determination by correlations of snapshot X-ray diffraction patterns, *Nature Communications* **3:1276**, 1–7, (2012).

[141] V. Elser, Strategies for processing diffraction data from randomly oriented particles, *Ultramicroscopy* **111**, 788–792, (2011).

Epilogue

This book is dedicated to a broad range of readers interested in the application of synchrotron radiation in various scientific fields where atomic arrangements are essential in research topics. Each chapter of this book tried to extract relevant characteristics of synchrotron radiation describing how they provide important information for typical applications.

Along with the progress in synchrotron radiation research, scientists realized that the recent trend of light source is seeking lower emittance and shorter pulse structure. For example, imaging and time-dependent experiments depend on lower emittance and shorter pulse, respectively, since higher brilliance and shorter pulse allows researchers to study smaller size samples, images with higher resolution and/or time-dependence with a narrower time-window. This trend of light source development led to the 4th generation light sources such as low emittance storage rings based on a novel multi-bend achlomat (MBA) lattice, and linac-based XFELs having superior peak brilliance, coherence and short pulse. Upgrades of the 3rd generation high energy (>6 GeV) storage rings and newly constructed rings with an MBA lattice such as MAXIV would achieve $\sim 10^{22}$ (photons/sec $\cdot$ mrad^2 $\cdot$ mm^2 $\cdot$ 0.1% band width) brilliance shifting a diffraction limit energy deeper into a higher energy region. It is difficult to describe the future of rapidly growing science but it is no doubt that the R&D efforts of the 4th generation light sources would continue allowing users experiments beyond the present limits.

But here we may need to note that tabletop compact light sources such as plasma (laser) wake field, inverse Compton scattering and higher harmonic generation are also under development which would allow researchers to perform their experiments in their laboratories.

Synchrotron radiation research would be particularly useful in energy science and related interdisciplinary fields that often utilize *operando* experiments. Not only due to the emission control and future energy shortage due to the depletion of fossil fuels expected in 30–40 years, efforts to utilize water splitting converting carbon dioxide into hydrocarbon fuels are becoming urgently needed. Synchrotron radiation research on catalysts to boost water splitting efficiency is an urgent demand for which *operando* experiments and time resolved experiments using high brilliance, short X-ray pulses is essential. We hope that this book would give opportunities to experienced researchers to widen their knowledge and to help beginners to step forward.

Hiroyuki Oyanagi
Research Institute of Instrumentation Frontier, AIST,
Tsukuba, Ibaraki 305-8568, Japan
h.oynagi@aist.go.jp

Xinyi Zhang
Department of Physics, Fudan University
xy-zhang@fudan.edu.cn

June 11, 2017, Shanghai, China

Index